A Million Zeros

A Million Zeros

Douglas Crockford

**virgule
solidus**

A Million Zeros
Douglas Crockford
Public Domain

978-1-949815-11-5 Hardcover

Books By Douglas Crockford

JavaScript: The Good Parts

How JavaScript Works

A Million And One Random Digits

A Million Nines

A Million Zeros

Forthcoming

Lower Mathematics

Misty System

Capability Security For Beginners

A Million Zeros

```
0000:  0000000000  0000000000  0000000000  0000000000  0000000000    0000000000  0000000000  0000000000  0000000000  0000000000
0001:  0000000000  0000000000  0000000000  0000000000  0000000000    0000000000  0000000000  0000000000  0000000000  0000000000
0002:  0000000000  0000000000  0000000000  0000000000  0000000000    0000000000  0000000000  0000000000  0000000000  0000000000
0003:  0000000000  0000000000  0000000000  0000000000  0000000000    0000000000  0000000000  0000000000  0000000000  0000000000
0004:  0000000000  0000000000  0000000000  0000000000  0000000000    0000000000  0000000000  0000000000  0000000000  0000000000
0005:  0000000000  0000000000  0000000000  0000000000  0000000000    0000000000  0000000000  0000000000  0000000000  0000000000
0006:  0000000000  0000000000  0000000000  0000000000  0000000000    0000000000  0000000000  0000000000  0000000000  0000000000
0007:  0000000000  0000000000  0000000000  0000000000  0000000000    0000000000  0000000000  0000000000  0000000000  0000000000
0008:  0000000000  0000000000  0000000000  0000000000  0000000000    0000000000  0000000000  0000000000  0000000000  0000000000
0009:  0000000000  0000000000  0000000000  0000000000  0000000000    0000000000  0000000000  0000000000  0000000000  0000000000
0010:  0000000000  0000000000  0000000000  0000000000  0000000000    0000000000  0000000000  0000000000  0000000000  0000000000
0011:  0000000000  0000000000  0000000000  0000000000  0000000000    0000000000  0000000000  0000000000  0000000000  0000000000
0012:  0000000000  0000000000  0000000000  0000000000  0000000000    0000000000  0000000000  0000000000  0000000000  0000000000
0013:  0000000000  0000000000  0000000000  0000000000  0000000000    0000000000  0000000000  0000000000  0000000000  0000000000
0014:  0000000000  0000000000  0000000000  0000000000  0000000000    0000000000  0000000000  0000000000  0000000000  0000000000
0015:  0000000000  0000000000  0000000000  0000000000  0000000000    0000000000  0000000000  0000000000  0000000000  0000000000
0016:  0000000000  0000000000  0000000000  0000000000  0000000000    0000000000  0000000000  0000000000  0000000000  0000000000
0017:  0000000000  0000000000  0000000000  0000000000  0000000000    0000000000  0000000000  0000000000  0000000000  0000000000
0018:  0000000000  0000000000  0000000000  0000000000  0000000000    0000000000  0000000000  0000000000  0000000000  0000000000
0019:  0000000000  0000000000  0000000000  0000000000  0000000000    0000000000  0000000000  0000000000  0000000000  0000000000
0020:  0000000000  0000000000  0000000000  0000000000  0000000000    0000000000  0000000000  0000000000  0000000000  0000000000
0021:  0000000000  0000000000  0000000000  0000000000  0000000000    0000000000  0000000000  0000000000  0000000000  0000000000
0022:  0000000000  0000000000  0000000000  0000000000  0000000000    0000000000  0000000000  0000000000  0000000000  0000000000
0023:  0000000000  0000000000  0000000000  0000000000  0000000000    0000000000  0000000000  0000000000  0000000000  0000000000
0024:  0000000000  0000000000  0000000000  0000000000  0000000000    0000000000  0000000000  0000000000  0000000000  0000000000
0025:  0000000000  0000000000  0000000000  0000000000  0000000000    0000000000  0000000000  0000000000  0000000000  0000000000
0026:  0000000000  0000000000  0000000000  0000000000  0000000000    0000000000  0000000000  0000000000  0000000000  0000000000
0027:  0000000000  0000000000  0000000000  0000000000  0000000000    0000000000  0000000000  0000000000  0000000000  0000000000
0028:  0000000000  0000000000  0000000000  0000000000  0000000000    0000000000  0000000000  0000000000  0000000000  0000000000
0029:  0000000000  0000000000  0000000000  0000000000  0000000000    0000000000  0000000000  0000000000  0000000000  0000000000
0030:  0000000000  0000000000  0000000000  0000000000  0000000000    0000000000  0000000000  0000000000  0000000000  0000000000
0031:  0000000000  0000000000  0000000000  0000000000  0000000000    0000000000  0000000000  0000000000  0000000000  0000000000
0032:  0000000000  0000000000  0000000000  0000000000  0000000000    0000000000  0000000000  0000000000  0000000000  0000000000
0033:  0000000000  0000000000  0000000000  0000000000  0000000000    0000000000  0000000000  0000000000  0000000000  0000000000
0034:  0000000000  0000000000  0000000000  0000000000  0000000000    0000000000  0000000000  0000000000  0000000000  0000000000
0035:  0000000000  0000000000  0000000000  0000000000  0000000000    0000000000  0000000000  0000000000  0000000000  0000000000
0036:  0000000000  0000000000  0000000000  0000000000  0000000000    0000000000  0000000000  0000000000  0000000000  0000000000
0037:  0000000000  0000000000  0000000000  0000000000  0000000000    0000000000  0000000000  0000000000  0000000000  0000000000
0038:  0000000000  0000000000  0000000000  0000000000  0000000000    0000000000  0000000000  0000000000  0000000000  0000000000
0039:  0000000000  0000000000  0000000000  0000000000  0000000000    0000000000  0000000000  0000000000  0000000000  0000000000
0040:  0000000000  0000000000  0000000000  0000000000  0000000000    0000000000  0000000000  0000000000  0000000000  0000000000
0041:  0000000000  0000000000  0000000000  0000000000  0000000000    0000000000  0000000000  0000000000  0000000000  0000000000
0042:  0000000000  0000000000  0000000000  0000000000  0000000000    0000000000  0000000000  0000000000  0000000000  0000000000
0043:  0000000000  0000000000  0000000000  0000000000  0000000000    0000000000  0000000000  0000000000  0000000000  0000000000
0044:  0000000000  0000000000  0000000000  0000000000  0000000000    0000000000  0000000000  0000000000  0000000000  0000000000
0045:  0000000000  0000000000  0000000000  0000000000  0000000000    0000000000  0000000000  0000000000  0000000000  0000000000
0046:  0000000000  0000000000  0000000000  0000000000  0000000000    0000000000  0000000000  0000000000  0000000000  0000000000
0047:  0000000000  0000000000  0000000000  0000000000  0000000000    0000000000  0000000000  0000000000  0000000000  0000000000
0048:  0000000000  0000000000  0000000000  0000000000  0000000000    0000000000  0000000000  0000000000  0000000000  0000000000
0049:  0000000000  0000000000  0000000000  0000000000  0000000000    0000000000  0000000000  0000000000  0000000000  0000000000
```

```
0050:  00000 00000  00000 00000  00000 00000  00000 00000  00000 00000    00000 00000  00000 00000  00000 00000  00000 00000  00000 00000
0051:  00000 00000  00000 00000  00000 00000  00000 00000  00000 00000    00000 00000  00000 00000  00000 00000  00000 00000  00000 00000
0052:  00000 00000  00000 00000  00000 00000  00000 00000  00000 00000    00000 00000  00000 00000  00000 00000  00000 00000  00000 00000
0053:  00000 00000  00000 00000  00000 00000  00000 00000  00000 00000    00000 00000  00000 00000  00000 00000  00000 00000  00000 00000
0054:  00000 00000  00000 00000  00000 00000  00000 00000  00000 00000    00000 00000  00000 00000  00000 00000  00000 00000  00000 00000
0055:  00000 00000  00000 00000  00000 00000  00000 00000  00000 00000    00000 00000  00000 00000  00000 00000  00000 00000  00000 00000
0056:  00000 00000  00000 00000  00000 00000  00000 00000  00000 00000    00000 00000  00000 00000  00000 00000  00000 00000  00000 00000
0057:  00000 00000  00000 00000  00000 00000  00000 00000  00000 00000    00000 00000  00000 00000  00000 00000  00000 00000  00000 00000
0058:  00000 00000  00000 00000  00000 00000  00000 00000  00000 00000    00000 00000  00000 00000  00000 00000  00000 00000  00000 00000
0059:  00000 00000  00000 00000  00000 00000  00000 00000  00000 00000    00000 00000  00000 00000  00000 00000  00000 00000  00000 00000
0060:  00000 00000  00000 00000  00000 00000  00000 00000  00000 00000    00000 00000  00000 00000  00000 00000  00000 00000  00000 00000
0061:  00000 00000  00000 00000  00000 00000  00000 00000  00000 00000    00000 00000  00000 00000  00000 00000  00000 00000  00000 00000
0062:  00000 00000  00000 00000  00000 00000  00000 00000  00000 00000    00000 00000  00000 00000  00000 00000  00000 00000  00000 00000
0063:  00000 00000  00000 00000  00000 00000  00000 00000  00000 00000    00000 00000  00000 00000  00000 00000  00000 00000  00000 00000
0064:  00000 00000  00000 00000  00000 00000  00000 00000  00000 00000    00000 00000  00000 00000  00000 00000  00000 00000  00000 00000
0065:  00000 00000  00000 00000  00000 00000  00000 00000  00000 00000    00000 00000  00000 00000  00000 00000  00000 00000  00000 00000
0066:  00000 00000  00000 00000  00000 00000  00000 00000  00000 00000    00000 00000  00000 00000  00000 00000  00000 00000  00000 00000
0067:  00000 00000  00000 00000  00000 00000  00000 00000  00000 00000    00000 00000  00000 00000  00000 00000  00000 00000  00000 00000
0068:  00000 00000  00000 00000  00000 00000  00000 00000  00000 00000    00000 00000  00000 00000  00000 00000  00000 00000  00000 00000
0069:  00000 00000  00000 00000  00000 00000  00000 00000  00000 00000    00000 00000  00000 00000  00000 00000  00000 00000  00000 00000
0070:  00000 00000  00000 00000  00000 00000  00000 00000  00000 00000    00000 00000  00000 00000  00000 00000  00000 00000  00000 00000
0071:  00000 00000  00000 00000  00000 00000  00000 00000  00000 00000    00000 00000  00000 00000  00000 00000  00000 00000  00000 00000
0072:  00000 00000  00000 00000  00000 00000  00000 00000  00000 00000    00000 00000  00000 00000  00000 00000  00000 00000  00000 00000
0073:  00000 00000  00000 00000  00000 00000  00000 00000  00000 00000    00000 00000  00000 00000  00000 00000  00000 00000  00000 00000
0074:  00000 00000  00000 00000  00000 00000  00000 00000  00000 00000    00000 00000  00000 00000  00000 00000  00000 00000  00000 00000
0075:  00000 00000  00000 00000  00000 00000  00000 00000  00000 00000    00000 00000  00000 00000  00000 00000  00000 00000  00000 00000
0076:  00000 00000  00000 00000  00000 00000  00000 00000  00000 00000    00000 00000  00000 00000  00000 00000  00000 00000  00000 00000
0077:  00000 00000  00000 00000  00000 00000  00000 00000  00000 00000    00000 00000  00000 00000  00000 00000  00000 00000  00000 00000
0078:  00000 00000  00000 00000  00000 00000  00000 00000  00000 00000    00000 00000  00000 00000  00000 00000  00000 00000  00000 00000
0079:  00000 00000  00000 00000  00000 00000  00000 00000  00000 00000    00000 00000  00000 00000  00000 00000  00000 00000  00000 00000
0080:  00000 00000  00000 00000  00000 00000  00000 00000  00000 00000    00000 00000  00000 00000  00000 00000  00000 00000  00000 00000
0081:  00000 00000  00000 00000  00000 00000  00000 00000  00000 00000    00000 00000  00000 00000  00000 00000  00000 00000  00000 00000
0082:  00000 00000  00000 00000  00000 00000  00000 00000  00000 00000    00000 00000  00000 00000  00000 00000  00000 00000  00000 00000
0083:  00000 00000  00000 00000  00000 00000  00000 00000  00000 00000    00000 00000  00000 00000  00000 00000  00000 00000  00000 00000
0084:  00000 00000  00000 00000  00000 00000  00000 00000  00000 00000    00000 00000  00000 00000  00000 00000  00000 00000  00000 00000
0085:  00000 00000  00000 00000  00000 00000  00000 00000  00000 00000    00000 00000  00000 00000  00000 00000  00000 00000  00000 00000
0086:  00000 00000  00000 00000  00000 00000  00000 00000  00000 00000    00000 00000  00000 00000  00000 00000  00000 00000  00000 00000
0087:  00000 00000  00000 00000  00000 00000  00000 00000  00000 00000    00000 00000  00000 00000  00000 00000  00000 00000  00000 00000
0088:  00000 00000  00000 00000  00000 00000  00000 00000  00000 00000    00000 00000  00000 00000  00000 00000  00000 00000  00000 00000
0089:  00000 00000  00000 00000  00000 00000  00000 00000  00000 00000    00000 00000  00000 00000  00000 00000  00000 00000  00000 00000
0090:  00000 00000  00000 00000  00000 00000  00000 00000  00000 00000    00000 00000  00000 00000  00000 00000  00000 00000  00000 00000
0091:  00000 00000  00000 00000  00000 00000  00000 00000  00000 00000    00000 00000  00000 00000  00000 00000  00000 00000  00000 00000
0092:  00000 00000  00000 00000  00000 00000  00000 00000  00000 00000    00000 00000  00000 00000  00000 00000  00000 00000  00000 00000
0093:  00000 00000  00000 00000  00000 00000  00000 00000  00000 00000    00000 00000  00000 00000  00000 00000  00000 00000  00000 00000
0094:  00000 00000  00000 00000  00000 00000  00000 00000  00000 00000    00000 00000  00000 00000  00000 00000  00000 00000  00000 00000
0095:  00000 00000  00000 00000  00000 00000  00000 00000  00000 00000    00000 00000  00000 00000  00000 00000  00000 00000  00000 00000
0096:  00000 00000  00000 00000  00000 00000  00000 00000  00000 00000    00000 00000  00000 00000  00000 00000  00000 00000  00000 00000
0097:  00000 00000  00000 00000  00000 00000  00000 00000  00000 00000    00000 00000  00000 00000  00000 00000  00000 00000  00000 00000
0098:  00000 00000  00000 00000  00000 00000  00000 00000  00000 00000    00000 00000  00000 00000  00000 00000  00000 00000  00000 00000
0099:  00000 00000  00000 00000  00000 00000  00000 00000  00000 00000    00000 00000  00000 00000  00000 00000  00000 00000  00000 00000
```

```
0100:  00000 00000  00000 00000  00000 00000  00000 00000  00000 00000    00000 00000  00000 00000  00000 00000  00000 00000  00000 00000
0101:  00000 00000  00000 00000  00000 00000  00000 00000  00000 00000    00000 00000  00000 00000  00000 00000  00000 00000  00000 00000
0102:  00000 00000  00000 00000  00000 00000  00000 00000  00000 00000    00000 00000  00000 00000  00000 00000  00000 00000  00000 00000
0103:  00000 00000  00000 00000  00000 00000  00000 00000  00000 00000    00000 00000  00000 00000  00000 00000  00000 00000  00000 00000
0104:  00000 00000  00000 00000  00000 00000  00000 00000  00000 00000    00000 00000  00000 00000  00000 00000  00000 00000  00000 00000
0105:  00000 00000  00000 00000  00000 00000  00000 00000  00000 00000    00000 00000  00000 00000  00000 00000  00000 00000  00000 00000
0106:  00000 00000  00000 00000  00000 00000  00000 00000  00000 00000    00000 00000  00000 00000  00000 00000  00000 00000  00000 00000
0107:  00000 00000  00000 00000  00000 00000  00000 00000  00000 00000    00000 00000  00000 00000  00000 00000  00000 00000  00000 00000
0108:  00000 00000  00000 00000  00000 00000  00000 00000  00000 00000    00000 00000  00000 00000  00000 00000  00000 00000  00000 00000
0109:  00000 00000  00000 00000  00000 00000  00000 00000  00000 00000    00000 00000  00000 00000  00000 00000  00000 00000  00000 00000
0110:  00000 00000  00000 00000  00000 00000  00000 00000  00000 00000    00000 00000  00000 00000  00000 00000  00000 00000  00000 00000
0111:  00000 00000  00000 00000  00000 00000  00000 00000  00000 00000    00000 00000  00000 00000  00000 00000  00000 00000  00000 00000
0112:  00000 00000  00000 00000  00000 00000  00000 00000  00000 00000    00000 00000  00000 00000  00000 00000  00000 00000  00000 00000
0113:  00000 00000  00000 00000  00000 00000  00000 00000  00000 00000    00000 00000  00000 00000  00000 00000  00000 00000  00000 00000
0114:  00000 00000  00000 00000  00000 00000  00000 00000  00000 00000    00000 00000  00000 00000  00000 00000  00000 00000  00000 00000
0115:  00000 00000  00000 00000  00000 00000  00000 00000  00000 00000    00000 00000  00000 00000  00000 00000  00000 00000  00000 00000
0116:  00000 00000  00000 00000  00000 00000  00000 00000  00000 00000    00000 00000  00000 00000  00000 00000  00000 00000  00000 00000
0117:  00000 00000  00000 00000  00000 00000  00000 00000  00000 00000    00000 00000  00000 00000  00000 00000  00000 00000  00000 00000
0118:  00000 00000  00000 00000  00000 00000  00000 00000  00000 00000    00000 00000  00000 00000  00000 00000  00000 00000  00000 00000
0119:  00000 00000  00000 00000  00000 00000  00000 00000  00000 00000    00000 00000  00000 00000  00000 00000  00000 00000  00000 00000
0120:  00000 00000  00000 00000  00000 00000  00000 00000  00000 00000    00000 00000  00000 00000  00000 00000  00000 00000  00000 00000
0121:  00000 00000  00000 00000  00000 00000  00000 00000  00000 00000    00000 00000  00000 00000  00000 00000  00000 00000  00000 00000
0122:  00000 00000  00000 00000  00000 00000  00000 00000  00000 00000    00000 00000  00000 00000  00000 00000  00000 00000  00000 00000
0123:  00000 00000  00000 00000  00000 00000  00000 00000  00000 00000    00000 00000  00000 00000  00000 00000  00000 00000  00000 00000
0124:  00000 00000  00000 00000  00000 00000  00000 00000  00000 00000    00000 00000  00000 00000  00000 00000  00000 00000  00000 00000
0125:  00000 00000  00000 00000  00000 00000  00000 00000  00000 00000    00000 00000  00000 00000  00000 00000  00000 00000  00000 00000
0126:  00000 00000  00000 00000  00000 00000  00000 00000  00000 00000    00000 00000  00000 00000  00000 00000  00000 00000  00000 00000
0127:  00000 00000  00000 00000  00000 00000  00000 00000  00000 00000    00000 00000  00000 00000  00000 00000  00000 00000  00000 00000
0128:  00000 00000  00000 00000  00000 00000  00000 00000  00000 00000    00000 00000  00000 00000  00000 00000  00000 00000  00000 00000
0129:  00000 00000  00000 00000  00000 00000  00000 00000  00000 00000    00000 00000  00000 00000  00000 00000  00000 00000  00000 00000
0130:  00000 00000  00000 00000  00000 00000  00000 00000  00000 00000    00000 00000  00000 00000  00000 00000  00000 00000  00000 00000
0131:  00000 00000  00000 00000  00000 00000  00000 00000  00000 00000    00000 00000  00000 00000  00000 00000  00000 00000  00000 00000
0132:  00000 00000  00000 00000  00000 00000  00000 00000  00000 00000    00000 00000  00000 00000  00000 00000  00000 00000  00000 00000
0133:  00000 00000  00000 00000  00000 00000  00000 00000  00000 00000    00000 00000  00000 00000  00000 00000  00000 00000  00000 00000
0134:  00000 00000  00000 00000  00000 00000  00000 00000  00000 00000    00000 00000  00000 00000  00000 00000  00000 00000  00000 00000
0135:  00000 00000  00000 00000  00000 00000  00000 00000  00000 00000    00000 00000  00000 00000  00000 00000  00000 00000  00000 00000
0136:  00000 00000  00000 00000  00000 00000  00000 00000  00000 00000    00000 00000  00000 00000  00000 00000  00000 00000  00000 00000
0137:  00000 00000  00000 00000  00000 00000  00000 00000  00000 00000    00000 00000  00000 00000  00000 00000  00000 00000  00000 00000
0138:  00000 00000  00000 00000  00000 00000  00000 00000  00000 00000    00000 00000  00000 00000  00000 00000  00000 00000  00000 00000
0139:  00000 00000  00000 00000  00000 00000  00000 00000  00000 00000    00000 00000  00000 00000  00000 00000  00000 00000  00000 00000
0140:  00000 00000  00000 00000  00000 00000  00000 00000  00000 00000    00000 00000  00000 00000  00000 00000  00000 00000  00000 00000
0141:  00000 00000  00000 00000  00000 00000  00000 00000  00000 00000    00000 00000  00000 00000  00000 00000  00000 00000  00000 00000
0142:  00000 00000  00000 00000  00000 00000  00000 00000  00000 00000    00000 00000  00000 00000  00000 00000  00000 00000  00000 00000
0143:  00000 00000  00000 00000  00000 00000  00000 00000  00000 00000    00000 00000  00000 00000  00000 00000  00000 00000  00000 00000
0144:  00000 00000  00000 00000  00000 00000  00000 00000  00000 00000    00000 00000  00000 00000  00000 00000  00000 00000  00000 00000
0145:  00000 00000  00000 00000  00000 00000  00000 00000  00000 00000    00000 00000  00000 00000  00000 00000  00000 00000  00000 00000
0146:  00000 00000  00000 00000  00000 00000  00000 00000  00000 00000    00000 00000  00000 00000  00000 00000  00000 00000  00000 00000
0147:  00000 00000  00000 00000  00000 00000  00000 00000  00000 00000    00000 00000  00000 00000  00000 00000  00000 00000  00000 00000
0148:  00000 00000  00000 00000  00000 00000  00000 00000  00000 00000    00000 00000  00000 00000  00000 00000  00000 00000  00000 00000
0149:  00000 00000  00000 00000  00000 00000  00000 00000  00000 00000    00000 00000  00000 00000  00000 00000  00000 00000  00000 00000
```

```
0150:  0000000000  0000000000  0000000000  0000000000  0000000000    0000000000  0000000000  0000000000  0000000000  0000000000
0151:  0000000000  0000000000  0000000000  0000000000  0000000000    0000000000  0000000000  0000000000  0000000000  0000000000
0152:  0000000000  0000000000  0000000000  0000000000  0000000000    0000000000  0000000000  0000000000  0000000000  0000000000
0153:  0000000000  0000000000  0000000000  0000000000  0000000000    0000000000  0000000000  0000000000  0000000000  0000000000
0154:  0000000000  0000000000  0000000000  0000000000  0000000000    0000000000  0000000000  0000000000  0000000000  0000000000
0155:  0000000000  0000000000  0000000000  0000000000  0000000000    0000000000  0000000000  0000000000  0000000000  0000000000
0156:  0000000000  0000000000  0000000000  0000000000  0000000000    0000000000  0000000000  0000000000  0000000000  0000000000
0157:  0000000000  0000000000  0000000000  0000000000  0000000000    0000000000  0000000000  0000000000  0000000000  0000000000
0158:  0000000000  0000000000  0000000000  0000000000  0000000000    0000000000  0000000000  0000000000  0000000000  0000000000
0159:  0000000000  0000000000  0000000000  0000000000  0000000000    0000000000  0000000000  0000000000  0000000000  0000000000
0160:  0000000000  0000000000  0000000000  0000000000  0000000000    0000000000  0000000000  0000000000  0000000000  0000000000
0161:  0000000000  0000000000  0000000000  0000000000  0000000000    0000000000  0000000000  0000000000  0000000000  0000000000
0162:  0000000000  0000000000  0000000000  0000000000  0000000000    0000000000  0000000000  0000000000  0000000000  0000000000
0163:  0000000000  0000000000  0000000000  0000000000  0000000000    0000000000  0000000000  0000000000  0000000000  0000000000
0164:  0000000000  0000000000  0000000000  0000000000  0000000000    0000000000  0000000000  0000000000  0000000000  0000000000
0165:  0000000000  0000000000  0000000000  0000000000  0000000000    0000000000  0000000000  0000000000  0000000000  0000000000
0166:  0000000000  0000000000  0000000000  0000000000  0000000000    0000000000  0000000000  0000000000  0000000000  0000000000
0167:  0000000000  0000000000  0000000000  0000000000  0000000000    0000000000  0000000000  0000000000  0000000000  0000000000
0168:  0000000000  0000000000  0000000000  0000000000  0000000000    0000000000  0000000000  0000000000  0000000000  0000000000
0169:  0000000000  0000000000  0000000000  0000000000  0000000000    0000000000  0000000000  0000000000  0000000000  0000000000
0170:  0000000000  0000000000  0000000000  0000000000  0000000000    0000000000  0000000000  0000000000  0000000000  0000000000
0171:  0000000000  0000000000  0000000000  0000000000  0000000000    0000000000  0000000000  0000000000  0000000000  0000000000
0172:  0000000000  0000000000  0000000000  0000000000  0000000000    0000000000  0000000000  0000000000  0000000000  0000000000
0173:  0000000000  0000000000  0000000000  0000000000  0000000000    0000000000  0000000000  0000000000  0000000000  0000000000
0174:  0000000000  0000000000  0000000000  0000000000  0000000000    0000000000  0000000000  0000000000  0000000000  0000000000
0175:  0000000000  0000000000  0000000000  0000000000  0000000000    0000000000  0000000000  0000000000  0000000000  0000000000
0176:  0000000000  0000000000  0000000000  0000000000  0000000000    0000000000  0000000000  0000000000  0000000000  0000000000
0177:  0000000000  0000000000  0000000000  0000000000  0000000000    0000000000  0000000000  0000000000  0000000000  0000000000
0178:  0000000000  0000000000  0000000000  0000000000  0000000000    0000000000  0000000000  0000000000  0000000000  0000000000
0179:  0000000000  0000000000  0000000000  0000000000  0000000000    0000000000  0000000000  0000000000  0000000000  0000000000
0180:  0000000000  0000000000  0000000000  0000000000  0000000000    0000000000  0000000000  0000000000  0000000000  0000000000
0181:  0000000000  0000000000  0000000000  0000000000  0000000000    0000000000  0000000000  0000000000  0000000000  0000000000
0182:  0000000000  0000000000  0000000000  0000000000  0000000000    0000000000  0000000000  0000000000  0000000000  0000000000
0183:  0000000000  0000000000  0000000000  0000000000  0000000000    0000000000  0000000000  0000000000  0000000000  0000000000
0184:  0000000000  0000000000  0000000000  0000000000  0000000000    0000000000  0000000000  0000000000  0000000000  0000000000
0185:  0000000000  0000000000  0000000000  0000000000  0000000000    0000000000  0000000000  0000000000  0000000000  0000000000
0186:  0000000000  0000000000  0000000000  0000000000  0000000000    0000000000  0000000000  0000000000  0000000000  0000000000
0187:  0000000000  0000000000  0000000000  0000000000  0000000000    0000000000  0000000000  0000000000  0000000000  0000000000
0188:  0000000000  0000000000  0000000000  0000000000  0000000000    0000000000  0000000000  0000000000  0000000000  0000000000
0189:  0000000000  0000000000  0000000000  0000000000  0000000000    0000000000  0000000000  0000000000  0000000000  0000000000
0190:  0000000000  0000000000  0000000000  0000000000  0000000000    0000000000  0000000000  0000000000  0000000000  0000000000
0191:  0000000000  0000000000  0000000000  0000000000  0000000000    0000000000  0000000000  0000000000  0000000000  0000000000
0192:  0000000000  0000000000  0000000000  0000000000  0000000000    0000000000  0000000000  0000000000  0000000000  0000000000
0193:  0000000000  0000000000  0000000000  0000000000  0000000000    0000000000  0000000000  0000000000  0000000000  0000000000
0194:  0000000000  0000000000  0000000000  0000000000  0000000000    0000000000  0000000000  0000000000  0000000000  0000000000
0195:  0000000000  0000000000  0000000000  0000000000  0000000000    0000000000  0000000000  0000000000  0000000000  0000000000
0196:  0000000000  0000000000  0000000000  0000000000  0000000000    0000000000  0000000000  0000000000  0000000000  0000000000
0197:  0000000000  0000000000  0000000000  0000000000  0000000000    0000000000  0000000000  0000000000  0000000000  0000000000
0198:  0000000000  0000000000  0000000000  0000000000  0000000000    0000000000  0000000000  0000000000  0000000000  0000000000
0199:  0000000000  0000000000  0000000000  0000000000  0000000000    0000000000  0000000000  0000000000  0000000000  0000000000
```

```
0200:  0000000000  0000000000  0000000000  0000000000  0000000000   0000000000  0000000000  0000000000  0000000000  0000000000
0201:  0000000000  0000000000  0000000000  0000000000  0000000000   0000000000  0000000000  0000000000  0000000000  0000000000
0202:  0000000000  0000000000  0000000000  0000000000  0000000000   0000000000  0000000000  0000000000  0000000000  0000000000
0203:  0000000000  0000000000  0000000000  0000000000  0000000000   0000000000  0000000000  0000000000  0000000000  0000000000
0204:  0000000000  0000000000  0000000000  0000000000  0000000000   0000000000  0000000000  0000000000  0000000000  0000000000
0205:  0000000000  0000000000  0000000000  0000000000  0000000000   0000000000  0000000000  0000000000  0000000000  0000000000
0206:  0000000000  0000000000  0000000000  0000000000  0000000000   0000000000  0000000000  0000000000  0000000000  0000000000
0207:  0000000000  0000000000  0000000000  0000000000  0000000000   0000000000  0000000000  0000000000  0000000000  0000000000
0208:  0000000000  0000000000  0000000000  0000000000  0000000000   0000000000  0000000000  0000000000  0000000000  0000000000
0209:  0000000000  0000000000  0000000000  0000000000  0000000000   0000000000  0000000000  0000000000  0000000000  0000000000
0210:  0000000000  0000000000  0000000000  0000000000  0000000000   0000000000  0000000000  0000000000  0000000000  0000000000
0211:  0000000000  0000000000  0000000000  0000000000  0000000000   0000000000  0000000000  0000000000  0000000000  0000000000
0212:  0000000000  0000000000  0000000000  0000000000  0000000000   0000000000  0000000000  0000000000  0000000000  0000000000
0213:  0000000000  0000000000  0000000000  0000000000  0000000000   0000000000  0000000000  0000000000  0000000000  0000000000
0214:  0000000000  0000000000  0000000000  0000000000  0000000000   0000000000  0000000000  0000000000  0000000000  0000000000
0215:  0000000000  0000000000  0000000000  0000000000  0000000000   0000000000  0000000000  0000000000  0000000000  0000000000
0216:  0000000000  0000000000  0000000000  0000000000  0000000000   0000000000  0000000000  0000000000  0000000000  0000000000
0217:  0000000000  0000000000  0000000000  0000000000  0000000000   0000000000  0000000000  0000000000  0000000000  0000000000
0218:  0000000000  0000000000  0000000000  0000000000  0000000000   0000000000  0000000000  0000000000  0000000000  0000000000
0219:  0000000000  0000000000  0000000000  0000000000  0000000000   0000000000  0000000000  0000000000  0000000000  0000000000
0220:  0000000000  0000000000  0000000000  0000000000  0000000000   0000000000  0000000000  0000000000  0000000000  0000000000
0221:  0000000000  0000000000  0000000000  0000000000  0000000000   0000000000  0000000000  0000000000  0000000000  0000000000
0222:  0000000000  0000000000  0000000000  0000000000  0000000000   0000000000  0000000000  0000000000  0000000000  0000000000
0223:  0000000000  0000000000  0000000000  0000000000  0000000000   0000000000  0000000000  0000000000  0000000000  0000000000
0224:  0000000000  0000000000  0000000000  0000000000  0000000000   0000000000  0000000000  0000000000  0000000000  0000000000
0225:  0000000000  0000000000  0000000000  0000000000  0000000000   0000000000  0000000000  0000000000  0000000000  0000000000
0226:  0000000000  0000000000  0000000000  0000000000  0000000000   0000000000  0000000000  0000000000  0000000000  0000000000
0227:  0000000000  0000000000  0000000000  0000000000  0000000000   0000000000  0000000000  0000000000  0000000000  0000000000
0228:  0000000000  0000000000  0000000000  0000000000  0000000000   0000000000  0000000000  0000000000  0000000000  0000000000
0229:  0000000000  0000000000  0000000000  0000000000  0000000000   0000000000  0000000000  0000000000  0000000000  0000000000
0230:  0000000000  0000000000  0000000000  0000000000  0000000000   0000000000  0000000000  0000000000  0000000000  0000000000
0231:  0000000000  0000000000  0000000000  0000000000  0000000000   0000000000  0000000000  0000000000  0000000000  0000000000
0232:  0000000000  0000000000  0000000000  0000000000  0000000000   0000000000  0000000000  0000000000  0000000000  0000000000
0233:  0000000000  0000000000  0000000000  0000000000  0000000000   0000000000  0000000000  0000000000  0000000000  0000000000
0234:  0000000000  0000000000  0000000000  0000000000  0000000000   0000000000  0000000000  0000000000  0000000000  0000000000
0235:  0000000000  0000000000  0000000000  0000000000  0000000000   0000000000  0000000000  0000000000  0000000000  0000000000
0236:  0000000000  0000000000  0000000000  0000000000  0000000000   0000000000  0000000000  0000000000  0000000000  0000000000
0237:  0000000000  0000000000  0000000000  0000000000  0000000000   0000000000  0000000000  0000000000  0000000000  0000000000
0238:  0000000000  0000000000  0000000000  0000000000  0000000000   0000000000  0000000000  0000000000  0000000000  0000000000
0239:  0000000000  0000000000  0000000000  0000000000  0000000000   0000000000  0000000000  0000000000  0000000000  0000000000
0240:  0000000000  0000000000  0000000000  0000000000  0000000000   0000000000  0000000000  0000000000  0000000000  0000000000
0241:  0000000000  0000000000  0000000000  0000000000  0000000000   0000000000  0000000000  0000000000  0000000000  0000000000
0242:  0000000000  0000000000  0000000000  0000000000  0000000000   0000000000  0000000000  0000000000  0000000000  0000000000
0243:  0000000000  0000000000  0000000000  0000000000  0000000000   0000000000  0000000000  0000000000  0000000000  0000000000
0244:  0000000000  0000000000  0000000000  0000000000  0000000000   0000000000  0000000000  0000000000  0000000000  0000000000
0245:  0000000000  0000000000  0000000000  0000000000  0000000000   0000000000  0000000000  0000000000  0000000000  0000000000
0246:  0000000000  0000000000  0000000000  0000000000  0000000000   0000000000  0000000000  0000000000  0000000000  0000000000
0247:  0000000000  0000000000  0000000000  0000000000  0000000000   0000000000  0000000000  0000000000  0000000000  0000000000
0248:  0000000000  0000000000  0000000000  0000000000  0000000000   0000000000  0000000000  0000000000  0000000000  0000000000
0249:  0000000000  0000000000  0000000000  0000000000  0000000000   0000000000  0000000000  0000000000  0000000000  0000000000
```

```
0250:  00000 00000  00000 00000  00000 00000  00000 00000  00000 00000    00000 00000  00000 00000  00000 00000  00000 00000  00000 00000
0251:  00000 00000  00000 00000  00000 00000  00000 00000  00000 00000    00000 00000  00000 00000  00000 00000  00000 00000  00000 00000
0252:  00000 00000  00000 00000  00000 00000  00000 00000  00000 00000    00000 00000  00000 00000  00000 00000  00000 00000  00000 00000
0253:  00000 00000  00000 00000  00000 00000  00000 00000  00000 00000    00000 00000  00000 00000  00000 00000  00000 00000  00000 00000
0254:  00000 00000  00000 00000  00000 00000  00000 00000  00000 00000    00000 00000  00000 00000  00000 00000  00000 00000  00000 00000
0255:  00000 00000  00000 00000  00000 00000  00000 00000  00000 00000    00000 00000  00000 00000  00000 00000  00000 00000  00000 00000
0256:  00000 00000  00000 00000  00000 00000  00000 00000  00000 00000    00000 00000  00000 00000  00000 00000  00000 00000  00000 00000
0257:  00000 00000  00000 00000  00000 00000  00000 00000  00000 00000    00000 00000  00000 00000  00000 00000  00000 00000  00000 00000
0258:  00000 00000  00000 00000  00000 00000  00000 00000  00000 00000    00000 00000  00000 00000  00000 00000  00000 00000  00000 00000
0259:  00000 00000  00000 00000  00000 00000  00000 00000  00000 00000    00000 00000  00000 00000  00000 00000  00000 00000  00000 00000
0260:  00000 00000  00000 00000  00000 00000  00000 00000  00000 00000    00000 00000  00000 00000  00000 00000  00000 00000  00000 00000
0261:  00000 00000  00000 00000  00000 00000  00000 00000  00000 00000    00000 00000  00000 00000  00000 00000  00000 00000  00000 00000
0262:  00000 00000  00000 00000  00000 00000  00000 00000  00000 00000    00000 00000  00000 00000  00000 00000  00000 00000  00000 00000
0263:  00000 00000  00000 00000  00000 00000  00000 00000  00000 00000    00000 00000  00000 00000  00000 00000  00000 00000  00000 00000
0264:  00000 00000  00000 00000  00000 00000  00000 00000  00000 00000    00000 00000  00000 00000  00000 00000  00000 00000  00000 00000
0265:  00000 00000  00000 00000  00000 00000  00000 00000  00000 00000    00000 00000  00000 00000  00000 00000  00000 00000  00000 00000
0266:  00000 00000  00000 00000  00000 00000  00000 00000  00000 00000    00000 00000  00000 00000  00000 00000  00000 00000  00000 00000
0267:  00000 00000  00000 00000  00000 00000  00000 00000  00000 00000    00000 00000  00000 00000  00000 00000  00000 00000  00000 00000
0268:  00000 00000  00000 00000  00000 00000  00000 00000  00000 00000    00000 00000  00000 00000  00000 00000  00000 00000  00000 00000
0269:  00000 00000  00000 00000  00000 00000  00000 00000  00000 00000    00000 00000  00000 00000  00000 00000  00000 00000  00000 00000
0270:  00000 00000  00000 00000  00000 00000  00000 00000  00000 00000    00000 00000  00000 00000  00000 00000  00000 00000  00000 00000
0271:  00000 00000  00000 00000  00000 00000  00000 00000  00000 00000    00000 00000  00000 00000  00000 00000  00000 00000  00000 00000
0272:  00000 00000  00000 00000  00000 00000  00000 00000  00000 00000    00000 00000  00000 00000  00000 00000  00000 00000  00000 00000
0273:  00000 00000  00000 00000  00000 00000  00000 00000  00000 00000    00000 00000  00000 00000  00000 00000  00000 00000  00000 00000
0274:  00000 00000  00000 00000  00000 00000  00000 00000  00000 00000    00000 00000  00000 00000  00000 00000  00000 00000  00000 00000
0275:  00000 00000  00000 00000  00000 00000  00000 00000  00000 00000    00000 00000  00000 00000  00000 00000  00000 00000  00000 00000
0276:  00000 00000  00000 00000  00000 00000  00000 00000  00000 00000    00000 00000  00000 00000  00000 00000  00000 00000  00000 00000
0277:  00000 00000  00000 00000  00000 00000  00000 00000  00000 00000    00000 00000  00000 00000  00000 00000  00000 00000  00000 00000
0278:  00000 00000  00000 00000  00000 00000  00000 00000  00000 00000    00000 00000  00000 00000  00000 00000  00000 00000  00000 00000
0279:  00000 00000  00000 00000  00000 00000  00000 00000  00000 00000    00000 00000  00000 00000  00000 00000  00000 00000  00000 00000
0280:  00000 00000  00000 00000  00000 00000  00000 00000  00000 00000    00000 00000  00000 00000  00000 00000  00000 00000  00000 00000
0281:  00000 00000  00000 00000  00000 00000  00000 00000  00000 00000    00000 00000  00000 00000  00000 00000  00000 00000  00000 00000
0282:  00000 00000  00000 00000  00000 00000  00000 00000  00000 00000    00000 00000  00000 00000  00000 00000  00000 00000  00000 00000
0283:  00000 00000  00000 00000  00000 00000  00000 00000  00000 00000    00000 00000  00000 00000  00000 00000  00000 00000  00000 00000
0284:  00000 00000  00000 00000  00000 00000  00000 00000  00000 00000    00000 00000  00000 00000  00000 00000  00000 00000  00000 00000
0285:  00000 00000  00000 00000  00000 00000  00000 00000  00000 00000    00000 00000  00000 00000  00000 00000  00000 00000  00000 00000
0286:  00000 00000  00000 00000  00000 00000  00000 00000  00000 00000    00000 00000  00000 00000  00000 00000  00000 00000  00000 00000
0287:  00000 00000  00000 00000  00000 00000  00000 00000  00000 00000    00000 00000  00000 00000  00000 00000  00000 00000  00000 00000
0288:  00000 00000  00000 00000  00000 00000  00000 00000  00000 00000    00000 00000  00000 00000  00000 00000  00000 00000  00000 00000
0289:  00000 00000  00000 00000  00000 00000  00000 00000  00000 00000    00000 00000  00000 00000  00000 00000  00000 00000  00000 00000
0290:  00000 00000  00000 00000  00000 00000  00000 00000  00000 00000    00000 00000  00000 00000  00000 00000  00000 00000  00000 00000
0291:  00000 00000  00000 00000  00000 00000  00000 00000  00000 00000    00000 00000  00000 00000  00000 00000  00000 00000  00000 00000
0292:  00000 00000  00000 00000  00000 00000  00000 00000  00000 00000    00000 00000  00000 00000  00000 00000  00000 00000  00000 00000
0293:  00000 00000  00000 00000  00000 00000  00000 00000  00000 00000    00000 00000  00000 00000  00000 00000  00000 00000  00000 00000
0294:  00000 00000  00000 00000  00000 00000  00000 00000  00000 00000    00000 00000  00000 00000  00000 00000  00000 00000  00000 00000
0295:  00000 00000  00000 00000  00000 00000  00000 00000  00000 00000    00000 00000  00000 00000  00000 00000  00000 00000  00000 00000
0296:  00000 00000  00000 00000  00000 00000  00000 00000  00000 00000    00000 00000  00000 00000  00000 00000  00000 00000  00000 00000
0297:  00000 00000  00000 00000  00000 00000  00000 00000  00000 00000    00000 00000  00000 00000  00000 00000  00000 00000  00000 00000
0298:  00000 00000  00000 00000  00000 00000  00000 00000  00000 00000    00000 00000  00000 00000  00000 00000  00000 00000  00000 00000
0299:  00000 00000  00000 00000  00000 00000  00000 00000  00000 00000    00000 00000  00000 00000  00000 00000  00000 00000  00000 00000
```

```
0300:  00000 00000  00000 00000  00000 00000  00000 00000  00000 00000     00000 00000  00000 00000  00000 00000  00000 00000  00000 00000
0301:  00000 00000  00000 00000  00000 00000  00000 00000  00000 00000     00000 00000  00000 00000  00000 00000  00000 00000  00000 00000
0302:  00000 00000  00000 00000  00000 00000  00000 00000  00000 00000     00000 00000  00000 00000  00000 00000  00000 00000  00000 00000
0303:  00000 00000  00000 00000  00000 00000  00000 00000  00000 00000     00000 00000  00000 00000  00000 00000  00000 00000  00000 00000
0304:  00000 00000  00000 00000  00000 00000  00000 00000  00000 00000     00000 00000  00000 00000  00000 00000  00000 00000  00000 00000
0305:  00000 00000  00000 00000  00000 00000  00000 00000  00000 00000     00000 00000  00000 00000  00000 00000  00000 00000  00000 00000
0306:  00000 00000  00000 00000  00000 00000  00000 00000  00000 00000     00000 00000  00000 00000  00000 00000  00000 00000  00000 00000
0307:  00000 00000  00000 00000  00000 00000  00000 00000  00000 00000     00000 00000  00000 00000  00000 00000  00000 00000  00000 00000
0308:  00000 00000  00000 00000  00000 00000  00000 00000  00000 00000     00000 00000  00000 00000  00000 00000  00000 00000  00000 00000
0309:  00000 00000  00000 00000  00000 00000  00000 00000  00000 00000     00000 00000  00000 00000  00000 00000  00000 00000  00000 00000
0310:  00000 00000  00000 00000  00000 00000  00000 00000  00000 00000     00000 00000  00000 00000  00000 00000  00000 00000  00000 00000
0311:  00000 00000  00000 00000  00000 00000  00000 00000  00000 00000     00000 00000  00000 00000  00000 00000  00000 00000  00000 00000
0312:  00000 00000  00000 00000  00000 00000  00000 00000  00000 00000     00000 00000  00000 00000  00000 00000  00000 00000  00000 00000
0313:  00000 00000  00000 00000  00000 00000  00000 00000  00000 00000     00000 00000  00000 00000  00000 00000  00000 00000  00000 00000
0314:  00000 00000  00000 00000  00000 00000  00000 00000  00000 00000     00000 00000  00000 00000  00000 00000  00000 00000  00000 00000
0315:  00000 00000  00000 00000  00000 00000  00000 00000  00000 00000     00000 00000  00000 00000  00000 00000  00000 00000  00000 00000
0316:  00000 00000  00000 00000  00000 00000  00000 00000  00000 00000     00000 00000  00000 00000  00000 00000  00000 00000  00000 00000
0317:  00000 00000  00000 00000  00000 00000  00000 00000  00000 00000     00000 00000  00000 00000  00000 00000  00000 00000  00000 00000
0318:  00000 00000  00000 00000  00000 00000  00000 00000  00000 00000     00000 00000  00000 00000  00000 00000  00000 00000  00000 00000
0319:  00000 00000  00000 00000  00000 00000  00000 00000  00000 00000     00000 00000  00000 00000  00000 00000  00000 00000  00000 00000
0320:  00000 00000  00000 00000  00000 00000  00000 00000  00000 00000     00000 00000  00000 00000  00000 00000  00000 00000  00000 00000
0321:  00000 00000  00000 00000  00000 00000  00000 00000  00000 00000     00000 00000  00000 00000  00000 00000  00000 00000  00000 00000
0322:  00000 00000  00000 00000  00000 00000  00000 00000  00000 00000     00000 00000  00000 00000  00000 00000  00000 00000  00000 00000
0323:  00000 00000  00000 00000  00000 00000  00000 00000  00000 00000     00000 00000  00000 00000  00000 00000  00000 00000  00000 00000
0324:  00000 00000  00000 00000  00000 00000  00000 00000  00000 00000     00000 00000  00000 00000  00000 00000  00000 00000  00000 00000
0325:  00000 00000  00000 00000  00000 00000  00000 00000  00000 00000     00000 00000  00000 00000  00000 00000  00000 00000  00000 00000
0326:  00000 00000  00000 00000  00000 00000  00000 00000  00000 00000     00000 00000  00000 00000  00000 00000  00000 00000  00000 00000
0327:  00000 00000  00000 00000  00000 00000  00000 00000  00000 00000     00000 00000  00000 00000  00000 00000  00000 00000  00000 00000
0328:  00000 00000  00000 00000  00000 00000  00000 00000  00000 00000     00000 00000  00000 00000  00000 00000  00000 00000  00000 00000
0329:  00000 00000  00000 00000  00000 00000  00000 00000  00000 00000     00000 00000  00000 00000  00000 00000  00000 00000  00000 00000
0330:  00000 00000  00000 00000  00000 00000  00000 00000  00000 00000     00000 00000  00000 00000  00000 00000  00000 00000  00000 00000
0331:  00000 00000  00000 00000  00000 00000  00000 00000  00000 00000     00000 00000  00000 00000  00000 00000  00000 00000  00000 00000
0332:  00000 00000  00000 00000  00000 00000  00000 00000  00000 00000     00000 00000  00000 00000  00000 00000  00000 00000  00000 00000
0333:  00000 00000  00000 00000  00000 00000  00000 00000  00000 00000     00000 00000  00000 00000  00000 00000  00000 00000  00000 00000
0334:  00000 00000  00000 00000  00000 00000  00000 00000  00000 00000     00000 00000  00000 00000  00000 00000  00000 00000  00000 00000
0335:  00000 00000  00000 00000  00000 00000  00000 00000  00000 00000     00000 00000  00000 00000  00000 00000  00000 00000  00000 00000
0336:  00000 00000  00000 00000  00000 00000  00000 00000  00000 00000     00000 00000  00000 00000  00000 00000  00000 00000  00000 00000
0337:  00000 00000  00000 00000  00000 00000  00000 00000  00000 00000     00000 00000  00000 00000  00000 00000  00000 00000  00000 00000
0338:  00000 00000  00000 00000  00000 00000  00000 00000  00000 00000     00000 00000  00000 00000  00000 00000  00000 00000  00000 00000
0339:  00000 00000  00000 00000  00000 00000  00000 00000  00000 00000     00000 00000  00000 00000  00000 00000  00000 00000  00000 00000
0340:  00000 00000  00000 00000  00000 00000  00000 00000  00000 00000     00000 00000  00000 00000  00000 00000  00000 00000  00000 00000
0341:  00000 00000  00000 00000  00000 00000  00000 00000  00000 00000     00000 00000  00000 00000  00000 00000  00000 00000  00000 00000
0342:  00000 00000  00000 00000  00000 00000  00000 00000  00000 00000     00000 00000  00000 00000  00000 00000  00000 00000  00000 00000
0343:  00000 00000  00000 00000  00000 00000  00000 00000  00000 00000     00000 00000  00000 00000  00000 00000  00000 00000  00000 00000
0344:  00000 00000  00000 00000  00000 00000  00000 00000  00000 00000     00000 00000  00000 00000  00000 00000  00000 00000  00000 00000
0345:  00000 00000  00000 00000  00000 00000  00000 00000  00000 00000     00000 00000  00000 00000  00000 00000  00000 00000  00000 00000
0346:  00000 00000  00000 00000  00000 00000  00000 00000  00000 00000     00000 00000  00000 00000  00000 00000  00000 00000  00000 00000
0347:  00000 00000  00000 00000  00000 00000  00000 00000  00000 00000     00000 00000  00000 00000  00000 00000  00000 00000  00000 00000
0348:  00000 00000  00000 00000  00000 00000  00000 00000  00000 00000     00000 00000  00000 00000  00000 00000  00000 00000  00000 00000
0349:  00000 00000  00000 00000  00000 00000  00000 00000  00000 00000     00000 00000  00000 00000  00000 00000  00000 00000  00000 00000
```

```
0350:  00000 00000  00000 00000  00000 00000  00000 00000  00000 00000    00000 00000  00000 00000  00000 00000  00000 00000  00000 00000
0351:  00000 00000  00000 00000  00000 00000  00000 00000  00000 00000    00000 00000  00000 00000  00000 00000  00000 00000  00000 00000
0352:  00000 00000  00000 00000  00000 00000  00000 00000  00000 00000    00000 00000  00000 00000  00000 00000  00000 00000  00000 00000
0353:  00000 00000  00000 00000  00000 00000  00000 00000  00000 00000    00000 00000  00000 00000  00000 00000  00000 00000  00000 00000
0354:  00000 00000  00000 00000  00000 00000  00000 00000  00000 00000    00000 00000  00000 00000  00000 00000  00000 00000  00000 00000
0355:  00000 00000  00000 00000  00000 00000  00000 00000  00000 00000    00000 00000  00000 00000  00000 00000  00000 00000  00000 00000
0356:  00000 00000  00000 00000  00000 00000  00000 00000  00000 00000    00000 00000  00000 00000  00000 00000  00000 00000  00000 00000
0357:  00000 00000  00000 00000  00000 00000  00000 00000  00000 00000    00000 00000  00000 00000  00000 00000  00000 00000  00000 00000
0358:  00000 00000  00000 00000  00000 00000  00000 00000  00000 00000    00000 00000  00000 00000  00000 00000  00000 00000  00000 00000
0359:  00000 00000  00000 00000  00000 00000  00000 00000  00000 00000    00000 00000  00000 00000  00000 00000  00000 00000  00000 00000
0360:  00000 00000  00000 00000  00000 00000  00000 00000  00000 00000    00000 00000  00000 00000  00000 00000  00000 00000  00000 00000
0361:  00000 00000  00000 00000  00000 00000  00000 00000  00000 00000    00000 00000  00000 00000  00000 00000  00000 00000  00000 00000
0362:  00000 00000  00000 00000  00000 00000  00000 00000  00000 00000    00000 00000  00000 00000  00000 00000  00000 00000  00000 00000
0363:  00000 00000  00000 00000  00000 00000  00000 00000  00000 00000    00000 00000  00000 00000  00000 00000  00000 00000  00000 00000
0364:  00000 00000  00000 00000  00000 00000  00000 00000  00000 00000    00000 00000  00000 00000  00000 00000  00000 00000  00000 00000
0365:  00000 00000  00000 00000  00000 00000  00000 00000  00000 00000    00000 00000  00000 00000  00000 00000  00000 00000  00000 00000
0366:  00000 00000  00000 00000  00000 00000  00000 00000  00000 00000    00000 00000  00000 00000  00000 00000  00000 00000  00000 00000
0367:  00000 00000  00000 00000  00000 00000  00000 00000  00000 00000    00000 00000  00000 00000  00000 00000  00000 00000  00000 00000
0368:  00000 00000  00000 00000  00000 00000  00000 00000  00000 00000    00000 00000  00000 00000  00000 00000  00000 00000  00000 00000
0369:  00000 00000  00000 00000  00000 00000  00000 00000  00000 00000    00000 00000  00000 00000  00000 00000  00000 00000  00000 00000
0370:  00000 00000  00000 00000  00000 00000  00000 00000  00000 00000    00000 00000  00000 00000  00000 00000  00000 00000  00000 00000
0371:  00000 00000  00000 00000  00000 00000  00000 00000  00000 00000    00000 00000  00000 00000  00000 00000  00000 00000  00000 00000
0372:  00000 00000  00000 00000  00000 00000  00000 00000  00000 00000    00000 00000  00000 00000  00000 00000  00000 00000  00000 00000
0373:  00000 00000  00000 00000  00000 00000  00000 00000  00000 00000    00000 00000  00000 00000  00000 00000  00000 00000  00000 00000
0374:  00000 00000  00000 00000  00000 00000  00000 00000  00000 00000    00000 00000  00000 00000  00000 00000  00000 00000  00000 00000
0375:  00000 00000  00000 00000  00000 00000  00000 00000  00000 00000    00000 00000  00000 00000  00000 00000  00000 00000  00000 00000
0376:  00000 00000  00000 00000  00000 00000  00000 00000  00000 00000    00000 00000  00000 00000  00000 00000  00000 00000  00000 00000
0377:  00000 00000  00000 00000  00000 00000  00000 00000  00000 00000    00000 00000  00000 00000  00000 00000  00000 00000  00000 00000
0378:  00000 00000  00000 00000  00000 00000  00000 00000  00000 00000    00000 00000  00000 00000  00000 00000  00000 00000  00000 00000
0379:  00000 00000  00000 00000  00000 00000  00000 00000  00000 00000    00000 00000  00000 00000  00000 00000  00000 00000  00000 00000
0380:  00000 00000  00000 00000  00000 00000  00000 00000  00000 00000    00000 00000  00000 00000  00000 00000  00000 00000  00000 00000
0381:  00000 00000  00000 00000  00000 00000  00000 00000  00000 00000    00000 00000  00000 00000  00000 00000  00000 00000  00000 00000
0382:  00000 00000  00000 00000  00000 00000  00000 00000  00000 00000    00000 00000  00000 00000  00000 00000  00000 00000  00000 00000
0383:  00000 00000  00000 00000  00000 00000  00000 00000  00000 00000    00000 00000  00000 00000  00000 00000  00000 00000  00000 00000
0384:  00000 00000  00000 00000  00000 00000  00000 00000  00000 00000    00000 00000  00000 00000  00000 00000  00000 00000  00000 00000
0385:  00000 00000  00000 00000  00000 00000  00000 00000  00000 00000    00000 00000  00000 00000  00000 00000  00000 00000  00000 00000
0386:  00000 00000  00000 00000  00000 00000  00000 00000  00000 00000    00000 00000  00000 00000  00000 00000  00000 00000  00000 00000
0387:  00000 00000  00000 00000  00000 00000  00000 00000  00000 00000    00000 00000  00000 00000  00000 00000  00000 00000  00000 00000
0388:  00000 00000  00000 00000  00000 00000  00000 00000  00000 00000    00000 00000  00000 00000  00000 00000  00000 00000  00000 00000
0389:  00000 00000  00000 00000  00000 00000  00000 00000  00000 00000    00000 00000  00000 00000  00000 00000  00000 00000  00000 00000
0390:  00000 00000  00000 00000  00000 00000  00000 00000  00000 00000    00000 00000  00000 00000  00000 00000  00000 00000  00000 00000
0391:  00000 00000  00000 00000  00000 00000  00000 00000  00000 00000    00000 00000  00000 00000  00000 00000  00000 00000  00000 00000
0392:  00000 00000  00000 00000  00000 00000  00000 00000  00000 00000    00000 00000  00000 00000  00000 00000  00000 00000  00000 00000
0393:  00000 00000  00000 00000  00000 00000  00000 00000  00000 00000    00000 00000  00000 00000  00000 00000  00000 00000  00000 00000
0394:  00000 00000  00000 00000  00000 00000  00000 00000  00000 00000    00000 00000  00000 00000  00000 00000  00000 00000  00000 00000
0395:  00000 00000  00000 00000  00000 00000  00000 00000  00000 00000    00000 00000  00000 00000  00000 00000  00000 00000  00000 00000
0396:  00000 00000  00000 00000  00000 00000  00000 00000  00000 00000    00000 00000  00000 00000  00000 00000  00000 00000  00000 00000
0397:  00000 00000  00000 00000  00000 00000  00000 00000  00000 00000    00000 00000  00000 00000  00000 00000  00000 00000  00000 00000
0398:  00000 00000  00000 00000  00000 00000  00000 00000  00000 00000    00000 00000  00000 00000  00000 00000  00000 00000  00000 00000
0399:  00000 00000  00000 00000  00000 00000  00000 00000  00000 00000    00000 00000  00000 00000  00000 00000  00000 00000  00000 00000
```

```
0400:   00000 00000   00000 00000   00000 00000   00000 00000   00000 00000     00000 00000   00000 00000   00000 00000   00000 00000   00000 00000
0401:   00000 00000   00000 00000   00000 00000   00000 00000   00000 00000     00000 00000   00000 00000   00000 00000   00000 00000   00000 00000
0402:   00000 00000   00000 00000   00000 00000   00000 00000   00000 00000     00000 00000   00000 00000   00000 00000   00000 00000   00000 00000
0403:   00000 00000   00000 00000   00000 00000   00000 00000   00000 00000     00000 00000   00000 00000   00000 00000   00000 00000   00000 00000
0404:   00000 00000   00000 00000   00000 00000   00000 00000   00000 00000     00000 00000   00000 00000   00000 00000   00000 00000   00000 00000
0405:   00000 00000   00000 00000   00000 00000   00000 00000   00000 00000     00000 00000   00000 00000   00000 00000   00000 00000   00000 00000
0406:   00000 00000   00000 00000   00000 00000   00000 00000   00000 00000     00000 00000   00000 00000   00000 00000   00000 00000   00000 00000
0407:   00000 00000   00000 00000   00000 00000   00000 00000   00000 00000     00000 00000   00000 00000   00000 00000   00000 00000   00000 00000
0408:   00000 00000   00000 00000   00000 00000   00000 00000   00000 00000     00000 00000   00000 00000   00000 00000   00000 00000   00000 00000
0409:   00000 00000   00000 00000   00000 00000   00000 00000   00000 00000     00000 00000   00000 00000   00000 00000   00000 00000   00000 00000
0410:   00000 00000   00000 00000   00000 00000   00000 00000   00000 00000     00000 00000   00000 00000   00000 00000   00000 00000   00000 00000
0411:   00000 00000   00000 00000   00000 00000   00000 00000   00000 00000     00000 00000   00000 00000   00000 00000   00000 00000   00000 00000
0412:   00000 00000   00000 00000   00000 00000   00000 00000   00000 00000     00000 00000   00000 00000   00000 00000   00000 00000   00000 00000
0413:   00000 00000   00000 00000   00000 00000   00000 00000   00000 00000     00000 00000   00000 00000   00000 00000   00000 00000   00000 00000
0414:   00000 00000   00000 00000   00000 00000   00000 00000   00000 00000     00000 00000   00000 00000   00000 00000   00000 00000   00000 00000
0415:   00000 00000   00000 00000   00000 00000   00000 00000   00000 00000     00000 00000   00000 00000   00000 00000   00000 00000   00000 00000
0416:   00000 00000   00000 00000   00000 00000   00000 00000   00000 00000     00000 00000   00000 00000   00000 00000   00000 00000   00000 00000
0417:   00000 00000   00000 00000   00000 00000   00000 00000   00000 00000     00000 00000   00000 00000   00000 00000   00000 00000   00000 00000
0418:   00000 00000   00000 00000   00000 00000   00000 00000   00000 00000     00000 00000   00000 00000   00000 00000   00000 00000   00000 00000
0419:   00000 00000   00000 00000   00000 00000   00000 00000   00000 00000     00000 00000   00000 00000   00000 00000   00000 00000   00000 00000
0420:   00000 00000   00000 00000   00000 00000   00000 00000   00000 00000     00000 00000   00000 00000   00000 00000   00000 00000   00000 00000
0421:   00000 00000   00000 00000   00000 00000   00000 00000   00000 00000     00000 00000   00000 00000   00000 00000   00000 00000   00000 00000
0422:   00000 00000   00000 00000   00000 00000   00000 00000   00000 00000     00000 00000   00000 00000   00000 00000   00000 00000   00000 00000
0423:   00000 00000   00000 00000   00000 00000   00000 00000   00000 00000     00000 00000   00000 00000   00000 00000   00000 00000   00000 00000
0424:   00000 00000   00000 00000   00000 00000   00000 00000   00000 00000     00000 00000   00000 00000   00000 00000   00000 00000   00000 00000
0425:   00000 00000   00000 00000   00000 00000   00000 00000   00000 00000     00000 00000   00000 00000   00000 00000   00000 00000   00000 00000
0426:   00000 00000   00000 00000   00000 00000   00000 00000   00000 00000     00000 00000   00000 00000   00000 00000   00000 00000   00000 00000
0427:   00000 00000   00000 00000   00000 00000   00000 00000   00000 00000     00000 00000   00000 00000   00000 00000   00000 00000   00000 00000
0428:   00000 00000   00000 00000   00000 00000   00000 00000   00000 00000     00000 00000   00000 00000   00000 00000   00000 00000   00000 00000
0429:   00000 00000   00000 00000   00000 00000   00000 00000   00000 00000     00000 00000   00000 00000   00000 00000   00000 00000   00000 00000
0430:   00000 00000   00000 00000   00000 00000   00000 00000   00000 00000     00000 00000   00000 00000   00000 00000   00000 00000   00000 00000
0431:   00000 00000   00000 00000   00000 00000   00000 00000   00000 00000     00000 00000   00000 00000   00000 00000   00000 00000   00000 00000
0432:   00000 00000   00000 00000   00000 00000   00000 00000   00000 00000     00000 00000   00000 00000   00000 00000   00000 00000   00000 00000
0433:   00000 00000   00000 00000   00000 00000   00000 00000   00000 00000     00000 00000   00000 00000   00000 00000   00000 00000   00000 00000
0434:   00000 00000   00000 00000   00000 00000   00000 00000   00000 00000     00000 00000   00000 00000   00000 00000   00000 00000   00000 00000
0435:   00000 00000   00000 00000   00000 00000   00000 00000   00000 00000     00000 00000   00000 00000   00000 00000   00000 00000   00000 00000
0436:   00000 00000   00000 00000   00000 00000   00000 00000   00000 00000     00000 00000   00000 00000   00000 00000   00000 00000   00000 00000
0437:   00000 00000   00000 00000   00000 00000   00000 00000   00000 00000     00000 00000   00000 00000   00000 00000   00000 00000   00000 00000
0438:   00000 00000   00000 00000   00000 00000   00000 00000   00000 00000     00000 00000   00000 00000   00000 00000   00000 00000   00000 00000
0439:   00000 00000   00000 00000   00000 00000   00000 00000   00000 00000     00000 00000   00000 00000   00000 00000   00000 00000   00000 00000
0440:   00000 00000   00000 00000   00000 00000   00000 00000   00000 00000     00000 00000   00000 00000   00000 00000   00000 00000   00000 00000
0441:   00000 00000   00000 00000   00000 00000   00000 00000   00000 00000     00000 00000   00000 00000   00000 00000   00000 00000   00000 00000
0442:   00000 00000   00000 00000   00000 00000   00000 00000   00000 00000     00000 00000   00000 00000   00000 00000   00000 00000   00000 00000
0443:   00000 00000   00000 00000   00000 00000   00000 00000   00000 00000     00000 00000   00000 00000   00000 00000   00000 00000   00000 00000
0444:   00000 00000   00000 00000   00000 00000   00000 00000   00000 00000     00000 00000   00000 00000   00000 00000   00000 00000   00000 00000
0445:   00000 00000   00000 00000   00000 00000   00000 00000   00000 00000     00000 00000   00000 00000   00000 00000   00000 00000   00000 00000
0446:   00000 00000   00000 00000   00000 00000   00000 00000   00000 00000     00000 00000   00000 00000   00000 00000   00000 00000   00000 00000
0447:   00000 00000   00000 00000   00000 00000   00000 00000   00000 00000     00000 00000   00000 00000   00000 00000   00000 00000   00000 00000
0448:   00000 00000   00000 00000   00000 00000   00000 00000   00000 00000     00000 00000   00000 00000   00000 00000   00000 00000   00000 00000
0449:   00000 00000   00000 00000   00000 00000   00000 00000   00000 00000     00000 00000   00000 00000   00000 00000   00000 00000   00000 00000
```

```
0450:  00000 00000  00000 00000  00000 00000  00000 00000  00000 00000    00000 00000  00000 00000  00000 00000  00000 00000  00000 00000
0451:  00000 00000  00000 00000  00000 00000  00000 00000  00000 00000    00000 00000  00000 00000  00000 00000  00000 00000  00000 00000
0452:  00000 00000  00000 00000  00000 00000  00000 00000  00000 00000    00000 00000  00000 00000  00000 00000  00000 00000  00000 00000
0453:  00000 00000  00000 00000  00000 00000  00000 00000  00000 00000    00000 00000  00000 00000  00000 00000  00000 00000  00000 00000
0454:  00000 00000  00000 00000  00000 00000  00000 00000  00000 00000    00000 00000  00000 00000  00000 00000  00000 00000  00000 00000
0455:  00000 00000  00000 00000  00000 00000  00000 00000  00000 00000    00000 00000  00000 00000  00000 00000  00000 00000  00000 00000
0456:  00000 00000  00000 00000  00000 00000  00000 00000  00000 00000    00000 00000  00000 00000  00000 00000  00000 00000  00000 00000
0457:  00000 00000  00000 00000  00000 00000  00000 00000  00000 00000    00000 00000  00000 00000  00000 00000  00000 00000  00000 00000
0458:  00000 00000  00000 00000  00000 00000  00000 00000  00000 00000    00000 00000  00000 00000  00000 00000  00000 00000  00000 00000
0459:  00000 00000  00000 00000  00000 00000  00000 00000  00000 00000    00000 00000  00000 00000  00000 00000  00000 00000  00000 00000
0460:  00000 00000  00000 00000  00000 00000  00000 00000  00000 00000    00000 00000  00000 00000  00000 00000  00000 00000  00000 00000
0461:  00000 00000  00000 00000  00000 00000  00000 00000  00000 00000    00000 00000  00000 00000  00000 00000  00000 00000  00000 00000
0462:  00000 00000  00000 00000  00000 00000  00000 00000  00000 00000    00000 00000  00000 00000  00000 00000  00000 00000  00000 00000
0463:  00000 00000  00000 00000  00000 00000  00000 00000  00000 00000    00000 00000  00000 00000  00000 00000  00000 00000  00000 00000
0464:  00000 00000  00000 00000  00000 00000  00000 00000  00000 00000    00000 00000  00000 00000  00000 00000  00000 00000  00000 00000
0465:  00000 00000  00000 00000  00000 00000  00000 00000  00000 00000    00000 00000  00000 00000  00000 00000  00000 00000  00000 00000
0466:  00000 00000  00000 00000  00000 00000  00000 00000  00000 00000    00000 00000  00000 00000  00000 00000  00000 00000  00000 00000
0467:  00000 00000  00000 00000  00000 00000  00000 00000  00000 00000    00000 00000  00000 00000  00000 00000  00000 00000  00000 00000
0468:  00000 00000  00000 00000  00000 00000  00000 00000  00000 00000    00000 00000  00000 00000  00000 00000  00000 00000  00000 00000
0469:  00000 00000  00000 00000  00000 00000  00000 00000  00000 00000    00000 00000  00000 00000  00000 00000  00000 00000  00000 00000
0470:  00000 00000  00000 00000  00000 00000  00000 00000  00000 00000    00000 00000  00000 00000  00000 00000  00000 00000  00000 00000
0471:  00000 00000  00000 00000  00000 00000  00000 00000  00000 00000    00000 00000  00000 00000  00000 00000  00000 00000  00000 00000
0472:  00000 00000  00000 00000  00000 00000  00000 00000  00000 00000    00000 00000  00000 00000  00000 00000  00000 00000  00000 00000
0473:  00000 00000  00000 00000  00000 00000  00000 00000  00000 00000    00000 00000  00000 00000  00000 00000  00000 00000  00000 00000
0474:  00000 00000  00000 00000  00000 00000  00000 00000  00000 00000    00000 00000  00000 00000  00000 00000  00000 00000  00000 00000
0475:  00000 00000  00000 00000  00000 00000  00000 00000  00000 00000    00000 00000  00000 00000  00000 00000  00000 00000  00000 00000
0476:  00000 00000  00000 00000  00000 00000  00000 00000  00000 00000    00000 00000  00000 00000  00000 00000  00000 00000  00000 00000
0477:  00000 00000  00000 00000  00000 00000  00000 00000  00000 00000    00000 00000  00000 00000  00000 00000  00000 00000  00000 00000
0478:  00000 00000  00000 00000  00000 00000  00000 00000  00000 00000    00000 00000  00000 00000  00000 00000  00000 00000  00000 00000
0479:  00000 00000  00000 00000  00000 00000  00000 00000  00000 00000    00000 00000  00000 00000  00000 00000  00000 00000  00000 00000
0480:  00000 00000  00000 00000  00000 00000  00000 00000  00000 00000    00000 00000  00000 00000  00000 00000  00000 00000  00000 00000
0481:  00000 00000  00000 00000  00000 00000  00000 00000  00000 00000    00000 00000  00000 00000  00000 00000  00000 00000  00000 00000
0482:  00000 00000  00000 00000  00000 00000  00000 00000  00000 00000    00000 00000  00000 00000  00000 00000  00000 00000  00000 00000
0483:  00000 00000  00000 00000  00000 00000  00000 00000  00000 00000    00000 00000  00000 00000  00000 00000  00000 00000  00000 00000
0484:  00000 00000  00000 00000  00000 00000  00000 00000  00000 00000    00000 00000  00000 00000  00000 00000  00000 00000  00000 00000
0485:  00000 00000  00000 00000  00000 00000  00000 00000  00000 00000    00000 00000  00000 00000  00000 00000  00000 00000  00000 00000
0486:  00000 00000  00000 00000  00000 00000  00000 00000  00000 00000    00000 00000  00000 00000  00000 00000  00000 00000  00000 00000
0487:  00000 00000  00000 00000  00000 00000  00000 00000  00000 00000    00000 00000  00000 00000  00000 00000  00000 00000  00000 00000
0488:  00000 00000  00000 00000  00000 00000  00000 00000  00000 00000    00000 00000  00000 00000  00000 00000  00000 00000  00000 00000
0489:  00000 00000  00000 00000  00000 00000  00000 00000  00000 00000    00000 00000  00000 00000  00000 00000  00000 00000  00000 00000
0490:  00000 00000  00000 00000  00000 00000  00000 00000  00000 00000    00000 00000  00000 00000  00000 00000  00000 00000  00000 00000
0491:  00000 00000  00000 00000  00000 00000  00000 00000  00000 00000    00000 00000  00000 00000  00000 00000  00000 00000  00000 00000
0492:  00000 00000  00000 00000  00000 00000  00000 00000  00000 00000    00000 00000  00000 00000  00000 00000  00000 00000  00000 00000
0493:  00000 00000  00000 00000  00000 00000  00000 00000  00000 00000    00000 00000  00000 00000  00000 00000  00000 00000  00000 00000
0494:  00000 00000  00000 00000  00000 00000  00000 00000  00000 00000    00000 00000  00000 00000  00000 00000  00000 00000  00000 00000
0495:  00000 00000  00000 00000  00000 00000  00000 00000  00000 00000    00000 00000  00000 00000  00000 00000  00000 00000  00000 00000
0496:  00000 00000  00000 00000  00000 00000  00000 00000  00000 00000    00000 00000  00000 00000  00000 00000  00000 00000  00000 00000
0497:  00000 00000  00000 00000  00000 00000  00000 00000  00000 00000    00000 00000  00000 00000  00000 00000  00000 00000  00000 00000
0498:  00000 00000  00000 00000  00000 00000  00000 00000  00000 00000    00000 00000  00000 00000  00000 00000  00000 00000  00000 00000
0499:  00000 00000  00000 00000  00000 00000  00000 00000  00000 00000    00000 00000  00000 00000  00000 00000  00000 00000  00000 00000
```

```
0500:  00000 00000  00000 00000  00000 00000  00000 00000  00000 00000    00000 00000  00000 00000  00000 00000  00000 00000  00000 00000
0501:  00000 00000  00000 00000  00000 00000  00000 00000  00000 00000    00000 00000  00000 00000  00000 00000  00000 00000  00000 00000
0502:  00000 00000  00000 00000  00000 00000  00000 00000  00000 00000    00000 00000  00000 00000  00000 00000  00000 00000  00000 00000
0503:  00000 00000  00000 00000  00000 00000  00000 00000  00000 00000    00000 00000  00000 00000  00000 00000  00000 00000  00000 00000
0504:  00000 00000  00000 00000  00000 00000  00000 00000  00000 00000    00000 00000  00000 00000  00000 00000  00000 00000  00000 00000
0505:  00000 00000  00000 00000  00000 00000  00000 00000  00000 00000    00000 00000  00000 00000  00000 00000  00000 00000  00000 00000
0506:  00000 00000  00000 00000  00000 00000  00000 00000  00000 00000    00000 00000  00000 00000  00000 00000  00000 00000  00000 00000
0507:  00000 00000  00000 00000  00000 00000  00000 00000  00000 00000    00000 00000  00000 00000  00000 00000  00000 00000  00000 00000
0508:  00000 00000  00000 00000  00000 00000  00000 00000  00000 00000    00000 00000  00000 00000  00000 00000  00000 00000  00000 00000
0509:  00000 00000  00000 00000  00000 00000  00000 00000  00000 00000    00000 00000  00000 00000  00000 00000  00000 00000  00000 00000
0510:  00000 00000  00000 00000  00000 00000  00000 00000  00000 00000    00000 00000  00000 00000  00000 00000  00000 00000  00000 00000
0511:  00000 00000  00000 00000  00000 00000  00000 00000  00000 00000    00000 00000  00000 00000  00000 00000  00000 00000  00000 00000
0512:  00000 00000  00000 00000  00000 00000  00000 00000  00000 00000    00000 00000  00000 00000  00000 00000  00000 00000  00000 00000
0513:  00000 00000  00000 00000  00000 00000  00000 00000  00000 00000    00000 00000  00000 00000  00000 00000  00000 00000  00000 00000
0514:  00000 00000  00000 00000  00000 00000  00000 00000  00000 00000    00000 00000  00000 00000  00000 00000  00000 00000  00000 00000
0515:  00000 00000  00000 00000  00000 00000  00000 00000  00000 00000    00000 00000  00000 00000  00000 00000  00000 00000  00000 00000
0516:  00000 00000  00000 00000  00000 00000  00000 00000  00000 00000    00000 00000  00000 00000  00000 00000  00000 00000  00000 00000
0517:  00000 00000  00000 00000  00000 00000  00000 00000  00000 00000    00000 00000  00000 00000  00000 00000  00000 00000  00000 00000
0518:  00000 00000  00000 00000  00000 00000  00000 00000  00000 00000    00000 00000  00000 00000  00000 00000  00000 00000  00000 00000
0519:  00000 00000  00000 00000  00000 00000  00000 00000  00000 00000    00000 00000  00000 00000  00000 00000  00000 00000  00000 00000
0520:  00000 00000  00000 00000  00000 00000  00000 00000  00000 00000    00000 00000  00000 00000  00000 00000  00000 00000  00000 00000
0521:  00000 00000  00000 00000  00000 00000  00000 00000  00000 00000    00000 00000  00000 00000  00000 00000  00000 00000  00000 00000
0522:  00000 00000  00000 00000  00000 00000  00000 00000  00000 00000    00000 00000  00000 00000  00000 00000  00000 00000  00000 00000
0523:  00000 00000  00000 00000  00000 00000  00000 00000  00000 00000    00000 00000  00000 00000  00000 00000  00000 00000  00000 00000
0524:  00000 00000  00000 00000  00000 00000  00000 00000  00000 00000    00000 00000  00000 00000  00000 00000  00000 00000  00000 00000
0525:  00000 00000  00000 00000  00000 00000  00000 00000  00000 00000    00000 00000  00000 00000  00000 00000  00000 00000  00000 00000
0526:  00000 00000  00000 00000  00000 00000  00000 00000  00000 00000    00000 00000  00000 00000  00000 00000  00000 00000  00000 00000
0527:  00000 00000  00000 00000  00000 00000  00000 00000  00000 00000    00000 00000  00000 00000  00000 00000  00000 00000  00000 00000
0528:  00000 00000  00000 00000  00000 00000  00000 00000  00000 00000    00000 00000  00000 00000  00000 00000  00000 00000  00000 00000
0529:  00000 00000  00000 00000  00000 00000  00000 00000  00000 00000    00000 00000  00000 00000  00000 00000  00000 00000  00000 00000
0530:  00000 00000  00000 00000  00000 00000  00000 00000  00000 00000    00000 00000  00000 00000  00000 00000  00000 00000  00000 00000
0531:  00000 00000  00000 00000  00000 00000  00000 00000  00000 00000    00000 00000  00000 00000  00000 00000  00000 00000  00000 00000
0532:  00000 00000  00000 00000  00000 00000  00000 00000  00000 00000    00000 00000  00000 00000  00000 00000  00000 00000  00000 00000
0533:  00000 00000  00000 00000  00000 00000  00000 00000  00000 00000    00000 00000  00000 00000  00000 00000  00000 00000  00000 00000
0534:  00000 00000  00000 00000  00000 00000  00000 00000  00000 00000    00000 00000  00000 00000  00000 00000  00000 00000  00000 00000
0535:  00000 00000  00000 00000  00000 00000  00000 00000  00000 00000    00000 00000  00000 00000  00000 00000  00000 00000  00000 00000
0536:  00000 00000  00000 00000  00000 00000  00000 00000  00000 00000    00000 00000  00000 00000  00000 00000  00000 00000  00000 00000
0537:  00000 00000  00000 00000  00000 00000  00000 00000  00000 00000    00000 00000  00000 00000  00000 00000  00000 00000  00000 00000
0538:  00000 00000  00000 00000  00000 00000  00000 00000  00000 00000    00000 00000  00000 00000  00000 00000  00000 00000  00000 00000
0539:  00000 00000  00000 00000  00000 00000  00000 00000  00000 00000    00000 00000  00000 00000  00000 00000  00000 00000  00000 00000
0540:  00000 00000  00000 00000  00000 00000  00000 00000  00000 00000    00000 00000  00000 00000  00000 00000  00000 00000  00000 00000
0541:  00000 00000  00000 00000  00000 00000  00000 00000  00000 00000    00000 00000  00000 00000  00000 00000  00000 00000  00000 00000
0542:  00000 00000  00000 00000  00000 00000  00000 00000  00000 00000    00000 00000  00000 00000  00000 00000  00000 00000  00000 00000
0543:  00000 00000  00000 00000  00000 00000  00000 00000  00000 00000    00000 00000  00000 00000  00000 00000  00000 00000  00000 00000
0544:  00000 00000  00000 00000  00000 00000  00000 00000  00000 00000    00000 00000  00000 00000  00000 00000  00000 00000  00000 00000
0545:  00000 00000  00000 00000  00000 00000  00000 00000  00000 00000    00000 00000  00000 00000  00000 00000  00000 00000  00000 00000
0546:  00000 00000  00000 00000  00000 00000  00000 00000  00000 00000    00000 00000  00000 00000  00000 00000  00000 00000  00000 00000
0547:  00000 00000  00000 00000  00000 00000  00000 00000  00000 00000    00000 00000  00000 00000  00000 00000  00000 00000  00000 00000
0548:  00000 00000  00000 00000  00000 00000  00000 00000  00000 00000    00000 00000  00000 00000  00000 00000  00000 00000  00000 00000
0549:  00000 00000  00000 00000  00000 00000  00000 00000  00000 00000    00000 00000  00000 00000  00000 00000  00000 00000  00000 00000
```

```
0550:   00000 00000   00000 00000   00000 00000   00000 00000   00000 00000     00000 00000   00000 00000   00000 00000   00000 00000   00000 00000
0551:   00000 00000   00000 00000   00000 00000   00000 00000   00000 00000     00000 00000   00000 00000   00000 00000   00000 00000   00000 00000
0552:   00000 00000   00000 00000   00000 00000   00000 00000   00000 00000     00000 00000   00000 00000   00000 00000   00000 00000   00000 00000
0553:   00000 00000   00000 00000   00000 00000   00000 00000   00000 00000     00000 00000   00000 00000   00000 00000   00000 00000   00000 00000
0554:   00000 00000   00000 00000   00000 00000   00000 00000   00000 00000     00000 00000   00000 00000   00000 00000   00000 00000   00000 00000
0555:   00000 00000   00000 00000   00000 00000   00000 00000   00000 00000     00000 00000   00000 00000   00000 00000   00000 00000   00000 00000
0556:   00000 00000   00000 00000   00000 00000   00000 00000   00000 00000     00000 00000   00000 00000   00000 00000   00000 00000   00000 00000
0557:   00000 00000   00000 00000   00000 00000   00000 00000   00000 00000     00000 00000   00000 00000   00000 00000   00000 00000   00000 00000
0558:   00000 00000   00000 00000   00000 00000   00000 00000   00000 00000     00000 00000   00000 00000   00000 00000   00000 00000   00000 00000
0559:   00000 00000   00000 00000   00000 00000   00000 00000   00000 00000     00000 00000   00000 00000   00000 00000   00000 00000   00000 00000
0560:   00000 00000   00000 00000   00000 00000   00000 00000   00000 00000     00000 00000   00000 00000   00000 00000   00000 00000   00000 00000
0561:   00000 00000   00000 00000   00000 00000   00000 00000   00000 00000     00000 00000   00000 00000   00000 00000   00000 00000   00000 00000
0562:   00000 00000   00000 00000   00000 00000   00000 00000   00000 00000     00000 00000   00000 00000   00000 00000   00000 00000   00000 00000
0563:   00000 00000   00000 00000   00000 00000   00000 00000   00000 00000     00000 00000   00000 00000   00000 00000   00000 00000   00000 00000
0564:   00000 00000   00000 00000   00000 00000   00000 00000   00000 00000     00000 00000   00000 00000   00000 00000   00000 00000   00000 00000
0565:   00000 00000   00000 00000   00000 00000   00000 00000   00000 00000     00000 00000   00000 00000   00000 00000   00000 00000   00000 00000
0566:   00000 00000   00000 00000   00000 00000   00000 00000   00000 00000     00000 00000   00000 00000   00000 00000   00000 00000   00000 00000
0567:   00000 00000   00000 00000   00000 00000   00000 00000   00000 00000     00000 00000   00000 00000   00000 00000   00000 00000   00000 00000
0568:   00000 00000   00000 00000   00000 00000   00000 00000   00000 00000     00000 00000   00000 00000   00000 00000   00000 00000   00000 00000
0569:   00000 00000   00000 00000   00000 00000   00000 00000   00000 00000     00000 00000   00000 00000   00000 00000   00000 00000   00000 00000
0570:   00000 00000   00000 00000   00000 00000   00000 00000   00000 00000     00000 00000   00000 00000   00000 00000   00000 00000   00000 00000
0571:   00000 00000   00000 00000   00000 00000   00000 00000   00000 00000     00000 00000   00000 00000   00000 00000   00000 00000   00000 00000
0572:   00000 00000   00000 00000   00000 00000   00000 00000   00000 00000     00000 00000   00000 00000   00000 00000   00000 00000   00000 00000
0573:   00000 00000   00000 00000   00000 00000   00000 00000   00000 00000     00000 00000   00000 00000   00000 00000   00000 00000   00000 00000
0574:   00000 00000   00000 00000   00000 00000   00000 00000   00000 00000     00000 00000   00000 00000   00000 00000   00000 00000   00000 00000
0575:   00000 00000   00000 00000   00000 00000   00000 00000   00000 00000     00000 00000   00000 00000   00000 00000   00000 00000   00000 00000
0576:   00000 00000   00000 00000   00000 00000   00000 00000   00000 00000     00000 00000   00000 00000   00000 00000   00000 00000   00000 00000
0577:   00000 00000   00000 00000   00000 00000   00000 00000   00000 00000     00000 00000   00000 00000   00000 00000   00000 00000   00000 00000
0578:   00000 00000   00000 00000   00000 00000   00000 00000   00000 00000     00000 00000   00000 00000   00000 00000   00000 00000   00000 00000
0579:   00000 00000   00000 00000   00000 00000   00000 00000   00000 00000     00000 00000   00000 00000   00000 00000   00000 00000   00000 00000
0580:   00000 00000   00000 00000   00000 00000   00000 00000   00000 00000     00000 00000   00000 00000   00000 00000   00000 00000   00000 00000
0581:   00000 00000   00000 00000   00000 00000   00000 00000   00000 00000     00000 00000   00000 00000   00000 00000   00000 00000   00000 00000
0582:   00000 00000   00000 00000   00000 00000   00000 00000   00000 00000     00000 00000   00000 00000   00000 00000   00000 00000   00000 00000
0583:   00000 00000   00000 00000   00000 00000   00000 00000   00000 00000     00000 00000   00000 00000   00000 00000   00000 00000   00000 00000
0584:   00000 00000   00000 00000   00000 00000   00000 00000   00000 00000     00000 00000   00000 00000   00000 00000   00000 00000   00000 00000
0585:   00000 00000   00000 00000   00000 00000   00000 00000   00000 00000     00000 00000   00000 00000   00000 00000   00000 00000   00000 00000
0586:   00000 00000   00000 00000   00000 00000   00000 00000   00000 00000     00000 00000   00000 00000   00000 00000   00000 00000   00000 00000
0587:   00000 00000   00000 00000   00000 00000   00000 00000   00000 00000     00000 00000   00000 00000   00000 00000   00000 00000   00000 00000
0588:   00000 00000   00000 00000   00000 00000   00000 00000   00000 00000     00000 00000   00000 00000   00000 00000   00000 00000   00000 00000
0589:   00000 00000   00000 00000   00000 00000   00000 00000   00000 00000     00000 00000   00000 00000   00000 00000   00000 00000   00000 00000
0590:   00000 00000   00000 00000   00000 00000   00000 00000   00000 00000     00000 00000   00000 00000   00000 00000   00000 00000   00000 00000
0591:   00000 00000   00000 00000   00000 00000   00000 00000   00000 00000     00000 00000   00000 00000   00000 00000   00000 00000   00000 00000
0592:   00000 00000   00000 00000   00000 00000   00000 00000   00000 00000     00000 00000   00000 00000   00000 00000   00000 00000   00000 00000
0593:   00000 00000   00000 00000   00000 00000   00000 00000   00000 00000     00000 00000   00000 00000   00000 00000   00000 00000   00000 00000
0594:   00000 00000   00000 00000   00000 00000   00000 00000   00000 00000     00000 00000   00000 00000   00000 00000   00000 00000   00000 00000
0595:   00000 00000   00000 00000   00000 00000   00000 00000   00000 00000     00000 00000   00000 00000   00000 00000   00000 00000   00000 00000
0596:   00000 00000   00000 00000   00000 00000   00000 00000   00000 00000     00000 00000   00000 00000   00000 00000   00000 00000   00000 00000
0597:   00000 00000   00000 00000   00000 00000   00000 00000   00000 00000     00000 00000   00000 00000   00000 00000   00000 00000   00000 00000
0598:   00000 00000   00000 00000   00000 00000   00000 00000   00000 00000     00000 00000   00000 00000   00000 00000   00000 00000   00000 00000
0599:   00000 00000   00000 00000   00000 00000   00000 00000   00000 00000     00000 00000   00000 00000   00000 00000   00000 00000   00000 00000
```

```
0600:   00000 00000   00000 00000   00000 00000   00000 00000   00000 00000     00000 00000   00000 00000   00000 00000   00000 00000   00000 00000
0601:   00000 00000   00000 00000   00000 00000   00000 00000   00000 00000     00000 00000   00000 00000   00000 00000   00000 00000   00000 00000
0602:   00000 00000   00000 00000   00000 00000   00000 00000   00000 00000     00000 00000   00000 00000   00000 00000   00000 00000   00000 00000
0603:   00000 00000   00000 00000   00000 00000   00000 00000   00000 00000     00000 00000   00000 00000   00000 00000   00000 00000   00000 00000
0604:   00000 00000   00000 00000   00000 00000   00000 00000   00000 00000     00000 00000   00000 00000   00000 00000   00000 00000   00000 00000
0605:   00000 00000   00000 00000   00000 00000   00000 00000   00000 00000     00000 00000   00000 00000   00000 00000   00000 00000   00000 00000
0606:   00000 00000   00000 00000   00000 00000   00000 00000   00000 00000     00000 00000   00000 00000   00000 00000   00000 00000   00000 00000
0607:   00000 00000   00000 00000   00000 00000   00000 00000   00000 00000     00000 00000   00000 00000   00000 00000   00000 00000   00000 00000
0608:   00000 00000   00000 00000   00000 00000   00000 00000   00000 00000     00000 00000   00000 00000   00000 00000   00000 00000   00000 00000
0609:   00000 00000   00000 00000   00000 00000   00000 00000   00000 00000     00000 00000   00000 00000   00000 00000   00000 00000   00000 00000
0610:   00000 00000   00000 00000   00000 00000   00000 00000   00000 00000     00000 00000   00000 00000   00000 00000   00000 00000   00000 00000
0611:   00000 00000   00000 00000   00000 00000   00000 00000   00000 00000     00000 00000   00000 00000   00000 00000   00000 00000   00000 00000
0612:   00000 00000   00000 00000   00000 00000   00000 00000   00000 00000     00000 00000   00000 00000   00000 00000   00000 00000   00000 00000
0613:   00000 00000   00000 00000   00000 00000   00000 00000   00000 00000     00000 00000   00000 00000   00000 00000   00000 00000   00000 00000
0614:   00000 00000   00000 00000   00000 00000   00000 00000   00000 00000     00000 00000   00000 00000   00000 00000   00000 00000   00000 00000
0615:   00000 00000   00000 00000   00000 00000   00000 00000   00000 00000     00000 00000   00000 00000   00000 00000   00000 00000   00000 00000
0616:   00000 00000   00000 00000   00000 00000   00000 00000   00000 00000     00000 00000   00000 00000   00000 00000   00000 00000   00000 00000
0617:   00000 00000   00000 00000   00000 00000   00000 00000   00000 00000     00000 00000   00000 00000   00000 00000   00000 00000   00000 00000
0618:   00000 00000   00000 00000   00000 00000   00000 00000   00000 00000     00000 00000   00000 00000   00000 00000   00000 00000   00000 00000
0619:   00000 00000   00000 00000   00000 00000   00000 00000   00000 00000     00000 00000   00000 00000   00000 00000   00000 00000   00000 00000
0620:   00000 00000   00000 00000   00000 00000   00000 00000   00000 00000     00000 00000   00000 00000   00000 00000   00000 00000   00000 00000
0621:   00000 00000   00000 00000   00000 00000   00000 00000   00000 00000     00000 00000   00000 00000   00000 00000   00000 00000   00000 00000
0622:   00000 00000   00000 00000   00000 00000   00000 00000   00000 00000     00000 00000   00000 00000   00000 00000   00000 00000   00000 00000
0623:   00000 00000   00000 00000   00000 00000   00000 00000   00000 00000     00000 00000   00000 00000   00000 00000   00000 00000   00000 00000
0624:   00000 00000   00000 00000   00000 00000   00000 00000   00000 00000     00000 00000   00000 00000   00000 00000   00000 00000   00000 00000
0625:   00000 00000   00000 00000   00000 00000   00000 00000   00000 00000     00000 00000   00000 00000   00000 00000   00000 00000   00000 00000
0626:   00000 00000   00000 00000   00000 00000   00000 00000   00000 00000     00000 00000   00000 00000   00000 00000   00000 00000   00000 00000
0627:   00000 00000   00000 00000   00000 00000   00000 00000   00000 00000     00000 00000   00000 00000   00000 00000   00000 00000   00000 00000
0628:   00000 00000   00000 00000   00000 00000   00000 00000   00000 00000     00000 00000   00000 00000   00000 00000   00000 00000   00000 00000
0629:   00000 00000   00000 00000   00000 00000   00000 00000   00000 00000     00000 00000   00000 00000   00000 00000   00000 00000   00000 00000
0630:   00000 00000   00000 00000   00000 00000   00000 00000   00000 00000     00000 00000   00000 00000   00000 00000   00000 00000   00000 00000
0631:   00000 00000   00000 00000   00000 00000   00000 00000   00000 00000     00000 00000   00000 00000   00000 00000   00000 00000   00000 00000
0632:   00000 00000   00000 00000   00000 00000   00000 00000   00000 00000     00000 00000   00000 00000   00000 00000   00000 00000   00000 00000
0633:   00000 00000   00000 00000   00000 00000   00000 00000   00000 00000     00000 00000   00000 00000   00000 00000   00000 00000   00000 00000
0634:   00000 00000   00000 00000   00000 00000   00000 00000   00000 00000     00000 00000   00000 00000   00000 00000   00000 00000   00000 00000
0635:   00000 00000   00000 00000   00000 00000   00000 00000   00000 00000     00000 00000   00000 00000   00000 00000   00000 00000   00000 00000
0636:   00000 00000   00000 00000   00000 00000   00000 00000   00000 00000     00000 00000   00000 00000   00000 00000   00000 00000   00000 00000
0637:   00000 00000   00000 00000   00000 00000   00000 00000   00000 00000     00000 00000   00000 00000   00000 00000   00000 00000   00000 00000
0638:   00000 00000   00000 00000   00000 00000   00000 00000   00000 00000     00000 00000   00000 00000   00000 00000   00000 00000   00000 00000
0639:   00000 00000   00000 00000   00000 00000   00000 00000   00000 00000     00000 00000   00000 00000   00000 00000   00000 00000   00000 00000
0640:   00000 00000   00000 00000   00000 00000   00000 00000   00000 00000     00000 00000   00000 00000   00000 00000   00000 00000   00000 00000
0641:   00000 00000   00000 00000   00000 00000   00000 00000   00000 00000     00000 00000   00000 00000   00000 00000   00000 00000   00000 00000
0642:   00000 00000   00000 00000   00000 00000   00000 00000   00000 00000     00000 00000   00000 00000   00000 00000   00000 00000   00000 00000
0643:   00000 00000   00000 00000   00000 00000   00000 00000   00000 00000     00000 00000   00000 00000   00000 00000   00000 00000   00000 00000
0644:   00000 00000   00000 00000   00000 00000   00000 00000   00000 00000     00000 00000   00000 00000   00000 00000   00000 00000   00000 00000
0645:   00000 00000   00000 00000   00000 00000   00000 00000   00000 00000     00000 00000   00000 00000   00000 00000   00000 00000   00000 00000
0646:   00000 00000   00000 00000   00000 00000   00000 00000   00000 00000     00000 00000   00000 00000   00000 00000   00000 00000   00000 00000
0647:   00000 00000   00000 00000   00000 00000   00000 00000   00000 00000     00000 00000   00000 00000   00000 00000   00000 00000   00000 00000
0648:   00000 00000   00000 00000   00000 00000   00000 00000   00000 00000     00000 00000   00000 00000   00000 00000   00000 00000   00000 00000
0649:   00000 00000   00000 00000   00000 00000   00000 00000   00000 00000     00000 00000   00000 00000   00000 00000   00000 00000   00000 00000
```

```
0650:   00000 00000   00000 00000   00000 00000   00000 00000   00000 00000      00000 00000   00000 00000   00000 00000   00000 00000   00000 00000
0651:   00000 00000   00000 00000   00000 00000   00000 00000   00000 00000      00000 00000   00000 00000   00000 00000   00000 00000   00000 00000
0652:   00000 00000   00000 00000   00000 00000   00000 00000   00000 00000      00000 00000   00000 00000   00000 00000   00000 00000   00000 00000
0653:   00000 00000   00000 00000   00000 00000   00000 00000   00000 00000      00000 00000   00000 00000   00000 00000   00000 00000   00000 00000
0654:   00000 00000   00000 00000   00000 00000   00000 00000   00000 00000      00000 00000   00000 00000   00000 00000   00000 00000   00000 00000
0655:   00000 00000   00000 00000   00000 00000   00000 00000   00000 00000      00000 00000   00000 00000   00000 00000   00000 00000   00000 00000
0656:   00000 00000   00000 00000   00000 00000   00000 00000   00000 00000      00000 00000   00000 00000   00000 00000   00000 00000   00000 00000
0657:   00000 00000   00000 00000   00000 00000   00000 00000   00000 00000      00000 00000   00000 00000   00000 00000   00000 00000   00000 00000
0658:   00000 00000   00000 00000   00000 00000   00000 00000   00000 00000      00000 00000   00000 00000   00000 00000   00000 00000   00000 00000
0659:   00000 00000   00000 00000   00000 00000   00000 00000   00000 00000      00000 00000   00000 00000   00000 00000   00000 00000   00000 00000
0660:   00000 00000   00000 00000   00000 00000   00000 00000   00000 00000      00000 00000   00000 00000   00000 00000   00000 00000   00000 00000
0661:   00000 00000   00000 00000   00000 00000   00000 00000   00000 00000      00000 00000   00000 00000   00000 00000   00000 00000   00000 00000
0662:   00000 00000   00000 00000   00000 00000   00000 00000   00000 00000      00000 00000   00000 00000   00000 00000   00000 00000   00000 00000
0663:   00000 00000   00000 00000   00000 00000   00000 00000   00000 00000      00000 00000   00000 00000   00000 00000   00000 00000   00000 00000
0664:   00000 00000   00000 00000   00000 00000   00000 00000   00000 00000      00000 00000   00000 00000   00000 00000   00000 00000   00000 00000
0665:   00000 00000   00000 00000   00000 00000   00000 00000   00000 00000      00000 00000   00000 00000   00000 00000   00000 00000   00000 00000
0666:   00000 00000   00000 00000   00000 00000   00000 00000   00000 00000      00000 00000   00000 00000   00000 00000   00000 00000   00000 00000
0667:   00000 00000   00000 00000   00000 00000   00000 00000   00000 00000      00000 00000   00000 00000   00000 00000   00000 00000   00000 00000
0668:   00000 00000   00000 00000   00000 00000   00000 00000   00000 00000      00000 00000   00000 00000   00000 00000   00000 00000   00000 00000
0669:   00000 00000   00000 00000   00000 00000   00000 00000   00000 00000      00000 00000   00000 00000   00000 00000   00000 00000   00000 00000
0670:   00000 00000   00000 00000   00000 00000   00000 00000   00000 00000      00000 00000   00000 00000   00000 00000   00000 00000   00000 00000
0671:   00000 00000   00000 00000   00000 00000   00000 00000   00000 00000      00000 00000   00000 00000   00000 00000   00000 00000   00000 00000
0672:   00000 00000   00000 00000   00000 00000   00000 00000   00000 00000      00000 00000   00000 00000   00000 00000   00000 00000   00000 00000
0673:   00000 00000   00000 00000   00000 00000   00000 00000   00000 00000      00000 00000   00000 00000   00000 00000   00000 00000   00000 00000
0674:   00000 00000   00000 00000   00000 00000   00000 00000   00000 00000      00000 00000   00000 00000   00000 00000   00000 00000   00000 00000
0675:   00000 00000   00000 00000   00000 00000   00000 00000   00000 00000      00000 00000   00000 00000   00000 00000   00000 00000   00000 00000
0676:   00000 00000   00000 00000   00000 00000   00000 00000   00000 00000      00000 00000   00000 00000   00000 00000   00000 00000   00000 00000
0677:   00000 00000   00000 00000   00000 00000   00000 00000   00000 00000      00000 00000   00000 00000   00000 00000   00000 00000   00000 00000
0678:   00000 00000   00000 00000   00000 00000   00000 00000   00000 00000      00000 00000   00000 00000   00000 00000   00000 00000   00000 00000
0679:   00000 00000   00000 00000   00000 00000   00000 00000   00000 00000      00000 00000   00000 00000   00000 00000   00000 00000   00000 00000
0680:   00000 00000   00000 00000   00000 00000   00000 00000   00000 00000      00000 00000   00000 00000   00000 00000   00000 00000   00000 00000
0681:   00000 00000   00000 00000   00000 00000   00000 00000   00000 00000      00000 00000   00000 00000   00000 00000   00000 00000   00000 00000
0682:   00000 00000   00000 00000   00000 00000   00000 00000   00000 00000      00000 00000   00000 00000   00000 00000   00000 00000   00000 00000
0683:   00000 00000   00000 00000   00000 00000   00000 00000   00000 00000      00000 00000   00000 00000   00000 00000   00000 00000   00000 00000
0684:   00000 00000   00000 00000   00000 00000   00000 00000   00000 00000      00000 00000   00000 00000   00000 00000   00000 00000   00000 00000
0685:   00000 00000   00000 00000   00000 00000   00000 00000   00000 00000      00000 00000   00000 00000   00000 00000   00000 00000   00000 00000
0686:   00000 00000   00000 00000   00000 00000   00000 00000   00000 00000      00000 00000   00000 00000   00000 00000   00000 00000   00000 00000
0687:   00000 00000   00000 00000   00000 00000   00000 00000   00000 00000      00000 00000   00000 00000   00000 00000   00000 00000   00000 00000
0688:   00000 00000   00000 00000   00000 00000   00000 00000   00000 00000      00000 00000   00000 00000   00000 00000   00000 00000   00000 00000
0689:   00000 00000   00000 00000   00000 00000   00000 00000   00000 00000      00000 00000   00000 00000   00000 00000   00000 00000   00000 00000
0690:   00000 00000   00000 00000   00000 00000   00000 00000   00000 00000      00000 00000   00000 00000   00000 00000   00000 00000   00000 00000
0691:   00000 00000   00000 00000   00000 00000   00000 00000   00000 00000      00000 00000   00000 00000   00000 00000   00000 00000   00000 00000
0692:   00000 00000   00000 00000   00000 00000   00000 00000   00000 00000      00000 00000   00000 00000   00000 00000   00000 00000   00000 00000
0693:   00000 00000   00000 00000   00000 00000   00000 00000   00000 00000      00000 00000   00000 00000   00000 00000   00000 00000   00000 00000
0694:   00000 00000   00000 00000   00000 00000   00000 00000   00000 00000      00000 00000   00000 00000   00000 00000   00000 00000   00000 00000
0695:   00000 00000   00000 00000   00000 00000   00000 00000   00000 00000      00000 00000   00000 00000   00000 00000   00000 00000   00000 00000
0696:   00000 00000   00000 00000   00000 00000   00000 00000   00000 00000      00000 00000   00000 00000   00000 00000   00000 00000   00000 00000
0697:   00000 00000   00000 00000   00000 00000   00000 00000   00000 00000      00000 00000   00000 00000   00000 00000   00000 00000   00000 00000
0698:   00000 00000   00000 00000   00000 00000   00000 00000   00000 00000      00000 00000   00000 00000   00000 00000   00000 00000   00000 00000
0699:   00000 00000   00000 00000   00000 00000   00000 00000   00000 00000      00000 00000   00000 00000   00000 00000   00000 00000   00000 00000
```

```
0700:  00000 00000  00000 00000  00000 00000  00000 00000  00000 00000    00000 00000  00000 00000  00000 00000  00000 00000  00000 00000
0701:  00000 00000  00000 00000  00000 00000  00000 00000  00000 00000    00000 00000  00000 00000  00000 00000  00000 00000  00000 00000
0702:  00000 00000  00000 00000  00000 00000  00000 00000  00000 00000    00000 00000  00000 00000  00000 00000  00000 00000  00000 00000
0703:  00000 00000  00000 00000  00000 00000  00000 00000  00000 00000    00000 00000  00000 00000  00000 00000  00000 00000  00000 00000
0704:  00000 00000  00000 00000  00000 00000  00000 00000  00000 00000    00000 00000  00000 00000  00000 00000  00000 00000  00000 00000
0705:  00000 00000  00000 00000  00000 00000  00000 00000  00000 00000    00000 00000  00000 00000  00000 00000  00000 00000  00000 00000
0706:  00000 00000  00000 00000  00000 00000  00000 00000  00000 00000    00000 00000  00000 00000  00000 00000  00000 00000  00000 00000
0707:  00000 00000  00000 00000  00000 00000  00000 00000  00000 00000    00000 00000  00000 00000  00000 00000  00000 00000  00000 00000
0708:  00000 00000  00000 00000  00000 00000  00000 00000  00000 00000    00000 00000  00000 00000  00000 00000  00000 00000  00000 00000
0709:  00000 00000  00000 00000  00000 00000  00000 00000  00000 00000    00000 00000  00000 00000  00000 00000  00000 00000  00000 00000
0710:  00000 00000  00000 00000  00000 00000  00000 00000  00000 00000    00000 00000  00000 00000  00000 00000  00000 00000  00000 00000
0711:  00000 00000  00000 00000  00000 00000  00000 00000  00000 00000    00000 00000  00000 00000  00000 00000  00000 00000  00000 00000
0712:  00000 00000  00000 00000  00000 00000  00000 00000  00000 00000    00000 00000  00000 00000  00000 00000  00000 00000  00000 00000
0713:  00000 00000  00000 00000  00000 00000  00000 00000  00000 00000    00000 00000  00000 00000  00000 00000  00000 00000  00000 00000
0714:  00000 00000  00000 00000  00000 00000  00000 00000  00000 00000    00000 00000  00000 00000  00000 00000  00000 00000  00000 00000
0715:  00000 00000  00000 00000  00000 00000  00000 00000  00000 00000    00000 00000  00000 00000  00000 00000  00000 00000  00000 00000
0716:  00000 00000  00000 00000  00000 00000  00000 00000  00000 00000    00000 00000  00000 00000  00000 00000  00000 00000  00000 00000
0717:  00000 00000  00000 00000  00000 00000  00000 00000  00000 00000    00000 00000  00000 00000  00000 00000  00000 00000  00000 00000
0718:  00000 00000  00000 00000  00000 00000  00000 00000  00000 00000    00000 00000  00000 00000  00000 00000  00000 00000  00000 00000
0719:  00000 00000  00000 00000  00000 00000  00000 00000  00000 00000    00000 00000  00000 00000  00000 00000  00000 00000  00000 00000
0720:  00000 00000  00000 00000  00000 00000  00000 00000  00000 00000    00000 00000  00000 00000  00000 00000  00000 00000  00000 00000
0721:  00000 00000  00000 00000  00000 00000  00000 00000  00000 00000    00000 00000  00000 00000  00000 00000  00000 00000  00000 00000
0722:  00000 00000  00000 00000  00000 00000  00000 00000  00000 00000    00000 00000  00000 00000  00000 00000  00000 00000  00000 00000
0723:  00000 00000  00000 00000  00000 00000  00000 00000  00000 00000    00000 00000  00000 00000  00000 00000  00000 00000  00000 00000
0724:  00000 00000  00000 00000  00000 00000  00000 00000  00000 00000    00000 00000  00000 00000  00000 00000  00000 00000  00000 00000
0725:  00000 00000  00000 00000  00000 00000  00000 00000  00000 00000    00000 00000  00000 00000  00000 00000  00000 00000  00000 00000
0726:  00000 00000  00000 00000  00000 00000  00000 00000  00000 00000    00000 00000  00000 00000  00000 00000  00000 00000  00000 00000
0727:  00000 00000  00000 00000  00000 00000  00000 00000  00000 00000    00000 00000  00000 00000  00000 00000  00000 00000  00000 00000
0728:  00000 00000  00000 00000  00000 00000  00000 00000  00000 00000    00000 00000  00000 00000  00000 00000  00000 00000  00000 00000
0729:  00000 00000  00000 00000  00000 00000  00000 00000  00000 00000    00000 00000  00000 00000  00000 00000  00000 00000  00000 00000
0730:  00000 00000  00000 00000  00000 00000  00000 00000  00000 00000    00000 00000  00000 00000  00000 00000  00000 00000  00000 00000
0731:  00000 00000  00000 00000  00000 00000  00000 00000  00000 00000    00000 00000  00000 00000  00000 00000  00000 00000  00000 00000
0732:  00000 00000  00000 00000  00000 00000  00000 00000  00000 00000    00000 00000  00000 00000  00000 00000  00000 00000  00000 00000
0733:  00000 00000  00000 00000  00000 00000  00000 00000  00000 00000    00000 00000  00000 00000  00000 00000  00000 00000  00000 00000
0734:  00000 00000  00000 00000  00000 00000  00000 00000  00000 00000    00000 00000  00000 00000  00000 00000  00000 00000  00000 00000
0735:  00000 00000  00000 00000  00000 00000  00000 00000  00000 00000    00000 00000  00000 00000  00000 00000  00000 00000  00000 00000
0736:  00000 00000  00000 00000  00000 00000  00000 00000  00000 00000    00000 00000  00000 00000  00000 00000  00000 00000  00000 00000
0737:  00000 00000  00000 00000  00000 00000  00000 00000  00000 00000    00000 00000  00000 00000  00000 00000  00000 00000  00000 00000
0738:  00000 00000  00000 00000  00000 00000  00000 00000  00000 00000    00000 00000  00000 00000  00000 00000  00000 00000  00000 00000
0739:  00000 00000  00000 00000  00000 00000  00000 00000  00000 00000    00000 00000  00000 00000  00000 00000  00000 00000  00000 00000
0740:  00000 00000  00000 00000  00000 00000  00000 00000  00000 00000    00000 00000  00000 00000  00000 00000  00000 00000  00000 00000
0741:  00000 00000  00000 00000  00000 00000  00000 00000  00000 00000    00000 00000  00000 00000  00000 00000  00000 00000  00000 00000
0742:  00000 00000  00000 00000  00000 00000  00000 00000  00000 00000    00000 00000  00000 00000  00000 00000  00000 00000  00000 00000
0743:  00000 00000  00000 00000  00000 00000  00000 00000  00000 00000    00000 00000  00000 00000  00000 00000  00000 00000  00000 00000
0744:  00000 00000  00000 00000  00000 00000  00000 00000  00000 00000    00000 00000  00000 00000  00000 00000  00000 00000  00000 00000
0745:  00000 00000  00000 00000  00000 00000  00000 00000  00000 00000    00000 00000  00000 00000  00000 00000  00000 00000  00000 00000
0746:  00000 00000  00000 00000  00000 00000  00000 00000  00000 00000    00000 00000  00000 00000  00000 00000  00000 00000  00000 00000
0747:  00000 00000  00000 00000  00000 00000  00000 00000  00000 00000    00000 00000  00000 00000  00000 00000  00000 00000  00000 00000
0748:  00000 00000  00000 00000  00000 00000  00000 00000  00000 00000    00000 00000  00000 00000  00000 00000  00000 00000  00000 00000
0749:  00000 00000  00000 00000  00000 00000  00000 00000  00000 00000    00000 00000  00000 00000  00000 00000  00000 00000  00000 00000
```

```
0750:  00000 00000  00000 00000  00000 00000  00000 00000  00000 00000    00000 00000  00000 00000  00000 00000  00000 00000  00000 00000
0751:  00000 00000  00000 00000  00000 00000  00000 00000  00000 00000    00000 00000  00000 00000  00000 00000  00000 00000  00000 00000
0752:  00000 00000  00000 00000  00000 00000  00000 00000  00000 00000    00000 00000  00000 00000  00000 00000  00000 00000  00000 00000
0753:  00000 00000  00000 00000  00000 00000  00000 00000  00000 00000    00000 00000  00000 00000  00000 00000  00000 00000  00000 00000
0754:  00000 00000  00000 00000  00000 00000  00000 00000  00000 00000    00000 00000  00000 00000  00000 00000  00000 00000  00000 00000
0755:  00000 00000  00000 00000  00000 00000  00000 00000  00000 00000    00000 00000  00000 00000  00000 00000  00000 00000  00000 00000
0756:  00000 00000  00000 00000  00000 00000  00000 00000  00000 00000    00000 00000  00000 00000  00000 00000  00000 00000  00000 00000
0757:  00000 00000  00000 00000  00000 00000  00000 00000  00000 00000    00000 00000  00000 00000  00000 00000  00000 00000  00000 00000
0758:  00000 00000  00000 00000  00000 00000  00000 00000  00000 00000    00000 00000  00000 00000  00000 00000  00000 00000  00000 00000
0759:  00000 00000  00000 00000  00000 00000  00000 00000  00000 00000    00000 00000  00000 00000  00000 00000  00000 00000  00000 00000
0760:  00000 00000  00000 00000  00000 00000  00000 00000  00000 00000    00000 00000  00000 00000  00000 00000  00000 00000  00000 00000
0761:  00000 00000  00000 00000  00000 00000  00000 00000  00000 00000    00000 00000  00000 00000  00000 00000  00000 00000  00000 00000
0762:  00000 00000  00000 00000  00000 00000  00000 00000  00000 00000    00000 00000  00000 00000  00000 00000  00000 00000  00000 00000
0763:  00000 00000  00000 00000  00000 00000  00000 00000  00000 00000    00000 00000  00000 00000  00000 00000  00000 00000  00000 00000
0764:  00000 00000  00000 00000  00000 00000  00000 00000  00000 00000    00000 00000  00000 00000  00000 00000  00000 00000  00000 00000
0765:  00000 00000  00000 00000  00000 00000  00000 00000  00000 00000    00000 00000  00000 00000  00000 00000  00000 00000  00000 00000
0766:  00000 00000  00000 00000  00000 00000  00000 00000  00000 00000    00000 00000  00000 00000  00000 00000  00000 00000  00000 00000
0767:  00000 00000  00000 00000  00000 00000  00000 00000  00000 00000    00000 00000  00000 00000  00000 00000  00000 00000  00000 00000
0768:  00000 00000  00000 00000  00000 00000  00000 00000  00000 00000    00000 00000  00000 00000  00000 00000  00000 00000  00000 00000
0769:  00000 00000  00000 00000  00000 00000  00000 00000  00000 00000    00000 00000  00000 00000  00000 00000  00000 00000  00000 00000
0770:  00000 00000  00000 00000  00000 00000  00000 00000  00000 00000    00000 00000  00000 00000  00000 00000  00000 00000  00000 00000
0771:  00000 00000  00000 00000  00000 00000  00000 00000  00000 00000    00000 00000  00000 00000  00000 00000  00000 00000  00000 00000
0772:  00000 00000  00000 00000  00000 00000  00000 00000  00000 00000    00000 00000  00000 00000  00000 00000  00000 00000  00000 00000
0773:  00000 00000  00000 00000  00000 00000  00000 00000  00000 00000    00000 00000  00000 00000  00000 00000  00000 00000  00000 00000
0774:  00000 00000  00000 00000  00000 00000  00000 00000  00000 00000    00000 00000  00000 00000  00000 00000  00000 00000  00000 00000
0775:  00000 00000  00000 00000  00000 00000  00000 00000  00000 00000    00000 00000  00000 00000  00000 00000  00000 00000  00000 00000
0776:  00000 00000  00000 00000  00000 00000  00000 00000  00000 00000    00000 00000  00000 00000  00000 00000  00000 00000  00000 00000
0777:  00000 00000  00000 00000  00000 00000  00000 00000  00000 00000    00000 00000  00000 00000  00000 00000  00000 00000  00000 00000
0778:  00000 00000  00000 00000  00000 00000  00000 00000  00000 00000    00000 00000  00000 00000  00000 00000  00000 00000  00000 00000
0779:  00000 00000  00000 00000  00000 00000  00000 00000  00000 00000    00000 00000  00000 00000  00000 00000  00000 00000  00000 00000
0780:  00000 00000  00000 00000  00000 00000  00000 00000  00000 00000    00000 00000  00000 00000  00000 00000  00000 00000  00000 00000
0781:  00000 00000  00000 00000  00000 00000  00000 00000  00000 00000    00000 00000  00000 00000  00000 00000  00000 00000  00000 00000
0782:  00000 00000  00000 00000  00000 00000  00000 00000  00000 00000    00000 00000  00000 00000  00000 00000  00000 00000  00000 00000
0783:  00000 00000  00000 00000  00000 00000  00000 00000  00000 00000    00000 00000  00000 00000  00000 00000  00000 00000  00000 00000
0784:  00000 00000  00000 00000  00000 00000  00000 00000  00000 00000    00000 00000  00000 00000  00000 00000  00000 00000  00000 00000
0785:  00000 00000  00000 00000  00000 00000  00000 00000  00000 00000    00000 00000  00000 00000  00000 00000  00000 00000  00000 00000
0786:  00000 00000  00000 00000  00000 00000  00000 00000  00000 00000    00000 00000  00000 00000  00000 00000  00000 00000  00000 00000
0787:  00000 00000  00000 00000  00000 00000  00000 00000  00000 00000    00000 00000  00000 00000  00000 00000  00000 00000  00000 00000
0788:  00000 00000  00000 00000  00000 00000  00000 00000  00000 00000    00000 00000  00000 00000  00000 00000  00000 00000  00000 00000
0789:  00000 00000  00000 00000  00000 00000  00000 00000  00000 00000    00000 00000  00000 00000  00000 00000  00000 00000  00000 00000
0790:  00000 00000  00000 00000  00000 00000  00000 00000  00000 00000    00000 00000  00000 00000  00000 00000  00000 00000  00000 00000
0791:  00000 00000  00000 00000  00000 00000  00000 00000  00000 00000    00000 00000  00000 00000  00000 00000  00000 00000  00000 00000
0792:  00000 00000  00000 00000  00000 00000  00000 00000  00000 00000    00000 00000  00000 00000  00000 00000  00000 00000  00000 00000
0793:  00000 00000  00000 00000  00000 00000  00000 00000  00000 00000    00000 00000  00000 00000  00000 00000  00000 00000  00000 00000
0794:  00000 00000  00000 00000  00000 00000  00000 00000  00000 00000    00000 00000  00000 00000  00000 00000  00000 00000  00000 00000
0795:  00000 00000  00000 00000  00000 00000  00000 00000  00000 00000    00000 00000  00000 00000  00000 00000  00000 00000  00000 00000
0796:  00000 00000  00000 00000  00000 00000  00000 00000  00000 00000    00000 00000  00000 00000  00000 00000  00000 00000  00000 00000
0797:  00000 00000  00000 00000  00000 00000  00000 00000  00000 00000    00000 00000  00000 00000  00000 00000  00000 00000  00000 00000
0798:  00000 00000  00000 00000  00000 00000  00000 00000  00000 00000    00000 00000  00000 00000  00000 00000  00000 00000  00000 00000
0799:  00000 00000  00000 00000  00000 00000  00000 00000  00000 00000    00000 00000  00000 00000  00000 00000  00000 00000  00000 00000
```

```
0800:  0000000000  0000000000  0000000000  0000000000  0000000000    0000000000  0000000000  0000000000  0000000000  0000000000
0801:  0000000000  0000000000  0000000000  0000000000  0000000000    0000000000  0000000000  0000000000  0000000000  0000000000
0802:  0000000000  0000000000  0000000000  0000000000  0000000000    0000000000  0000000000  0000000000  0000000000  0000000000
0803:  0000000000  0000000000  0000000000  0000000000  0000000000    0000000000  0000000000  0000000000  0000000000  0000000000
0804:  0000000000  0000000000  0000000000  0000000000  0000000000    0000000000  0000000000  0000000000  0000000000  0000000000
0805:  0000000000  0000000000  0000000000  0000000000  0000000000    0000000000  0000000000  0000000000  0000000000  0000000000
0806:  0000000000  0000000000  0000000000  0000000000  0000000000    0000000000  0000000000  0000000000  0000000000  0000000000
0807:  0000000000  0000000000  0000000000  0000000000  0000000000    0000000000  0000000000  0000000000  0000000000  0000000000
0808:  0000000000  0000000000  0000000000  0000000000  0000000000    0000000000  0000000000  0000000000  0000000000  0000000000
0809:  0000000000  0000000000  0000000000  0000000000  0000000000    0000000000  0000000000  0000000000  0000000000  0000000000
0810:  0000000000  0000000000  0000000000  0000000000  0000000000    0000000000  0000000000  0000000000  0000000000  0000000000
0811:  0000000000  0000000000  0000000000  0000000000  0000000000    0000000000  0000000000  0000000000  0000000000  0000000000
0812:  0000000000  0000000000  0000000000  0000000000  0000000000    0000000000  0000000000  0000000000  0000000000  0000000000
0813:  0000000000  0000000000  0000000000  0000000000  0000000000    0000000000  0000000000  0000000000  0000000000  0000000000
0814:  0000000000  0000000000  0000000000  0000000000  0000000000    0000000000  0000000000  0000000000  0000000000  0000000000
0815:  0000000000  0000000000  0000000000  0000000000  0000000000    0000000000  0000000000  0000000000  0000000000  0000000000
0816:  0000000000  0000000000  0000000000  0000000000  0000000000    0000000000  0000000000  0000000000  0000000000  0000000000
0817:  0000000000  0000000000  0000000000  0000000000  0000000000    0000000000  0000000000  0000000000  0000000000  0000000000
0818:  0000000000  0000000000  0000000000  0000000000  0000000000    0000000000  0000000000  0000000000  0000000000  0000000000
0819:  0000000000  0000000000  0000000000  0000000000  0000000000    0000000000  0000000000  0000000000  0000000000  0000000000
0820:  0000000000  0000000000  0000000000  0000000000  0000000000    0000000000  0000000000  0000000000  0000000000  0000000000
0821:  0000000000  0000000000  0000000000  0000000000  0000000000    0000000000  0000000000  0000000000  0000000000  0000000000
0822:  0000000000  0000000000  0000000000  0000000000  0000000000    0000000000  0000000000  0000000000  0000000000  0000000000
0823:  0000000000  0000000000  0000000000  0000000000  0000000000    0000000000  0000000000  0000000000  0000000000  0000000000
0824:  0000000000  0000000000  0000000000  0000000000  0000000000    0000000000  0000000000  0000000000  0000000000  0000000000
0825:  0000000000  0000000000  0000000000  0000000000  0000000000    0000000000  0000000000  0000000000  0000000000  0000000000
0826:  0000000000  0000000000  0000000000  0000000000  0000000000    0000000000  0000000000  0000000000  0000000000  0000000000
0827:  0000000000  0000000000  0000000000  0000000000  0000000000    0000000000  0000000000  0000000000  0000000000  0000000000
0828:  0000000000  0000000000  0000000000  0000000000  0000000000    0000000000  0000000000  0000000000  0000000000  0000000000
0829:  0000000000  0000000000  0000000000  0000000000  0000000000    0000000000  0000000000  0000000000  0000000000  0000000000
0830:  0000000000  0000000000  0000000000  0000000000  0000000000    0000000000  0000000000  0000000000  0000000000  0000000000
0831:  0000000000  0000000000  0000000000  0000000000  0000000000    0000000000  0000000000  0000000000  0000000000  0000000000
0832:  0000000000  0000000000  0000000000  0000000000  0000000000    0000000000  0000000000  0000000000  0000000000  0000000000
0833:  0000000000  0000000000  0000000000  0000000000  0000000000    0000000000  0000000000  0000000000  0000000000  0000000000
0834:  0000000000  0000000000  0000000000  0000000000  0000000000    0000000000  0000000000  0000000000  0000000000  0000000000
0835:  0000000000  0000000000  0000000000  0000000000  0000000000    0000000000  0000000000  0000000000  0000000000  0000000000
0836:  0000000000  0000000000  0000000000  0000000000  0000000000    0000000000  0000000000  0000000000  0000000000  0000000000
0837:  0000000000  0000000000  0000000000  0000000000  0000000000    0000000000  0000000000  0000000000  0000000000  0000000000
0838:  0000000000  0000000000  0000000000  0000000000  0000000000    0000000000  0000000000  0000000000  0000000000  0000000000
0839:  0000000000  0000000000  0000000000  0000000000  0000000000    0000000000  0000000000  0000000000  0000000000  0000000000
0840:  0000000000  0000000000  0000000000  0000000000  0000000000    0000000000  0000000000  0000000000  0000000000  0000000000
0841:  0000000000  0000000000  0000000000  0000000000  0000000000    0000000000  0000000000  0000000000  0000000000  0000000000
0842:  0000000000  0000000000  0000000000  0000000000  0000000000    0000000000  0000000000  0000000000  0000000000  0000000000
0843:  0000000000  0000000000  0000000000  0000000000  0000000000    0000000000  0000000000  0000000000  0000000000  0000000000
0844:  0000000000  0000000000  0000000000  0000000000  0000000000    0000000000  0000000000  0000000000  0000000000  0000000000
0845:  0000000000  0000000000  0000000000  0000000000  0000000000    0000000000  0000000000  0000000000  0000000000  0000000000
0846:  0000000000  0000000000  0000000000  0000000000  0000000000    0000000000  0000000000  0000000000  0000000000  0000000000
0847:  0000000000  0000000000  0000000000  0000000000  0000000000    0000000000  0000000000  0000000000  0000000000  0000000000
0848:  0000000000  0000000000  0000000000  0000000000  0000000000    0000000000  0000000000  0000000000  0000000000  0000000000
0849:  0000000000  0000000000  0000000000  0000000000  0000000000    0000000000  0000000000  0000000000  0000000000  0000000000
```

```
0850:  00000 00000  00000 00000  00000 00000  00000 00000  00000 00000    00000 00000  00000 00000  00000 00000  00000 00000  00000 00000
0851:  00000 00000  00000 00000  00000 00000  00000 00000  00000 00000    00000 00000  00000 00000  00000 00000  00000 00000  00000 00000
0852:  00000 00000  00000 00000  00000 00000  00000 00000  00000 00000    00000 00000  00000 00000  00000 00000  00000 00000  00000 00000
0853:  00000 00000  00000 00000  00000 00000  00000 00000  00000 00000    00000 00000  00000 00000  00000 00000  00000 00000  00000 00000
0854:  00000 00000  00000 00000  00000 00000  00000 00000  00000 00000    00000 00000  00000 00000  00000 00000  00000 00000  00000 00000
0855:  00000 00000  00000 00000  00000 00000  00000 00000  00000 00000    00000 00000  00000 00000  00000 00000  00000 00000  00000 00000
0856:  00000 00000  00000 00000  00000 00000  00000 00000  00000 00000    00000 00000  00000 00000  00000 00000  00000 00000  00000 00000
0857:  00000 00000  00000 00000  00000 00000  00000 00000  00000 00000    00000 00000  00000 00000  00000 00000  00000 00000  00000 00000
0858:  00000 00000  00000 00000  00000 00000  00000 00000  00000 00000    00000 00000  00000 00000  00000 00000  00000 00000  00000 00000
0859:  00000 00000  00000 00000  00000 00000  00000 00000  00000 00000    00000 00000  00000 00000  00000 00000  00000 00000  00000 00000
0860:  00000 00000  00000 00000  00000 00000  00000 00000  00000 00000    00000 00000  00000 00000  00000 00000  00000 00000  00000 00000
0861:  00000 00000  00000 00000  00000 00000  00000 00000  00000 00000    00000 00000  00000 00000  00000 00000  00000 00000  00000 00000
0862:  00000 00000  00000 00000  00000 00000  00000 00000  00000 00000    00000 00000  00000 00000  00000 00000  00000 00000  00000 00000
0863:  00000 00000  00000 00000  00000 00000  00000 00000  00000 00000    00000 00000  00000 00000  00000 00000  00000 00000  00000 00000
0864:  00000 00000  00000 00000  00000 00000  00000 00000  00000 00000    00000 00000  00000 00000  00000 00000  00000 00000  00000 00000
0865:  00000 00000  00000 00000  00000 00000  00000 00000  00000 00000    00000 00000  00000 00000  00000 00000  00000 00000  00000 00000
0866:  00000 00000  00000 00000  00000 00000  00000 00000  00000 00000    00000 00000  00000 00000  00000 00000  00000 00000  00000 00000
0867:  00000 00000  00000 00000  00000 00000  00000 00000  00000 00000    00000 00000  00000 00000  00000 00000  00000 00000  00000 00000
0868:  00000 00000  00000 00000  00000 00000  00000 00000  00000 00000    00000 00000  00000 00000  00000 00000  00000 00000  00000 00000
0869:  00000 00000  00000 00000  00000 00000  00000 00000  00000 00000    00000 00000  00000 00000  00000 00000  00000 00000  00000 00000
0870:  00000 00000  00000 00000  00000 00000  00000 00000  00000 00000    00000 00000  00000 00000  00000 00000  00000 00000  00000 00000
0871:  00000 00000  00000 00000  00000 00000  00000 00000  00000 00000    00000 00000  00000 00000  00000 00000  00000 00000  00000 00000
0872:  00000 00000  00000 00000  00000 00000  00000 00000  00000 00000    00000 00000  00000 00000  00000 00000  00000 00000  00000 00000
0873:  00000 00000  00000 00000  00000 00000  00000 00000  00000 00000    00000 00000  00000 00000  00000 00000  00000 00000  00000 00000
0874:  00000 00000  00000 00000  00000 00000  00000 00000  00000 00000    00000 00000  00000 00000  00000 00000  00000 00000  00000 00000
0875:  00000 00000  00000 00000  00000 00000  00000 00000  00000 00000    00000 00000  00000 00000  00000 00000  00000 00000  00000 00000
0876:  00000 00000  00000 00000  00000 00000  00000 00000  00000 00000    00000 00000  00000 00000  00000 00000  00000 00000  00000 00000
0877:  00000 00000  00000 00000  00000 00000  00000 00000  00000 00000    00000 00000  00000 00000  00000 00000  00000 00000  00000 00000
0878:  00000 00000  00000 00000  00000 00000  00000 00000  00000 00000    00000 00000  00000 00000  00000 00000  00000 00000  00000 00000
0879:  00000 00000  00000 00000  00000 00000  00000 00000  00000 00000    00000 00000  00000 00000  00000 00000  00000 00000  00000 00000
0880:  00000 00000  00000 00000  00000 00000  00000 00000  00000 00000    00000 00000  00000 00000  00000 00000  00000 00000  00000 00000
0881:  00000 00000  00000 00000  00000 00000  00000 00000  00000 00000    00000 00000  00000 00000  00000 00000  00000 00000  00000 00000
0882:  00000 00000  00000 00000  00000 00000  00000 00000  00000 00000    00000 00000  00000 00000  00000 00000  00000 00000  00000 00000
0883:  00000 00000  00000 00000  00000 00000  00000 00000  00000 00000    00000 00000  00000 00000  00000 00000  00000 00000  00000 00000
0884:  00000 00000  00000 00000  00000 00000  00000 00000  00000 00000    00000 00000  00000 00000  00000 00000  00000 00000  00000 00000
0885:  00000 00000  00000 00000  00000 00000  00000 00000  00000 00000    00000 00000  00000 00000  00000 00000  00000 00000  00000 00000
0886:  00000 00000  00000 00000  00000 00000  00000 00000  00000 00000    00000 00000  00000 00000  00000 00000  00000 00000  00000 00000
0887:  00000 00000  00000 00000  00000 00000  00000 00000  00000 00000    00000 00000  00000 00000  00000 00000  00000 00000  00000 00000
0888:  00000 00000  00000 00000  00000 00000  00000 00000  00000 00000    00000 00000  00000 00000  00000 00000  00000 00000  00000 00000
0889:  00000 00000  00000 00000  00000 00000  00000 00000  00000 00000    00000 00000  00000 00000  00000 00000  00000 00000  00000 00000
0890:  00000 00000  00000 00000  00000 00000  00000 00000  00000 00000    00000 00000  00000 00000  00000 00000  00000 00000  00000 00000
0891:  00000 00000  00000 00000  00000 00000  00000 00000  00000 00000    00000 00000  00000 00000  00000 00000  00000 00000  00000 00000
0892:  00000 00000  00000 00000  00000 00000  00000 00000  00000 00000    00000 00000  00000 00000  00000 00000  00000 00000  00000 00000
0893:  00000 00000  00000 00000  00000 00000  00000 00000  00000 00000    00000 00000  00000 00000  00000 00000  00000 00000  00000 00000
0894:  00000 00000  00000 00000  00000 00000  00000 00000  00000 00000    00000 00000  00000 00000  00000 00000  00000 00000  00000 00000
0895:  00000 00000  00000 00000  00000 00000  00000 00000  00000 00000    00000 00000  00000 00000  00000 00000  00000 00000  00000 00000
0896:  00000 00000  00000 00000  00000 00000  00000 00000  00000 00000    00000 00000  00000 00000  00000 00000  00000 00000  00000 00000
0897:  00000 00000  00000 00000  00000 00000  00000 00000  00000 00000    00000 00000  00000 00000  00000 00000  00000 00000  00000 00000
0898:  00000 00000  00000 00000  00000 00000  00000 00000  00000 00000    00000 00000  00000 00000  00000 00000  00000 00000  00000 00000
0899:  00000 00000  00000 00000  00000 00000  00000 00000  00000 00000    00000 00000  00000 00000  00000 00000  00000 00000  00000 00000
```

```
0900:  00000 00000  00000 00000  00000 00000  00000 00000  00000 00000   00000 00000  00000 00000  00000 00000  00000 00000  00000 00000
0901:  00000 00000  00000 00000  00000 00000  00000 00000  00000 00000   00000 00000  00000 00000  00000 00000  00000 00000  00000 00000
0902:  00000 00000  00000 00000  00000 00000  00000 00000  00000 00000   00000 00000  00000 00000  00000 00000  00000 00000  00000 00000
0903:  00000 00000  00000 00000  00000 00000  00000 00000  00000 00000   00000 00000  00000 00000  00000 00000  00000 00000  00000 00000
0904:  00000 00000  00000 00000  00000 00000  00000 00000  00000 00000   00000 00000  00000 00000  00000 00000  00000 00000  00000 00000
0905:  00000 00000  00000 00000  00000 00000  00000 00000  00000 00000   00000 00000  00000 00000  00000 00000  00000 00000  00000 00000
0906:  00000 00000  00000 00000  00000 00000  00000 00000  00000 00000   00000 00000  00000 00000  00000 00000  00000 00000  00000 00000
0907:  00000 00000  00000 00000  00000 00000  00000 00000  00000 00000   00000 00000  00000 00000  00000 00000  00000 00000  00000 00000
0908:  00000 00000  00000 00000  00000 00000  00000 00000  00000 00000   00000 00000  00000 00000  00000 00000  00000 00000  00000 00000
0909:  00000 00000  00000 00000  00000 00000  00000 00000  00000 00000   00000 00000  00000 00000  00000 00000  00000 00000  00000 00000
0910:  00000 00000  00000 00000  00000 00000  00000 00000  00000 00000   00000 00000  00000 00000  00000 00000  00000 00000  00000 00000
0911:  00000 00000  00000 00000  00000 00000  00000 00000  00000 00000   00000 00000  00000 00000  00000 00000  00000 00000  00000 00000
0912:  00000 00000  00000 00000  00000 00000  00000 00000  00000 00000   00000 00000  00000 00000  00000 00000  00000 00000  00000 00000
0913:  00000 00000  00000 00000  00000 00000  00000 00000  00000 00000   00000 00000  00000 00000  00000 00000  00000 00000  00000 00000
0914:  00000 00000  00000 00000  00000 00000  00000 00000  00000 00000   00000 00000  00000 00000  00000 00000  00000 00000  00000 00000
0915:  00000 00000  00000 00000  00000 00000  00000 00000  00000 00000   00000 00000  00000 00000  00000 00000  00000 00000  00000 00000
0916:  00000 00000  00000 00000  00000 00000  00000 00000  00000 00000   00000 00000  00000 00000  00000 00000  00000 00000  00000 00000
0917:  00000 00000  00000 00000  00000 00000  00000 00000  00000 00000   00000 00000  00000 00000  00000 00000  00000 00000  00000 00000
0918:  00000 00000  00000 00000  00000 00000  00000 00000  00000 00000   00000 00000  00000 00000  00000 00000  00000 00000  00000 00000
0919:  00000 00000  00000 00000  00000 00000  00000 00000  00000 00000   00000 00000  00000 00000  00000 00000  00000 00000  00000 00000
0920:  00000 00000  00000 00000  00000 00000  00000 00000  00000 00000   00000 00000  00000 00000  00000 00000  00000 00000  00000 00000
0921:  00000 00000  00000 00000  00000 00000  00000 00000  00000 00000   00000 00000  00000 00000  00000 00000  00000 00000  00000 00000
0922:  00000 00000  00000 00000  00000 00000  00000 00000  00000 00000   00000 00000  00000 00000  00000 00000  00000 00000  00000 00000
0923:  00000 00000  00000 00000  00000 00000  00000 00000  00000 00000   00000 00000  00000 00000  00000 00000  00000 00000  00000 00000
0924:  00000 00000  00000 00000  00000 00000  00000 00000  00000 00000   00000 00000  00000 00000  00000 00000  00000 00000  00000 00000
0925:  00000 00000  00000 00000  00000 00000  00000 00000  00000 00000   00000 00000  00000 00000  00000 00000  00000 00000  00000 00000
0926:  00000 00000  00000 00000  00000 00000  00000 00000  00000 00000   00000 00000  00000 00000  00000 00000  00000 00000  00000 00000
0927:  00000 00000  00000 00000  00000 00000  00000 00000  00000 00000   00000 00000  00000 00000  00000 00000  00000 00000  00000 00000
0928:  00000 00000  00000 00000  00000 00000  00000 00000  00000 00000   00000 00000  00000 00000  00000 00000  00000 00000  00000 00000
0929:  00000 00000  00000 00000  00000 00000  00000 00000  00000 00000   00000 00000  00000 00000  00000 00000  00000 00000  00000 00000
0930:  00000 00000  00000 00000  00000 00000  00000 00000  00000 00000   00000 00000  00000 00000  00000 00000  00000 00000  00000 00000
0931:  00000 00000  00000 00000  00000 00000  00000 00000  00000 00000   00000 00000  00000 00000  00000 00000  00000 00000  00000 00000
0932:  00000 00000  00000 00000  00000 00000  00000 00000  00000 00000   00000 00000  00000 00000  00000 00000  00000 00000  00000 00000
0933:  00000 00000  00000 00000  00000 00000  00000 00000  00000 00000   00000 00000  00000 00000  00000 00000  00000 00000  00000 00000
0934:  00000 00000  00000 00000  00000 00000  00000 00000  00000 00000   00000 00000  00000 00000  00000 00000  00000 00000  00000 00000
0935:  00000 00000  00000 00000  00000 00000  00000 00000  00000 00000   00000 00000  00000 00000  00000 00000  00000 00000  00000 00000
0936:  00000 00000  00000 00000  00000 00000  00000 00000  00000 00000   00000 00000  00000 00000  00000 00000  00000 00000  00000 00000
0937:  00000 00000  00000 00000  00000 00000  00000 00000  00000 00000   00000 00000  00000 00000  00000 00000  00000 00000  00000 00000
0938:  00000 00000  00000 00000  00000 00000  00000 00000  00000 00000   00000 00000  00000 00000  00000 00000  00000 00000  00000 00000
0939:  00000 00000  00000 00000  00000 00000  00000 00000  00000 00000   00000 00000  00000 00000  00000 00000  00000 00000  00000 00000
0940:  00000 00000  00000 00000  00000 00000  00000 00000  00000 00000   00000 00000  00000 00000  00000 00000  00000 00000  00000 00000
0941:  00000 00000  00000 00000  00000 00000  00000 00000  00000 00000   00000 00000  00000 00000  00000 00000  00000 00000  00000 00000
0942:  00000 00000  00000 00000  00000 00000  00000 00000  00000 00000   00000 00000  00000 00000  00000 00000  00000 00000  00000 00000
0943:  00000 00000  00000 00000  00000 00000  00000 00000  00000 00000   00000 00000  00000 00000  00000 00000  00000 00000  00000 00000
0944:  00000 00000  00000 00000  00000 00000  00000 00000  00000 00000   00000 00000  00000 00000  00000 00000  00000 00000  00000 00000
0945:  00000 00000  00000 00000  00000 00000  00000 00000  00000 00000   00000 00000  00000 00000  00000 00000  00000 00000  00000 00000
0946:  00000 00000  00000 00000  00000 00000  00000 00000  00000 00000   00000 00000  00000 00000  00000 00000  00000 00000  00000 00000
0947:  00000 00000  00000 00000  00000 00000  00000 00000  00000 00000   00000 00000  00000 00000  00000 00000  00000 00000  00000 00000
0948:  00000 00000  00000 00000  00000 00000  00000 00000  00000 00000   00000 00000  00000 00000  00000 00000  00000 00000  00000 00000
0949:  00000 00000  00000 00000  00000 00000  00000 00000  00000 00000   00000 00000  00000 00000  00000 00000  00000 00000  00000 00000
```

```
0950:  00000 00000  00000 00000  00000 00000  00000 00000  00000 00000    00000 00000  00000 00000  00000 00000  00000 00000  00000 00000
0951:  00000 00000  00000 00000  00000 00000  00000 00000  00000 00000    00000 00000  00000 00000  00000 00000  00000 00000  00000 00000
0952:  00000 00000  00000 00000  00000 00000  00000 00000  00000 00000    00000 00000  00000 00000  00000 00000  00000 00000  00000 00000
0953:  00000 00000  00000 00000  00000 00000  00000 00000  00000 00000    00000 00000  00000 00000  00000 00000  00000 00000  00000 00000
0954:  00000 00000  00000 00000  00000 00000  00000 00000  00000 00000    00000 00000  00000 00000  00000 00000  00000 00000  00000 00000
0955:  00000 00000  00000 00000  00000 00000  00000 00000  00000 00000    00000 00000  00000 00000  00000 00000  00000 00000  00000 00000
0956:  00000 00000  00000 00000  00000 00000  00000 00000  00000 00000    00000 00000  00000 00000  00000 00000  00000 00000  00000 00000
0957:  00000 00000  00000 00000  00000 00000  00000 00000  00000 00000    00000 00000  00000 00000  00000 00000  00000 00000  00000 00000
0958:  00000 00000  00000 00000  00000 00000  00000 00000  00000 00000    00000 00000  00000 00000  00000 00000  00000 00000  00000 00000
0959:  00000 00000  00000 00000  00000 00000  00000 00000  00000 00000    00000 00000  00000 00000  00000 00000  00000 00000  00000 00000
0960:  00000 00000  00000 00000  00000 00000  00000 00000  00000 00000    00000 00000  00000 00000  00000 00000  00000 00000  00000 00000
0961:  00000 00000  00000 00000  00000 00000  00000 00000  00000 00000    00000 00000  00000 00000  00000 00000  00000 00000  00000 00000
0962:  00000 00000  00000 00000  00000 00000  00000 00000  00000 00000    00000 00000  00000 00000  00000 00000  00000 00000  00000 00000
0963:  00000 00000  00000 00000  00000 00000  00000 00000  00000 00000    00000 00000  00000 00000  00000 00000  00000 00000  00000 00000
0964:  00000 00000  00000 00000  00000 00000  00000 00000  00000 00000    00000 00000  00000 00000  00000 00000  00000 00000  00000 00000
0965:  00000 00000  00000 00000  00000 00000  00000 00000  00000 00000    00000 00000  00000 00000  00000 00000  00000 00000  00000 00000
0966:  00000 00000  00000 00000  00000 00000  00000 00000  00000 00000    00000 00000  00000 00000  00000 00000  00000 00000  00000 00000
0967:  00000 00000  00000 00000  00000 00000  00000 00000  00000 00000    00000 00000  00000 00000  00000 00000  00000 00000  00000 00000
0968:  00000 00000  00000 00000  00000 00000  00000 00000  00000 00000    00000 00000  00000 00000  00000 00000  00000 00000  00000 00000
0969:  00000 00000  00000 00000  00000 00000  00000 00000  00000 00000    00000 00000  00000 00000  00000 00000  00000 00000  00000 00000
0970:  00000 00000  00000 00000  00000 00000  00000 00000  00000 00000    00000 00000  00000 00000  00000 00000  00000 00000  00000 00000
0971:  00000 00000  00000 00000  00000 00000  00000 00000  00000 00000    00000 00000  00000 00000  00000 00000  00000 00000  00000 00000
0972:  00000 00000  00000 00000  00000 00000  00000 00000  00000 00000    00000 00000  00000 00000  00000 00000  00000 00000  00000 00000
0973:  00000 00000  00000 00000  00000 00000  00000 00000  00000 00000    00000 00000  00000 00000  00000 00000  00000 00000  00000 00000
0974:  00000 00000  00000 00000  00000 00000  00000 00000  00000 00000    00000 00000  00000 00000  00000 00000  00000 00000  00000 00000
0975:  00000 00000  00000 00000  00000 00000  00000 00000  00000 00000    00000 00000  00000 00000  00000 00000  00000 00000  00000 00000
0976:  00000 00000  00000 00000  00000 00000  00000 00000  00000 00000    00000 00000  00000 00000  00000 00000  00000 00000  00000 00000
0977:  00000 00000  00000 00000  00000 00000  00000 00000  00000 00000    00000 00000  00000 00000  00000 00000  00000 00000  00000 00000
0978:  00000 00000  00000 00000  00000 00000  00000 00000  00000 00000    00000 00000  00000 00000  00000 00000  00000 00000  00000 00000
0979:  00000 00000  00000 00000  00000 00000  00000 00000  00000 00000    00000 00000  00000 00000  00000 00000  00000 00000  00000 00000
0980:  00000 00000  00000 00000  00000 00000  00000 00000  00000 00000    00000 00000  00000 00000  00000 00000  00000 00000  00000 00000
0981:  00000 00000  00000 00000  00000 00000  00000 00000  00000 00000    00000 00000  00000 00000  00000 00000  00000 00000  00000 00000
0982:  00000 00000  00000 00000  00000 00000  00000 00000  00000 00000    00000 00000  00000 00000  00000 00000  00000 00000  00000 00000
0983:  00000 00000  00000 00000  00000 00000  00000 00000  00000 00000    00000 00000  00000 00000  00000 00000  00000 00000  00000 00000
0984:  00000 00000  00000 00000  00000 00000  00000 00000  00000 00000    00000 00000  00000 00000  00000 00000  00000 00000  00000 00000
0985:  00000 00000  00000 00000  00000 00000  00000 00000  00000 00000    00000 00000  00000 00000  00000 00000  00000 00000  00000 00000
0986:  00000 00000  00000 00000  00000 00000  00000 00000  00000 00000    00000 00000  00000 00000  00000 00000  00000 00000  00000 00000
0987:  00000 00000  00000 00000  00000 00000  00000 00000  00000 00000    00000 00000  00000 00000  00000 00000  00000 00000  00000 00000
0988:  00000 00000  00000 00000  00000 00000  00000 00000  00000 00000    00000 00000  00000 00000  00000 00000  00000 00000  00000 00000
0989:  00000 00000  00000 00000  00000 00000  00000 00000  00000 00000    00000 00000  00000 00000  00000 00000  00000 00000  00000 00000
0990:  00000 00000  00000 00000  00000 00000  00000 00000  00000 00000    00000 00000  00000 00000  00000 00000  00000 00000  00000 00000
0991:  00000 00000  00000 00000  00000 00000  00000 00000  00000 00000    00000 00000  00000 00000  00000 00000  00000 00000  00000 00000
0992:  00000 00000  00000 00000  00000 00000  00000 00000  00000 00000    00000 00000  00000 00000  00000 00000  00000 00000  00000 00000
0993:  00000 00000  00000 00000  00000 00000  00000 00000  00000 00000    00000 00000  00000 00000  00000 00000  00000 00000  00000 00000
0994:  00000 00000  00000 00000  00000 00000  00000 00000  00000 00000    00000 00000  00000 00000  00000 00000  00000 00000  00000 00000
0995:  00000 00000  00000 00000  00000 00000  00000 00000  00000 00000    00000 00000  00000 00000  00000 00000  00000 00000  00000 00000
0996:  00000 00000  00000 00000  00000 00000  00000 00000  00000 00000    00000 00000  00000 00000  00000 00000  00000 00000  00000 00000
0997:  00000 00000  00000 00000  00000 00000  00000 00000  00000 00000    00000 00000  00000 00000  00000 00000  00000 00000  00000 00000
0998:  00000 00000  00000 00000  00000 00000  00000 00000  00000 00000    00000 00000  00000 00000  00000 00000  00000 00000  00000 00000
0999:  00000 00000  00000 00000  00000 00000  00000 00000  00000 00000    00000 00000  00000 00000  00000 00000  00000 00000  00000 00000
```

```
1000:   00000 00000   00000 00000   00000 00000   00000 00000   00000 00000      00000 00000   00000 00000   00000 00000   00000 00000   00000 00000
1001:   00000 00000   00000 00000   00000 00000   00000 00000   00000 00000      00000 00000   00000 00000   00000 00000   00000 00000   00000 00000
1002:   00000 00000   00000 00000   00000 00000   00000 00000   00000 00000      00000 00000   00000 00000   00000 00000   00000 00000   00000 00000
1003:   00000 00000   00000 00000   00000 00000   00000 00000   00000 00000      00000 00000   00000 00000   00000 00000   00000 00000   00000 00000
1004:   00000 00000   00000 00000   00000 00000   00000 00000   00000 00000      00000 00000   00000 00000   00000 00000   00000 00000   00000 00000
1005:   00000 00000   00000 00000   00000 00000   00000 00000   00000 00000      00000 00000   00000 00000   00000 00000   00000 00000   00000 00000
1006:   00000 00000   00000 00000   00000 00000   00000 00000   00000 00000      00000 00000   00000 00000   00000 00000   00000 00000   00000 00000
1007:   00000 00000   00000 00000   00000 00000   00000 00000   00000 00000      00000 00000   00000 00000   00000 00000   00000 00000   00000 00000
1008:   00000 00000   00000 00000   00000 00000   00000 00000   00000 00000      00000 00000   00000 00000   00000 00000   00000 00000   00000 00000
1009:   00000 00000   00000 00000   00000 00000   00000 00000   00000 00000      00000 00000   00000 00000   00000 00000   00000 00000   00000 00000
1010:   00000 00000   00000 00000   00000 00000   00000 00000   00000 00000      00000 00000   00000 00000   00000 00000   00000 00000   00000 00000
1011:   00000 00000   00000 00000   00000 00000   00000 00000   00000 00000      00000 00000   00000 00000   00000 00000   00000 00000   00000 00000
1012:   00000 00000   00000 00000   00000 00000   00000 00000   00000 00000      00000 00000   00000 00000   00000 00000   00000 00000   00000 00000
1013:   00000 00000   00000 00000   00000 00000   00000 00000   00000 00000      00000 00000   00000 00000   00000 00000   00000 00000   00000 00000
1014:   00000 00000   00000 00000   00000 00000   00000 00000   00000 00000      00000 00000   00000 00000   00000 00000   00000 00000   00000 00000
1015:   00000 00000   00000 00000   00000 00000   00000 00000   00000 00000      00000 00000   00000 00000   00000 00000   00000 00000   00000 00000
1016:   00000 00000   00000 00000   00000 00000   00000 00000   00000 00000      00000 00000   00000 00000   00000 00000   00000 00000   00000 00000
1017:   00000 00000   00000 00000   00000 00000   00000 00000   00000 00000      00000 00000   00000 00000   00000 00000   00000 00000   00000 00000
1018:   00000 00000   00000 00000   00000 00000   00000 00000   00000 00000      00000 00000   00000 00000   00000 00000   00000 00000   00000 00000
1019:   00000 00000   00000 00000   00000 00000   00000 00000   00000 00000      00000 00000   00000 00000   00000 00000   00000 00000   00000 00000
1020:   00000 00000   00000 00000   00000 00000   00000 00000   00000 00000      00000 00000   00000 00000   00000 00000   00000 00000   00000 00000
1021:   00000 00000   00000 00000   00000 00000   00000 00000   00000 00000      00000 00000   00000 00000   00000 00000   00000 00000   00000 00000
1022:   00000 00000   00000 00000   00000 00000   00000 00000   00000 00000      00000 00000   00000 00000   00000 00000   00000 00000   00000 00000
1023:   00000 00000   00000 00000   00000 00000   00000 00000   00000 00000      00000 00000   00000 00000   00000 00000   00000 00000   00000 00000
1024:   00000 00000   00000 00000   00000 00000   00000 00000   00000 00000      00000 00000   00000 00000   00000 00000   00000 00000   00000 00000
1025:   00000 00000   00000 00000   00000 00000   00000 00000   00000 00000      00000 00000   00000 00000   00000 00000   00000 00000   00000 00000
1026:   00000 00000   00000 00000   00000 00000   00000 00000   00000 00000      00000 00000   00000 00000   00000 00000   00000 00000   00000 00000
1027:   00000 00000   00000 00000   00000 00000   00000 00000   00000 00000      00000 00000   00000 00000   00000 00000   00000 00000   00000 00000
1028:   00000 00000   00000 00000   00000 00000   00000 00000   00000 00000      00000 00000   00000 00000   00000 00000   00000 00000   00000 00000
1029:   00000 00000   00000 00000   00000 00000   00000 00000   00000 00000      00000 00000   00000 00000   00000 00000   00000 00000   00000 00000
1030:   00000 00000   00000 00000   00000 00000   00000 00000   00000 00000      00000 00000   00000 00000   00000 00000   00000 00000   00000 00000
1031:   00000 00000   00000 00000   00000 00000   00000 00000   00000 00000      00000 00000   00000 00000   00000 00000   00000 00000   00000 00000
1032:   00000 00000   00000 00000   00000 00000   00000 00000   00000 00000      00000 00000   00000 00000   00000 00000   00000 00000   00000 00000
1033:   00000 00000   00000 00000   00000 00000   00000 00000   00000 00000      00000 00000   00000 00000   00000 00000   00000 00000   00000 00000
1034:   00000 00000   00000 00000   00000 00000   00000 00000   00000 00000      00000 00000   00000 00000   00000 00000   00000 00000   00000 00000
1035:   00000 00000   00000 00000   00000 00000   00000 00000   00000 00000      00000 00000   00000 00000   00000 00000   00000 00000   00000 00000
1036:   00000 00000   00000 00000   00000 00000   00000 00000   00000 00000      00000 00000   00000 00000   00000 00000   00000 00000   00000 00000
1037:   00000 00000   00000 00000   00000 00000   00000 00000   00000 00000      00000 00000   00000 00000   00000 00000   00000 00000   00000 00000
1038:   00000 00000   00000 00000   00000 00000   00000 00000   00000 00000      00000 00000   00000 00000   00000 00000   00000 00000   00000 00000
1039:   00000 00000   00000 00000   00000 00000   00000 00000   00000 00000      00000 00000   00000 00000   00000 00000   00000 00000   00000 00000
1040:   00000 00000   00000 00000   00000 00000   00000 00000   00000 00000      00000 00000   00000 00000   00000 00000   00000 00000   00000 00000
1041:   00000 00000   00000 00000   00000 00000   00000 00000   00000 00000      00000 00000   00000 00000   00000 00000   00000 00000   00000 00000
1042:   00000 00000   00000 00000   00000 00000   00000 00000   00000 00000      00000 00000   00000 00000   00000 00000   00000 00000   00000 00000
1043:   00000 00000   00000 00000   00000 00000   00000 00000   00000 00000      00000 00000   00000 00000   00000 00000   00000 00000   00000 00000
1044:   00000 00000   00000 00000   00000 00000   00000 00000   00000 00000      00000 00000   00000 00000   00000 00000   00000 00000   00000 00000
1045:   00000 00000   00000 00000   00000 00000   00000 00000   00000 00000      00000 00000   00000 00000   00000 00000   00000 00000   00000 00000
1046:   00000 00000   00000 00000   00000 00000   00000 00000   00000 00000      00000 00000   00000 00000   00000 00000   00000 00000   00000 00000
1047:   00000 00000   00000 00000   00000 00000   00000 00000   00000 00000      00000 00000   00000 00000   00000 00000   00000 00000   00000 00000
1048:   00000 00000   00000 00000   00000 00000   00000 00000   00000 00000      00000 00000   00000 00000   00000 00000   00000 00000   00000 00000
1049:   00000 00000   00000 00000   00000 00000   00000 00000   00000 00000      00000 00000   00000 00000   00000 00000   00000 00000   00000 00000
```

```
1050:   00000 00000   00000 00000   00000 00000   00000 00000   00000 00000     00000 00000   00000 00000   00000 00000   00000 00000   00000 00000
1051:   00000 00000   00000 00000   00000 00000   00000 00000   00000 00000     00000 00000   00000 00000   00000 00000   00000 00000   00000 00000
1052:   00000 00000   00000 00000   00000 00000   00000 00000   00000 00000     00000 00000   00000 00000   00000 00000   00000 00000   00000 00000
1053:   00000 00000   00000 00000   00000 00000   00000 00000   00000 00000     00000 00000   00000 00000   00000 00000   00000 00000   00000 00000
1054:   00000 00000   00000 00000   00000 00000   00000 00000   00000 00000     00000 00000   00000 00000   00000 00000   00000 00000   00000 00000
1055:   00000 00000   00000 00000   00000 00000   00000 00000   00000 00000     00000 00000   00000 00000   00000 00000   00000 00000   00000 00000
1056:   00000 00000   00000 00000   00000 00000   00000 00000   00000 00000     00000 00000   00000 00000   00000 00000   00000 00000   00000 00000
1057:   00000 00000   00000 00000   00000 00000   00000 00000   00000 00000     00000 00000   00000 00000   00000 00000   00000 00000   00000 00000
1058:   00000 00000   00000 00000   00000 00000   00000 00000   00000 00000     00000 00000   00000 00000   00000 00000   00000 00000   00000 00000
1059:   00000 00000   00000 00000   00000 00000   00000 00000   00000 00000     00000 00000   00000 00000   00000 00000   00000 00000   00000 00000
1060:   00000 00000   00000 00000   00000 00000   00000 00000   00000 00000     00000 00000   00000 00000   00000 00000   00000 00000   00000 00000
1061:   00000 00000   00000 00000   00000 00000   00000 00000   00000 00000     00000 00000   00000 00000   00000 00000   00000 00000   00000 00000
1062:   00000 00000   00000 00000   00000 00000   00000 00000   00000 00000     00000 00000   00000 00000   00000 00000   00000 00000   00000 00000
1063:   00000 00000   00000 00000   00000 00000   00000 00000   00000 00000     00000 00000   00000 00000   00000 00000   00000 00000   00000 00000
1064:   00000 00000   00000 00000   00000 00000   00000 00000   00000 00000     00000 00000   00000 00000   00000 00000   00000 00000   00000 00000
1065:   00000 00000   00000 00000   00000 00000   00000 00000   00000 00000     00000 00000   00000 00000   00000 00000   00000 00000   00000 00000
1066:   00000 00000   00000 00000   00000 00000   00000 00000   00000 00000     00000 00000   00000 00000   00000 00000   00000 00000   00000 00000
1067:   00000 00000   00000 00000   00000 00000   00000 00000   00000 00000     00000 00000   00000 00000   00000 00000   00000 00000   00000 00000
1068:   00000 00000   00000 00000   00000 00000   00000 00000   00000 00000     00000 00000   00000 00000   00000 00000   00000 00000   00000 00000
1069:   00000 00000   00000 00000   00000 00000   00000 00000   00000 00000     00000 00000   00000 00000   00000 00000   00000 00000   00000 00000
1070:   00000 00000   00000 00000   00000 00000   00000 00000   00000 00000     00000 00000   00000 00000   00000 00000   00000 00000   00000 00000
1071:   00000 00000   00000 00000   00000 00000   00000 00000   00000 00000     00000 00000   00000 00000   00000 00000   00000 00000   00000 00000
1072:   00000 00000   00000 00000   00000 00000   00000 00000   00000 00000     00000 00000   00000 00000   00000 00000   00000 00000   00000 00000
1073:   00000 00000   00000 00000   00000 00000   00000 00000   00000 00000     00000 00000   00000 00000   00000 00000   00000 00000   00000 00000
1074:   00000 00000   00000 00000   00000 00000   00000 00000   00000 00000     00000 00000   00000 00000   00000 00000   00000 00000   00000 00000
1075:   00000 00000   00000 00000   00000 00000   00000 00000   00000 00000     00000 00000   00000 00000   00000 00000   00000 00000   00000 00000
1076:   00000 00000   00000 00000   00000 00000   00000 00000   00000 00000     00000 00000   00000 00000   00000 00000   00000 00000   00000 00000
1077:   00000 00000   00000 00000   00000 00000   00000 00000   00000 00000     00000 00000   00000 00000   00000 00000   00000 00000   00000 00000
1078:   00000 00000   00000 00000   00000 00000   00000 00000   00000 00000     00000 00000   00000 00000   00000 00000   00000 00000   00000 00000
1079:   00000 00000   00000 00000   00000 00000   00000 00000   00000 00000     00000 00000   00000 00000   00000 00000   00000 00000   00000 00000
1080:   00000 00000   00000 00000   00000 00000   00000 00000   00000 00000     00000 00000   00000 00000   00000 00000   00000 00000   00000 00000
1081:   00000 00000   00000 00000   00000 00000   00000 00000   00000 00000     00000 00000   00000 00000   00000 00000   00000 00000   00000 00000
1082:   00000 00000   00000 00000   00000 00000   00000 00000   00000 00000     00000 00000   00000 00000   00000 00000   00000 00000   00000 00000
1083:   00000 00000   00000 00000   00000 00000   00000 00000   00000 00000     00000 00000   00000 00000   00000 00000   00000 00000   00000 00000
1084:   00000 00000   00000 00000   00000 00000   00000 00000   00000 00000     00000 00000   00000 00000   00000 00000   00000 00000   00000 00000
1085:   00000 00000   00000 00000   00000 00000   00000 00000   00000 00000     00000 00000   00000 00000   00000 00000   00000 00000   00000 00000
1086:   00000 00000   00000 00000   00000 00000   00000 00000   00000 00000     00000 00000   00000 00000   00000 00000   00000 00000   00000 00000
1087:   00000 00000   00000 00000   00000 00000   00000 00000   00000 00000     00000 00000   00000 00000   00000 00000   00000 00000   00000 00000
1088:   00000 00000   00000 00000   00000 00000   00000 00000   00000 00000     00000 00000   00000 00000   00000 00000   00000 00000   00000 00000
1089:   00000 00000   00000 00000   00000 00000   00000 00000   00000 00000     00000 00000   00000 00000   00000 00000   00000 00000   00000 00000
1090:   00000 00000   00000 00000   00000 00000   00000 00000   00000 00000     00000 00000   00000 00000   00000 00000   00000 00000   00000 00000
1091:   00000 00000   00000 00000   00000 00000   00000 00000   00000 00000     00000 00000   00000 00000   00000 00000   00000 00000   00000 00000
1092:   00000 00000   00000 00000   00000 00000   00000 00000   00000 00000     00000 00000   00000 00000   00000 00000   00000 00000   00000 00000
1093:   00000 00000   00000 00000   00000 00000   00000 00000   00000 00000     00000 00000   00000 00000   00000 00000   00000 00000   00000 00000
1094:   00000 00000   00000 00000   00000 00000   00000 00000   00000 00000     00000 00000   00000 00000   00000 00000   00000 00000   00000 00000
1095:   00000 00000   00000 00000   00000 00000   00000 00000   00000 00000     00000 00000   00000 00000   00000 00000   00000 00000   00000 00000
1096:   00000 00000   00000 00000   00000 00000   00000 00000   00000 00000     00000 00000   00000 00000   00000 00000   00000 00000   00000 00000
1097:   00000 00000   00000 00000   00000 00000   00000 00000   00000 00000     00000 00000   00000 00000   00000 00000   00000 00000   00000 00000
1098:   00000 00000   00000 00000   00000 00000   00000 00000   00000 00000     00000 00000   00000 00000   00000 00000   00000 00000   00000 00000
1099:   00000 00000   00000 00000   00000 00000   00000 00000   00000 00000     00000 00000   00000 00000   00000 00000   00000 00000   00000 00000
```

```
1100:  00000 00000  00000 00000  00000 00000  00000 00000  00000 00000    00000 00000  00000 00000  00000 00000  00000 00000  00000 00000
1101:  00000 00000  00000 00000  00000 00000  00000 00000  00000 00000    00000 00000  00000 00000  00000 00000  00000 00000  00000 00000
1102:  00000 00000  00000 00000  00000 00000  00000 00000  00000 00000    00000 00000  00000 00000  00000 00000  00000 00000  00000 00000
1103:  00000 00000  00000 00000  00000 00000  00000 00000  00000 00000    00000 00000  00000 00000  00000 00000  00000 00000  00000 00000
1104:  00000 00000  00000 00000  00000 00000  00000 00000  00000 00000    00000 00000  00000 00000  00000 00000  00000 00000  00000 00000
1105:  00000 00000  00000 00000  00000 00000  00000 00000  00000 00000    00000 00000  00000 00000  00000 00000  00000 00000  00000 00000
1106:  00000 00000  00000 00000  00000 00000  00000 00000  00000 00000    00000 00000  00000 00000  00000 00000  00000 00000  00000 00000
1107:  00000 00000  00000 00000  00000 00000  00000 00000  00000 00000    00000 00000  00000 00000  00000 00000  00000 00000  00000 00000
1108:  00000 00000  00000 00000  00000 00000  00000 00000  00000 00000    00000 00000  00000 00000  00000 00000  00000 00000  00000 00000
1109:  00000 00000  00000 00000  00000 00000  00000 00000  00000 00000    00000 00000  00000 00000  00000 00000  00000 00000  00000 00000
1110:  00000 00000  00000 00000  00000 00000  00000 00000  00000 00000    00000 00000  00000 00000  00000 00000  00000 00000  00000 00000
1111:  00000 00000  00000 00000  00000 00000  00000 00000  00000 00000    00000 00000  00000 00000  00000 00000  00000 00000  00000 00000
1112:  00000 00000  00000 00000  00000 00000  00000 00000  00000 00000    00000 00000  00000 00000  00000 00000  00000 00000  00000 00000
1113:  00000 00000  00000 00000  00000 00000  00000 00000  00000 00000    00000 00000  00000 00000  00000 00000  00000 00000  00000 00000
1114:  00000 00000  00000 00000  00000 00000  00000 00000  00000 00000    00000 00000  00000 00000  00000 00000  00000 00000  00000 00000
1115:  00000 00000  00000 00000  00000 00000  00000 00000  00000 00000    00000 00000  00000 00000  00000 00000  00000 00000  00000 00000
1116:  00000 00000  00000 00000  00000 00000  00000 00000  00000 00000    00000 00000  00000 00000  00000 00000  00000 00000  00000 00000
1117:  00000 00000  00000 00000  00000 00000  00000 00000  00000 00000    00000 00000  00000 00000  00000 00000  00000 00000  00000 00000
1118:  00000 00000  00000 00000  00000 00000  00000 00000  00000 00000    00000 00000  00000 00000  00000 00000  00000 00000  00000 00000
1119:  00000 00000  00000 00000  00000 00000  00000 00000  00000 00000    00000 00000  00000 00000  00000 00000  00000 00000  00000 00000
1120:  00000 00000  00000 00000  00000 00000  00000 00000  00000 00000    00000 00000  00000 00000  00000 00000  00000 00000  00000 00000
1121:  00000 00000  00000 00000  00000 00000  00000 00000  00000 00000    00000 00000  00000 00000  00000 00000  00000 00000  00000 00000
1122:  00000 00000  00000 00000  00000 00000  00000 00000  00000 00000    00000 00000  00000 00000  00000 00000  00000 00000  00000 00000
1123:  00000 00000  00000 00000  00000 00000  00000 00000  00000 00000    00000 00000  00000 00000  00000 00000  00000 00000  00000 00000
1124:  00000 00000  00000 00000  00000 00000  00000 00000  00000 00000    00000 00000  00000 00000  00000 00000  00000 00000  00000 00000
1125:  00000 00000  00000 00000  00000 00000  00000 00000  00000 00000    00000 00000  00000 00000  00000 00000  00000 00000  00000 00000
1126:  00000 00000  00000 00000  00000 00000  00000 00000  00000 00000    00000 00000  00000 00000  00000 00000  00000 00000  00000 00000
1127:  00000 00000  00000 00000  00000 00000  00000 00000  00000 00000    00000 00000  00000 00000  00000 00000  00000 00000  00000 00000
1128:  00000 00000  00000 00000  00000 00000  00000 00000  00000 00000    00000 00000  00000 00000  00000 00000  00000 00000  00000 00000
1129:  00000 00000  00000 00000  00000 00000  00000 00000  00000 00000    00000 00000  00000 00000  00000 00000  00000 00000  00000 00000
1130:  00000 00000  00000 00000  00000 00000  00000 00000  00000 00000    00000 00000  00000 00000  00000 00000  00000 00000  00000 00000
1131:  00000 00000  00000 00000  00000 00000  00000 00000  00000 00000    00000 00000  00000 00000  00000 00000  00000 00000  00000 00000
1132:  00000 00000  00000 00000  00000 00000  00000 00000  00000 00000    00000 00000  00000 00000  00000 00000  00000 00000  00000 00000
1133:  00000 00000  00000 00000  00000 00000  00000 00000  00000 00000    00000 00000  00000 00000  00000 00000  00000 00000  00000 00000
1134:  00000 00000  00000 00000  00000 00000  00000 00000  00000 00000    00000 00000  00000 00000  00000 00000  00000 00000  00000 00000
1135:  00000 00000  00000 00000  00000 00000  00000 00000  00000 00000    00000 00000  00000 00000  00000 00000  00000 00000  00000 00000
1136:  00000 00000  00000 00000  00000 00000  00000 00000  00000 00000    00000 00000  00000 00000  00000 00000  00000 00000  00000 00000
1137:  00000 00000  00000 00000  00000 00000  00000 00000  00000 00000    00000 00000  00000 00000  00000 00000  00000 00000  00000 00000
1138:  00000 00000  00000 00000  00000 00000  00000 00000  00000 00000    00000 00000  00000 00000  00000 00000  00000 00000  00000 00000
1139:  00000 00000  00000 00000  00000 00000  00000 00000  00000 00000    00000 00000  00000 00000  00000 00000  00000 00000  00000 00000
1140:  00000 00000  00000 00000  00000 00000  00000 00000  00000 00000    00000 00000  00000 00000  00000 00000  00000 00000  00000 00000
1141:  00000 00000  00000 00000  00000 00000  00000 00000  00000 00000    00000 00000  00000 00000  00000 00000  00000 00000  00000 00000
1142:  00000 00000  00000 00000  00000 00000  00000 00000  00000 00000    00000 00000  00000 00000  00000 00000  00000 00000  00000 00000
1143:  00000 00000  00000 00000  00000 00000  00000 00000  00000 00000    00000 00000  00000 00000  00000 00000  00000 00000  00000 00000
1144:  00000 00000  00000 00000  00000 00000  00000 00000  00000 00000    00000 00000  00000 00000  00000 00000  00000 00000  00000 00000
1145:  00000 00000  00000 00000  00000 00000  00000 00000  00000 00000    00000 00000  00000 00000  00000 00000  00000 00000  00000 00000
1146:  00000 00000  00000 00000  00000 00000  00000 00000  00000 00000    00000 00000  00000 00000  00000 00000  00000 00000  00000 00000
1147:  00000 00000  00000 00000  00000 00000  00000 00000  00000 00000    00000 00000  00000 00000  00000 00000  00000 00000  00000 00000
1148:  00000 00000  00000 00000  00000 00000  00000 00000  00000 00000    00000 00000  00000 00000  00000 00000  00000 00000  00000 00000
1149:  00000 00000  00000 00000  00000 00000  00000 00000  00000 00000    00000 00000  00000 00000  00000 00000  00000 00000  00000 00000
```

```
1150:  00000 00000  00000 00000  00000 00000  00000 00000  00000 00000    00000 00000  00000 00000  00000 00000  00000 00000  00000 00000
1151:  00000 00000  00000 00000  00000 00000  00000 00000  00000 00000    00000 00000  00000 00000  00000 00000  00000 00000  00000 00000
1152:  00000 00000  00000 00000  00000 00000  00000 00000  00000 00000    00000 00000  00000 00000  00000 00000  00000 00000  00000 00000
1153:  00000 00000  00000 00000  00000 00000  00000 00000  00000 00000    00000 00000  00000 00000  00000 00000  00000 00000  00000 00000
1154:  00000 00000  00000 00000  00000 00000  00000 00000  00000 00000    00000 00000  00000 00000  00000 00000  00000 00000  00000 00000
1155:  00000 00000  00000 00000  00000 00000  00000 00000  00000 00000    00000 00000  00000 00000  00000 00000  00000 00000  00000 00000
1156:  00000 00000  00000 00000  00000 00000  00000 00000  00000 00000    00000 00000  00000 00000  00000 00000  00000 00000  00000 00000
1157:  00000 00000  00000 00000  00000 00000  00000 00000  00000 00000    00000 00000  00000 00000  00000 00000  00000 00000  00000 00000
1158:  00000 00000  00000 00000  00000 00000  00000 00000  00000 00000    00000 00000  00000 00000  00000 00000  00000 00000  00000 00000
1159:  00000 00000  00000 00000  00000 00000  00000 00000  00000 00000    00000 00000  00000 00000  00000 00000  00000 00000  00000 00000
1160:  00000 00000  00000 00000  00000 00000  00000 00000  00000 00000    00000 00000  00000 00000  00000 00000  00000 00000  00000 00000
1161:  00000 00000  00000 00000  00000 00000  00000 00000  00000 00000    00000 00000  00000 00000  00000 00000  00000 00000  00000 00000
1162:  00000 00000  00000 00000  00000 00000  00000 00000  00000 00000    00000 00000  00000 00000  00000 00000  00000 00000  00000 00000
1163:  00000 00000  00000 00000  00000 00000  00000 00000  00000 00000    00000 00000  00000 00000  00000 00000  00000 00000  00000 00000
1164:  00000 00000  00000 00000  00000 00000  00000 00000  00000 00000    00000 00000  00000 00000  00000 00000  00000 00000  00000 00000
1165:  00000 00000  00000 00000  00000 00000  00000 00000  00000 00000    00000 00000  00000 00000  00000 00000  00000 00000  00000 00000
1166:  00000 00000  00000 00000  00000 00000  00000 00000  00000 00000    00000 00000  00000 00000  00000 00000  00000 00000  00000 00000
1167:  00000 00000  00000 00000  00000 00000  00000 00000  00000 00000    00000 00000  00000 00000  00000 00000  00000 00000  00000 00000
1168:  00000 00000  00000 00000  00000 00000  00000 00000  00000 00000    00000 00000  00000 00000  00000 00000  00000 00000  00000 00000
1169:  00000 00000  00000 00000  00000 00000  00000 00000  00000 00000    00000 00000  00000 00000  00000 00000  00000 00000  00000 00000
1170:  00000 00000  00000 00000  00000 00000  00000 00000  00000 00000    00000 00000  00000 00000  00000 00000  00000 00000  00000 00000
1171:  00000 00000  00000 00000  00000 00000  00000 00000  00000 00000    00000 00000  00000 00000  00000 00000  00000 00000  00000 00000
1172:  00000 00000  00000 00000  00000 00000  00000 00000  00000 00000    00000 00000  00000 00000  00000 00000  00000 00000  00000 00000
1173:  00000 00000  00000 00000  00000 00000  00000 00000  00000 00000    00000 00000  00000 00000  00000 00000  00000 00000  00000 00000
1174:  00000 00000  00000 00000  00000 00000  00000 00000  00000 00000    00000 00000  00000 00000  00000 00000  00000 00000  00000 00000
1175:  00000 00000  00000 00000  00000 00000  00000 00000  00000 00000    00000 00000  00000 00000  00000 00000  00000 00000  00000 00000
1176:  00000 00000  00000 00000  00000 00000  00000 00000  00000 00000    00000 00000  00000 00000  00000 00000  00000 00000  00000 00000
1177:  00000 00000  00000 00000  00000 00000  00000 00000  00000 00000    00000 00000  00000 00000  00000 00000  00000 00000  00000 00000
1178:  00000 00000  00000 00000  00000 00000  00000 00000  00000 00000    00000 00000  00000 00000  00000 00000  00000 00000  00000 00000
1179:  00000 00000  00000 00000  00000 00000  00000 00000  00000 00000    00000 00000  00000 00000  00000 00000  00000 00000  00000 00000
1180:  00000 00000  00000 00000  00000 00000  00000 00000  00000 00000    00000 00000  00000 00000  00000 00000  00000 00000  00000 00000
1181:  00000 00000  00000 00000  00000 00000  00000 00000  00000 00000    00000 00000  00000 00000  00000 00000  00000 00000  00000 00000
1182:  00000 00000  00000 00000  00000 00000  00000 00000  00000 00000    00000 00000  00000 00000  00000 00000  00000 00000  00000 00000
1183:  00000 00000  00000 00000  00000 00000  00000 00000  00000 00000    00000 00000  00000 00000  00000 00000  00000 00000  00000 00000
1184:  00000 00000  00000 00000  00000 00000  00000 00000  00000 00000    00000 00000  00000 00000  00000 00000  00000 00000  00000 00000
1185:  00000 00000  00000 00000  00000 00000  00000 00000  00000 00000    00000 00000  00000 00000  00000 00000  00000 00000  00000 00000
1186:  00000 00000  00000 00000  00000 00000  00000 00000  00000 00000    00000 00000  00000 00000  00000 00000  00000 00000  00000 00000
1187:  00000 00000  00000 00000  00000 00000  00000 00000  00000 00000    00000 00000  00000 00000  00000 00000  00000 00000  00000 00000
1188:  00000 00000  00000 00000  00000 00000  00000 00000  00000 00000    00000 00000  00000 00000  00000 00000  00000 00000  00000 00000
1189:  00000 00000  00000 00000  00000 00000  00000 00000  00000 00000    00000 00000  00000 00000  00000 00000  00000 00000  00000 00000
1190:  00000 00000  00000 00000  00000 00000  00000 00000  00000 00000    00000 00000  00000 00000  00000 00000  00000 00000  00000 00000
1191:  00000 00000  00000 00000  00000 00000  00000 00000  00000 00000    00000 00000  00000 00000  00000 00000  00000 00000  00000 00000
1192:  00000 00000  00000 00000  00000 00000  00000 00000  00000 00000    00000 00000  00000 00000  00000 00000  00000 00000  00000 00000
1193:  00000 00000  00000 00000  00000 00000  00000 00000  00000 00000    00000 00000  00000 00000  00000 00000  00000 00000  00000 00000
1194:  00000 00000  00000 00000  00000 00000  00000 00000  00000 00000    00000 00000  00000 00000  00000 00000  00000 00000  00000 00000
1195:  00000 00000  00000 00000  00000 00000  00000 00000  00000 00000    00000 00000  00000 00000  00000 00000  00000 00000  00000 00000
1196:  00000 00000  00000 00000  00000 00000  00000 00000  00000 00000    00000 00000  00000 00000  00000 00000  00000 00000  00000 00000
1197:  00000 00000  00000 00000  00000 00000  00000 00000  00000 00000    00000 00000  00000 00000  00000 00000  00000 00000  00000 00000
1198:  00000 00000  00000 00000  00000 00000  00000 00000  00000 00000    00000 00000  00000 00000  00000 00000  00000 00000  00000 00000
1199:  00000 00000  00000 00000  00000 00000  00000 00000  00000 00000    00000 00000  00000 00000  00000 00000  00000 00000  00000 00000
```

```
1200:  0000000000  0000000000  0000000000  0000000000  0000000000    0000000000  0000000000  0000000000  0000000000  0000000000
1201:  0000000000  0000000000  0000000000  0000000000  0000000000    0000000000  0000000000  0000000000  0000000000  0000000000
1202:  0000000000  0000000000  0000000000  0000000000  0000000000    0000000000  0000000000  0000000000  0000000000  0000000000
1203:  0000000000  0000000000  0000000000  0000000000  0000000000    0000000000  0000000000  0000000000  0000000000  0000000000
1204:  0000000000  0000000000  0000000000  0000000000  0000000000    0000000000  0000000000  0000000000  0000000000  0000000000
1205:  0000000000  0000000000  0000000000  0000000000  0000000000    0000000000  0000000000  0000000000  0000000000  0000000000
1206:  0000000000  0000000000  0000000000  0000000000  0000000000    0000000000  0000000000  0000000000  0000000000  0000000000
1207:  0000000000  0000000000  0000000000  0000000000  0000000000    0000000000  0000000000  0000000000  0000000000  0000000000
1208:  0000000000  0000000000  0000000000  0000000000  0000000000    0000000000  0000000000  0000000000  0000000000  0000000000
1209:  0000000000  0000000000  0000000000  0000000000  0000000000    0000000000  0000000000  0000000000  0000000000  0000000000
1210:  0000000000  0000000000  0000000000  0000000000  0000000000    0000000000  0000000000  0000000000  0000000000  0000000000
1211:  0000000000  0000000000  0000000000  0000000000  0000000000    0000000000  0000000000  0000000000  0000000000  0000000000
1212:  0000000000  0000000000  0000000000  0000000000  0000000000    0000000000  0000000000  0000000000  0000000000  0000000000
1213:  0000000000  0000000000  0000000000  0000000000  0000000000    0000000000  0000000000  0000000000  0000000000  0000000000
1214:  0000000000  0000000000  0000000000  0000000000  0000000000    0000000000  0000000000  0000000000  0000000000  0000000000
1215:  0000000000  0000000000  0000000000  0000000000  0000000000    0000000000  0000000000  0000000000  0000000000  0000000000
1216:  0000000000  0000000000  0000000000  0000000000  0000000000    0000000000  0000000000  0000000000  0000000000  0000000000
1217:  0000000000  0000000000  0000000000  0000000000  0000000000    0000000000  0000000000  0000000000  0000000000  0000000000
1218:  0000000000  0000000000  0000000000  0000000000  0000000000    0000000000  0000000000  0000000000  0000000000  0000000000
1219:  0000000000  0000000000  0000000000  0000000000  0000000000    0000000000  0000000000  0000000000  0000000000  0000000000
1220:  0000000000  0000000000  0000000000  0000000000  0000000000    0000000000  0000000000  0000000000  0000000000  0000000000
1221:  0000000000  0000000000  0000000000  0000000000  0000000000    0000000000  0000000000  0000000000  0000000000  0000000000
1222:  0000000000  0000000000  0000000000  0000000000  0000000000    0000000000  0000000000  0000000000  0000000000  0000000000
1223:  0000000000  0000000000  0000000000  0000000000  0000000000    0000000000  0000000000  0000000000  0000000000  0000000000
1224:  0000000000  0000000000  0000000000  0000000000  0000000000    0000000000  0000000000  0000000000  0000000000  0000000000
1225:  0000000000  0000000000  0000000000  0000000000  0000000000    0000000000  0000000000  0000000000  0000000000  0000000000
1226:  0000000000  0000000000  0000000000  0000000000  0000000000    0000000000  0000000000  0000000000  0000000000  0000000000
1227:  0000000000  0000000000  0000000000  0000000000  0000000000    0000000000  0000000000  0000000000  0000000000  0000000000
1228:  0000000000  0000000000  0000000000  0000000000  0000000000    0000000000  0000000000  0000000000  0000000000  0000000000
1229:  0000000000  0000000000  0000000000  0000000000  0000000000    0000000000  0000000000  0000000000  0000000000  0000000000
1230:  0000000000  0000000000  0000000000  0000000000  0000000000    0000000000  0000000000  0000000000  0000000000  0000000000
1231:  0000000000  0000000000  0000000000  0000000000  0000000000    0000000000  0000000000  0000000000  0000000000  0000000000
1232:  0000000000  0000000000  0000000000  0000000000  0000000000    0000000000  0000000000  0000000000  0000000000  0000000000
1233:  0000000000  0000000000  0000000000  0000000000  0000000000    0000000000  0000000000  0000000000  0000000000  0000000000
1234:  0000000000  0000000000  0000000000  0000000000  0000000000    0000000000  0000000000  0000000000  0000000000  0000000000
1235:  0000000000  0000000000  0000000000  0000000000  0000000000    0000000000  0000000000  0000000000  0000000000  0000000000
1236:  0000000000  0000000000  0000000000  0000000000  0000000000    0000000000  0000000000  0000000000  0000000000  0000000000
1237:  0000000000  0000000000  0000000000  0000000000  0000000000    0000000000  0000000000  0000000000  0000000000  0000000000
1238:  0000000000  0000000000  0000000000  0000000000  0000000000    0000000000  0000000000  0000000000  0000000000  0000000000
1239:  0000000000  0000000000  0000000000  0000000000  0000000000    0000000000  0000000000  0000000000  0000000000  0000000000
1240:  0000000000  0000000000  0000000000  0000000000  0000000000    0000000000  0000000000  0000000000  0000000000  0000000000
1241:  0000000000  0000000000  0000000000  0000000000  0000000000    0000000000  0000000000  0000000000  0000000000  0000000000
1242:  0000000000  0000000000  0000000000  0000000000  0000000000    0000000000  0000000000  0000000000  0000000000  0000000000
1243:  0000000000  0000000000  0000000000  0000000000  0000000000    0000000000  0000000000  0000000000  0000000000  0000000000
1244:  0000000000  0000000000  0000000000  0000000000  0000000000    0000000000  0000000000  0000000000  0000000000  0000000000
1245:  0000000000  0000000000  0000000000  0000000000  0000000000    0000000000  0000000000  0000000000  0000000000  0000000000
1246:  0000000000  0000000000  0000000000  0000000000  0000000000    0000000000  0000000000  0000000000  0000000000  0000000000
1247:  0000000000  0000000000  0000000000  0000000000  0000000000    0000000000  0000000000  0000000000  0000000000  0000000000
1248:  0000000000  0000000000  0000000000  0000000000  0000000000    0000000000  0000000000  0000000000  0000000000  0000000000
1249:  0000000000  0000000000  0000000000  0000000000  0000000000    0000000000  0000000000  0000000000  0000000000  0000000000
```

```
1250:  00000 00000  00000 00000  00000 00000  00000 00000  00000 00000    00000 00000  00000 00000  00000 00000  00000 00000  00000 00000
1251:  00000 00000  00000 00000  00000 00000  00000 00000  00000 00000    00000 00000  00000 00000  00000 00000  00000 00000  00000 00000
1252:  00000 00000  00000 00000  00000 00000  00000 00000  00000 00000    00000 00000  00000 00000  00000 00000  00000 00000  00000 00000
1253:  00000 00000  00000 00000  00000 00000  00000 00000  00000 00000    00000 00000  00000 00000  00000 00000  00000 00000  00000 00000
1254:  00000 00000  00000 00000  00000 00000  00000 00000  00000 00000    00000 00000  00000 00000  00000 00000  00000 00000  00000 00000
1255:  00000 00000  00000 00000  00000 00000  00000 00000  00000 00000    00000 00000  00000 00000  00000 00000  00000 00000  00000 00000
1256:  00000 00000  00000 00000  00000 00000  00000 00000  00000 00000    00000 00000  00000 00000  00000 00000  00000 00000  00000 00000
1257:  00000 00000  00000 00000  00000 00000  00000 00000  00000 00000    00000 00000  00000 00000  00000 00000  00000 00000  00000 00000
1258:  00000 00000  00000 00000  00000 00000  00000 00000  00000 00000    00000 00000  00000 00000  00000 00000  00000 00000  00000 00000
1259:  00000 00000  00000 00000  00000 00000  00000 00000  00000 00000    00000 00000  00000 00000  00000 00000  00000 00000  00000 00000
1260:  00000 00000  00000 00000  00000 00000  00000 00000  00000 00000    00000 00000  00000 00000  00000 00000  00000 00000  00000 00000
1261:  00000 00000  00000 00000  00000 00000  00000 00000  00000 00000    00000 00000  00000 00000  00000 00000  00000 00000  00000 00000
1262:  00000 00000  00000 00000  00000 00000  00000 00000  00000 00000    00000 00000  00000 00000  00000 00000  00000 00000  00000 00000
1263:  00000 00000  00000 00000  00000 00000  00000 00000  00000 00000    00000 00000  00000 00000  00000 00000  00000 00000  00000 00000
1264:  00000 00000  00000 00000  00000 00000  00000 00000  00000 00000    00000 00000  00000 00000  00000 00000  00000 00000  00000 00000
1265:  00000 00000  00000 00000  00000 00000  00000 00000  00000 00000    00000 00000  00000 00000  00000 00000  00000 00000  00000 00000
1266:  00000 00000  00000 00000  00000 00000  00000 00000  00000 00000    00000 00000  00000 00000  00000 00000  00000 00000  00000 00000
1267:  00000 00000  00000 00000  00000 00000  00000 00000  00000 00000    00000 00000  00000 00000  00000 00000  00000 00000  00000 00000
1268:  00000 00000  00000 00000  00000 00000  00000 00000  00000 00000    00000 00000  00000 00000  00000 00000  00000 00000  00000 00000
1269:  00000 00000  00000 00000  00000 00000  00000 00000  00000 00000    00000 00000  00000 00000  00000 00000  00000 00000  00000 00000
1270:  00000 00000  00000 00000  00000 00000  00000 00000  00000 00000    00000 00000  00000 00000  00000 00000  00000 00000  00000 00000
1271:  00000 00000  00000 00000  00000 00000  00000 00000  00000 00000    00000 00000  00000 00000  00000 00000  00000 00000  00000 00000
1272:  00000 00000  00000 00000  00000 00000  00000 00000  00000 00000    00000 00000  00000 00000  00000 00000  00000 00000  00000 00000
1273:  00000 00000  00000 00000  00000 00000  00000 00000  00000 00000    00000 00000  00000 00000  00000 00000  00000 00000  00000 00000
1274:  00000 00000  00000 00000  00000 00000  00000 00000  00000 00000    00000 00000  00000 00000  00000 00000  00000 00000  00000 00000
1275:  00000 00000  00000 00000  00000 00000  00000 00000  00000 00000    00000 00000  00000 00000  00000 00000  00000 00000  00000 00000
1276:  00000 00000  00000 00000  00000 00000  00000 00000  00000 00000    00000 00000  00000 00000  00000 00000  00000 00000  00000 00000
1277:  00000 00000  00000 00000  00000 00000  00000 00000  00000 00000    00000 00000  00000 00000  00000 00000  00000 00000  00000 00000
1278:  00000 00000  00000 00000  00000 00000  00000 00000  00000 00000    00000 00000  00000 00000  00000 00000  00000 00000  00000 00000
1279:  00000 00000  00000 00000  00000 00000  00000 00000  00000 00000    00000 00000  00000 00000  00000 00000  00000 00000  00000 00000
1280:  00000 00000  00000 00000  00000 00000  00000 00000  00000 00000    00000 00000  00000 00000  00000 00000  00000 00000  00000 00000
1281:  00000 00000  00000 00000  00000 00000  00000 00000  00000 00000    00000 00000  00000 00000  00000 00000  00000 00000  00000 00000
1282:  00000 00000  00000 00000  00000 00000  00000 00000  00000 00000    00000 00000  00000 00000  00000 00000  00000 00000  00000 00000
1283:  00000 00000  00000 00000  00000 00000  00000 00000  00000 00000    00000 00000  00000 00000  00000 00000  00000 00000  00000 00000
1284:  00000 00000  00000 00000  00000 00000  00000 00000  00000 00000    00000 00000  00000 00000  00000 00000  00000 00000  00000 00000
1285:  00000 00000  00000 00000  00000 00000  00000 00000  00000 00000    00000 00000  00000 00000  00000 00000  00000 00000  00000 00000
1286:  00000 00000  00000 00000  00000 00000  00000 00000  00000 00000    00000 00000  00000 00000  00000 00000  00000 00000  00000 00000
1287:  00000 00000  00000 00000  00000 00000  00000 00000  00000 00000    00000 00000  00000 00000  00000 00000  00000 00000  00000 00000
1288:  00000 00000  00000 00000  00000 00000  00000 00000  00000 00000    00000 00000  00000 00000  00000 00000  00000 00000  00000 00000
1289:  00000 00000  00000 00000  00000 00000  00000 00000  00000 00000    00000 00000  00000 00000  00000 00000  00000 00000  00000 00000
1290:  00000 00000  00000 00000  00000 00000  00000 00000  00000 00000    00000 00000  00000 00000  00000 00000  00000 00000  00000 00000
1291:  00000 00000  00000 00000  00000 00000  00000 00000  00000 00000    00000 00000  00000 00000  00000 00000  00000 00000  00000 00000
1292:  00000 00000  00000 00000  00000 00000  00000 00000  00000 00000    00000 00000  00000 00000  00000 00000  00000 00000  00000 00000
1293:  00000 00000  00000 00000  00000 00000  00000 00000  00000 00000    00000 00000  00000 00000  00000 00000  00000 00000  00000 00000
1294:  00000 00000  00000 00000  00000 00000  00000 00000  00000 00000    00000 00000  00000 00000  00000 00000  00000 00000  00000 00000
1295:  00000 00000  00000 00000  00000 00000  00000 00000  00000 00000    00000 00000  00000 00000  00000 00000  00000 00000  00000 00000
1296:  00000 00000  00000 00000  00000 00000  00000 00000  00000 00000    00000 00000  00000 00000  00000 00000  00000 00000  00000 00000
1297:  00000 00000  00000 00000  00000 00000  00000 00000  00000 00000    00000 00000  00000 00000  00000 00000  00000 00000  00000 00000
1298:  00000 00000  00000 00000  00000 00000  00000 00000  00000 00000    00000 00000  00000 00000  00000 00000  00000 00000  00000 00000
1299:  00000 00000  00000 00000  00000 00000  00000 00000  00000 00000    00000 00000  00000 00000  00000 00000  00000 00000  00000 00000
```

```
1300:  00000 00000  00000 00000  00000 00000  00000 00000  00000 00000    00000 00000  00000 00000  00000 00000  00000 00000  00000 00000
1301:  00000 00000  00000 00000  00000 00000  00000 00000  00000 00000    00000 00000  00000 00000  00000 00000  00000 00000  00000 00000
1302:  00000 00000  00000 00000  00000 00000  00000 00000  00000 00000    00000 00000  00000 00000  00000 00000  00000 00000  00000 00000
1303:  00000 00000  00000 00000  00000 00000  00000 00000  00000 00000    00000 00000  00000 00000  00000 00000  00000 00000  00000 00000
1304:  00000 00000  00000 00000  00000 00000  00000 00000  00000 00000    00000 00000  00000 00000  00000 00000  00000 00000  00000 00000
1305:  00000 00000  00000 00000  00000 00000  00000 00000  00000 00000    00000 00000  00000 00000  00000 00000  00000 00000  00000 00000
1306:  00000 00000  00000 00000  00000 00000  00000 00000  00000 00000    00000 00000  00000 00000  00000 00000  00000 00000  00000 00000
1307:  00000 00000  00000 00000  00000 00000  00000 00000  00000 00000    00000 00000  00000 00000  00000 00000  00000 00000  00000 00000
1308:  00000 00000  00000 00000  00000 00000  00000 00000  00000 00000    00000 00000  00000 00000  00000 00000  00000 00000  00000 00000
1309:  00000 00000  00000 00000  00000 00000  00000 00000  00000 00000    00000 00000  00000 00000  00000 00000  00000 00000  00000 00000
1310:  00000 00000  00000 00000  00000 00000  00000 00000  00000 00000    00000 00000  00000 00000  00000 00000  00000 00000  00000 00000
1311:  00000 00000  00000 00000  00000 00000  00000 00000  00000 00000    00000 00000  00000 00000  00000 00000  00000 00000  00000 00000
1312:  00000 00000  00000 00000  00000 00000  00000 00000  00000 00000    00000 00000  00000 00000  00000 00000  00000 00000  00000 00000
1313:  00000 00000  00000 00000  00000 00000  00000 00000  00000 00000    00000 00000  00000 00000  00000 00000  00000 00000  00000 00000
1314:  00000 00000  00000 00000  00000 00000  00000 00000  00000 00000    00000 00000  00000 00000  00000 00000  00000 00000  00000 00000
1315:  00000 00000  00000 00000  00000 00000  00000 00000  00000 00000    00000 00000  00000 00000  00000 00000  00000 00000  00000 00000
1316:  00000 00000  00000 00000  00000 00000  00000 00000  00000 00000    00000 00000  00000 00000  00000 00000  00000 00000  00000 00000
1317:  00000 00000  00000 00000  00000 00000  00000 00000  00000 00000    00000 00000  00000 00000  00000 00000  00000 00000  00000 00000
1318:  00000 00000  00000 00000  00000 00000  00000 00000  00000 00000    00000 00000  00000 00000  00000 00000  00000 00000  00000 00000
1319:  00000 00000  00000 00000  00000 00000  00000 00000  00000 00000    00000 00000  00000 00000  00000 00000  00000 00000  00000 00000
1320:  00000 00000  00000 00000  00000 00000  00000 00000  00000 00000    00000 00000  00000 00000  00000 00000  00000 00000  00000 00000
1321:  00000 00000  00000 00000  00000 00000  00000 00000  00000 00000    00000 00000  00000 00000  00000 00000  00000 00000  00000 00000
1322:  00000 00000  00000 00000  00000 00000  00000 00000  00000 00000    00000 00000  00000 00000  00000 00000  00000 00000  00000 00000
1323:  00000 00000  00000 00000  00000 00000  00000 00000  00000 00000    00000 00000  00000 00000  00000 00000  00000 00000  00000 00000
1324:  00000 00000  00000 00000  00000 00000  00000 00000  00000 00000    00000 00000  00000 00000  00000 00000  00000 00000  00000 00000
1325:  00000 00000  00000 00000  00000 00000  00000 00000  00000 00000    00000 00000  00000 00000  00000 00000  00000 00000  00000 00000
1326:  00000 00000  00000 00000  00000 00000  00000 00000  00000 00000    00000 00000  00000 00000  00000 00000  00000 00000  00000 00000
1327:  00000 00000  00000 00000  00000 00000  00000 00000  00000 00000    00000 00000  00000 00000  00000 00000  00000 00000  00000 00000
1328:  00000 00000  00000 00000  00000 00000  00000 00000  00000 00000    00000 00000  00000 00000  00000 00000  00000 00000  00000 00000
1329:  00000 00000  00000 00000  00000 00000  00000 00000  00000 00000    00000 00000  00000 00000  00000 00000  00000 00000  00000 00000
1330:  00000 00000  00000 00000  00000 00000  00000 00000  00000 00000    00000 00000  00000 00000  00000 00000  00000 00000  00000 00000
1331:  00000 00000  00000 00000  00000 00000  00000 00000  00000 00000    00000 00000  00000 00000  00000 00000  00000 00000  00000 00000
1332:  00000 00000  00000 00000  00000 00000  00000 00000  00000 00000    00000 00000  00000 00000  00000 00000  00000 00000  00000 00000
1333:  00000 00000  00000 00000  00000 00000  00000 00000  00000 00000    00000 00000  00000 00000  00000 00000  00000 00000  00000 00000
1334:  00000 00000  00000 00000  00000 00000  00000 00000  00000 00000    00000 00000  00000 00000  00000 00000  00000 00000  00000 00000
1335:  00000 00000  00000 00000  00000 00000  00000 00000  00000 00000    00000 00000  00000 00000  00000 00000  00000 00000  00000 00000
1336:  00000 00000  00000 00000  00000 00000  00000 00000  00000 00000    00000 00000  00000 00000  00000 00000  00000 00000  00000 00000
1337:  00000 00000  00000 00000  00000 00000  00000 00000  00000 00000    00000 00000  00000 00000  00000 00000  00000 00000  00000 00000
1338:  00000 00000  00000 00000  00000 00000  00000 00000  00000 00000    00000 00000  00000 00000  00000 00000  00000 00000  00000 00000
1339:  00000 00000  00000 00000  00000 00000  00000 00000  00000 00000    00000 00000  00000 00000  00000 00000  00000 00000  00000 00000
1340:  00000 00000  00000 00000  00000 00000  00000 00000  00000 00000    00000 00000  00000 00000  00000 00000  00000 00000  00000 00000
1341:  00000 00000  00000 00000  00000 00000  00000 00000  00000 00000    00000 00000  00000 00000  00000 00000  00000 00000  00000 00000
1342:  00000 00000  00000 00000  00000 00000  00000 00000  00000 00000    00000 00000  00000 00000  00000 00000  00000 00000  00000 00000
1343:  00000 00000  00000 00000  00000 00000  00000 00000  00000 00000    00000 00000  00000 00000  00000 00000  00000 00000  00000 00000
1344:  00000 00000  00000 00000  00000 00000  00000 00000  00000 00000    00000 00000  00000 00000  00000 00000  00000 00000  00000 00000
1345:  00000 00000  00000 00000  00000 00000  00000 00000  00000 00000    00000 00000  00000 00000  00000 00000  00000 00000  00000 00000
1346:  00000 00000  00000 00000  00000 00000  00000 00000  00000 00000    00000 00000  00000 00000  00000 00000  00000 00000  00000 00000
1347:  00000 00000  00000 00000  00000 00000  00000 00000  00000 00000    00000 00000  00000 00000  00000 00000  00000 00000  00000 00000
1348:  00000 00000  00000 00000  00000 00000  00000 00000  00000 00000    00000 00000  00000 00000  00000 00000  00000 00000  00000 00000
1349:  00000 00000  00000 00000  00000 00000  00000 00000  00000 00000    00000 00000  00000 00000  00000 00000  00000 00000  00000 00000
```

```
1350:  00000 00000  00000 00000  00000 00000  00000 00000  00000 00000    00000 00000  00000 00000  00000 00000  00000 00000  00000 00000
1351:  00000 00000  00000 00000  00000 00000  00000 00000  00000 00000    00000 00000  00000 00000  00000 00000  00000 00000  00000 00000
1352:  00000 00000  00000 00000  00000 00000  00000 00000  00000 00000    00000 00000  00000 00000  00000 00000  00000 00000  00000 00000
1353:  00000 00000  00000 00000  00000 00000  00000 00000  00000 00000    00000 00000  00000 00000  00000 00000  00000 00000  00000 00000
1354:  00000 00000  00000 00000  00000 00000  00000 00000  00000 00000    00000 00000  00000 00000  00000 00000  00000 00000  00000 00000
1355:  00000 00000  00000 00000  00000 00000  00000 00000  00000 00000    00000 00000  00000 00000  00000 00000  00000 00000  00000 00000
1356:  00000 00000  00000 00000  00000 00000  00000 00000  00000 00000    00000 00000  00000 00000  00000 00000  00000 00000  00000 00000
1357:  00000 00000  00000 00000  00000 00000  00000 00000  00000 00000    00000 00000  00000 00000  00000 00000  00000 00000  00000 00000
1358:  00000 00000  00000 00000  00000 00000  00000 00000  00000 00000    00000 00000  00000 00000  00000 00000  00000 00000  00000 00000
1359:  00000 00000  00000 00000  00000 00000  00000 00000  00000 00000    00000 00000  00000 00000  00000 00000  00000 00000  00000 00000
1360:  00000 00000  00000 00000  00000 00000  00000 00000  00000 00000    00000 00000  00000 00000  00000 00000  00000 00000  00000 00000
1361:  00000 00000  00000 00000  00000 00000  00000 00000  00000 00000    00000 00000  00000 00000  00000 00000  00000 00000  00000 00000
1362:  00000 00000  00000 00000  00000 00000  00000 00000  00000 00000    00000 00000  00000 00000  00000 00000  00000 00000  00000 00000
1363:  00000 00000  00000 00000  00000 00000  00000 00000  00000 00000    00000 00000  00000 00000  00000 00000  00000 00000  00000 00000
1364:  00000 00000  00000 00000  00000 00000  00000 00000  00000 00000    00000 00000  00000 00000  00000 00000  00000 00000  00000 00000
1365:  00000 00000  00000 00000  00000 00000  00000 00000  00000 00000    00000 00000  00000 00000  00000 00000  00000 00000  00000 00000
1366:  00000 00000  00000 00000  00000 00000  00000 00000  00000 00000    00000 00000  00000 00000  00000 00000  00000 00000  00000 00000
1367:  00000 00000  00000 00000  00000 00000  00000 00000  00000 00000    00000 00000  00000 00000  00000 00000  00000 00000  00000 00000
1368:  00000 00000  00000 00000  00000 00000  00000 00000  00000 00000    00000 00000  00000 00000  00000 00000  00000 00000  00000 00000
1369:  00000 00000  00000 00000  00000 00000  00000 00000  00000 00000    00000 00000  00000 00000  00000 00000  00000 00000  00000 00000
1370:  00000 00000  00000 00000  00000 00000  00000 00000  00000 00000    00000 00000  00000 00000  00000 00000  00000 00000  00000 00000
1371:  00000 00000  00000 00000  00000 00000  00000 00000  00000 00000    00000 00000  00000 00000  00000 00000  00000 00000  00000 00000
1372:  00000 00000  00000 00000  00000 00000  00000 00000  00000 00000    00000 00000  00000 00000  00000 00000  00000 00000  00000 00000
1373:  00000 00000  00000 00000  00000 00000  00000 00000  00000 00000    00000 00000  00000 00000  00000 00000  00000 00000  00000 00000
1374:  00000 00000  00000 00000  00000 00000  00000 00000  00000 00000    00000 00000  00000 00000  00000 00000  00000 00000  00000 00000
1375:  00000 00000  00000 00000  00000 00000  00000 00000  00000 00000    00000 00000  00000 00000  00000 00000  00000 00000  00000 00000
1376:  00000 00000  00000 00000  00000 00000  00000 00000  00000 00000    00000 00000  00000 00000  00000 00000  00000 00000  00000 00000
1377:  00000 00000  00000 00000  00000 00000  00000 00000  00000 00000    00000 00000  00000 00000  00000 00000  00000 00000  00000 00000
1378:  00000 00000  00000 00000  00000 00000  00000 00000  00000 00000    00000 00000  00000 00000  00000 00000  00000 00000  00000 00000
1379:  00000 00000  00000 00000  00000 00000  00000 00000  00000 00000    00000 00000  00000 00000  00000 00000  00000 00000  00000 00000
1380:  00000 00000  00000 00000  00000 00000  00000 00000  00000 00000    00000 00000  00000 00000  00000 00000  00000 00000  00000 00000
1381:  00000 00000  00000 00000  00000 00000  00000 00000  00000 00000    00000 00000  00000 00000  00000 00000  00000 00000  00000 00000
1382:  00000 00000  00000 00000  00000 00000  00000 00000  00000 00000    00000 00000  00000 00000  00000 00000  00000 00000  00000 00000
1383:  00000 00000  00000 00000  00000 00000  00000 00000  00000 00000    00000 00000  00000 00000  00000 00000  00000 00000  00000 00000
1384:  00000 00000  00000 00000  00000 00000  00000 00000  00000 00000    00000 00000  00000 00000  00000 00000  00000 00000  00000 00000
1385:  00000 00000  00000 00000  00000 00000  00000 00000  00000 00000    00000 00000  00000 00000  00000 00000  00000 00000  00000 00000
1386:  00000 00000  00000 00000  00000 00000  00000 00000  00000 00000    00000 00000  00000 00000  00000 00000  00000 00000  00000 00000
1387:  00000 00000  00000 00000  00000 00000  00000 00000  00000 00000    00000 00000  00000 00000  00000 00000  00000 00000  00000 00000
1388:  00000 00000  00000 00000  00000 00000  00000 00000  00000 00000    00000 00000  00000 00000  00000 00000  00000 00000  00000 00000
1389:  00000 00000  00000 00000  00000 00000  00000 00000  00000 00000    00000 00000  00000 00000  00000 00000  00000 00000  00000 00000
1390:  00000 00000  00000 00000  00000 00000  00000 00000  00000 00000    00000 00000  00000 00000  00000 00000  00000 00000  00000 00000
1391:  00000 00000  00000 00000  00000 00000  00000 00000  00000 00000    00000 00000  00000 00000  00000 00000  00000 00000  00000 00000
1392:  00000 00000  00000 00000  00000 00000  00000 00000  00000 00000    00000 00000  00000 00000  00000 00000  00000 00000  00000 00000
1393:  00000 00000  00000 00000  00000 00000  00000 00000  00000 00000    00000 00000  00000 00000  00000 00000  00000 00000  00000 00000
1394:  00000 00000  00000 00000  00000 00000  00000 00000  00000 00000    00000 00000  00000 00000  00000 00000  00000 00000  00000 00000
1395:  00000 00000  00000 00000  00000 00000  00000 00000  00000 00000    00000 00000  00000 00000  00000 00000  00000 00000  00000 00000
1396:  00000 00000  00000 00000  00000 00000  00000 00000  00000 00000    00000 00000  00000 00000  00000 00000  00000 00000  00000 00000
1397:  00000 00000  00000 00000  00000 00000  00000 00000  00000 00000    00000 00000  00000 00000  00000 00000  00000 00000  00000 00000
1398:  00000 00000  00000 00000  00000 00000  00000 00000  00000 00000    00000 00000  00000 00000  00000 00000  00000 00000  00000 00000
1399:  00000 00000  00000 00000  00000 00000  00000 00000  00000 00000    00000 00000  00000 00000  00000 00000  00000 00000  00000 00000
```

```
1400:   00000 00000   00000 00000   00000 00000   00000 00000   00000 00000     00000 00000   00000 00000   00000 00000   00000 00000   00000 00000
1401:   00000 00000   00000 00000   00000 00000   00000 00000   00000 00000     00000 00000   00000 00000   00000 00000   00000 00000   00000 00000
1402:   00000 00000   00000 00000   00000 00000   00000 00000   00000 00000     00000 00000   00000 00000   00000 00000   00000 00000   00000 00000
1403:   00000 00000   00000 00000   00000 00000   00000 00000   00000 00000     00000 00000   00000 00000   00000 00000   00000 00000   00000 00000
1404:   00000 00000   00000 00000   00000 00000   00000 00000   00000 00000     00000 00000   00000 00000   00000 00000   00000 00000   00000 00000
1405:   00000 00000   00000 00000   00000 00000   00000 00000   00000 00000     00000 00000   00000 00000   00000 00000   00000 00000   00000 00000
1406:   00000 00000   00000 00000   00000 00000   00000 00000   00000 00000     00000 00000   00000 00000   00000 00000   00000 00000   00000 00000
1407:   00000 00000   00000 00000   00000 00000   00000 00000   00000 00000     00000 00000   00000 00000   00000 00000   00000 00000   00000 00000
1408:   00000 00000   00000 00000   00000 00000   00000 00000   00000 00000     00000 00000   00000 00000   00000 00000   00000 00000   00000 00000
1409:   00000 00000   00000 00000   00000 00000   00000 00000   00000 00000     00000 00000   00000 00000   00000 00000   00000 00000   00000 00000
1410:   00000 00000   00000 00000   00000 00000   00000 00000   00000 00000     00000 00000   00000 00000   00000 00000   00000 00000   00000 00000
1411:   00000 00000   00000 00000   00000 00000   00000 00000   00000 00000     00000 00000   00000 00000   00000 00000   00000 00000   00000 00000
1412:   00000 00000   00000 00000   00000 00000   00000 00000   00000 00000     00000 00000   00000 00000   00000 00000   00000 00000   00000 00000
1413:   00000 00000   00000 00000   00000 00000   00000 00000   00000 00000     00000 00000   00000 00000   00000 00000   00000 00000   00000 00000
1414:   00000 00000   00000 00000   00000 00000   00000 00000   00000 00000     00000 00000   00000 00000   00000 00000   00000 00000   00000 00000
1415:   00000 00000   00000 00000   00000 00000   00000 00000   00000 00000     00000 00000   00000 00000   00000 00000   00000 00000   00000 00000
1416:   00000 00000   00000 00000   00000 00000   00000 00000   00000 00000     00000 00000   00000 00000   00000 00000   00000 00000   00000 00000
1417:   00000 00000   00000 00000   00000 00000   00000 00000   00000 00000     00000 00000   00000 00000   00000 00000   00000 00000   00000 00000
1418:   00000 00000   00000 00000   00000 00000   00000 00000   00000 00000     00000 00000   00000 00000   00000 00000   00000 00000   00000 00000
1419:   00000 00000   00000 00000   00000 00000   00000 00000   00000 00000     00000 00000   00000 00000   00000 00000   00000 00000   00000 00000
1420:   00000 00000   00000 00000   00000 00000   00000 00000   00000 00000     00000 00000   00000 00000   00000 00000   00000 00000   00000 00000
1421:   00000 00000   00000 00000   00000 00000   00000 00000   00000 00000     00000 00000   00000 00000   00000 00000   00000 00000   00000 00000
1422:   00000 00000   00000 00000   00000 00000   00000 00000   00000 00000     00000 00000   00000 00000   00000 00000   00000 00000   00000 00000
1423:   00000 00000   00000 00000   00000 00000   00000 00000   00000 00000     00000 00000   00000 00000   00000 00000   00000 00000   00000 00000
1424:   00000 00000   00000 00000   00000 00000   00000 00000   00000 00000     00000 00000   00000 00000   00000 00000   00000 00000   00000 00000
1425:   00000 00000   00000 00000   00000 00000   00000 00000   00000 00000     00000 00000   00000 00000   00000 00000   00000 00000   00000 00000
1426:   00000 00000   00000 00000   00000 00000   00000 00000   00000 00000     00000 00000   00000 00000   00000 00000   00000 00000   00000 00000
1427:   00000 00000   00000 00000   00000 00000   00000 00000   00000 00000     00000 00000   00000 00000   00000 00000   00000 00000   00000 00000
1428:   00000 00000   00000 00000   00000 00000   00000 00000   00000 00000     00000 00000   00000 00000   00000 00000   00000 00000   00000 00000
1429:   00000 00000   00000 00000   00000 00000   00000 00000   00000 00000     00000 00000   00000 00000   00000 00000   00000 00000   00000 00000
1430:   00000 00000   00000 00000   00000 00000   00000 00000   00000 00000     00000 00000   00000 00000   00000 00000   00000 00000   00000 00000
1431:   00000 00000   00000 00000   00000 00000   00000 00000   00000 00000     00000 00000   00000 00000   00000 00000   00000 00000   00000 00000
1432:   00000 00000   00000 00000   00000 00000   00000 00000   00000 00000     00000 00000   00000 00000   00000 00000   00000 00000   00000 00000
1433:   00000 00000   00000 00000   00000 00000   00000 00000   00000 00000     00000 00000   00000 00000   00000 00000   00000 00000   00000 00000
1434:   00000 00000   00000 00000   00000 00000   00000 00000   00000 00000     00000 00000   00000 00000   00000 00000   00000 00000   00000 00000
1435:   00000 00000   00000 00000   00000 00000   00000 00000   00000 00000     00000 00000   00000 00000   00000 00000   00000 00000   00000 00000
1436:   00000 00000   00000 00000   00000 00000   00000 00000   00000 00000     00000 00000   00000 00000   00000 00000   00000 00000   00000 00000
1437:   00000 00000   00000 00000   00000 00000   00000 00000   00000 00000     00000 00000   00000 00000   00000 00000   00000 00000   00000 00000
1438:   00000 00000   00000 00000   00000 00000   00000 00000   00000 00000     00000 00000   00000 00000   00000 00000   00000 00000   00000 00000
1439:   00000 00000   00000 00000   00000 00000   00000 00000   00000 00000     00000 00000   00000 00000   00000 00000   00000 00000   00000 00000
1440:   00000 00000   00000 00000   00000 00000   00000 00000   00000 00000     00000 00000   00000 00000   00000 00000   00000 00000   00000 00000
1441:   00000 00000   00000 00000   00000 00000   00000 00000   00000 00000     00000 00000   00000 00000   00000 00000   00000 00000   00000 00000
1442:   00000 00000   00000 00000   00000 00000   00000 00000   00000 00000     00000 00000   00000 00000   00000 00000   00000 00000   00000 00000
1443:   00000 00000   00000 00000   00000 00000   00000 00000   00000 00000     00000 00000   00000 00000   00000 00000   00000 00000   00000 00000
1444:   00000 00000   00000 00000   00000 00000   00000 00000   00000 00000     00000 00000   00000 00000   00000 00000   00000 00000   00000 00000
1445:   00000 00000   00000 00000   00000 00000   00000 00000   00000 00000     00000 00000   00000 00000   00000 00000   00000 00000   00000 00000
1446:   00000 00000   00000 00000   00000 00000   00000 00000   00000 00000     00000 00000   00000 00000   00000 00000   00000 00000   00000 00000
1447:   00000 00000   00000 00000   00000 00000   00000 00000   00000 00000     00000 00000   00000 00000   00000 00000   00000 00000   00000 00000
1448:   00000 00000   00000 00000   00000 00000   00000 00000   00000 00000     00000 00000   00000 00000   00000 00000   00000 00000   00000 00000
1449:   00000 00000   00000 00000   00000 00000   00000 00000   00000 00000     00000 00000   00000 00000   00000 00000   00000 00000   00000 00000
```

```
1450:  00000 00000  00000 00000  00000 00000  00000 00000  00000 00000    00000 00000  00000 00000  00000 00000  00000 00000  00000 00000
1451:  00000 00000  00000 00000  00000 00000  00000 00000  00000 00000    00000 00000  00000 00000  00000 00000  00000 00000  00000 00000
1452:  00000 00000  00000 00000  00000 00000  00000 00000  00000 00000    00000 00000  00000 00000  00000 00000  00000 00000  00000 00000
1453:  00000 00000  00000 00000  00000 00000  00000 00000  00000 00000    00000 00000  00000 00000  00000 00000  00000 00000  00000 00000
1454:  00000 00000  00000 00000  00000 00000  00000 00000  00000 00000    00000 00000  00000 00000  00000 00000  00000 00000  00000 00000
1455:  00000 00000  00000 00000  00000 00000  00000 00000  00000 00000    00000 00000  00000 00000  00000 00000  00000 00000  00000 00000
1456:  00000 00000  00000 00000  00000 00000  00000 00000  00000 00000    00000 00000  00000 00000  00000 00000  00000 00000  00000 00000
1457:  00000 00000  00000 00000  00000 00000  00000 00000  00000 00000    00000 00000  00000 00000  00000 00000  00000 00000  00000 00000
1458:  00000 00000  00000 00000  00000 00000  00000 00000  00000 00000    00000 00000  00000 00000  00000 00000  00000 00000  00000 00000
1459:  00000 00000  00000 00000  00000 00000  00000 00000  00000 00000    00000 00000  00000 00000  00000 00000  00000 00000  00000 00000
1460:  00000 00000  00000 00000  00000 00000  00000 00000  00000 00000    00000 00000  00000 00000  00000 00000  00000 00000  00000 00000
1461:  00000 00000  00000 00000  00000 00000  00000 00000  00000 00000    00000 00000  00000 00000  00000 00000  00000 00000  00000 00000
1462:  00000 00000  00000 00000  00000 00000  00000 00000  00000 00000    00000 00000  00000 00000  00000 00000  00000 00000  00000 00000
1463:  00000 00000  00000 00000  00000 00000  00000 00000  00000 00000    00000 00000  00000 00000  00000 00000  00000 00000  00000 00000
1464:  00000 00000  00000 00000  00000 00000  00000 00000  00000 00000    00000 00000  00000 00000  00000 00000  00000 00000  00000 00000
1465:  00000 00000  00000 00000  00000 00000  00000 00000  00000 00000    00000 00000  00000 00000  00000 00000  00000 00000  00000 00000
1466:  00000 00000  00000 00000  00000 00000  00000 00000  00000 00000    00000 00000  00000 00000  00000 00000  00000 00000  00000 00000
1467:  00000 00000  00000 00000  00000 00000  00000 00000  00000 00000    00000 00000  00000 00000  00000 00000  00000 00000  00000 00000
1468:  00000 00000  00000 00000  00000 00000  00000 00000  00000 00000    00000 00000  00000 00000  00000 00000  00000 00000  00000 00000
1469:  00000 00000  00000 00000  00000 00000  00000 00000  00000 00000    00000 00000  00000 00000  00000 00000  00000 00000  00000 00000
1470:  00000 00000  00000 00000  00000 00000  00000 00000  00000 00000    00000 00000  00000 00000  00000 00000  00000 00000  00000 00000
1471:  00000 00000  00000 00000  00000 00000  00000 00000  00000 00000    00000 00000  00000 00000  00000 00000  00000 00000  00000 00000
1472:  00000 00000  00000 00000  00000 00000  00000 00000  00000 00000    00000 00000  00000 00000  00000 00000  00000 00000  00000 00000
1473:  00000 00000  00000 00000  00000 00000  00000 00000  00000 00000    00000 00000  00000 00000  00000 00000  00000 00000  00000 00000
1474:  00000 00000  00000 00000  00000 00000  00000 00000  00000 00000    00000 00000  00000 00000  00000 00000  00000 00000  00000 00000
1475:  00000 00000  00000 00000  00000 00000  00000 00000  00000 00000    00000 00000  00000 00000  00000 00000  00000 00000  00000 00000
1476:  00000 00000  00000 00000  00000 00000  00000 00000  00000 00000    00000 00000  00000 00000  00000 00000  00000 00000  00000 00000
1477:  00000 00000  00000 00000  00000 00000  00000 00000  00000 00000    00000 00000  00000 00000  00000 00000  00000 00000  00000 00000
1478:  00000 00000  00000 00000  00000 00000  00000 00000  00000 00000    00000 00000  00000 00000  00000 00000  00000 00000  00000 00000
1479:  00000 00000  00000 00000  00000 00000  00000 00000  00000 00000    00000 00000  00000 00000  00000 00000  00000 00000  00000 00000
1480:  00000 00000  00000 00000  00000 00000  00000 00000  00000 00000    00000 00000  00000 00000  00000 00000  00000 00000  00000 00000
1481:  00000 00000  00000 00000  00000 00000  00000 00000  00000 00000    00000 00000  00000 00000  00000 00000  00000 00000  00000 00000
1482:  00000 00000  00000 00000  00000 00000  00000 00000  00000 00000    00000 00000  00000 00000  00000 00000  00000 00000  00000 00000
1483:  00000 00000  00000 00000  00000 00000  00000 00000  00000 00000    00000 00000  00000 00000  00000 00000  00000 00000  00000 00000
1484:  00000 00000  00000 00000  00000 00000  00000 00000  00000 00000    00000 00000  00000 00000  00000 00000  00000 00000  00000 00000
1485:  00000 00000  00000 00000  00000 00000  00000 00000  00000 00000    00000 00000  00000 00000  00000 00000  00000 00000  00000 00000
1486:  00000 00000  00000 00000  00000 00000  00000 00000  00000 00000    00000 00000  00000 00000  00000 00000  00000 00000  00000 00000
1487:  00000 00000  00000 00000  00000 00000  00000 00000  00000 00000    00000 00000  00000 00000  00000 00000  00000 00000  00000 00000
1488:  00000 00000  00000 00000  00000 00000  00000 00000  00000 00000    00000 00000  00000 00000  00000 00000  00000 00000  00000 00000
1489:  00000 00000  00000 00000  00000 00000  00000 00000  00000 00000    00000 00000  00000 00000  00000 00000  00000 00000  00000 00000
1490:  00000 00000  00000 00000  00000 00000  00000 00000  00000 00000    00000 00000  00000 00000  00000 00000  00000 00000  00000 00000
1491:  00000 00000  00000 00000  00000 00000  00000 00000  00000 00000    00000 00000  00000 00000  00000 00000  00000 00000  00000 00000
1492:  00000 00000  00000 00000  00000 00000  00000 00000  00000 00000    00000 00000  00000 00000  00000 00000  00000 00000  00000 00000
1493:  00000 00000  00000 00000  00000 00000  00000 00000  00000 00000    00000 00000  00000 00000  00000 00000  00000 00000  00000 00000
1494:  00000 00000  00000 00000  00000 00000  00000 00000  00000 00000    00000 00000  00000 00000  00000 00000  00000 00000  00000 00000
1495:  00000 00000  00000 00000  00000 00000  00000 00000  00000 00000    00000 00000  00000 00000  00000 00000  00000 00000  00000 00000
1496:  00000 00000  00000 00000  00000 00000  00000 00000  00000 00000    00000 00000  00000 00000  00000 00000  00000 00000  00000 00000
1497:  00000 00000  00000 00000  00000 00000  00000 00000  00000 00000    00000 00000  00000 00000  00000 00000  00000 00000  00000 00000
1498:  00000 00000  00000 00000  00000 00000  00000 00000  00000 00000    00000 00000  00000 00000  00000 00000  00000 00000  00000 00000
1499:  00000 00000  00000 00000  00000 00000  00000 00000  00000 00000    00000 00000  00000 00000  00000 00000  00000 00000  00000 00000
```

```
1500:  00000 00000  00000 00000  00000 00000  00000 00000  00000 00000    00000 00000  00000 00000  00000 00000  00000 00000  00000 00000
1501:  00000 00000  00000 00000  00000 00000  00000 00000  00000 00000    00000 00000  00000 00000  00000 00000  00000 00000  00000 00000
1502:  00000 00000  00000 00000  00000 00000  00000 00000  00000 00000    00000 00000  00000 00000  00000 00000  00000 00000  00000 00000
1503:  00000 00000  00000 00000  00000 00000  00000 00000  00000 00000    00000 00000  00000 00000  00000 00000  00000 00000  00000 00000
1504:  00000 00000  00000 00000  00000 00000  00000 00000  00000 00000    00000 00000  00000 00000  00000 00000  00000 00000  00000 00000
1505:  00000 00000  00000 00000  00000 00000  00000 00000  00000 00000    00000 00000  00000 00000  00000 00000  00000 00000  00000 00000
1506:  00000 00000  00000 00000  00000 00000  00000 00000  00000 00000    00000 00000  00000 00000  00000 00000  00000 00000  00000 00000
1507:  00000 00000  00000 00000  00000 00000  00000 00000  00000 00000    00000 00000  00000 00000  00000 00000  00000 00000  00000 00000
1508:  00000 00000  00000 00000  00000 00000  00000 00000  00000 00000    00000 00000  00000 00000  00000 00000  00000 00000  00000 00000
1509:  00000 00000  00000 00000  00000 00000  00000 00000  00000 00000    00000 00000  00000 00000  00000 00000  00000 00000  00000 00000
1510:  00000 00000  00000 00000  00000 00000  00000 00000  00000 00000    00000 00000  00000 00000  00000 00000  00000 00000  00000 00000
1511:  00000 00000  00000 00000  00000 00000  00000 00000  00000 00000    00000 00000  00000 00000  00000 00000  00000 00000  00000 00000
1512:  00000 00000  00000 00000  00000 00000  00000 00000  00000 00000    00000 00000  00000 00000  00000 00000  00000 00000  00000 00000
1513:  00000 00000  00000 00000  00000 00000  00000 00000  00000 00000    00000 00000  00000 00000  00000 00000  00000 00000  00000 00000
1514:  00000 00000  00000 00000  00000 00000  00000 00000  00000 00000    00000 00000  00000 00000  00000 00000  00000 00000  00000 00000
1515:  00000 00000  00000 00000  00000 00000  00000 00000  00000 00000    00000 00000  00000 00000  00000 00000  00000 00000  00000 00000
1516:  00000 00000  00000 00000  00000 00000  00000 00000  00000 00000    00000 00000  00000 00000  00000 00000  00000 00000  00000 00000
1517:  00000 00000  00000 00000  00000 00000  00000 00000  00000 00000    00000 00000  00000 00000  00000 00000  00000 00000  00000 00000
1518:  00000 00000  00000 00000  00000 00000  00000 00000  00000 00000    00000 00000  00000 00000  00000 00000  00000 00000  00000 00000
1519:  00000 00000  00000 00000  00000 00000  00000 00000  00000 00000    00000 00000  00000 00000  00000 00000  00000 00000  00000 00000
1520:  00000 00000  00000 00000  00000 00000  00000 00000  00000 00000    00000 00000  00000 00000  00000 00000  00000 00000  00000 00000
1521:  00000 00000  00000 00000  00000 00000  00000 00000  00000 00000    00000 00000  00000 00000  00000 00000  00000 00000  00000 00000
1522:  00000 00000  00000 00000  00000 00000  00000 00000  00000 00000    00000 00000  00000 00000  00000 00000  00000 00000  00000 00000
1523:  00000 00000  00000 00000  00000 00000  00000 00000  00000 00000    00000 00000  00000 00000  00000 00000  00000 00000  00000 00000
1524:  00000 00000  00000 00000  00000 00000  00000 00000  00000 00000    00000 00000  00000 00000  00000 00000  00000 00000  00000 00000
1525:  00000 00000  00000 00000  00000 00000  00000 00000  00000 00000    00000 00000  00000 00000  00000 00000  00000 00000  00000 00000
1526:  00000 00000  00000 00000  00000 00000  00000 00000  00000 00000    00000 00000  00000 00000  00000 00000  00000 00000  00000 00000
1527:  00000 00000  00000 00000  00000 00000  00000 00000  00000 00000    00000 00000  00000 00000  00000 00000  00000 00000  00000 00000
1528:  00000 00000  00000 00000  00000 00000  00000 00000  00000 00000    00000 00000  00000 00000  00000 00000  00000 00000  00000 00000
1529:  00000 00000  00000 00000  00000 00000  00000 00000  00000 00000    00000 00000  00000 00000  00000 00000  00000 00000  00000 00000
1530:  00000 00000  00000 00000  00000 00000  00000 00000  00000 00000    00000 00000  00000 00000  00000 00000  00000 00000  00000 00000
1531:  00000 00000  00000 00000  00000 00000  00000 00000  00000 00000    00000 00000  00000 00000  00000 00000  00000 00000  00000 00000
1532:  00000 00000  00000 00000  00000 00000  00000 00000  00000 00000    00000 00000  00000 00000  00000 00000  00000 00000  00000 00000
1533:  00000 00000  00000 00000  00000 00000  00000 00000  00000 00000    00000 00000  00000 00000  00000 00000  00000 00000  00000 00000
1534:  00000 00000  00000 00000  00000 00000  00000 00000  00000 00000    00000 00000  00000 00000  00000 00000  00000 00000  00000 00000
1535:  00000 00000  00000 00000  00000 00000  00000 00000  00000 00000    00000 00000  00000 00000  00000 00000  00000 00000  00000 00000
1536:  00000 00000  00000 00000  00000 00000  00000 00000  00000 00000    00000 00000  00000 00000  00000 00000  00000 00000  00000 00000
1537:  00000 00000  00000 00000  00000 00000  00000 00000  00000 00000    00000 00000  00000 00000  00000 00000  00000 00000  00000 00000
1538:  00000 00000  00000 00000  00000 00000  00000 00000  00000 00000    00000 00000  00000 00000  00000 00000  00000 00000  00000 00000
1539:  00000 00000  00000 00000  00000 00000  00000 00000  00000 00000    00000 00000  00000 00000  00000 00000  00000 00000  00000 00000
1540:  00000 00000  00000 00000  00000 00000  00000 00000  00000 00000    00000 00000  00000 00000  00000 00000  00000 00000  00000 00000
1541:  00000 00000  00000 00000  00000 00000  00000 00000  00000 00000    00000 00000  00000 00000  00000 00000  00000 00000  00000 00000
1542:  00000 00000  00000 00000  00000 00000  00000 00000  00000 00000    00000 00000  00000 00000  00000 00000  00000 00000  00000 00000
1543:  00000 00000  00000 00000  00000 00000  00000 00000  00000 00000    00000 00000  00000 00000  00000 00000  00000 00000  00000 00000
1544:  00000 00000  00000 00000  00000 00000  00000 00000  00000 00000    00000 00000  00000 00000  00000 00000  00000 00000  00000 00000
1545:  00000 00000  00000 00000  00000 00000  00000 00000  00000 00000    00000 00000  00000 00000  00000 00000  00000 00000  00000 00000
1546:  00000 00000  00000 00000  00000 00000  00000 00000  00000 00000    00000 00000  00000 00000  00000 00000  00000 00000  00000 00000
1547:  00000 00000  00000 00000  00000 00000  00000 00000  00000 00000    00000 00000  00000 00000  00000 00000  00000 00000  00000 00000
1548:  00000 00000  00000 00000  00000 00000  00000 00000  00000 00000    00000 00000  00000 00000  00000 00000  00000 00000  00000 00000
1549:  00000 00000  00000 00000  00000 00000  00000 00000  00000 00000    00000 00000  00000 00000  00000 00000  00000 00000  00000 00000
```

```
1550:  00000 00000  00000 00000  00000 00000  00000 00000  00000 00000   00000 00000  00000 00000  00000 00000  00000 00000  00000 00000
1551:  00000 00000  00000 00000  00000 00000  00000 00000  00000 00000   00000 00000  00000 00000  00000 00000  00000 00000  00000 00000
1552:  00000 00000  00000 00000  00000 00000  00000 00000  00000 00000   00000 00000  00000 00000  00000 00000  00000 00000  00000 00000
1553:  00000 00000  00000 00000  00000 00000  00000 00000  00000 00000   00000 00000  00000 00000  00000 00000  00000 00000  00000 00000
1554:  00000 00000  00000 00000  00000 00000  00000 00000  00000 00000   00000 00000  00000 00000  00000 00000  00000 00000  00000 00000
1555:  00000 00000  00000 00000  00000 00000  00000 00000  00000 00000   00000 00000  00000 00000  00000 00000  00000 00000  00000 00000
1556:  00000 00000  00000 00000  00000 00000  00000 00000  00000 00000   00000 00000  00000 00000  00000 00000  00000 00000  00000 00000
1557:  00000 00000  00000 00000  00000 00000  00000 00000  00000 00000   00000 00000  00000 00000  00000 00000  00000 00000  00000 00000
1558:  00000 00000  00000 00000  00000 00000  00000 00000  00000 00000   00000 00000  00000 00000  00000 00000  00000 00000  00000 00000
1559:  00000 00000  00000 00000  00000 00000  00000 00000  00000 00000   00000 00000  00000 00000  00000 00000  00000 00000  00000 00000
1560:  00000 00000  00000 00000  00000 00000  00000 00000  00000 00000   00000 00000  00000 00000  00000 00000  00000 00000  00000 00000
1561:  00000 00000  00000 00000  00000 00000  00000 00000  00000 00000   00000 00000  00000 00000  00000 00000  00000 00000  00000 00000
1562:  00000 00000  00000 00000  00000 00000  00000 00000  00000 00000   00000 00000  00000 00000  00000 00000  00000 00000  00000 00000
1563:  00000 00000  00000 00000  00000 00000  00000 00000  00000 00000   00000 00000  00000 00000  00000 00000  00000 00000  00000 00000
1564:  00000 00000  00000 00000  00000 00000  00000 00000  00000 00000   00000 00000  00000 00000  00000 00000  00000 00000  00000 00000
1565:  00000 00000  00000 00000  00000 00000  00000 00000  00000 00000   00000 00000  00000 00000  00000 00000  00000 00000  00000 00000
1566:  00000 00000  00000 00000  00000 00000  00000 00000  00000 00000   00000 00000  00000 00000  00000 00000  00000 00000  00000 00000
1567:  00000 00000  00000 00000  00000 00000  00000 00000  00000 00000   00000 00000  00000 00000  00000 00000  00000 00000  00000 00000
1568:  00000 00000  00000 00000  00000 00000  00000 00000  00000 00000   00000 00000  00000 00000  00000 00000  00000 00000  00000 00000
1569:  00000 00000  00000 00000  00000 00000  00000 00000  00000 00000   00000 00000  00000 00000  00000 00000  00000 00000  00000 00000
1570:  00000 00000  00000 00000  00000 00000  00000 00000  00000 00000   00000 00000  00000 00000  00000 00000  00000 00000  00000 00000
1571:  00000 00000  00000 00000  00000 00000  00000 00000  00000 00000   00000 00000  00000 00000  00000 00000  00000 00000  00000 00000
1572:  00000 00000  00000 00000  00000 00000  00000 00000  00000 00000   00000 00000  00000 00000  00000 00000  00000 00000  00000 00000
1573:  00000 00000  00000 00000  00000 00000  00000 00000  00000 00000   00000 00000  00000 00000  00000 00000  00000 00000  00000 00000
1574:  00000 00000  00000 00000  00000 00000  00000 00000  00000 00000   00000 00000  00000 00000  00000 00000  00000 00000  00000 00000
1575:  00000 00000  00000 00000  00000 00000  00000 00000  00000 00000   00000 00000  00000 00000  00000 00000  00000 00000  00000 00000
1576:  00000 00000  00000 00000  00000 00000  00000 00000  00000 00000   00000 00000  00000 00000  00000 00000  00000 00000  00000 00000
1577:  00000 00000  00000 00000  00000 00000  00000 00000  00000 00000   00000 00000  00000 00000  00000 00000  00000 00000  00000 00000
1578:  00000 00000  00000 00000  00000 00000  00000 00000  00000 00000   00000 00000  00000 00000  00000 00000  00000 00000  00000 00000
1579:  00000 00000  00000 00000  00000 00000  00000 00000  00000 00000   00000 00000  00000 00000  00000 00000  00000 00000  00000 00000
1580:  00000 00000  00000 00000  00000 00000  00000 00000  00000 00000   00000 00000  00000 00000  00000 00000  00000 00000  00000 00000
1581:  00000 00000  00000 00000  00000 00000  00000 00000  00000 00000   00000 00000  00000 00000  00000 00000  00000 00000  00000 00000
1582:  00000 00000  00000 00000  00000 00000  00000 00000  00000 00000   00000 00000  00000 00000  00000 00000  00000 00000  00000 00000
1583:  00000 00000  00000 00000  00000 00000  00000 00000  00000 00000   00000 00000  00000 00000  00000 00000  00000 00000  00000 00000
1584:  00000 00000  00000 00000  00000 00000  00000 00000  00000 00000   00000 00000  00000 00000  00000 00000  00000 00000  00000 00000
1585:  00000 00000  00000 00000  00000 00000  00000 00000  00000 00000   00000 00000  00000 00000  00000 00000  00000 00000  00000 00000
1586:  00000 00000  00000 00000  00000 00000  00000 00000  00000 00000   00000 00000  00000 00000  00000 00000  00000 00000  00000 00000
1587:  00000 00000  00000 00000  00000 00000  00000 00000  00000 00000   00000 00000  00000 00000  00000 00000  00000 00000  00000 00000
1588:  00000 00000  00000 00000  00000 00000  00000 00000  00000 00000   00000 00000  00000 00000  00000 00000  00000 00000  00000 00000
1589:  00000 00000  00000 00000  00000 00000  00000 00000  00000 00000   00000 00000  00000 00000  00000 00000  00000 00000  00000 00000
1590:  00000 00000  00000 00000  00000 00000  00000 00000  00000 00000   00000 00000  00000 00000  00000 00000  00000 00000  00000 00000
1591:  00000 00000  00000 00000  00000 00000  00000 00000  00000 00000   00000 00000  00000 00000  00000 00000  00000 00000  00000 00000
1592:  00000 00000  00000 00000  00000 00000  00000 00000  00000 00000   00000 00000  00000 00000  00000 00000  00000 00000  00000 00000
1593:  00000 00000  00000 00000  00000 00000  00000 00000  00000 00000   00000 00000  00000 00000  00000 00000  00000 00000  00000 00000
1594:  00000 00000  00000 00000  00000 00000  00000 00000  00000 00000   00000 00000  00000 00000  00000 00000  00000 00000  00000 00000
1595:  00000 00000  00000 00000  00000 00000  00000 00000  00000 00000   00000 00000  00000 00000  00000 00000  00000 00000  00000 00000
1596:  00000 00000  00000 00000  00000 00000  00000 00000  00000 00000   00000 00000  00000 00000  00000 00000  00000 00000  00000 00000
1597:  00000 00000  00000 00000  00000 00000  00000 00000  00000 00000   00000 00000  00000 00000  00000 00000  00000 00000  00000 00000
1598:  00000 00000  00000 00000  00000 00000  00000 00000  00000 00000   00000 00000  00000 00000  00000 00000  00000 00000  00000 00000
1599:  00000 00000  00000 00000  00000 00000  00000 00000  00000 00000   00000 00000  00000 00000  00000 00000  00000 00000  00000 00000
```

```
1600:  00000 00000  00000 00000  00000 00000  00000 00000  00000 00000    00000 00000  00000 00000  00000 00000  00000 00000  00000 00000
1601:  00000 00000  00000 00000  00000 00000  00000 00000  00000 00000    00000 00000  00000 00000  00000 00000  00000 00000  00000 00000
1602:  00000 00000  00000 00000  00000 00000  00000 00000  00000 00000    00000 00000  00000 00000  00000 00000  00000 00000  00000 00000
1603:  00000 00000  00000 00000  00000 00000  00000 00000  00000 00000    00000 00000  00000 00000  00000 00000  00000 00000  00000 00000
1604:  00000 00000  00000 00000  00000 00000  00000 00000  00000 00000    00000 00000  00000 00000  00000 00000  00000 00000  00000 00000
1605:  00000 00000  00000 00000  00000 00000  00000 00000  00000 00000    00000 00000  00000 00000  00000 00000  00000 00000  00000 00000
1606:  00000 00000  00000 00000  00000 00000  00000 00000  00000 00000    00000 00000  00000 00000  00000 00000  00000 00000  00000 00000
1607:  00000 00000  00000 00000  00000 00000  00000 00000  00000 00000    00000 00000  00000 00000  00000 00000  00000 00000  00000 00000
1608:  00000 00000  00000 00000  00000 00000  00000 00000  00000 00000    00000 00000  00000 00000  00000 00000  00000 00000  00000 00000
1609:  00000 00000  00000 00000  00000 00000  00000 00000  00000 00000    00000 00000  00000 00000  00000 00000  00000 00000  00000 00000
1610:  00000 00000  00000 00000  00000 00000  00000 00000  00000 00000    00000 00000  00000 00000  00000 00000  00000 00000  00000 00000
1611:  00000 00000  00000 00000  00000 00000  00000 00000  00000 00000    00000 00000  00000 00000  00000 00000  00000 00000  00000 00000
1612:  00000 00000  00000 00000  00000 00000  00000 00000  00000 00000    00000 00000  00000 00000  00000 00000  00000 00000  00000 00000
1613:  00000 00000  00000 00000  00000 00000  00000 00000  00000 00000    00000 00000  00000 00000  00000 00000  00000 00000  00000 00000
1614:  00000 00000  00000 00000  00000 00000  00000 00000  00000 00000    00000 00000  00000 00000  00000 00000  00000 00000  00000 00000
1615:  00000 00000  00000 00000  00000 00000  00000 00000  00000 00000    00000 00000  00000 00000  00000 00000  00000 00000  00000 00000
1616:  00000 00000  00000 00000  00000 00000  00000 00000  00000 00000    00000 00000  00000 00000  00000 00000  00000 00000  00000 00000
1617:  00000 00000  00000 00000  00000 00000  00000 00000  00000 00000    00000 00000  00000 00000  00000 00000  00000 00000  00000 00000
1618:  00000 00000  00000 00000  00000 00000  00000 00000  00000 00000    00000 00000  00000 00000  00000 00000  00000 00000  00000 00000
1619:  00000 00000  00000 00000  00000 00000  00000 00000  00000 00000    00000 00000  00000 00000  00000 00000  00000 00000  00000 00000
1620:  00000 00000  00000 00000  00000 00000  00000 00000  00000 00000    00000 00000  00000 00000  00000 00000  00000 00000  00000 00000
1621:  00000 00000  00000 00000  00000 00000  00000 00000  00000 00000    00000 00000  00000 00000  00000 00000  00000 00000  00000 00000
1622:  00000 00000  00000 00000  00000 00000  00000 00000  00000 00000    00000 00000  00000 00000  00000 00000  00000 00000  00000 00000
1623:  00000 00000  00000 00000  00000 00000  00000 00000  00000 00000    00000 00000  00000 00000  00000 00000  00000 00000  00000 00000
1624:  00000 00000  00000 00000  00000 00000  00000 00000  00000 00000    00000 00000  00000 00000  00000 00000  00000 00000  00000 00000
1625:  00000 00000  00000 00000  00000 00000  00000 00000  00000 00000    00000 00000  00000 00000  00000 00000  00000 00000  00000 00000
1626:  00000 00000  00000 00000  00000 00000  00000 00000  00000 00000    00000 00000  00000 00000  00000 00000  00000 00000  00000 00000
1627:  00000 00000  00000 00000  00000 00000  00000 00000  00000 00000    00000 00000  00000 00000  00000 00000  00000 00000  00000 00000
1628:  00000 00000  00000 00000  00000 00000  00000 00000  00000 00000    00000 00000  00000 00000  00000 00000  00000 00000  00000 00000
1629:  00000 00000  00000 00000  00000 00000  00000 00000  00000 00000    00000 00000  00000 00000  00000 00000  00000 00000  00000 00000
1630:  00000 00000  00000 00000  00000 00000  00000 00000  00000 00000    00000 00000  00000 00000  00000 00000  00000 00000  00000 00000
1631:  00000 00000  00000 00000  00000 00000  00000 00000  00000 00000    00000 00000  00000 00000  00000 00000  00000 00000  00000 00000
1632:  00000 00000  00000 00000  00000 00000  00000 00000  00000 00000    00000 00000  00000 00000  00000 00000  00000 00000  00000 00000
1633:  00000 00000  00000 00000  00000 00000  00000 00000  00000 00000    00000 00000  00000 00000  00000 00000  00000 00000  00000 00000
1634:  00000 00000  00000 00000  00000 00000  00000 00000  00000 00000    00000 00000  00000 00000  00000 00000  00000 00000  00000 00000
1635:  00000 00000  00000 00000  00000 00000  00000 00000  00000 00000    00000 00000  00000 00000  00000 00000  00000 00000  00000 00000
1636:  00000 00000  00000 00000  00000 00000  00000 00000  00000 00000    00000 00000  00000 00000  00000 00000  00000 00000  00000 00000
1637:  00000 00000  00000 00000  00000 00000  00000 00000  00000 00000    00000 00000  00000 00000  00000 00000  00000 00000  00000 00000
1638:  00000 00000  00000 00000  00000 00000  00000 00000  00000 00000    00000 00000  00000 00000  00000 00000  00000 00000  00000 00000
1639:  00000 00000  00000 00000  00000 00000  00000 00000  00000 00000    00000 00000  00000 00000  00000 00000  00000 00000  00000 00000
1640:  00000 00000  00000 00000  00000 00000  00000 00000  00000 00000    00000 00000  00000 00000  00000 00000  00000 00000  00000 00000
1641:  00000 00000  00000 00000  00000 00000  00000 00000  00000 00000    00000 00000  00000 00000  00000 00000  00000 00000  00000 00000
1642:  00000 00000  00000 00000  00000 00000  00000 00000  00000 00000    00000 00000  00000 00000  00000 00000  00000 00000  00000 00000
1643:  00000 00000  00000 00000  00000 00000  00000 00000  00000 00000    00000 00000  00000 00000  00000 00000  00000 00000  00000 00000
1644:  00000 00000  00000 00000  00000 00000  00000 00000  00000 00000    00000 00000  00000 00000  00000 00000  00000 00000  00000 00000
1645:  00000 00000  00000 00000  00000 00000  00000 00000  00000 00000    00000 00000  00000 00000  00000 00000  00000 00000  00000 00000
1646:  00000 00000  00000 00000  00000 00000  00000 00000  00000 00000    00000 00000  00000 00000  00000 00000  00000 00000  00000 00000
1647:  00000 00000  00000 00000  00000 00000  00000 00000  00000 00000    00000 00000  00000 00000  00000 00000  00000 00000  00000 00000
1648:  00000 00000  00000 00000  00000 00000  00000 00000  00000 00000    00000 00000  00000 00000  00000 00000  00000 00000  00000 00000
1649:  00000 00000  00000 00000  00000 00000  00000 00000  00000 00000    00000 00000  00000 00000  00000 00000  00000 00000  00000 00000
```

```
1650:  00000 00000  00000 00000  00000 00000  00000 00000  00000 00000   00000 00000  00000 00000  00000 00000  00000 00000  00000 00000
1651:  00000 00000  00000 00000  00000 00000  00000 00000  00000 00000   00000 00000  00000 00000  00000 00000  00000 00000  00000 00000
1652:  00000 00000  00000 00000  00000 00000  00000 00000  00000 00000   00000 00000  00000 00000  00000 00000  00000 00000  00000 00000
1653:  00000 00000  00000 00000  00000 00000  00000 00000  00000 00000   00000 00000  00000 00000  00000 00000  00000 00000  00000 00000
1654:  00000 00000  00000 00000  00000 00000  00000 00000  00000 00000   00000 00000  00000 00000  00000 00000  00000 00000  00000 00000
1655:  00000 00000  00000 00000  00000 00000  00000 00000  00000 00000   00000 00000  00000 00000  00000 00000  00000 00000  00000 00000
1656:  00000 00000  00000 00000  00000 00000  00000 00000  00000 00000   00000 00000  00000 00000  00000 00000  00000 00000  00000 00000
1657:  00000 00000  00000 00000  00000 00000  00000 00000  00000 00000   00000 00000  00000 00000  00000 00000  00000 00000  00000 00000
1658:  00000 00000  00000 00000  00000 00000  00000 00000  00000 00000   00000 00000  00000 00000  00000 00000  00000 00000  00000 00000
1659:  00000 00000  00000 00000  00000 00000  00000 00000  00000 00000   00000 00000  00000 00000  00000 00000  00000 00000  00000 00000
1660:  00000 00000  00000 00000  00000 00000  00000 00000  00000 00000   00000 00000  00000 00000  00000 00000  00000 00000  00000 00000
1661:  00000 00000  00000 00000  00000 00000  00000 00000  00000 00000   00000 00000  00000 00000  00000 00000  00000 00000  00000 00000
1662:  00000 00000  00000 00000  00000 00000  00000 00000  00000 00000   00000 00000  00000 00000  00000 00000  00000 00000  00000 00000
1663:  00000 00000  00000 00000  00000 00000  00000 00000  00000 00000   00000 00000  00000 00000  00000 00000  00000 00000  00000 00000
1664:  00000 00000  00000 00000  00000 00000  00000 00000  00000 00000   00000 00000  00000 00000  00000 00000  00000 00000  00000 00000
1665:  00000 00000  00000 00000  00000 00000  00000 00000  00000 00000   00000 00000  00000 00000  00000 00000  00000 00000  00000 00000
1666:  00000 00000  00000 00000  00000 00000  00000 00000  00000 00000   00000 00000  00000 00000  00000 00000  00000 00000  00000 00000
1667:  00000 00000  00000 00000  00000 00000  00000 00000  00000 00000   00000 00000  00000 00000  00000 00000  00000 00000  00000 00000
1668:  00000 00000  00000 00000  00000 00000  00000 00000  00000 00000   00000 00000  00000 00000  00000 00000  00000 00000  00000 00000
1669:  00000 00000  00000 00000  00000 00000  00000 00000  00000 00000   00000 00000  00000 00000  00000 00000  00000 00000  00000 00000
1670:  00000 00000  00000 00000  00000 00000  00000 00000  00000 00000   00000 00000  00000 00000  00000 00000  00000 00000  00000 00000
1671:  00000 00000  00000 00000  00000 00000  00000 00000  00000 00000   00000 00000  00000 00000  00000 00000  00000 00000  00000 00000
1672:  00000 00000  00000 00000  00000 00000  00000 00000  00000 00000   00000 00000  00000 00000  00000 00000  00000 00000  00000 00000
1673:  00000 00000  00000 00000  00000 00000  00000 00000  00000 00000   00000 00000  00000 00000  00000 00000  00000 00000  00000 00000
1674:  00000 00000  00000 00000  00000 00000  00000 00000  00000 00000   00000 00000  00000 00000  00000 00000  00000 00000  00000 00000
1675:  00000 00000  00000 00000  00000 00000  00000 00000  00000 00000   00000 00000  00000 00000  00000 00000  00000 00000  00000 00000
1676:  00000 00000  00000 00000  00000 00000  00000 00000  00000 00000   00000 00000  00000 00000  00000 00000  00000 00000  00000 00000
1677:  00000 00000  00000 00000  00000 00000  00000 00000  00000 00000   00000 00000  00000 00000  00000 00000  00000 00000  00000 00000
1678:  00000 00000  00000 00000  00000 00000  00000 00000  00000 00000   00000 00000  00000 00000  00000 00000  00000 00000  00000 00000
1679:  00000 00000  00000 00000  00000 00000  00000 00000  00000 00000   00000 00000  00000 00000  00000 00000  00000 00000  00000 00000
1680:  00000 00000  00000 00000  00000 00000  00000 00000  00000 00000   00000 00000  00000 00000  00000 00000  00000 00000  00000 00000
1681:  00000 00000  00000 00000  00000 00000  00000 00000  00000 00000   00000 00000  00000 00000  00000 00000  00000 00000  00000 00000
1682:  00000 00000  00000 00000  00000 00000  00000 00000  00000 00000   00000 00000  00000 00000  00000 00000  00000 00000  00000 00000
1683:  00000 00000  00000 00000  00000 00000  00000 00000  00000 00000   00000 00000  00000 00000  00000 00000  00000 00000  00000 00000
1684:  00000 00000  00000 00000  00000 00000  00000 00000  00000 00000   00000 00000  00000 00000  00000 00000  00000 00000  00000 00000
1685:  00000 00000  00000 00000  00000 00000  00000 00000  00000 00000   00000 00000  00000 00000  00000 00000  00000 00000  00000 00000
1686:  00000 00000  00000 00000  00000 00000  00000 00000  00000 00000   00000 00000  00000 00000  00000 00000  00000 00000  00000 00000
1687:  00000 00000  00000 00000  00000 00000  00000 00000  00000 00000   00000 00000  00000 00000  00000 00000  00000 00000  00000 00000
1688:  00000 00000  00000 00000  00000 00000  00000 00000  00000 00000   00000 00000  00000 00000  00000 00000  00000 00000  00000 00000
1689:  00000 00000  00000 00000  00000 00000  00000 00000  00000 00000   00000 00000  00000 00000  00000 00000  00000 00000  00000 00000
1690:  00000 00000  00000 00000  00000 00000  00000 00000  00000 00000   00000 00000  00000 00000  00000 00000  00000 00000  00000 00000
1691:  00000 00000  00000 00000  00000 00000  00000 00000  00000 00000   00000 00000  00000 00000  00000 00000  00000 00000  00000 00000
1692:  00000 00000  00000 00000  00000 00000  00000 00000  00000 00000   00000 00000  00000 00000  00000 00000  00000 00000  00000 00000
1693:  00000 00000  00000 00000  00000 00000  00000 00000  00000 00000   00000 00000  00000 00000  00000 00000  00000 00000  00000 00000
1694:  00000 00000  00000 00000  00000 00000  00000 00000  00000 00000   00000 00000  00000 00000  00000 00000  00000 00000  00000 00000
1695:  00000 00000  00000 00000  00000 00000  00000 00000  00000 00000   00000 00000  00000 00000  00000 00000  00000 00000  00000 00000
1696:  00000 00000  00000 00000  00000 00000  00000 00000  00000 00000   00000 00000  00000 00000  00000 00000  00000 00000  00000 00000
1697:  00000 00000  00000 00000  00000 00000  00000 00000  00000 00000   00000 00000  00000 00000  00000 00000  00000 00000  00000 00000
1698:  00000 00000  00000 00000  00000 00000  00000 00000  00000 00000   00000 00000  00000 00000  00000 00000  00000 00000  00000 00000
1699:  00000 00000  00000 00000  00000 00000  00000 00000  00000 00000   00000 00000  00000 00000  00000 00000  00000 00000  00000 00000
```

```
1700:   00000 00000   00000 00000   00000 00000   00000 00000   00000 00000      00000 00000   00000 00000   00000 00000   00000 00000   00000 00000
1701:   00000 00000   00000 00000   00000 00000   00000 00000   00000 00000      00000 00000   00000 00000   00000 00000   00000 00000   00000 00000
1702:   00000 00000   00000 00000   00000 00000   00000 00000   00000 00000      00000 00000   00000 00000   00000 00000   00000 00000   00000 00000
1703:   00000 00000   00000 00000   00000 00000   00000 00000   00000 00000      00000 00000   00000 00000   00000 00000   00000 00000   00000 00000
1704:   00000 00000   00000 00000   00000 00000   00000 00000   00000 00000      00000 00000   00000 00000   00000 00000   00000 00000   00000 00000
1705:   00000 00000   00000 00000   00000 00000   00000 00000   00000 00000      00000 00000   00000 00000   00000 00000   00000 00000   00000 00000
1706:   00000 00000   00000 00000   00000 00000   00000 00000   00000 00000      00000 00000   00000 00000   00000 00000   00000 00000   00000 00000
1707:   00000 00000   00000 00000   00000 00000   00000 00000   00000 00000      00000 00000   00000 00000   00000 00000   00000 00000   00000 00000
1708:   00000 00000   00000 00000   00000 00000   00000 00000   00000 00000      00000 00000   00000 00000   00000 00000   00000 00000   00000 00000
1709:   00000 00000   00000 00000   00000 00000   00000 00000   00000 00000      00000 00000   00000 00000   00000 00000   00000 00000   00000 00000
1710:   00000 00000   00000 00000   00000 00000   00000 00000   00000 00000      00000 00000   00000 00000   00000 00000   00000 00000   00000 00000
1711:   00000 00000   00000 00000   00000 00000   00000 00000   00000 00000      00000 00000   00000 00000   00000 00000   00000 00000   00000 00000
1712:   00000 00000   00000 00000   00000 00000   00000 00000   00000 00000      00000 00000   00000 00000   00000 00000   00000 00000   00000 00000
1713:   00000 00000   00000 00000   00000 00000   00000 00000   00000 00000      00000 00000   00000 00000   00000 00000   00000 00000   00000 00000
1714:   00000 00000   00000 00000   00000 00000   00000 00000   00000 00000      00000 00000   00000 00000   00000 00000   00000 00000   00000 00000
1715:   00000 00000   00000 00000   00000 00000   00000 00000   00000 00000      00000 00000   00000 00000   00000 00000   00000 00000   00000 00000
1716:   00000 00000   00000 00000   00000 00000   00000 00000   00000 00000      00000 00000   00000 00000   00000 00000   00000 00000   00000 00000
1717:   00000 00000   00000 00000   00000 00000   00000 00000   00000 00000      00000 00000   00000 00000   00000 00000   00000 00000   00000 00000
1718:   00000 00000   00000 00000   00000 00000   00000 00000   00000 00000      00000 00000   00000 00000   00000 00000   00000 00000   00000 00000
1719:   00000 00000   00000 00000   00000 00000   00000 00000   00000 00000      00000 00000   00000 00000   00000 00000   00000 00000   00000 00000
1720:   00000 00000   00000 00000   00000 00000   00000 00000   00000 00000      00000 00000   00000 00000   00000 00000   00000 00000   00000 00000
1721:   00000 00000   00000 00000   00000 00000   00000 00000   00000 00000      00000 00000   00000 00000   00000 00000   00000 00000   00000 00000
1722:   00000 00000   00000 00000   00000 00000   00000 00000   00000 00000      00000 00000   00000 00000   00000 00000   00000 00000   00000 00000
1723:   00000 00000   00000 00000   00000 00000   00000 00000   00000 00000      00000 00000   00000 00000   00000 00000   00000 00000   00000 00000
1724:   00000 00000   00000 00000   00000 00000   00000 00000   00000 00000      00000 00000   00000 00000   00000 00000   00000 00000   00000 00000
1725:   00000 00000   00000 00000   00000 00000   00000 00000   00000 00000      00000 00000   00000 00000   00000 00000   00000 00000   00000 00000
1726:   00000 00000   00000 00000   00000 00000   00000 00000   00000 00000      00000 00000   00000 00000   00000 00000   00000 00000   00000 00000
1727:   00000 00000   00000 00000   00000 00000   00000 00000   00000 00000      00000 00000   00000 00000   00000 00000   00000 00000   00000 00000
1728:   00000 00000   00000 00000   00000 00000   00000 00000   00000 00000      00000 00000   00000 00000   00000 00000   00000 00000   00000 00000
1729:   00000 00000   00000 00000   00000 00000   00000 00000   00000 00000      00000 00000   00000 00000   00000 00000   00000 00000   00000 00000
1730:   00000 00000   00000 00000   00000 00000   00000 00000   00000 00000      00000 00000   00000 00000   00000 00000   00000 00000   00000 00000
1731:   00000 00000   00000 00000   00000 00000   00000 00000   00000 00000      00000 00000   00000 00000   00000 00000   00000 00000   00000 00000
1732:   00000 00000   00000 00000   00000 00000   00000 00000   00000 00000      00000 00000   00000 00000   00000 00000   00000 00000   00000 00000
1733:   00000 00000   00000 00000   00000 00000   00000 00000   00000 00000      00000 00000   00000 00000   00000 00000   00000 00000   00000 00000
1734:   00000 00000   00000 00000   00000 00000   00000 00000   00000 00000      00000 00000   00000 00000   00000 00000   00000 00000   00000 00000
1735:   00000 00000   00000 00000   00000 00000   00000 00000   00000 00000      00000 00000   00000 00000   00000 00000   00000 00000   00000 00000
1736:   00000 00000   00000 00000   00000 00000   00000 00000   00000 00000      00000 00000   00000 00000   00000 00000   00000 00000   00000 00000
1737:   00000 00000   00000 00000   00000 00000   00000 00000   00000 00000      00000 00000   00000 00000   00000 00000   00000 00000   00000 00000
1738:   00000 00000   00000 00000   00000 00000   00000 00000   00000 00000      00000 00000   00000 00000   00000 00000   00000 00000   00000 00000
1739:   00000 00000   00000 00000   00000 00000   00000 00000   00000 00000      00000 00000   00000 00000   00000 00000   00000 00000   00000 00000
1740:   00000 00000   00000 00000   00000 00000   00000 00000   00000 00000      00000 00000   00000 00000   00000 00000   00000 00000   00000 00000
1741:   00000 00000   00000 00000   00000 00000   00000 00000   00000 00000      00000 00000   00000 00000   00000 00000   00000 00000   00000 00000
1742:   00000 00000   00000 00000   00000 00000   00000 00000   00000 00000      00000 00000   00000 00000   00000 00000   00000 00000   00000 00000
1743:   00000 00000   00000 00000   00000 00000   00000 00000   00000 00000      00000 00000   00000 00000   00000 00000   00000 00000   00000 00000
1744:   00000 00000   00000 00000   00000 00000   00000 00000   00000 00000      00000 00000   00000 00000   00000 00000   00000 00000   00000 00000
1745:   00000 00000   00000 00000   00000 00000   00000 00000   00000 00000      00000 00000   00000 00000   00000 00000   00000 00000   00000 00000
1746:   00000 00000   00000 00000   00000 00000   00000 00000   00000 00000      00000 00000   00000 00000   00000 00000   00000 00000   00000 00000
1747:   00000 00000   00000 00000   00000 00000   00000 00000   00000 00000      00000 00000   00000 00000   00000 00000   00000 00000   00000 00000
1748:   00000 00000   00000 00000   00000 00000   00000 00000   00000 00000      00000 00000   00000 00000   00000 00000   00000 00000   00000 00000
1749:   00000 00000   00000 00000   00000 00000   00000 00000   00000 00000      00000 00000   00000 00000   00000 00000   00000 00000   00000 00000
```

```
1750:  00000 00000  00000 00000  00000 00000  00000 00000  00000 00000    00000 00000  00000 00000  00000 00000  00000 00000  00000 00000
1751:  00000 00000  00000 00000  00000 00000  00000 00000  00000 00000    00000 00000  00000 00000  00000 00000  00000 00000  00000 00000
1752:  00000 00000  00000 00000  00000 00000  00000 00000  00000 00000    00000 00000  00000 00000  00000 00000  00000 00000  00000 00000
1753:  00000 00000  00000 00000  00000 00000  00000 00000  00000 00000    00000 00000  00000 00000  00000 00000  00000 00000  00000 00000
1754:  00000 00000  00000 00000  00000 00000  00000 00000  00000 00000    00000 00000  00000 00000  00000 00000  00000 00000  00000 00000
1755:  00000 00000  00000 00000  00000 00000  00000 00000  00000 00000    00000 00000  00000 00000  00000 00000  00000 00000  00000 00000
1756:  00000 00000  00000 00000  00000 00000  00000 00000  00000 00000    00000 00000  00000 00000  00000 00000  00000 00000  00000 00000
1757:  00000 00000  00000 00000  00000 00000  00000 00000  00000 00000    00000 00000  00000 00000  00000 00000  00000 00000  00000 00000
1758:  00000 00000  00000 00000  00000 00000  00000 00000  00000 00000    00000 00000  00000 00000  00000 00000  00000 00000  00000 00000
1759:  00000 00000  00000 00000  00000 00000  00000 00000  00000 00000    00000 00000  00000 00000  00000 00000  00000 00000  00000 00000
1760:  00000 00000  00000 00000  00000 00000  00000 00000  00000 00000    00000 00000  00000 00000  00000 00000  00000 00000  00000 00000
1761:  00000 00000  00000 00000  00000 00000  00000 00000  00000 00000    00000 00000  00000 00000  00000 00000  00000 00000  00000 00000
1762:  00000 00000  00000 00000  00000 00000  00000 00000  00000 00000    00000 00000  00000 00000  00000 00000  00000 00000  00000 00000
1763:  00000 00000  00000 00000  00000 00000  00000 00000  00000 00000    00000 00000  00000 00000  00000 00000  00000 00000  00000 00000
1764:  00000 00000  00000 00000  00000 00000  00000 00000  00000 00000    00000 00000  00000 00000  00000 00000  00000 00000  00000 00000
1765:  00000 00000  00000 00000  00000 00000  00000 00000  00000 00000    00000 00000  00000 00000  00000 00000  00000 00000  00000 00000
1766:  00000 00000  00000 00000  00000 00000  00000 00000  00000 00000    00000 00000  00000 00000  00000 00000  00000 00000  00000 00000
1767:  00000 00000  00000 00000  00000 00000  00000 00000  00000 00000    00000 00000  00000 00000  00000 00000  00000 00000  00000 00000
1768:  00000 00000  00000 00000  00000 00000  00000 00000  00000 00000    00000 00000  00000 00000  00000 00000  00000 00000  00000 00000
1769:  00000 00000  00000 00000  00000 00000  00000 00000  00000 00000    00000 00000  00000 00000  00000 00000  00000 00000  00000 00000
1770:  00000 00000  00000 00000  00000 00000  00000 00000  00000 00000    00000 00000  00000 00000  00000 00000  00000 00000  00000 00000
1771:  00000 00000  00000 00000  00000 00000  00000 00000  00000 00000    00000 00000  00000 00000  00000 00000  00000 00000  00000 00000
1772:  00000 00000  00000 00000  00000 00000  00000 00000  00000 00000    00000 00000  00000 00000  00000 00000  00000 00000  00000 00000
1773:  00000 00000  00000 00000  00000 00000  00000 00000  00000 00000    00000 00000  00000 00000  00000 00000  00000 00000  00000 00000
1774:  00000 00000  00000 00000  00000 00000  00000 00000  00000 00000    00000 00000  00000 00000  00000 00000  00000 00000  00000 00000
1775:  00000 00000  00000 00000  00000 00000  00000 00000  00000 00000    00000 00000  00000 00000  00000 00000  00000 00000  00000 00000
1776:  00000 00000  00000 00000  00000 00000  00000 00000  00000 00000    00000 00000  00000 00000  00000 00000  00000 00000  00000 00000
1777:  00000 00000  00000 00000  00000 00000  00000 00000  00000 00000    00000 00000  00000 00000  00000 00000  00000 00000  00000 00000
1778:  00000 00000  00000 00000  00000 00000  00000 00000  00000 00000    00000 00000  00000 00000  00000 00000  00000 00000  00000 00000
1779:  00000 00000  00000 00000  00000 00000  00000 00000  00000 00000    00000 00000  00000 00000  00000 00000  00000 00000  00000 00000
1780:  00000 00000  00000 00000  00000 00000  00000 00000  00000 00000    00000 00000  00000 00000  00000 00000  00000 00000  00000 00000
1781:  00000 00000  00000 00000  00000 00000  00000 00000  00000 00000    00000 00000  00000 00000  00000 00000  00000 00000  00000 00000
1782:  00000 00000  00000 00000  00000 00000  00000 00000  00000 00000    00000 00000  00000 00000  00000 00000  00000 00000  00000 00000
1783:  00000 00000  00000 00000  00000 00000  00000 00000  00000 00000    00000 00000  00000 00000  00000 00000  00000 00000  00000 00000
1784:  00000 00000  00000 00000  00000 00000  00000 00000  00000 00000    00000 00000  00000 00000  00000 00000  00000 00000  00000 00000
1785:  00000 00000  00000 00000  00000 00000  00000 00000  00000 00000    00000 00000  00000 00000  00000 00000  00000 00000  00000 00000
1786:  00000 00000  00000 00000  00000 00000  00000 00000  00000 00000    00000 00000  00000 00000  00000 00000  00000 00000  00000 00000
1787:  00000 00000  00000 00000  00000 00000  00000 00000  00000 00000    00000 00000  00000 00000  00000 00000  00000 00000  00000 00000
1788:  00000 00000  00000 00000  00000 00000  00000 00000  00000 00000    00000 00000  00000 00000  00000 00000  00000 00000  00000 00000
1789:  00000 00000  00000 00000  00000 00000  00000 00000  00000 00000    00000 00000  00000 00000  00000 00000  00000 00000  00000 00000
1790:  00000 00000  00000 00000  00000 00000  00000 00000  00000 00000    00000 00000  00000 00000  00000 00000  00000 00000  00000 00000
1791:  00000 00000  00000 00000  00000 00000  00000 00000  00000 00000    00000 00000  00000 00000  00000 00000  00000 00000  00000 00000
1792:  00000 00000  00000 00000  00000 00000  00000 00000  00000 00000    00000 00000  00000 00000  00000 00000  00000 00000  00000 00000
1793:  00000 00000  00000 00000  00000 00000  00000 00000  00000 00000    00000 00000  00000 00000  00000 00000  00000 00000  00000 00000
1794:  00000 00000  00000 00000  00000 00000  00000 00000  00000 00000    00000 00000  00000 00000  00000 00000  00000 00000  00000 00000
1795:  00000 00000  00000 00000  00000 00000  00000 00000  00000 00000    00000 00000  00000 00000  00000 00000  00000 00000  00000 00000
1796:  00000 00000  00000 00000  00000 00000  00000 00000  00000 00000    00000 00000  00000 00000  00000 00000  00000 00000  00000 00000
1797:  00000 00000  00000 00000  00000 00000  00000 00000  00000 00000    00000 00000  00000 00000  00000 00000  00000 00000  00000 00000
1798:  00000 00000  00000 00000  00000 00000  00000 00000  00000 00000    00000 00000  00000 00000  00000 00000  00000 00000  00000 00000
1799:  00000 00000  00000 00000  00000 00000  00000 00000  00000 00000    00000 00000  00000 00000  00000 00000  00000 00000  00000 00000
```

```
1800:  00000 00000   00000 00000   00000 00000   00000 00000   00000 00000      00000 00000   00000 00000   00000 00000   00000 00000   00000 00000
1801:  00000 00000   00000 00000   00000 00000   00000 00000   00000 00000      00000 00000   00000 00000   00000 00000   00000 00000   00000 00000
1802:  00000 00000   00000 00000   00000 00000   00000 00000   00000 00000      00000 00000   00000 00000   00000 00000   00000 00000   00000 00000
1803:  00000 00000   00000 00000   00000 00000   00000 00000   00000 00000      00000 00000   00000 00000   00000 00000   00000 00000   00000 00000
1804:  00000 00000   00000 00000   00000 00000   00000 00000   00000 00000      00000 00000   00000 00000   00000 00000   00000 00000   00000 00000
1805:  00000 00000   00000 00000   00000 00000   00000 00000   00000 00000      00000 00000   00000 00000   00000 00000   00000 00000   00000 00000
1806:  00000 00000   00000 00000   00000 00000   00000 00000   00000 00000      00000 00000   00000 00000   00000 00000   00000 00000   00000 00000
1807:  00000 00000   00000 00000   00000 00000   00000 00000   00000 00000      00000 00000   00000 00000   00000 00000   00000 00000   00000 00000
1808:  00000 00000   00000 00000   00000 00000   00000 00000   00000 00000      00000 00000   00000 00000   00000 00000   00000 00000   00000 00000
1809:  00000 00000   00000 00000   00000 00000   00000 00000   00000 00000      00000 00000   00000 00000   00000 00000   00000 00000   00000 00000
1810:  00000 00000   00000 00000   00000 00000   00000 00000   00000 00000      00000 00000   00000 00000   00000 00000   00000 00000   00000 00000
1811:  00000 00000   00000 00000   00000 00000   00000 00000   00000 00000      00000 00000   00000 00000   00000 00000   00000 00000   00000 00000
1812:  00000 00000   00000 00000   00000 00000   00000 00000   00000 00000      00000 00000   00000 00000   00000 00000   00000 00000   00000 00000
1813:  00000 00000   00000 00000   00000 00000   00000 00000   00000 00000      00000 00000   00000 00000   00000 00000   00000 00000   00000 00000
1814:  00000 00000   00000 00000   00000 00000   00000 00000   00000 00000      00000 00000   00000 00000   00000 00000   00000 00000   00000 00000
1815:  00000 00000   00000 00000   00000 00000   00000 00000   00000 00000      00000 00000   00000 00000   00000 00000   00000 00000   00000 00000
1816:  00000 00000   00000 00000   00000 00000   00000 00000   00000 00000      00000 00000   00000 00000   00000 00000   00000 00000   00000 00000
1817:  00000 00000   00000 00000   00000 00000   00000 00000   00000 00000      00000 00000   00000 00000   00000 00000   00000 00000   00000 00000
1818:  00000 00000   00000 00000   00000 00000   00000 00000   00000 00000      00000 00000   00000 00000   00000 00000   00000 00000   00000 00000
1819:  00000 00000   00000 00000   00000 00000   00000 00000   00000 00000      00000 00000   00000 00000   00000 00000   00000 00000   00000 00000
1820:  00000 00000   00000 00000   00000 00000   00000 00000   00000 00000      00000 00000   00000 00000   00000 00000   00000 00000   00000 00000
1821:  00000 00000   00000 00000   00000 00000   00000 00000   00000 00000      00000 00000   00000 00000   00000 00000   00000 00000   00000 00000
1822:  00000 00000   00000 00000   00000 00000   00000 00000   00000 00000      00000 00000   00000 00000   00000 00000   00000 00000   00000 00000
1823:  00000 00000   00000 00000   00000 00000   00000 00000   00000 00000      00000 00000   00000 00000   00000 00000   00000 00000   00000 00000
1824:  00000 00000   00000 00000   00000 00000   00000 00000   00000 00000      00000 00000   00000 00000   00000 00000   00000 00000   00000 00000
1825:  00000 00000   00000 00000   00000 00000   00000 00000   00000 00000      00000 00000   00000 00000   00000 00000   00000 00000   00000 00000
1826:  00000 00000   00000 00000   00000 00000   00000 00000   00000 00000      00000 00000   00000 00000   00000 00000   00000 00000   00000 00000
1827:  00000 00000   00000 00000   00000 00000   00000 00000   00000 00000      00000 00000   00000 00000   00000 00000   00000 00000   00000 00000
1828:  00000 00000   00000 00000   00000 00000   00000 00000   00000 00000      00000 00000   00000 00000   00000 00000   00000 00000   00000 00000
1829:  00000 00000   00000 00000   00000 00000   00000 00000   00000 00000      00000 00000   00000 00000   00000 00000   00000 00000   00000 00000
1830:  00000 00000   00000 00000   00000 00000   00000 00000   00000 00000      00000 00000   00000 00000   00000 00000   00000 00000   00000 00000
1831:  00000 00000   00000 00000   00000 00000   00000 00000   00000 00000      00000 00000   00000 00000   00000 00000   00000 00000   00000 00000
1832:  00000 00000   00000 00000   00000 00000   00000 00000   00000 00000      00000 00000   00000 00000   00000 00000   00000 00000   00000 00000
1833:  00000 00000   00000 00000   00000 00000   00000 00000   00000 00000      00000 00000   00000 00000   00000 00000   00000 00000   00000 00000
1834:  00000 00000   00000 00000   00000 00000   00000 00000   00000 00000      00000 00000   00000 00000   00000 00000   00000 00000   00000 00000
1835:  00000 00000   00000 00000   00000 00000   00000 00000   00000 00000      00000 00000   00000 00000   00000 00000   00000 00000   00000 00000
1836:  00000 00000   00000 00000   00000 00000   00000 00000   00000 00000      00000 00000   00000 00000   00000 00000   00000 00000   00000 00000
1837:  00000 00000   00000 00000   00000 00000   00000 00000   00000 00000      00000 00000   00000 00000   00000 00000   00000 00000   00000 00000
1838:  00000 00000   00000 00000   00000 00000   00000 00000   00000 00000      00000 00000   00000 00000   00000 00000   00000 00000   00000 00000
1839:  00000 00000   00000 00000   00000 00000   00000 00000   00000 00000      00000 00000   00000 00000   00000 00000   00000 00000   00000 00000
1840:  00000 00000   00000 00000   00000 00000   00000 00000   00000 00000      00000 00000   00000 00000   00000 00000   00000 00000   00000 00000
1841:  00000 00000   00000 00000   00000 00000   00000 00000   00000 00000      00000 00000   00000 00000   00000 00000   00000 00000   00000 00000
1842:  00000 00000   00000 00000   00000 00000   00000 00000   00000 00000      00000 00000   00000 00000   00000 00000   00000 00000   00000 00000
1843:  00000 00000   00000 00000   00000 00000   00000 00000   00000 00000      00000 00000   00000 00000   00000 00000   00000 00000   00000 00000
1844:  00000 00000   00000 00000   00000 00000   00000 00000   00000 00000      00000 00000   00000 00000   00000 00000   00000 00000   00000 00000
1845:  00000 00000   00000 00000   00000 00000   00000 00000   00000 00000      00000 00000   00000 00000   00000 00000   00000 00000   00000 00000
1846:  00000 00000   00000 00000   00000 00000   00000 00000   00000 00000      00000 00000   00000 00000   00000 00000   00000 00000   00000 00000
1847:  00000 00000   00000 00000   00000 00000   00000 00000   00000 00000      00000 00000   00000 00000   00000 00000   00000 00000   00000 00000
1848:  00000 00000   00000 00000   00000 00000   00000 00000   00000 00000      00000 00000   00000 00000   00000 00000   00000 00000   00000 00000
1849:  00000 00000   00000 00000   00000 00000   00000 00000   00000 00000      00000 00000   00000 00000   00000 00000   00000 00000   00000 00000
```

```
1850:   00000 00000   00000 00000   00000 00000   00000 00000   00000 00000      00000 00000   00000 00000   00000 00000   00000 00000   00000 00000
1851:   00000 00000   00000 00000   00000 00000   00000 00000   00000 00000      00000 00000   00000 00000   00000 00000   00000 00000   00000 00000
1852:   00000 00000   00000 00000   00000 00000   00000 00000   00000 00000      00000 00000   00000 00000   00000 00000   00000 00000   00000 00000
1853:   00000 00000   00000 00000   00000 00000   00000 00000   00000 00000      00000 00000   00000 00000   00000 00000   00000 00000   00000 00000
1854:   00000 00000   00000 00000   00000 00000   00000 00000   00000 00000      00000 00000   00000 00000   00000 00000   00000 00000   00000 00000
1855:   00000 00000   00000 00000   00000 00000   00000 00000   00000 00000      00000 00000   00000 00000   00000 00000   00000 00000   00000 00000
1856:   00000 00000   00000 00000   00000 00000   00000 00000   00000 00000      00000 00000   00000 00000   00000 00000   00000 00000   00000 00000
1857:   00000 00000   00000 00000   00000 00000   00000 00000   00000 00000      00000 00000   00000 00000   00000 00000   00000 00000   00000 00000
1858:   00000 00000   00000 00000   00000 00000   00000 00000   00000 00000      00000 00000   00000 00000   00000 00000   00000 00000   00000 00000
1859:   00000 00000   00000 00000   00000 00000   00000 00000   00000 00000      00000 00000   00000 00000   00000 00000   00000 00000   00000 00000
1860:   00000 00000   00000 00000   00000 00000   00000 00000   00000 00000      00000 00000   00000 00000   00000 00000   00000 00000   00000 00000
1861:   00000 00000   00000 00000   00000 00000   00000 00000   00000 00000      00000 00000   00000 00000   00000 00000   00000 00000   00000 00000
1862:   00000 00000   00000 00000   00000 00000   00000 00000   00000 00000      00000 00000   00000 00000   00000 00000   00000 00000   00000 00000
1863:   00000 00000   00000 00000   00000 00000   00000 00000   00000 00000      00000 00000   00000 00000   00000 00000   00000 00000   00000 00000
1864:   00000 00000   00000 00000   00000 00000   00000 00000   00000 00000      00000 00000   00000 00000   00000 00000   00000 00000   00000 00000
1865:   00000 00000   00000 00000   00000 00000   00000 00000   00000 00000      00000 00000   00000 00000   00000 00000   00000 00000   00000 00000
1866:   00000 00000   00000 00000   00000 00000   00000 00000   00000 00000      00000 00000   00000 00000   00000 00000   00000 00000   00000 00000
1867:   00000 00000   00000 00000   00000 00000   00000 00000   00000 00000      00000 00000   00000 00000   00000 00000   00000 00000   00000 00000
1868:   00000 00000   00000 00000   00000 00000   00000 00000   00000 00000      00000 00000   00000 00000   00000 00000   00000 00000   00000 00000
1869:   00000 00000   00000 00000   00000 00000   00000 00000   00000 00000      00000 00000   00000 00000   00000 00000   00000 00000   00000 00000
1870:   00000 00000   00000 00000   00000 00000   00000 00000   00000 00000      00000 00000   00000 00000   00000 00000   00000 00000   00000 00000
1871:   00000 00000   00000 00000   00000 00000   00000 00000   00000 00000      00000 00000   00000 00000   00000 00000   00000 00000   00000 00000
1872:   00000 00000   00000 00000   00000 00000   00000 00000   00000 00000      00000 00000   00000 00000   00000 00000   00000 00000   00000 00000
1873:   00000 00000   00000 00000   00000 00000   00000 00000   00000 00000      00000 00000   00000 00000   00000 00000   00000 00000   00000 00000
1874:   00000 00000   00000 00000   00000 00000   00000 00000   00000 00000      00000 00000   00000 00000   00000 00000   00000 00000   00000 00000
1875:   00000 00000   00000 00000   00000 00000   00000 00000   00000 00000      00000 00000   00000 00000   00000 00000   00000 00000   00000 00000
1876:   00000 00000   00000 00000   00000 00000   00000 00000   00000 00000      00000 00000   00000 00000   00000 00000   00000 00000   00000 00000
1877:   00000 00000   00000 00000   00000 00000   00000 00000   00000 00000      00000 00000   00000 00000   00000 00000   00000 00000   00000 00000
1878:   00000 00000   00000 00000   00000 00000   00000 00000   00000 00000      00000 00000   00000 00000   00000 00000   00000 00000   00000 00000
1879:   00000 00000   00000 00000   00000 00000   00000 00000   00000 00000      00000 00000   00000 00000   00000 00000   00000 00000   00000 00000
1880:   00000 00000   00000 00000   00000 00000   00000 00000   00000 00000      00000 00000   00000 00000   00000 00000   00000 00000   00000 00000
1881:   00000 00000   00000 00000   00000 00000   00000 00000   00000 00000      00000 00000   00000 00000   00000 00000   00000 00000   00000 00000
1882:   00000 00000   00000 00000   00000 00000   00000 00000   00000 00000      00000 00000   00000 00000   00000 00000   00000 00000   00000 00000
1883:   00000 00000   00000 00000   00000 00000   00000 00000   00000 00000      00000 00000   00000 00000   00000 00000   00000 00000   00000 00000
1884:   00000 00000   00000 00000   00000 00000   00000 00000   00000 00000      00000 00000   00000 00000   00000 00000   00000 00000   00000 00000
1885:   00000 00000   00000 00000   00000 00000   00000 00000   00000 00000      00000 00000   00000 00000   00000 00000   00000 00000   00000 00000
1886:   00000 00000   00000 00000   00000 00000   00000 00000   00000 00000      00000 00000   00000 00000   00000 00000   00000 00000   00000 00000
1887:   00000 00000   00000 00000   00000 00000   00000 00000   00000 00000      00000 00000   00000 00000   00000 00000   00000 00000   00000 00000
1888:   00000 00000   00000 00000   00000 00000   00000 00000   00000 00000      00000 00000   00000 00000   00000 00000   00000 00000   00000 00000
1889:   00000 00000   00000 00000   00000 00000   00000 00000   00000 00000      00000 00000   00000 00000   00000 00000   00000 00000   00000 00000
1890:   00000 00000   00000 00000   00000 00000   00000 00000   00000 00000      00000 00000   00000 00000   00000 00000   00000 00000   00000 00000
1891:   00000 00000   00000 00000   00000 00000   00000 00000   00000 00000      00000 00000   00000 00000   00000 00000   00000 00000   00000 00000
1892:   00000 00000   00000 00000   00000 00000   00000 00000   00000 00000      00000 00000   00000 00000   00000 00000   00000 00000   00000 00000
1893:   00000 00000   00000 00000   00000 00000   00000 00000   00000 00000      00000 00000   00000 00000   00000 00000   00000 00000   00000 00000
1894:   00000 00000   00000 00000   00000 00000   00000 00000   00000 00000      00000 00000   00000 00000   00000 00000   00000 00000   00000 00000
1895:   00000 00000   00000 00000   00000 00000   00000 00000   00000 00000      00000 00000   00000 00000   00000 00000   00000 00000   00000 00000
1896:   00000 00000   00000 00000   00000 00000   00000 00000   00000 00000      00000 00000   00000 00000   00000 00000   00000 00000   00000 00000
1897:   00000 00000   00000 00000   00000 00000   00000 00000   00000 00000      00000 00000   00000 00000   00000 00000   00000 00000   00000 00000
1898:   00000 00000   00000 00000   00000 00000   00000 00000   00000 00000      00000 00000   00000 00000   00000 00000   00000 00000   00000 00000
1899:   00000 00000   00000 00000   00000 00000   00000 00000   00000 00000      00000 00000   00000 00000   00000 00000   00000 00000   00000 00000
```

```
1900:  00000 00000  00000 00000  00000 00000  00000 00000  00000 00000   00000 00000  00000 00000  00000 00000  00000 00000  00000 00000
1901:  00000 00000  00000 00000  00000 00000  00000 00000  00000 00000   00000 00000  00000 00000  00000 00000  00000 00000  00000 00000
1902:  00000 00000  00000 00000  00000 00000  00000 00000  00000 00000   00000 00000  00000 00000  00000 00000  00000 00000  00000 00000
1903:  00000 00000  00000 00000  00000 00000  00000 00000  00000 00000   00000 00000  00000 00000  00000 00000  00000 00000  00000 00000
1904:  00000 00000  00000 00000  00000 00000  00000 00000  00000 00000   00000 00000  00000 00000  00000 00000  00000 00000  00000 00000
1905:  00000 00000  00000 00000  00000 00000  00000 00000  00000 00000   00000 00000  00000 00000  00000 00000  00000 00000  00000 00000
1906:  00000 00000  00000 00000  00000 00000  00000 00000  00000 00000   00000 00000  00000 00000  00000 00000  00000 00000  00000 00000
1907:  00000 00000  00000 00000  00000 00000  00000 00000  00000 00000   00000 00000  00000 00000  00000 00000  00000 00000  00000 00000
1908:  00000 00000  00000 00000  00000 00000  00000 00000  00000 00000   00000 00000  00000 00000  00000 00000  00000 00000  00000 00000
1909:  00000 00000  00000 00000  00000 00000  00000 00000  00000 00000   00000 00000  00000 00000  00000 00000  00000 00000  00000 00000
1910:  00000 00000  00000 00000  00000 00000  00000 00000  00000 00000   00000 00000  00000 00000  00000 00000  00000 00000  00000 00000
1911:  00000 00000  00000 00000  00000 00000  00000 00000  00000 00000   00000 00000  00000 00000  00000 00000  00000 00000  00000 00000
1912:  00000 00000  00000 00000  00000 00000  00000 00000  00000 00000   00000 00000  00000 00000  00000 00000  00000 00000  00000 00000
1913:  00000 00000  00000 00000  00000 00000  00000 00000  00000 00000   00000 00000  00000 00000  00000 00000  00000 00000  00000 00000
1914:  00000 00000  00000 00000  00000 00000  00000 00000  00000 00000   00000 00000  00000 00000  00000 00000  00000 00000  00000 00000
1915:  00000 00000  00000 00000  00000 00000  00000 00000  00000 00000   00000 00000  00000 00000  00000 00000  00000 00000  00000 00000
1916:  00000 00000  00000 00000  00000 00000  00000 00000  00000 00000   00000 00000  00000 00000  00000 00000  00000 00000  00000 00000
1917:  00000 00000  00000 00000  00000 00000  00000 00000  00000 00000   00000 00000  00000 00000  00000 00000  00000 00000  00000 00000
1918:  00000 00000  00000 00000  00000 00000  00000 00000  00000 00000   00000 00000  00000 00000  00000 00000  00000 00000  00000 00000
1919:  00000 00000  00000 00000  00000 00000  00000 00000  00000 00000   00000 00000  00000 00000  00000 00000  00000 00000  00000 00000
1920:  00000 00000  00000 00000  00000 00000  00000 00000  00000 00000   00000 00000  00000 00000  00000 00000  00000 00000  00000 00000
1921:  00000 00000  00000 00000  00000 00000  00000 00000  00000 00000   00000 00000  00000 00000  00000 00000  00000 00000  00000 00000
1922:  00000 00000  00000 00000  00000 00000  00000 00000  00000 00000   00000 00000  00000 00000  00000 00000  00000 00000  00000 00000
1923:  00000 00000  00000 00000  00000 00000  00000 00000  00000 00000   00000 00000  00000 00000  00000 00000  00000 00000  00000 00000
1924:  00000 00000  00000 00000  00000 00000  00000 00000  00000 00000   00000 00000  00000 00000  00000 00000  00000 00000  00000 00000
1925:  00000 00000  00000 00000  00000 00000  00000 00000  00000 00000   00000 00000  00000 00000  00000 00000  00000 00000  00000 00000
1926:  00000 00000  00000 00000  00000 00000  00000 00000  00000 00000   00000 00000  00000 00000  00000 00000  00000 00000  00000 00000
1927:  00000 00000  00000 00000  00000 00000  00000 00000  00000 00000   00000 00000  00000 00000  00000 00000  00000 00000  00000 00000
1928:  00000 00000  00000 00000  00000 00000  00000 00000  00000 00000   00000 00000  00000 00000  00000 00000  00000 00000  00000 00000
1929:  00000 00000  00000 00000  00000 00000  00000 00000  00000 00000   00000 00000  00000 00000  00000 00000  00000 00000  00000 00000
1930:  00000 00000  00000 00000  00000 00000  00000 00000  00000 00000   00000 00000  00000 00000  00000 00000  00000 00000  00000 00000
1931:  00000 00000  00000 00000  00000 00000  00000 00000  00000 00000   00000 00000  00000 00000  00000 00000  00000 00000  00000 00000
1932:  00000 00000  00000 00000  00000 00000  00000 00000  00000 00000   00000 00000  00000 00000  00000 00000  00000 00000  00000 00000
1933:  00000 00000  00000 00000  00000 00000  00000 00000  00000 00000   00000 00000  00000 00000  00000 00000  00000 00000  00000 00000
1934:  00000 00000  00000 00000  00000 00000  00000 00000  00000 00000   00000 00000  00000 00000  00000 00000  00000 00000  00000 00000
1935:  00000 00000  00000 00000  00000 00000  00000 00000  00000 00000   00000 00000  00000 00000  00000 00000  00000 00000  00000 00000
1936:  00000 00000  00000 00000  00000 00000  00000 00000  00000 00000   00000 00000  00000 00000  00000 00000  00000 00000  00000 00000
1937:  00000 00000  00000 00000  00000 00000  00000 00000  00000 00000   00000 00000  00000 00000  00000 00000  00000 00000  00000 00000
1938:  00000 00000  00000 00000  00000 00000  00000 00000  00000 00000   00000 00000  00000 00000  00000 00000  00000 00000  00000 00000
1939:  00000 00000  00000 00000  00000 00000  00000 00000  00000 00000   00000 00000  00000 00000  00000 00000  00000 00000  00000 00000
1940:  00000 00000  00000 00000  00000 00000  00000 00000  00000 00000   00000 00000  00000 00000  00000 00000  00000 00000  00000 00000
1941:  00000 00000  00000 00000  00000 00000  00000 00000  00000 00000   00000 00000  00000 00000  00000 00000  00000 00000  00000 00000
1942:  00000 00000  00000 00000  00000 00000  00000 00000  00000 00000   00000 00000  00000 00000  00000 00000  00000 00000  00000 00000
1943:  00000 00000  00000 00000  00000 00000  00000 00000  00000 00000   00000 00000  00000 00000  00000 00000  00000 00000  00000 00000
1944:  00000 00000  00000 00000  00000 00000  00000 00000  00000 00000   00000 00000  00000 00000  00000 00000  00000 00000  00000 00000
1945:  00000 00000  00000 00000  00000 00000  00000 00000  00000 00000   00000 00000  00000 00000  00000 00000  00000 00000  00000 00000
1946:  00000 00000  00000 00000  00000 00000  00000 00000  00000 00000   00000 00000  00000 00000  00000 00000  00000 00000  00000 00000
1947:  00000 00000  00000 00000  00000 00000  00000 00000  00000 00000   00000 00000  00000 00000  00000 00000  00000 00000  00000 00000
1948:  00000 00000  00000 00000  00000 00000  00000 00000  00000 00000   00000 00000  00000 00000  00000 00000  00000 00000  00000 00000
1949:  00000 00000  00000 00000  00000 00000  00000 00000  00000 00000   00000 00000  00000 00000  00000 00000  00000 00000  00000 00000
```

```
1950:  00000 00000  00000 00000  00000 00000  00000 00000  00000 00000    00000 00000  00000 00000  00000 00000  00000 00000  00000 00000
1951:  00000 00000  00000 00000  00000 00000  00000 00000  00000 00000    00000 00000  00000 00000  00000 00000  00000 00000  00000 00000
1952:  00000 00000  00000 00000  00000 00000  00000 00000  00000 00000    00000 00000  00000 00000  00000 00000  00000 00000  00000 00000
1953:  00000 00000  00000 00000  00000 00000  00000 00000  00000 00000    00000 00000  00000 00000  00000 00000  00000 00000  00000 00000
1954:  00000 00000  00000 00000  00000 00000  00000 00000  00000 00000    00000 00000  00000 00000  00000 00000  00000 00000  00000 00000
1955:  00000 00000  00000 00000  00000 00000  00000 00000  00000 00000    00000 00000  00000 00000  00000 00000  00000 00000  00000 00000
1956:  00000 00000  00000 00000  00000 00000  00000 00000  00000 00000    00000 00000  00000 00000  00000 00000  00000 00000  00000 00000
1957:  00000 00000  00000 00000  00000 00000  00000 00000  00000 00000    00000 00000  00000 00000  00000 00000  00000 00000  00000 00000
1958:  00000 00000  00000 00000  00000 00000  00000 00000  00000 00000    00000 00000  00000 00000  00000 00000  00000 00000  00000 00000
1959:  00000 00000  00000 00000  00000 00000  00000 00000  00000 00000    00000 00000  00000 00000  00000 00000  00000 00000  00000 00000
1960:  00000 00000  00000 00000  00000 00000  00000 00000  00000 00000    00000 00000  00000 00000  00000 00000  00000 00000  00000 00000
1961:  00000 00000  00000 00000  00000 00000  00000 00000  00000 00000    00000 00000  00000 00000  00000 00000  00000 00000  00000 00000
1962:  00000 00000  00000 00000  00000 00000  00000 00000  00000 00000    00000 00000  00000 00000  00000 00000  00000 00000  00000 00000
1963:  00000 00000  00000 00000  00000 00000  00000 00000  00000 00000    00000 00000  00000 00000  00000 00000  00000 00000  00000 00000
1964:  00000 00000  00000 00000  00000 00000  00000 00000  00000 00000    00000 00000  00000 00000  00000 00000  00000 00000  00000 00000
1965:  00000 00000  00000 00000  00000 00000  00000 00000  00000 00000    00000 00000  00000 00000  00000 00000  00000 00000  00000 00000
1966:  00000 00000  00000 00000  00000 00000  00000 00000  00000 00000    00000 00000  00000 00000  00000 00000  00000 00000  00000 00000
1967:  00000 00000  00000 00000  00000 00000  00000 00000  00000 00000    00000 00000  00000 00000  00000 00000  00000 00000  00000 00000
1968:  00000 00000  00000 00000  00000 00000  00000 00000  00000 00000    00000 00000  00000 00000  00000 00000  00000 00000  00000 00000
1969:  00000 00000  00000 00000  00000 00000  00000 00000  00000 00000    00000 00000  00000 00000  00000 00000  00000 00000  00000 00000
1970:  00000 00000  00000 00000  00000 00000  00000 00000  00000 00000    00000 00000  00000 00000  00000 00000  00000 00000  00000 00000
1971:  00000 00000  00000 00000  00000 00000  00000 00000  00000 00000    00000 00000  00000 00000  00000 00000  00000 00000  00000 00000
1972:  00000 00000  00000 00000  00000 00000  00000 00000  00000 00000    00000 00000  00000 00000  00000 00000  00000 00000  00000 00000
1973:  00000 00000  00000 00000  00000 00000  00000 00000  00000 00000    00000 00000  00000 00000  00000 00000  00000 00000  00000 00000
1974:  00000 00000  00000 00000  00000 00000  00000 00000  00000 00000    00000 00000  00000 00000  00000 00000  00000 00000  00000 00000
1975:  00000 00000  00000 00000  00000 00000  00000 00000  00000 00000    00000 00000  00000 00000  00000 00000  00000 00000  00000 00000
1976:  00000 00000  00000 00000  00000 00000  00000 00000  00000 00000    00000 00000  00000 00000  00000 00000  00000 00000  00000 00000
1977:  00000 00000  00000 00000  00000 00000  00000 00000  00000 00000    00000 00000  00000 00000  00000 00000  00000 00000  00000 00000
1978:  00000 00000  00000 00000  00000 00000  00000 00000  00000 00000    00000 00000  00000 00000  00000 00000  00000 00000  00000 00000
1979:  00000 00000  00000 00000  00000 00000  00000 00000  00000 00000    00000 00000  00000 00000  00000 00000  00000 00000  00000 00000
1980:  00000 00000  00000 00000  00000 00000  00000 00000  00000 00000    00000 00000  00000 00000  00000 00000  00000 00000  00000 00000
1981:  00000 00000  00000 00000  00000 00000  00000 00000  00000 00000    00000 00000  00000 00000  00000 00000  00000 00000  00000 00000
1982:  00000 00000  00000 00000  00000 00000  00000 00000  00000 00000    00000 00000  00000 00000  00000 00000  00000 00000  00000 00000
1983:  00000 00000  00000 00000  00000 00000  00000 00000  00000 00000    00000 00000  00000 00000  00000 00000  00000 00000  00000 00000
1984:  00000 00000  00000 00000  00000 00000  00000 00000  00000 00000    00000 00000  00000 00000  00000 00000  00000 00000  00000 00000
1985:  00000 00000  00000 00000  00000 00000  00000 00000  00000 00000    00000 00000  00000 00000  00000 00000  00000 00000  00000 00000
1986:  00000 00000  00000 00000  00000 00000  00000 00000  00000 00000    00000 00000  00000 00000  00000 00000  00000 00000  00000 00000
1987:  00000 00000  00000 00000  00000 00000  00000 00000  00000 00000    00000 00000  00000 00000  00000 00000  00000 00000  00000 00000
1988:  00000 00000  00000 00000  00000 00000  00000 00000  00000 00000    00000 00000  00000 00000  00000 00000  00000 00000  00000 00000
1989:  00000 00000  00000 00000  00000 00000  00000 00000  00000 00000    00000 00000  00000 00000  00000 00000  00000 00000  00000 00000
1990:  00000 00000  00000 00000  00000 00000  00000 00000  00000 00000    00000 00000  00000 00000  00000 00000  00000 00000  00000 00000
1991:  00000 00000  00000 00000  00000 00000  00000 00000  00000 00000    00000 00000  00000 00000  00000 00000  00000 00000  00000 00000
1992:  00000 00000  00000 00000  00000 00000  00000 00000  00000 00000    00000 00000  00000 00000  00000 00000  00000 00000  00000 00000
1993:  00000 00000  00000 00000  00000 00000  00000 00000  00000 00000    00000 00000  00000 00000  00000 00000  00000 00000  00000 00000
1994:  00000 00000  00000 00000  00000 00000  00000 00000  00000 00000    00000 00000  00000 00000  00000 00000  00000 00000  00000 00000
1995:  00000 00000  00000 00000  00000 00000  00000 00000  00000 00000    00000 00000  00000 00000  00000 00000  00000 00000  00000 00000
1996:  00000 00000  00000 00000  00000 00000  00000 00000  00000 00000    00000 00000  00000 00000  00000 00000  00000 00000  00000 00000
1997:  00000 00000  00000 00000  00000 00000  00000 00000  00000 00000    00000 00000  00000 00000  00000 00000  00000 00000  00000 00000
1998:  00000 00000  00000 00000  00000 00000  00000 00000  00000 00000    00000 00000  00000 00000  00000 00000  00000 00000  00000 00000
1999:  00000 00000  00000 00000  00000 00000  00000 00000  00000 00000    00000 00000  00000 00000  00000 00000  00000 00000  00000 00000
```

```
2000:   00000 00000   00000 00000   00000 00000   00000 00000   00000 00000      00000 00000   00000 00000   00000 00000   00000 00000   00000 00000
2001:   00000 00000   00000 00000   00000 00000   00000 00000   00000 00000      00000 00000   00000 00000   00000 00000   00000 00000   00000 00000
2002:   00000 00000   00000 00000   00000 00000   00000 00000   00000 00000      00000 00000   00000 00000   00000 00000   00000 00000   00000 00000
2003:   00000 00000   00000 00000   00000 00000   00000 00000   00000 00000      00000 00000   00000 00000   00000 00000   00000 00000   00000 00000
2004:   00000 00000   00000 00000   00000 00000   00000 00000   00000 00000      00000 00000   00000 00000   00000 00000   00000 00000   00000 00000
2005:   00000 00000   00000 00000   00000 00000   00000 00000   00000 00000      00000 00000   00000 00000   00000 00000   00000 00000   00000 00000
2006:   00000 00000   00000 00000   00000 00000   00000 00000   00000 00000      00000 00000   00000 00000   00000 00000   00000 00000   00000 00000
2007:   00000 00000   00000 00000   00000 00000   00000 00000   00000 00000      00000 00000   00000 00000   00000 00000   00000 00000   00000 00000
2008:   00000 00000   00000 00000   00000 00000   00000 00000   00000 00000      00000 00000   00000 00000   00000 00000   00000 00000   00000 00000
2009:   00000 00000   00000 00000   00000 00000   00000 00000   00000 00000      00000 00000   00000 00000   00000 00000   00000 00000   00000 00000
2010:   00000 00000   00000 00000   00000 00000   00000 00000   00000 00000      00000 00000   00000 00000   00000 00000   00000 00000   00000 00000
2011:   00000 00000   00000 00000   00000 00000   00000 00000   00000 00000      00000 00000   00000 00000   00000 00000   00000 00000   00000 00000
2012:   00000 00000   00000 00000   00000 00000   00000 00000   00000 00000      00000 00000   00000 00000   00000 00000   00000 00000   00000 00000
2013:   00000 00000   00000 00000   00000 00000   00000 00000   00000 00000      00000 00000   00000 00000   00000 00000   00000 00000   00000 00000
2014:   00000 00000   00000 00000   00000 00000   00000 00000   00000 00000      00000 00000   00000 00000   00000 00000   00000 00000   00000 00000
2015:   00000 00000   00000 00000   00000 00000   00000 00000   00000 00000      00000 00000   00000 00000   00000 00000   00000 00000   00000 00000
2016:   00000 00000   00000 00000   00000 00000   00000 00000   00000 00000      00000 00000   00000 00000   00000 00000   00000 00000   00000 00000
2017:   00000 00000   00000 00000   00000 00000   00000 00000   00000 00000      00000 00000   00000 00000   00000 00000   00000 00000   00000 00000
2018:   00000 00000   00000 00000   00000 00000   00000 00000   00000 00000      00000 00000   00000 00000   00000 00000   00000 00000   00000 00000
2019:   00000 00000   00000 00000   00000 00000   00000 00000   00000 00000      00000 00000   00000 00000   00000 00000   00000 00000   00000 00000
2020:   00000 00000   00000 00000   00000 00000   00000 00000   00000 00000      00000 00000   00000 00000   00000 00000   00000 00000   00000 00000
2021:   00000 00000   00000 00000   00000 00000   00000 00000   00000 00000      00000 00000   00000 00000   00000 00000   00000 00000   00000 00000
2022:   00000 00000   00000 00000   00000 00000   00000 00000   00000 00000      00000 00000   00000 00000   00000 00000   00000 00000   00000 00000
2023:   00000 00000   00000 00000   00000 00000   00000 00000   00000 00000      00000 00000   00000 00000   00000 00000   00000 00000   00000 00000
2024:   00000 00000   00000 00000   00000 00000   00000 00000   00000 00000      00000 00000   00000 00000   00000 00000   00000 00000   00000 00000
2025:   00000 00000   00000 00000   00000 00000   00000 00000   00000 00000      00000 00000   00000 00000   00000 00000   00000 00000   00000 00000
2026:   00000 00000   00000 00000   00000 00000   00000 00000   00000 00000      00000 00000   00000 00000   00000 00000   00000 00000   00000 00000
2027:   00000 00000   00000 00000   00000 00000   00000 00000   00000 00000      00000 00000   00000 00000   00000 00000   00000 00000   00000 00000
2028:   00000 00000   00000 00000   00000 00000   00000 00000   00000 00000      00000 00000   00000 00000   00000 00000   00000 00000   00000 00000
2029:   00000 00000   00000 00000   00000 00000   00000 00000   00000 00000      00000 00000   00000 00000   00000 00000   00000 00000   00000 00000
2030:   00000 00000   00000 00000   00000 00000   00000 00000   00000 00000      00000 00000   00000 00000   00000 00000   00000 00000   00000 00000
2031:   00000 00000   00000 00000   00000 00000   00000 00000   00000 00000      00000 00000   00000 00000   00000 00000   00000 00000   00000 00000
2032:   00000 00000   00000 00000   00000 00000   00000 00000   00000 00000      00000 00000   00000 00000   00000 00000   00000 00000   00000 00000
2033:   00000 00000   00000 00000   00000 00000   00000 00000   00000 00000      00000 00000   00000 00000   00000 00000   00000 00000   00000 00000
2034:   00000 00000   00000 00000   00000 00000   00000 00000   00000 00000      00000 00000   00000 00000   00000 00000   00000 00000   00000 00000
2035:   00000 00000   00000 00000   00000 00000   00000 00000   00000 00000      00000 00000   00000 00000   00000 00000   00000 00000   00000 00000
2036:   00000 00000   00000 00000   00000 00000   00000 00000   00000 00000      00000 00000   00000 00000   00000 00000   00000 00000   00000 00000
2037:   00000 00000   00000 00000   00000 00000   00000 00000   00000 00000      00000 00000   00000 00000   00000 00000   00000 00000   00000 00000
2038:   00000 00000   00000 00000   00000 00000   00000 00000   00000 00000      00000 00000   00000 00000   00000 00000   00000 00000   00000 00000
2039:   00000 00000   00000 00000   00000 00000   00000 00000   00000 00000      00000 00000   00000 00000   00000 00000   00000 00000   00000 00000
2040:   00000 00000   00000 00000   00000 00000   00000 00000   00000 00000      00000 00000   00000 00000   00000 00000   00000 00000   00000 00000
2041:   00000 00000   00000 00000   00000 00000   00000 00000   00000 00000      00000 00000   00000 00000   00000 00000   00000 00000   00000 00000
2042:   00000 00000   00000 00000   00000 00000   00000 00000   00000 00000      00000 00000   00000 00000   00000 00000   00000 00000   00000 00000
2043:   00000 00000   00000 00000   00000 00000   00000 00000   00000 00000      00000 00000   00000 00000   00000 00000   00000 00000   00000 00000
2044:   00000 00000   00000 00000   00000 00000   00000 00000   00000 00000      00000 00000   00000 00000   00000 00000   00000 00000   00000 00000
2045:   00000 00000   00000 00000   00000 00000   00000 00000   00000 00000      00000 00000   00000 00000   00000 00000   00000 00000   00000 00000
2046:   00000 00000   00000 00000   00000 00000   00000 00000   00000 00000      00000 00000   00000 00000   00000 00000   00000 00000   00000 00000
2047:   00000 00000   00000 00000   00000 00000   00000 00000   00000 00000      00000 00000   00000 00000   00000 00000   00000 00000   00000 00000
2048:   00000 00000   00000 00000   00000 00000   00000 00000   00000 00000      00000 00000   00000 00000   00000 00000   00000 00000   00000 00000
2049:   00000 00000   00000 00000   00000 00000   00000 00000   00000 00000      00000 00000   00000 00000   00000 00000   00000 00000   00000 00000
```

```
2050: 00000 00000  00000 00000  00000 00000  00000 00000  00000 00000    00000 00000  00000 00000  00000 00000  00000 00000  00000 00000
2051: 00000 00000  00000 00000  00000 00000  00000 00000  00000 00000    00000 00000  00000 00000  00000 00000  00000 00000  00000 00000
2052: 00000 00000  00000 00000  00000 00000  00000 00000  00000 00000    00000 00000  00000 00000  00000 00000  00000 00000  00000 00000
2053: 00000 00000  00000 00000  00000 00000  00000 00000  00000 00000    00000 00000  00000 00000  00000 00000  00000 00000  00000 00000
2054: 00000 00000  00000 00000  00000 00000  00000 00000  00000 00000    00000 00000  00000 00000  00000 00000  00000 00000  00000 00000
2055: 00000 00000  00000 00000  00000 00000  00000 00000  00000 00000    00000 00000  00000 00000  00000 00000  00000 00000  00000 00000
2056: 00000 00000  00000 00000  00000 00000  00000 00000  00000 00000    00000 00000  00000 00000  00000 00000  00000 00000  00000 00000
2057: 00000 00000  00000 00000  00000 00000  00000 00000  00000 00000    00000 00000  00000 00000  00000 00000  00000 00000  00000 00000
2058: 00000 00000  00000 00000  00000 00000  00000 00000  00000 00000    00000 00000  00000 00000  00000 00000  00000 00000  00000 00000
2059: 00000 00000  00000 00000  00000 00000  00000 00000  00000 00000    00000 00000  00000 00000  00000 00000  00000 00000  00000 00000
2060: 00000 00000  00000 00000  00000 00000  00000 00000  00000 00000    00000 00000  00000 00000  00000 00000  00000 00000  00000 00000
2061: 00000 00000  00000 00000  00000 00000  00000 00000  00000 00000    00000 00000  00000 00000  00000 00000  00000 00000  00000 00000
2062: 00000 00000  00000 00000  00000 00000  00000 00000  00000 00000    00000 00000  00000 00000  00000 00000  00000 00000  00000 00000
2063: 00000 00000  00000 00000  00000 00000  00000 00000  00000 00000    00000 00000  00000 00000  00000 00000  00000 00000  00000 00000
2064: 00000 00000  00000 00000  00000 00000  00000 00000  00000 00000    00000 00000  00000 00000  00000 00000  00000 00000  00000 00000
2065: 00000 00000  00000 00000  00000 00000  00000 00000  00000 00000    00000 00000  00000 00000  00000 00000  00000 00000  00000 00000
2066: 00000 00000  00000 00000  00000 00000  00000 00000  00000 00000    00000 00000  00000 00000  00000 00000  00000 00000  00000 00000
2067: 00000 00000  00000 00000  00000 00000  00000 00000  00000 00000    00000 00000  00000 00000  00000 00000  00000 00000  00000 00000
2068: 00000 00000  00000 00000  00000 00000  00000 00000  00000 00000    00000 00000  00000 00000  00000 00000  00000 00000  00000 00000
2069: 00000 00000  00000 00000  00000 00000  00000 00000  00000 00000    00000 00000  00000 00000  00000 00000  00000 00000  00000 00000
2070: 00000 00000  00000 00000  00000 00000  00000 00000  00000 00000    00000 00000  00000 00000  00000 00000  00000 00000  00000 00000
2071: 00000 00000  00000 00000  00000 00000  00000 00000  00000 00000    00000 00000  00000 00000  00000 00000  00000 00000  00000 00000
2072: 00000 00000  00000 00000  00000 00000  00000 00000  00000 00000    00000 00000  00000 00000  00000 00000  00000 00000  00000 00000
2073: 00000 00000  00000 00000  00000 00000  00000 00000  00000 00000    00000 00000  00000 00000  00000 00000  00000 00000  00000 00000
2074: 00000 00000  00000 00000  00000 00000  00000 00000  00000 00000    00000 00000  00000 00000  00000 00000  00000 00000  00000 00000
2075: 00000 00000  00000 00000  00000 00000  00000 00000  00000 00000    00000 00000  00000 00000  00000 00000  00000 00000  00000 00000
2076: 00000 00000  00000 00000  00000 00000  00000 00000  00000 00000    00000 00000  00000 00000  00000 00000  00000 00000  00000 00000
2077: 00000 00000  00000 00000  00000 00000  00000 00000  00000 00000    00000 00000  00000 00000  00000 00000  00000 00000  00000 00000
2078: 00000 00000  00000 00000  00000 00000  00000 00000  00000 00000    00000 00000  00000 00000  00000 00000  00000 00000  00000 00000
2079: 00000 00000  00000 00000  00000 00000  00000 00000  00000 00000    00000 00000  00000 00000  00000 00000  00000 00000  00000 00000
2080: 00000 00000  00000 00000  00000 00000  00000 00000  00000 00000    00000 00000  00000 00000  00000 00000  00000 00000  00000 00000
2081: 00000 00000  00000 00000  00000 00000  00000 00000  00000 00000    00000 00000  00000 00000  00000 00000  00000 00000  00000 00000
2082: 00000 00000  00000 00000  00000 00000  00000 00000  00000 00000    00000 00000  00000 00000  00000 00000  00000 00000  00000 00000
2083: 00000 00000  00000 00000  00000 00000  00000 00000  00000 00000    00000 00000  00000 00000  00000 00000  00000 00000  00000 00000
2084: 00000 00000  00000 00000  00000 00000  00000 00000  00000 00000    00000 00000  00000 00000  00000 00000  00000 00000  00000 00000
2085: 00000 00000  00000 00000  00000 00000  00000 00000  00000 00000    00000 00000  00000 00000  00000 00000  00000 00000  00000 00000
2086: 00000 00000  00000 00000  00000 00000  00000 00000  00000 00000    00000 00000  00000 00000  00000 00000  00000 00000  00000 00000
2087: 00000 00000  00000 00000  00000 00000  00000 00000  00000 00000    00000 00000  00000 00000  00000 00000  00000 00000  00000 00000
2088: 00000 00000  00000 00000  00000 00000  00000 00000  00000 00000    00000 00000  00000 00000  00000 00000  00000 00000  00000 00000
2089: 00000 00000  00000 00000  00000 00000  00000 00000  00000 00000    00000 00000  00000 00000  00000 00000  00000 00000  00000 00000
2090: 00000 00000  00000 00000  00000 00000  00000 00000  00000 00000    00000 00000  00000 00000  00000 00000  00000 00000  00000 00000
2091: 00000 00000  00000 00000  00000 00000  00000 00000  00000 00000    00000 00000  00000 00000  00000 00000  00000 00000  00000 00000
2092: 00000 00000  00000 00000  00000 00000  00000 00000  00000 00000    00000 00000  00000 00000  00000 00000  00000 00000  00000 00000
2093: 00000 00000  00000 00000  00000 00000  00000 00000  00000 00000    00000 00000  00000 00000  00000 00000  00000 00000  00000 00000
2094: 00000 00000  00000 00000  00000 00000  00000 00000  00000 00000    00000 00000  00000 00000  00000 00000  00000 00000  00000 00000
2095: 00000 00000  00000 00000  00000 00000  00000 00000  00000 00000    00000 00000  00000 00000  00000 00000  00000 00000  00000 00000
2096: 00000 00000  00000 00000  00000 00000  00000 00000  00000 00000    00000 00000  00000 00000  00000 00000  00000 00000  00000 00000
2097: 00000 00000  00000 00000  00000 00000  00000 00000  00000 00000    00000 00000  00000 00000  00000 00000  00000 00000  00000 00000
2098: 00000 00000  00000 00000  00000 00000  00000 00000  00000 00000    00000 00000  00000 00000  00000 00000  00000 00000  00000 00000
2099: 00000 00000  00000 00000  00000 00000  00000 00000  00000 00000    00000 00000  00000 00000  00000 00000  00000 00000  00000 00000
```

```
2100:  00000 00000  00000 00000  00000 00000  00000 00000  00000 00000   00000 00000  00000 00000  00000 00000  00000 00000  00000 00000
2101:  00000 00000  00000 00000  00000 00000  00000 00000  00000 00000   00000 00000  00000 00000  00000 00000  00000 00000  00000 00000
2102:  00000 00000  00000 00000  00000 00000  00000 00000  00000 00000   00000 00000  00000 00000  00000 00000  00000 00000  00000 00000
2103:  00000 00000  00000 00000  00000 00000  00000 00000  00000 00000   00000 00000  00000 00000  00000 00000  00000 00000  00000 00000
2104:  00000 00000  00000 00000  00000 00000  00000 00000  00000 00000   00000 00000  00000 00000  00000 00000  00000 00000  00000 00000
2105:  00000 00000  00000 00000  00000 00000  00000 00000  00000 00000   00000 00000  00000 00000  00000 00000  00000 00000  00000 00000
2106:  00000 00000  00000 00000  00000 00000  00000 00000  00000 00000   00000 00000  00000 00000  00000 00000  00000 00000  00000 00000
2107:  00000 00000  00000 00000  00000 00000  00000 00000  00000 00000   00000 00000  00000 00000  00000 00000  00000 00000  00000 00000
2108:  00000 00000  00000 00000  00000 00000  00000 00000  00000 00000   00000 00000  00000 00000  00000 00000  00000 00000  00000 00000
2109:  00000 00000  00000 00000  00000 00000  00000 00000  00000 00000   00000 00000  00000 00000  00000 00000  00000 00000  00000 00000
2110:  00000 00000  00000 00000  00000 00000  00000 00000  00000 00000   00000 00000  00000 00000  00000 00000  00000 00000  00000 00000
2111:  00000 00000  00000 00000  00000 00000  00000 00000  00000 00000   00000 00000  00000 00000  00000 00000  00000 00000  00000 00000
2112:  00000 00000  00000 00000  00000 00000  00000 00000  00000 00000   00000 00000  00000 00000  00000 00000  00000 00000  00000 00000
2113:  00000 00000  00000 00000  00000 00000  00000 00000  00000 00000   00000 00000  00000 00000  00000 00000  00000 00000  00000 00000
2114:  00000 00000  00000 00000  00000 00000  00000 00000  00000 00000   00000 00000  00000 00000  00000 00000  00000 00000  00000 00000
2115:  00000 00000  00000 00000  00000 00000  00000 00000  00000 00000   00000 00000  00000 00000  00000 00000  00000 00000  00000 00000
2116:  00000 00000  00000 00000  00000 00000  00000 00000  00000 00000   00000 00000  00000 00000  00000 00000  00000 00000  00000 00000
2117:  00000 00000  00000 00000  00000 00000  00000 00000  00000 00000   00000 00000  00000 00000  00000 00000  00000 00000  00000 00000
2118:  00000 00000  00000 00000  00000 00000  00000 00000  00000 00000   00000 00000  00000 00000  00000 00000  00000 00000  00000 00000
2119:  00000 00000  00000 00000  00000 00000  00000 00000  00000 00000   00000 00000  00000 00000  00000 00000  00000 00000  00000 00000
2120:  00000 00000  00000 00000  00000 00000  00000 00000  00000 00000   00000 00000  00000 00000  00000 00000  00000 00000  00000 00000
2121:  00000 00000  00000 00000  00000 00000  00000 00000  00000 00000   00000 00000  00000 00000  00000 00000  00000 00000  00000 00000
2122:  00000 00000  00000 00000  00000 00000  00000 00000  00000 00000   00000 00000  00000 00000  00000 00000  00000 00000  00000 00000
2123:  00000 00000  00000 00000  00000 00000  00000 00000  00000 00000   00000 00000  00000 00000  00000 00000  00000 00000  00000 00000
2124:  00000 00000  00000 00000  00000 00000  00000 00000  00000 00000   00000 00000  00000 00000  00000 00000  00000 00000  00000 00000
2125:  00000 00000  00000 00000  00000 00000  00000 00000  00000 00000   00000 00000  00000 00000  00000 00000  00000 00000  00000 00000
2126:  00000 00000  00000 00000  00000 00000  00000 00000  00000 00000   00000 00000  00000 00000  00000 00000  00000 00000  00000 00000
2127:  00000 00000  00000 00000  00000 00000  00000 00000  00000 00000   00000 00000  00000 00000  00000 00000  00000 00000  00000 00000
2128:  00000 00000  00000 00000  00000 00000  00000 00000  00000 00000   00000 00000  00000 00000  00000 00000  00000 00000  00000 00000
2129:  00000 00000  00000 00000  00000 00000  00000 00000  00000 00000   00000 00000  00000 00000  00000 00000  00000 00000  00000 00000
2130:  00000 00000  00000 00000  00000 00000  00000 00000  00000 00000   00000 00000  00000 00000  00000 00000  00000 00000  00000 00000
2131:  00000 00000  00000 00000  00000 00000  00000 00000  00000 00000   00000 00000  00000 00000  00000 00000  00000 00000  00000 00000
2132:  00000 00000  00000 00000  00000 00000  00000 00000  00000 00000   00000 00000  00000 00000  00000 00000  00000 00000  00000 00000
2133:  00000 00000  00000 00000  00000 00000  00000 00000  00000 00000   00000 00000  00000 00000  00000 00000  00000 00000  00000 00000
2134:  00000 00000  00000 00000  00000 00000  00000 00000  00000 00000   00000 00000  00000 00000  00000 00000  00000 00000  00000 00000
2135:  00000 00000  00000 00000  00000 00000  00000 00000  00000 00000   00000 00000  00000 00000  00000 00000  00000 00000  00000 00000
2136:  00000 00000  00000 00000  00000 00000  00000 00000  00000 00000   00000 00000  00000 00000  00000 00000  00000 00000  00000 00000
2137:  00000 00000  00000 00000  00000 00000  00000 00000  00000 00000   00000 00000  00000 00000  00000 00000  00000 00000  00000 00000
2138:  00000 00000  00000 00000  00000 00000  00000 00000  00000 00000   00000 00000  00000 00000  00000 00000  00000 00000  00000 00000
2139:  00000 00000  00000 00000  00000 00000  00000 00000  00000 00000   00000 00000  00000 00000  00000 00000  00000 00000  00000 00000
2140:  00000 00000  00000 00000  00000 00000  00000 00000  00000 00000   00000 00000  00000 00000  00000 00000  00000 00000  00000 00000
2141:  00000 00000  00000 00000  00000 00000  00000 00000  00000 00000   00000 00000  00000 00000  00000 00000  00000 00000  00000 00000
2142:  00000 00000  00000 00000  00000 00000  00000 00000  00000 00000   00000 00000  00000 00000  00000 00000  00000 00000  00000 00000
2143:  00000 00000  00000 00000  00000 00000  00000 00000  00000 00000   00000 00000  00000 00000  00000 00000  00000 00000  00000 00000
2144:  00000 00000  00000 00000  00000 00000  00000 00000  00000 00000   00000 00000  00000 00000  00000 00000  00000 00000  00000 00000
2145:  00000 00000  00000 00000  00000 00000  00000 00000  00000 00000   00000 00000  00000 00000  00000 00000  00000 00000  00000 00000
2146:  00000 00000  00000 00000  00000 00000  00000 00000  00000 00000   00000 00000  00000 00000  00000 00000  00000 00000  00000 00000
2147:  00000 00000  00000 00000  00000 00000  00000 00000  00000 00000   00000 00000  00000 00000  00000 00000  00000 00000  00000 00000
2148:  00000 00000  00000 00000  00000 00000  00000 00000  00000 00000   00000 00000  00000 00000  00000 00000  00000 00000  00000 00000
2149:  00000 00000  00000 00000  00000 00000  00000 00000  00000 00000   00000 00000  00000 00000  00000 00000  00000 00000  00000 00000
```

```
2150:   00000 00000   00000 00000   00000 00000   00000 00000   00000 00000     00000 00000   00000 00000   00000 00000   00000 00000   00000 00000
2151:   00000 00000   00000 00000   00000 00000   00000 00000   00000 00000     00000 00000   00000 00000   00000 00000   00000 00000   00000 00000
2152:   00000 00000   00000 00000   00000 00000   00000 00000   00000 00000     00000 00000   00000 00000   00000 00000   00000 00000   00000 00000
2153:   00000 00000   00000 00000   00000 00000   00000 00000   00000 00000     00000 00000   00000 00000   00000 00000   00000 00000   00000 00000
2154:   00000 00000   00000 00000   00000 00000   00000 00000   00000 00000     00000 00000   00000 00000   00000 00000   00000 00000   00000 00000
2155:   00000 00000   00000 00000   00000 00000   00000 00000   00000 00000     00000 00000   00000 00000   00000 00000   00000 00000   00000 00000
2156:   00000 00000   00000 00000   00000 00000   00000 00000   00000 00000     00000 00000   00000 00000   00000 00000   00000 00000   00000 00000
2157:   00000 00000   00000 00000   00000 00000   00000 00000   00000 00000     00000 00000   00000 00000   00000 00000   00000 00000   00000 00000
2158:   00000 00000   00000 00000   00000 00000   00000 00000   00000 00000     00000 00000   00000 00000   00000 00000   00000 00000   00000 00000
2159:   00000 00000   00000 00000   00000 00000   00000 00000   00000 00000     00000 00000   00000 00000   00000 00000   00000 00000   00000 00000
2160:   00000 00000   00000 00000   00000 00000   00000 00000   00000 00000     00000 00000   00000 00000   00000 00000   00000 00000   00000 00000
2161:   00000 00000   00000 00000   00000 00000   00000 00000   00000 00000     00000 00000   00000 00000   00000 00000   00000 00000   00000 00000
2162:   00000 00000   00000 00000   00000 00000   00000 00000   00000 00000     00000 00000   00000 00000   00000 00000   00000 00000   00000 00000
2163:   00000 00000   00000 00000   00000 00000   00000 00000   00000 00000     00000 00000   00000 00000   00000 00000   00000 00000   00000 00000
2164:   00000 00000   00000 00000   00000 00000   00000 00000   00000 00000     00000 00000   00000 00000   00000 00000   00000 00000   00000 00000
2165:   00000 00000   00000 00000   00000 00000   00000 00000   00000 00000     00000 00000   00000 00000   00000 00000   00000 00000   00000 00000
2166:   00000 00000   00000 00000   00000 00000   00000 00000   00000 00000     00000 00000   00000 00000   00000 00000   00000 00000   00000 00000
2167:   00000 00000   00000 00000   00000 00000   00000 00000   00000 00000     00000 00000   00000 00000   00000 00000   00000 00000   00000 00000
2168:   00000 00000   00000 00000   00000 00000   00000 00000   00000 00000     00000 00000   00000 00000   00000 00000   00000 00000   00000 00000
2169:   00000 00000   00000 00000   00000 00000   00000 00000   00000 00000     00000 00000   00000 00000   00000 00000   00000 00000   00000 00000
2170:   00000 00000   00000 00000   00000 00000   00000 00000   00000 00000     00000 00000   00000 00000   00000 00000   00000 00000   00000 00000
2171:   00000 00000   00000 00000   00000 00000   00000 00000   00000 00000     00000 00000   00000 00000   00000 00000   00000 00000   00000 00000
2172:   00000 00000   00000 00000   00000 00000   00000 00000   00000 00000     00000 00000   00000 00000   00000 00000   00000 00000   00000 00000
2173:   00000 00000   00000 00000   00000 00000   00000 00000   00000 00000     00000 00000   00000 00000   00000 00000   00000 00000   00000 00000
2174:   00000 00000   00000 00000   00000 00000   00000 00000   00000 00000     00000 00000   00000 00000   00000 00000   00000 00000   00000 00000
2175:   00000 00000   00000 00000   00000 00000   00000 00000   00000 00000     00000 00000   00000 00000   00000 00000   00000 00000   00000 00000
2176:   00000 00000   00000 00000   00000 00000   00000 00000   00000 00000     00000 00000   00000 00000   00000 00000   00000 00000   00000 00000
2177:   00000 00000   00000 00000   00000 00000   00000 00000   00000 00000     00000 00000   00000 00000   00000 00000   00000 00000   00000 00000
2178:   00000 00000   00000 00000   00000 00000   00000 00000   00000 00000     00000 00000   00000 00000   00000 00000   00000 00000   00000 00000
2179:   00000 00000   00000 00000   00000 00000   00000 00000   00000 00000     00000 00000   00000 00000   00000 00000   00000 00000   00000 00000
2180:   00000 00000   00000 00000   00000 00000   00000 00000   00000 00000     00000 00000   00000 00000   00000 00000   00000 00000   00000 00000
2181:   00000 00000   00000 00000   00000 00000   00000 00000   00000 00000     00000 00000   00000 00000   00000 00000   00000 00000   00000 00000
2182:   00000 00000   00000 00000   00000 00000   00000 00000   00000 00000     00000 00000   00000 00000   00000 00000   00000 00000   00000 00000
2183:   00000 00000   00000 00000   00000 00000   00000 00000   00000 00000     00000 00000   00000 00000   00000 00000   00000 00000   00000 00000
2184:   00000 00000   00000 00000   00000 00000   00000 00000   00000 00000     00000 00000   00000 00000   00000 00000   00000 00000   00000 00000
2185:   00000 00000   00000 00000   00000 00000   00000 00000   00000 00000     00000 00000   00000 00000   00000 00000   00000 00000   00000 00000
2186:   00000 00000   00000 00000   00000 00000   00000 00000   00000 00000     00000 00000   00000 00000   00000 00000   00000 00000   00000 00000
2187:   00000 00000   00000 00000   00000 00000   00000 00000   00000 00000     00000 00000   00000 00000   00000 00000   00000 00000   00000 00000
2188:   00000 00000   00000 00000   00000 00000   00000 00000   00000 00000     00000 00000   00000 00000   00000 00000   00000 00000   00000 00000
2189:   00000 00000   00000 00000   00000 00000   00000 00000   00000 00000     00000 00000   00000 00000   00000 00000   00000 00000   00000 00000
2190:   00000 00000   00000 00000   00000 00000   00000 00000   00000 00000     00000 00000   00000 00000   00000 00000   00000 00000   00000 00000
2191:   00000 00000   00000 00000   00000 00000   00000 00000   00000 00000     00000 00000   00000 00000   00000 00000   00000 00000   00000 00000
2192:   00000 00000   00000 00000   00000 00000   00000 00000   00000 00000     00000 00000   00000 00000   00000 00000   00000 00000   00000 00000
2193:   00000 00000   00000 00000   00000 00000   00000 00000   00000 00000     00000 00000   00000 00000   00000 00000   00000 00000   00000 00000
2194:   00000 00000   00000 00000   00000 00000   00000 00000   00000 00000     00000 00000   00000 00000   00000 00000   00000 00000   00000 00000
2195:   00000 00000   00000 00000   00000 00000   00000 00000   00000 00000     00000 00000   00000 00000   00000 00000   00000 00000   00000 00000
2196:   00000 00000   00000 00000   00000 00000   00000 00000   00000 00000     00000 00000   00000 00000   00000 00000   00000 00000   00000 00000
2197:   00000 00000   00000 00000   00000 00000   00000 00000   00000 00000     00000 00000   00000 00000   00000 00000   00000 00000   00000 00000
2198:   00000 00000   00000 00000   00000 00000   00000 00000   00000 00000     00000 00000   00000 00000   00000 00000   00000 00000   00000 00000
2199:   00000 00000   00000 00000   00000 00000   00000 00000   00000 00000     00000 00000   00000 00000   00000 00000   00000 00000   00000 00000
```

```
2200:  00000 00000  00000 00000  00000 00000  00000 00000  00000 00000    00000 00000  00000 00000  00000 00000  00000 00000  00000 00000
2201:  00000 00000  00000 00000  00000 00000  00000 00000  00000 00000    00000 00000  00000 00000  00000 00000  00000 00000  00000 00000
2202:  00000 00000  00000 00000  00000 00000  00000 00000  00000 00000    00000 00000  00000 00000  00000 00000  00000 00000  00000 00000
2203:  00000 00000  00000 00000  00000 00000  00000 00000  00000 00000    00000 00000  00000 00000  00000 00000  00000 00000  00000 00000
2204:  00000 00000  00000 00000  00000 00000  00000 00000  00000 00000    00000 00000  00000 00000  00000 00000  00000 00000  00000 00000
2205:  00000 00000  00000 00000  00000 00000  00000 00000  00000 00000    00000 00000  00000 00000  00000 00000  00000 00000  00000 00000
2206:  00000 00000  00000 00000  00000 00000  00000 00000  00000 00000    00000 00000  00000 00000  00000 00000  00000 00000  00000 00000
2207:  00000 00000  00000 00000  00000 00000  00000 00000  00000 00000    00000 00000  00000 00000  00000 00000  00000 00000  00000 00000
2208:  00000 00000  00000 00000  00000 00000  00000 00000  00000 00000    00000 00000  00000 00000  00000 00000  00000 00000  00000 00000
2209:  00000 00000  00000 00000  00000 00000  00000 00000  00000 00000    00000 00000  00000 00000  00000 00000  00000 00000  00000 00000
2210:  00000 00000  00000 00000  00000 00000  00000 00000  00000 00000    00000 00000  00000 00000  00000 00000  00000 00000  00000 00000
2211:  00000 00000  00000 00000  00000 00000  00000 00000  00000 00000    00000 00000  00000 00000  00000 00000  00000 00000  00000 00000
2212:  00000 00000  00000 00000  00000 00000  00000 00000  00000 00000    00000 00000  00000 00000  00000 00000  00000 00000  00000 00000
2213:  00000 00000  00000 00000  00000 00000  00000 00000  00000 00000    00000 00000  00000 00000  00000 00000  00000 00000  00000 00000
2214:  00000 00000  00000 00000  00000 00000  00000 00000  00000 00000    00000 00000  00000 00000  00000 00000  00000 00000  00000 00000
2215:  00000 00000  00000 00000  00000 00000  00000 00000  00000 00000    00000 00000  00000 00000  00000 00000  00000 00000  00000 00000
2216:  00000 00000  00000 00000  00000 00000  00000 00000  00000 00000    00000 00000  00000 00000  00000 00000  00000 00000  00000 00000
2217:  00000 00000  00000 00000  00000 00000  00000 00000  00000 00000    00000 00000  00000 00000  00000 00000  00000 00000  00000 00000
2218:  00000 00000  00000 00000  00000 00000  00000 00000  00000 00000    00000 00000  00000 00000  00000 00000  00000 00000  00000 00000
2219:  00000 00000  00000 00000  00000 00000  00000 00000  00000 00000    00000 00000  00000 00000  00000 00000  00000 00000  00000 00000
2220:  00000 00000  00000 00000  00000 00000  00000 00000  00000 00000    00000 00000  00000 00000  00000 00000  00000 00000  00000 00000
2221:  00000 00000  00000 00000  00000 00000  00000 00000  00000 00000    00000 00000  00000 00000  00000 00000  00000 00000  00000 00000
2222:  00000 00000  00000 00000  00000 00000  00000 00000  00000 00000    00000 00000  00000 00000  00000 00000  00000 00000  00000 00000
2223:  00000 00000  00000 00000  00000 00000  00000 00000  00000 00000    00000 00000  00000 00000  00000 00000  00000 00000  00000 00000
2224:  00000 00000  00000 00000  00000 00000  00000 00000  00000 00000    00000 00000  00000 00000  00000 00000  00000 00000  00000 00000
2225:  00000 00000  00000 00000  00000 00000  00000 00000  00000 00000    00000 00000  00000 00000  00000 00000  00000 00000  00000 00000
2226:  00000 00000  00000 00000  00000 00000  00000 00000  00000 00000    00000 00000  00000 00000  00000 00000  00000 00000  00000 00000
2227:  00000 00000  00000 00000  00000 00000  00000 00000  00000 00000    00000 00000  00000 00000  00000 00000  00000 00000  00000 00000
2228:  00000 00000  00000 00000  00000 00000  00000 00000  00000 00000    00000 00000  00000 00000  00000 00000  00000 00000  00000 00000
2229:  00000 00000  00000 00000  00000 00000  00000 00000  00000 00000    00000 00000  00000 00000  00000 00000  00000 00000  00000 00000
2230:  00000 00000  00000 00000  00000 00000  00000 00000  00000 00000    00000 00000  00000 00000  00000 00000  00000 00000  00000 00000
2231:  00000 00000  00000 00000  00000 00000  00000 00000  00000 00000    00000 00000  00000 00000  00000 00000  00000 00000  00000 00000
2232:  00000 00000  00000 00000  00000 00000  00000 00000  00000 00000    00000 00000  00000 00000  00000 00000  00000 00000  00000 00000
2233:  00000 00000  00000 00000  00000 00000  00000 00000  00000 00000    00000 00000  00000 00000  00000 00000  00000 00000  00000 00000
2234:  00000 00000  00000 00000  00000 00000  00000 00000  00000 00000    00000 00000  00000 00000  00000 00000  00000 00000  00000 00000
2235:  00000 00000  00000 00000  00000 00000  00000 00000  00000 00000    00000 00000  00000 00000  00000 00000  00000 00000  00000 00000
2236:  00000 00000  00000 00000  00000 00000  00000 00000  00000 00000    00000 00000  00000 00000  00000 00000  00000 00000  00000 00000
2237:  00000 00000  00000 00000  00000 00000  00000 00000  00000 00000    00000 00000  00000 00000  00000 00000  00000 00000  00000 00000
2238:  00000 00000  00000 00000  00000 00000  00000 00000  00000 00000    00000 00000  00000 00000  00000 00000  00000 00000  00000 00000
2239:  00000 00000  00000 00000  00000 00000  00000 00000  00000 00000    00000 00000  00000 00000  00000 00000  00000 00000  00000 00000
2240:  00000 00000  00000 00000  00000 00000  00000 00000  00000 00000    00000 00000  00000 00000  00000 00000  00000 00000  00000 00000
2241:  00000 00000  00000 00000  00000 00000  00000 00000  00000 00000    00000 00000  00000 00000  00000 00000  00000 00000  00000 00000
2242:  00000 00000  00000 00000  00000 00000  00000 00000  00000 00000    00000 00000  00000 00000  00000 00000  00000 00000  00000 00000
2243:  00000 00000  00000 00000  00000 00000  00000 00000  00000 00000    00000 00000  00000 00000  00000 00000  00000 00000  00000 00000
2244:  00000 00000  00000 00000  00000 00000  00000 00000  00000 00000    00000 00000  00000 00000  00000 00000  00000 00000  00000 00000
2245:  00000 00000  00000 00000  00000 00000  00000 00000  00000 00000    00000 00000  00000 00000  00000 00000  00000 00000  00000 00000
2246:  00000 00000  00000 00000  00000 00000  00000 00000  00000 00000    00000 00000  00000 00000  00000 00000  00000 00000  00000 00000
2247:  00000 00000  00000 00000  00000 00000  00000 00000  00000 00000    00000 00000  00000 00000  00000 00000  00000 00000  00000 00000
2248:  00000 00000  00000 00000  00000 00000  00000 00000  00000 00000    00000 00000  00000 00000  00000 00000  00000 00000  00000 00000
2249:  00000 00000  00000 00000  00000 00000  00000 00000  00000 00000    00000 00000  00000 00000  00000 00000  00000 00000  00000 00000
```

```
2250:  00000 00000  00000 00000  00000 00000  00000 00000  00000 00000    00000 00000  00000 00000  00000 00000  00000 00000  00000 00000
2251:  00000 00000  00000 00000  00000 00000  00000 00000  00000 00000    00000 00000  00000 00000  00000 00000  00000 00000  00000 00000
2252:  00000 00000  00000 00000  00000 00000  00000 00000  00000 00000    00000 00000  00000 00000  00000 00000  00000 00000  00000 00000
2253:  00000 00000  00000 00000  00000 00000  00000 00000  00000 00000    00000 00000  00000 00000  00000 00000  00000 00000  00000 00000
2254:  00000 00000  00000 00000  00000 00000  00000 00000  00000 00000    00000 00000  00000 00000  00000 00000  00000 00000  00000 00000
2255:  00000 00000  00000 00000  00000 00000  00000 00000  00000 00000    00000 00000  00000 00000  00000 00000  00000 00000  00000 00000
2256:  00000 00000  00000 00000  00000 00000  00000 00000  00000 00000    00000 00000  00000 00000  00000 00000  00000 00000  00000 00000
2257:  00000 00000  00000 00000  00000 00000  00000 00000  00000 00000    00000 00000  00000 00000  00000 00000  00000 00000  00000 00000
2258:  00000 00000  00000 00000  00000 00000  00000 00000  00000 00000    00000 00000  00000 00000  00000 00000  00000 00000  00000 00000
2259:  00000 00000  00000 00000  00000 00000  00000 00000  00000 00000    00000 00000  00000 00000  00000 00000  00000 00000  00000 00000
2260:  00000 00000  00000 00000  00000 00000  00000 00000  00000 00000    00000 00000  00000 00000  00000 00000  00000 00000  00000 00000
2261:  00000 00000  00000 00000  00000 00000  00000 00000  00000 00000    00000 00000  00000 00000  00000 00000  00000 00000  00000 00000
2262:  00000 00000  00000 00000  00000 00000  00000 00000  00000 00000    00000 00000  00000 00000  00000 00000  00000 00000  00000 00000
2263:  00000 00000  00000 00000  00000 00000  00000 00000  00000 00000    00000 00000  00000 00000  00000 00000  00000 00000  00000 00000
2264:  00000 00000  00000 00000  00000 00000  00000 00000  00000 00000    00000 00000  00000 00000  00000 00000  00000 00000  00000 00000
2265:  00000 00000  00000 00000  00000 00000  00000 00000  00000 00000    00000 00000  00000 00000  00000 00000  00000 00000  00000 00000
2266:  00000 00000  00000 00000  00000 00000  00000 00000  00000 00000    00000 00000  00000 00000  00000 00000  00000 00000  00000 00000
2267:  00000 00000  00000 00000  00000 00000  00000 00000  00000 00000    00000 00000  00000 00000  00000 00000  00000 00000  00000 00000
2268:  00000 00000  00000 00000  00000 00000  00000 00000  00000 00000    00000 00000  00000 00000  00000 00000  00000 00000  00000 00000
2269:  00000 00000  00000 00000  00000 00000  00000 00000  00000 00000    00000 00000  00000 00000  00000 00000  00000 00000  00000 00000
2270:  00000 00000  00000 00000  00000 00000  00000 00000  00000 00000    00000 00000  00000 00000  00000 00000  00000 00000  00000 00000
2271:  00000 00000  00000 00000  00000 00000  00000 00000  00000 00000    00000 00000  00000 00000  00000 00000  00000 00000  00000 00000
2272:  00000 00000  00000 00000  00000 00000  00000 00000  00000 00000    00000 00000  00000 00000  00000 00000  00000 00000  00000 00000
2273:  00000 00000  00000 00000  00000 00000  00000 00000  00000 00000    00000 00000  00000 00000  00000 00000  00000 00000  00000 00000
2274:  00000 00000  00000 00000  00000 00000  00000 00000  00000 00000    00000 00000  00000 00000  00000 00000  00000 00000  00000 00000
2275:  00000 00000  00000 00000  00000 00000  00000 00000  00000 00000    00000 00000  00000 00000  00000 00000  00000 00000  00000 00000
2276:  00000 00000  00000 00000  00000 00000  00000 00000  00000 00000    00000 00000  00000 00000  00000 00000  00000 00000  00000 00000
2277:  00000 00000  00000 00000  00000 00000  00000 00000  00000 00000    00000 00000  00000 00000  00000 00000  00000 00000  00000 00000
2278:  00000 00000  00000 00000  00000 00000  00000 00000  00000 00000    00000 00000  00000 00000  00000 00000  00000 00000  00000 00000
2279:  00000 00000  00000 00000  00000 00000  00000 00000  00000 00000    00000 00000  00000 00000  00000 00000  00000 00000  00000 00000
2280:  00000 00000  00000 00000  00000 00000  00000 00000  00000 00000    00000 00000  00000 00000  00000 00000  00000 00000  00000 00000
2281:  00000 00000  00000 00000  00000 00000  00000 00000  00000 00000    00000 00000  00000 00000  00000 00000  00000 00000  00000 00000
2282:  00000 00000  00000 00000  00000 00000  00000 00000  00000 00000    00000 00000  00000 00000  00000 00000  00000 00000  00000 00000
2283:  00000 00000  00000 00000  00000 00000  00000 00000  00000 00000    00000 00000  00000 00000  00000 00000  00000 00000  00000 00000
2284:  00000 00000  00000 00000  00000 00000  00000 00000  00000 00000    00000 00000  00000 00000  00000 00000  00000 00000  00000 00000
2285:  00000 00000  00000 00000  00000 00000  00000 00000  00000 00000    00000 00000  00000 00000  00000 00000  00000 00000  00000 00000
2286:  00000 00000  00000 00000  00000 00000  00000 00000  00000 00000    00000 00000  00000 00000  00000 00000  00000 00000  00000 00000
2287:  00000 00000  00000 00000  00000 00000  00000 00000  00000 00000    00000 00000  00000 00000  00000 00000  00000 00000  00000 00000
2288:  00000 00000  00000 00000  00000 00000  00000 00000  00000 00000    00000 00000  00000 00000  00000 00000  00000 00000  00000 00000
2289:  00000 00000  00000 00000  00000 00000  00000 00000  00000 00000    00000 00000  00000 00000  00000 00000  00000 00000  00000 00000
2290:  00000 00000  00000 00000  00000 00000  00000 00000  00000 00000    00000 00000  00000 00000  00000 00000  00000 00000  00000 00000
2291:  00000 00000  00000 00000  00000 00000  00000 00000  00000 00000    00000 00000  00000 00000  00000 00000  00000 00000  00000 00000
2292:  00000 00000  00000 00000  00000 00000  00000 00000  00000 00000    00000 00000  00000 00000  00000 00000  00000 00000  00000 00000
2293:  00000 00000  00000 00000  00000 00000  00000 00000  00000 00000    00000 00000  00000 00000  00000 00000  00000 00000  00000 00000
2294:  00000 00000  00000 00000  00000 00000  00000 00000  00000 00000    00000 00000  00000 00000  00000 00000  00000 00000  00000 00000
2295:  00000 00000  00000 00000  00000 00000  00000 00000  00000 00000    00000 00000  00000 00000  00000 00000  00000 00000  00000 00000
2296:  00000 00000  00000 00000  00000 00000  00000 00000  00000 00000    00000 00000  00000 00000  00000 00000  00000 00000  00000 00000
2297:  00000 00000  00000 00000  00000 00000  00000 00000  00000 00000    00000 00000  00000 00000  00000 00000  00000 00000  00000 00000
2298:  00000 00000  00000 00000  00000 00000  00000 00000  00000 00000    00000 00000  00000 00000  00000 00000  00000 00000  00000 00000
2299:  00000 00000  00000 00000  00000 00000  00000 00000  00000 00000    00000 00000  00000 00000  00000 00000  00000 00000  00000 00000
```

```
2300:  00000 00000  00000 00000  00000 00000  00000 00000  00000 00000    00000 00000  00000 00000  00000 00000  00000 00000  00000 00000
2301:  00000 00000  00000 00000  00000 00000  00000 00000  00000 00000    00000 00000  00000 00000  00000 00000  00000 00000  00000 00000
2302:  00000 00000  00000 00000  00000 00000  00000 00000  00000 00000    00000 00000  00000 00000  00000 00000  00000 00000  00000 00000
2303:  00000 00000  00000 00000  00000 00000  00000 00000  00000 00000    00000 00000  00000 00000  00000 00000  00000 00000  00000 00000
2304:  00000 00000  00000 00000  00000 00000  00000 00000  00000 00000    00000 00000  00000 00000  00000 00000  00000 00000  00000 00000
2305:  00000 00000  00000 00000  00000 00000  00000 00000  00000 00000    00000 00000  00000 00000  00000 00000  00000 00000  00000 00000
2306:  00000 00000  00000 00000  00000 00000  00000 00000  00000 00000    00000 00000  00000 00000  00000 00000  00000 00000  00000 00000
2307:  00000 00000  00000 00000  00000 00000  00000 00000  00000 00000    00000 00000  00000 00000  00000 00000  00000 00000  00000 00000
2308:  00000 00000  00000 00000  00000 00000  00000 00000  00000 00000    00000 00000  00000 00000  00000 00000  00000 00000  00000 00000
2309:  00000 00000  00000 00000  00000 00000  00000 00000  00000 00000    00000 00000  00000 00000  00000 00000  00000 00000  00000 00000
2310:  00000 00000  00000 00000  00000 00000  00000 00000  00000 00000    00000 00000  00000 00000  00000 00000  00000 00000  00000 00000
2311:  00000 00000  00000 00000  00000 00000  00000 00000  00000 00000    00000 00000  00000 00000  00000 00000  00000 00000  00000 00000
2312:  00000 00000  00000 00000  00000 00000  00000 00000  00000 00000    00000 00000  00000 00000  00000 00000  00000 00000  00000 00000
2313:  00000 00000  00000 00000  00000 00000  00000 00000  00000 00000    00000 00000  00000 00000  00000 00000  00000 00000  00000 00000
2314:  00000 00000  00000 00000  00000 00000  00000 00000  00000 00000    00000 00000  00000 00000  00000 00000  00000 00000  00000 00000
2315:  00000 00000  00000 00000  00000 00000  00000 00000  00000 00000    00000 00000  00000 00000  00000 00000  00000 00000  00000 00000
2316:  00000 00000  00000 00000  00000 00000  00000 00000  00000 00000    00000 00000  00000 00000  00000 00000  00000 00000  00000 00000
2317:  00000 00000  00000 00000  00000 00000  00000 00000  00000 00000    00000 00000  00000 00000  00000 00000  00000 00000  00000 00000
2318:  00000 00000  00000 00000  00000 00000  00000 00000  00000 00000    00000 00000  00000 00000  00000 00000  00000 00000  00000 00000
2319:  00000 00000  00000 00000  00000 00000  00000 00000  00000 00000    00000 00000  00000 00000  00000 00000  00000 00000  00000 00000
2320:  00000 00000  00000 00000  00000 00000  00000 00000  00000 00000    00000 00000  00000 00000  00000 00000  00000 00000  00000 00000
2321:  00000 00000  00000 00000  00000 00000  00000 00000  00000 00000    00000 00000  00000 00000  00000 00000  00000 00000  00000 00000
2322:  00000 00000  00000 00000  00000 00000  00000 00000  00000 00000    00000 00000  00000 00000  00000 00000  00000 00000  00000 00000
2323:  00000 00000  00000 00000  00000 00000  00000 00000  00000 00000    00000 00000  00000 00000  00000 00000  00000 00000  00000 00000
2324:  00000 00000  00000 00000  00000 00000  00000 00000  00000 00000    00000 00000  00000 00000  00000 00000  00000 00000  00000 00000
2325:  00000 00000  00000 00000  00000 00000  00000 00000  00000 00000    00000 00000  00000 00000  00000 00000  00000 00000  00000 00000
2326:  00000 00000  00000 00000  00000 00000  00000 00000  00000 00000    00000 00000  00000 00000  00000 00000  00000 00000  00000 00000
2327:  00000 00000  00000 00000  00000 00000  00000 00000  00000 00000    00000 00000  00000 00000  00000 00000  00000 00000  00000 00000
2328:  00000 00000  00000 00000  00000 00000  00000 00000  00000 00000    00000 00000  00000 00000  00000 00000  00000 00000  00000 00000
2329:  00000 00000  00000 00000  00000 00000  00000 00000  00000 00000    00000 00000  00000 00000  00000 00000  00000 00000  00000 00000
2330:  00000 00000  00000 00000  00000 00000  00000 00000  00000 00000    00000 00000  00000 00000  00000 00000  00000 00000  00000 00000
2331:  00000 00000  00000 00000  00000 00000  00000 00000  00000 00000    00000 00000  00000 00000  00000 00000  00000 00000  00000 00000
2332:  00000 00000  00000 00000  00000 00000  00000 00000  00000 00000    00000 00000  00000 00000  00000 00000  00000 00000  00000 00000
2333:  00000 00000  00000 00000  00000 00000  00000 00000  00000 00000    00000 00000  00000 00000  00000 00000  00000 00000  00000 00000
2334:  00000 00000  00000 00000  00000 00000  00000 00000  00000 00000    00000 00000  00000 00000  00000 00000  00000 00000  00000 00000
2335:  00000 00000  00000 00000  00000 00000  00000 00000  00000 00000    00000 00000  00000 00000  00000 00000  00000 00000  00000 00000
2336:  00000 00000  00000 00000  00000 00000  00000 00000  00000 00000    00000 00000  00000 00000  00000 00000  00000 00000  00000 00000
2337:  00000 00000  00000 00000  00000 00000  00000 00000  00000 00000    00000 00000  00000 00000  00000 00000  00000 00000  00000 00000
2338:  00000 00000  00000 00000  00000 00000  00000 00000  00000 00000    00000 00000  00000 00000  00000 00000  00000 00000  00000 00000
2339:  00000 00000  00000 00000  00000 00000  00000 00000  00000 00000    00000 00000  00000 00000  00000 00000  00000 00000  00000 00000
2340:  00000 00000  00000 00000  00000 00000  00000 00000  00000 00000    00000 00000  00000 00000  00000 00000  00000 00000  00000 00000
2341:  00000 00000  00000 00000  00000 00000  00000 00000  00000 00000    00000 00000  00000 00000  00000 00000  00000 00000  00000 00000
2342:  00000 00000  00000 00000  00000 00000  00000 00000  00000 00000    00000 00000  00000 00000  00000 00000  00000 00000  00000 00000
2343:  00000 00000  00000 00000  00000 00000  00000 00000  00000 00000    00000 00000  00000 00000  00000 00000  00000 00000  00000 00000
2344:  00000 00000  00000 00000  00000 00000  00000 00000  00000 00000    00000 00000  00000 00000  00000 00000  00000 00000  00000 00000
2345:  00000 00000  00000 00000  00000 00000  00000 00000  00000 00000    00000 00000  00000 00000  00000 00000  00000 00000  00000 00000
2346:  00000 00000  00000 00000  00000 00000  00000 00000  00000 00000    00000 00000  00000 00000  00000 00000  00000 00000  00000 00000
2347:  00000 00000  00000 00000  00000 00000  00000 00000  00000 00000    00000 00000  00000 00000  00000 00000  00000 00000  00000 00000
2348:  00000 00000  00000 00000  00000 00000  00000 00000  00000 00000    00000 00000  00000 00000  00000 00000  00000 00000  00000 00000
2349:  00000 00000  00000 00000  00000 00000  00000 00000  00000 00000    00000 00000  00000 00000  00000 00000  00000 00000  00000 00000
```

```
2350:  00000 00000   00000 00000   00000 00000   00000 00000   00000 00000     00000 00000   00000 00000   00000 00000   00000 00000   00000 00000
2351:  00000 00000   00000 00000   00000 00000   00000 00000   00000 00000     00000 00000   00000 00000   00000 00000   00000 00000   00000 00000
2352:  00000 00000   00000 00000   00000 00000   00000 00000   00000 00000     00000 00000   00000 00000   00000 00000   00000 00000   00000 00000
2353:  00000 00000   00000 00000   00000 00000   00000 00000   00000 00000     00000 00000   00000 00000   00000 00000   00000 00000   00000 00000
2354:  00000 00000   00000 00000   00000 00000   00000 00000   00000 00000     00000 00000   00000 00000   00000 00000   00000 00000   00000 00000
2355:  00000 00000   00000 00000   00000 00000   00000 00000   00000 00000     00000 00000   00000 00000   00000 00000   00000 00000   00000 00000
2356:  00000 00000   00000 00000   00000 00000   00000 00000   00000 00000     00000 00000   00000 00000   00000 00000   00000 00000   00000 00000
2357:  00000 00000   00000 00000   00000 00000   00000 00000   00000 00000     00000 00000   00000 00000   00000 00000   00000 00000   00000 00000
2358:  00000 00000   00000 00000   00000 00000   00000 00000   00000 00000     00000 00000   00000 00000   00000 00000   00000 00000   00000 00000
2359:  00000 00000   00000 00000   00000 00000   00000 00000   00000 00000     00000 00000   00000 00000   00000 00000   00000 00000   00000 00000
2360:  00000 00000   00000 00000   00000 00000   00000 00000   00000 00000     00000 00000   00000 00000   00000 00000   00000 00000   00000 00000
2361:  00000 00000   00000 00000   00000 00000   00000 00000   00000 00000     00000 00000   00000 00000   00000 00000   00000 00000   00000 00000
2362:  00000 00000   00000 00000   00000 00000   00000 00000   00000 00000     00000 00000   00000 00000   00000 00000   00000 00000   00000 00000
2363:  00000 00000   00000 00000   00000 00000   00000 00000   00000 00000     00000 00000   00000 00000   00000 00000   00000 00000   00000 00000
2364:  00000 00000   00000 00000   00000 00000   00000 00000   00000 00000     00000 00000   00000 00000   00000 00000   00000 00000   00000 00000
2365:  00000 00000   00000 00000   00000 00000   00000 00000   00000 00000     00000 00000   00000 00000   00000 00000   00000 00000   00000 00000
2366:  00000 00000   00000 00000   00000 00000   00000 00000   00000 00000     00000 00000   00000 00000   00000 00000   00000 00000   00000 00000
2367:  00000 00000   00000 00000   00000 00000   00000 00000   00000 00000     00000 00000   00000 00000   00000 00000   00000 00000   00000 00000
2368:  00000 00000   00000 00000   00000 00000   00000 00000   00000 00000     00000 00000   00000 00000   00000 00000   00000 00000   00000 00000
2369:  00000 00000   00000 00000   00000 00000   00000 00000   00000 00000     00000 00000   00000 00000   00000 00000   00000 00000   00000 00000
2370:  00000 00000   00000 00000   00000 00000   00000 00000   00000 00000     00000 00000   00000 00000   00000 00000   00000 00000   00000 00000
2371:  00000 00000   00000 00000   00000 00000   00000 00000   00000 00000     00000 00000   00000 00000   00000 00000   00000 00000   00000 00000
2372:  00000 00000   00000 00000   00000 00000   00000 00000   00000 00000     00000 00000   00000 00000   00000 00000   00000 00000   00000 00000
2373:  00000 00000   00000 00000   00000 00000   00000 00000   00000 00000     00000 00000   00000 00000   00000 00000   00000 00000   00000 00000
2374:  00000 00000   00000 00000   00000 00000   00000 00000   00000 00000     00000 00000   00000 00000   00000 00000   00000 00000   00000 00000
2375:  00000 00000   00000 00000   00000 00000   00000 00000   00000 00000     00000 00000   00000 00000   00000 00000   00000 00000   00000 00000
2376:  00000 00000   00000 00000   00000 00000   00000 00000   00000 00000     00000 00000   00000 00000   00000 00000   00000 00000   00000 00000
2377:  00000 00000   00000 00000   00000 00000   00000 00000   00000 00000     00000 00000   00000 00000   00000 00000   00000 00000   00000 00000
2378:  00000 00000   00000 00000   00000 00000   00000 00000   00000 00000     00000 00000   00000 00000   00000 00000   00000 00000   00000 00000
2379:  00000 00000   00000 00000   00000 00000   00000 00000   00000 00000     00000 00000   00000 00000   00000 00000   00000 00000   00000 00000
2380:  00000 00000   00000 00000   00000 00000   00000 00000   00000 00000     00000 00000   00000 00000   00000 00000   00000 00000   00000 00000
2381:  00000 00000   00000 00000   00000 00000   00000 00000   00000 00000     00000 00000   00000 00000   00000 00000   00000 00000   00000 00000
2382:  00000 00000   00000 00000   00000 00000   00000 00000   00000 00000     00000 00000   00000 00000   00000 00000   00000 00000   00000 00000
2383:  00000 00000   00000 00000   00000 00000   00000 00000   00000 00000     00000 00000   00000 00000   00000 00000   00000 00000   00000 00000
2384:  00000 00000   00000 00000   00000 00000   00000 00000   00000 00000     00000 00000   00000 00000   00000 00000   00000 00000   00000 00000
2385:  00000 00000   00000 00000   00000 00000   00000 00000   00000 00000     00000 00000   00000 00000   00000 00000   00000 00000   00000 00000
2386:  00000 00000   00000 00000   00000 00000   00000 00000   00000 00000     00000 00000   00000 00000   00000 00000   00000 00000   00000 00000
2387:  00000 00000   00000 00000   00000 00000   00000 00000   00000 00000     00000 00000   00000 00000   00000 00000   00000 00000   00000 00000
2388:  00000 00000   00000 00000   00000 00000   00000 00000   00000 00000     00000 00000   00000 00000   00000 00000   00000 00000   00000 00000
2389:  00000 00000   00000 00000   00000 00000   00000 00000   00000 00000     00000 00000   00000 00000   00000 00000   00000 00000   00000 00000
2390:  00000 00000   00000 00000   00000 00000   00000 00000   00000 00000     00000 00000   00000 00000   00000 00000   00000 00000   00000 00000
2391:  00000 00000   00000 00000   00000 00000   00000 00000   00000 00000     00000 00000   00000 00000   00000 00000   00000 00000   00000 00000
2392:  00000 00000   00000 00000   00000 00000   00000 00000   00000 00000     00000 00000   00000 00000   00000 00000   00000 00000   00000 00000
2393:  00000 00000   00000 00000   00000 00000   00000 00000   00000 00000     00000 00000   00000 00000   00000 00000   00000 00000   00000 00000
2394:  00000 00000   00000 00000   00000 00000   00000 00000   00000 00000     00000 00000   00000 00000   00000 00000   00000 00000   00000 00000
2395:  00000 00000   00000 00000   00000 00000   00000 00000   00000 00000     00000 00000   00000 00000   00000 00000   00000 00000   00000 00000
2396:  00000 00000   00000 00000   00000 00000   00000 00000   00000 00000     00000 00000   00000 00000   00000 00000   00000 00000   00000 00000
2397:  00000 00000   00000 00000   00000 00000   00000 00000   00000 00000     00000 00000   00000 00000   00000 00000   00000 00000   00000 00000
2398:  00000 00000   00000 00000   00000 00000   00000 00000   00000 00000     00000 00000   00000 00000   00000 00000   00000 00000   00000 00000
2399:  00000 00000   00000 00000   00000 00000   00000 00000   00000 00000     00000 00000   00000 00000   00000 00000   00000 00000   00000 00000
```

```
2400:  00000 00000  00000 00000  00000 00000  00000 00000  00000 00000    00000 00000  00000 00000  00000 00000  00000 00000  00000 00000
2401:  00000 00000  00000 00000  00000 00000  00000 00000  00000 00000    00000 00000  00000 00000  00000 00000  00000 00000  00000 00000
2402:  00000 00000  00000 00000  00000 00000  00000 00000  00000 00000    00000 00000  00000 00000  00000 00000  00000 00000  00000 00000
2403:  00000 00000  00000 00000  00000 00000  00000 00000  00000 00000    00000 00000  00000 00000  00000 00000  00000 00000  00000 00000
2404:  00000 00000  00000 00000  00000 00000  00000 00000  00000 00000    00000 00000  00000 00000  00000 00000  00000 00000  00000 00000
2405:  00000 00000  00000 00000  00000 00000  00000 00000  00000 00000    00000 00000  00000 00000  00000 00000  00000 00000  00000 00000
2406:  00000 00000  00000 00000  00000 00000  00000 00000  00000 00000    00000 00000  00000 00000  00000 00000  00000 00000  00000 00000
2407:  00000 00000  00000 00000  00000 00000  00000 00000  00000 00000    00000 00000  00000 00000  00000 00000  00000 00000  00000 00000
2408:  00000 00000  00000 00000  00000 00000  00000 00000  00000 00000    00000 00000  00000 00000  00000 00000  00000 00000  00000 00000
2409:  00000 00000  00000 00000  00000 00000  00000 00000  00000 00000    00000 00000  00000 00000  00000 00000  00000 00000  00000 00000
2410:  00000 00000  00000 00000  00000 00000  00000 00000  00000 00000    00000 00000  00000 00000  00000 00000  00000 00000  00000 00000
2411:  00000 00000  00000 00000  00000 00000  00000 00000  00000 00000    00000 00000  00000 00000  00000 00000  00000 00000  00000 00000
2412:  00000 00000  00000 00000  00000 00000  00000 00000  00000 00000    00000 00000  00000 00000  00000 00000  00000 00000  00000 00000
2413:  00000 00000  00000 00000  00000 00000  00000 00000  00000 00000    00000 00000  00000 00000  00000 00000  00000 00000  00000 00000
2414:  00000 00000  00000 00000  00000 00000  00000 00000  00000 00000    00000 00000  00000 00000  00000 00000  00000 00000  00000 00000
2415:  00000 00000  00000 00000  00000 00000  00000 00000  00000 00000    00000 00000  00000 00000  00000 00000  00000 00000  00000 00000
2416:  00000 00000  00000 00000  00000 00000  00000 00000  00000 00000    00000 00000  00000 00000  00000 00000  00000 00000  00000 00000
2417:  00000 00000  00000 00000  00000 00000  00000 00000  00000 00000    00000 00000  00000 00000  00000 00000  00000 00000  00000 00000
2418:  00000 00000  00000 00000  00000 00000  00000 00000  00000 00000    00000 00000  00000 00000  00000 00000  00000 00000  00000 00000
2419:  00000 00000  00000 00000  00000 00000  00000 00000  00000 00000    00000 00000  00000 00000  00000 00000  00000 00000  00000 00000
2420:  00000 00000  00000 00000  00000 00000  00000 00000  00000 00000    00000 00000  00000 00000  00000 00000  00000 00000  00000 00000
2421:  00000 00000  00000 00000  00000 00000  00000 00000  00000 00000    00000 00000  00000 00000  00000 00000  00000 00000  00000 00000
2422:  00000 00000  00000 00000  00000 00000  00000 00000  00000 00000    00000 00000  00000 00000  00000 00000  00000 00000  00000 00000
2423:  00000 00000  00000 00000  00000 00000  00000 00000  00000 00000    00000 00000  00000 00000  00000 00000  00000 00000  00000 00000
2424:  00000 00000  00000 00000  00000 00000  00000 00000  00000 00000    00000 00000  00000 00000  00000 00000  00000 00000  00000 00000
2425:  00000 00000  00000 00000  00000 00000  00000 00000  00000 00000    00000 00000  00000 00000  00000 00000  00000 00000  00000 00000
2426:  00000 00000  00000 00000  00000 00000  00000 00000  00000 00000    00000 00000  00000 00000  00000 00000  00000 00000  00000 00000
2427:  00000 00000  00000 00000  00000 00000  00000 00000  00000 00000    00000 00000  00000 00000  00000 00000  00000 00000  00000 00000
2428:  00000 00000  00000 00000  00000 00000  00000 00000  00000 00000    00000 00000  00000 00000  00000 00000  00000 00000  00000 00000
2429:  00000 00000  00000 00000  00000 00000  00000 00000  00000 00000    00000 00000  00000 00000  00000 00000  00000 00000  00000 00000
2430:  00000 00000  00000 00000  00000 00000  00000 00000  00000 00000    00000 00000  00000 00000  00000 00000  00000 00000  00000 00000
2431:  00000 00000  00000 00000  00000 00000  00000 00000  00000 00000    00000 00000  00000 00000  00000 00000  00000 00000  00000 00000
2432:  00000 00000  00000 00000  00000 00000  00000 00000  00000 00000    00000 00000  00000 00000  00000 00000  00000 00000  00000 00000
2433:  00000 00000  00000 00000  00000 00000  00000 00000  00000 00000    00000 00000  00000 00000  00000 00000  00000 00000  00000 00000
2434:  00000 00000  00000 00000  00000 00000  00000 00000  00000 00000    00000 00000  00000 00000  00000 00000  00000 00000  00000 00000
2435:  00000 00000  00000 00000  00000 00000  00000 00000  00000 00000    00000 00000  00000 00000  00000 00000  00000 00000  00000 00000
2436:  00000 00000  00000 00000  00000 00000  00000 00000  00000 00000    00000 00000  00000 00000  00000 00000  00000 00000  00000 00000
2437:  00000 00000  00000 00000  00000 00000  00000 00000  00000 00000    00000 00000  00000 00000  00000 00000  00000 00000  00000 00000
2438:  00000 00000  00000 00000  00000 00000  00000 00000  00000 00000    00000 00000  00000 00000  00000 00000  00000 00000  00000 00000
2439:  00000 00000  00000 00000  00000 00000  00000 00000  00000 00000    00000 00000  00000 00000  00000 00000  00000 00000  00000 00000
2440:  00000 00000  00000 00000  00000 00000  00000 00000  00000 00000    00000 00000  00000 00000  00000 00000  00000 00000  00000 00000
2441:  00000 00000  00000 00000  00000 00000  00000 00000  00000 00000    00000 00000  00000 00000  00000 00000  00000 00000  00000 00000
2442:  00000 00000  00000 00000  00000 00000  00000 00000  00000 00000    00000 00000  00000 00000  00000 00000  00000 00000  00000 00000
2443:  00000 00000  00000 00000  00000 00000  00000 00000  00000 00000    00000 00000  00000 00000  00000 00000  00000 00000  00000 00000
2444:  00000 00000  00000 00000  00000 00000  00000 00000  00000 00000    00000 00000  00000 00000  00000 00000  00000 00000  00000 00000
2445:  00000 00000  00000 00000  00000 00000  00000 00000  00000 00000    00000 00000  00000 00000  00000 00000  00000 00000  00000 00000
2446:  00000 00000  00000 00000  00000 00000  00000 00000  00000 00000    00000 00000  00000 00000  00000 00000  00000 00000  00000 00000
2447:  00000 00000  00000 00000  00000 00000  00000 00000  00000 00000    00000 00000  00000 00000  00000 00000  00000 00000  00000 00000
2448:  00000 00000  00000 00000  00000 00000  00000 00000  00000 00000    00000 00000  00000 00000  00000 00000  00000 00000  00000 00000
2449:  00000 00000  00000 00000  00000 00000  00000 00000  00000 00000    00000 00000  00000 00000  00000 00000  00000 00000  00000 00000
```

```
2450:   00000 00000   00000 00000   00000 00000   00000 00000   00000 00000     00000 00000   00000 00000   00000 00000   00000 00000   00000 00000
2451:   00000 00000   00000 00000   00000 00000   00000 00000   00000 00000     00000 00000   00000 00000   00000 00000   00000 00000   00000 00000
2452:   00000 00000   00000 00000   00000 00000   00000 00000   00000 00000     00000 00000   00000 00000   00000 00000   00000 00000   00000 00000
2453:   00000 00000   00000 00000   00000 00000   00000 00000   00000 00000     00000 00000   00000 00000   00000 00000   00000 00000   00000 00000
2454:   00000 00000   00000 00000   00000 00000   00000 00000   00000 00000     00000 00000   00000 00000   00000 00000   00000 00000   00000 00000
2455:   00000 00000   00000 00000   00000 00000   00000 00000   00000 00000     00000 00000   00000 00000   00000 00000   00000 00000   00000 00000
2456:   00000 00000   00000 00000   00000 00000   00000 00000   00000 00000     00000 00000   00000 00000   00000 00000   00000 00000   00000 00000
2457:   00000 00000   00000 00000   00000 00000   00000 00000   00000 00000     00000 00000   00000 00000   00000 00000   00000 00000   00000 00000
2458:   00000 00000   00000 00000   00000 00000   00000 00000   00000 00000     00000 00000   00000 00000   00000 00000   00000 00000   00000 00000
2459:   00000 00000   00000 00000   00000 00000   00000 00000   00000 00000     00000 00000   00000 00000   00000 00000   00000 00000   00000 00000
2460:   00000 00000   00000 00000   00000 00000   00000 00000   00000 00000     00000 00000   00000 00000   00000 00000   00000 00000   00000 00000
2461:   00000 00000   00000 00000   00000 00000   00000 00000   00000 00000     00000 00000   00000 00000   00000 00000   00000 00000   00000 00000
2462:   00000 00000   00000 00000   00000 00000   00000 00000   00000 00000     00000 00000   00000 00000   00000 00000   00000 00000   00000 00000
2463:   00000 00000   00000 00000   00000 00000   00000 00000   00000 00000     00000 00000   00000 00000   00000 00000   00000 00000   00000 00000
2464:   00000 00000   00000 00000   00000 00000   00000 00000   00000 00000     00000 00000   00000 00000   00000 00000   00000 00000   00000 00000
2465:   00000 00000   00000 00000   00000 00000   00000 00000   00000 00000     00000 00000   00000 00000   00000 00000   00000 00000   00000 00000
2466:   00000 00000   00000 00000   00000 00000   00000 00000   00000 00000     00000 00000   00000 00000   00000 00000   00000 00000   00000 00000
2467:   00000 00000   00000 00000   00000 00000   00000 00000   00000 00000     00000 00000   00000 00000   00000 00000   00000 00000   00000 00000
2468:   00000 00000   00000 00000   00000 00000   00000 00000   00000 00000     00000 00000   00000 00000   00000 00000   00000 00000   00000 00000
2469:   00000 00000   00000 00000   00000 00000   00000 00000   00000 00000     00000 00000   00000 00000   00000 00000   00000 00000   00000 00000
2470:   00000 00000   00000 00000   00000 00000   00000 00000   00000 00000     00000 00000   00000 00000   00000 00000   00000 00000   00000 00000
2471:   00000 00000   00000 00000   00000 00000   00000 00000   00000 00000     00000 00000   00000 00000   00000 00000   00000 00000   00000 00000
2472:   00000 00000   00000 00000   00000 00000   00000 00000   00000 00000     00000 00000   00000 00000   00000 00000   00000 00000   00000 00000
2473:   00000 00000   00000 00000   00000 00000   00000 00000   00000 00000     00000 00000   00000 00000   00000 00000   00000 00000   00000 00000
2474:   00000 00000   00000 00000   00000 00000   00000 00000   00000 00000     00000 00000   00000 00000   00000 00000   00000 00000   00000 00000
2475:   00000 00000   00000 00000   00000 00000   00000 00000   00000 00000     00000 00000   00000 00000   00000 00000   00000 00000   00000 00000
2476:   00000 00000   00000 00000   00000 00000   00000 00000   00000 00000     00000 00000   00000 00000   00000 00000   00000 00000   00000 00000
2477:   00000 00000   00000 00000   00000 00000   00000 00000   00000 00000     00000 00000   00000 00000   00000 00000   00000 00000   00000 00000
2478:   00000 00000   00000 00000   00000 00000   00000 00000   00000 00000     00000 00000   00000 00000   00000 00000   00000 00000   00000 00000
2479:   00000 00000   00000 00000   00000 00000   00000 00000   00000 00000     00000 00000   00000 00000   00000 00000   00000 00000   00000 00000
2480:   00000 00000   00000 00000   00000 00000   00000 00000   00000 00000     00000 00000   00000 00000   00000 00000   00000 00000   00000 00000
2481:   00000 00000   00000 00000   00000 00000   00000 00000   00000 00000     00000 00000   00000 00000   00000 00000   00000 00000   00000 00000
2482:   00000 00000   00000 00000   00000 00000   00000 00000   00000 00000     00000 00000   00000 00000   00000 00000   00000 00000   00000 00000
2483:   00000 00000   00000 00000   00000 00000   00000 00000   00000 00000     00000 00000   00000 00000   00000 00000   00000 00000   00000 00000
2484:   00000 00000   00000 00000   00000 00000   00000 00000   00000 00000     00000 00000   00000 00000   00000 00000   00000 00000   00000 00000
2485:   00000 00000   00000 00000   00000 00000   00000 00000   00000 00000     00000 00000   00000 00000   00000 00000   00000 00000   00000 00000
2486:   00000 00000   00000 00000   00000 00000   00000 00000   00000 00000     00000 00000   00000 00000   00000 00000   00000 00000   00000 00000
2487:   00000 00000   00000 00000   00000 00000   00000 00000   00000 00000     00000 00000   00000 00000   00000 00000   00000 00000   00000 00000
2488:   00000 00000   00000 00000   00000 00000   00000 00000   00000 00000     00000 00000   00000 00000   00000 00000   00000 00000   00000 00000
2489:   00000 00000   00000 00000   00000 00000   00000 00000   00000 00000     00000 00000   00000 00000   00000 00000   00000 00000   00000 00000
2490:   00000 00000   00000 00000   00000 00000   00000 00000   00000 00000     00000 00000   00000 00000   00000 00000   00000 00000   00000 00000
2491:   00000 00000   00000 00000   00000 00000   00000 00000   00000 00000     00000 00000   00000 00000   00000 00000   00000 00000   00000 00000
2492:   00000 00000   00000 00000   00000 00000   00000 00000   00000 00000     00000 00000   00000 00000   00000 00000   00000 00000   00000 00000
2493:   00000 00000   00000 00000   00000 00000   00000 00000   00000 00000     00000 00000   00000 00000   00000 00000   00000 00000   00000 00000
2494:   00000 00000   00000 00000   00000 00000   00000 00000   00000 00000     00000 00000   00000 00000   00000 00000   00000 00000   00000 00000
2495:   00000 00000   00000 00000   00000 00000   00000 00000   00000 00000     00000 00000   00000 00000   00000 00000   00000 00000   00000 00000
2496:   00000 00000   00000 00000   00000 00000   00000 00000   00000 00000     00000 00000   00000 00000   00000 00000   00000 00000   00000 00000
2497:   00000 00000   00000 00000   00000 00000   00000 00000   00000 00000     00000 00000   00000 00000   00000 00000   00000 00000   00000 00000
2498:   00000 00000   00000 00000   00000 00000   00000 00000   00000 00000     00000 00000   00000 00000   00000 00000   00000 00000   00000 00000
2499:   00000 00000   00000 00000   00000 00000   00000 00000   00000 00000     00000 00000   00000 00000   00000 00000   00000 00000   00000 00000
```

```
2500:  00000 00000  00000 00000  00000 00000  00000 00000  00000 00000    00000 00000  00000 00000  00000 00000  00000 00000  00000 00000
2501:  00000 00000  00000 00000  00000 00000  00000 00000  00000 00000    00000 00000  00000 00000  00000 00000  00000 00000  00000 00000
2502:  00000 00000  00000 00000  00000 00000  00000 00000  00000 00000    00000 00000  00000 00000  00000 00000  00000 00000  00000 00000
2503:  00000 00000  00000 00000  00000 00000  00000 00000  00000 00000    00000 00000  00000 00000  00000 00000  00000 00000  00000 00000
2504:  00000 00000  00000 00000  00000 00000  00000 00000  00000 00000    00000 00000  00000 00000  00000 00000  00000 00000  00000 00000
2505:  00000 00000  00000 00000  00000 00000  00000 00000  00000 00000    00000 00000  00000 00000  00000 00000  00000 00000  00000 00000
2506:  00000 00000  00000 00000  00000 00000  00000 00000  00000 00000    00000 00000  00000 00000  00000 00000  00000 00000  00000 00000
2507:  00000 00000  00000 00000  00000 00000  00000 00000  00000 00000    00000 00000  00000 00000  00000 00000  00000 00000  00000 00000
2508:  00000 00000  00000 00000  00000 00000  00000 00000  00000 00000    00000 00000  00000 00000  00000 00000  00000 00000  00000 00000
2509:  00000 00000  00000 00000  00000 00000  00000 00000  00000 00000    00000 00000  00000 00000  00000 00000  00000 00000  00000 00000
2510:  00000 00000  00000 00000  00000 00000  00000 00000  00000 00000    00000 00000  00000 00000  00000 00000  00000 00000  00000 00000
2511:  00000 00000  00000 00000  00000 00000  00000 00000  00000 00000    00000 00000  00000 00000  00000 00000  00000 00000  00000 00000
2512:  00000 00000  00000 00000  00000 00000  00000 00000  00000 00000    00000 00000  00000 00000  00000 00000  00000 00000  00000 00000
2513:  00000 00000  00000 00000  00000 00000  00000 00000  00000 00000    00000 00000  00000 00000  00000 00000  00000 00000  00000 00000
2514:  00000 00000  00000 00000  00000 00000  00000 00000  00000 00000    00000 00000  00000 00000  00000 00000  00000 00000  00000 00000
2515:  00000 00000  00000 00000  00000 00000  00000 00000  00000 00000    00000 00000  00000 00000  00000 00000  00000 00000  00000 00000
2516:  00000 00000  00000 00000  00000 00000  00000 00000  00000 00000    00000 00000  00000 00000  00000 00000  00000 00000  00000 00000
2517:  00000 00000  00000 00000  00000 00000  00000 00000  00000 00000    00000 00000  00000 00000  00000 00000  00000 00000  00000 00000
2518:  00000 00000  00000 00000  00000 00000  00000 00000  00000 00000    00000 00000  00000 00000  00000 00000  00000 00000  00000 00000
2519:  00000 00000  00000 00000  00000 00000  00000 00000  00000 00000    00000 00000  00000 00000  00000 00000  00000 00000  00000 00000
2520:  00000 00000  00000 00000  00000 00000  00000 00000  00000 00000    00000 00000  00000 00000  00000 00000  00000 00000  00000 00000
2521:  00000 00000  00000 00000  00000 00000  00000 00000  00000 00000    00000 00000  00000 00000  00000 00000  00000 00000  00000 00000
2522:  00000 00000  00000 00000  00000 00000  00000 00000  00000 00000    00000 00000  00000 00000  00000 00000  00000 00000  00000 00000
2523:  00000 00000  00000 00000  00000 00000  00000 00000  00000 00000    00000 00000  00000 00000  00000 00000  00000 00000  00000 00000
2524:  00000 00000  00000 00000  00000 00000  00000 00000  00000 00000    00000 00000  00000 00000  00000 00000  00000 00000  00000 00000
2525:  00000 00000  00000 00000  00000 00000  00000 00000  00000 00000    00000 00000  00000 00000  00000 00000  00000 00000  00000 00000
2526:  00000 00000  00000 00000  00000 00000  00000 00000  00000 00000    00000 00000  00000 00000  00000 00000  00000 00000  00000 00000
2527:  00000 00000  00000 00000  00000 00000  00000 00000  00000 00000    00000 00000  00000 00000  00000 00000  00000 00000  00000 00000
2528:  00000 00000  00000 00000  00000 00000  00000 00000  00000 00000    00000 00000  00000 00000  00000 00000  00000 00000  00000 00000
2529:  00000 00000  00000 00000  00000 00000  00000 00000  00000 00000    00000 00000  00000 00000  00000 00000  00000 00000  00000 00000
2530:  00000 00000  00000 00000  00000 00000  00000 00000  00000 00000    00000 00000  00000 00000  00000 00000  00000 00000  00000 00000
2531:  00000 00000  00000 00000  00000 00000  00000 00000  00000 00000    00000 00000  00000 00000  00000 00000  00000 00000  00000 00000
2532:  00000 00000  00000 00000  00000 00000  00000 00000  00000 00000    00000 00000  00000 00000  00000 00000  00000 00000  00000 00000
2533:  00000 00000  00000 00000  00000 00000  00000 00000  00000 00000    00000 00000  00000 00000  00000 00000  00000 00000  00000 00000
2534:  00000 00000  00000 00000  00000 00000  00000 00000  00000 00000    00000 00000  00000 00000  00000 00000  00000 00000  00000 00000
2535:  00000 00000  00000 00000  00000 00000  00000 00000  00000 00000    00000 00000  00000 00000  00000 00000  00000 00000  00000 00000
2536:  00000 00000  00000 00000  00000 00000  00000 00000  00000 00000    00000 00000  00000 00000  00000 00000  00000 00000  00000 00000
2537:  00000 00000  00000 00000  00000 00000  00000 00000  00000 00000    00000 00000  00000 00000  00000 00000  00000 00000  00000 00000
2538:  00000 00000  00000 00000  00000 00000  00000 00000  00000 00000    00000 00000  00000 00000  00000 00000  00000 00000  00000 00000
2539:  00000 00000  00000 00000  00000 00000  00000 00000  00000 00000    00000 00000  00000 00000  00000 00000  00000 00000  00000 00000
2540:  00000 00000  00000 00000  00000 00000  00000 00000  00000 00000    00000 00000  00000 00000  00000 00000  00000 00000  00000 00000
2541:  00000 00000  00000 00000  00000 00000  00000 00000  00000 00000    00000 00000  00000 00000  00000 00000  00000 00000  00000 00000
2542:  00000 00000  00000 00000  00000 00000  00000 00000  00000 00000    00000 00000  00000 00000  00000 00000  00000 00000  00000 00000
2543:  00000 00000  00000 00000  00000 00000  00000 00000  00000 00000    00000 00000  00000 00000  00000 00000  00000 00000  00000 00000
2544:  00000 00000  00000 00000  00000 00000  00000 00000  00000 00000    00000 00000  00000 00000  00000 00000  00000 00000  00000 00000
2545:  00000 00000  00000 00000  00000 00000  00000 00000  00000 00000    00000 00000  00000 00000  00000 00000  00000 00000  00000 00000
2546:  00000 00000  00000 00000  00000 00000  00000 00000  00000 00000    00000 00000  00000 00000  00000 00000  00000 00000  00000 00000
2547:  00000 00000  00000 00000  00000 00000  00000 00000  00000 00000    00000 00000  00000 00000  00000 00000  00000 00000  00000 00000
2548:  00000 00000  00000 00000  00000 00000  00000 00000  00000 00000    00000 00000  00000 00000  00000 00000  00000 00000  00000 00000
2549:  00000 00000  00000 00000  00000 00000  00000 00000  00000 00000    00000 00000  00000 00000  00000 00000  00000 00000  00000 00000
```

```
2550:  00000 00000  00000 00000  00000 00000  00000 00000  00000 00000    00000 00000  00000 00000  00000 00000  00000 00000  00000 00000
2551:  00000 00000  00000 00000  00000 00000  00000 00000  00000 00000    00000 00000  00000 00000  00000 00000  00000 00000  00000 00000
2552:  00000 00000  00000 00000  00000 00000  00000 00000  00000 00000    00000 00000  00000 00000  00000 00000  00000 00000  00000 00000
2553:  00000 00000  00000 00000  00000 00000  00000 00000  00000 00000    00000 00000  00000 00000  00000 00000  00000 00000  00000 00000
2554:  00000 00000  00000 00000  00000 00000  00000 00000  00000 00000    00000 00000  00000 00000  00000 00000  00000 00000  00000 00000
2555:  00000 00000  00000 00000  00000 00000  00000 00000  00000 00000    00000 00000  00000 00000  00000 00000  00000 00000  00000 00000
2556:  00000 00000  00000 00000  00000 00000  00000 00000  00000 00000    00000 00000  00000 00000  00000 00000  00000 00000  00000 00000
2557:  00000 00000  00000 00000  00000 00000  00000 00000  00000 00000    00000 00000  00000 00000  00000 00000  00000 00000  00000 00000
2558:  00000 00000  00000 00000  00000 00000  00000 00000  00000 00000    00000 00000  00000 00000  00000 00000  00000 00000  00000 00000
2559:  00000 00000  00000 00000  00000 00000  00000 00000  00000 00000    00000 00000  00000 00000  00000 00000  00000 00000  00000 00000
2560:  00000 00000  00000 00000  00000 00000  00000 00000  00000 00000    00000 00000  00000 00000  00000 00000  00000 00000  00000 00000
2561:  00000 00000  00000 00000  00000 00000  00000 00000  00000 00000    00000 00000  00000 00000  00000 00000  00000 00000  00000 00000
2562:  00000 00000  00000 00000  00000 00000  00000 00000  00000 00000    00000 00000  00000 00000  00000 00000  00000 00000  00000 00000
2563:  00000 00000  00000 00000  00000 00000  00000 00000  00000 00000    00000 00000  00000 00000  00000 00000  00000 00000  00000 00000
2564:  00000 00000  00000 00000  00000 00000  00000 00000  00000 00000    00000 00000  00000 00000  00000 00000  00000 00000  00000 00000
2565:  00000 00000  00000 00000  00000 00000  00000 00000  00000 00000    00000 00000  00000 00000  00000 00000  00000 00000  00000 00000
2566:  00000 00000  00000 00000  00000 00000  00000 00000  00000 00000    00000 00000  00000 00000  00000 00000  00000 00000  00000 00000
2567:  00000 00000  00000 00000  00000 00000  00000 00000  00000 00000    00000 00000  00000 00000  00000 00000  00000 00000  00000 00000
2568:  00000 00000  00000 00000  00000 00000  00000 00000  00000 00000    00000 00000  00000 00000  00000 00000  00000 00000  00000 00000
2569:  00000 00000  00000 00000  00000 00000  00000 00000  00000 00000    00000 00000  00000 00000  00000 00000  00000 00000  00000 00000
2570:  00000 00000  00000 00000  00000 00000  00000 00000  00000 00000    00000 00000  00000 00000  00000 00000  00000 00000  00000 00000
2571:  00000 00000  00000 00000  00000 00000  00000 00000  00000 00000    00000 00000  00000 00000  00000 00000  00000 00000  00000 00000
2572:  00000 00000  00000 00000  00000 00000  00000 00000  00000 00000    00000 00000  00000 00000  00000 00000  00000 00000  00000 00000
2573:  00000 00000  00000 00000  00000 00000  00000 00000  00000 00000    00000 00000  00000 00000  00000 00000  00000 00000  00000 00000
2574:  00000 00000  00000 00000  00000 00000  00000 00000  00000 00000    00000 00000  00000 00000  00000 00000  00000 00000  00000 00000
2575:  00000 00000  00000 00000  00000 00000  00000 00000  00000 00000    00000 00000  00000 00000  00000 00000  00000 00000  00000 00000
2576:  00000 00000  00000 00000  00000 00000  00000 00000  00000 00000    00000 00000  00000 00000  00000 00000  00000 00000  00000 00000
2577:  00000 00000  00000 00000  00000 00000  00000 00000  00000 00000    00000 00000  00000 00000  00000 00000  00000 00000  00000 00000
2578:  00000 00000  00000 00000  00000 00000  00000 00000  00000 00000    00000 00000  00000 00000  00000 00000  00000 00000  00000 00000
2579:  00000 00000  00000 00000  00000 00000  00000 00000  00000 00000    00000 00000  00000 00000  00000 00000  00000 00000  00000 00000
2580:  00000 00000  00000 00000  00000 00000  00000 00000  00000 00000    00000 00000  00000 00000  00000 00000  00000 00000  00000 00000
2581:  00000 00000  00000 00000  00000 00000  00000 00000  00000 00000    00000 00000  00000 00000  00000 00000  00000 00000  00000 00000
2582:  00000 00000  00000 00000  00000 00000  00000 00000  00000 00000    00000 00000  00000 00000  00000 00000  00000 00000  00000 00000
2583:  00000 00000  00000 00000  00000 00000  00000 00000  00000 00000    00000 00000  00000 00000  00000 00000  00000 00000  00000 00000
2584:  00000 00000  00000 00000  00000 00000  00000 00000  00000 00000    00000 00000  00000 00000  00000 00000  00000 00000  00000 00000
2585:  00000 00000  00000 00000  00000 00000  00000 00000  00000 00000    00000 00000  00000 00000  00000 00000  00000 00000  00000 00000
2586:  00000 00000  00000 00000  00000 00000  00000 00000  00000 00000    00000 00000  00000 00000  00000 00000  00000 00000  00000 00000
2587:  00000 00000  00000 00000  00000 00000  00000 00000  00000 00000    00000 00000  00000 00000  00000 00000  00000 00000  00000 00000
2588:  00000 00000  00000 00000  00000 00000  00000 00000  00000 00000    00000 00000  00000 00000  00000 00000  00000 00000  00000 00000
2589:  00000 00000  00000 00000  00000 00000  00000 00000  00000 00000    00000 00000  00000 00000  00000 00000  00000 00000  00000 00000
2590:  00000 00000  00000 00000  00000 00000  00000 00000  00000 00000    00000 00000  00000 00000  00000 00000  00000 00000  00000 00000
2591:  00000 00000  00000 00000  00000 00000  00000 00000  00000 00000    00000 00000  00000 00000  00000 00000  00000 00000  00000 00000
2592:  00000 00000  00000 00000  00000 00000  00000 00000  00000 00000    00000 00000  00000 00000  00000 00000  00000 00000  00000 00000
2593:  00000 00000  00000 00000  00000 00000  00000 00000  00000 00000    00000 00000  00000 00000  00000 00000  00000 00000  00000 00000
2594:  00000 00000  00000 00000  00000 00000  00000 00000  00000 00000    00000 00000  00000 00000  00000 00000  00000 00000  00000 00000
2595:  00000 00000  00000 00000  00000 00000  00000 00000  00000 00000    00000 00000  00000 00000  00000 00000  00000 00000  00000 00000
2596:  00000 00000  00000 00000  00000 00000  00000 00000  00000 00000    00000 00000  00000 00000  00000 00000  00000 00000  00000 00000
2597:  00000 00000  00000 00000  00000 00000  00000 00000  00000 00000    00000 00000  00000 00000  00000 00000  00000 00000  00000 00000
2598:  00000 00000  00000 00000  00000 00000  00000 00000  00000 00000    00000 00000  00000 00000  00000 00000  00000 00000  00000 00000
2599:  00000 00000  00000 00000  00000 00000  00000 00000  00000 00000    00000 00000  00000 00000  00000 00000  00000 00000  00000 00000
```

```
2600:  0000000000  0000000000  0000000000  0000000000  0000000000    0000000000  0000000000  0000000000  0000000000  0000000000
2601:  0000000000  0000000000  0000000000  0000000000  0000000000    0000000000  0000000000  0000000000  0000000000  0000000000
2602:  0000000000  0000000000  0000000000  0000000000  0000000000    0000000000  0000000000  0000000000  0000000000  0000000000
2603:  0000000000  0000000000  0000000000  0000000000  0000000000    0000000000  0000000000  0000000000  0000000000  0000000000
2604:  0000000000  0000000000  0000000000  0000000000  0000000000    0000000000  0000000000  0000000000  0000000000  0000000000
2605:  0000000000  0000000000  0000000000  0000000000  0000000000    0000000000  0000000000  0000000000  0000000000  0000000000
2606:  0000000000  0000000000  0000000000  0000000000  0000000000    0000000000  0000000000  0000000000  0000000000  0000000000
2607:  0000000000  0000000000  0000000000  0000000000  0000000000    0000000000  0000000000  0000000000  0000000000  0000000000
2608:  0000000000  0000000000  0000000000  0000000000  0000000000    0000000000  0000000000  0000000000  0000000000  0000000000
2609:  0000000000  0000000000  0000000000  0000000000  0000000000    0000000000  0000000000  0000000000  0000000000  0000000000
2610:  0000000000  0000000000  0000000000  0000000000  0000000000    0000000000  0000000000  0000000000  0000000000  0000000000
2611:  0000000000  0000000000  0000000000  0000000000  0000000000    0000000000  0000000000  0000000000  0000000000  0000000000
2612:  0000000000  0000000000  0000000000  0000000000  0000000000    0000000000  0000000000  0000000000  0000000000  0000000000
2613:  0000000000  0000000000  0000000000  0000000000  0000000000    0000000000  0000000000  0000000000  0000000000  0000000000
2614:  0000000000  0000000000  0000000000  0000000000  0000000000    0000000000  0000000000  0000000000  0000000000  0000000000
2615:  0000000000  0000000000  0000000000  0000000000  0000000000    0000000000  0000000000  0000000000  0000000000  0000000000
2616:  0000000000  0000000000  0000000000  0000000000  0000000000    0000000000  0000000000  0000000000  0000000000  0000000000
2617:  0000000000  0000000000  0000000000  0000000000  0000000000    0000000000  0000000000  0000000000  0000000000  0000000000
2618:  0000000000  0000000000  0000000000  0000000000  0000000000    0000000000  0000000000  0000000000  0000000000  0000000000
2619:  0000000000  0000000000  0000000000  0000000000  0000000000    0000000000  0000000000  0000000000  0000000000  0000000000
2620:  0000000000  0000000000  0000000000  0000000000  0000000000    0000000000  0000000000  0000000000  0000000000  0000000000
2621:  0000000000  0000000000  0000000000  0000000000  0000000000    0000000000  0000000000  0000000000  0000000000  0000000000
2622:  0000000000  0000000000  0000000000  0000000000  0000000000    0000000000  0000000000  0000000000  0000000000  0000000000
2623:  0000000000  0000000000  0000000000  0000000000  0000000000    0000000000  0000000000  0000000000  0000000000  0000000000
2624:  0000000000  0000000000  0000000000  0000000000  0000000000    0000000000  0000000000  0000000000  0000000000  0000000000
2625:  0000000000  0000000000  0000000000  0000000000  0000000000    0000000000  0000000000  0000000000  0000000000  0000000000
2626:  0000000000  0000000000  0000000000  0000000000  0000000000    0000000000  0000000000  0000000000  0000000000  0000000000
2627:  0000000000  0000000000  0000000000  0000000000  0000000000    0000000000  0000000000  0000000000  0000000000  0000000000
2628:  0000000000  0000000000  0000000000  0000000000  0000000000    0000000000  0000000000  0000000000  0000000000  0000000000
2629:  0000000000  0000000000  0000000000  0000000000  0000000000    0000000000  0000000000  0000000000  0000000000  0000000000
2630:  0000000000  0000000000  0000000000  0000000000  0000000000    0000000000  0000000000  0000000000  0000000000  0000000000
2631:  0000000000  0000000000  0000000000  0000000000  0000000000    0000000000  0000000000  0000000000  0000000000  0000000000
2632:  0000000000  0000000000  0000000000  0000000000  0000000000    0000000000  0000000000  0000000000  0000000000  0000000000
2633:  0000000000  0000000000  0000000000  0000000000  0000000000    0000000000  0000000000  0000000000  0000000000  0000000000
2634:  0000000000  0000000000  0000000000  0000000000  0000000000    0000000000  0000000000  0000000000  0000000000  0000000000
2635:  0000000000  0000000000  0000000000  0000000000  0000000000    0000000000  0000000000  0000000000  0000000000  0000000000
2636:  0000000000  0000000000  0000000000  0000000000  0000000000    0000000000  0000000000  0000000000  0000000000  0000000000
2637:  0000000000  0000000000  0000000000  0000000000  0000000000    0000000000  0000000000  0000000000  0000000000  0000000000
2638:  0000000000  0000000000  0000000000  0000000000  0000000000    0000000000  0000000000  0000000000  0000000000  0000000000
2639:  0000000000  0000000000  0000000000  0000000000  0000000000    0000000000  0000000000  0000000000  0000000000  0000000000
2640:  0000000000  0000000000  0000000000  0000000000  0000000000    0000000000  0000000000  0000000000  0000000000  0000000000
2641:  0000000000  0000000000  0000000000  0000000000  0000000000    0000000000  0000000000  0000000000  0000000000  0000000000
2642:  0000000000  0000000000  0000000000  0000000000  0000000000    0000000000  0000000000  0000000000  0000000000  0000000000
2643:  0000000000  0000000000  0000000000  0000000000  0000000000    0000000000  0000000000  0000000000  0000000000  0000000000
2644:  0000000000  0000000000  0000000000  0000000000  0000000000    0000000000  0000000000  0000000000  0000000000  0000000000
2645:  0000000000  0000000000  0000000000  0000000000  0000000000    0000000000  0000000000  0000000000  0000000000  0000000000
2646:  0000000000  0000000000  0000000000  0000000000  0000000000    0000000000  0000000000  0000000000  0000000000  0000000000
2647:  0000000000  0000000000  0000000000  0000000000  0000000000    0000000000  0000000000  0000000000  0000000000  0000000000
2648:  0000000000  0000000000  0000000000  0000000000  0000000000    0000000000  0000000000  0000000000  0000000000  0000000000
2649:  0000000000  0000000000  0000000000  0000000000  0000000000    0000000000  0000000000  0000000000  0000000000  0000000000
```

```
2650:  00000 00000  00000 00000  00000 00000  00000 00000  00000 00000    00000 00000  00000 00000  00000 00000  00000 00000  00000 00000
2651:  00000 00000  00000 00000  00000 00000  00000 00000  00000 00000    00000 00000  00000 00000  00000 00000  00000 00000  00000 00000
2652:  00000 00000  00000 00000  00000 00000  00000 00000  00000 00000    00000 00000  00000 00000  00000 00000  00000 00000  00000 00000
2653:  00000 00000  00000 00000  00000 00000  00000 00000  00000 00000    00000 00000  00000 00000  00000 00000  00000 00000  00000 00000
2654:  00000 00000  00000 00000  00000 00000  00000 00000  00000 00000    00000 00000  00000 00000  00000 00000  00000 00000  00000 00000
2655:  00000 00000  00000 00000  00000 00000  00000 00000  00000 00000    00000 00000  00000 00000  00000 00000  00000 00000  00000 00000
2656:  00000 00000  00000 00000  00000 00000  00000 00000  00000 00000    00000 00000  00000 00000  00000 00000  00000 00000  00000 00000
2657:  00000 00000  00000 00000  00000 00000  00000 00000  00000 00000    00000 00000  00000 00000  00000 00000  00000 00000  00000 00000
2658:  00000 00000  00000 00000  00000 00000  00000 00000  00000 00000    00000 00000  00000 00000  00000 00000  00000 00000  00000 00000
2659:  00000 00000  00000 00000  00000 00000  00000 00000  00000 00000    00000 00000  00000 00000  00000 00000  00000 00000  00000 00000
2660:  00000 00000  00000 00000  00000 00000  00000 00000  00000 00000    00000 00000  00000 00000  00000 00000  00000 00000  00000 00000
2661:  00000 00000  00000 00000  00000 00000  00000 00000  00000 00000    00000 00000  00000 00000  00000 00000  00000 00000  00000 00000
2662:  00000 00000  00000 00000  00000 00000  00000 00000  00000 00000    00000 00000  00000 00000  00000 00000  00000 00000  00000 00000
2663:  00000 00000  00000 00000  00000 00000  00000 00000  00000 00000    00000 00000  00000 00000  00000 00000  00000 00000  00000 00000
2664:  00000 00000  00000 00000  00000 00000  00000 00000  00000 00000    00000 00000  00000 00000  00000 00000  00000 00000  00000 00000
2665:  00000 00000  00000 00000  00000 00000  00000 00000  00000 00000    00000 00000  00000 00000  00000 00000  00000 00000  00000 00000
2666:  00000 00000  00000 00000  00000 00000  00000 00000  00000 00000    00000 00000  00000 00000  00000 00000  00000 00000  00000 00000
2667:  00000 00000  00000 00000  00000 00000  00000 00000  00000 00000    00000 00000  00000 00000  00000 00000  00000 00000  00000 00000
2668:  00000 00000  00000 00000  00000 00000  00000 00000  00000 00000    00000 00000  00000 00000  00000 00000  00000 00000  00000 00000
2669:  00000 00000  00000 00000  00000 00000  00000 00000  00000 00000    00000 00000  00000 00000  00000 00000  00000 00000  00000 00000
2670:  00000 00000  00000 00000  00000 00000  00000 00000  00000 00000    00000 00000  00000 00000  00000 00000  00000 00000  00000 00000
2671:  00000 00000  00000 00000  00000 00000  00000 00000  00000 00000    00000 00000  00000 00000  00000 00000  00000 00000  00000 00000
2672:  00000 00000  00000 00000  00000 00000  00000 00000  00000 00000    00000 00000  00000 00000  00000 00000  00000 00000  00000 00000
2673:  00000 00000  00000 00000  00000 00000  00000 00000  00000 00000    00000 00000  00000 00000  00000 00000  00000 00000  00000 00000
2674:  00000 00000  00000 00000  00000 00000  00000 00000  00000 00000    00000 00000  00000 00000  00000 00000  00000 00000  00000 00000
2675:  00000 00000  00000 00000  00000 00000  00000 00000  00000 00000    00000 00000  00000 00000  00000 00000  00000 00000  00000 00000
2676:  00000 00000  00000 00000  00000 00000  00000 00000  00000 00000    00000 00000  00000 00000  00000 00000  00000 00000  00000 00000
2677:  00000 00000  00000 00000  00000 00000  00000 00000  00000 00000    00000 00000  00000 00000  00000 00000  00000 00000  00000 00000
2678:  00000 00000  00000 00000  00000 00000  00000 00000  00000 00000    00000 00000  00000 00000  00000 00000  00000 00000  00000 00000
2679:  00000 00000  00000 00000  00000 00000  00000 00000  00000 00000    00000 00000  00000 00000  00000 00000  00000 00000  00000 00000
2680:  00000 00000  00000 00000  00000 00000  00000 00000  00000 00000    00000 00000  00000 00000  00000 00000  00000 00000  00000 00000
2681:  00000 00000  00000 00000  00000 00000  00000 00000  00000 00000    00000 00000  00000 00000  00000 00000  00000 00000  00000 00000
2682:  00000 00000  00000 00000  00000 00000  00000 00000  00000 00000    00000 00000  00000 00000  00000 00000  00000 00000  00000 00000
2683:  00000 00000  00000 00000  00000 00000  00000 00000  00000 00000    00000 00000  00000 00000  00000 00000  00000 00000  00000 00000
2684:  00000 00000  00000 00000  00000 00000  00000 00000  00000 00000    00000 00000  00000 00000  00000 00000  00000 00000  00000 00000
2685:  00000 00000  00000 00000  00000 00000  00000 00000  00000 00000    00000 00000  00000 00000  00000 00000  00000 00000  00000 00000
2686:  00000 00000  00000 00000  00000 00000  00000 00000  00000 00000    00000 00000  00000 00000  00000 00000  00000 00000  00000 00000
2687:  00000 00000  00000 00000  00000 00000  00000 00000  00000 00000    00000 00000  00000 00000  00000 00000  00000 00000  00000 00000
2688:  00000 00000  00000 00000  00000 00000  00000 00000  00000 00000    00000 00000  00000 00000  00000 00000  00000 00000  00000 00000
2689:  00000 00000  00000 00000  00000 00000  00000 00000  00000 00000    00000 00000  00000 00000  00000 00000  00000 00000  00000 00000
2690:  00000 00000  00000 00000  00000 00000  00000 00000  00000 00000    00000 00000  00000 00000  00000 00000  00000 00000  00000 00000
2691:  00000 00000  00000 00000  00000 00000  00000 00000  00000 00000    00000 00000  00000 00000  00000 00000  00000 00000  00000 00000
2692:  00000 00000  00000 00000  00000 00000  00000 00000  00000 00000    00000 00000  00000 00000  00000 00000  00000 00000  00000 00000
2693:  00000 00000  00000 00000  00000 00000  00000 00000  00000 00000    00000 00000  00000 00000  00000 00000  00000 00000  00000 00000
2694:  00000 00000  00000 00000  00000 00000  00000 00000  00000 00000    00000 00000  00000 00000  00000 00000  00000 00000  00000 00000
2695:  00000 00000  00000 00000  00000 00000  00000 00000  00000 00000    00000 00000  00000 00000  00000 00000  00000 00000  00000 00000
2696:  00000 00000  00000 00000  00000 00000  00000 00000  00000 00000    00000 00000  00000 00000  00000 00000  00000 00000  00000 00000
2697:  00000 00000  00000 00000  00000 00000  00000 00000  00000 00000    00000 00000  00000 00000  00000 00000  00000 00000  00000 00000
2698:  00000 00000  00000 00000  00000 00000  00000 00000  00000 00000    00000 00000  00000 00000  00000 00000  00000 00000  00000 00000
2699:  00000 00000  00000 00000  00000 00000  00000 00000  00000 00000    00000 00000  00000 00000  00000 00000  00000 00000  00000 00000
```

```
2700:  00000 00000  00000 00000  00000 00000  00000 00000  00000 00000    00000 00000  00000 00000  00000 00000  00000 00000  00000 00000
2701:  00000 00000  00000 00000  00000 00000  00000 00000  00000 00000    00000 00000  00000 00000  00000 00000  00000 00000  00000 00000
2702:  00000 00000  00000 00000  00000 00000  00000 00000  00000 00000    00000 00000  00000 00000  00000 00000  00000 00000  00000 00000
2703:  00000 00000  00000 00000  00000 00000  00000 00000  00000 00000    00000 00000  00000 00000  00000 00000  00000 00000  00000 00000
2704:  00000 00000  00000 00000  00000 00000  00000 00000  00000 00000    00000 00000  00000 00000  00000 00000  00000 00000  00000 00000
2705:  00000 00000  00000 00000  00000 00000  00000 00000  00000 00000    00000 00000  00000 00000  00000 00000  00000 00000  00000 00000
2706:  00000 00000  00000 00000  00000 00000  00000 00000  00000 00000    00000 00000  00000 00000  00000 00000  00000 00000  00000 00000
2707:  00000 00000  00000 00000  00000 00000  00000 00000  00000 00000    00000 00000  00000 00000  00000 00000  00000 00000  00000 00000
2708:  00000 00000  00000 00000  00000 00000  00000 00000  00000 00000    00000 00000  00000 00000  00000 00000  00000 00000  00000 00000
2709:  00000 00000  00000 00000  00000 00000  00000 00000  00000 00000    00000 00000  00000 00000  00000 00000  00000 00000  00000 00000
2710:  00000 00000  00000 00000  00000 00000  00000 00000  00000 00000    00000 00000  00000 00000  00000 00000  00000 00000  00000 00000
2711:  00000 00000  00000 00000  00000 00000  00000 00000  00000 00000    00000 00000  00000 00000  00000 00000  00000 00000  00000 00000
2712:  00000 00000  00000 00000  00000 00000  00000 00000  00000 00000    00000 00000  00000 00000  00000 00000  00000 00000  00000 00000
2713:  00000 00000  00000 00000  00000 00000  00000 00000  00000 00000    00000 00000  00000 00000  00000 00000  00000 00000  00000 00000
2714:  00000 00000  00000 00000  00000 00000  00000 00000  00000 00000    00000 00000  00000 00000  00000 00000  00000 00000  00000 00000
2715:  00000 00000  00000 00000  00000 00000  00000 00000  00000 00000    00000 00000  00000 00000  00000 00000  00000 00000  00000 00000
2716:  00000 00000  00000 00000  00000 00000  00000 00000  00000 00000    00000 00000  00000 00000  00000 00000  00000 00000  00000 00000
2717:  00000 00000  00000 00000  00000 00000  00000 00000  00000 00000    00000 00000  00000 00000  00000 00000  00000 00000  00000 00000
2718:  00000 00000  00000 00000  00000 00000  00000 00000  00000 00000    00000 00000  00000 00000  00000 00000  00000 00000  00000 00000
2719:  00000 00000  00000 00000  00000 00000  00000 00000  00000 00000    00000 00000  00000 00000  00000 00000  00000 00000  00000 00000
2720:  00000 00000  00000 00000  00000 00000  00000 00000  00000 00000    00000 00000  00000 00000  00000 00000  00000 00000  00000 00000
2721:  00000 00000  00000 00000  00000 00000  00000 00000  00000 00000    00000 00000  00000 00000  00000 00000  00000 00000  00000 00000
2722:  00000 00000  00000 00000  00000 00000  00000 00000  00000 00000    00000 00000  00000 00000  00000 00000  00000 00000  00000 00000
2723:  00000 00000  00000 00000  00000 00000  00000 00000  00000 00000    00000 00000  00000 00000  00000 00000  00000 00000  00000 00000
2724:  00000 00000  00000 00000  00000 00000  00000 00000  00000 00000    00000 00000  00000 00000  00000 00000  00000 00000  00000 00000
2725:  00000 00000  00000 00000  00000 00000  00000 00000  00000 00000    00000 00000  00000 00000  00000 00000  00000 00000  00000 00000
2726:  00000 00000  00000 00000  00000 00000  00000 00000  00000 00000    00000 00000  00000 00000  00000 00000  00000 00000  00000 00000
2727:  00000 00000  00000 00000  00000 00000  00000 00000  00000 00000    00000 00000  00000 00000  00000 00000  00000 00000  00000 00000
2728:  00000 00000  00000 00000  00000 00000  00000 00000  00000 00000    00000 00000  00000 00000  00000 00000  00000 00000  00000 00000
2729:  00000 00000  00000 00000  00000 00000  00000 00000  00000 00000    00000 00000  00000 00000  00000 00000  00000 00000  00000 00000
2730:  00000 00000  00000 00000  00000 00000  00000 00000  00000 00000    00000 00000  00000 00000  00000 00000  00000 00000  00000 00000
2731:  00000 00000  00000 00000  00000 00000  00000 00000  00000 00000    00000 00000  00000 00000  00000 00000  00000 00000  00000 00000
2732:  00000 00000  00000 00000  00000 00000  00000 00000  00000 00000    00000 00000  00000 00000  00000 00000  00000 00000  00000 00000
2733:  00000 00000  00000 00000  00000 00000  00000 00000  00000 00000    00000 00000  00000 00000  00000 00000  00000 00000  00000 00000
2734:  00000 00000  00000 00000  00000 00000  00000 00000  00000 00000    00000 00000  00000 00000  00000 00000  00000 00000  00000 00000
2735:  00000 00000  00000 00000  00000 00000  00000 00000  00000 00000    00000 00000  00000 00000  00000 00000  00000 00000  00000 00000
2736:  00000 00000  00000 00000  00000 00000  00000 00000  00000 00000    00000 00000  00000 00000  00000 00000  00000 00000  00000 00000
2737:  00000 00000  00000 00000  00000 00000  00000 00000  00000 00000    00000 00000  00000 00000  00000 00000  00000 00000  00000 00000
2738:  00000 00000  00000 00000  00000 00000  00000 00000  00000 00000    00000 00000  00000 00000  00000 00000  00000 00000  00000 00000
2739:  00000 00000  00000 00000  00000 00000  00000 00000  00000 00000    00000 00000  00000 00000  00000 00000  00000 00000  00000 00000
2740:  00000 00000  00000 00000  00000 00000  00000 00000  00000 00000    00000 00000  00000 00000  00000 00000  00000 00000  00000 00000
2741:  00000 00000  00000 00000  00000 00000  00000 00000  00000 00000    00000 00000  00000 00000  00000 00000  00000 00000  00000 00000
2742:  00000 00000  00000 00000  00000 00000  00000 00000  00000 00000    00000 00000  00000 00000  00000 00000  00000 00000  00000 00000
2743:  00000 00000  00000 00000  00000 00000  00000 00000  00000 00000    00000 00000  00000 00000  00000 00000  00000 00000  00000 00000
2744:  00000 00000  00000 00000  00000 00000  00000 00000  00000 00000    00000 00000  00000 00000  00000 00000  00000 00000  00000 00000
2745:  00000 00000  00000 00000  00000 00000  00000 00000  00000 00000    00000 00000  00000 00000  00000 00000  00000 00000  00000 00000
2746:  00000 00000  00000 00000  00000 00000  00000 00000  00000 00000    00000 00000  00000 00000  00000 00000  00000 00000  00000 00000
2747:  00000 00000  00000 00000  00000 00000  00000 00000  00000 00000    00000 00000  00000 00000  00000 00000  00000 00000  00000 00000
2748:  00000 00000  00000 00000  00000 00000  00000 00000  00000 00000    00000 00000  00000 00000  00000 00000  00000 00000  00000 00000
2749:  00000 00000  00000 00000  00000 00000  00000 00000  00000 00000    00000 00000  00000 00000  00000 00000  00000 00000  00000 00000
```

```
2750: 00000 00000  00000 00000  00000 00000  00000 00000  00000 00000    00000 00000  00000 00000  00000 00000  00000 00000  00000 00000
2751: 00000 00000  00000 00000  00000 00000  00000 00000  00000 00000    00000 00000  00000 00000  00000 00000  00000 00000  00000 00000
2752: 00000 00000  00000 00000  00000 00000  00000 00000  00000 00000    00000 00000  00000 00000  00000 00000  00000 00000  00000 00000
2753: 00000 00000  00000 00000  00000 00000  00000 00000  00000 00000    00000 00000  00000 00000  00000 00000  00000 00000  00000 00000
2754: 00000 00000  00000 00000  00000 00000  00000 00000  00000 00000    00000 00000  00000 00000  00000 00000  00000 00000  00000 00000
2755: 00000 00000  00000 00000  00000 00000  00000 00000  00000 00000    00000 00000  00000 00000  00000 00000  00000 00000  00000 00000
2756: 00000 00000  00000 00000  00000 00000  00000 00000  00000 00000    00000 00000  00000 00000  00000 00000  00000 00000  00000 00000
2757: 00000 00000  00000 00000  00000 00000  00000 00000  00000 00000    00000 00000  00000 00000  00000 00000  00000 00000  00000 00000
2758: 00000 00000  00000 00000  00000 00000  00000 00000  00000 00000    00000 00000  00000 00000  00000 00000  00000 00000  00000 00000
2759: 00000 00000  00000 00000  00000 00000  00000 00000  00000 00000    00000 00000  00000 00000  00000 00000  00000 00000  00000 00000
2760: 00000 00000  00000 00000  00000 00000  00000 00000  00000 00000    00000 00000  00000 00000  00000 00000  00000 00000  00000 00000
2761: 00000 00000  00000 00000  00000 00000  00000 00000  00000 00000    00000 00000  00000 00000  00000 00000  00000 00000  00000 00000
2762: 00000 00000  00000 00000  00000 00000  00000 00000  00000 00000    00000 00000  00000 00000  00000 00000  00000 00000  00000 00000
2763: 00000 00000  00000 00000  00000 00000  00000 00000  00000 00000    00000 00000  00000 00000  00000 00000  00000 00000  00000 00000
2764: 00000 00000  00000 00000  00000 00000  00000 00000  00000 00000    00000 00000  00000 00000  00000 00000  00000 00000  00000 00000
2765: 00000 00000  00000 00000  00000 00000  00000 00000  00000 00000    00000 00000  00000 00000  00000 00000  00000 00000  00000 00000
2766: 00000 00000  00000 00000  00000 00000  00000 00000  00000 00000    00000 00000  00000 00000  00000 00000  00000 00000  00000 00000
2767: 00000 00000  00000 00000  00000 00000  00000 00000  00000 00000    00000 00000  00000 00000  00000 00000  00000 00000  00000 00000
2768: 00000 00000  00000 00000  00000 00000  00000 00000  00000 00000    00000 00000  00000 00000  00000 00000  00000 00000  00000 00000
2769: 00000 00000  00000 00000  00000 00000  00000 00000  00000 00000    00000 00000  00000 00000  00000 00000  00000 00000  00000 00000
2770: 00000 00000  00000 00000  00000 00000  00000 00000  00000 00000    00000 00000  00000 00000  00000 00000  00000 00000  00000 00000
2771: 00000 00000  00000 00000  00000 00000  00000 00000  00000 00000    00000 00000  00000 00000  00000 00000  00000 00000  00000 00000
2772: 00000 00000  00000 00000  00000 00000  00000 00000  00000 00000    00000 00000  00000 00000  00000 00000  00000 00000  00000 00000
2773: 00000 00000  00000 00000  00000 00000  00000 00000  00000 00000    00000 00000  00000 00000  00000 00000  00000 00000  00000 00000
2774: 00000 00000  00000 00000  00000 00000  00000 00000  00000 00000    00000 00000  00000 00000  00000 00000  00000 00000  00000 00000
2775: 00000 00000  00000 00000  00000 00000  00000 00000  00000 00000    00000 00000  00000 00000  00000 00000  00000 00000  00000 00000
2776: 00000 00000  00000 00000  00000 00000  00000 00000  00000 00000    00000 00000  00000 00000  00000 00000  00000 00000  00000 00000
2777: 00000 00000  00000 00000  00000 00000  00000 00000  00000 00000    00000 00000  00000 00000  00000 00000  00000 00000  00000 00000
2778: 00000 00000  00000 00000  00000 00000  00000 00000  00000 00000    00000 00000  00000 00000  00000 00000  00000 00000  00000 00000
2779: 00000 00000  00000 00000  00000 00000  00000 00000  00000 00000    00000 00000  00000 00000  00000 00000  00000 00000  00000 00000
2780: 00000 00000  00000 00000  00000 00000  00000 00000  00000 00000    00000 00000  00000 00000  00000 00000  00000 00000  00000 00000
2781: 00000 00000  00000 00000  00000 00000  00000 00000  00000 00000    00000 00000  00000 00000  00000 00000  00000 00000  00000 00000
2782: 00000 00000  00000 00000  00000 00000  00000 00000  00000 00000    00000 00000  00000 00000  00000 00000  00000 00000  00000 00000
2783: 00000 00000  00000 00000  00000 00000  00000 00000  00000 00000    00000 00000  00000 00000  00000 00000  00000 00000  00000 00000
2784: 00000 00000  00000 00000  00000 00000  00000 00000  00000 00000    00000 00000  00000 00000  00000 00000  00000 00000  00000 00000
2785: 00000 00000  00000 00000  00000 00000  00000 00000  00000 00000    00000 00000  00000 00000  00000 00000  00000 00000  00000 00000
2786: 00000 00000  00000 00000  00000 00000  00000 00000  00000 00000    00000 00000  00000 00000  00000 00000  00000 00000  00000 00000
2787: 00000 00000  00000 00000  00000 00000  00000 00000  00000 00000    00000 00000  00000 00000  00000 00000  00000 00000  00000 00000
2788: 00000 00000  00000 00000  00000 00000  00000 00000  00000 00000    00000 00000  00000 00000  00000 00000  00000 00000  00000 00000
2789: 00000 00000  00000 00000  00000 00000  00000 00000  00000 00000    00000 00000  00000 00000  00000 00000  00000 00000  00000 00000
2790: 00000 00000  00000 00000  00000 00000  00000 00000  00000 00000    00000 00000  00000 00000  00000 00000  00000 00000  00000 00000
2791: 00000 00000  00000 00000  00000 00000  00000 00000  00000 00000    00000 00000  00000 00000  00000 00000  00000 00000  00000 00000
2792: 00000 00000  00000 00000  00000 00000  00000 00000  00000 00000    00000 00000  00000 00000  00000 00000  00000 00000  00000 00000
2793: 00000 00000  00000 00000  00000 00000  00000 00000  00000 00000    00000 00000  00000 00000  00000 00000  00000 00000  00000 00000
2794: 00000 00000  00000 00000  00000 00000  00000 00000  00000 00000    00000 00000  00000 00000  00000 00000  00000 00000  00000 00000
2795: 00000 00000  00000 00000  00000 00000  00000 00000  00000 00000    00000 00000  00000 00000  00000 00000  00000 00000  00000 00000
2796: 00000 00000  00000 00000  00000 00000  00000 00000  00000 00000    00000 00000  00000 00000  00000 00000  00000 00000  00000 00000
2797: 00000 00000  00000 00000  00000 00000  00000 00000  00000 00000    00000 00000  00000 00000  00000 00000  00000 00000  00000 00000
2798: 00000 00000  00000 00000  00000 00000  00000 00000  00000 00000    00000 00000  00000 00000  00000 00000  00000 00000  00000 00000
2799: 00000 00000  00000 00000  00000 00000  00000 00000  00000 00000    00000 00000  00000 00000  00000 00000  00000 00000  00000 00000
```

```
2800:   00000 00000   00000 00000   00000 00000   00000 00000   00000 00000    00000 00000   00000 00000   00000 00000   00000 00000   00000 00000
2801:   00000 00000   00000 00000   00000 00000   00000 00000   00000 00000    00000 00000   00000 00000   00000 00000   00000 00000   00000 00000
2802:   00000 00000   00000 00000   00000 00000   00000 00000   00000 00000    00000 00000   00000 00000   00000 00000   00000 00000   00000 00000
2803:   00000 00000   00000 00000   00000 00000   00000 00000   00000 00000    00000 00000   00000 00000   00000 00000   00000 00000   00000 00000
2804:   00000 00000   00000 00000   00000 00000   00000 00000   00000 00000    00000 00000   00000 00000   00000 00000   00000 00000   00000 00000
2805:   00000 00000   00000 00000   00000 00000   00000 00000   00000 00000    00000 00000   00000 00000   00000 00000   00000 00000   00000 00000
2806:   00000 00000   00000 00000   00000 00000   00000 00000   00000 00000    00000 00000   00000 00000   00000 00000   00000 00000   00000 00000
2807:   00000 00000   00000 00000   00000 00000   00000 00000   00000 00000    00000 00000   00000 00000   00000 00000   00000 00000   00000 00000
2808:   00000 00000   00000 00000   00000 00000   00000 00000   00000 00000    00000 00000   00000 00000   00000 00000   00000 00000   00000 00000
2809:   00000 00000   00000 00000   00000 00000   00000 00000   00000 00000    00000 00000   00000 00000   00000 00000   00000 00000   00000 00000
2810:   00000 00000   00000 00000   00000 00000   00000 00000   00000 00000    00000 00000   00000 00000   00000 00000   00000 00000   00000 00000
2811:   00000 00000   00000 00000   00000 00000   00000 00000   00000 00000    00000 00000   00000 00000   00000 00000   00000 00000   00000 00000
2812:   00000 00000   00000 00000   00000 00000   00000 00000   00000 00000    00000 00000   00000 00000   00000 00000   00000 00000   00000 00000
2813:   00000 00000   00000 00000   00000 00000   00000 00000   00000 00000    00000 00000   00000 00000   00000 00000   00000 00000   00000 00000
2814:   00000 00000   00000 00000   00000 00000   00000 00000   00000 00000    00000 00000   00000 00000   00000 00000   00000 00000   00000 00000
2815:   00000 00000   00000 00000   00000 00000   00000 00000   00000 00000    00000 00000   00000 00000   00000 00000   00000 00000   00000 00000
2816:   00000 00000   00000 00000   00000 00000   00000 00000   00000 00000    00000 00000   00000 00000   00000 00000   00000 00000   00000 00000
2817:   00000 00000   00000 00000   00000 00000   00000 00000   00000 00000    00000 00000   00000 00000   00000 00000   00000 00000   00000 00000
2818:   00000 00000   00000 00000   00000 00000   00000 00000   00000 00000    00000 00000   00000 00000   00000 00000   00000 00000   00000 00000
2819:   00000 00000   00000 00000   00000 00000   00000 00000   00000 00000    00000 00000   00000 00000   00000 00000   00000 00000   00000 00000
2820:   00000 00000   00000 00000   00000 00000   00000 00000   00000 00000    00000 00000   00000 00000   00000 00000   00000 00000   00000 00000
2821:   00000 00000   00000 00000   00000 00000   00000 00000   00000 00000    00000 00000   00000 00000   00000 00000   00000 00000   00000 00000
2822:   00000 00000   00000 00000   00000 00000   00000 00000   00000 00000    00000 00000   00000 00000   00000 00000   00000 00000   00000 00000
2823:   00000 00000   00000 00000   00000 00000   00000 00000   00000 00000    00000 00000   00000 00000   00000 00000   00000 00000   00000 00000
2824:   00000 00000   00000 00000   00000 00000   00000 00000   00000 00000    00000 00000   00000 00000   00000 00000   00000 00000   00000 00000
2825:   00000 00000   00000 00000   00000 00000   00000 00000   00000 00000    00000 00000   00000 00000   00000 00000   00000 00000   00000 00000
2826:   00000 00000   00000 00000   00000 00000   00000 00000   00000 00000    00000 00000   00000 00000   00000 00000   00000 00000   00000 00000
2827:   00000 00000   00000 00000   00000 00000   00000 00000   00000 00000    00000 00000   00000 00000   00000 00000   00000 00000   00000 00000
2828:   00000 00000   00000 00000   00000 00000   00000 00000   00000 00000    00000 00000   00000 00000   00000 00000   00000 00000   00000 00000
2829:   00000 00000   00000 00000   00000 00000   00000 00000   00000 00000    00000 00000   00000 00000   00000 00000   00000 00000   00000 00000
2830:   00000 00000   00000 00000   00000 00000   00000 00000   00000 00000    00000 00000   00000 00000   00000 00000   00000 00000   00000 00000
2831:   00000 00000   00000 00000   00000 00000   00000 00000   00000 00000    00000 00000   00000 00000   00000 00000   00000 00000   00000 00000
2832:   00000 00000   00000 00000   00000 00000   00000 00000   00000 00000    00000 00000   00000 00000   00000 00000   00000 00000   00000 00000
2833:   00000 00000   00000 00000   00000 00000   00000 00000   00000 00000    00000 00000   00000 00000   00000 00000   00000 00000   00000 00000
2834:   00000 00000   00000 00000   00000 00000   00000 00000   00000 00000    00000 00000   00000 00000   00000 00000   00000 00000   00000 00000
2835:   00000 00000   00000 00000   00000 00000   00000 00000   00000 00000    00000 00000   00000 00000   00000 00000   00000 00000   00000 00000
2836:   00000 00000   00000 00000   00000 00000   00000 00000   00000 00000    00000 00000   00000 00000   00000 00000   00000 00000   00000 00000
2837:   00000 00000   00000 00000   00000 00000   00000 00000   00000 00000    00000 00000   00000 00000   00000 00000   00000 00000   00000 00000
2838:   00000 00000   00000 00000   00000 00000   00000 00000   00000 00000    00000 00000   00000 00000   00000 00000   00000 00000   00000 00000
2839:   00000 00000   00000 00000   00000 00000   00000 00000   00000 00000    00000 00000   00000 00000   00000 00000   00000 00000   00000 00000
2840:   00000 00000   00000 00000   00000 00000   00000 00000   00000 00000    00000 00000   00000 00000   00000 00000   00000 00000   00000 00000
2841:   00000 00000   00000 00000   00000 00000   00000 00000   00000 00000    00000 00000   00000 00000   00000 00000   00000 00000   00000 00000
2842:   00000 00000   00000 00000   00000 00000   00000 00000   00000 00000    00000 00000   00000 00000   00000 00000   00000 00000   00000 00000
2843:   00000 00000   00000 00000   00000 00000   00000 00000   00000 00000    00000 00000   00000 00000   00000 00000   00000 00000   00000 00000
2844:   00000 00000   00000 00000   00000 00000   00000 00000   00000 00000    00000 00000   00000 00000   00000 00000   00000 00000   00000 00000
2845:   00000 00000   00000 00000   00000 00000   00000 00000   00000 00000    00000 00000   00000 00000   00000 00000   00000 00000   00000 00000
2846:   00000 00000   00000 00000   00000 00000   00000 00000   00000 00000    00000 00000   00000 00000   00000 00000   00000 00000   00000 00000
2847:   00000 00000   00000 00000   00000 00000   00000 00000   00000 00000    00000 00000   00000 00000   00000 00000   00000 00000   00000 00000
2848:   00000 00000   00000 00000   00000 00000   00000 00000   00000 00000    00000 00000   00000 00000   00000 00000   00000 00000   00000 00000
2849:   00000 00000   00000 00000   00000 00000   00000 00000   00000 00000    00000 00000   00000 00000   00000 00000   00000 00000   00000 00000
```

```
2850:  00000 00000   00000 00000   00000 00000   00000 00000   00000 00000      00000 00000   00000 00000   00000 00000   00000 00000   00000 00000
2851:  00000 00000   00000 00000   00000 00000   00000 00000   00000 00000      00000 00000   00000 00000   00000 00000   00000 00000   00000 00000
2852:  00000 00000   00000 00000   00000 00000   00000 00000   00000 00000      00000 00000   00000 00000   00000 00000   00000 00000   00000 00000
2853:  00000 00000   00000 00000   00000 00000   00000 00000   00000 00000      00000 00000   00000 00000   00000 00000   00000 00000   00000 00000
2854:  00000 00000   00000 00000   00000 00000   00000 00000   00000 00000      00000 00000   00000 00000   00000 00000   00000 00000   00000 00000
2855:  00000 00000   00000 00000   00000 00000   00000 00000   00000 00000      00000 00000   00000 00000   00000 00000   00000 00000   00000 00000
2856:  00000 00000   00000 00000   00000 00000   00000 00000   00000 00000      00000 00000   00000 00000   00000 00000   00000 00000   00000 00000
2857:  00000 00000   00000 00000   00000 00000   00000 00000   00000 00000      00000 00000   00000 00000   00000 00000   00000 00000   00000 00000
2858:  00000 00000   00000 00000   00000 00000   00000 00000   00000 00000      00000 00000   00000 00000   00000 00000   00000 00000   00000 00000
2859:  00000 00000   00000 00000   00000 00000   00000 00000   00000 00000      00000 00000   00000 00000   00000 00000   00000 00000   00000 00000
2860:  00000 00000   00000 00000   00000 00000   00000 00000   00000 00000      00000 00000   00000 00000   00000 00000   00000 00000   00000 00000
2861:  00000 00000   00000 00000   00000 00000   00000 00000   00000 00000      00000 00000   00000 00000   00000 00000   00000 00000   00000 00000
2862:  00000 00000   00000 00000   00000 00000   00000 00000   00000 00000      00000 00000   00000 00000   00000 00000   00000 00000   00000 00000
2863:  00000 00000   00000 00000   00000 00000   00000 00000   00000 00000      00000 00000   00000 00000   00000 00000   00000 00000   00000 00000
2864:  00000 00000   00000 00000   00000 00000   00000 00000   00000 00000      00000 00000   00000 00000   00000 00000   00000 00000   00000 00000
2865:  00000 00000   00000 00000   00000 00000   00000 00000   00000 00000      00000 00000   00000 00000   00000 00000   00000 00000   00000 00000
2866:  00000 00000   00000 00000   00000 00000   00000 00000   00000 00000      00000 00000   00000 00000   00000 00000   00000 00000   00000 00000
2867:  00000 00000   00000 00000   00000 00000   00000 00000   00000 00000      00000 00000   00000 00000   00000 00000   00000 00000   00000 00000
2868:  00000 00000   00000 00000   00000 00000   00000 00000   00000 00000      00000 00000   00000 00000   00000 00000   00000 00000   00000 00000
2869:  00000 00000   00000 00000   00000 00000   00000 00000   00000 00000      00000 00000   00000 00000   00000 00000   00000 00000   00000 00000
2870:  00000 00000   00000 00000   00000 00000   00000 00000   00000 00000      00000 00000   00000 00000   00000 00000   00000 00000   00000 00000
2871:  00000 00000   00000 00000   00000 00000   00000 00000   00000 00000      00000 00000   00000 00000   00000 00000   00000 00000   00000 00000
2872:  00000 00000   00000 00000   00000 00000   00000 00000   00000 00000      00000 00000   00000 00000   00000 00000   00000 00000   00000 00000
2873:  00000 00000   00000 00000   00000 00000   00000 00000   00000 00000      00000 00000   00000 00000   00000 00000   00000 00000   00000 00000
2874:  00000 00000   00000 00000   00000 00000   00000 00000   00000 00000      00000 00000   00000 00000   00000 00000   00000 00000   00000 00000
2875:  00000 00000   00000 00000   00000 00000   00000 00000   00000 00000      00000 00000   00000 00000   00000 00000   00000 00000   00000 00000
2876:  00000 00000   00000 00000   00000 00000   00000 00000   00000 00000      00000 00000   00000 00000   00000 00000   00000 00000   00000 00000
2877:  00000 00000   00000 00000   00000 00000   00000 00000   00000 00000      00000 00000   00000 00000   00000 00000   00000 00000   00000 00000
2878:  00000 00000   00000 00000   00000 00000   00000 00000   00000 00000      00000 00000   00000 00000   00000 00000   00000 00000   00000 00000
2879:  00000 00000   00000 00000   00000 00000   00000 00000   00000 00000      00000 00000   00000 00000   00000 00000   00000 00000   00000 00000
2880:  00000 00000   00000 00000   00000 00000   00000 00000   00000 00000      00000 00000   00000 00000   00000 00000   00000 00000   00000 00000
2881:  00000 00000   00000 00000   00000 00000   00000 00000   00000 00000      00000 00000   00000 00000   00000 00000   00000 00000   00000 00000
2882:  00000 00000   00000 00000   00000 00000   00000 00000   00000 00000      00000 00000   00000 00000   00000 00000   00000 00000   00000 00000
2883:  00000 00000   00000 00000   00000 00000   00000 00000   00000 00000      00000 00000   00000 00000   00000 00000   00000 00000   00000 00000
2884:  00000 00000   00000 00000   00000 00000   00000 00000   00000 00000      00000 00000   00000 00000   00000 00000   00000 00000   00000 00000
2885:  00000 00000   00000 00000   00000 00000   00000 00000   00000 00000      00000 00000   00000 00000   00000 00000   00000 00000   00000 00000
2886:  00000 00000   00000 00000   00000 00000   00000 00000   00000 00000      00000 00000   00000 00000   00000 00000   00000 00000   00000 00000
2887:  00000 00000   00000 00000   00000 00000   00000 00000   00000 00000      00000 00000   00000 00000   00000 00000   00000 00000   00000 00000
2888:  00000 00000   00000 00000   00000 00000   00000 00000   00000 00000      00000 00000   00000 00000   00000 00000   00000 00000   00000 00000
2889:  00000 00000   00000 00000   00000 00000   00000 00000   00000 00000      00000 00000   00000 00000   00000 00000   00000 00000   00000 00000
2890:  00000 00000   00000 00000   00000 00000   00000 00000   00000 00000      00000 00000   00000 00000   00000 00000   00000 00000   00000 00000
2891:  00000 00000   00000 00000   00000 00000   00000 00000   00000 00000      00000 00000   00000 00000   00000 00000   00000 00000   00000 00000
2892:  00000 00000   00000 00000   00000 00000   00000 00000   00000 00000      00000 00000   00000 00000   00000 00000   00000 00000   00000 00000
2893:  00000 00000   00000 00000   00000 00000   00000 00000   00000 00000      00000 00000   00000 00000   00000 00000   00000 00000   00000 00000
2894:  00000 00000   00000 00000   00000 00000   00000 00000   00000 00000      00000 00000   00000 00000   00000 00000   00000 00000   00000 00000
2895:  00000 00000   00000 00000   00000 00000   00000 00000   00000 00000      00000 00000   00000 00000   00000 00000   00000 00000   00000 00000
2896:  00000 00000   00000 00000   00000 00000   00000 00000   00000 00000      00000 00000   00000 00000   00000 00000   00000 00000   00000 00000
2897:  00000 00000   00000 00000   00000 00000   00000 00000   00000 00000      00000 00000   00000 00000   00000 00000   00000 00000   00000 00000
2898:  00000 00000   00000 00000   00000 00000   00000 00000   00000 00000      00000 00000   00000 00000   00000 00000   00000 00000   00000 00000
2899:  00000 00000   00000 00000   00000 00000   00000 00000   00000 00000      00000 00000   00000 00000   00000 00000   00000 00000   00000 00000
```

```
2900:  00000 00000  00000 00000  00000 00000  00000 00000  00000 00000    00000 00000  00000 00000  00000 00000  00000 00000  00000 00000
2901:  00000 00000  00000 00000  00000 00000  00000 00000  00000 00000    00000 00000  00000 00000  00000 00000  00000 00000  00000 00000
2902:  00000 00000  00000 00000  00000 00000  00000 00000  00000 00000    00000 00000  00000 00000  00000 00000  00000 00000  00000 00000
2903:  00000 00000  00000 00000  00000 00000  00000 00000  00000 00000    00000 00000  00000 00000  00000 00000  00000 00000  00000 00000
2904:  00000 00000  00000 00000  00000 00000  00000 00000  00000 00000    00000 00000  00000 00000  00000 00000  00000 00000  00000 00000
2905:  00000 00000  00000 00000  00000 00000  00000 00000  00000 00000    00000 00000  00000 00000  00000 00000  00000 00000  00000 00000
2906:  00000 00000  00000 00000  00000 00000  00000 00000  00000 00000    00000 00000  00000 00000  00000 00000  00000 00000  00000 00000
2907:  00000 00000  00000 00000  00000 00000  00000 00000  00000 00000    00000 00000  00000 00000  00000 00000  00000 00000  00000 00000
2908:  00000 00000  00000 00000  00000 00000  00000 00000  00000 00000    00000 00000  00000 00000  00000 00000  00000 00000  00000 00000
2909:  00000 00000  00000 00000  00000 00000  00000 00000  00000 00000    00000 00000  00000 00000  00000 00000  00000 00000  00000 00000
2910:  00000 00000  00000 00000  00000 00000  00000 00000  00000 00000    00000 00000  00000 00000  00000 00000  00000 00000  00000 00000
2911:  00000 00000  00000 00000  00000 00000  00000 00000  00000 00000    00000 00000  00000 00000  00000 00000  00000 00000  00000 00000
2912:  00000 00000  00000 00000  00000 00000  00000 00000  00000 00000    00000 00000  00000 00000  00000 00000  00000 00000  00000 00000
2913:  00000 00000  00000 00000  00000 00000  00000 00000  00000 00000    00000 00000  00000 00000  00000 00000  00000 00000  00000 00000
2914:  00000 00000  00000 00000  00000 00000  00000 00000  00000 00000    00000 00000  00000 00000  00000 00000  00000 00000  00000 00000
2915:  00000 00000  00000 00000  00000 00000  00000 00000  00000 00000    00000 00000  00000 00000  00000 00000  00000 00000  00000 00000
2916:  00000 00000  00000 00000  00000 00000  00000 00000  00000 00000    00000 00000  00000 00000  00000 00000  00000 00000  00000 00000
2917:  00000 00000  00000 00000  00000 00000  00000 00000  00000 00000    00000 00000  00000 00000  00000 00000  00000 00000  00000 00000
2918:  00000 00000  00000 00000  00000 00000  00000 00000  00000 00000    00000 00000  00000 00000  00000 00000  00000 00000  00000 00000
2919:  00000 00000  00000 00000  00000 00000  00000 00000  00000 00000    00000 00000  00000 00000  00000 00000  00000 00000  00000 00000
2920:  00000 00000  00000 00000  00000 00000  00000 00000  00000 00000    00000 00000  00000 00000  00000 00000  00000 00000  00000 00000
2921:  00000 00000  00000 00000  00000 00000  00000 00000  00000 00000    00000 00000  00000 00000  00000 00000  00000 00000  00000 00000
2922:  00000 00000  00000 00000  00000 00000  00000 00000  00000 00000    00000 00000  00000 00000  00000 00000  00000 00000  00000 00000
2923:  00000 00000  00000 00000  00000 00000  00000 00000  00000 00000    00000 00000  00000 00000  00000 00000  00000 00000  00000 00000
2924:  00000 00000  00000 00000  00000 00000  00000 00000  00000 00000    00000 00000  00000 00000  00000 00000  00000 00000  00000 00000
2925:  00000 00000  00000 00000  00000 00000  00000 00000  00000 00000    00000 00000  00000 00000  00000 00000  00000 00000  00000 00000
2926:  00000 00000  00000 00000  00000 00000  00000 00000  00000 00000    00000 00000  00000 00000  00000 00000  00000 00000  00000 00000
2927:  00000 00000  00000 00000  00000 00000  00000 00000  00000 00000    00000 00000  00000 00000  00000 00000  00000 00000  00000 00000
2928:  00000 00000  00000 00000  00000 00000  00000 00000  00000 00000    00000 00000  00000 00000  00000 00000  00000 00000  00000 00000
2929:  00000 00000  00000 00000  00000 00000  00000 00000  00000 00000    00000 00000  00000 00000  00000 00000  00000 00000  00000 00000
2930:  00000 00000  00000 00000  00000 00000  00000 00000  00000 00000    00000 00000  00000 00000  00000 00000  00000 00000  00000 00000
2931:  00000 00000  00000 00000  00000 00000  00000 00000  00000 00000    00000 00000  00000 00000  00000 00000  00000 00000  00000 00000
2932:  00000 00000  00000 00000  00000 00000  00000 00000  00000 00000    00000 00000  00000 00000  00000 00000  00000 00000  00000 00000
2933:  00000 00000  00000 00000  00000 00000  00000 00000  00000 00000    00000 00000  00000 00000  00000 00000  00000 00000  00000 00000
2934:  00000 00000  00000 00000  00000 00000  00000 00000  00000 00000    00000 00000  00000 00000  00000 00000  00000 00000  00000 00000
2935:  00000 00000  00000 00000  00000 00000  00000 00000  00000 00000    00000 00000  00000 00000  00000 00000  00000 00000  00000 00000
2936:  00000 00000  00000 00000  00000 00000  00000 00000  00000 00000    00000 00000  00000 00000  00000 00000  00000 00000  00000 00000
2937:  00000 00000  00000 00000  00000 00000  00000 00000  00000 00000    00000 00000  00000 00000  00000 00000  00000 00000  00000 00000
2938:  00000 00000  00000 00000  00000 00000  00000 00000  00000 00000    00000 00000  00000 00000  00000 00000  00000 00000  00000 00000
2939:  00000 00000  00000 00000  00000 00000  00000 00000  00000 00000    00000 00000  00000 00000  00000 00000  00000 00000  00000 00000
2940:  00000 00000  00000 00000  00000 00000  00000 00000  00000 00000    00000 00000  00000 00000  00000 00000  00000 00000  00000 00000
2941:  00000 00000  00000 00000  00000 00000  00000 00000  00000 00000    00000 00000  00000 00000  00000 00000  00000 00000  00000 00000
2942:  00000 00000  00000 00000  00000 00000  00000 00000  00000 00000    00000 00000  00000 00000  00000 00000  00000 00000  00000 00000
2943:  00000 00000  00000 00000  00000 00000  00000 00000  00000 00000    00000 00000  00000 00000  00000 00000  00000 00000  00000 00000
2944:  00000 00000  00000 00000  00000 00000  00000 00000  00000 00000    00000 00000  00000 00000  00000 00000  00000 00000  00000 00000
2945:  00000 00000  00000 00000  00000 00000  00000 00000  00000 00000    00000 00000  00000 00000  00000 00000  00000 00000  00000 00000
2946:  00000 00000  00000 00000  00000 00000  00000 00000  00000 00000    00000 00000  00000 00000  00000 00000  00000 00000  00000 00000
2947:  00000 00000  00000 00000  00000 00000  00000 00000  00000 00000    00000 00000  00000 00000  00000 00000  00000 00000  00000 00000
2948:  00000 00000  00000 00000  00000 00000  00000 00000  00000 00000    00000 00000  00000 00000  00000 00000  00000 00000  00000 00000
2949:  00000 00000  00000 00000  00000 00000  00000 00000  00000 00000    00000 00000  00000 00000  00000 00000  00000 00000  00000 00000
```

```
2950:  00000 00000  00000 00000  00000 00000  00000 00000  00000 00000    00000 00000  00000 00000  00000 00000  00000 00000  00000 00000
2951:  00000 00000  00000 00000  00000 00000  00000 00000  00000 00000    00000 00000  00000 00000  00000 00000  00000 00000  00000 00000
2952:  00000 00000  00000 00000  00000 00000  00000 00000  00000 00000    00000 00000  00000 00000  00000 00000  00000 00000  00000 00000
2953:  00000 00000  00000 00000  00000 00000  00000 00000  00000 00000    00000 00000  00000 00000  00000 00000  00000 00000  00000 00000
2954:  00000 00000  00000 00000  00000 00000  00000 00000  00000 00000    00000 00000  00000 00000  00000 00000  00000 00000  00000 00000
2955:  00000 00000  00000 00000  00000 00000  00000 00000  00000 00000    00000 00000  00000 00000  00000 00000  00000 00000  00000 00000
2956:  00000 00000  00000 00000  00000 00000  00000 00000  00000 00000    00000 00000  00000 00000  00000 00000  00000 00000  00000 00000
2957:  00000 00000  00000 00000  00000 00000  00000 00000  00000 00000    00000 00000  00000 00000  00000 00000  00000 00000  00000 00000
2958:  00000 00000  00000 00000  00000 00000  00000 00000  00000 00000    00000 00000  00000 00000  00000 00000  00000 00000  00000 00000
2959:  00000 00000  00000 00000  00000 00000  00000 00000  00000 00000    00000 00000  00000 00000  00000 00000  00000 00000  00000 00000
2960:  00000 00000  00000 00000  00000 00000  00000 00000  00000 00000    00000 00000  00000 00000  00000 00000  00000 00000  00000 00000
2961:  00000 00000  00000 00000  00000 00000  00000 00000  00000 00000    00000 00000  00000 00000  00000 00000  00000 00000  00000 00000
2962:  00000 00000  00000 00000  00000 00000  00000 00000  00000 00000    00000 00000  00000 00000  00000 00000  00000 00000  00000 00000
2963:  00000 00000  00000 00000  00000 00000  00000 00000  00000 00000    00000 00000  00000 00000  00000 00000  00000 00000  00000 00000
2964:  00000 00000  00000 00000  00000 00000  00000 00000  00000 00000    00000 00000  00000 00000  00000 00000  00000 00000  00000 00000
2965:  00000 00000  00000 00000  00000 00000  00000 00000  00000 00000    00000 00000  00000 00000  00000 00000  00000 00000  00000 00000
2966:  00000 00000  00000 00000  00000 00000  00000 00000  00000 00000    00000 00000  00000 00000  00000 00000  00000 00000  00000 00000
2967:  00000 00000  00000 00000  00000 00000  00000 00000  00000 00000    00000 00000  00000 00000  00000 00000  00000 00000  00000 00000
2968:  00000 00000  00000 00000  00000 00000  00000 00000  00000 00000    00000 00000  00000 00000  00000 00000  00000 00000  00000 00000
2969:  00000 00000  00000 00000  00000 00000  00000 00000  00000 00000    00000 00000  00000 00000  00000 00000  00000 00000  00000 00000
2970:  00000 00000  00000 00000  00000 00000  00000 00000  00000 00000    00000 00000  00000 00000  00000 00000  00000 00000  00000 00000
2971:  00000 00000  00000 00000  00000 00000  00000 00000  00000 00000    00000 00000  00000 00000  00000 00000  00000 00000  00000 00000
2972:  00000 00000  00000 00000  00000 00000  00000 00000  00000 00000    00000 00000  00000 00000  00000 00000  00000 00000  00000 00000
2973:  00000 00000  00000 00000  00000 00000  00000 00000  00000 00000    00000 00000  00000 00000  00000 00000  00000 00000  00000 00000
2974:  00000 00000  00000 00000  00000 00000  00000 00000  00000 00000    00000 00000  00000 00000  00000 00000  00000 00000  00000 00000
2975:  00000 00000  00000 00000  00000 00000  00000 00000  00000 00000    00000 00000  00000 00000  00000 00000  00000 00000  00000 00000
2976:  00000 00000  00000 00000  00000 00000  00000 00000  00000 00000    00000 00000  00000 00000  00000 00000  00000 00000  00000 00000
2977:  00000 00000  00000 00000  00000 00000  00000 00000  00000 00000    00000 00000  00000 00000  00000 00000  00000 00000  00000 00000
2978:  00000 00000  00000 00000  00000 00000  00000 00000  00000 00000    00000 00000  00000 00000  00000 00000  00000 00000  00000 00000
2979:  00000 00000  00000 00000  00000 00000  00000 00000  00000 00000    00000 00000  00000 00000  00000 00000  00000 00000  00000 00000
2980:  00000 00000  00000 00000  00000 00000  00000 00000  00000 00000    00000 00000  00000 00000  00000 00000  00000 00000  00000 00000
2981:  00000 00000  00000 00000  00000 00000  00000 00000  00000 00000    00000 00000  00000 00000  00000 00000  00000 00000  00000 00000
2982:  00000 00000  00000 00000  00000 00000  00000 00000  00000 00000    00000 00000  00000 00000  00000 00000  00000 00000  00000 00000
2983:  00000 00000  00000 00000  00000 00000  00000 00000  00000 00000    00000 00000  00000 00000  00000 00000  00000 00000  00000 00000
2984:  00000 00000  00000 00000  00000 00000  00000 00000  00000 00000    00000 00000  00000 00000  00000 00000  00000 00000  00000 00000
2985:  00000 00000  00000 00000  00000 00000  00000 00000  00000 00000    00000 00000  00000 00000  00000 00000  00000 00000  00000 00000
2986:  00000 00000  00000 00000  00000 00000  00000 00000  00000 00000    00000 00000  00000 00000  00000 00000  00000 00000  00000 00000
2987:  00000 00000  00000 00000  00000 00000  00000 00000  00000 00000    00000 00000  00000 00000  00000 00000  00000 00000  00000 00000
2988:  00000 00000  00000 00000  00000 00000  00000 00000  00000 00000    00000 00000  00000 00000  00000 00000  00000 00000  00000 00000
2989:  00000 00000  00000 00000  00000 00000  00000 00000  00000 00000    00000 00000  00000 00000  00000 00000  00000 00000  00000 00000
2990:  00000 00000  00000 00000  00000 00000  00000 00000  00000 00000    00000 00000  00000 00000  00000 00000  00000 00000  00000 00000
2991:  00000 00000  00000 00000  00000 00000  00000 00000  00000 00000    00000 00000  00000 00000  00000 00000  00000 00000  00000 00000
2992:  00000 00000  00000 00000  00000 00000  00000 00000  00000 00000    00000 00000  00000 00000  00000 00000  00000 00000  00000 00000
2993:  00000 00000  00000 00000  00000 00000  00000 00000  00000 00000    00000 00000  00000 00000  00000 00000  00000 00000  00000 00000
2994:  00000 00000  00000 00000  00000 00000  00000 00000  00000 00000    00000 00000  00000 00000  00000 00000  00000 00000  00000 00000
2995:  00000 00000  00000 00000  00000 00000  00000 00000  00000 00000    00000 00000  00000 00000  00000 00000  00000 00000  00000 00000
2996:  00000 00000  00000 00000  00000 00000  00000 00000  00000 00000    00000 00000  00000 00000  00000 00000  00000 00000  00000 00000
2997:  00000 00000  00000 00000  00000 00000  00000 00000  00000 00000    00000 00000  00000 00000  00000 00000  00000 00000  00000 00000
2998:  00000 00000  00000 00000  00000 00000  00000 00000  00000 00000    00000 00000  00000 00000  00000 00000  00000 00000  00000 00000
2999:  00000 00000  00000 00000  00000 00000  00000 00000  00000 00000    00000 00000  00000 00000  00000 00000  00000 00000  00000 00000
```

```
3000:  00000 00000   00000 00000   00000 00000   00000 00000   00000 00000     00000 00000   00000 00000   00000 00000   00000 00000   00000 00000
3001:  00000 00000   00000 00000   00000 00000   00000 00000   00000 00000     00000 00000   00000 00000   00000 00000   00000 00000   00000 00000
3002:  00000 00000   00000 00000   00000 00000   00000 00000   00000 00000     00000 00000   00000 00000   00000 00000   00000 00000   00000 00000
3003:  00000 00000   00000 00000   00000 00000   00000 00000   00000 00000     00000 00000   00000 00000   00000 00000   00000 00000   00000 00000
3004:  00000 00000   00000 00000   00000 00000   00000 00000   00000 00000     00000 00000   00000 00000   00000 00000   00000 00000   00000 00000
3005:  00000 00000   00000 00000   00000 00000   00000 00000   00000 00000     00000 00000   00000 00000   00000 00000   00000 00000   00000 00000
3006:  00000 00000   00000 00000   00000 00000   00000 00000   00000 00000     00000 00000   00000 00000   00000 00000   00000 00000   00000 00000
3007:  00000 00000   00000 00000   00000 00000   00000 00000   00000 00000     00000 00000   00000 00000   00000 00000   00000 00000   00000 00000
3008:  00000 00000   00000 00000   00000 00000   00000 00000   00000 00000     00000 00000   00000 00000   00000 00000   00000 00000   00000 00000
3009:  00000 00000   00000 00000   00000 00000   00000 00000   00000 00000     00000 00000   00000 00000   00000 00000   00000 00000   00000 00000
3010:  00000 00000   00000 00000   00000 00000   00000 00000   00000 00000     00000 00000   00000 00000   00000 00000   00000 00000   00000 00000
3011:  00000 00000   00000 00000   00000 00000   00000 00000   00000 00000     00000 00000   00000 00000   00000 00000   00000 00000   00000 00000
3012:  00000 00000   00000 00000   00000 00000   00000 00000   00000 00000     00000 00000   00000 00000   00000 00000   00000 00000   00000 00000
3013:  00000 00000   00000 00000   00000 00000   00000 00000   00000 00000     00000 00000   00000 00000   00000 00000   00000 00000   00000 00000
3014:  00000 00000   00000 00000   00000 00000   00000 00000   00000 00000     00000 00000   00000 00000   00000 00000   00000 00000   00000 00000
3015:  00000 00000   00000 00000   00000 00000   00000 00000   00000 00000     00000 00000   00000 00000   00000 00000   00000 00000   00000 00000
3016:  00000 00000   00000 00000   00000 00000   00000 00000   00000 00000     00000 00000   00000 00000   00000 00000   00000 00000   00000 00000
3017:  00000 00000   00000 00000   00000 00000   00000 00000   00000 00000     00000 00000   00000 00000   00000 00000   00000 00000   00000 00000
3018:  00000 00000   00000 00000   00000 00000   00000 00000   00000 00000     00000 00000   00000 00000   00000 00000   00000 00000   00000 00000
3019:  00000 00000   00000 00000   00000 00000   00000 00000   00000 00000     00000 00000   00000 00000   00000 00000   00000 00000   00000 00000
3020:  00000 00000   00000 00000   00000 00000   00000 00000   00000 00000     00000 00000   00000 00000   00000 00000   00000 00000   00000 00000
3021:  00000 00000   00000 00000   00000 00000   00000 00000   00000 00000     00000 00000   00000 00000   00000 00000   00000 00000   00000 00000
3022:  00000 00000   00000 00000   00000 00000   00000 00000   00000 00000     00000 00000   00000 00000   00000 00000   00000 00000   00000 00000
3023:  00000 00000   00000 00000   00000 00000   00000 00000   00000 00000     00000 00000   00000 00000   00000 00000   00000 00000   00000 00000
3024:  00000 00000   00000 00000   00000 00000   00000 00000   00000 00000     00000 00000   00000 00000   00000 00000   00000 00000   00000 00000
3025:  00000 00000   00000 00000   00000 00000   00000 00000   00000 00000     00000 00000   00000 00000   00000 00000   00000 00000   00000 00000
3026:  00000 00000   00000 00000   00000 00000   00000 00000   00000 00000     00000 00000   00000 00000   00000 00000   00000 00000   00000 00000
3027:  00000 00000   00000 00000   00000 00000   00000 00000   00000 00000     00000 00000   00000 00000   00000 00000   00000 00000   00000 00000
3028:  00000 00000   00000 00000   00000 00000   00000 00000   00000 00000     00000 00000   00000 00000   00000 00000   00000 00000   00000 00000
3029:  00000 00000   00000 00000   00000 00000   00000 00000   00000 00000     00000 00000   00000 00000   00000 00000   00000 00000   00000 00000
3030:  00000 00000   00000 00000   00000 00000   00000 00000   00000 00000     00000 00000   00000 00000   00000 00000   00000 00000   00000 00000
3031:  00000 00000   00000 00000   00000 00000   00000 00000   00000 00000     00000 00000   00000 00000   00000 00000   00000 00000   00000 00000
3032:  00000 00000   00000 00000   00000 00000   00000 00000   00000 00000     00000 00000   00000 00000   00000 00000   00000 00000   00000 00000
3033:  00000 00000   00000 00000   00000 00000   00000 00000   00000 00000     00000 00000   00000 00000   00000 00000   00000 00000   00000 00000
3034:  00000 00000   00000 00000   00000 00000   00000 00000   00000 00000     00000 00000   00000 00000   00000 00000   00000 00000   00000 00000
3035:  00000 00000   00000 00000   00000 00000   00000 00000   00000 00000     00000 00000   00000 00000   00000 00000   00000 00000   00000 00000
3036:  00000 00000   00000 00000   00000 00000   00000 00000   00000 00000     00000 00000   00000 00000   00000 00000   00000 00000   00000 00000
3037:  00000 00000   00000 00000   00000 00000   00000 00000   00000 00000     00000 00000   00000 00000   00000 00000   00000 00000   00000 00000
3038:  00000 00000   00000 00000   00000 00000   00000 00000   00000 00000     00000 00000   00000 00000   00000 00000   00000 00000   00000 00000
3039:  00000 00000   00000 00000   00000 00000   00000 00000   00000 00000     00000 00000   00000 00000   00000 00000   00000 00000   00000 00000
3040:  00000 00000   00000 00000   00000 00000   00000 00000   00000 00000     00000 00000   00000 00000   00000 00000   00000 00000   00000 00000
3041:  00000 00000   00000 00000   00000 00000   00000 00000   00000 00000     00000 00000   00000 00000   00000 00000   00000 00000   00000 00000
3042:  00000 00000   00000 00000   00000 00000   00000 00000   00000 00000     00000 00000   00000 00000   00000 00000   00000 00000   00000 00000
3043:  00000 00000   00000 00000   00000 00000   00000 00000   00000 00000     00000 00000   00000 00000   00000 00000   00000 00000   00000 00000
3044:  00000 00000   00000 00000   00000 00000   00000 00000   00000 00000     00000 00000   00000 00000   00000 00000   00000 00000   00000 00000
3045:  00000 00000   00000 00000   00000 00000   00000 00000   00000 00000     00000 00000   00000 00000   00000 00000   00000 00000   00000 00000
3046:  00000 00000   00000 00000   00000 00000   00000 00000   00000 00000     00000 00000   00000 00000   00000 00000   00000 00000   00000 00000
3047:  00000 00000   00000 00000   00000 00000   00000 00000   00000 00000     00000 00000   00000 00000   00000 00000   00000 00000   00000 00000
3048:  00000 00000   00000 00000   00000 00000   00000 00000   00000 00000     00000 00000   00000 00000   00000 00000   00000 00000   00000 00000
3049:  00000 00000   00000 00000   00000 00000   00000 00000   00000 00000     00000 00000   00000 00000   00000 00000   00000 00000   00000 00000
```

```
3050:  00000 00000   00000 00000   00000 00000   00000 00000   00000 00000     00000 00000   00000 00000   00000 00000   00000 00000   00000 00000
3051:  00000 00000   00000 00000   00000 00000   00000 00000   00000 00000     00000 00000   00000 00000   00000 00000   00000 00000   00000 00000
3052:  00000 00000   00000 00000   00000 00000   00000 00000   00000 00000     00000 00000   00000 00000   00000 00000   00000 00000   00000 00000
3053:  00000 00000   00000 00000   00000 00000   00000 00000   00000 00000     00000 00000   00000 00000   00000 00000   00000 00000   00000 00000
3054:  00000 00000   00000 00000   00000 00000   00000 00000   00000 00000     00000 00000   00000 00000   00000 00000   00000 00000   00000 00000
3055:  00000 00000   00000 00000   00000 00000   00000 00000   00000 00000     00000 00000   00000 00000   00000 00000   00000 00000   00000 00000
3056:  00000 00000   00000 00000   00000 00000   00000 00000   00000 00000     00000 00000   00000 00000   00000 00000   00000 00000   00000 00000
3057:  00000 00000   00000 00000   00000 00000   00000 00000   00000 00000     00000 00000   00000 00000   00000 00000   00000 00000   00000 00000
3058:  00000 00000   00000 00000   00000 00000   00000 00000   00000 00000     00000 00000   00000 00000   00000 00000   00000 00000   00000 00000
3059:  00000 00000   00000 00000   00000 00000   00000 00000   00000 00000     00000 00000   00000 00000   00000 00000   00000 00000   00000 00000
3060:  00000 00000   00000 00000   00000 00000   00000 00000   00000 00000     00000 00000   00000 00000   00000 00000   00000 00000   00000 00000
3061:  00000 00000   00000 00000   00000 00000   00000 00000   00000 00000     00000 00000   00000 00000   00000 00000   00000 00000   00000 00000
3062:  00000 00000   00000 00000   00000 00000   00000 00000   00000 00000     00000 00000   00000 00000   00000 00000   00000 00000   00000 00000
3063:  00000 00000   00000 00000   00000 00000   00000 00000   00000 00000     00000 00000   00000 00000   00000 00000   00000 00000   00000 00000
3064:  00000 00000   00000 00000   00000 00000   00000 00000   00000 00000     00000 00000   00000 00000   00000 00000   00000 00000   00000 00000
3065:  00000 00000   00000 00000   00000 00000   00000 00000   00000 00000     00000 00000   00000 00000   00000 00000   00000 00000   00000 00000
3066:  00000 00000   00000 00000   00000 00000   00000 00000   00000 00000     00000 00000   00000 00000   00000 00000   00000 00000   00000 00000
3067:  00000 00000   00000 00000   00000 00000   00000 00000   00000 00000     00000 00000   00000 00000   00000 00000   00000 00000   00000 00000
3068:  00000 00000   00000 00000   00000 00000   00000 00000   00000 00000     00000 00000   00000 00000   00000 00000   00000 00000   00000 00000
3069:  00000 00000   00000 00000   00000 00000   00000 00000   00000 00000     00000 00000   00000 00000   00000 00000   00000 00000   00000 00000
3070:  00000 00000   00000 00000   00000 00000   00000 00000   00000 00000     00000 00000   00000 00000   00000 00000   00000 00000   00000 00000
3071:  00000 00000   00000 00000   00000 00000   00000 00000   00000 00000     00000 00000   00000 00000   00000 00000   00000 00000   00000 00000
3072:  00000 00000   00000 00000   00000 00000   00000 00000   00000 00000     00000 00000   00000 00000   00000 00000   00000 00000   00000 00000
3073:  00000 00000   00000 00000   00000 00000   00000 00000   00000 00000     00000 00000   00000 00000   00000 00000   00000 00000   00000 00000
3074:  00000 00000   00000 00000   00000 00000   00000 00000   00000 00000     00000 00000   00000 00000   00000 00000   00000 00000   00000 00000
3075:  00000 00000   00000 00000   00000 00000   00000 00000   00000 00000     00000 00000   00000 00000   00000 00000   00000 00000   00000 00000
3076:  00000 00000   00000 00000   00000 00000   00000 00000   00000 00000     00000 00000   00000 00000   00000 00000   00000 00000   00000 00000
3077:  00000 00000   00000 00000   00000 00000   00000 00000   00000 00000     00000 00000   00000 00000   00000 00000   00000 00000   00000 00000
3078:  00000 00000   00000 00000   00000 00000   00000 00000   00000 00000     00000 00000   00000 00000   00000 00000   00000 00000   00000 00000
3079:  00000 00000   00000 00000   00000 00000   00000 00000   00000 00000     00000 00000   00000 00000   00000 00000   00000 00000   00000 00000
3080:  00000 00000   00000 00000   00000 00000   00000 00000   00000 00000     00000 00000   00000 00000   00000 00000   00000 00000   00000 00000
3081:  00000 00000   00000 00000   00000 00000   00000 00000   00000 00000     00000 00000   00000 00000   00000 00000   00000 00000   00000 00000
3082:  00000 00000   00000 00000   00000 00000   00000 00000   00000 00000     00000 00000   00000 00000   00000 00000   00000 00000   00000 00000
3083:  00000 00000   00000 00000   00000 00000   00000 00000   00000 00000     00000 00000   00000 00000   00000 00000   00000 00000   00000 00000
3084:  00000 00000   00000 00000   00000 00000   00000 00000   00000 00000     00000 00000   00000 00000   00000 00000   00000 00000   00000 00000
3085:  00000 00000   00000 00000   00000 00000   00000 00000   00000 00000     00000 00000   00000 00000   00000 00000   00000 00000   00000 00000
3086:  00000 00000   00000 00000   00000 00000   00000 00000   00000 00000     00000 00000   00000 00000   00000 00000   00000 00000   00000 00000
3087:  00000 00000   00000 00000   00000 00000   00000 00000   00000 00000     00000 00000   00000 00000   00000 00000   00000 00000   00000 00000
3088:  00000 00000   00000 00000   00000 00000   00000 00000   00000 00000     00000 00000   00000 00000   00000 00000   00000 00000   00000 00000
3089:  00000 00000   00000 00000   00000 00000   00000 00000   00000 00000     00000 00000   00000 00000   00000 00000   00000 00000   00000 00000
3090:  00000 00000   00000 00000   00000 00000   00000 00000   00000 00000     00000 00000   00000 00000   00000 00000   00000 00000   00000 00000
3091:  00000 00000   00000 00000   00000 00000   00000 00000   00000 00000     00000 00000   00000 00000   00000 00000   00000 00000   00000 00000
3092:  00000 00000   00000 00000   00000 00000   00000 00000   00000 00000     00000 00000   00000 00000   00000 00000   00000 00000   00000 00000
3093:  00000 00000   00000 00000   00000 00000   00000 00000   00000 00000     00000 00000   00000 00000   00000 00000   00000 00000   00000 00000
3094:  00000 00000   00000 00000   00000 00000   00000 00000   00000 00000     00000 00000   00000 00000   00000 00000   00000 00000   00000 00000
3095:  00000 00000   00000 00000   00000 00000   00000 00000   00000 00000     00000 00000   00000 00000   00000 00000   00000 00000   00000 00000
3096:  00000 00000   00000 00000   00000 00000   00000 00000   00000 00000     00000 00000   00000 00000   00000 00000   00000 00000   00000 00000
3097:  00000 00000   00000 00000   00000 00000   00000 00000   00000 00000     00000 00000   00000 00000   00000 00000   00000 00000   00000 00000
3098:  00000 00000   00000 00000   00000 00000   00000 00000   00000 00000     00000 00000   00000 00000   00000 00000   00000 00000   00000 00000
3099:  00000 00000   00000 00000   00000 00000   00000 00000   00000 00000     00000 00000   00000 00000   00000 00000   00000 00000   00000 00000
```

```
3100:   00000 00000   00000 00000   00000 00000   00000 00000   00000 00000     00000 00000   00000 00000   00000 00000   00000 00000   00000 00000
3101:   00000 00000   00000 00000   00000 00000   00000 00000   00000 00000     00000 00000   00000 00000   00000 00000   00000 00000   00000 00000
3102:   00000 00000   00000 00000   00000 00000   00000 00000   00000 00000     00000 00000   00000 00000   00000 00000   00000 00000   00000 00000
3103:   00000 00000   00000 00000   00000 00000   00000 00000   00000 00000     00000 00000   00000 00000   00000 00000   00000 00000   00000 00000
3104:   00000 00000   00000 00000   00000 00000   00000 00000   00000 00000     00000 00000   00000 00000   00000 00000   00000 00000   00000 00000
3105:   00000 00000   00000 00000   00000 00000   00000 00000   00000 00000     00000 00000   00000 00000   00000 00000   00000 00000   00000 00000
3106:   00000 00000   00000 00000   00000 00000   00000 00000   00000 00000     00000 00000   00000 00000   00000 00000   00000 00000   00000 00000
3107:   00000 00000   00000 00000   00000 00000   00000 00000   00000 00000     00000 00000   00000 00000   00000 00000   00000 00000   00000 00000
3108:   00000 00000   00000 00000   00000 00000   00000 00000   00000 00000     00000 00000   00000 00000   00000 00000   00000 00000   00000 00000
3109:   00000 00000   00000 00000   00000 00000   00000 00000   00000 00000     00000 00000   00000 00000   00000 00000   00000 00000   00000 00000
3110:   00000 00000   00000 00000   00000 00000   00000 00000   00000 00000     00000 00000   00000 00000   00000 00000   00000 00000   00000 00000
3111:   00000 00000   00000 00000   00000 00000   00000 00000   00000 00000     00000 00000   00000 00000   00000 00000   00000 00000   00000 00000
3112:   00000 00000   00000 00000   00000 00000   00000 00000   00000 00000     00000 00000   00000 00000   00000 00000   00000 00000   00000 00000
3113:   00000 00000   00000 00000   00000 00000   00000 00000   00000 00000     00000 00000   00000 00000   00000 00000   00000 00000   00000 00000
3114:   00000 00000   00000 00000   00000 00000   00000 00000   00000 00000     00000 00000   00000 00000   00000 00000   00000 00000   00000 00000
3115:   00000 00000   00000 00000   00000 00000   00000 00000   00000 00000     00000 00000   00000 00000   00000 00000   00000 00000   00000 00000
3116:   00000 00000   00000 00000   00000 00000   00000 00000   00000 00000     00000 00000   00000 00000   00000 00000   00000 00000   00000 00000
3117:   00000 00000   00000 00000   00000 00000   00000 00000   00000 00000     00000 00000   00000 00000   00000 00000   00000 00000   00000 00000
3118:   00000 00000   00000 00000   00000 00000   00000 00000   00000 00000     00000 00000   00000 00000   00000 00000   00000 00000   00000 00000
3119:   00000 00000   00000 00000   00000 00000   00000 00000   00000 00000     00000 00000   00000 00000   00000 00000   00000 00000   00000 00000
3120:   00000 00000   00000 00000   00000 00000   00000 00000   00000 00000     00000 00000   00000 00000   00000 00000   00000 00000   00000 00000
3121:   00000 00000   00000 00000   00000 00000   00000 00000   00000 00000     00000 00000   00000 00000   00000 00000   00000 00000   00000 00000
3122:   00000 00000   00000 00000   00000 00000   00000 00000   00000 00000     00000 00000   00000 00000   00000 00000   00000 00000   00000 00000
3123:   00000 00000   00000 00000   00000 00000   00000 00000   00000 00000     00000 00000   00000 00000   00000 00000   00000 00000   00000 00000
3124:   00000 00000   00000 00000   00000 00000   00000 00000   00000 00000     00000 00000   00000 00000   00000 00000   00000 00000   00000 00000
3125:   00000 00000   00000 00000   00000 00000   00000 00000   00000 00000     00000 00000   00000 00000   00000 00000   00000 00000   00000 00000
3126:   00000 00000   00000 00000   00000 00000   00000 00000   00000 00000     00000 00000   00000 00000   00000 00000   00000 00000   00000 00000
3127:   00000 00000   00000 00000   00000 00000   00000 00000   00000 00000     00000 00000   00000 00000   00000 00000   00000 00000   00000 00000
3128:   00000 00000   00000 00000   00000 00000   00000 00000   00000 00000     00000 00000   00000 00000   00000 00000   00000 00000   00000 00000
3129:   00000 00000   00000 00000   00000 00000   00000 00000   00000 00000     00000 00000   00000 00000   00000 00000   00000 00000   00000 00000
3130:   00000 00000   00000 00000   00000 00000   00000 00000   00000 00000     00000 00000   00000 00000   00000 00000   00000 00000   00000 00000
3131:   00000 00000   00000 00000   00000 00000   00000 00000   00000 00000     00000 00000   00000 00000   00000 00000   00000 00000   00000 00000
3132:   00000 00000   00000 00000   00000 00000   00000 00000   00000 00000     00000 00000   00000 00000   00000 00000   00000 00000   00000 00000
3133:   00000 00000   00000 00000   00000 00000   00000 00000   00000 00000     00000 00000   00000 00000   00000 00000   00000 00000   00000 00000
3134:   00000 00000   00000 00000   00000 00000   00000 00000   00000 00000     00000 00000   00000 00000   00000 00000   00000 00000   00000 00000
3135:   00000 00000   00000 00000   00000 00000   00000 00000   00000 00000     00000 00000   00000 00000   00000 00000   00000 00000   00000 00000
3136:   00000 00000   00000 00000   00000 00000   00000 00000   00000 00000     00000 00000   00000 00000   00000 00000   00000 00000   00000 00000
3137:   00000 00000   00000 00000   00000 00000   00000 00000   00000 00000     00000 00000   00000 00000   00000 00000   00000 00000   00000 00000
3138:   00000 00000   00000 00000   00000 00000   00000 00000   00000 00000     00000 00000   00000 00000   00000 00000   00000 00000   00000 00000
3139:   00000 00000   00000 00000   00000 00000   00000 00000   00000 00000     00000 00000   00000 00000   00000 00000   00000 00000   00000 00000
3140:   00000 00000   00000 00000   00000 00000   00000 00000   00000 00000     00000 00000   00000 00000   00000 00000   00000 00000   00000 00000
3141:   00000 00000   00000 00000   00000 00000   00000 00000   00000 00000     00000 00000   00000 00000   00000 00000   00000 00000   00000 00000
3142:   00000 00000   00000 00000   00000 00000   00000 00000   00000 00000     00000 00000   00000 00000   00000 00000   00000 00000   00000 00000
3143:   00000 00000   00000 00000   00000 00000   00000 00000   00000 00000     00000 00000   00000 00000   00000 00000   00000 00000   00000 00000
3144:   00000 00000   00000 00000   00000 00000   00000 00000   00000 00000     00000 00000   00000 00000   00000 00000   00000 00000   00000 00000
3145:   00000 00000   00000 00000   00000 00000   00000 00000   00000 00000     00000 00000   00000 00000   00000 00000   00000 00000   00000 00000
3146:   00000 00000   00000 00000   00000 00000   00000 00000   00000 00000     00000 00000   00000 00000   00000 00000   00000 00000   00000 00000
3147:   00000 00000   00000 00000   00000 00000   00000 00000   00000 00000     00000 00000   00000 00000   00000 00000   00000 00000   00000 00000
3148:   00000 00000   00000 00000   00000 00000   00000 00000   00000 00000     00000 00000   00000 00000   00000 00000   00000 00000   00000 00000
3149:   00000 00000   00000 00000   00000 00000   00000 00000   00000 00000     00000 00000   00000 00000   00000 00000   00000 00000   00000 00000
```

```
3150:  00000 00000   00000 00000   00000 00000   00000 00000   00000 00000     00000 00000   00000 00000   00000 00000   00000 00000   00000 00000
3151:  00000 00000   00000 00000   00000 00000   00000 00000   00000 00000     00000 00000   00000 00000   00000 00000   00000 00000   00000 00000
3152:  00000 00000   00000 00000   00000 00000   00000 00000   00000 00000     00000 00000   00000 00000   00000 00000   00000 00000   00000 00000
3153:  00000 00000   00000 00000   00000 00000   00000 00000   00000 00000     00000 00000   00000 00000   00000 00000   00000 00000   00000 00000
3154:  00000 00000   00000 00000   00000 00000   00000 00000   00000 00000     00000 00000   00000 00000   00000 00000   00000 00000   00000 00000
3155:  00000 00000   00000 00000   00000 00000   00000 00000   00000 00000     00000 00000   00000 00000   00000 00000   00000 00000   00000 00000
3156:  00000 00000   00000 00000   00000 00000   00000 00000   00000 00000     00000 00000   00000 00000   00000 00000   00000 00000   00000 00000
3157:  00000 00000   00000 00000   00000 00000   00000 00000   00000 00000     00000 00000   00000 00000   00000 00000   00000 00000   00000 00000
3158:  00000 00000   00000 00000   00000 00000   00000 00000   00000 00000     00000 00000   00000 00000   00000 00000   00000 00000   00000 00000
3159:  00000 00000   00000 00000   00000 00000   00000 00000   00000 00000     00000 00000   00000 00000   00000 00000   00000 00000   00000 00000
3160:  00000 00000   00000 00000   00000 00000   00000 00000   00000 00000     00000 00000   00000 00000   00000 00000   00000 00000   00000 00000
3161:  00000 00000   00000 00000   00000 00000   00000 00000   00000 00000     00000 00000   00000 00000   00000 00000   00000 00000   00000 00000
3162:  00000 00000   00000 00000   00000 00000   00000 00000   00000 00000     00000 00000   00000 00000   00000 00000   00000 00000   00000 00000
3163:  00000 00000   00000 00000   00000 00000   00000 00000   00000 00000     00000 00000   00000 00000   00000 00000   00000 00000   00000 00000
3164:  00000 00000   00000 00000   00000 00000   00000 00000   00000 00000     00000 00000   00000 00000   00000 00000   00000 00000   00000 00000
3165:  00000 00000   00000 00000   00000 00000   00000 00000   00000 00000     00000 00000   00000 00000   00000 00000   00000 00000   00000 00000
3166:  00000 00000   00000 00000   00000 00000   00000 00000   00000 00000     00000 00000   00000 00000   00000 00000   00000 00000   00000 00000
3167:  00000 00000   00000 00000   00000 00000   00000 00000   00000 00000     00000 00000   00000 00000   00000 00000   00000 00000   00000 00000
3168:  00000 00000   00000 00000   00000 00000   00000 00000   00000 00000     00000 00000   00000 00000   00000 00000   00000 00000   00000 00000
3169:  00000 00000   00000 00000   00000 00000   00000 00000   00000 00000     00000 00000   00000 00000   00000 00000   00000 00000   00000 00000
3170:  00000 00000   00000 00000   00000 00000   00000 00000   00000 00000     00000 00000   00000 00000   00000 00000   00000 00000   00000 00000
3171:  00000 00000   00000 00000   00000 00000   00000 00000   00000 00000     00000 00000   00000 00000   00000 00000   00000 00000   00000 00000
3172:  00000 00000   00000 00000   00000 00000   00000 00000   00000 00000     00000 00000   00000 00000   00000 00000   00000 00000   00000 00000
3173:  00000 00000   00000 00000   00000 00000   00000 00000   00000 00000     00000 00000   00000 00000   00000 00000   00000 00000   00000 00000
3174:  00000 00000   00000 00000   00000 00000   00000 00000   00000 00000     00000 00000   00000 00000   00000 00000   00000 00000   00000 00000
3175:  00000 00000   00000 00000   00000 00000   00000 00000   00000 00000     00000 00000   00000 00000   00000 00000   00000 00000   00000 00000
3176:  00000 00000   00000 00000   00000 00000   00000 00000   00000 00000     00000 00000   00000 00000   00000 00000   00000 00000   00000 00000
3177:  00000 00000   00000 00000   00000 00000   00000 00000   00000 00000     00000 00000   00000 00000   00000 00000   00000 00000   00000 00000
3178:  00000 00000   00000 00000   00000 00000   00000 00000   00000 00000     00000 00000   00000 00000   00000 00000   00000 00000   00000 00000
3179:  00000 00000   00000 00000   00000 00000   00000 00000   00000 00000     00000 00000   00000 00000   00000 00000   00000 00000   00000 00000
3180:  00000 00000   00000 00000   00000 00000   00000 00000   00000 00000     00000 00000   00000 00000   00000 00000   00000 00000   00000 00000
3181:  00000 00000   00000 00000   00000 00000   00000 00000   00000 00000     00000 00000   00000 00000   00000 00000   00000 00000   00000 00000
3182:  00000 00000   00000 00000   00000 00000   00000 00000   00000 00000     00000 00000   00000 00000   00000 00000   00000 00000   00000 00000
3183:  00000 00000   00000 00000   00000 00000   00000 00000   00000 00000     00000 00000   00000 00000   00000 00000   00000 00000   00000 00000
3184:  00000 00000   00000 00000   00000 00000   00000 00000   00000 00000     00000 00000   00000 00000   00000 00000   00000 00000   00000 00000
3185:  00000 00000   00000 00000   00000 00000   00000 00000   00000 00000     00000 00000   00000 00000   00000 00000   00000 00000   00000 00000
3186:  00000 00000   00000 00000   00000 00000   00000 00000   00000 00000     00000 00000   00000 00000   00000 00000   00000 00000   00000 00000
3187:  00000 00000   00000 00000   00000 00000   00000 00000   00000 00000     00000 00000   00000 00000   00000 00000   00000 00000   00000 00000
3188:  00000 00000   00000 00000   00000 00000   00000 00000   00000 00000     00000 00000   00000 00000   00000 00000   00000 00000   00000 00000
3189:  00000 00000   00000 00000   00000 00000   00000 00000   00000 00000     00000 00000   00000 00000   00000 00000   00000 00000   00000 00000
3190:  00000 00000   00000 00000   00000 00000   00000 00000   00000 00000     00000 00000   00000 00000   00000 00000   00000 00000   00000 00000
3191:  00000 00000   00000 00000   00000 00000   00000 00000   00000 00000     00000 00000   00000 00000   00000 00000   00000 00000   00000 00000
3192:  00000 00000   00000 00000   00000 00000   00000 00000   00000 00000     00000 00000   00000 00000   00000 00000   00000 00000   00000 00000
3193:  00000 00000   00000 00000   00000 00000   00000 00000   00000 00000     00000 00000   00000 00000   00000 00000   00000 00000   00000 00000
3194:  00000 00000   00000 00000   00000 00000   00000 00000   00000 00000     00000 00000   00000 00000   00000 00000   00000 00000   00000 00000
3195:  00000 00000   00000 00000   00000 00000   00000 00000   00000 00000     00000 00000   00000 00000   00000 00000   00000 00000   00000 00000
3196:  00000 00000   00000 00000   00000 00000   00000 00000   00000 00000     00000 00000   00000 00000   00000 00000   00000 00000   00000 00000
3197:  00000 00000   00000 00000   00000 00000   00000 00000   00000 00000     00000 00000   00000 00000   00000 00000   00000 00000   00000 00000
3198:  00000 00000   00000 00000   00000 00000   00000 00000   00000 00000     00000 00000   00000 00000   00000 00000   00000 00000   00000 00000
3199:  00000 00000   00000 00000   00000 00000   00000 00000   00000 00000     00000 00000   00000 00000   00000 00000   00000 00000   00000 00000
```

```
3200:  00000 00000  00000 00000  00000 00000  00000 00000  00000 00000     00000 00000  00000 00000  00000 00000  00000 00000  00000 00000
3201:  00000 00000  00000 00000  00000 00000  00000 00000  00000 00000     00000 00000  00000 00000  00000 00000  00000 00000  00000 00000
3202:  00000 00000  00000 00000  00000 00000  00000 00000  00000 00000     00000 00000  00000 00000  00000 00000  00000 00000  00000 00000
3203:  00000 00000  00000 00000  00000 00000  00000 00000  00000 00000     00000 00000  00000 00000  00000 00000  00000 00000  00000 00000
3204:  00000 00000  00000 00000  00000 00000  00000 00000  00000 00000     00000 00000  00000 00000  00000 00000  00000 00000  00000 00000
3205:  00000 00000  00000 00000  00000 00000  00000 00000  00000 00000     00000 00000  00000 00000  00000 00000  00000 00000  00000 00000
3206:  00000 00000  00000 00000  00000 00000  00000 00000  00000 00000     00000 00000  00000 00000  00000 00000  00000 00000  00000 00000
3207:  00000 00000  00000 00000  00000 00000  00000 00000  00000 00000     00000 00000  00000 00000  00000 00000  00000 00000  00000 00000
3208:  00000 00000  00000 00000  00000 00000  00000 00000  00000 00000     00000 00000  00000 00000  00000 00000  00000 00000  00000 00000
3209:  00000 00000  00000 00000  00000 00000  00000 00000  00000 00000     00000 00000  00000 00000  00000 00000  00000 00000  00000 00000
3210:  00000 00000  00000 00000  00000 00000  00000 00000  00000 00000     00000 00000  00000 00000  00000 00000  00000 00000  00000 00000
3211:  00000 00000  00000 00000  00000 00000  00000 00000  00000 00000     00000 00000  00000 00000  00000 00000  00000 00000  00000 00000
3212:  00000 00000  00000 00000  00000 00000  00000 00000  00000 00000     00000 00000  00000 00000  00000 00000  00000 00000  00000 00000
3213:  00000 00000  00000 00000  00000 00000  00000 00000  00000 00000     00000 00000  00000 00000  00000 00000  00000 00000  00000 00000
3214:  00000 00000  00000 00000  00000 00000  00000 00000  00000 00000     00000 00000  00000 00000  00000 00000  00000 00000  00000 00000
3215:  00000 00000  00000 00000  00000 00000  00000 00000  00000 00000     00000 00000  00000 00000  00000 00000  00000 00000  00000 00000
3216:  00000 00000  00000 00000  00000 00000  00000 00000  00000 00000     00000 00000  00000 00000  00000 00000  00000 00000  00000 00000
3217:  00000 00000  00000 00000  00000 00000  00000 00000  00000 00000     00000 00000  00000 00000  00000 00000  00000 00000  00000 00000
3218:  00000 00000  00000 00000  00000 00000  00000 00000  00000 00000     00000 00000  00000 00000  00000 00000  00000 00000  00000 00000
3219:  00000 00000  00000 00000  00000 00000  00000 00000  00000 00000     00000 00000  00000 00000  00000 00000  00000 00000  00000 00000
3220:  00000 00000  00000 00000  00000 00000  00000 00000  00000 00000     00000 00000  00000 00000  00000 00000  00000 00000  00000 00000
3221:  00000 00000  00000 00000  00000 00000  00000 00000  00000 00000     00000 00000  00000 00000  00000 00000  00000 00000  00000 00000
3222:  00000 00000  00000 00000  00000 00000  00000 00000  00000 00000     00000 00000  00000 00000  00000 00000  00000 00000  00000 00000
3223:  00000 00000  00000 00000  00000 00000  00000 00000  00000 00000     00000 00000  00000 00000  00000 00000  00000 00000  00000 00000
3224:  00000 00000  00000 00000  00000 00000  00000 00000  00000 00000     00000 00000  00000 00000  00000 00000  00000 00000  00000 00000
3225:  00000 00000  00000 00000  00000 00000  00000 00000  00000 00000     00000 00000  00000 00000  00000 00000  00000 00000  00000 00000
3226:  00000 00000  00000 00000  00000 00000  00000 00000  00000 00000     00000 00000  00000 00000  00000 00000  00000 00000  00000 00000
3227:  00000 00000  00000 00000  00000 00000  00000 00000  00000 00000     00000 00000  00000 00000  00000 00000  00000 00000  00000 00000
3228:  00000 00000  00000 00000  00000 00000  00000 00000  00000 00000     00000 00000  00000 00000  00000 00000  00000 00000  00000 00000
3229:  00000 00000  00000 00000  00000 00000  00000 00000  00000 00000     00000 00000  00000 00000  00000 00000  00000 00000  00000 00000
3230:  00000 00000  00000 00000  00000 00000  00000 00000  00000 00000     00000 00000  00000 00000  00000 00000  00000 00000  00000 00000
3231:  00000 00000  00000 00000  00000 00000  00000 00000  00000 00000     00000 00000  00000 00000  00000 00000  00000 00000  00000 00000
3232:  00000 00000  00000 00000  00000 00000  00000 00000  00000 00000     00000 00000  00000 00000  00000 00000  00000 00000  00000 00000
3233:  00000 00000  00000 00000  00000 00000  00000 00000  00000 00000     00000 00000  00000 00000  00000 00000  00000 00000  00000 00000
3234:  00000 00000  00000 00000  00000 00000  00000 00000  00000 00000     00000 00000  00000 00000  00000 00000  00000 00000  00000 00000
3235:  00000 00000  00000 00000  00000 00000  00000 00000  00000 00000     00000 00000  00000 00000  00000 00000  00000 00000  00000 00000
3236:  00000 00000  00000 00000  00000 00000  00000 00000  00000 00000     00000 00000  00000 00000  00000 00000  00000 00000  00000 00000
3237:  00000 00000  00000 00000  00000 00000  00000 00000  00000 00000     00000 00000  00000 00000  00000 00000  00000 00000  00000 00000
3238:  00000 00000  00000 00000  00000 00000  00000 00000  00000 00000     00000 00000  00000 00000  00000 00000  00000 00000  00000 00000
3239:  00000 00000  00000 00000  00000 00000  00000 00000  00000 00000     00000 00000  00000 00000  00000 00000  00000 00000  00000 00000
3240:  00000 00000  00000 00000  00000 00000  00000 00000  00000 00000     00000 00000  00000 00000  00000 00000  00000 00000  00000 00000
3241:  00000 00000  00000 00000  00000 00000  00000 00000  00000 00000     00000 00000  00000 00000  00000 00000  00000 00000  00000 00000
3242:  00000 00000  00000 00000  00000 00000  00000 00000  00000 00000     00000 00000  00000 00000  00000 00000  00000 00000  00000 00000
3243:  00000 00000  00000 00000  00000 00000  00000 00000  00000 00000     00000 00000  00000 00000  00000 00000  00000 00000  00000 00000
3244:  00000 00000  00000 00000  00000 00000  00000 00000  00000 00000     00000 00000  00000 00000  00000 00000  00000 00000  00000 00000
3245:  00000 00000  00000 00000  00000 00000  00000 00000  00000 00000     00000 00000  00000 00000  00000 00000  00000 00000  00000 00000
3246:  00000 00000  00000 00000  00000 00000  00000 00000  00000 00000     00000 00000  00000 00000  00000 00000  00000 00000  00000 00000
3247:  00000 00000  00000 00000  00000 00000  00000 00000  00000 00000     00000 00000  00000 00000  00000 00000  00000 00000  00000 00000
3248:  00000 00000  00000 00000  00000 00000  00000 00000  00000 00000     00000 00000  00000 00000  00000 00000  00000 00000  00000 00000
3249:  00000 00000  00000 00000  00000 00000  00000 00000  00000 00000     00000 00000  00000 00000  00000 00000  00000 00000  00000 00000
```

```
3250: 00000 00000  00000 00000  00000 00000  00000 00000  00000 00000    00000 00000  00000 00000  00000 00000  00000 00000  00000 00000
3251: 00000 00000  00000 00000  00000 00000  00000 00000  00000 00000    00000 00000  00000 00000  00000 00000  00000 00000  00000 00000
3252: 00000 00000  00000 00000  00000 00000  00000 00000  00000 00000    00000 00000  00000 00000  00000 00000  00000 00000  00000 00000
3253: 00000 00000  00000 00000  00000 00000  00000 00000  00000 00000    00000 00000  00000 00000  00000 00000  00000 00000  00000 00000
3254: 00000 00000  00000 00000  00000 00000  00000 00000  00000 00000    00000 00000  00000 00000  00000 00000  00000 00000  00000 00000
3255: 00000 00000  00000 00000  00000 00000  00000 00000  00000 00000    00000 00000  00000 00000  00000 00000  00000 00000  00000 00000
3256: 00000 00000  00000 00000  00000 00000  00000 00000  00000 00000    00000 00000  00000 00000  00000 00000  00000 00000  00000 00000
3257: 00000 00000  00000 00000  00000 00000  00000 00000  00000 00000    00000 00000  00000 00000  00000 00000  00000 00000  00000 00000
3258: 00000 00000  00000 00000  00000 00000  00000 00000  00000 00000    00000 00000  00000 00000  00000 00000  00000 00000  00000 00000
3259: 00000 00000  00000 00000  00000 00000  00000 00000  00000 00000    00000 00000  00000 00000  00000 00000  00000 00000  00000 00000
3260: 00000 00000  00000 00000  00000 00000  00000 00000  00000 00000    00000 00000  00000 00000  00000 00000  00000 00000  00000 00000
3261: 00000 00000  00000 00000  00000 00000  00000 00000  00000 00000    00000 00000  00000 00000  00000 00000  00000 00000  00000 00000
3262: 00000 00000  00000 00000  00000 00000  00000 00000  00000 00000    00000 00000  00000 00000  00000 00000  00000 00000  00000 00000
3263: 00000 00000  00000 00000  00000 00000  00000 00000  00000 00000    00000 00000  00000 00000  00000 00000  00000 00000  00000 00000
3264: 00000 00000  00000 00000  00000 00000  00000 00000  00000 00000    00000 00000  00000 00000  00000 00000  00000 00000  00000 00000
3265: 00000 00000  00000 00000  00000 00000  00000 00000  00000 00000    00000 00000  00000 00000  00000 00000  00000 00000  00000 00000
3266: 00000 00000  00000 00000  00000 00000  00000 00000  00000 00000    00000 00000  00000 00000  00000 00000  00000 00000  00000 00000
3267: 00000 00000  00000 00000  00000 00000  00000 00000  00000 00000    00000 00000  00000 00000  00000 00000  00000 00000  00000 00000
3268: 00000 00000  00000 00000  00000 00000  00000 00000  00000 00000    00000 00000  00000 00000  00000 00000  00000 00000  00000 00000
3269: 00000 00000  00000 00000  00000 00000  00000 00000  00000 00000    00000 00000  00000 00000  00000 00000  00000 00000  00000 00000
3270: 00000 00000  00000 00000  00000 00000  00000 00000  00000 00000    00000 00000  00000 00000  00000 00000  00000 00000  00000 00000
3271: 00000 00000  00000 00000  00000 00000  00000 00000  00000 00000    00000 00000  00000 00000  00000 00000  00000 00000  00000 00000
3272: 00000 00000  00000 00000  00000 00000  00000 00000  00000 00000    00000 00000  00000 00000  00000 00000  00000 00000  00000 00000
3273: 00000 00000  00000 00000  00000 00000  00000 00000  00000 00000    00000 00000  00000 00000  00000 00000  00000 00000  00000 00000
3274: 00000 00000  00000 00000  00000 00000  00000 00000  00000 00000    00000 00000  00000 00000  00000 00000  00000 00000  00000 00000
3275: 00000 00000  00000 00000  00000 00000  00000 00000  00000 00000    00000 00000  00000 00000  00000 00000  00000 00000  00000 00000
3276: 00000 00000  00000 00000  00000 00000  00000 00000  00000 00000    00000 00000  00000 00000  00000 00000  00000 00000  00000 00000
3277: 00000 00000  00000 00000  00000 00000  00000 00000  00000 00000    00000 00000  00000 00000  00000 00000  00000 00000  00000 00000
3278: 00000 00000  00000 00000  00000 00000  00000 00000  00000 00000    00000 00000  00000 00000  00000 00000  00000 00000  00000 00000
3279: 00000 00000  00000 00000  00000 00000  00000 00000  00000 00000    00000 00000  00000 00000  00000 00000  00000 00000  00000 00000
3280: 00000 00000  00000 00000  00000 00000  00000 00000  00000 00000    00000 00000  00000 00000  00000 00000  00000 00000  00000 00000
3281: 00000 00000  00000 00000  00000 00000  00000 00000  00000 00000    00000 00000  00000 00000  00000 00000  00000 00000  00000 00000
3282: 00000 00000  00000 00000  00000 00000  00000 00000  00000 00000    00000 00000  00000 00000  00000 00000  00000 00000  00000 00000
3283: 00000 00000  00000 00000  00000 00000  00000 00000  00000 00000    00000 00000  00000 00000  00000 00000  00000 00000  00000 00000
3284: 00000 00000  00000 00000  00000 00000  00000 00000  00000 00000    00000 00000  00000 00000  00000 00000  00000 00000  00000 00000
3285: 00000 00000  00000 00000  00000 00000  00000 00000  00000 00000    00000 00000  00000 00000  00000 00000  00000 00000  00000 00000
3286: 00000 00000  00000 00000  00000 00000  00000 00000  00000 00000    00000 00000  00000 00000  00000 00000  00000 00000  00000 00000
3287: 00000 00000  00000 00000  00000 00000  00000 00000  00000 00000    00000 00000  00000 00000  00000 00000  00000 00000  00000 00000
3288: 00000 00000  00000 00000  00000 00000  00000 00000  00000 00000    00000 00000  00000 00000  00000 00000  00000 00000  00000 00000
3289: 00000 00000  00000 00000  00000 00000  00000 00000  00000 00000    00000 00000  00000 00000  00000 00000  00000 00000  00000 00000
3290: 00000 00000  00000 00000  00000 00000  00000 00000  00000 00000    00000 00000  00000 00000  00000 00000  00000 00000  00000 00000
3291: 00000 00000  00000 00000  00000 00000  00000 00000  00000 00000    00000 00000  00000 00000  00000 00000  00000 00000  00000 00000
3292: 00000 00000  00000 00000  00000 00000  00000 00000  00000 00000    00000 00000  00000 00000  00000 00000  00000 00000  00000 00000
3293: 00000 00000  00000 00000  00000 00000  00000 00000  00000 00000    00000 00000  00000 00000  00000 00000  00000 00000  00000 00000
3294: 00000 00000  00000 00000  00000 00000  00000 00000  00000 00000    00000 00000  00000 00000  00000 00000  00000 00000  00000 00000
3295: 00000 00000  00000 00000  00000 00000  00000 00000  00000 00000    00000 00000  00000 00000  00000 00000  00000 00000  00000 00000
3296: 00000 00000  00000 00000  00000 00000  00000 00000  00000 00000    00000 00000  00000 00000  00000 00000  00000 00000  00000 00000
3297: 00000 00000  00000 00000  00000 00000  00000 00000  00000 00000    00000 00000  00000 00000  00000 00000  00000 00000  00000 00000
3298: 00000 00000  00000 00000  00000 00000  00000 00000  00000 00000    00000 00000  00000 00000  00000 00000  00000 00000  00000 00000
3299: 00000 00000  00000 00000  00000 00000  00000 00000  00000 00000    00000 00000  00000 00000  00000 00000  00000 00000  00000 00000
```

```
3300:  00000 00000  00000 00000  00000 00000  00000 00000  00000 00000   00000 00000  00000 00000  00000 00000  00000 00000  00000 00000
3301:  00000 00000  00000 00000  00000 00000  00000 00000  00000 00000   00000 00000  00000 00000  00000 00000  00000 00000  00000 00000
3302:  00000 00000  00000 00000  00000 00000  00000 00000  00000 00000   00000 00000  00000 00000  00000 00000  00000 00000  00000 00000
3303:  00000 00000  00000 00000  00000 00000  00000 00000  00000 00000   00000 00000  00000 00000  00000 00000  00000 00000  00000 00000
3304:  00000 00000  00000 00000  00000 00000  00000 00000  00000 00000   00000 00000  00000 00000  00000 00000  00000 00000  00000 00000
3305:  00000 00000  00000 00000  00000 00000  00000 00000  00000 00000   00000 00000  00000 00000  00000 00000  00000 00000  00000 00000
3306:  00000 00000  00000 00000  00000 00000  00000 00000  00000 00000   00000 00000  00000 00000  00000 00000  00000 00000  00000 00000
3307:  00000 00000  00000 00000  00000 00000  00000 00000  00000 00000   00000 00000  00000 00000  00000 00000  00000 00000  00000 00000
3308:  00000 00000  00000 00000  00000 00000  00000 00000  00000 00000   00000 00000  00000 00000  00000 00000  00000 00000  00000 00000
3309:  00000 00000  00000 00000  00000 00000  00000 00000  00000 00000   00000 00000  00000 00000  00000 00000  00000 00000  00000 00000
3310:  00000 00000  00000 00000  00000 00000  00000 00000  00000 00000   00000 00000  00000 00000  00000 00000  00000 00000  00000 00000
3311:  00000 00000  00000 00000  00000 00000  00000 00000  00000 00000   00000 00000  00000 00000  00000 00000  00000 00000  00000 00000
3312:  00000 00000  00000 00000  00000 00000  00000 00000  00000 00000   00000 00000  00000 00000  00000 00000  00000 00000  00000 00000
3313:  00000 00000  00000 00000  00000 00000  00000 00000  00000 00000   00000 00000  00000 00000  00000 00000  00000 00000  00000 00000
3314:  00000 00000  00000 00000  00000 00000  00000 00000  00000 00000   00000 00000  00000 00000  00000 00000  00000 00000  00000 00000
3315:  00000 00000  00000 00000  00000 00000  00000 00000  00000 00000   00000 00000  00000 00000  00000 00000  00000 00000  00000 00000
3316:  00000 00000  00000 00000  00000 00000  00000 00000  00000 00000   00000 00000  00000 00000  00000 00000  00000 00000  00000 00000
3317:  00000 00000  00000 00000  00000 00000  00000 00000  00000 00000   00000 00000  00000 00000  00000 00000  00000 00000  00000 00000
3318:  00000 00000  00000 00000  00000 00000  00000 00000  00000 00000   00000 00000  00000 00000  00000 00000  00000 00000  00000 00000
3319:  00000 00000  00000 00000  00000 00000  00000 00000  00000 00000   00000 00000  00000 00000  00000 00000  00000 00000  00000 00000
3320:  00000 00000  00000 00000  00000 00000  00000 00000  00000 00000   00000 00000  00000 00000  00000 00000  00000 00000  00000 00000
3321:  00000 00000  00000 00000  00000 00000  00000 00000  00000 00000   00000 00000  00000 00000  00000 00000  00000 00000  00000 00000
3322:  00000 00000  00000 00000  00000 00000  00000 00000  00000 00000   00000 00000  00000 00000  00000 00000  00000 00000  00000 00000
3323:  00000 00000  00000 00000  00000 00000  00000 00000  00000 00000   00000 00000  00000 00000  00000 00000  00000 00000  00000 00000
3324:  00000 00000  00000 00000  00000 00000  00000 00000  00000 00000   00000 00000  00000 00000  00000 00000  00000 00000  00000 00000
3325:  00000 00000  00000 00000  00000 00000  00000 00000  00000 00000   00000 00000  00000 00000  00000 00000  00000 00000  00000 00000
3326:  00000 00000  00000 00000  00000 00000  00000 00000  00000 00000   00000 00000  00000 00000  00000 00000  00000 00000  00000 00000
3327:  00000 00000  00000 00000  00000 00000  00000 00000  00000 00000   00000 00000  00000 00000  00000 00000  00000 00000  00000 00000
3328:  00000 00000  00000 00000  00000 00000  00000 00000  00000 00000   00000 00000  00000 00000  00000 00000  00000 00000  00000 00000
3329:  00000 00000  00000 00000  00000 00000  00000 00000  00000 00000   00000 00000  00000 00000  00000 00000  00000 00000  00000 00000
3330:  00000 00000  00000 00000  00000 00000  00000 00000  00000 00000   00000 00000  00000 00000  00000 00000  00000 00000  00000 00000
3331:  00000 00000  00000 00000  00000 00000  00000 00000  00000 00000   00000 00000  00000 00000  00000 00000  00000 00000  00000 00000
3332:  00000 00000  00000 00000  00000 00000  00000 00000  00000 00000   00000 00000  00000 00000  00000 00000  00000 00000  00000 00000
3333:  00000 00000  00000 00000  00000 00000  00000 00000  00000 00000   00000 00000  00000 00000  00000 00000  00000 00000  00000 00000
3334:  00000 00000  00000 00000  00000 00000  00000 00000  00000 00000   00000 00000  00000 00000  00000 00000  00000 00000  00000 00000
3335:  00000 00000  00000 00000  00000 00000  00000 00000  00000 00000   00000 00000  00000 00000  00000 00000  00000 00000  00000 00000
3336:  00000 00000  00000 00000  00000 00000  00000 00000  00000 00000   00000 00000  00000 00000  00000 00000  00000 00000  00000 00000
3337:  00000 00000  00000 00000  00000 00000  00000 00000  00000 00000   00000 00000  00000 00000  00000 00000  00000 00000  00000 00000
3338:  00000 00000  00000 00000  00000 00000  00000 00000  00000 00000   00000 00000  00000 00000  00000 00000  00000 00000  00000 00000
3339:  00000 00000  00000 00000  00000 00000  00000 00000  00000 00000   00000 00000  00000 00000  00000 00000  00000 00000  00000 00000
3340:  00000 00000  00000 00000  00000 00000  00000 00000  00000 00000   00000 00000  00000 00000  00000 00000  00000 00000  00000 00000
3341:  00000 00000  00000 00000  00000 00000  00000 00000  00000 00000   00000 00000  00000 00000  00000 00000  00000 00000  00000 00000
3342:  00000 00000  00000 00000  00000 00000  00000 00000  00000 00000   00000 00000  00000 00000  00000 00000  00000 00000  00000 00000
3343:  00000 00000  00000 00000  00000 00000  00000 00000  00000 00000   00000 00000  00000 00000  00000 00000  00000 00000  00000 00000
3344:  00000 00000  00000 00000  00000 00000  00000 00000  00000 00000   00000 00000  00000 00000  00000 00000  00000 00000  00000 00000
3345:  00000 00000  00000 00000  00000 00000  00000 00000  00000 00000   00000 00000  00000 00000  00000 00000  00000 00000  00000 00000
3346:  00000 00000  00000 00000  00000 00000  00000 00000  00000 00000   00000 00000  00000 00000  00000 00000  00000 00000  00000 00000
3347:  00000 00000  00000 00000  00000 00000  00000 00000  00000 00000   00000 00000  00000 00000  00000 00000  00000 00000  00000 00000
3348:  00000 00000  00000 00000  00000 00000  00000 00000  00000 00000   00000 00000  00000 00000  00000 00000  00000 00000  00000 00000
3349:  00000 00000  00000 00000  00000 00000  00000 00000  00000 00000   00000 00000  00000 00000  00000 00000  00000 00000  00000 00000
```

```
3350:  00000 00000  00000 00000  00000 00000  00000 00000  00000 00000    00000 00000  00000 00000  00000 00000  00000 00000  00000 00000
3351:  00000 00000  00000 00000  00000 00000  00000 00000  00000 00000    00000 00000  00000 00000  00000 00000  00000 00000  00000 00000
3352:  00000 00000  00000 00000  00000 00000  00000 00000  00000 00000    00000 00000  00000 00000  00000 00000  00000 00000  00000 00000
3353:  00000 00000  00000 00000  00000 00000  00000 00000  00000 00000    00000 00000  00000 00000  00000 00000  00000 00000  00000 00000
3354:  00000 00000  00000 00000  00000 00000  00000 00000  00000 00000    00000 00000  00000 00000  00000 00000  00000 00000  00000 00000
3355:  00000 00000  00000 00000  00000 00000  00000 00000  00000 00000    00000 00000  00000 00000  00000 00000  00000 00000  00000 00000
3356:  00000 00000  00000 00000  00000 00000  00000 00000  00000 00000    00000 00000  00000 00000  00000 00000  00000 00000  00000 00000
3357:  00000 00000  00000 00000  00000 00000  00000 00000  00000 00000    00000 00000  00000 00000  00000 00000  00000 00000  00000 00000
3358:  00000 00000  00000 00000  00000 00000  00000 00000  00000 00000    00000 00000  00000 00000  00000 00000  00000 00000  00000 00000
3359:  00000 00000  00000 00000  00000 00000  00000 00000  00000 00000    00000 00000  00000 00000  00000 00000  00000 00000  00000 00000
3360:  00000 00000  00000 00000  00000 00000  00000 00000  00000 00000    00000 00000  00000 00000  00000 00000  00000 00000  00000 00000
3361:  00000 00000  00000 00000  00000 00000  00000 00000  00000 00000    00000 00000  00000 00000  00000 00000  00000 00000  00000 00000
3362:  00000 00000  00000 00000  00000 00000  00000 00000  00000 00000    00000 00000  00000 00000  00000 00000  00000 00000  00000 00000
3363:  00000 00000  00000 00000  00000 00000  00000 00000  00000 00000    00000 00000  00000 00000  00000 00000  00000 00000  00000 00000
3364:  00000 00000  00000 00000  00000 00000  00000 00000  00000 00000    00000 00000  00000 00000  00000 00000  00000 00000  00000 00000
3365:  00000 00000  00000 00000  00000 00000  00000 00000  00000 00000    00000 00000  00000 00000  00000 00000  00000 00000  00000 00000
3366:  00000 00000  00000 00000  00000 00000  00000 00000  00000 00000    00000 00000  00000 00000  00000 00000  00000 00000  00000 00000
3367:  00000 00000  00000 00000  00000 00000  00000 00000  00000 00000    00000 00000  00000 00000  00000 00000  00000 00000  00000 00000
3368:  00000 00000  00000 00000  00000 00000  00000 00000  00000 00000    00000 00000  00000 00000  00000 00000  00000 00000  00000 00000
3369:  00000 00000  00000 00000  00000 00000  00000 00000  00000 00000    00000 00000  00000 00000  00000 00000  00000 00000  00000 00000
3370:  00000 00000  00000 00000  00000 00000  00000 00000  00000 00000    00000 00000  00000 00000  00000 00000  00000 00000  00000 00000
3371:  00000 00000  00000 00000  00000 00000  00000 00000  00000 00000    00000 00000  00000 00000  00000 00000  00000 00000  00000 00000
3372:  00000 00000  00000 00000  00000 00000  00000 00000  00000 00000    00000 00000  00000 00000  00000 00000  00000 00000  00000 00000
3373:  00000 00000  00000 00000  00000 00000  00000 00000  00000 00000    00000 00000  00000 00000  00000 00000  00000 00000  00000 00000
3374:  00000 00000  00000 00000  00000 00000  00000 00000  00000 00000    00000 00000  00000 00000  00000 00000  00000 00000  00000 00000
3375:  00000 00000  00000 00000  00000 00000  00000 00000  00000 00000    00000 00000  00000 00000  00000 00000  00000 00000  00000 00000
3376:  00000 00000  00000 00000  00000 00000  00000 00000  00000 00000    00000 00000  00000 00000  00000 00000  00000 00000  00000 00000
3377:  00000 00000  00000 00000  00000 00000  00000 00000  00000 00000    00000 00000  00000 00000  00000 00000  00000 00000  00000 00000
3378:  00000 00000  00000 00000  00000 00000  00000 00000  00000 00000    00000 00000  00000 00000  00000 00000  00000 00000  00000 00000
3379:  00000 00000  00000 00000  00000 00000  00000 00000  00000 00000    00000 00000  00000 00000  00000 00000  00000 00000  00000 00000
3380:  00000 00000  00000 00000  00000 00000  00000 00000  00000 00000    00000 00000  00000 00000  00000 00000  00000 00000  00000 00000
3381:  00000 00000  00000 00000  00000 00000  00000 00000  00000 00000    00000 00000  00000 00000  00000 00000  00000 00000  00000 00000
3382:  00000 00000  00000 00000  00000 00000  00000 00000  00000 00000    00000 00000  00000 00000  00000 00000  00000 00000  00000 00000
3383:  00000 00000  00000 00000  00000 00000  00000 00000  00000 00000    00000 00000  00000 00000  00000 00000  00000 00000  00000 00000
3384:  00000 00000  00000 00000  00000 00000  00000 00000  00000 00000    00000 00000  00000 00000  00000 00000  00000 00000  00000 00000
3385:  00000 00000  00000 00000  00000 00000  00000 00000  00000 00000    00000 00000  00000 00000  00000 00000  00000 00000  00000 00000
3386:  00000 00000  00000 00000  00000 00000  00000 00000  00000 00000    00000 00000  00000 00000  00000 00000  00000 00000  00000 00000
3387:  00000 00000  00000 00000  00000 00000  00000 00000  00000 00000    00000 00000  00000 00000  00000 00000  00000 00000  00000 00000
3388:  00000 00000  00000 00000  00000 00000  00000 00000  00000 00000    00000 00000  00000 00000  00000 00000  00000 00000  00000 00000
3389:  00000 00000  00000 00000  00000 00000  00000 00000  00000 00000    00000 00000  00000 00000  00000 00000  00000 00000  00000 00000
3390:  00000 00000  00000 00000  00000 00000  00000 00000  00000 00000    00000 00000  00000 00000  00000 00000  00000 00000  00000 00000
3391:  00000 00000  00000 00000  00000 00000  00000 00000  00000 00000    00000 00000  00000 00000  00000 00000  00000 00000  00000 00000
3392:  00000 00000  00000 00000  00000 00000  00000 00000  00000 00000    00000 00000  00000 00000  00000 00000  00000 00000  00000 00000
3393:  00000 00000  00000 00000  00000 00000  00000 00000  00000 00000    00000 00000  00000 00000  00000 00000  00000 00000  00000 00000
3394:  00000 00000  00000 00000  00000 00000  00000 00000  00000 00000    00000 00000  00000 00000  00000 00000  00000 00000  00000 00000
3395:  00000 00000  00000 00000  00000 00000  00000 00000  00000 00000    00000 00000  00000 00000  00000 00000  00000 00000  00000 00000
3396:  00000 00000  00000 00000  00000 00000  00000 00000  00000 00000    00000 00000  00000 00000  00000 00000  00000 00000  00000 00000
3397:  00000 00000  00000 00000  00000 00000  00000 00000  00000 00000    00000 00000  00000 00000  00000 00000  00000 00000  00000 00000
3398:  00000 00000  00000 00000  00000 00000  00000 00000  00000 00000    00000 00000  00000 00000  00000 00000  00000 00000  00000 00000
3399:  00000 00000  00000 00000  00000 00000  00000 00000  00000 00000    00000 00000  00000 00000  00000 00000  00000 00000  00000 00000
```

```
3400:  00000 00000  00000 00000  00000 00000  00000 00000  00000 00000    00000 00000  00000 00000  00000 00000  00000 00000  00000 00000
3401:  00000 00000  00000 00000  00000 00000  00000 00000  00000 00000    00000 00000  00000 00000  00000 00000  00000 00000  00000 00000
3402:  00000 00000  00000 00000  00000 00000  00000 00000  00000 00000    00000 00000  00000 00000  00000 00000  00000 00000  00000 00000
3403:  00000 00000  00000 00000  00000 00000  00000 00000  00000 00000    00000 00000  00000 00000  00000 00000  00000 00000  00000 00000
3404:  00000 00000  00000 00000  00000 00000  00000 00000  00000 00000    00000 00000  00000 00000  00000 00000  00000 00000  00000 00000
3405:  00000 00000  00000 00000  00000 00000  00000 00000  00000 00000    00000 00000  00000 00000  00000 00000  00000 00000  00000 00000
3406:  00000 00000  00000 00000  00000 00000  00000 00000  00000 00000    00000 00000  00000 00000  00000 00000  00000 00000  00000 00000
3407:  00000 00000  00000 00000  00000 00000  00000 00000  00000 00000    00000 00000  00000 00000  00000 00000  00000 00000  00000 00000
3408:  00000 00000  00000 00000  00000 00000  00000 00000  00000 00000    00000 00000  00000 00000  00000 00000  00000 00000  00000 00000
3409:  00000 00000  00000 00000  00000 00000  00000 00000  00000 00000    00000 00000  00000 00000  00000 00000  00000 00000  00000 00000
3410:  00000 00000  00000 00000  00000 00000  00000 00000  00000 00000    00000 00000  00000 00000  00000 00000  00000 00000  00000 00000
3411:  00000 00000  00000 00000  00000 00000  00000 00000  00000 00000    00000 00000  00000 00000  00000 00000  00000 00000  00000 00000
3412:  00000 00000  00000 00000  00000 00000  00000 00000  00000 00000    00000 00000  00000 00000  00000 00000  00000 00000  00000 00000
3413:  00000 00000  00000 00000  00000 00000  00000 00000  00000 00000    00000 00000  00000 00000  00000 00000  00000 00000  00000 00000
3414:  00000 00000  00000 00000  00000 00000  00000 00000  00000 00000    00000 00000  00000 00000  00000 00000  00000 00000  00000 00000
3415:  00000 00000  00000 00000  00000 00000  00000 00000  00000 00000    00000 00000  00000 00000  00000 00000  00000 00000  00000 00000
3416:  00000 00000  00000 00000  00000 00000  00000 00000  00000 00000    00000 00000  00000 00000  00000 00000  00000 00000  00000 00000
3417:  00000 00000  00000 00000  00000 00000  00000 00000  00000 00000    00000 00000  00000 00000  00000 00000  00000 00000  00000 00000
3418:  00000 00000  00000 00000  00000 00000  00000 00000  00000 00000    00000 00000  00000 00000  00000 00000  00000 00000  00000 00000
3419:  00000 00000  00000 00000  00000 00000  00000 00000  00000 00000    00000 00000  00000 00000  00000 00000  00000 00000  00000 00000
3420:  00000 00000  00000 00000  00000 00000  00000 00000  00000 00000    00000 00000  00000 00000  00000 00000  00000 00000  00000 00000
3421:  00000 00000  00000 00000  00000 00000  00000 00000  00000 00000    00000 00000  00000 00000  00000 00000  00000 00000  00000 00000
3422:  00000 00000  00000 00000  00000 00000  00000 00000  00000 00000    00000 00000  00000 00000  00000 00000  00000 00000  00000 00000
3423:  00000 00000  00000 00000  00000 00000  00000 00000  00000 00000    00000 00000  00000 00000  00000 00000  00000 00000  00000 00000
3424:  00000 00000  00000 00000  00000 00000  00000 00000  00000 00000    00000 00000  00000 00000  00000 00000  00000 00000  00000 00000
3425:  00000 00000  00000 00000  00000 00000  00000 00000  00000 00000    00000 00000  00000 00000  00000 00000  00000 00000  00000 00000
3426:  00000 00000  00000 00000  00000 00000  00000 00000  00000 00000    00000 00000  00000 00000  00000 00000  00000 00000  00000 00000
3427:  00000 00000  00000 00000  00000 00000  00000 00000  00000 00000    00000 00000  00000 00000  00000 00000  00000 00000  00000 00000
3428:  00000 00000  00000 00000  00000 00000  00000 00000  00000 00000    00000 00000  00000 00000  00000 00000  00000 00000  00000 00000
3429:  00000 00000  00000 00000  00000 00000  00000 00000  00000 00000    00000 00000  00000 00000  00000 00000  00000 00000  00000 00000
3430:  00000 00000  00000 00000  00000 00000  00000 00000  00000 00000    00000 00000  00000 00000  00000 00000  00000 00000  00000 00000
3431:  00000 00000  00000 00000  00000 00000  00000 00000  00000 00000    00000 00000  00000 00000  00000 00000  00000 00000  00000 00000
3432:  00000 00000  00000 00000  00000 00000  00000 00000  00000 00000    00000 00000  00000 00000  00000 00000  00000 00000  00000 00000
3433:  00000 00000  00000 00000  00000 00000  00000 00000  00000 00000    00000 00000  00000 00000  00000 00000  00000 00000  00000 00000
3434:  00000 00000  00000 00000  00000 00000  00000 00000  00000 00000    00000 00000  00000 00000  00000 00000  00000 00000  00000 00000
3435:  00000 00000  00000 00000  00000 00000  00000 00000  00000 00000    00000 00000  00000 00000  00000 00000  00000 00000  00000 00000
3436:  00000 00000  00000 00000  00000 00000  00000 00000  00000 00000    00000 00000  00000 00000  00000 00000  00000 00000  00000 00000
3437:  00000 00000  00000 00000  00000 00000  00000 00000  00000 00000    00000 00000  00000 00000  00000 00000  00000 00000  00000 00000
3438:  00000 00000  00000 00000  00000 00000  00000 00000  00000 00000    00000 00000  00000 00000  00000 00000  00000 00000  00000 00000
3439:  00000 00000  00000 00000  00000 00000  00000 00000  00000 00000    00000 00000  00000 00000  00000 00000  00000 00000  00000 00000
3440:  00000 00000  00000 00000  00000 00000  00000 00000  00000 00000    00000 00000  00000 00000  00000 00000  00000 00000  00000 00000
3441:  00000 00000  00000 00000  00000 00000  00000 00000  00000 00000    00000 00000  00000 00000  00000 00000  00000 00000  00000 00000
3442:  00000 00000  00000 00000  00000 00000  00000 00000  00000 00000    00000 00000  00000 00000  00000 00000  00000 00000  00000 00000
3443:  00000 00000  00000 00000  00000 00000  00000 00000  00000 00000    00000 00000  00000 00000  00000 00000  00000 00000  00000 00000
3444:  00000 00000  00000 00000  00000 00000  00000 00000  00000 00000    00000 00000  00000 00000  00000 00000  00000 00000  00000 00000
3445:  00000 00000  00000 00000  00000 00000  00000 00000  00000 00000    00000 00000  00000 00000  00000 00000  00000 00000  00000 00000
3446:  00000 00000  00000 00000  00000 00000  00000 00000  00000 00000    00000 00000  00000 00000  00000 00000  00000 00000  00000 00000
3447:  00000 00000  00000 00000  00000 00000  00000 00000  00000 00000    00000 00000  00000 00000  00000 00000  00000 00000  00000 00000
3448:  00000 00000  00000 00000  00000 00000  00000 00000  00000 00000    00000 00000  00000 00000  00000 00000  00000 00000  00000 00000
3449:  00000 00000  00000 00000  00000 00000  00000 00000  00000 00000    00000 00000  00000 00000  00000 00000  00000 00000  00000 00000
```

```
3450:  00000 00000  00000 00000  00000 00000  00000 00000  00000 00000    00000 00000  00000 00000  00000 00000  00000 00000  00000 00000
3451:  00000 00000  00000 00000  00000 00000  00000 00000  00000 00000    00000 00000  00000 00000  00000 00000  00000 00000  00000 00000
3452:  00000 00000  00000 00000  00000 00000  00000 00000  00000 00000    00000 00000  00000 00000  00000 00000  00000 00000  00000 00000
3453:  00000 00000  00000 00000  00000 00000  00000 00000  00000 00000    00000 00000  00000 00000  00000 00000  00000 00000  00000 00000
3454:  00000 00000  00000 00000  00000 00000  00000 00000  00000 00000    00000 00000  00000 00000  00000 00000  00000 00000  00000 00000
3455:  00000 00000  00000 00000  00000 00000  00000 00000  00000 00000    00000 00000  00000 00000  00000 00000  00000 00000  00000 00000
3456:  00000 00000  00000 00000  00000 00000  00000 00000  00000 00000    00000 00000  00000 00000  00000 00000  00000 00000  00000 00000
3457:  00000 00000  00000 00000  00000 00000  00000 00000  00000 00000    00000 00000  00000 00000  00000 00000  00000 00000  00000 00000
3458:  00000 00000  00000 00000  00000 00000  00000 00000  00000 00000    00000 00000  00000 00000  00000 00000  00000 00000  00000 00000
3459:  00000 00000  00000 00000  00000 00000  00000 00000  00000 00000    00000 00000  00000 00000  00000 00000  00000 00000  00000 00000
3460:  00000 00000  00000 00000  00000 00000  00000 00000  00000 00000    00000 00000  00000 00000  00000 00000  00000 00000  00000 00000
3461:  00000 00000  00000 00000  00000 00000  00000 00000  00000 00000    00000 00000  00000 00000  00000 00000  00000 00000  00000 00000
3462:  00000 00000  00000 00000  00000 00000  00000 00000  00000 00000    00000 00000  00000 00000  00000 00000  00000 00000  00000 00000
3463:  00000 00000  00000 00000  00000 00000  00000 00000  00000 00000    00000 00000  00000 00000  00000 00000  00000 00000  00000 00000
3464:  00000 00000  00000 00000  00000 00000  00000 00000  00000 00000    00000 00000  00000 00000  00000 00000  00000 00000  00000 00000
3465:  00000 00000  00000 00000  00000 00000  00000 00000  00000 00000    00000 00000  00000 00000  00000 00000  00000 00000  00000 00000
3466:  00000 00000  00000 00000  00000 00000  00000 00000  00000 00000    00000 00000  00000 00000  00000 00000  00000 00000  00000 00000
3467:  00000 00000  00000 00000  00000 00000  00000 00000  00000 00000    00000 00000  00000 00000  00000 00000  00000 00000  00000 00000
3468:  00000 00000  00000 00000  00000 00000  00000 00000  00000 00000    00000 00000  00000 00000  00000 00000  00000 00000  00000 00000
3469:  00000 00000  00000 00000  00000 00000  00000 00000  00000 00000    00000 00000  00000 00000  00000 00000  00000 00000  00000 00000
3470:  00000 00000  00000 00000  00000 00000  00000 00000  00000 00000    00000 00000  00000 00000  00000 00000  00000 00000  00000 00000
3471:  00000 00000  00000 00000  00000 00000  00000 00000  00000 00000    00000 00000  00000 00000  00000 00000  00000 00000  00000 00000
3472:  00000 00000  00000 00000  00000 00000  00000 00000  00000 00000    00000 00000  00000 00000  00000 00000  00000 00000  00000 00000
3473:  00000 00000  00000 00000  00000 00000  00000 00000  00000 00000    00000 00000  00000 00000  00000 00000  00000 00000  00000 00000
3474:  00000 00000  00000 00000  00000 00000  00000 00000  00000 00000    00000 00000  00000 00000  00000 00000  00000 00000  00000 00000
3475:  00000 00000  00000 00000  00000 00000  00000 00000  00000 00000    00000 00000  00000 00000  00000 00000  00000 00000  00000 00000
3476:  00000 00000  00000 00000  00000 00000  00000 00000  00000 00000    00000 00000  00000 00000  00000 00000  00000 00000  00000 00000
3477:  00000 00000  00000 00000  00000 00000  00000 00000  00000 00000    00000 00000  00000 00000  00000 00000  00000 00000  00000 00000
3478:  00000 00000  00000 00000  00000 00000  00000 00000  00000 00000    00000 00000  00000 00000  00000 00000  00000 00000  00000 00000
3479:  00000 00000  00000 00000  00000 00000  00000 00000  00000 00000    00000 00000  00000 00000  00000 00000  00000 00000  00000 00000
3480:  00000 00000  00000 00000  00000 00000  00000 00000  00000 00000    00000 00000  00000 00000  00000 00000  00000 00000  00000 00000
3481:  00000 00000  00000 00000  00000 00000  00000 00000  00000 00000    00000 00000  00000 00000  00000 00000  00000 00000  00000 00000
3482:  00000 00000  00000 00000  00000 00000  00000 00000  00000 00000    00000 00000  00000 00000  00000 00000  00000 00000  00000 00000
3483:  00000 00000  00000 00000  00000 00000  00000 00000  00000 00000    00000 00000  00000 00000  00000 00000  00000 00000  00000 00000
3484:  00000 00000  00000 00000  00000 00000  00000 00000  00000 00000    00000 00000  00000 00000  00000 00000  00000 00000  00000 00000
3485:  00000 00000  00000 00000  00000 00000  00000 00000  00000 00000    00000 00000  00000 00000  00000 00000  00000 00000  00000 00000
3486:  00000 00000  00000 00000  00000 00000  00000 00000  00000 00000    00000 00000  00000 00000  00000 00000  00000 00000  00000 00000
3487:  00000 00000  00000 00000  00000 00000  00000 00000  00000 00000    00000 00000  00000 00000  00000 00000  00000 00000  00000 00000
3488:  00000 00000  00000 00000  00000 00000  00000 00000  00000 00000    00000 00000  00000 00000  00000 00000  00000 00000  00000 00000
3489:  00000 00000  00000 00000  00000 00000  00000 00000  00000 00000    00000 00000  00000 00000  00000 00000  00000 00000  00000 00000
3490:  00000 00000  00000 00000  00000 00000  00000 00000  00000 00000    00000 00000  00000 00000  00000 00000  00000 00000  00000 00000
3491:  00000 00000  00000 00000  00000 00000  00000 00000  00000 00000    00000 00000  00000 00000  00000 00000  00000 00000  00000 00000
3492:  00000 00000  00000 00000  00000 00000  00000 00000  00000 00000    00000 00000  00000 00000  00000 00000  00000 00000  00000 00000
3493:  00000 00000  00000 00000  00000 00000  00000 00000  00000 00000    00000 00000  00000 00000  00000 00000  00000 00000  00000 00000
3494:  00000 00000  00000 00000  00000 00000  00000 00000  00000 00000    00000 00000  00000 00000  00000 00000  00000 00000  00000 00000
3495:  00000 00000  00000 00000  00000 00000  00000 00000  00000 00000    00000 00000  00000 00000  00000 00000  00000 00000  00000 00000
3496:  00000 00000  00000 00000  00000 00000  00000 00000  00000 00000    00000 00000  00000 00000  00000 00000  00000 00000  00000 00000
3497:  00000 00000  00000 00000  00000 00000  00000 00000  00000 00000    00000 00000  00000 00000  00000 00000  00000 00000  00000 00000
3498:  00000 00000  00000 00000  00000 00000  00000 00000  00000 00000    00000 00000  00000 00000  00000 00000  00000 00000  00000 00000
3499:  00000 00000  00000 00000  00000 00000  00000 00000  00000 00000    00000 00000  00000 00000  00000 00000  00000 00000  00000 00000
```

```
3500:  00000 00000  00000 00000  00000 00000  00000 00000  00000 00000   00000 00000  00000 00000  00000 00000  00000 00000  00000 00000
3501:  00000 00000  00000 00000  00000 00000  00000 00000  00000 00000   00000 00000  00000 00000  00000 00000  00000 00000  00000 00000
3502:  00000 00000  00000 00000  00000 00000  00000 00000  00000 00000   00000 00000  00000 00000  00000 00000  00000 00000  00000 00000
3503:  00000 00000  00000 00000  00000 00000  00000 00000  00000 00000   00000 00000  00000 00000  00000 00000  00000 00000  00000 00000
3504:  00000 00000  00000 00000  00000 00000  00000 00000  00000 00000   00000 00000  00000 00000  00000 00000  00000 00000  00000 00000
3505:  00000 00000  00000 00000  00000 00000  00000 00000  00000 00000   00000 00000  00000 00000  00000 00000  00000 00000  00000 00000
3506:  00000 00000  00000 00000  00000 00000  00000 00000  00000 00000   00000 00000  00000 00000  00000 00000  00000 00000  00000 00000
3507:  00000 00000  00000 00000  00000 00000  00000 00000  00000 00000   00000 00000  00000 00000  00000 00000  00000 00000  00000 00000
3508:  00000 00000  00000 00000  00000 00000  00000 00000  00000 00000   00000 00000  00000 00000  00000 00000  00000 00000  00000 00000
3509:  00000 00000  00000 00000  00000 00000  00000 00000  00000 00000   00000 00000  00000 00000  00000 00000  00000 00000  00000 00000
3510:  00000 00000  00000 00000  00000 00000  00000 00000  00000 00000   00000 00000  00000 00000  00000 00000  00000 00000  00000 00000
3511:  00000 00000  00000 00000  00000 00000  00000 00000  00000 00000   00000 00000  00000 00000  00000 00000  00000 00000  00000 00000
3512:  00000 00000  00000 00000  00000 00000  00000 00000  00000 00000   00000 00000  00000 00000  00000 00000  00000 00000  00000 00000
3513:  00000 00000  00000 00000  00000 00000  00000 00000  00000 00000   00000 00000  00000 00000  00000 00000  00000 00000  00000 00000
3514:  00000 00000  00000 00000  00000 00000  00000 00000  00000 00000   00000 00000  00000 00000  00000 00000  00000 00000  00000 00000
3515:  00000 00000  00000 00000  00000 00000  00000 00000  00000 00000   00000 00000  00000 00000  00000 00000  00000 00000  00000 00000
3516:  00000 00000  00000 00000  00000 00000  00000 00000  00000 00000   00000 00000  00000 00000  00000 00000  00000 00000  00000 00000
3517:  00000 00000  00000 00000  00000 00000  00000 00000  00000 00000   00000 00000  00000 00000  00000 00000  00000 00000  00000 00000
3518:  00000 00000  00000 00000  00000 00000  00000 00000  00000 00000   00000 00000  00000 00000  00000 00000  00000 00000  00000 00000
3519:  00000 00000  00000 00000  00000 00000  00000 00000  00000 00000   00000 00000  00000 00000  00000 00000  00000 00000  00000 00000
3520:  00000 00000  00000 00000  00000 00000  00000 00000  00000 00000   00000 00000  00000 00000  00000 00000  00000 00000  00000 00000
3521:  00000 00000  00000 00000  00000 00000  00000 00000  00000 00000   00000 00000  00000 00000  00000 00000  00000 00000  00000 00000
3522:  00000 00000  00000 00000  00000 00000  00000 00000  00000 00000   00000 00000  00000 00000  00000 00000  00000 00000  00000 00000
3523:  00000 00000  00000 00000  00000 00000  00000 00000  00000 00000   00000 00000  00000 00000  00000 00000  00000 00000  00000 00000
3524:  00000 00000  00000 00000  00000 00000  00000 00000  00000 00000   00000 00000  00000 00000  00000 00000  00000 00000  00000 00000
3525:  00000 00000  00000 00000  00000 00000  00000 00000  00000 00000   00000 00000  00000 00000  00000 00000  00000 00000  00000 00000
3526:  00000 00000  00000 00000  00000 00000  00000 00000  00000 00000   00000 00000  00000 00000  00000 00000  00000 00000  00000 00000
3527:  00000 00000  00000 00000  00000 00000  00000 00000  00000 00000   00000 00000  00000 00000  00000 00000  00000 00000  00000 00000
3528:  00000 00000  00000 00000  00000 00000  00000 00000  00000 00000   00000 00000  00000 00000  00000 00000  00000 00000  00000 00000
3529:  00000 00000  00000 00000  00000 00000  00000 00000  00000 00000   00000 00000  00000 00000  00000 00000  00000 00000  00000 00000
3530:  00000 00000  00000 00000  00000 00000  00000 00000  00000 00000   00000 00000  00000 00000  00000 00000  00000 00000  00000 00000
3531:  00000 00000  00000 00000  00000 00000  00000 00000  00000 00000   00000 00000  00000 00000  00000 00000  00000 00000  00000 00000
3532:  00000 00000  00000 00000  00000 00000  00000 00000  00000 00000   00000 00000  00000 00000  00000 00000  00000 00000  00000 00000
3533:  00000 00000  00000 00000  00000 00000  00000 00000  00000 00000   00000 00000  00000 00000  00000 00000  00000 00000  00000 00000
3534:  00000 00000  00000 00000  00000 00000  00000 00000  00000 00000   00000 00000  00000 00000  00000 00000  00000 00000  00000 00000
3535:  00000 00000  00000 00000  00000 00000  00000 00000  00000 00000   00000 00000  00000 00000  00000 00000  00000 00000  00000 00000
3536:  00000 00000  00000 00000  00000 00000  00000 00000  00000 00000   00000 00000  00000 00000  00000 00000  00000 00000  00000 00000
3537:  00000 00000  00000 00000  00000 00000  00000 00000  00000 00000   00000 00000  00000 00000  00000 00000  00000 00000  00000 00000
3538:  00000 00000  00000 00000  00000 00000  00000 00000  00000 00000   00000 00000  00000 00000  00000 00000  00000 00000  00000 00000
3539:  00000 00000  00000 00000  00000 00000  00000 00000  00000 00000   00000 00000  00000 00000  00000 00000  00000 00000  00000 00000
3540:  00000 00000  00000 00000  00000 00000  00000 00000  00000 00000   00000 00000  00000 00000  00000 00000  00000 00000  00000 00000
3541:  00000 00000  00000 00000  00000 00000  00000 00000  00000 00000   00000 00000  00000 00000  00000 00000  00000 00000  00000 00000
3542:  00000 00000  00000 00000  00000 00000  00000 00000  00000 00000   00000 00000  00000 00000  00000 00000  00000 00000  00000 00000
3543:  00000 00000  00000 00000  00000 00000  00000 00000  00000 00000   00000 00000  00000 00000  00000 00000  00000 00000  00000 00000
3544:  00000 00000  00000 00000  00000 00000  00000 00000  00000 00000   00000 00000  00000 00000  00000 00000  00000 00000  00000 00000
3545:  00000 00000  00000 00000  00000 00000  00000 00000  00000 00000   00000 00000  00000 00000  00000 00000  00000 00000  00000 00000
3546:  00000 00000  00000 00000  00000 00000  00000 00000  00000 00000   00000 00000  00000 00000  00000 00000  00000 00000  00000 00000
3547:  00000 00000  00000 00000  00000 00000  00000 00000  00000 00000   00000 00000  00000 00000  00000 00000  00000 00000  00000 00000
3548:  00000 00000  00000 00000  00000 00000  00000 00000  00000 00000   00000 00000  00000 00000  00000 00000  00000 00000  00000 00000
3549:  00000 00000  00000 00000  00000 00000  00000 00000  00000 00000   00000 00000  00000 00000  00000 00000  00000 00000  00000 00000
```

```
3550:  00000 00000  00000 00000  00000 00000  00000 00000  00000 00000    00000 00000  00000 00000  00000 00000  00000 00000  00000 00000
3551:  00000 00000  00000 00000  00000 00000  00000 00000  00000 00000    00000 00000  00000 00000  00000 00000  00000 00000  00000 00000
3552:  00000 00000  00000 00000  00000 00000  00000 00000  00000 00000    00000 00000  00000 00000  00000 00000  00000 00000  00000 00000
3553:  00000 00000  00000 00000  00000 00000  00000 00000  00000 00000    00000 00000  00000 00000  00000 00000  00000 00000  00000 00000
3554:  00000 00000  00000 00000  00000 00000  00000 00000  00000 00000    00000 00000  00000 00000  00000 00000  00000 00000  00000 00000
3555:  00000 00000  00000 00000  00000 00000  00000 00000  00000 00000    00000 00000  00000 00000  00000 00000  00000 00000  00000 00000
3556:  00000 00000  00000 00000  00000 00000  00000 00000  00000 00000    00000 00000  00000 00000  00000 00000  00000 00000  00000 00000
3557:  00000 00000  00000 00000  00000 00000  00000 00000  00000 00000    00000 00000  00000 00000  00000 00000  00000 00000  00000 00000
3558:  00000 00000  00000 00000  00000 00000  00000 00000  00000 00000    00000 00000  00000 00000  00000 00000  00000 00000  00000 00000
3559:  00000 00000  00000 00000  00000 00000  00000 00000  00000 00000    00000 00000  00000 00000  00000 00000  00000 00000  00000 00000
3560:  00000 00000  00000 00000  00000 00000  00000 00000  00000 00000    00000 00000  00000 00000  00000 00000  00000 00000  00000 00000
3561:  00000 00000  00000 00000  00000 00000  00000 00000  00000 00000    00000 00000  00000 00000  00000 00000  00000 00000  00000 00000
3562:  00000 00000  00000 00000  00000 00000  00000 00000  00000 00000    00000 00000  00000 00000  00000 00000  00000 00000  00000 00000
3563:  00000 00000  00000 00000  00000 00000  00000 00000  00000 00000    00000 00000  00000 00000  00000 00000  00000 00000  00000 00000
3564:  00000 00000  00000 00000  00000 00000  00000 00000  00000 00000    00000 00000  00000 00000  00000 00000  00000 00000  00000 00000
3565:  00000 00000  00000 00000  00000 00000  00000 00000  00000 00000    00000 00000  00000 00000  00000 00000  00000 00000  00000 00000
3566:  00000 00000  00000 00000  00000 00000  00000 00000  00000 00000    00000 00000  00000 00000  00000 00000  00000 00000  00000 00000
3567:  00000 00000  00000 00000  00000 00000  00000 00000  00000 00000    00000 00000  00000 00000  00000 00000  00000 00000  00000 00000
3568:  00000 00000  00000 00000  00000 00000  00000 00000  00000 00000    00000 00000  00000 00000  00000 00000  00000 00000  00000 00000
3569:  00000 00000  00000 00000  00000 00000  00000 00000  00000 00000    00000 00000  00000 00000  00000 00000  00000 00000  00000 00000
3570:  00000 00000  00000 00000  00000 00000  00000 00000  00000 00000    00000 00000  00000 00000  00000 00000  00000 00000  00000 00000
3571:  00000 00000  00000 00000  00000 00000  00000 00000  00000 00000    00000 00000  00000 00000  00000 00000  00000 00000  00000 00000
3572:  00000 00000  00000 00000  00000 00000  00000 00000  00000 00000    00000 00000  00000 00000  00000 00000  00000 00000  00000 00000
3573:  00000 00000  00000 00000  00000 00000  00000 00000  00000 00000    00000 00000  00000 00000  00000 00000  00000 00000  00000 00000
3574:  00000 00000  00000 00000  00000 00000  00000 00000  00000 00000    00000 00000  00000 00000  00000 00000  00000 00000  00000 00000
3575:  00000 00000  00000 00000  00000 00000  00000 00000  00000 00000    00000 00000  00000 00000  00000 00000  00000 00000  00000 00000
3576:  00000 00000  00000 00000  00000 00000  00000 00000  00000 00000    00000 00000  00000 00000  00000 00000  00000 00000  00000 00000
3577:  00000 00000  00000 00000  00000 00000  00000 00000  00000 00000    00000 00000  00000 00000  00000 00000  00000 00000  00000 00000
3578:  00000 00000  00000 00000  00000 00000  00000 00000  00000 00000    00000 00000  00000 00000  00000 00000  00000 00000  00000 00000
3579:  00000 00000  00000 00000  00000 00000  00000 00000  00000 00000    00000 00000  00000 00000  00000 00000  00000 00000  00000 00000
3580:  00000 00000  00000 00000  00000 00000  00000 00000  00000 00000    00000 00000  00000 00000  00000 00000  00000 00000  00000 00000
3581:  00000 00000  00000 00000  00000 00000  00000 00000  00000 00000    00000 00000  00000 00000  00000 00000  00000 00000  00000 00000
3582:  00000 00000  00000 00000  00000 00000  00000 00000  00000 00000    00000 00000  00000 00000  00000 00000  00000 00000  00000 00000
3583:  00000 00000  00000 00000  00000 00000  00000 00000  00000 00000    00000 00000  00000 00000  00000 00000  00000 00000  00000 00000
3584:  00000 00000  00000 00000  00000 00000  00000 00000  00000 00000    00000 00000  00000 00000  00000 00000  00000 00000  00000 00000
3585:  00000 00000  00000 00000  00000 00000  00000 00000  00000 00000    00000 00000  00000 00000  00000 00000  00000 00000  00000 00000
3586:  00000 00000  00000 00000  00000 00000  00000 00000  00000 00000    00000 00000  00000 00000  00000 00000  00000 00000  00000 00000
3587:  00000 00000  00000 00000  00000 00000  00000 00000  00000 00000    00000 00000  00000 00000  00000 00000  00000 00000  00000 00000
3588:  00000 00000  00000 00000  00000 00000  00000 00000  00000 00000    00000 00000  00000 00000  00000 00000  00000 00000  00000 00000
3589:  00000 00000  00000 00000  00000 00000  00000 00000  00000 00000    00000 00000  00000 00000  00000 00000  00000 00000  00000 00000
3590:  00000 00000  00000 00000  00000 00000  00000 00000  00000 00000    00000 00000  00000 00000  00000 00000  00000 00000  00000 00000
3591:  00000 00000  00000 00000  00000 00000  00000 00000  00000 00000    00000 00000  00000 00000  00000 00000  00000 00000  00000 00000
3592:  00000 00000  00000 00000  00000 00000  00000 00000  00000 00000    00000 00000  00000 00000  00000 00000  00000 00000  00000 00000
3593:  00000 00000  00000 00000  00000 00000  00000 00000  00000 00000    00000 00000  00000 00000  00000 00000  00000 00000  00000 00000
3594:  00000 00000  00000 00000  00000 00000  00000 00000  00000 00000    00000 00000  00000 00000  00000 00000  00000 00000  00000 00000
3595:  00000 00000  00000 00000  00000 00000  00000 00000  00000 00000    00000 00000  00000 00000  00000 00000  00000 00000  00000 00000
3596:  00000 00000  00000 00000  00000 00000  00000 00000  00000 00000    00000 00000  00000 00000  00000 00000  00000 00000  00000 00000
3597:  00000 00000  00000 00000  00000 00000  00000 00000  00000 00000    00000 00000  00000 00000  00000 00000  00000 00000  00000 00000
3598:  00000 00000  00000 00000  00000 00000  00000 00000  00000 00000    00000 00000  00000 00000  00000 00000  00000 00000  00000 00000
3599:  00000 00000  00000 00000  00000 00000  00000 00000  00000 00000    00000 00000  00000 00000  00000 00000  00000 00000  00000 00000
```

```
3600:  00000 00000  00000 00000  00000 00000  00000 00000  00000 00000   00000 00000  00000 00000  00000 00000  00000 00000  00000 00000
3601:  00000 00000  00000 00000  00000 00000  00000 00000  00000 00000   00000 00000  00000 00000  00000 00000  00000 00000  00000 00000
3602:  00000 00000  00000 00000  00000 00000  00000 00000  00000 00000   00000 00000  00000 00000  00000 00000  00000 00000  00000 00000
3603:  00000 00000  00000 00000  00000 00000  00000 00000  00000 00000   00000 00000  00000 00000  00000 00000  00000 00000  00000 00000
3604:  00000 00000  00000 00000  00000 00000  00000 00000  00000 00000   00000 00000  00000 00000  00000 00000  00000 00000  00000 00000
3605:  00000 00000  00000 00000  00000 00000  00000 00000  00000 00000   00000 00000  00000 00000  00000 00000  00000 00000  00000 00000
3606:  00000 00000  00000 00000  00000 00000  00000 00000  00000 00000   00000 00000  00000 00000  00000 00000  00000 00000  00000 00000
3607:  00000 00000  00000 00000  00000 00000  00000 00000  00000 00000   00000 00000  00000 00000  00000 00000  00000 00000  00000 00000
3608:  00000 00000  00000 00000  00000 00000  00000 00000  00000 00000   00000 00000  00000 00000  00000 00000  00000 00000  00000 00000
3609:  00000 00000  00000 00000  00000 00000  00000 00000  00000 00000   00000 00000  00000 00000  00000 00000  00000 00000  00000 00000
3610:  00000 00000  00000 00000  00000 00000  00000 00000  00000 00000   00000 00000  00000 00000  00000 00000  00000 00000  00000 00000
3611:  00000 00000  00000 00000  00000 00000  00000 00000  00000 00000   00000 00000  00000 00000  00000 00000  00000 00000  00000 00000
3612:  00000 00000  00000 00000  00000 00000  00000 00000  00000 00000   00000 00000  00000 00000  00000 00000  00000 00000  00000 00000
3613:  00000 00000  00000 00000  00000 00000  00000 00000  00000 00000   00000 00000  00000 00000  00000 00000  00000 00000  00000 00000
3614:  00000 00000  00000 00000  00000 00000  00000 00000  00000 00000   00000 00000  00000 00000  00000 00000  00000 00000  00000 00000
3615:  00000 00000  00000 00000  00000 00000  00000 00000  00000 00000   00000 00000  00000 00000  00000 00000  00000 00000  00000 00000
3616:  00000 00000  00000 00000  00000 00000  00000 00000  00000 00000   00000 00000  00000 00000  00000 00000  00000 00000  00000 00000
3617:  00000 00000  00000 00000  00000 00000  00000 00000  00000 00000   00000 00000  00000 00000  00000 00000  00000 00000  00000 00000
3618:  00000 00000  00000 00000  00000 00000  00000 00000  00000 00000   00000 00000  00000 00000  00000 00000  00000 00000  00000 00000
3619:  00000 00000  00000 00000  00000 00000  00000 00000  00000 00000   00000 00000  00000 00000  00000 00000  00000 00000  00000 00000
3620:  00000 00000  00000 00000  00000 00000  00000 00000  00000 00000   00000 00000  00000 00000  00000 00000  00000 00000  00000 00000
3621:  00000 00000  00000 00000  00000 00000  00000 00000  00000 00000   00000 00000  00000 00000  00000 00000  00000 00000  00000 00000
3622:  00000 00000  00000 00000  00000 00000  00000 00000  00000 00000   00000 00000  00000 00000  00000 00000  00000 00000  00000 00000
3623:  00000 00000  00000 00000  00000 00000  00000 00000  00000 00000   00000 00000  00000 00000  00000 00000  00000 00000  00000 00000
3624:  00000 00000  00000 00000  00000 00000  00000 00000  00000 00000   00000 00000  00000 00000  00000 00000  00000 00000  00000 00000
3625:  00000 00000  00000 00000  00000 00000  00000 00000  00000 00000   00000 00000  00000 00000  00000 00000  00000 00000  00000 00000
3626:  00000 00000  00000 00000  00000 00000  00000 00000  00000 00000   00000 00000  00000 00000  00000 00000  00000 00000  00000 00000
3627:  00000 00000  00000 00000  00000 00000  00000 00000  00000 00000   00000 00000  00000 00000  00000 00000  00000 00000  00000 00000
3628:  00000 00000  00000 00000  00000 00000  00000 00000  00000 00000   00000 00000  00000 00000  00000 00000  00000 00000  00000 00000
3629:  00000 00000  00000 00000  00000 00000  00000 00000  00000 00000   00000 00000  00000 00000  00000 00000  00000 00000  00000 00000
3630:  00000 00000  00000 00000  00000 00000  00000 00000  00000 00000   00000 00000  00000 00000  00000 00000  00000 00000  00000 00000
3631:  00000 00000  00000 00000  00000 00000  00000 00000  00000 00000   00000 00000  00000 00000  00000 00000  00000 00000  00000 00000
3632:  00000 00000  00000 00000  00000 00000  00000 00000  00000 00000   00000 00000  00000 00000  00000 00000  00000 00000  00000 00000
3633:  00000 00000  00000 00000  00000 00000  00000 00000  00000 00000   00000 00000  00000 00000  00000 00000  00000 00000  00000 00000
3634:  00000 00000  00000 00000  00000 00000  00000 00000  00000 00000   00000 00000  00000 00000  00000 00000  00000 00000  00000 00000
3635:  00000 00000  00000 00000  00000 00000  00000 00000  00000 00000   00000 00000  00000 00000  00000 00000  00000 00000  00000 00000
3636:  00000 00000  00000 00000  00000 00000  00000 00000  00000 00000   00000 00000  00000 00000  00000 00000  00000 00000  00000 00000
3637:  00000 00000  00000 00000  00000 00000  00000 00000  00000 00000   00000 00000  00000 00000  00000 00000  00000 00000  00000 00000
3638:  00000 00000  00000 00000  00000 00000  00000 00000  00000 00000   00000 00000  00000 00000  00000 00000  00000 00000  00000 00000
3639:  00000 00000  00000 00000  00000 00000  00000 00000  00000 00000   00000 00000  00000 00000  00000 00000  00000 00000  00000 00000
3640:  00000 00000  00000 00000  00000 00000  00000 00000  00000 00000   00000 00000  00000 00000  00000 00000  00000 00000  00000 00000
3641:  00000 00000  00000 00000  00000 00000  00000 00000  00000 00000   00000 00000  00000 00000  00000 00000  00000 00000  00000 00000
3642:  00000 00000  00000 00000  00000 00000  00000 00000  00000 00000   00000 00000  00000 00000  00000 00000  00000 00000  00000 00000
3643:  00000 00000  00000 00000  00000 00000  00000 00000  00000 00000   00000 00000  00000 00000  00000 00000  00000 00000  00000 00000
3644:  00000 00000  00000 00000  00000 00000  00000 00000  00000 00000   00000 00000  00000 00000  00000 00000  00000 00000  00000 00000
3645:  00000 00000  00000 00000  00000 00000  00000 00000  00000 00000   00000 00000  00000 00000  00000 00000  00000 00000  00000 00000
3646:  00000 00000  00000 00000  00000 00000  00000 00000  00000 00000   00000 00000  00000 00000  00000 00000  00000 00000  00000 00000
3647:  00000 00000  00000 00000  00000 00000  00000 00000  00000 00000   00000 00000  00000 00000  00000 00000  00000 00000  00000 00000
3648:  00000 00000  00000 00000  00000 00000  00000 00000  00000 00000   00000 00000  00000 00000  00000 00000  00000 00000  00000 00000
3649:  00000 00000  00000 00000  00000 00000  00000 00000  00000 00000   00000 00000  00000 00000  00000 00000  00000 00000  00000 00000
```

```
3650:  00000 00000  00000 00000  00000 00000  00000 00000  00000 00000    00000 00000  00000 00000  00000 00000  00000 00000  00000 00000
3651:  00000 00000  00000 00000  00000 00000  00000 00000  00000 00000    00000 00000  00000 00000  00000 00000  00000 00000  00000 00000
3652:  00000 00000  00000 00000  00000 00000  00000 00000  00000 00000    00000 00000  00000 00000  00000 00000  00000 00000  00000 00000
3653:  00000 00000  00000 00000  00000 00000  00000 00000  00000 00000    00000 00000  00000 00000  00000 00000  00000 00000  00000 00000
3654:  00000 00000  00000 00000  00000 00000  00000 00000  00000 00000    00000 00000  00000 00000  00000 00000  00000 00000  00000 00000
3655:  00000 00000  00000 00000  00000 00000  00000 00000  00000 00000    00000 00000  00000 00000  00000 00000  00000 00000  00000 00000
3656:  00000 00000  00000 00000  00000 00000  00000 00000  00000 00000    00000 00000  00000 00000  00000 00000  00000 00000  00000 00000
3657:  00000 00000  00000 00000  00000 00000  00000 00000  00000 00000    00000 00000  00000 00000  00000 00000  00000 00000  00000 00000
3658:  00000 00000  00000 00000  00000 00000  00000 00000  00000 00000    00000 00000  00000 00000  00000 00000  00000 00000  00000 00000
3659:  00000 00000  00000 00000  00000 00000  00000 00000  00000 00000    00000 00000  00000 00000  00000 00000  00000 00000  00000 00000
3660:  00000 00000  00000 00000  00000 00000  00000 00000  00000 00000    00000 00000  00000 00000  00000 00000  00000 00000  00000 00000
3661:  00000 00000  00000 00000  00000 00000  00000 00000  00000 00000    00000 00000  00000 00000  00000 00000  00000 00000  00000 00000
3662:  00000 00000  00000 00000  00000 00000  00000 00000  00000 00000    00000 00000  00000 00000  00000 00000  00000 00000  00000 00000
3663:  00000 00000  00000 00000  00000 00000  00000 00000  00000 00000    00000 00000  00000 00000  00000 00000  00000 00000  00000 00000
3664:  00000 00000  00000 00000  00000 00000  00000 00000  00000 00000    00000 00000  00000 00000  00000 00000  00000 00000  00000 00000
3665:  00000 00000  00000 00000  00000 00000  00000 00000  00000 00000    00000 00000  00000 00000  00000 00000  00000 00000  00000 00000
3666:  00000 00000  00000 00000  00000 00000  00000 00000  00000 00000    00000 00000  00000 00000  00000 00000  00000 00000  00000 00000
3667:  00000 00000  00000 00000  00000 00000  00000 00000  00000 00000    00000 00000  00000 00000  00000 00000  00000 00000  00000 00000
3668:  00000 00000  00000 00000  00000 00000  00000 00000  00000 00000    00000 00000  00000 00000  00000 00000  00000 00000  00000 00000
3669:  00000 00000  00000 00000  00000 00000  00000 00000  00000 00000    00000 00000  00000 00000  00000 00000  00000 00000  00000 00000
3670:  00000 00000  00000 00000  00000 00000  00000 00000  00000 00000    00000 00000  00000 00000  00000 00000  00000 00000  00000 00000
3671:  00000 00000  00000 00000  00000 00000  00000 00000  00000 00000    00000 00000  00000 00000  00000 00000  00000 00000  00000 00000
3672:  00000 00000  00000 00000  00000 00000  00000 00000  00000 00000    00000 00000  00000 00000  00000 00000  00000 00000  00000 00000
3673:  00000 00000  00000 00000  00000 00000  00000 00000  00000 00000    00000 00000  00000 00000  00000 00000  00000 00000  00000 00000
3674:  00000 00000  00000 00000  00000 00000  00000 00000  00000 00000    00000 00000  00000 00000  00000 00000  00000 00000  00000 00000
3675:  00000 00000  00000 00000  00000 00000  00000 00000  00000 00000    00000 00000  00000 00000  00000 00000  00000 00000  00000 00000
3676:  00000 00000  00000 00000  00000 00000  00000 00000  00000 00000    00000 00000  00000 00000  00000 00000  00000 00000  00000 00000
3677:  00000 00000  00000 00000  00000 00000  00000 00000  00000 00000    00000 00000  00000 00000  00000 00000  00000 00000  00000 00000
3678:  00000 00000  00000 00000  00000 00000  00000 00000  00000 00000    00000 00000  00000 00000  00000 00000  00000 00000  00000 00000
3679:  00000 00000  00000 00000  00000 00000  00000 00000  00000 00000    00000 00000  00000 00000  00000 00000  00000 00000  00000 00000
3680:  00000 00000  00000 00000  00000 00000  00000 00000  00000 00000    00000 00000  00000 00000  00000 00000  00000 00000  00000 00000
3681:  00000 00000  00000 00000  00000 00000  00000 00000  00000 00000    00000 00000  00000 00000  00000 00000  00000 00000  00000 00000
3682:  00000 00000  00000 00000  00000 00000  00000 00000  00000 00000    00000 00000  00000 00000  00000 00000  00000 00000  00000 00000
3683:  00000 00000  00000 00000  00000 00000  00000 00000  00000 00000    00000 00000  00000 00000  00000 00000  00000 00000  00000 00000
3684:  00000 00000  00000 00000  00000 00000  00000 00000  00000 00000    00000 00000  00000 00000  00000 00000  00000 00000  00000 00000
3685:  00000 00000  00000 00000  00000 00000  00000 00000  00000 00000    00000 00000  00000 00000  00000 00000  00000 00000  00000 00000
3686:  00000 00000  00000 00000  00000 00000  00000 00000  00000 00000    00000 00000  00000 00000  00000 00000  00000 00000  00000 00000
3687:  00000 00000  00000 00000  00000 00000  00000 00000  00000 00000    00000 00000  00000 00000  00000 00000  00000 00000  00000 00000
3688:  00000 00000  00000 00000  00000 00000  00000 00000  00000 00000    00000 00000  00000 00000  00000 00000  00000 00000  00000 00000
3689:  00000 00000  00000 00000  00000 00000  00000 00000  00000 00000    00000 00000  00000 00000  00000 00000  00000 00000  00000 00000
3690:  00000 00000  00000 00000  00000 00000  00000 00000  00000 00000    00000 00000  00000 00000  00000 00000  00000 00000  00000 00000
3691:  00000 00000  00000 00000  00000 00000  00000 00000  00000 00000    00000 00000  00000 00000  00000 00000  00000 00000  00000 00000
3692:  00000 00000  00000 00000  00000 00000  00000 00000  00000 00000    00000 00000  00000 00000  00000 00000  00000 00000  00000 00000
3693:  00000 00000  00000 00000  00000 00000  00000 00000  00000 00000    00000 00000  00000 00000  00000 00000  00000 00000  00000 00000
3694:  00000 00000  00000 00000  00000 00000  00000 00000  00000 00000    00000 00000  00000 00000  00000 00000  00000 00000  00000 00000
3695:  00000 00000  00000 00000  00000 00000  00000 00000  00000 00000    00000 00000  00000 00000  00000 00000  00000 00000  00000 00000
3696:  00000 00000  00000 00000  00000 00000  00000 00000  00000 00000    00000 00000  00000 00000  00000 00000  00000 00000  00000 00000
3697:  00000 00000  00000 00000  00000 00000  00000 00000  00000 00000    00000 00000  00000 00000  00000 00000  00000 00000  00000 00000
3698:  00000 00000  00000 00000  00000 00000  00000 00000  00000 00000    00000 00000  00000 00000  00000 00000  00000 00000  00000 00000
3699:  00000 00000  00000 00000  00000 00000  00000 00000  00000 00000    00000 00000  00000 00000  00000 00000  00000 00000  00000 00000
```

```
3700:  00000 00000  00000 00000  00000 00000  00000 00000  00000 00000   00000 00000  00000 00000  00000 00000  00000 00000  00000 00000
3701:  00000 00000  00000 00000  00000 00000  00000 00000  00000 00000   00000 00000  00000 00000  00000 00000  00000 00000  00000 00000
3702:  00000 00000  00000 00000  00000 00000  00000 00000  00000 00000   00000 00000  00000 00000  00000 00000  00000 00000  00000 00000
3703:  00000 00000  00000 00000  00000 00000  00000 00000  00000 00000   00000 00000  00000 00000  00000 00000  00000 00000  00000 00000
3704:  00000 00000  00000 00000  00000 00000  00000 00000  00000 00000   00000 00000  00000 00000  00000 00000  00000 00000  00000 00000
3705:  00000 00000  00000 00000  00000 00000  00000 00000  00000 00000   00000 00000  00000 00000  00000 00000  00000 00000  00000 00000
3706:  00000 00000  00000 00000  00000 00000  00000 00000  00000 00000   00000 00000  00000 00000  00000 00000  00000 00000  00000 00000
3707:  00000 00000  00000 00000  00000 00000  00000 00000  00000 00000   00000 00000  00000 00000  00000 00000  00000 00000  00000 00000
3708:  00000 00000  00000 00000  00000 00000  00000 00000  00000 00000   00000 00000  00000 00000  00000 00000  00000 00000  00000 00000
3709:  00000 00000  00000 00000  00000 00000  00000 00000  00000 00000   00000 00000  00000 00000  00000 00000  00000 00000  00000 00000
3710:  00000 00000  00000 00000  00000 00000  00000 00000  00000 00000   00000 00000  00000 00000  00000 00000  00000 00000  00000 00000
3711:  00000 00000  00000 00000  00000 00000  00000 00000  00000 00000   00000 00000  00000 00000  00000 00000  00000 00000  00000 00000
3712:  00000 00000  00000 00000  00000 00000  00000 00000  00000 00000   00000 00000  00000 00000  00000 00000  00000 00000  00000 00000
3713:  00000 00000  00000 00000  00000 00000  00000 00000  00000 00000   00000 00000  00000 00000  00000 00000  00000 00000  00000 00000
3714:  00000 00000  00000 00000  00000 00000  00000 00000  00000 00000   00000 00000  00000 00000  00000 00000  00000 00000  00000 00000
3715:  00000 00000  00000 00000  00000 00000  00000 00000  00000 00000   00000 00000  00000 00000  00000 00000  00000 00000  00000 00000
3716:  00000 00000  00000 00000  00000 00000  00000 00000  00000 00000   00000 00000  00000 00000  00000 00000  00000 00000  00000 00000
3717:  00000 00000  00000 00000  00000 00000  00000 00000  00000 00000   00000 00000  00000 00000  00000 00000  00000 00000  00000 00000
3718:  00000 00000  00000 00000  00000 00000  00000 00000  00000 00000   00000 00000  00000 00000  00000 00000  00000 00000  00000 00000
3719:  00000 00000  00000 00000  00000 00000  00000 00000  00000 00000   00000 00000  00000 00000  00000 00000  00000 00000  00000 00000
3720:  00000 00000  00000 00000  00000 00000  00000 00000  00000 00000   00000 00000  00000 00000  00000 00000  00000 00000  00000 00000
3721:  00000 00000  00000 00000  00000 00000  00000 00000  00000 00000   00000 00000  00000 00000  00000 00000  00000 00000  00000 00000
3722:  00000 00000  00000 00000  00000 00000  00000 00000  00000 00000   00000 00000  00000 00000  00000 00000  00000 00000  00000 00000
3723:  00000 00000  00000 00000  00000 00000  00000 00000  00000 00000   00000 00000  00000 00000  00000 00000  00000 00000  00000 00000
3724:  00000 00000  00000 00000  00000 00000  00000 00000  00000 00000   00000 00000  00000 00000  00000 00000  00000 00000  00000 00000
3725:  00000 00000  00000 00000  00000 00000  00000 00000  00000 00000   00000 00000  00000 00000  00000 00000  00000 00000  00000 00000
3726:  00000 00000  00000 00000  00000 00000  00000 00000  00000 00000   00000 00000  00000 00000  00000 00000  00000 00000  00000 00000
3727:  00000 00000  00000 00000  00000 00000  00000 00000  00000 00000   00000 00000  00000 00000  00000 00000  00000 00000  00000 00000
3728:  00000 00000  00000 00000  00000 00000  00000 00000  00000 00000   00000 00000  00000 00000  00000 00000  00000 00000  00000 00000
3729:  00000 00000  00000 00000  00000 00000  00000 00000  00000 00000   00000 00000  00000 00000  00000 00000  00000 00000  00000 00000
3730:  00000 00000  00000 00000  00000 00000  00000 00000  00000 00000   00000 00000  00000 00000  00000 00000  00000 00000  00000 00000
3731:  00000 00000  00000 00000  00000 00000  00000 00000  00000 00000   00000 00000  00000 00000  00000 00000  00000 00000  00000 00000
3732:  00000 00000  00000 00000  00000 00000  00000 00000  00000 00000   00000 00000  00000 00000  00000 00000  00000 00000  00000 00000
3733:  00000 00000  00000 00000  00000 00000  00000 00000  00000 00000   00000 00000  00000 00000  00000 00000  00000 00000  00000 00000
3734:  00000 00000  00000 00000  00000 00000  00000 00000  00000 00000   00000 00000  00000 00000  00000 00000  00000 00000  00000 00000
3735:  00000 00000  00000 00000  00000 00000  00000 00000  00000 00000   00000 00000  00000 00000  00000 00000  00000 00000  00000 00000
3736:  00000 00000  00000 00000  00000 00000  00000 00000  00000 00000   00000 00000  00000 00000  00000 00000  00000 00000  00000 00000
3737:  00000 00000  00000 00000  00000 00000  00000 00000  00000 00000   00000 00000  00000 00000  00000 00000  00000 00000  00000 00000
3738:  00000 00000  00000 00000  00000 00000  00000 00000  00000 00000   00000 00000  00000 00000  00000 00000  00000 00000  00000 00000
3739:  00000 00000  00000 00000  00000 00000  00000 00000  00000 00000   00000 00000  00000 00000  00000 00000  00000 00000  00000 00000
3740:  00000 00000  00000 00000  00000 00000  00000 00000  00000 00000   00000 00000  00000 00000  00000 00000  00000 00000  00000 00000
3741:  00000 00000  00000 00000  00000 00000  00000 00000  00000 00000   00000 00000  00000 00000  00000 00000  00000 00000  00000 00000
3742:  00000 00000  00000 00000  00000 00000  00000 00000  00000 00000   00000 00000  00000 00000  00000 00000  00000 00000  00000 00000
3743:  00000 00000  00000 00000  00000 00000  00000 00000  00000 00000   00000 00000  00000 00000  00000 00000  00000 00000  00000 00000
3744:  00000 00000  00000 00000  00000 00000  00000 00000  00000 00000   00000 00000  00000 00000  00000 00000  00000 00000  00000 00000
3745:  00000 00000  00000 00000  00000 00000  00000 00000  00000 00000   00000 00000  00000 00000  00000 00000  00000 00000  00000 00000
3746:  00000 00000  00000 00000  00000 00000  00000 00000  00000 00000   00000 00000  00000 00000  00000 00000  00000 00000  00000 00000
3747:  00000 00000  00000 00000  00000 00000  00000 00000  00000 00000   00000 00000  00000 00000  00000 00000  00000 00000  00000 00000
3748:  00000 00000  00000 00000  00000 00000  00000 00000  00000 00000   00000 00000  00000 00000  00000 00000  00000 00000  00000 00000
3749:  00000 00000  00000 00000  00000 00000  00000 00000  00000 00000   00000 00000  00000 00000  00000 00000  00000 00000  00000 00000
```

```
3750:  00000 00000  00000 00000  00000 00000  00000 00000  00000 00000    00000 00000  00000 00000  00000 00000  00000 00000  00000 00000
3751:  00000 00000  00000 00000  00000 00000  00000 00000  00000 00000    00000 00000  00000 00000  00000 00000  00000 00000  00000 00000
3752:  00000 00000  00000 00000  00000 00000  00000 00000  00000 00000    00000 00000  00000 00000  00000 00000  00000 00000  00000 00000
3753:  00000 00000  00000 00000  00000 00000  00000 00000  00000 00000    00000 00000  00000 00000  00000 00000  00000 00000  00000 00000
3754:  00000 00000  00000 00000  00000 00000  00000 00000  00000 00000    00000 00000  00000 00000  00000 00000  00000 00000  00000 00000
3755:  00000 00000  00000 00000  00000 00000  00000 00000  00000 00000    00000 00000  00000 00000  00000 00000  00000 00000  00000 00000
3756:  00000 00000  00000 00000  00000 00000  00000 00000  00000 00000    00000 00000  00000 00000  00000 00000  00000 00000  00000 00000
3757:  00000 00000  00000 00000  00000 00000  00000 00000  00000 00000    00000 00000  00000 00000  00000 00000  00000 00000  00000 00000
3758:  00000 00000  00000 00000  00000 00000  00000 00000  00000 00000    00000 00000  00000 00000  00000 00000  00000 00000  00000 00000
3759:  00000 00000  00000 00000  00000 00000  00000 00000  00000 00000    00000 00000  00000 00000  00000 00000  00000 00000  00000 00000
3760:  00000 00000  00000 00000  00000 00000  00000 00000  00000 00000    00000 00000  00000 00000  00000 00000  00000 00000  00000 00000
3761:  00000 00000  00000 00000  00000 00000  00000 00000  00000 00000    00000 00000  00000 00000  00000 00000  00000 00000  00000 00000
3762:  00000 00000  00000 00000  00000 00000  00000 00000  00000 00000    00000 00000  00000 00000  00000 00000  00000 00000  00000 00000
3763:  00000 00000  00000 00000  00000 00000  00000 00000  00000 00000    00000 00000  00000 00000  00000 00000  00000 00000  00000 00000
3764:  00000 00000  00000 00000  00000 00000  00000 00000  00000 00000    00000 00000  00000 00000  00000 00000  00000 00000  00000 00000
3765:  00000 00000  00000 00000  00000 00000  00000 00000  00000 00000    00000 00000  00000 00000  00000 00000  00000 00000  00000 00000
3766:  00000 00000  00000 00000  00000 00000  00000 00000  00000 00000    00000 00000  00000 00000  00000 00000  00000 00000  00000 00000
3767:  00000 00000  00000 00000  00000 00000  00000 00000  00000 00000    00000 00000  00000 00000  00000 00000  00000 00000  00000 00000
3768:  00000 00000  00000 00000  00000 00000  00000 00000  00000 00000    00000 00000  00000 00000  00000 00000  00000 00000  00000 00000
3769:  00000 00000  00000 00000  00000 00000  00000 00000  00000 00000    00000 00000  00000 00000  00000 00000  00000 00000  00000 00000
3770:  00000 00000  00000 00000  00000 00000  00000 00000  00000 00000    00000 00000  00000 00000  00000 00000  00000 00000  00000 00000
3771:  00000 00000  00000 00000  00000 00000  00000 00000  00000 00000    00000 00000  00000 00000  00000 00000  00000 00000  00000 00000
3772:  00000 00000  00000 00000  00000 00000  00000 00000  00000 00000    00000 00000  00000 00000  00000 00000  00000 00000  00000 00000
3773:  00000 00000  00000 00000  00000 00000  00000 00000  00000 00000    00000 00000  00000 00000  00000 00000  00000 00000  00000 00000
3774:  00000 00000  00000 00000  00000 00000  00000 00000  00000 00000    00000 00000  00000 00000  00000 00000  00000 00000  00000 00000
3775:  00000 00000  00000 00000  00000 00000  00000 00000  00000 00000    00000 00000  00000 00000  00000 00000  00000 00000  00000 00000
3776:  00000 00000  00000 00000  00000 00000  00000 00000  00000 00000    00000 00000  00000 00000  00000 00000  00000 00000  00000 00000
3777:  00000 00000  00000 00000  00000 00000  00000 00000  00000 00000    00000 00000  00000 00000  00000 00000  00000 00000  00000 00000
3778:  00000 00000  00000 00000  00000 00000  00000 00000  00000 00000    00000 00000  00000 00000  00000 00000  00000 00000  00000 00000
3779:  00000 00000  00000 00000  00000 00000  00000 00000  00000 00000    00000 00000  00000 00000  00000 00000  00000 00000  00000 00000
3780:  00000 00000  00000 00000  00000 00000  00000 00000  00000 00000    00000 00000  00000 00000  00000 00000  00000 00000  00000 00000
3781:  00000 00000  00000 00000  00000 00000  00000 00000  00000 00000    00000 00000  00000 00000  00000 00000  00000 00000  00000 00000
3782:  00000 00000  00000 00000  00000 00000  00000 00000  00000 00000    00000 00000  00000 00000  00000 00000  00000 00000  00000 00000
3783:  00000 00000  00000 00000  00000 00000  00000 00000  00000 00000    00000 00000  00000 00000  00000 00000  00000 00000  00000 00000
3784:  00000 00000  00000 00000  00000 00000  00000 00000  00000 00000    00000 00000  00000 00000  00000 00000  00000 00000  00000 00000
3785:  00000 00000  00000 00000  00000 00000  00000 00000  00000 00000    00000 00000  00000 00000  00000 00000  00000 00000  00000 00000
3786:  00000 00000  00000 00000  00000 00000  00000 00000  00000 00000    00000 00000  00000 00000  00000 00000  00000 00000  00000 00000
3787:  00000 00000  00000 00000  00000 00000  00000 00000  00000 00000    00000 00000  00000 00000  00000 00000  00000 00000  00000 00000
3788:  00000 00000  00000 00000  00000 00000  00000 00000  00000 00000    00000 00000  00000 00000  00000 00000  00000 00000  00000 00000
3789:  00000 00000  00000 00000  00000 00000  00000 00000  00000 00000    00000 00000  00000 00000  00000 00000  00000 00000  00000 00000
3790:  00000 00000  00000 00000  00000 00000  00000 00000  00000 00000    00000 00000  00000 00000  00000 00000  00000 00000  00000 00000
3791:  00000 00000  00000 00000  00000 00000  00000 00000  00000 00000    00000 00000  00000 00000  00000 00000  00000 00000  00000 00000
3792:  00000 00000  00000 00000  00000 00000  00000 00000  00000 00000    00000 00000  00000 00000  00000 00000  00000 00000  00000 00000
3793:  00000 00000  00000 00000  00000 00000  00000 00000  00000 00000    00000 00000  00000 00000  00000 00000  00000 00000  00000 00000
3794:  00000 00000  00000 00000  00000 00000  00000 00000  00000 00000    00000 00000  00000 00000  00000 00000  00000 00000  00000 00000
3795:  00000 00000  00000 00000  00000 00000  00000 00000  00000 00000    00000 00000  00000 00000  00000 00000  00000 00000  00000 00000
3796:  00000 00000  00000 00000  00000 00000  00000 00000  00000 00000    00000 00000  00000 00000  00000 00000  00000 00000  00000 00000
3797:  00000 00000  00000 00000  00000 00000  00000 00000  00000 00000    00000 00000  00000 00000  00000 00000  00000 00000  00000 00000
3798:  00000 00000  00000 00000  00000 00000  00000 00000  00000 00000    00000 00000  00000 00000  00000 00000  00000 00000  00000 00000
3799:  00000 00000  00000 00000  00000 00000  00000 00000  00000 00000    00000 00000  00000 00000  00000 00000  00000 00000  00000 00000
```

```
3800:  00000 00000  00000 00000  00000 00000  00000 00000  00000 00000    00000 00000  00000 00000  00000 00000  00000 00000  00000 00000
3801:  00000 00000  00000 00000  00000 00000  00000 00000  00000 00000    00000 00000  00000 00000  00000 00000  00000 00000  00000 00000
3802:  00000 00000  00000 00000  00000 00000  00000 00000  00000 00000    00000 00000  00000 00000  00000 00000  00000 00000  00000 00000
3803:  00000 00000  00000 00000  00000 00000  00000 00000  00000 00000    00000 00000  00000 00000  00000 00000  00000 00000  00000 00000
3804:  00000 00000  00000 00000  00000 00000  00000 00000  00000 00000    00000 00000  00000 00000  00000 00000  00000 00000  00000 00000
3805:  00000 00000  00000 00000  00000 00000  00000 00000  00000 00000    00000 00000  00000 00000  00000 00000  00000 00000  00000 00000
3806:  00000 00000  00000 00000  00000 00000  00000 00000  00000 00000    00000 00000  00000 00000  00000 00000  00000 00000  00000 00000
3807:  00000 00000  00000 00000  00000 00000  00000 00000  00000 00000    00000 00000  00000 00000  00000 00000  00000 00000  00000 00000
3808:  00000 00000  00000 00000  00000 00000  00000 00000  00000 00000    00000 00000  00000 00000  00000 00000  00000 00000  00000 00000
3809:  00000 00000  00000 00000  00000 00000  00000 00000  00000 00000    00000 00000  00000 00000  00000 00000  00000 00000  00000 00000
3810:  00000 00000  00000 00000  00000 00000  00000 00000  00000 00000    00000 00000  00000 00000  00000 00000  00000 00000  00000 00000
3811:  00000 00000  00000 00000  00000 00000  00000 00000  00000 00000    00000 00000  00000 00000  00000 00000  00000 00000  00000 00000
3812:  00000 00000  00000 00000  00000 00000  00000 00000  00000 00000    00000 00000  00000 00000  00000 00000  00000 00000  00000 00000
3813:  00000 00000  00000 00000  00000 00000  00000 00000  00000 00000    00000 00000  00000 00000  00000 00000  00000 00000  00000 00000
3814:  00000 00000  00000 00000  00000 00000  00000 00000  00000 00000    00000 00000  00000 00000  00000 00000  00000 00000  00000 00000
3815:  00000 00000  00000 00000  00000 00000  00000 00000  00000 00000    00000 00000  00000 00000  00000 00000  00000 00000  00000 00000
3816:  00000 00000  00000 00000  00000 00000  00000 00000  00000 00000    00000 00000  00000 00000  00000 00000  00000 00000  00000 00000
3817:  00000 00000  00000 00000  00000 00000  00000 00000  00000 00000    00000 00000  00000 00000  00000 00000  00000 00000  00000 00000
3818:  00000 00000  00000 00000  00000 00000  00000 00000  00000 00000    00000 00000  00000 00000  00000 00000  00000 00000  00000 00000
3819:  00000 00000  00000 00000  00000 00000  00000 00000  00000 00000    00000 00000  00000 00000  00000 00000  00000 00000  00000 00000
3820:  00000 00000  00000 00000  00000 00000  00000 00000  00000 00000    00000 00000  00000 00000  00000 00000  00000 00000  00000 00000
3821:  00000 00000  00000 00000  00000 00000  00000 00000  00000 00000    00000 00000  00000 00000  00000 00000  00000 00000  00000 00000
3822:  00000 00000  00000 00000  00000 00000  00000 00000  00000 00000    00000 00000  00000 00000  00000 00000  00000 00000  00000 00000
3823:  00000 00000  00000 00000  00000 00000  00000 00000  00000 00000    00000 00000  00000 00000  00000 00000  00000 00000  00000 00000
3824:  00000 00000  00000 00000  00000 00000  00000 00000  00000 00000    00000 00000  00000 00000  00000 00000  00000 00000  00000 00000
3825:  00000 00000  00000 00000  00000 00000  00000 00000  00000 00000    00000 00000  00000 00000  00000 00000  00000 00000  00000 00000
3826:  00000 00000  00000 00000  00000 00000  00000 00000  00000 00000    00000 00000  00000 00000  00000 00000  00000 00000  00000 00000
3827:  00000 00000  00000 00000  00000 00000  00000 00000  00000 00000    00000 00000  00000 00000  00000 00000  00000 00000  00000 00000
3828:  00000 00000  00000 00000  00000 00000  00000 00000  00000 00000    00000 00000  00000 00000  00000 00000  00000 00000  00000 00000
3829:  00000 00000  00000 00000  00000 00000  00000 00000  00000 00000    00000 00000  00000 00000  00000 00000  00000 00000  00000 00000
3830:  00000 00000  00000 00000  00000 00000  00000 00000  00000 00000    00000 00000  00000 00000  00000 00000  00000 00000  00000 00000
3831:  00000 00000  00000 00000  00000 00000  00000 00000  00000 00000    00000 00000  00000 00000  00000 00000  00000 00000  00000 00000
3832:  00000 00000  00000 00000  00000 00000  00000 00000  00000 00000    00000 00000  00000 00000  00000 00000  00000 00000  00000 00000
3833:  00000 00000  00000 00000  00000 00000  00000 00000  00000 00000    00000 00000  00000 00000  00000 00000  00000 00000  00000 00000
3834:  00000 00000  00000 00000  00000 00000  00000 00000  00000 00000    00000 00000  00000 00000  00000 00000  00000 00000  00000 00000
3835:  00000 00000  00000 00000  00000 00000  00000 00000  00000 00000    00000 00000  00000 00000  00000 00000  00000 00000  00000 00000
3836:  00000 00000  00000 00000  00000 00000  00000 00000  00000 00000    00000 00000  00000 00000  00000 00000  00000 00000  00000 00000
3837:  00000 00000  00000 00000  00000 00000  00000 00000  00000 00000    00000 00000  00000 00000  00000 00000  00000 00000  00000 00000
3838:  00000 00000  00000 00000  00000 00000  00000 00000  00000 00000    00000 00000  00000 00000  00000 00000  00000 00000  00000 00000
3839:  00000 00000  00000 00000  00000 00000  00000 00000  00000 00000    00000 00000  00000 00000  00000 00000  00000 00000  00000 00000
3840:  00000 00000  00000 00000  00000 00000  00000 00000  00000 00000    00000 00000  00000 00000  00000 00000  00000 00000  00000 00000
3841:  00000 00000  00000 00000  00000 00000  00000 00000  00000 00000    00000 00000  00000 00000  00000 00000  00000 00000  00000 00000
3842:  00000 00000  00000 00000  00000 00000  00000 00000  00000 00000    00000 00000  00000 00000  00000 00000  00000 00000  00000 00000
3843:  00000 00000  00000 00000  00000 00000  00000 00000  00000 00000    00000 00000  00000 00000  00000 00000  00000 00000  00000 00000
3844:  00000 00000  00000 00000  00000 00000  00000 00000  00000 00000    00000 00000  00000 00000  00000 00000  00000 00000  00000 00000
3845:  00000 00000  00000 00000  00000 00000  00000 00000  00000 00000    00000 00000  00000 00000  00000 00000  00000 00000  00000 00000
3846:  00000 00000  00000 00000  00000 00000  00000 00000  00000 00000    00000 00000  00000 00000  00000 00000  00000 00000  00000 00000
3847:  00000 00000  00000 00000  00000 00000  00000 00000  00000 00000    00000 00000  00000 00000  00000 00000  00000 00000  00000 00000
3848:  00000 00000  00000 00000  00000 00000  00000 00000  00000 00000    00000 00000  00000 00000  00000 00000  00000 00000  00000 00000
3849:  00000 00000  00000 00000  00000 00000  00000 00000  00000 00000    00000 00000  00000 00000  00000 00000  00000 00000  00000 00000
```

```
3850:   00000 00000   00000 00000   00000 00000   00000 00000   00000 00000     00000 00000   00000 00000   00000 00000   00000 00000   00000 00000
3851:   00000 00000   00000 00000   00000 00000   00000 00000   00000 00000     00000 00000   00000 00000   00000 00000   00000 00000   00000 00000
3852:   00000 00000   00000 00000   00000 00000   00000 00000   00000 00000     00000 00000   00000 00000   00000 00000   00000 00000   00000 00000
3853:   00000 00000   00000 00000   00000 00000   00000 00000   00000 00000     00000 00000   00000 00000   00000 00000   00000 00000   00000 00000
3854:   00000 00000   00000 00000   00000 00000   00000 00000   00000 00000     00000 00000   00000 00000   00000 00000   00000 00000   00000 00000
3855:   00000 00000   00000 00000   00000 00000   00000 00000   00000 00000     00000 00000   00000 00000   00000 00000   00000 00000   00000 00000
3856:   00000 00000   00000 00000   00000 00000   00000 00000   00000 00000     00000 00000   00000 00000   00000 00000   00000 00000   00000 00000
3857:   00000 00000   00000 00000   00000 00000   00000 00000   00000 00000     00000 00000   00000 00000   00000 00000   00000 00000   00000 00000
3858:   00000 00000   00000 00000   00000 00000   00000 00000   00000 00000     00000 00000   00000 00000   00000 00000   00000 00000   00000 00000
3859:   00000 00000   00000 00000   00000 00000   00000 00000   00000 00000     00000 00000   00000 00000   00000 00000   00000 00000   00000 00000
3860:   00000 00000   00000 00000   00000 00000   00000 00000   00000 00000     00000 00000   00000 00000   00000 00000   00000 00000   00000 00000
3861:   00000 00000   00000 00000   00000 00000   00000 00000   00000 00000     00000 00000   00000 00000   00000 00000   00000 00000   00000 00000
3862:   00000 00000   00000 00000   00000 00000   00000 00000   00000 00000     00000 00000   00000 00000   00000 00000   00000 00000   00000 00000
3863:   00000 00000   00000 00000   00000 00000   00000 00000   00000 00000     00000 00000   00000 00000   00000 00000   00000 00000   00000 00000
3864:   00000 00000   00000 00000   00000 00000   00000 00000   00000 00000     00000 00000   00000 00000   00000 00000   00000 00000   00000 00000
3865:   00000 00000   00000 00000   00000 00000   00000 00000   00000 00000     00000 00000   00000 00000   00000 00000   00000 00000   00000 00000
3866:   00000 00000   00000 00000   00000 00000   00000 00000   00000 00000     00000 00000   00000 00000   00000 00000   00000 00000   00000 00000
3867:   00000 00000   00000 00000   00000 00000   00000 00000   00000 00000     00000 00000   00000 00000   00000 00000   00000 00000   00000 00000
3868:   00000 00000   00000 00000   00000 00000   00000 00000   00000 00000     00000 00000   00000 00000   00000 00000   00000 00000   00000 00000
3869:   00000 00000   00000 00000   00000 00000   00000 00000   00000 00000     00000 00000   00000 00000   00000 00000   00000 00000   00000 00000
3870:   00000 00000   00000 00000   00000 00000   00000 00000   00000 00000     00000 00000   00000 00000   00000 00000   00000 00000   00000 00000
3871:   00000 00000   00000 00000   00000 00000   00000 00000   00000 00000     00000 00000   00000 00000   00000 00000   00000 00000   00000 00000
3872:   00000 00000   00000 00000   00000 00000   00000 00000   00000 00000     00000 00000   00000 00000   00000 00000   00000 00000   00000 00000
3873:   00000 00000   00000 00000   00000 00000   00000 00000   00000 00000     00000 00000   00000 00000   00000 00000   00000 00000   00000 00000
3874:   00000 00000   00000 00000   00000 00000   00000 00000   00000 00000     00000 00000   00000 00000   00000 00000   00000 00000   00000 00000
3875:   00000 00000   00000 00000   00000 00000   00000 00000   00000 00000     00000 00000   00000 00000   00000 00000   00000 00000   00000 00000
3876:   00000 00000   00000 00000   00000 00000   00000 00000   00000 00000     00000 00000   00000 00000   00000 00000   00000 00000   00000 00000
3877:   00000 00000   00000 00000   00000 00000   00000 00000   00000 00000     00000 00000   00000 00000   00000 00000   00000 00000   00000 00000
3878:   00000 00000   00000 00000   00000 00000   00000 00000   00000 00000     00000 00000   00000 00000   00000 00000   00000 00000   00000 00000
3879:   00000 00000   00000 00000   00000 00000   00000 00000   00000 00000     00000 00000   00000 00000   00000 00000   00000 00000   00000 00000
3880:   00000 00000   00000 00000   00000 00000   00000 00000   00000 00000     00000 00000   00000 00000   00000 00000   00000 00000   00000 00000
3881:   00000 00000   00000 00000   00000 00000   00000 00000   00000 00000     00000 00000   00000 00000   00000 00000   00000 00000   00000 00000
3882:   00000 00000   00000 00000   00000 00000   00000 00000   00000 00000     00000 00000   00000 00000   00000 00000   00000 00000   00000 00000
3883:   00000 00000   00000 00000   00000 00000   00000 00000   00000 00000     00000 00000   00000 00000   00000 00000   00000 00000   00000 00000
3884:   00000 00000   00000 00000   00000 00000   00000 00000   00000 00000     00000 00000   00000 00000   00000 00000   00000 00000   00000 00000
3885:   00000 00000   00000 00000   00000 00000   00000 00000   00000 00000     00000 00000   00000 00000   00000 00000   00000 00000   00000 00000
3886:   00000 00000   00000 00000   00000 00000   00000 00000   00000 00000     00000 00000   00000 00000   00000 00000   00000 00000   00000 00000
3887:   00000 00000   00000 00000   00000 00000   00000 00000   00000 00000     00000 00000   00000 00000   00000 00000   00000 00000   00000 00000
3888:   00000 00000   00000 00000   00000 00000   00000 00000   00000 00000     00000 00000   00000 00000   00000 00000   00000 00000   00000 00000
3889:   00000 00000   00000 00000   00000 00000   00000 00000   00000 00000     00000 00000   00000 00000   00000 00000   00000 00000   00000 00000
3890:   00000 00000   00000 00000   00000 00000   00000 00000   00000 00000     00000 00000   00000 00000   00000 00000   00000 00000   00000 00000
3891:   00000 00000   00000 00000   00000 00000   00000 00000   00000 00000     00000 00000   00000 00000   00000 00000   00000 00000   00000 00000
3892:   00000 00000   00000 00000   00000 00000   00000 00000   00000 00000     00000 00000   00000 00000   00000 00000   00000 00000   00000 00000
3893:   00000 00000   00000 00000   00000 00000   00000 00000   00000 00000     00000 00000   00000 00000   00000 00000   00000 00000   00000 00000
3894:   00000 00000   00000 00000   00000 00000   00000 00000   00000 00000     00000 00000   00000 00000   00000 00000   00000 00000   00000 00000
3895:   00000 00000   00000 00000   00000 00000   00000 00000   00000 00000     00000 00000   00000 00000   00000 00000   00000 00000   00000 00000
3896:   00000 00000   00000 00000   00000 00000   00000 00000   00000 00000     00000 00000   00000 00000   00000 00000   00000 00000   00000 00000
3897:   00000 00000   00000 00000   00000 00000   00000 00000   00000 00000     00000 00000   00000 00000   00000 00000   00000 00000   00000 00000
3898:   00000 00000   00000 00000   00000 00000   00000 00000   00000 00000     00000 00000   00000 00000   00000 00000   00000 00000   00000 00000
3899:   00000 00000   00000 00000   00000 00000   00000 00000   00000 00000     00000 00000   00000 00000   00000 00000   00000 00000   00000 00000
```

```
3900:  00000 00000  00000 00000  00000 00000  00000 00000  00000 00000   00000 00000  00000 00000  00000 00000  00000 00000  00000 00000
3901:  00000 00000  00000 00000  00000 00000  00000 00000  00000 00000   00000 00000  00000 00000  00000 00000  00000 00000  00000 00000
3902:  00000 00000  00000 00000  00000 00000  00000 00000  00000 00000   00000 00000  00000 00000  00000 00000  00000 00000  00000 00000
3903:  00000 00000  00000 00000  00000 00000  00000 00000  00000 00000   00000 00000  00000 00000  00000 00000  00000 00000  00000 00000
3904:  00000 00000  00000 00000  00000 00000  00000 00000  00000 00000   00000 00000  00000 00000  00000 00000  00000 00000  00000 00000
3905:  00000 00000  00000 00000  00000 00000  00000 00000  00000 00000   00000 00000  00000 00000  00000 00000  00000 00000  00000 00000
3906:  00000 00000  00000 00000  00000 00000  00000 00000  00000 00000   00000 00000  00000 00000  00000 00000  00000 00000  00000 00000
3907:  00000 00000  00000 00000  00000 00000  00000 00000  00000 00000   00000 00000  00000 00000  00000 00000  00000 00000  00000 00000
3908:  00000 00000  00000 00000  00000 00000  00000 00000  00000 00000   00000 00000  00000 00000  00000 00000  00000 00000  00000 00000
3909:  00000 00000  00000 00000  00000 00000  00000 00000  00000 00000   00000 00000  00000 00000  00000 00000  00000 00000  00000 00000
3910:  00000 00000  00000 00000  00000 00000  00000 00000  00000 00000   00000 00000  00000 00000  00000 00000  00000 00000  00000 00000
3911:  00000 00000  00000 00000  00000 00000  00000 00000  00000 00000   00000 00000  00000 00000  00000 00000  00000 00000  00000 00000
3912:  00000 00000  00000 00000  00000 00000  00000 00000  00000 00000   00000 00000  00000 00000  00000 00000  00000 00000  00000 00000
3913:  00000 00000  00000 00000  00000 00000  00000 00000  00000 00000   00000 00000  00000 00000  00000 00000  00000 00000  00000 00000
3914:  00000 00000  00000 00000  00000 00000  00000 00000  00000 00000   00000 00000  00000 00000  00000 00000  00000 00000  00000 00000
3915:  00000 00000  00000 00000  00000 00000  00000 00000  00000 00000   00000 00000  00000 00000  00000 00000  00000 00000  00000 00000
3916:  00000 00000  00000 00000  00000 00000  00000 00000  00000 00000   00000 00000  00000 00000  00000 00000  00000 00000  00000 00000
3917:  00000 00000  00000 00000  00000 00000  00000 00000  00000 00000   00000 00000  00000 00000  00000 00000  00000 00000  00000 00000
3918:  00000 00000  00000 00000  00000 00000  00000 00000  00000 00000   00000 00000  00000 00000  00000 00000  00000 00000  00000 00000
3919:  00000 00000  00000 00000  00000 00000  00000 00000  00000 00000   00000 00000  00000 00000  00000 00000  00000 00000  00000 00000
3920:  00000 00000  00000 00000  00000 00000  00000 00000  00000 00000   00000 00000  00000 00000  00000 00000  00000 00000  00000 00000
3921:  00000 00000  00000 00000  00000 00000  00000 00000  00000 00000   00000 00000  00000 00000  00000 00000  00000 00000  00000 00000
3922:  00000 00000  00000 00000  00000 00000  00000 00000  00000 00000   00000 00000  00000 00000  00000 00000  00000 00000  00000 00000
3923:  00000 00000  00000 00000  00000 00000  00000 00000  00000 00000   00000 00000  00000 00000  00000 00000  00000 00000  00000 00000
3924:  00000 00000  00000 00000  00000 00000  00000 00000  00000 00000   00000 00000  00000 00000  00000 00000  00000 00000  00000 00000
3925:  00000 00000  00000 00000  00000 00000  00000 00000  00000 00000   00000 00000  00000 00000  00000 00000  00000 00000  00000 00000
3926:  00000 00000  00000 00000  00000 00000  00000 00000  00000 00000   00000 00000  00000 00000  00000 00000  00000 00000  00000 00000
3927:  00000 00000  00000 00000  00000 00000  00000 00000  00000 00000   00000 00000  00000 00000  00000 00000  00000 00000  00000 00000
3928:  00000 00000  00000 00000  00000 00000  00000 00000  00000 00000   00000 00000  00000 00000  00000 00000  00000 00000  00000 00000
3929:  00000 00000  00000 00000  00000 00000  00000 00000  00000 00000   00000 00000  00000 00000  00000 00000  00000 00000  00000 00000
3930:  00000 00000  00000 00000  00000 00000  00000 00000  00000 00000   00000 00000  00000 00000  00000 00000  00000 00000  00000 00000
3931:  00000 00000  00000 00000  00000 00000  00000 00000  00000 00000   00000 00000  00000 00000  00000 00000  00000 00000  00000 00000
3932:  00000 00000  00000 00000  00000 00000  00000 00000  00000 00000   00000 00000  00000 00000  00000 00000  00000 00000  00000 00000
3933:  00000 00000  00000 00000  00000 00000  00000 00000  00000 00000   00000 00000  00000 00000  00000 00000  00000 00000  00000 00000
3934:  00000 00000  00000 00000  00000 00000  00000 00000  00000 00000   00000 00000  00000 00000  00000 00000  00000 00000  00000 00000
3935:  00000 00000  00000 00000  00000 00000  00000 00000  00000 00000   00000 00000  00000 00000  00000 00000  00000 00000  00000 00000
3936:  00000 00000  00000 00000  00000 00000  00000 00000  00000 00000   00000 00000  00000 00000  00000 00000  00000 00000  00000 00000
3937:  00000 00000  00000 00000  00000 00000  00000 00000  00000 00000   00000 00000  00000 00000  00000 00000  00000 00000  00000 00000
3938:  00000 00000  00000 00000  00000 00000  00000 00000  00000 00000   00000 00000  00000 00000  00000 00000  00000 00000  00000 00000
3939:  00000 00000  00000 00000  00000 00000  00000 00000  00000 00000   00000 00000  00000 00000  00000 00000  00000 00000  00000 00000
3940:  00000 00000  00000 00000  00000 00000  00000 00000  00000 00000   00000 00000  00000 00000  00000 00000  00000 00000  00000 00000
3941:  00000 00000  00000 00000  00000 00000  00000 00000  00000 00000   00000 00000  00000 00000  00000 00000  00000 00000  00000 00000
3942:  00000 00000  00000 00000  00000 00000  00000 00000  00000 00000   00000 00000  00000 00000  00000 00000  00000 00000  00000 00000
3943:  00000 00000  00000 00000  00000 00000  00000 00000  00000 00000   00000 00000  00000 00000  00000 00000  00000 00000  00000 00000
3944:  00000 00000  00000 00000  00000 00000  00000 00000  00000 00000   00000 00000  00000 00000  00000 00000  00000 00000  00000 00000
3945:  00000 00000  00000 00000  00000 00000  00000 00000  00000 00000   00000 00000  00000 00000  00000 00000  00000 00000  00000 00000
3946:  00000 00000  00000 00000  00000 00000  00000 00000  00000 00000   00000 00000  00000 00000  00000 00000  00000 00000  00000 00000
3947:  00000 00000  00000 00000  00000 00000  00000 00000  00000 00000   00000 00000  00000 00000  00000 00000  00000 00000  00000 00000
3948:  00000 00000  00000 00000  00000 00000  00000 00000  00000 00000   00000 00000  00000 00000  00000 00000  00000 00000  00000 00000
3949:  00000 00000  00000 00000  00000 00000  00000 00000  00000 00000   00000 00000  00000 00000  00000 00000  00000 00000  00000 00000
```

```
3950:  00000 00000  00000 00000  00000 00000  00000 00000  00000 00000    00000 00000  00000 00000  00000 00000  00000 00000  00000 00000
3951:  00000 00000  00000 00000  00000 00000  00000 00000  00000 00000    00000 00000  00000 00000  00000 00000  00000 00000  00000 00000
3952:  00000 00000  00000 00000  00000 00000  00000 00000  00000 00000    00000 00000  00000 00000  00000 00000  00000 00000  00000 00000
3953:  00000 00000  00000 00000  00000 00000  00000 00000  00000 00000    00000 00000  00000 00000  00000 00000  00000 00000  00000 00000
3954:  00000 00000  00000 00000  00000 00000  00000 00000  00000 00000    00000 00000  00000 00000  00000 00000  00000 00000  00000 00000
3955:  00000 00000  00000 00000  00000 00000  00000 00000  00000 00000    00000 00000  00000 00000  00000 00000  00000 00000  00000 00000
3956:  00000 00000  00000 00000  00000 00000  00000 00000  00000 00000    00000 00000  00000 00000  00000 00000  00000 00000  00000 00000
3957:  00000 00000  00000 00000  00000 00000  00000 00000  00000 00000    00000 00000  00000 00000  00000 00000  00000 00000  00000 00000
3958:  00000 00000  00000 00000  00000 00000  00000 00000  00000 00000    00000 00000  00000 00000  00000 00000  00000 00000  00000 00000
3959:  00000 00000  00000 00000  00000 00000  00000 00000  00000 00000    00000 00000  00000 00000  00000 00000  00000 00000  00000 00000
3960:  00000 00000  00000 00000  00000 00000  00000 00000  00000 00000    00000 00000  00000 00000  00000 00000  00000 00000  00000 00000
3961:  00000 00000  00000 00000  00000 00000  00000 00000  00000 00000    00000 00000  00000 00000  00000 00000  00000 00000  00000 00000
3962:  00000 00000  00000 00000  00000 00000  00000 00000  00000 00000    00000 00000  00000 00000  00000 00000  00000 00000  00000 00000
3963:  00000 00000  00000 00000  00000 00000  00000 00000  00000 00000    00000 00000  00000 00000  00000 00000  00000 00000  00000 00000
3964:  00000 00000  00000 00000  00000 00000  00000 00000  00000 00000    00000 00000  00000 00000  00000 00000  00000 00000  00000 00000
3965:  00000 00000  00000 00000  00000 00000  00000 00000  00000 00000    00000 00000  00000 00000  00000 00000  00000 00000  00000 00000
3966:  00000 00000  00000 00000  00000 00000  00000 00000  00000 00000    00000 00000  00000 00000  00000 00000  00000 00000  00000 00000
3967:  00000 00000  00000 00000  00000 00000  00000 00000  00000 00000    00000 00000  00000 00000  00000 00000  00000 00000  00000 00000
3968:  00000 00000  00000 00000  00000 00000  00000 00000  00000 00000    00000 00000  00000 00000  00000 00000  00000 00000  00000 00000
3969:  00000 00000  00000 00000  00000 00000  00000 00000  00000 00000    00000 00000  00000 00000  00000 00000  00000 00000  00000 00000
3970:  00000 00000  00000 00000  00000 00000  00000 00000  00000 00000    00000 00000  00000 00000  00000 00000  00000 00000  00000 00000
3971:  00000 00000  00000 00000  00000 00000  00000 00000  00000 00000    00000 00000  00000 00000  00000 00000  00000 00000  00000 00000
3972:  00000 00000  00000 00000  00000 00000  00000 00000  00000 00000    00000 00000  00000 00000  00000 00000  00000 00000  00000 00000
3973:  00000 00000  00000 00000  00000 00000  00000 00000  00000 00000    00000 00000  00000 00000  00000 00000  00000 00000  00000 00000
3974:  00000 00000  00000 00000  00000 00000  00000 00000  00000 00000    00000 00000  00000 00000  00000 00000  00000 00000  00000 00000
3975:  00000 00000  00000 00000  00000 00000  00000 00000  00000 00000    00000 00000  00000 00000  00000 00000  00000 00000  00000 00000
3976:  00000 00000  00000 00000  00000 00000  00000 00000  00000 00000    00000 00000  00000 00000  00000 00000  00000 00000  00000 00000
3977:  00000 00000  00000 00000  00000 00000  00000 00000  00000 00000    00000 00000  00000 00000  00000 00000  00000 00000  00000 00000
3978:  00000 00000  00000 00000  00000 00000  00000 00000  00000 00000    00000 00000  00000 00000  00000 00000  00000 00000  00000 00000
3979:  00000 00000  00000 00000  00000 00000  00000 00000  00000 00000    00000 00000  00000 00000  00000 00000  00000 00000  00000 00000
3980:  00000 00000  00000 00000  00000 00000  00000 00000  00000 00000    00000 00000  00000 00000  00000 00000  00000 00000  00000 00000
3981:  00000 00000  00000 00000  00000 00000  00000 00000  00000 00000    00000 00000  00000 00000  00000 00000  00000 00000  00000 00000
3982:  00000 00000  00000 00000  00000 00000  00000 00000  00000 00000    00000 00000  00000 00000  00000 00000  00000 00000  00000 00000
3983:  00000 00000  00000 00000  00000 00000  00000 00000  00000 00000    00000 00000  00000 00000  00000 00000  00000 00000  00000 00000
3984:  00000 00000  00000 00000  00000 00000  00000 00000  00000 00000    00000 00000  00000 00000  00000 00000  00000 00000  00000 00000
3985:  00000 00000  00000 00000  00000 00000  00000 00000  00000 00000    00000 00000  00000 00000  00000 00000  00000 00000  00000 00000
3986:  00000 00000  00000 00000  00000 00000  00000 00000  00000 00000    00000 00000  00000 00000  00000 00000  00000 00000  00000 00000
3987:  00000 00000  00000 00000  00000 00000  00000 00000  00000 00000    00000 00000  00000 00000  00000 00000  00000 00000  00000 00000
3988:  00000 00000  00000 00000  00000 00000  00000 00000  00000 00000    00000 00000  00000 00000  00000 00000  00000 00000  00000 00000
3989:  00000 00000  00000 00000  00000 00000  00000 00000  00000 00000    00000 00000  00000 00000  00000 00000  00000 00000  00000 00000
3990:  00000 00000  00000 00000  00000 00000  00000 00000  00000 00000    00000 00000  00000 00000  00000 00000  00000 00000  00000 00000
3991:  00000 00000  00000 00000  00000 00000  00000 00000  00000 00000    00000 00000  00000 00000  00000 00000  00000 00000  00000 00000
3992:  00000 00000  00000 00000  00000 00000  00000 00000  00000 00000    00000 00000  00000 00000  00000 00000  00000 00000  00000 00000
3993:  00000 00000  00000 00000  00000 00000  00000 00000  00000 00000    00000 00000  00000 00000  00000 00000  00000 00000  00000 00000
3994:  00000 00000  00000 00000  00000 00000  00000 00000  00000 00000    00000 00000  00000 00000  00000 00000  00000 00000  00000 00000
3995:  00000 00000  00000 00000  00000 00000  00000 00000  00000 00000    00000 00000  00000 00000  00000 00000  00000 00000  00000 00000
3996:  00000 00000  00000 00000  00000 00000  00000 00000  00000 00000    00000 00000  00000 00000  00000 00000  00000 00000  00000 00000
3997:  00000 00000  00000 00000  00000 00000  00000 00000  00000 00000    00000 00000  00000 00000  00000 00000  00000 00000  00000 00000
3998:  00000 00000  00000 00000  00000 00000  00000 00000  00000 00000    00000 00000  00000 00000  00000 00000  00000 00000  00000 00000
3999:  00000 00000  00000 00000  00000 00000  00000 00000  00000 00000    00000 00000  00000 00000  00000 00000  00000 00000  00000 00000
```

```
4000:   00000 00000   00000 00000   00000 00000   00000 00000   00000 00000      00000 00000   00000 00000   00000 00000   00000 00000   00000 00000
4001:   00000 00000   00000 00000   00000 00000   00000 00000   00000 00000      00000 00000   00000 00000   00000 00000   00000 00000   00000 00000
4002:   00000 00000   00000 00000   00000 00000   00000 00000   00000 00000      00000 00000   00000 00000   00000 00000   00000 00000   00000 00000
4003:   00000 00000   00000 00000   00000 00000   00000 00000   00000 00000      00000 00000   00000 00000   00000 00000   00000 00000   00000 00000
4004:   00000 00000   00000 00000   00000 00000   00000 00000   00000 00000      00000 00000   00000 00000   00000 00000   00000 00000   00000 00000
4005:   00000 00000   00000 00000   00000 00000   00000 00000   00000 00000      00000 00000   00000 00000   00000 00000   00000 00000   00000 00000
4006:   00000 00000   00000 00000   00000 00000   00000 00000   00000 00000      00000 00000   00000 00000   00000 00000   00000 00000   00000 00000
4007:   00000 00000   00000 00000   00000 00000   00000 00000   00000 00000      00000 00000   00000 00000   00000 00000   00000 00000   00000 00000
4008:   00000 00000   00000 00000   00000 00000   00000 00000   00000 00000      00000 00000   00000 00000   00000 00000   00000 00000   00000 00000
4009:   00000 00000   00000 00000   00000 00000   00000 00000   00000 00000      00000 00000   00000 00000   00000 00000   00000 00000   00000 00000
4010:   00000 00000   00000 00000   00000 00000   00000 00000   00000 00000      00000 00000   00000 00000   00000 00000   00000 00000   00000 00000
4011:   00000 00000   00000 00000   00000 00000   00000 00000   00000 00000      00000 00000   00000 00000   00000 00000   00000 00000   00000 00000
4012:   00000 00000   00000 00000   00000 00000   00000 00000   00000 00000      00000 00000   00000 00000   00000 00000   00000 00000   00000 00000
4013:   00000 00000   00000 00000   00000 00000   00000 00000   00000 00000      00000 00000   00000 00000   00000 00000   00000 00000   00000 00000
4014:   00000 00000   00000 00000   00000 00000   00000 00000   00000 00000      00000 00000   00000 00000   00000 00000   00000 00000   00000 00000
4015:   00000 00000   00000 00000   00000 00000   00000 00000   00000 00000      00000 00000   00000 00000   00000 00000   00000 00000   00000 00000
4016:   00000 00000   00000 00000   00000 00000   00000 00000   00000 00000      00000 00000   00000 00000   00000 00000   00000 00000   00000 00000
4017:   00000 00000   00000 00000   00000 00000   00000 00000   00000 00000      00000 00000   00000 00000   00000 00000   00000 00000   00000 00000
4018:   00000 00000   00000 00000   00000 00000   00000 00000   00000 00000      00000 00000   00000 00000   00000 00000   00000 00000   00000 00000
4019:   00000 00000   00000 00000   00000 00000   00000 00000   00000 00000      00000 00000   00000 00000   00000 00000   00000 00000   00000 00000
4020:   00000 00000   00000 00000   00000 00000   00000 00000   00000 00000      00000 00000   00000 00000   00000 00000   00000 00000   00000 00000
4021:   00000 00000   00000 00000   00000 00000   00000 00000   00000 00000      00000 00000   00000 00000   00000 00000   00000 00000   00000 00000
4022:   00000 00000   00000 00000   00000 00000   00000 00000   00000 00000      00000 00000   00000 00000   00000 00000   00000 00000   00000 00000
4023:   00000 00000   00000 00000   00000 00000   00000 00000   00000 00000      00000 00000   00000 00000   00000 00000   00000 00000   00000 00000
4024:   00000 00000   00000 00000   00000 00000   00000 00000   00000 00000      00000 00000   00000 00000   00000 00000   00000 00000   00000 00000
4025:   00000 00000   00000 00000   00000 00000   00000 00000   00000 00000      00000 00000   00000 00000   00000 00000   00000 00000   00000 00000
4026:   00000 00000   00000 00000   00000 00000   00000 00000   00000 00000      00000 00000   00000 00000   00000 00000   00000 00000   00000 00000
4027:   00000 00000   00000 00000   00000 00000   00000 00000   00000 00000      00000 00000   00000 00000   00000 00000   00000 00000   00000 00000
4028:   00000 00000   00000 00000   00000 00000   00000 00000   00000 00000      00000 00000   00000 00000   00000 00000   00000 00000   00000 00000
4029:   00000 00000   00000 00000   00000 00000   00000 00000   00000 00000      00000 00000   00000 00000   00000 00000   00000 00000   00000 00000
4030:   00000 00000   00000 00000   00000 00000   00000 00000   00000 00000      00000 00000   00000 00000   00000 00000   00000 00000   00000 00000
4031:   00000 00000   00000 00000   00000 00000   00000 00000   00000 00000      00000 00000   00000 00000   00000 00000   00000 00000   00000 00000
4032:   00000 00000   00000 00000   00000 00000   00000 00000   00000 00000      00000 00000   00000 00000   00000 00000   00000 00000   00000 00000
4033:   00000 00000   00000 00000   00000 00000   00000 00000   00000 00000      00000 00000   00000 00000   00000 00000   00000 00000   00000 00000
4034:   00000 00000   00000 00000   00000 00000   00000 00000   00000 00000      00000 00000   00000 00000   00000 00000   00000 00000   00000 00000
4035:   00000 00000   00000 00000   00000 00000   00000 00000   00000 00000      00000 00000   00000 00000   00000 00000   00000 00000   00000 00000
4036:   00000 00000   00000 00000   00000 00000   00000 00000   00000 00000      00000 00000   00000 00000   00000 00000   00000 00000   00000 00000
4037:   00000 00000   00000 00000   00000 00000   00000 00000   00000 00000      00000 00000   00000 00000   00000 00000   00000 00000   00000 00000
4038:   00000 00000   00000 00000   00000 00000   00000 00000   00000 00000      00000 00000   00000 00000   00000 00000   00000 00000   00000 00000
4039:   00000 00000   00000 00000   00000 00000   00000 00000   00000 00000      00000 00000   00000 00000   00000 00000   00000 00000   00000 00000
4040:   00000 00000   00000 00000   00000 00000   00000 00000   00000 00000      00000 00000   00000 00000   00000 00000   00000 00000   00000 00000
4041:   00000 00000   00000 00000   00000 00000   00000 00000   00000 00000      00000 00000   00000 00000   00000 00000   00000 00000   00000 00000
4042:   00000 00000   00000 00000   00000 00000   00000 00000   00000 00000      00000 00000   00000 00000   00000 00000   00000 00000   00000 00000
4043:   00000 00000   00000 00000   00000 00000   00000 00000   00000 00000      00000 00000   00000 00000   00000 00000   00000 00000   00000 00000
4044:   00000 00000   00000 00000   00000 00000   00000 00000   00000 00000      00000 00000   00000 00000   00000 00000   00000 00000   00000 00000
4045:   00000 00000   00000 00000   00000 00000   00000 00000   00000 00000      00000 00000   00000 00000   00000 00000   00000 00000   00000 00000
4046:   00000 00000   00000 00000   00000 00000   00000 00000   00000 00000      00000 00000   00000 00000   00000 00000   00000 00000   00000 00000
4047:   00000 00000   00000 00000   00000 00000   00000 00000   00000 00000      00000 00000   00000 00000   00000 00000   00000 00000   00000 00000
4048:   00000 00000   00000 00000   00000 00000   00000 00000   00000 00000      00000 00000   00000 00000   00000 00000   00000 00000   00000 00000
4049:   00000 00000   00000 00000   00000 00000   00000 00000   00000 00000      00000 00000   00000 00000   00000 00000   00000 00000   00000 00000
```

```
4050:  00000 00000  00000 00000  00000 00000  00000 00000  00000 00000    00000 00000  00000 00000  00000 00000  00000 00000  00000 00000
4051:  00000 00000  00000 00000  00000 00000  00000 00000  00000 00000    00000 00000  00000 00000  00000 00000  00000 00000  00000 00000
4052:  00000 00000  00000 00000  00000 00000  00000 00000  00000 00000    00000 00000  00000 00000  00000 00000  00000 00000  00000 00000
4053:  00000 00000  00000 00000  00000 00000  00000 00000  00000 00000    00000 00000  00000 00000  00000 00000  00000 00000  00000 00000
4054:  00000 00000  00000 00000  00000 00000  00000 00000  00000 00000    00000 00000  00000 00000  00000 00000  00000 00000  00000 00000
4055:  00000 00000  00000 00000  00000 00000  00000 00000  00000 00000    00000 00000  00000 00000  00000 00000  00000 00000  00000 00000
4056:  00000 00000  00000 00000  00000 00000  00000 00000  00000 00000    00000 00000  00000 00000  00000 00000  00000 00000  00000 00000
4057:  00000 00000  00000 00000  00000 00000  00000 00000  00000 00000    00000 00000  00000 00000  00000 00000  00000 00000  00000 00000
4058:  00000 00000  00000 00000  00000 00000  00000 00000  00000 00000    00000 00000  00000 00000  00000 00000  00000 00000  00000 00000
4059:  00000 00000  00000 00000  00000 00000  00000 00000  00000 00000    00000 00000  00000 00000  00000 00000  00000 00000  00000 00000
4060:  00000 00000  00000 00000  00000 00000  00000 00000  00000 00000    00000 00000  00000 00000  00000 00000  00000 00000  00000 00000
4061:  00000 00000  00000 00000  00000 00000  00000 00000  00000 00000    00000 00000  00000 00000  00000 00000  00000 00000  00000 00000
4062:  00000 00000  00000 00000  00000 00000  00000 00000  00000 00000    00000 00000  00000 00000  00000 00000  00000 00000  00000 00000
4063:  00000 00000  00000 00000  00000 00000  00000 00000  00000 00000    00000 00000  00000 00000  00000 00000  00000 00000  00000 00000
4064:  00000 00000  00000 00000  00000 00000  00000 00000  00000 00000    00000 00000  00000 00000  00000 00000  00000 00000  00000 00000
4065:  00000 00000  00000 00000  00000 00000  00000 00000  00000 00000    00000 00000  00000 00000  00000 00000  00000 00000  00000 00000
4066:  00000 00000  00000 00000  00000 00000  00000 00000  00000 00000    00000 00000  00000 00000  00000 00000  00000 00000  00000 00000
4067:  00000 00000  00000 00000  00000 00000  00000 00000  00000 00000    00000 00000  00000 00000  00000 00000  00000 00000  00000 00000
4068:  00000 00000  00000 00000  00000 00000  00000 00000  00000 00000    00000 00000  00000 00000  00000 00000  00000 00000  00000 00000
4069:  00000 00000  00000 00000  00000 00000  00000 00000  00000 00000    00000 00000  00000 00000  00000 00000  00000 00000  00000 00000
4070:  00000 00000  00000 00000  00000 00000  00000 00000  00000 00000    00000 00000  00000 00000  00000 00000  00000 00000  00000 00000
4071:  00000 00000  00000 00000  00000 00000  00000 00000  00000 00000    00000 00000  00000 00000  00000 00000  00000 00000  00000 00000
4072:  00000 00000  00000 00000  00000 00000  00000 00000  00000 00000    00000 00000  00000 00000  00000 00000  00000 00000  00000 00000
4073:  00000 00000  00000 00000  00000 00000  00000 00000  00000 00000    00000 00000  00000 00000  00000 00000  00000 00000  00000 00000
4074:  00000 00000  00000 00000  00000 00000  00000 00000  00000 00000    00000 00000  00000 00000  00000 00000  00000 00000  00000 00000
4075:  00000 00000  00000 00000  00000 00000  00000 00000  00000 00000    00000 00000  00000 00000  00000 00000  00000 00000  00000 00000
4076:  00000 00000  00000 00000  00000 00000  00000 00000  00000 00000    00000 00000  00000 00000  00000 00000  00000 00000  00000 00000
4077:  00000 00000  00000 00000  00000 00000  00000 00000  00000 00000    00000 00000  00000 00000  00000 00000  00000 00000  00000 00000
4078:  00000 00000  00000 00000  00000 00000  00000 00000  00000 00000    00000 00000  00000 00000  00000 00000  00000 00000  00000 00000
4079:  00000 00000  00000 00000  00000 00000  00000 00000  00000 00000    00000 00000  00000 00000  00000 00000  00000 00000  00000 00000
4080:  00000 00000  00000 00000  00000 00000  00000 00000  00000 00000    00000 00000  00000 00000  00000 00000  00000 00000  00000 00000
4081:  00000 00000  00000 00000  00000 00000  00000 00000  00000 00000    00000 00000  00000 00000  00000 00000  00000 00000  00000 00000
4082:  00000 00000  00000 00000  00000 00000  00000 00000  00000 00000    00000 00000  00000 00000  00000 00000  00000 00000  00000 00000
4083:  00000 00000  00000 00000  00000 00000  00000 00000  00000 00000    00000 00000  00000 00000  00000 00000  00000 00000  00000 00000
4084:  00000 00000  00000 00000  00000 00000  00000 00000  00000 00000    00000 00000  00000 00000  00000 00000  00000 00000  00000 00000
4085:  00000 00000  00000 00000  00000 00000  00000 00000  00000 00000    00000 00000  00000 00000  00000 00000  00000 00000  00000 00000
4086:  00000 00000  00000 00000  00000 00000  00000 00000  00000 00000    00000 00000  00000 00000  00000 00000  00000 00000  00000 00000
4087:  00000 00000  00000 00000  00000 00000  00000 00000  00000 00000    00000 00000  00000 00000  00000 00000  00000 00000  00000 00000
4088:  00000 00000  00000 00000  00000 00000  00000 00000  00000 00000    00000 00000  00000 00000  00000 00000  00000 00000  00000 00000
4089:  00000 00000  00000 00000  00000 00000  00000 00000  00000 00000    00000 00000  00000 00000  00000 00000  00000 00000  00000 00000
4090:  00000 00000  00000 00000  00000 00000  00000 00000  00000 00000    00000 00000  00000 00000  00000 00000  00000 00000  00000 00000
4091:  00000 00000  00000 00000  00000 00000  00000 00000  00000 00000    00000 00000  00000 00000  00000 00000  00000 00000  00000 00000
4092:  00000 00000  00000 00000  00000 00000  00000 00000  00000 00000    00000 00000  00000 00000  00000 00000  00000 00000  00000 00000
4093:  00000 00000  00000 00000  00000 00000  00000 00000  00000 00000    00000 00000  00000 00000  00000 00000  00000 00000  00000 00000
4094:  00000 00000  00000 00000  00000 00000  00000 00000  00000 00000    00000 00000  00000 00000  00000 00000  00000 00000  00000 00000
4095:  00000 00000  00000 00000  00000 00000  00000 00000  00000 00000    00000 00000  00000 00000  00000 00000  00000 00000  00000 00000
4096:  00000 00000  00000 00000  00000 00000  00000 00000  00000 00000    00000 00000  00000 00000  00000 00000  00000 00000  00000 00000
4097:  00000 00000  00000 00000  00000 00000  00000 00000  00000 00000    00000 00000  00000 00000  00000 00000  00000 00000  00000 00000
4098:  00000 00000  00000 00000  00000 00000  00000 00000  00000 00000    00000 00000  00000 00000  00000 00000  00000 00000  00000 00000
4099:  00000 00000  00000 00000  00000 00000  00000 00000  00000 00000    00000 00000  00000 00000  00000 00000  00000 00000  00000 00000
```

```
4100:   00000 00000   00000 00000   00000 00000   00000 00000   00000 00000     00000 00000   00000 00000   00000 00000   00000 00000   00000 00000
4101:   00000 00000   00000 00000   00000 00000   00000 00000   00000 00000     00000 00000   00000 00000   00000 00000   00000 00000   00000 00000
4102:   00000 00000   00000 00000   00000 00000   00000 00000   00000 00000     00000 00000   00000 00000   00000 00000   00000 00000   00000 00000
4103:   00000 00000   00000 00000   00000 00000   00000 00000   00000 00000     00000 00000   00000 00000   00000 00000   00000 00000   00000 00000
4104:   00000 00000   00000 00000   00000 00000   00000 00000   00000 00000     00000 00000   00000 00000   00000 00000   00000 00000   00000 00000
4105:   00000 00000   00000 00000   00000 00000   00000 00000   00000 00000     00000 00000   00000 00000   00000 00000   00000 00000   00000 00000
4106:   00000 00000   00000 00000   00000 00000   00000 00000   00000 00000     00000 00000   00000 00000   00000 00000   00000 00000   00000 00000
4107:   00000 00000   00000 00000   00000 00000   00000 00000   00000 00000     00000 00000   00000 00000   00000 00000   00000 00000   00000 00000
4108:   00000 00000   00000 00000   00000 00000   00000 00000   00000 00000     00000 00000   00000 00000   00000 00000   00000 00000   00000 00000
4109:   00000 00000   00000 00000   00000 00000   00000 00000   00000 00000     00000 00000   00000 00000   00000 00000   00000 00000   00000 00000
4110:   00000 00000   00000 00000   00000 00000   00000 00000   00000 00000     00000 00000   00000 00000   00000 00000   00000 00000   00000 00000
4111:   00000 00000   00000 00000   00000 00000   00000 00000   00000 00000     00000 00000   00000 00000   00000 00000   00000 00000   00000 00000
4112:   00000 00000   00000 00000   00000 00000   00000 00000   00000 00000     00000 00000   00000 00000   00000 00000   00000 00000   00000 00000
4113:   00000 00000   00000 00000   00000 00000   00000 00000   00000 00000     00000 00000   00000 00000   00000 00000   00000 00000   00000 00000
4114:   00000 00000   00000 00000   00000 00000   00000 00000   00000 00000     00000 00000   00000 00000   00000 00000   00000 00000   00000 00000
4115:   00000 00000   00000 00000   00000 00000   00000 00000   00000 00000     00000 00000   00000 00000   00000 00000   00000 00000   00000 00000
4116:   00000 00000   00000 00000   00000 00000   00000 00000   00000 00000     00000 00000   00000 00000   00000 00000   00000 00000   00000 00000
4117:   00000 00000   00000 00000   00000 00000   00000 00000   00000 00000     00000 00000   00000 00000   00000 00000   00000 00000   00000 00000
4118:   00000 00000   00000 00000   00000 00000   00000 00000   00000 00000     00000 00000   00000 00000   00000 00000   00000 00000   00000 00000
4119:   00000 00000   00000 00000   00000 00000   00000 00000   00000 00000     00000 00000   00000 00000   00000 00000   00000 00000   00000 00000
4120:   00000 00000   00000 00000   00000 00000   00000 00000   00000 00000     00000 00000   00000 00000   00000 00000   00000 00000   00000 00000
4121:   00000 00000   00000 00000   00000 00000   00000 00000   00000 00000     00000 00000   00000 00000   00000 00000   00000 00000   00000 00000
4122:   00000 00000   00000 00000   00000 00000   00000 00000   00000 00000     00000 00000   00000 00000   00000 00000   00000 00000   00000 00000
4123:   00000 00000   00000 00000   00000 00000   00000 00000   00000 00000     00000 00000   00000 00000   00000 00000   00000 00000   00000 00000
4124:   00000 00000   00000 00000   00000 00000   00000 00000   00000 00000     00000 00000   00000 00000   00000 00000   00000 00000   00000 00000
4125:   00000 00000   00000 00000   00000 00000   00000 00000   00000 00000     00000 00000   00000 00000   00000 00000   00000 00000   00000 00000
4126:   00000 00000   00000 00000   00000 00000   00000 00000   00000 00000     00000 00000   00000 00000   00000 00000   00000 00000   00000 00000
4127:   00000 00000   00000 00000   00000 00000   00000 00000   00000 00000     00000 00000   00000 00000   00000 00000   00000 00000   00000 00000
4128:   00000 00000   00000 00000   00000 00000   00000 00000   00000 00000     00000 00000   00000 00000   00000 00000   00000 00000   00000 00000
4129:   00000 00000   00000 00000   00000 00000   00000 00000   00000 00000     00000 00000   00000 00000   00000 00000   00000 00000   00000 00000
4130:   00000 00000   00000 00000   00000 00000   00000 00000   00000 00000     00000 00000   00000 00000   00000 00000   00000 00000   00000 00000
4131:   00000 00000   00000 00000   00000 00000   00000 00000   00000 00000     00000 00000   00000 00000   00000 00000   00000 00000   00000 00000
4132:   00000 00000   00000 00000   00000 00000   00000 00000   00000 00000     00000 00000   00000 00000   00000 00000   00000 00000   00000 00000
4133:   00000 00000   00000 00000   00000 00000   00000 00000   00000 00000     00000 00000   00000 00000   00000 00000   00000 00000   00000 00000
4134:   00000 00000   00000 00000   00000 00000   00000 00000   00000 00000     00000 00000   00000 00000   00000 00000   00000 00000   00000 00000
4135:   00000 00000   00000 00000   00000 00000   00000 00000   00000 00000     00000 00000   00000 00000   00000 00000   00000 00000   00000 00000
4136:   00000 00000   00000 00000   00000 00000   00000 00000   00000 00000     00000 00000   00000 00000   00000 00000   00000 00000   00000 00000
4137:   00000 00000   00000 00000   00000 00000   00000 00000   00000 00000     00000 00000   00000 00000   00000 00000   00000 00000   00000 00000
4138:   00000 00000   00000 00000   00000 00000   00000 00000   00000 00000     00000 00000   00000 00000   00000 00000   00000 00000   00000 00000
4139:   00000 00000   00000 00000   00000 00000   00000 00000   00000 00000     00000 00000   00000 00000   00000 00000   00000 00000   00000 00000
4140:   00000 00000   00000 00000   00000 00000   00000 00000   00000 00000     00000 00000   00000 00000   00000 00000   00000 00000   00000 00000
4141:   00000 00000   00000 00000   00000 00000   00000 00000   00000 00000     00000 00000   00000 00000   00000 00000   00000 00000   00000 00000
4142:   00000 00000   00000 00000   00000 00000   00000 00000   00000 00000     00000 00000   00000 00000   00000 00000   00000 00000   00000 00000
4143:   00000 00000   00000 00000   00000 00000   00000 00000   00000 00000     00000 00000   00000 00000   00000 00000   00000 00000   00000 00000
4144:   00000 00000   00000 00000   00000 00000   00000 00000   00000 00000     00000 00000   00000 00000   00000 00000   00000 00000   00000 00000
4145:   00000 00000   00000 00000   00000 00000   00000 00000   00000 00000     00000 00000   00000 00000   00000 00000   00000 00000   00000 00000
4146:   00000 00000   00000 00000   00000 00000   00000 00000   00000 00000     00000 00000   00000 00000   00000 00000   00000 00000   00000 00000
4147:   00000 00000   00000 00000   00000 00000   00000 00000   00000 00000     00000 00000   00000 00000   00000 00000   00000 00000   00000 00000
4148:   00000 00000   00000 00000   00000 00000   00000 00000   00000 00000     00000 00000   00000 00000   00000 00000   00000 00000   00000 00000
4149:   00000 00000   00000 00000   00000 00000   00000 00000   00000 00000     00000 00000   00000 00000   00000 00000   00000 00000   00000 00000
```

```
4150:  00000 00000  00000 00000  00000 00000  00000 00000  00000 00000   00000 00000  00000 00000  00000 00000  00000 00000  00000 00000
4151:  00000 00000  00000 00000  00000 00000  00000 00000  00000 00000   00000 00000  00000 00000  00000 00000  00000 00000  00000 00000
4152:  00000 00000  00000 00000  00000 00000  00000 00000  00000 00000   00000 00000  00000 00000  00000 00000  00000 00000  00000 00000
4153:  00000 00000  00000 00000  00000 00000  00000 00000  00000 00000   00000 00000  00000 00000  00000 00000  00000 00000  00000 00000
4154:  00000 00000  00000 00000  00000 00000  00000 00000  00000 00000   00000 00000  00000 00000  00000 00000  00000 00000  00000 00000
4155:  00000 00000  00000 00000  00000 00000  00000 00000  00000 00000   00000 00000  00000 00000  00000 00000  00000 00000  00000 00000
4156:  00000 00000  00000 00000  00000 00000  00000 00000  00000 00000   00000 00000  00000 00000  00000 00000  00000 00000  00000 00000
4157:  00000 00000  00000 00000  00000 00000  00000 00000  00000 00000   00000 00000  00000 00000  00000 00000  00000 00000  00000 00000
4158:  00000 00000  00000 00000  00000 00000  00000 00000  00000 00000   00000 00000  00000 00000  00000 00000  00000 00000  00000 00000
4159:  00000 00000  00000 00000  00000 00000  00000 00000  00000 00000   00000 00000  00000 00000  00000 00000  00000 00000  00000 00000
4160:  00000 00000  00000 00000  00000 00000  00000 00000  00000 00000   00000 00000  00000 00000  00000 00000  00000 00000  00000 00000
4161:  00000 00000  00000 00000  00000 00000  00000 00000  00000 00000   00000 00000  00000 00000  00000 00000  00000 00000  00000 00000
4162:  00000 00000  00000 00000  00000 00000  00000 00000  00000 00000   00000 00000  00000 00000  00000 00000  00000 00000  00000 00000
4163:  00000 00000  00000 00000  00000 00000  00000 00000  00000 00000   00000 00000  00000 00000  00000 00000  00000 00000  00000 00000
4164:  00000 00000  00000 00000  00000 00000  00000 00000  00000 00000   00000 00000  00000 00000  00000 00000  00000 00000  00000 00000
4165:  00000 00000  00000 00000  00000 00000  00000 00000  00000 00000   00000 00000  00000 00000  00000 00000  00000 00000  00000 00000
4166:  00000 00000  00000 00000  00000 00000  00000 00000  00000 00000   00000 00000  00000 00000  00000 00000  00000 00000  00000 00000
4167:  00000 00000  00000 00000  00000 00000  00000 00000  00000 00000   00000 00000  00000 00000  00000 00000  00000 00000  00000 00000
4168:  00000 00000  00000 00000  00000 00000  00000 00000  00000 00000   00000 00000  00000 00000  00000 00000  00000 00000  00000 00000
4169:  00000 00000  00000 00000  00000 00000  00000 00000  00000 00000   00000 00000  00000 00000  00000 00000  00000 00000  00000 00000
4170:  00000 00000  00000 00000  00000 00000  00000 00000  00000 00000   00000 00000  00000 00000  00000 00000  00000 00000  00000 00000
4171:  00000 00000  00000 00000  00000 00000  00000 00000  00000 00000   00000 00000  00000 00000  00000 00000  00000 00000  00000 00000
4172:  00000 00000  00000 00000  00000 00000  00000 00000  00000 00000   00000 00000  00000 00000  00000 00000  00000 00000  00000 00000
4173:  00000 00000  00000 00000  00000 00000  00000 00000  00000 00000   00000 00000  00000 00000  00000 00000  00000 00000  00000 00000
4174:  00000 00000  00000 00000  00000 00000  00000 00000  00000 00000   00000 00000  00000 00000  00000 00000  00000 00000  00000 00000
4175:  00000 00000  00000 00000  00000 00000  00000 00000  00000 00000   00000 00000  00000 00000  00000 00000  00000 00000  00000 00000
4176:  00000 00000  00000 00000  00000 00000  00000 00000  00000 00000   00000 00000  00000 00000  00000 00000  00000 00000  00000 00000
4177:  00000 00000  00000 00000  00000 00000  00000 00000  00000 00000   00000 00000  00000 00000  00000 00000  00000 00000  00000 00000
4178:  00000 00000  00000 00000  00000 00000  00000 00000  00000 00000   00000 00000  00000 00000  00000 00000  00000 00000  00000 00000
4179:  00000 00000  00000 00000  00000 00000  00000 00000  00000 00000   00000 00000  00000 00000  00000 00000  00000 00000  00000 00000
4180:  00000 00000  00000 00000  00000 00000  00000 00000  00000 00000   00000 00000  00000 00000  00000 00000  00000 00000  00000 00000
4181:  00000 00000  00000 00000  00000 00000  00000 00000  00000 00000   00000 00000  00000 00000  00000 00000  00000 00000  00000 00000
4182:  00000 00000  00000 00000  00000 00000  00000 00000  00000 00000   00000 00000  00000 00000  00000 00000  00000 00000  00000 00000
4183:  00000 00000  00000 00000  00000 00000  00000 00000  00000 00000   00000 00000  00000 00000  00000 00000  00000 00000  00000 00000
4184:  00000 00000  00000 00000  00000 00000  00000 00000  00000 00000   00000 00000  00000 00000  00000 00000  00000 00000  00000 00000
4185:  00000 00000  00000 00000  00000 00000  00000 00000  00000 00000   00000 00000  00000 00000  00000 00000  00000 00000  00000 00000
4186:  00000 00000  00000 00000  00000 00000  00000 00000  00000 00000   00000 00000  00000 00000  00000 00000  00000 00000  00000 00000
4187:  00000 00000  00000 00000  00000 00000  00000 00000  00000 00000   00000 00000  00000 00000  00000 00000  00000 00000  00000 00000
4188:  00000 00000  00000 00000  00000 00000  00000 00000  00000 00000   00000 00000  00000 00000  00000 00000  00000 00000  00000 00000
4189:  00000 00000  00000 00000  00000 00000  00000 00000  00000 00000   00000 00000  00000 00000  00000 00000  00000 00000  00000 00000
4190:  00000 00000  00000 00000  00000 00000  00000 00000  00000 00000   00000 00000  00000 00000  00000 00000  00000 00000  00000 00000
4191:  00000 00000  00000 00000  00000 00000  00000 00000  00000 00000   00000 00000  00000 00000  00000 00000  00000 00000  00000 00000
4192:  00000 00000  00000 00000  00000 00000  00000 00000  00000 00000   00000 00000  00000 00000  00000 00000  00000 00000  00000 00000
4193:  00000 00000  00000 00000  00000 00000  00000 00000  00000 00000   00000 00000  00000 00000  00000 00000  00000 00000  00000 00000
4194:  00000 00000  00000 00000  00000 00000  00000 00000  00000 00000   00000 00000  00000 00000  00000 00000  00000 00000  00000 00000
4195:  00000 00000  00000 00000  00000 00000  00000 00000  00000 00000   00000 00000  00000 00000  00000 00000  00000 00000  00000 00000
4196:  00000 00000  00000 00000  00000 00000  00000 00000  00000 00000   00000 00000  00000 00000  00000 00000  00000 00000  00000 00000
4197:  00000 00000  00000 00000  00000 00000  00000 00000  00000 00000   00000 00000  00000 00000  00000 00000  00000 00000  00000 00000
4198:  00000 00000  00000 00000  00000 00000  00000 00000  00000 00000   00000 00000  00000 00000  00000 00000  00000 00000  00000 00000
4199:  00000 00000  00000 00000  00000 00000  00000 00000  00000 00000   00000 00000  00000 00000  00000 00000  00000 00000  00000 00000
```

```
4200:   00000 00000   00000 00000   00000 00000   00000 00000   00000 00000     00000 00000   00000 00000   00000 00000   00000 00000   00000 00000
4201:   00000 00000   00000 00000   00000 00000   00000 00000   00000 00000     00000 00000   00000 00000   00000 00000   00000 00000   00000 00000
4202:   00000 00000   00000 00000   00000 00000   00000 00000   00000 00000     00000 00000   00000 00000   00000 00000   00000 00000   00000 00000
4203:   00000 00000   00000 00000   00000 00000   00000 00000   00000 00000     00000 00000   00000 00000   00000 00000   00000 00000   00000 00000
4204:   00000 00000   00000 00000   00000 00000   00000 00000   00000 00000     00000 00000   00000 00000   00000 00000   00000 00000   00000 00000
4205:   00000 00000   00000 00000   00000 00000   00000 00000   00000 00000     00000 00000   00000 00000   00000 00000   00000 00000   00000 00000
4206:   00000 00000   00000 00000   00000 00000   00000 00000   00000 00000     00000 00000   00000 00000   00000 00000   00000 00000   00000 00000
4207:   00000 00000   00000 00000   00000 00000   00000 00000   00000 00000     00000 00000   00000 00000   00000 00000   00000 00000   00000 00000
4208:   00000 00000   00000 00000   00000 00000   00000 00000   00000 00000     00000 00000   00000 00000   00000 00000   00000 00000   00000 00000
4209:   00000 00000   00000 00000   00000 00000   00000 00000   00000 00000     00000 00000   00000 00000   00000 00000   00000 00000   00000 00000
4210:   00000 00000   00000 00000   00000 00000   00000 00000   00000 00000     00000 00000   00000 00000   00000 00000   00000 00000   00000 00000
4211:   00000 00000   00000 00000   00000 00000   00000 00000   00000 00000     00000 00000   00000 00000   00000 00000   00000 00000   00000 00000
4212:   00000 00000   00000 00000   00000 00000   00000 00000   00000 00000     00000 00000   00000 00000   00000 00000   00000 00000   00000 00000
4213:   00000 00000   00000 00000   00000 00000   00000 00000   00000 00000     00000 00000   00000 00000   00000 00000   00000 00000   00000 00000
4214:   00000 00000   00000 00000   00000 00000   00000 00000   00000 00000     00000 00000   00000 00000   00000 00000   00000 00000   00000 00000
4215:   00000 00000   00000 00000   00000 00000   00000 00000   00000 00000     00000 00000   00000 00000   00000 00000   00000 00000   00000 00000
4216:   00000 00000   00000 00000   00000 00000   00000 00000   00000 00000     00000 00000   00000 00000   00000 00000   00000 00000   00000 00000
4217:   00000 00000   00000 00000   00000 00000   00000 00000   00000 00000     00000 00000   00000 00000   00000 00000   00000 00000   00000 00000
4218:   00000 00000   00000 00000   00000 00000   00000 00000   00000 00000     00000 00000   00000 00000   00000 00000   00000 00000   00000 00000
4219:   00000 00000   00000 00000   00000 00000   00000 00000   00000 00000     00000 00000   00000 00000   00000 00000   00000 00000   00000 00000
4220:   00000 00000   00000 00000   00000 00000   00000 00000   00000 00000     00000 00000   00000 00000   00000 00000   00000 00000   00000 00000
4221:   00000 00000   00000 00000   00000 00000   00000 00000   00000 00000     00000 00000   00000 00000   00000 00000   00000 00000   00000 00000
4222:   00000 00000   00000 00000   00000 00000   00000 00000   00000 00000     00000 00000   00000 00000   00000 00000   00000 00000   00000 00000
4223:   00000 00000   00000 00000   00000 00000   00000 00000   00000 00000     00000 00000   00000 00000   00000 00000   00000 00000   00000 00000
4224:   00000 00000   00000 00000   00000 00000   00000 00000   00000 00000     00000 00000   00000 00000   00000 00000   00000 00000   00000 00000
4225:   00000 00000   00000 00000   00000 00000   00000 00000   00000 00000     00000 00000   00000 00000   00000 00000   00000 00000   00000 00000
4226:   00000 00000   00000 00000   00000 00000   00000 00000   00000 00000     00000 00000   00000 00000   00000 00000   00000 00000   00000 00000
4227:   00000 00000   00000 00000   00000 00000   00000 00000   00000 00000     00000 00000   00000 00000   00000 00000   00000 00000   00000 00000
4228:   00000 00000   00000 00000   00000 00000   00000 00000   00000 00000     00000 00000   00000 00000   00000 00000   00000 00000   00000 00000
4229:   00000 00000   00000 00000   00000 00000   00000 00000   00000 00000     00000 00000   00000 00000   00000 00000   00000 00000   00000 00000
4230:   00000 00000   00000 00000   00000 00000   00000 00000   00000 00000     00000 00000   00000 00000   00000 00000   00000 00000   00000 00000
4231:   00000 00000   00000 00000   00000 00000   00000 00000   00000 00000     00000 00000   00000 00000   00000 00000   00000 00000   00000 00000
4232:   00000 00000   00000 00000   00000 00000   00000 00000   00000 00000     00000 00000   00000 00000   00000 00000   00000 00000   00000 00000
4233:   00000 00000   00000 00000   00000 00000   00000 00000   00000 00000     00000 00000   00000 00000   00000 00000   00000 00000   00000 00000
4234:   00000 00000   00000 00000   00000 00000   00000 00000   00000 00000     00000 00000   00000 00000   00000 00000   00000 00000   00000 00000
4235:   00000 00000   00000 00000   00000 00000   00000 00000   00000 00000     00000 00000   00000 00000   00000 00000   00000 00000   00000 00000
4236:   00000 00000   00000 00000   00000 00000   00000 00000   00000 00000     00000 00000   00000 00000   00000 00000   00000 00000   00000 00000
4237:   00000 00000   00000 00000   00000 00000   00000 00000   00000 00000     00000 00000   00000 00000   00000 00000   00000 00000   00000 00000
4238:   00000 00000   00000 00000   00000 00000   00000 00000   00000 00000     00000 00000   00000 00000   00000 00000   00000 00000   00000 00000
4239:   00000 00000   00000 00000   00000 00000   00000 00000   00000 00000     00000 00000   00000 00000   00000 00000   00000 00000   00000 00000
4240:   00000 00000   00000 00000   00000 00000   00000 00000   00000 00000     00000 00000   00000 00000   00000 00000   00000 00000   00000 00000
4241:   00000 00000   00000 00000   00000 00000   00000 00000   00000 00000     00000 00000   00000 00000   00000 00000   00000 00000   00000 00000
4242:   00000 00000   00000 00000   00000 00000   00000 00000   00000 00000     00000 00000   00000 00000   00000 00000   00000 00000   00000 00000
4243:   00000 00000   00000 00000   00000 00000   00000 00000   00000 00000     00000 00000   00000 00000   00000 00000   00000 00000   00000 00000
4244:   00000 00000   00000 00000   00000 00000   00000 00000   00000 00000     00000 00000   00000 00000   00000 00000   00000 00000   00000 00000
4245:   00000 00000   00000 00000   00000 00000   00000 00000   00000 00000     00000 00000   00000 00000   00000 00000   00000 00000   00000 00000
4246:   00000 00000   00000 00000   00000 00000   00000 00000   00000 00000     00000 00000   00000 00000   00000 00000   00000 00000   00000 00000
4247:   00000 00000   00000 00000   00000 00000   00000 00000   00000 00000     00000 00000   00000 00000   00000 00000   00000 00000   00000 00000
4248:   00000 00000   00000 00000   00000 00000   00000 00000   00000 00000     00000 00000   00000 00000   00000 00000   00000 00000   00000 00000
4249:   00000 00000   00000 00000   00000 00000   00000 00000   00000 00000     00000 00000   00000 00000   00000 00000   00000 00000   00000 00000
```

```
4250: 00000 00000  00000 00000  00000 00000  00000 00000  00000 00000    00000 00000  00000 00000  00000 00000  00000 00000  00000 00000
4251: 00000 00000  00000 00000  00000 00000  00000 00000  00000 00000    00000 00000  00000 00000  00000 00000  00000 00000  00000 00000
4252: 00000 00000  00000 00000  00000 00000  00000 00000  00000 00000    00000 00000  00000 00000  00000 00000  00000 00000  00000 00000
4253: 00000 00000  00000 00000  00000 00000  00000 00000  00000 00000    00000 00000  00000 00000  00000 00000  00000 00000  00000 00000
4254: 00000 00000  00000 00000  00000 00000  00000 00000  00000 00000    00000 00000  00000 00000  00000 00000  00000 00000  00000 00000
4255: 00000 00000  00000 00000  00000 00000  00000 00000  00000 00000    00000 00000  00000 00000  00000 00000  00000 00000  00000 00000
4256: 00000 00000  00000 00000  00000 00000  00000 00000  00000 00000    00000 00000  00000 00000  00000 00000  00000 00000  00000 00000
4257: 00000 00000  00000 00000  00000 00000  00000 00000  00000 00000    00000 00000  00000 00000  00000 00000  00000 00000  00000 00000
4258: 00000 00000  00000 00000  00000 00000  00000 00000  00000 00000    00000 00000  00000 00000  00000 00000  00000 00000  00000 00000
4259: 00000 00000  00000 00000  00000 00000  00000 00000  00000 00000    00000 00000  00000 00000  00000 00000  00000 00000  00000 00000
4260: 00000 00000  00000 00000  00000 00000  00000 00000  00000 00000    00000 00000  00000 00000  00000 00000  00000 00000  00000 00000
4261: 00000 00000  00000 00000  00000 00000  00000 00000  00000 00000    00000 00000  00000 00000  00000 00000  00000 00000  00000 00000
4262: 00000 00000  00000 00000  00000 00000  00000 00000  00000 00000    00000 00000  00000 00000  00000 00000  00000 00000  00000 00000
4263: 00000 00000  00000 00000  00000 00000  00000 00000  00000 00000    00000 00000  00000 00000  00000 00000  00000 00000  00000 00000
4264: 00000 00000  00000 00000  00000 00000  00000 00000  00000 00000    00000 00000  00000 00000  00000 00000  00000 00000  00000 00000
4265: 00000 00000  00000 00000  00000 00000  00000 00000  00000 00000    00000 00000  00000 00000  00000 00000  00000 00000  00000 00000
4266: 00000 00000  00000 00000  00000 00000  00000 00000  00000 00000    00000 00000  00000 00000  00000 00000  00000 00000  00000 00000
4267: 00000 00000  00000 00000  00000 00000  00000 00000  00000 00000    00000 00000  00000 00000  00000 00000  00000 00000  00000 00000
4268: 00000 00000  00000 00000  00000 00000  00000 00000  00000 00000    00000 00000  00000 00000  00000 00000  00000 00000  00000 00000
4269: 00000 00000  00000 00000  00000 00000  00000 00000  00000 00000    00000 00000  00000 00000  00000 00000  00000 00000  00000 00000
4270: 00000 00000  00000 00000  00000 00000  00000 00000  00000 00000    00000 00000  00000 00000  00000 00000  00000 00000  00000 00000
4271: 00000 00000  00000 00000  00000 00000  00000 00000  00000 00000    00000 00000  00000 00000  00000 00000  00000 00000  00000 00000
4272: 00000 00000  00000 00000  00000 00000  00000 00000  00000 00000    00000 00000  00000 00000  00000 00000  00000 00000  00000 00000
4273: 00000 00000  00000 00000  00000 00000  00000 00000  00000 00000    00000 00000  00000 00000  00000 00000  00000 00000  00000 00000
4274: 00000 00000  00000 00000  00000 00000  00000 00000  00000 00000    00000 00000  00000 00000  00000 00000  00000 00000  00000 00000
4275: 00000 00000  00000 00000  00000 00000  00000 00000  00000 00000    00000 00000  00000 00000  00000 00000  00000 00000  00000 00000
4276: 00000 00000  00000 00000  00000 00000  00000 00000  00000 00000    00000 00000  00000 00000  00000 00000  00000 00000  00000 00000
4277: 00000 00000  00000 00000  00000 00000  00000 00000  00000 00000    00000 00000  00000 00000  00000 00000  00000 00000  00000 00000
4278: 00000 00000  00000 00000  00000 00000  00000 00000  00000 00000    00000 00000  00000 00000  00000 00000  00000 00000  00000 00000
4279: 00000 00000  00000 00000  00000 00000  00000 00000  00000 00000    00000 00000  00000 00000  00000 00000  00000 00000  00000 00000
4280: 00000 00000  00000 00000  00000 00000  00000 00000  00000 00000    00000 00000  00000 00000  00000 00000  00000 00000  00000 00000
4281: 00000 00000  00000 00000  00000 00000  00000 00000  00000 00000    00000 00000  00000 00000  00000 00000  00000 00000  00000 00000
4282: 00000 00000  00000 00000  00000 00000  00000 00000  00000 00000    00000 00000  00000 00000  00000 00000  00000 00000  00000 00000
4283: 00000 00000  00000 00000  00000 00000  00000 00000  00000 00000    00000 00000  00000 00000  00000 00000  00000 00000  00000 00000
4284: 00000 00000  00000 00000  00000 00000  00000 00000  00000 00000    00000 00000  00000 00000  00000 00000  00000 00000  00000 00000
4285: 00000 00000  00000 00000  00000 00000  00000 00000  00000 00000    00000 00000  00000 00000  00000 00000  00000 00000  00000 00000
4286: 00000 00000  00000 00000  00000 00000  00000 00000  00000 00000    00000 00000  00000 00000  00000 00000  00000 00000  00000 00000
4287: 00000 00000  00000 00000  00000 00000  00000 00000  00000 00000    00000 00000  00000 00000  00000 00000  00000 00000  00000 00000
4288: 00000 00000  00000 00000  00000 00000  00000 00000  00000 00000    00000 00000  00000 00000  00000 00000  00000 00000  00000 00000
4289: 00000 00000  00000 00000  00000 00000  00000 00000  00000 00000    00000 00000  00000 00000  00000 00000  00000 00000  00000 00000
4290: 00000 00000  00000 00000  00000 00000  00000 00000  00000 00000    00000 00000  00000 00000  00000 00000  00000 00000  00000 00000
4291: 00000 00000  00000 00000  00000 00000  00000 00000  00000 00000    00000 00000  00000 00000  00000 00000  00000 00000  00000 00000
4292: 00000 00000  00000 00000  00000 00000  00000 00000  00000 00000    00000 00000  00000 00000  00000 00000  00000 00000  00000 00000
4293: 00000 00000  00000 00000  00000 00000  00000 00000  00000 00000    00000 00000  00000 00000  00000 00000  00000 00000  00000 00000
4294: 00000 00000  00000 00000  00000 00000  00000 00000  00000 00000    00000 00000  00000 00000  00000 00000  00000 00000  00000 00000
4295: 00000 00000  00000 00000  00000 00000  00000 00000  00000 00000    00000 00000  00000 00000  00000 00000  00000 00000  00000 00000
4296: 00000 00000  00000 00000  00000 00000  00000 00000  00000 00000    00000 00000  00000 00000  00000 00000  00000 00000  00000 00000
4297: 00000 00000  00000 00000  00000 00000  00000 00000  00000 00000    00000 00000  00000 00000  00000 00000  00000 00000  00000 00000
4298: 00000 00000  00000 00000  00000 00000  00000 00000  00000 00000    00000 00000  00000 00000  00000 00000  00000 00000  00000 00000
4299: 00000 00000  00000 00000  00000 00000  00000 00000  00000 00000    00000 00000  00000 00000  00000 00000  00000 00000  00000 00000
```

```
4300:   00000 00000   00000 00000   00000 00000   00000 00000   00000 00000      00000 00000   00000 00000   00000 00000   00000 00000   00000 00000
4301:   00000 00000   00000 00000   00000 00000   00000 00000   00000 00000      00000 00000   00000 00000   00000 00000   00000 00000   00000 00000
4302:   00000 00000   00000 00000   00000 00000   00000 00000   00000 00000      00000 00000   00000 00000   00000 00000   00000 00000   00000 00000
4303:   00000 00000   00000 00000   00000 00000   00000 00000   00000 00000      00000 00000   00000 00000   00000 00000   00000 00000   00000 00000
4304:   00000 00000   00000 00000   00000 00000   00000 00000   00000 00000      00000 00000   00000 00000   00000 00000   00000 00000   00000 00000
4305:   00000 00000   00000 00000   00000 00000   00000 00000   00000 00000      00000 00000   00000 00000   00000 00000   00000 00000   00000 00000
4306:   00000 00000   00000 00000   00000 00000   00000 00000   00000 00000      00000 00000   00000 00000   00000 00000   00000 00000   00000 00000
4307:   00000 00000   00000 00000   00000 00000   00000 00000   00000 00000      00000 00000   00000 00000   00000 00000   00000 00000   00000 00000
4308:   00000 00000   00000 00000   00000 00000   00000 00000   00000 00000      00000 00000   00000 00000   00000 00000   00000 00000   00000 00000
4309:   00000 00000   00000 00000   00000 00000   00000 00000   00000 00000      00000 00000   00000 00000   00000 00000   00000 00000   00000 00000
4310:   00000 00000   00000 00000   00000 00000   00000 00000   00000 00000      00000 00000   00000 00000   00000 00000   00000 00000   00000 00000
4311:   00000 00000   00000 00000   00000 00000   00000 00000   00000 00000      00000 00000   00000 00000   00000 00000   00000 00000   00000 00000
4312:   00000 00000   00000 00000   00000 00000   00000 00000   00000 00000      00000 00000   00000 00000   00000 00000   00000 00000   00000 00000
4313:   00000 00000   00000 00000   00000 00000   00000 00000   00000 00000      00000 00000   00000 00000   00000 00000   00000 00000   00000 00000
4314:   00000 00000   00000 00000   00000 00000   00000 00000   00000 00000      00000 00000   00000 00000   00000 00000   00000 00000   00000 00000
4315:   00000 00000   00000 00000   00000 00000   00000 00000   00000 00000      00000 00000   00000 00000   00000 00000   00000 00000   00000 00000
4316:   00000 00000   00000 00000   00000 00000   00000 00000   00000 00000      00000 00000   00000 00000   00000 00000   00000 00000   00000 00000
4317:   00000 00000   00000 00000   00000 00000   00000 00000   00000 00000      00000 00000   00000 00000   00000 00000   00000 00000   00000 00000
4318:   00000 00000   00000 00000   00000 00000   00000 00000   00000 00000      00000 00000   00000 00000   00000 00000   00000 00000   00000 00000
4319:   00000 00000   00000 00000   00000 00000   00000 00000   00000 00000      00000 00000   00000 00000   00000 00000   00000 00000   00000 00000
4320:   00000 00000   00000 00000   00000 00000   00000 00000   00000 00000      00000 00000   00000 00000   00000 00000   00000 00000   00000 00000
4321:   00000 00000   00000 00000   00000 00000   00000 00000   00000 00000      00000 00000   00000 00000   00000 00000   00000 00000   00000 00000
4322:   00000 00000   00000 00000   00000 00000   00000 00000   00000 00000      00000 00000   00000 00000   00000 00000   00000 00000   00000 00000
4323:   00000 00000   00000 00000   00000 00000   00000 00000   00000 00000      00000 00000   00000 00000   00000 00000   00000 00000   00000 00000
4324:   00000 00000   00000 00000   00000 00000   00000 00000   00000 00000      00000 00000   00000 00000   00000 00000   00000 00000   00000 00000
4325:   00000 00000   00000 00000   00000 00000   00000 00000   00000 00000      00000 00000   00000 00000   00000 00000   00000 00000   00000 00000
4326:   00000 00000   00000 00000   00000 00000   00000 00000   00000 00000      00000 00000   00000 00000   00000 00000   00000 00000   00000 00000
4327:   00000 00000   00000 00000   00000 00000   00000 00000   00000 00000      00000 00000   00000 00000   00000 00000   00000 00000   00000 00000
4328:   00000 00000   00000 00000   00000 00000   00000 00000   00000 00000      00000 00000   00000 00000   00000 00000   00000 00000   00000 00000
4329:   00000 00000   00000 00000   00000 00000   00000 00000   00000 00000      00000 00000   00000 00000   00000 00000   00000 00000   00000 00000
4330:   00000 00000   00000 00000   00000 00000   00000 00000   00000 00000      00000 00000   00000 00000   00000 00000   00000 00000   00000 00000
4331:   00000 00000   00000 00000   00000 00000   00000 00000   00000 00000      00000 00000   00000 00000   00000 00000   00000 00000   00000 00000
4332:   00000 00000   00000 00000   00000 00000   00000 00000   00000 00000      00000 00000   00000 00000   00000 00000   00000 00000   00000 00000
4333:   00000 00000   00000 00000   00000 00000   00000 00000   00000 00000      00000 00000   00000 00000   00000 00000   00000 00000   00000 00000
4334:   00000 00000   00000 00000   00000 00000   00000 00000   00000 00000      00000 00000   00000 00000   00000 00000   00000 00000   00000 00000
4335:   00000 00000   00000 00000   00000 00000   00000 00000   00000 00000      00000 00000   00000 00000   00000 00000   00000 00000   00000 00000
4336:   00000 00000   00000 00000   00000 00000   00000 00000   00000 00000      00000 00000   00000 00000   00000 00000   00000 00000   00000 00000
4337:   00000 00000   00000 00000   00000 00000   00000 00000   00000 00000      00000 00000   00000 00000   00000 00000   00000 00000   00000 00000
4338:   00000 00000   00000 00000   00000 00000   00000 00000   00000 00000      00000 00000   00000 00000   00000 00000   00000 00000   00000 00000
4339:   00000 00000   00000 00000   00000 00000   00000 00000   00000 00000      00000 00000   00000 00000   00000 00000   00000 00000   00000 00000
4340:   00000 00000   00000 00000   00000 00000   00000 00000   00000 00000      00000 00000   00000 00000   00000 00000   00000 00000   00000 00000
4341:   00000 00000   00000 00000   00000 00000   00000 00000   00000 00000      00000 00000   00000 00000   00000 00000   00000 00000   00000 00000
4342:   00000 00000   00000 00000   00000 00000   00000 00000   00000 00000      00000 00000   00000 00000   00000 00000   00000 00000   00000 00000
4343:   00000 00000   00000 00000   00000 00000   00000 00000   00000 00000      00000 00000   00000 00000   00000 00000   00000 00000   00000 00000
4344:   00000 00000   00000 00000   00000 00000   00000 00000   00000 00000      00000 00000   00000 00000   00000 00000   00000 00000   00000 00000
4345:   00000 00000   00000 00000   00000 00000   00000 00000   00000 00000      00000 00000   00000 00000   00000 00000   00000 00000   00000 00000
4346:   00000 00000   00000 00000   00000 00000   00000 00000   00000 00000      00000 00000   00000 00000   00000 00000   00000 00000   00000 00000
4347:   00000 00000   00000 00000   00000 00000   00000 00000   00000 00000      00000 00000   00000 00000   00000 00000   00000 00000   00000 00000
4348:   00000 00000   00000 00000   00000 00000   00000 00000   00000 00000      00000 00000   00000 00000   00000 00000   00000 00000   00000 00000
4349:   00000 00000   00000 00000   00000 00000   00000 00000   00000 00000      00000 00000   00000 00000   00000 00000   00000 00000   00000 00000
```

```
4350:  00000 00000  00000 00000  00000 00000  00000 00000  00000 00000    00000 00000  00000 00000  00000 00000  00000 00000  00000 00000
4351:  00000 00000  00000 00000  00000 00000  00000 00000  00000 00000    00000 00000  00000 00000  00000 00000  00000 00000  00000 00000
4352:  00000 00000  00000 00000  00000 00000  00000 00000  00000 00000    00000 00000  00000 00000  00000 00000  00000 00000  00000 00000
4353:  00000 00000  00000 00000  00000 00000  00000 00000  00000 00000    00000 00000  00000 00000  00000 00000  00000 00000  00000 00000
4354:  00000 00000  00000 00000  00000 00000  00000 00000  00000 00000    00000 00000  00000 00000  00000 00000  00000 00000  00000 00000
4355:  00000 00000  00000 00000  00000 00000  00000 00000  00000 00000    00000 00000  00000 00000  00000 00000  00000 00000  00000 00000
4356:  00000 00000  00000 00000  00000 00000  00000 00000  00000 00000    00000 00000  00000 00000  00000 00000  00000 00000  00000 00000
4357:  00000 00000  00000 00000  00000 00000  00000 00000  00000 00000    00000 00000  00000 00000  00000 00000  00000 00000  00000 00000
4358:  00000 00000  00000 00000  00000 00000  00000 00000  00000 00000    00000 00000  00000 00000  00000 00000  00000 00000  00000 00000
4359:  00000 00000  00000 00000  00000 00000  00000 00000  00000 00000    00000 00000  00000 00000  00000 00000  00000 00000  00000 00000
4360:  00000 00000  00000 00000  00000 00000  00000 00000  00000 00000    00000 00000  00000 00000  00000 00000  00000 00000  00000 00000
4361:  00000 00000  00000 00000  00000 00000  00000 00000  00000 00000    00000 00000  00000 00000  00000 00000  00000 00000  00000 00000
4362:  00000 00000  00000 00000  00000 00000  00000 00000  00000 00000    00000 00000  00000 00000  00000 00000  00000 00000  00000 00000
4363:  00000 00000  00000 00000  00000 00000  00000 00000  00000 00000    00000 00000  00000 00000  00000 00000  00000 00000  00000 00000
4364:  00000 00000  00000 00000  00000 00000  00000 00000  00000 00000    00000 00000  00000 00000  00000 00000  00000 00000  00000 00000
4365:  00000 00000  00000 00000  00000 00000  00000 00000  00000 00000    00000 00000  00000 00000  00000 00000  00000 00000  00000 00000
4366:  00000 00000  00000 00000  00000 00000  00000 00000  00000 00000    00000 00000  00000 00000  00000 00000  00000 00000  00000 00000
4367:  00000 00000  00000 00000  00000 00000  00000 00000  00000 00000    00000 00000  00000 00000  00000 00000  00000 00000  00000 00000
4368:  00000 00000  00000 00000  00000 00000  00000 00000  00000 00000    00000 00000  00000 00000  00000 00000  00000 00000  00000 00000
4369:  00000 00000  00000 00000  00000 00000  00000 00000  00000 00000    00000 00000  00000 00000  00000 00000  00000 00000  00000 00000
4370:  00000 00000  00000 00000  00000 00000  00000 00000  00000 00000    00000 00000  00000 00000  00000 00000  00000 00000  00000 00000
4371:  00000 00000  00000 00000  00000 00000  00000 00000  00000 00000    00000 00000  00000 00000  00000 00000  00000 00000  00000 00000
4372:  00000 00000  00000 00000  00000 00000  00000 00000  00000 00000    00000 00000  00000 00000  00000 00000  00000 00000  00000 00000
4373:  00000 00000  00000 00000  00000 00000  00000 00000  00000 00000    00000 00000  00000 00000  00000 00000  00000 00000  00000 00000
4374:  00000 00000  00000 00000  00000 00000  00000 00000  00000 00000    00000 00000  00000 00000  00000 00000  00000 00000  00000 00000
4375:  00000 00000  00000 00000  00000 00000  00000 00000  00000 00000    00000 00000  00000 00000  00000 00000  00000 00000  00000 00000
4376:  00000 00000  00000 00000  00000 00000  00000 00000  00000 00000    00000 00000  00000 00000  00000 00000  00000 00000  00000 00000
4377:  00000 00000  00000 00000  00000 00000  00000 00000  00000 00000    00000 00000  00000 00000  00000 00000  00000 00000  00000 00000
4378:  00000 00000  00000 00000  00000 00000  00000 00000  00000 00000    00000 00000  00000 00000  00000 00000  00000 00000  00000 00000
4379:  00000 00000  00000 00000  00000 00000  00000 00000  00000 00000    00000 00000  00000 00000  00000 00000  00000 00000  00000 00000
4380:  00000 00000  00000 00000  00000 00000  00000 00000  00000 00000    00000 00000  00000 00000  00000 00000  00000 00000  00000 00000
4381:  00000 00000  00000 00000  00000 00000  00000 00000  00000 00000    00000 00000  00000 00000  00000 00000  00000 00000  00000 00000
4382:  00000 00000  00000 00000  00000 00000  00000 00000  00000 00000    00000 00000  00000 00000  00000 00000  00000 00000  00000 00000
4383:  00000 00000  00000 00000  00000 00000  00000 00000  00000 00000    00000 00000  00000 00000  00000 00000  00000 00000  00000 00000
4384:  00000 00000  00000 00000  00000 00000  00000 00000  00000 00000    00000 00000  00000 00000  00000 00000  00000 00000  00000 00000
4385:  00000 00000  00000 00000  00000 00000  00000 00000  00000 00000    00000 00000  00000 00000  00000 00000  00000 00000  00000 00000
4386:  00000 00000  00000 00000  00000 00000  00000 00000  00000 00000    00000 00000  00000 00000  00000 00000  00000 00000  00000 00000
4387:  00000 00000  00000 00000  00000 00000  00000 00000  00000 00000    00000 00000  00000 00000  00000 00000  00000 00000  00000 00000
4388:  00000 00000  00000 00000  00000 00000  00000 00000  00000 00000    00000 00000  00000 00000  00000 00000  00000 00000  00000 00000
4389:  00000 00000  00000 00000  00000 00000  00000 00000  00000 00000    00000 00000  00000 00000  00000 00000  00000 00000  00000 00000
4390:  00000 00000  00000 00000  00000 00000  00000 00000  00000 00000    00000 00000  00000 00000  00000 00000  00000 00000  00000 00000
4391:  00000 00000  00000 00000  00000 00000  00000 00000  00000 00000    00000 00000  00000 00000  00000 00000  00000 00000  00000 00000
4392:  00000 00000  00000 00000  00000 00000  00000 00000  00000 00000    00000 00000  00000 00000  00000 00000  00000 00000  00000 00000
4393:  00000 00000  00000 00000  00000 00000  00000 00000  00000 00000    00000 00000  00000 00000  00000 00000  00000 00000  00000 00000
4394:  00000 00000  00000 00000  00000 00000  00000 00000  00000 00000    00000 00000  00000 00000  00000 00000  00000 00000  00000 00000
4395:  00000 00000  00000 00000  00000 00000  00000 00000  00000 00000    00000 00000  00000 00000  00000 00000  00000 00000  00000 00000
4396:  00000 00000  00000 00000  00000 00000  00000 00000  00000 00000    00000 00000  00000 00000  00000 00000  00000 00000  00000 00000
4397:  00000 00000  00000 00000  00000 00000  00000 00000  00000 00000    00000 00000  00000 00000  00000 00000  00000 00000  00000 00000
4398:  00000 00000  00000 00000  00000 00000  00000 00000  00000 00000    00000 00000  00000 00000  00000 00000  00000 00000  00000 00000
4399:  00000 00000  00000 00000  00000 00000  00000 00000  00000 00000    00000 00000  00000 00000  00000 00000  00000 00000  00000 00000
```

```
4400:   00000 00000   00000 00000   00000 00000   00000 00000   00000 00000    00000 00000   00000 00000   00000 00000   00000 00000   00000 00000
4401:   00000 00000   00000 00000   00000 00000   00000 00000   00000 00000    00000 00000   00000 00000   00000 00000   00000 00000   00000 00000
4402:   00000 00000   00000 00000   00000 00000   00000 00000   00000 00000    00000 00000   00000 00000   00000 00000   00000 00000   00000 00000
4403:   00000 00000   00000 00000   00000 00000   00000 00000   00000 00000    00000 00000   00000 00000   00000 00000   00000 00000   00000 00000
4404:   00000 00000   00000 00000   00000 00000   00000 00000   00000 00000    00000 00000   00000 00000   00000 00000   00000 00000   00000 00000
4405:   00000 00000   00000 00000   00000 00000   00000 00000   00000 00000    00000 00000   00000 00000   00000 00000   00000 00000   00000 00000
4406:   00000 00000   00000 00000   00000 00000   00000 00000   00000 00000    00000 00000   00000 00000   00000 00000   00000 00000   00000 00000
4407:   00000 00000   00000 00000   00000 00000   00000 00000   00000 00000    00000 00000   00000 00000   00000 00000   00000 00000   00000 00000
4408:   00000 00000   00000 00000   00000 00000   00000 00000   00000 00000    00000 00000   00000 00000   00000 00000   00000 00000   00000 00000
4409:   00000 00000   00000 00000   00000 00000   00000 00000   00000 00000    00000 00000   00000 00000   00000 00000   00000 00000   00000 00000
4410:   00000 00000   00000 00000   00000 00000   00000 00000   00000 00000    00000 00000   00000 00000   00000 00000   00000 00000   00000 00000
4411:   00000 00000   00000 00000   00000 00000   00000 00000   00000 00000    00000 00000   00000 00000   00000 00000   00000 00000   00000 00000
4412:   00000 00000   00000 00000   00000 00000   00000 00000   00000 00000    00000 00000   00000 00000   00000 00000   00000 00000   00000 00000
4413:   00000 00000   00000 00000   00000 00000   00000 00000   00000 00000    00000 00000   00000 00000   00000 00000   00000 00000   00000 00000
4414:   00000 00000   00000 00000   00000 00000   00000 00000   00000 00000    00000 00000   00000 00000   00000 00000   00000 00000   00000 00000
4415:   00000 00000   00000 00000   00000 00000   00000 00000   00000 00000    00000 00000   00000 00000   00000 00000   00000 00000   00000 00000
4416:   00000 00000   00000 00000   00000 00000   00000 00000   00000 00000    00000 00000   00000 00000   00000 00000   00000 00000   00000 00000
4417:   00000 00000   00000 00000   00000 00000   00000 00000   00000 00000    00000 00000   00000 00000   00000 00000   00000 00000   00000 00000
4418:   00000 00000   00000 00000   00000 00000   00000 00000   00000 00000    00000 00000   00000 00000   00000 00000   00000 00000   00000 00000
4419:   00000 00000   00000 00000   00000 00000   00000 00000   00000 00000    00000 00000   00000 00000   00000 00000   00000 00000   00000 00000
4420:   00000 00000   00000 00000   00000 00000   00000 00000   00000 00000    00000 00000   00000 00000   00000 00000   00000 00000   00000 00000
4421:   00000 00000   00000 00000   00000 00000   00000 00000   00000 00000    00000 00000   00000 00000   00000 00000   00000 00000   00000 00000
4422:   00000 00000   00000 00000   00000 00000   00000 00000   00000 00000    00000 00000   00000 00000   00000 00000   00000 00000   00000 00000
4423:   00000 00000   00000 00000   00000 00000   00000 00000   00000 00000    00000 00000   00000 00000   00000 00000   00000 00000   00000 00000
4424:   00000 00000   00000 00000   00000 00000   00000 00000   00000 00000    00000 00000   00000 00000   00000 00000   00000 00000   00000 00000
4425:   00000 00000   00000 00000   00000 00000   00000 00000   00000 00000    00000 00000   00000 00000   00000 00000   00000 00000   00000 00000
4426:   00000 00000   00000 00000   00000 00000   00000 00000   00000 00000    00000 00000   00000 00000   00000 00000   00000 00000   00000 00000
4427:   00000 00000   00000 00000   00000 00000   00000 00000   00000 00000    00000 00000   00000 00000   00000 00000   00000 00000   00000 00000
4428:   00000 00000   00000 00000   00000 00000   00000 00000   00000 00000    00000 00000   00000 00000   00000 00000   00000 00000   00000 00000
4429:   00000 00000   00000 00000   00000 00000   00000 00000   00000 00000    00000 00000   00000 00000   00000 00000   00000 00000   00000 00000
4430:   00000 00000   00000 00000   00000 00000   00000 00000   00000 00000    00000 00000   00000 00000   00000 00000   00000 00000   00000 00000
4431:   00000 00000   00000 00000   00000 00000   00000 00000   00000 00000    00000 00000   00000 00000   00000 00000   00000 00000   00000 00000
4432:   00000 00000   00000 00000   00000 00000   00000 00000   00000 00000    00000 00000   00000 00000   00000 00000   00000 00000   00000 00000
4433:   00000 00000   00000 00000   00000 00000   00000 00000   00000 00000    00000 00000   00000 00000   00000 00000   00000 00000   00000 00000
4434:   00000 00000   00000 00000   00000 00000   00000 00000   00000 00000    00000 00000   00000 00000   00000 00000   00000 00000   00000 00000
4435:   00000 00000   00000 00000   00000 00000   00000 00000   00000 00000    00000 00000   00000 00000   00000 00000   00000 00000   00000 00000
4436:   00000 00000   00000 00000   00000 00000   00000 00000   00000 00000    00000 00000   00000 00000   00000 00000   00000 00000   00000 00000
4437:   00000 00000   00000 00000   00000 00000   00000 00000   00000 00000    00000 00000   00000 00000   00000 00000   00000 00000   00000 00000
4438:   00000 00000   00000 00000   00000 00000   00000 00000   00000 00000    00000 00000   00000 00000   00000 00000   00000 00000   00000 00000
4439:   00000 00000   00000 00000   00000 00000   00000 00000   00000 00000    00000 00000   00000 00000   00000 00000   00000 00000   00000 00000
4440:   00000 00000   00000 00000   00000 00000   00000 00000   00000 00000    00000 00000   00000 00000   00000 00000   00000 00000   00000 00000
4441:   00000 00000   00000 00000   00000 00000   00000 00000   00000 00000    00000 00000   00000 00000   00000 00000   00000 00000   00000 00000
4442:   00000 00000   00000 00000   00000 00000   00000 00000   00000 00000    00000 00000   00000 00000   00000 00000   00000 00000   00000 00000
4443:   00000 00000   00000 00000   00000 00000   00000 00000   00000 00000    00000 00000   00000 00000   00000 00000   00000 00000   00000 00000
4444:   00000 00000   00000 00000   00000 00000   00000 00000   00000 00000    00000 00000   00000 00000   00000 00000   00000 00000   00000 00000
4445:   00000 00000   00000 00000   00000 00000   00000 00000   00000 00000    00000 00000   00000 00000   00000 00000   00000 00000   00000 00000
4446:   00000 00000   00000 00000   00000 00000   00000 00000   00000 00000    00000 00000   00000 00000   00000 00000   00000 00000   00000 00000
4447:   00000 00000   00000 00000   00000 00000   00000 00000   00000 00000    00000 00000   00000 00000   00000 00000   00000 00000   00000 00000
4448:   00000 00000   00000 00000   00000 00000   00000 00000   00000 00000    00000 00000   00000 00000   00000 00000   00000 00000   00000 00000
4449:   00000 00000   00000 00000   00000 00000   00000 00000   00000 00000    00000 00000   00000 00000   00000 00000   00000 00000   00000 00000
```

```
4450:  00000 00000  00000 00000  00000 00000  00000 00000  00000 00000    00000 00000  00000 00000  00000 00000  00000 00000  00000 00000
4451:  00000 00000  00000 00000  00000 00000  00000 00000  00000 00000    00000 00000  00000 00000  00000 00000  00000 00000  00000 00000
4452:  00000 00000  00000 00000  00000 00000  00000 00000  00000 00000    00000 00000  00000 00000  00000 00000  00000 00000  00000 00000
4453:  00000 00000  00000 00000  00000 00000  00000 00000  00000 00000    00000 00000  00000 00000  00000 00000  00000 00000  00000 00000
4454:  00000 00000  00000 00000  00000 00000  00000 00000  00000 00000    00000 00000  00000 00000  00000 00000  00000 00000  00000 00000
4455:  00000 00000  00000 00000  00000 00000  00000 00000  00000 00000    00000 00000  00000 00000  00000 00000  00000 00000  00000 00000
4456:  00000 00000  00000 00000  00000 00000  00000 00000  00000 00000    00000 00000  00000 00000  00000 00000  00000 00000  00000 00000
4457:  00000 00000  00000 00000  00000 00000  00000 00000  00000 00000    00000 00000  00000 00000  00000 00000  00000 00000  00000 00000
4458:  00000 00000  00000 00000  00000 00000  00000 00000  00000 00000    00000 00000  00000 00000  00000 00000  00000 00000  00000 00000
4459:  00000 00000  00000 00000  00000 00000  00000 00000  00000 00000    00000 00000  00000 00000  00000 00000  00000 00000  00000 00000
4460:  00000 00000  00000 00000  00000 00000  00000 00000  00000 00000    00000 00000  00000 00000  00000 00000  00000 00000  00000 00000
4461:  00000 00000  00000 00000  00000 00000  00000 00000  00000 00000    00000 00000  00000 00000  00000 00000  00000 00000  00000 00000
4462:  00000 00000  00000 00000  00000 00000  00000 00000  00000 00000    00000 00000  00000 00000  00000 00000  00000 00000  00000 00000
4463:  00000 00000  00000 00000  00000 00000  00000 00000  00000 00000    00000 00000  00000 00000  00000 00000  00000 00000  00000 00000
4464:  00000 00000  00000 00000  00000 00000  00000 00000  00000 00000    00000 00000  00000 00000  00000 00000  00000 00000  00000 00000
4465:  00000 00000  00000 00000  00000 00000  00000 00000  00000 00000    00000 00000  00000 00000  00000 00000  00000 00000  00000 00000
4466:  00000 00000  00000 00000  00000 00000  00000 00000  00000 00000    00000 00000  00000 00000  00000 00000  00000 00000  00000 00000
4467:  00000 00000  00000 00000  00000 00000  00000 00000  00000 00000    00000 00000  00000 00000  00000 00000  00000 00000  00000 00000
4468:  00000 00000  00000 00000  00000 00000  00000 00000  00000 00000    00000 00000  00000 00000  00000 00000  00000 00000  00000 00000
4469:  00000 00000  00000 00000  00000 00000  00000 00000  00000 00000    00000 00000  00000 00000  00000 00000  00000 00000  00000 00000
4470:  00000 00000  00000 00000  00000 00000  00000 00000  00000 00000    00000 00000  00000 00000  00000 00000  00000 00000  00000 00000
4471:  00000 00000  00000 00000  00000 00000  00000 00000  00000 00000    00000 00000  00000 00000  00000 00000  00000 00000  00000 00000
4472:  00000 00000  00000 00000  00000 00000  00000 00000  00000 00000    00000 00000  00000 00000  00000 00000  00000 00000  00000 00000
4473:  00000 00000  00000 00000  00000 00000  00000 00000  00000 00000    00000 00000  00000 00000  00000 00000  00000 00000  00000 00000
4474:  00000 00000  00000 00000  00000 00000  00000 00000  00000 00000    00000 00000  00000 00000  00000 00000  00000 00000  00000 00000
4475:  00000 00000  00000 00000  00000 00000  00000 00000  00000 00000    00000 00000  00000 00000  00000 00000  00000 00000  00000 00000
4476:  00000 00000  00000 00000  00000 00000  00000 00000  00000 00000    00000 00000  00000 00000  00000 00000  00000 00000  00000 00000
4477:  00000 00000  00000 00000  00000 00000  00000 00000  00000 00000    00000 00000  00000 00000  00000 00000  00000 00000  00000 00000
4478:  00000 00000  00000 00000  00000 00000  00000 00000  00000 00000    00000 00000  00000 00000  00000 00000  00000 00000  00000 00000
4479:  00000 00000  00000 00000  00000 00000  00000 00000  00000 00000    00000 00000  00000 00000  00000 00000  00000 00000  00000 00000
4480:  00000 00000  00000 00000  00000 00000  00000 00000  00000 00000    00000 00000  00000 00000  00000 00000  00000 00000  00000 00000
4481:  00000 00000  00000 00000  00000 00000  00000 00000  00000 00000    00000 00000  00000 00000  00000 00000  00000 00000  00000 00000
4482:  00000 00000  00000 00000  00000 00000  00000 00000  00000 00000    00000 00000  00000 00000  00000 00000  00000 00000  00000 00000
4483:  00000 00000  00000 00000  00000 00000  00000 00000  00000 00000    00000 00000  00000 00000  00000 00000  00000 00000  00000 00000
4484:  00000 00000  00000 00000  00000 00000  00000 00000  00000 00000    00000 00000  00000 00000  00000 00000  00000 00000  00000 00000
4485:  00000 00000  00000 00000  00000 00000  00000 00000  00000 00000    00000 00000  00000 00000  00000 00000  00000 00000  00000 00000
4486:  00000 00000  00000 00000  00000 00000  00000 00000  00000 00000    00000 00000  00000 00000  00000 00000  00000 00000  00000 00000
4487:  00000 00000  00000 00000  00000 00000  00000 00000  00000 00000    00000 00000  00000 00000  00000 00000  00000 00000  00000 00000
4488:  00000 00000  00000 00000  00000 00000  00000 00000  00000 00000    00000 00000  00000 00000  00000 00000  00000 00000  00000 00000
4489:  00000 00000  00000 00000  00000 00000  00000 00000  00000 00000    00000 00000  00000 00000  00000 00000  00000 00000  00000 00000
4490:  00000 00000  00000 00000  00000 00000  00000 00000  00000 00000    00000 00000  00000 00000  00000 00000  00000 00000  00000 00000
4491:  00000 00000  00000 00000  00000 00000  00000 00000  00000 00000    00000 00000  00000 00000  00000 00000  00000 00000  00000 00000
4492:  00000 00000  00000 00000  00000 00000  00000 00000  00000 00000    00000 00000  00000 00000  00000 00000  00000 00000  00000 00000
4493:  00000 00000  00000 00000  00000 00000  00000 00000  00000 00000    00000 00000  00000 00000  00000 00000  00000 00000  00000 00000
4494:  00000 00000  00000 00000  00000 00000  00000 00000  00000 00000    00000 00000  00000 00000  00000 00000  00000 00000  00000 00000
4495:  00000 00000  00000 00000  00000 00000  00000 00000  00000 00000    00000 00000  00000 00000  00000 00000  00000 00000  00000 00000
4496:  00000 00000  00000 00000  00000 00000  00000 00000  00000 00000    00000 00000  00000 00000  00000 00000  00000 00000  00000 00000
4497:  00000 00000  00000 00000  00000 00000  00000 00000  00000 00000    00000 00000  00000 00000  00000 00000  00000 00000  00000 00000
4498:  00000 00000  00000 00000  00000 00000  00000 00000  00000 00000    00000 00000  00000 00000  00000 00000  00000 00000  00000 00000
4499:  00000 00000  00000 00000  00000 00000  00000 00000  00000 00000    00000 00000  00000 00000  00000 00000  00000 00000  00000 00000
```

```
4500:  00000 00000  00000 00000  00000 00000  00000 00000  00000 00000    00000 00000  00000 00000  00000 00000  00000 00000  00000 00000
4501:  00000 00000  00000 00000  00000 00000  00000 00000  00000 00000    00000 00000  00000 00000  00000 00000  00000 00000  00000 00000
4502:  00000 00000  00000 00000  00000 00000  00000 00000  00000 00000    00000 00000  00000 00000  00000 00000  00000 00000  00000 00000
4503:  00000 00000  00000 00000  00000 00000  00000 00000  00000 00000    00000 00000  00000 00000  00000 00000  00000 00000  00000 00000
4504:  00000 00000  00000 00000  00000 00000  00000 00000  00000 00000    00000 00000  00000 00000  00000 00000  00000 00000  00000 00000
4505:  00000 00000  00000 00000  00000 00000  00000 00000  00000 00000    00000 00000  00000 00000  00000 00000  00000 00000  00000 00000
4506:  00000 00000  00000 00000  00000 00000  00000 00000  00000 00000    00000 00000  00000 00000  00000 00000  00000 00000  00000 00000
4507:  00000 00000  00000 00000  00000 00000  00000 00000  00000 00000    00000 00000  00000 00000  00000 00000  00000 00000  00000 00000
4508:  00000 00000  00000 00000  00000 00000  00000 00000  00000 00000    00000 00000  00000 00000  00000 00000  00000 00000  00000 00000
4509:  00000 00000  00000 00000  00000 00000  00000 00000  00000 00000    00000 00000  00000 00000  00000 00000  00000 00000  00000 00000
4510:  00000 00000  00000 00000  00000 00000  00000 00000  00000 00000    00000 00000  00000 00000  00000 00000  00000 00000  00000 00000
4511:  00000 00000  00000 00000  00000 00000  00000 00000  00000 00000    00000 00000  00000 00000  00000 00000  00000 00000  00000 00000
4512:  00000 00000  00000 00000  00000 00000  00000 00000  00000 00000    00000 00000  00000 00000  00000 00000  00000 00000  00000 00000
4513:  00000 00000  00000 00000  00000 00000  00000 00000  00000 00000    00000 00000  00000 00000  00000 00000  00000 00000  00000 00000
4514:  00000 00000  00000 00000  00000 00000  00000 00000  00000 00000    00000 00000  00000 00000  00000 00000  00000 00000  00000 00000
4515:  00000 00000  00000 00000  00000 00000  00000 00000  00000 00000    00000 00000  00000 00000  00000 00000  00000 00000  00000 00000
4516:  00000 00000  00000 00000  00000 00000  00000 00000  00000 00000    00000 00000  00000 00000  00000 00000  00000 00000  00000 00000
4517:  00000 00000  00000 00000  00000 00000  00000 00000  00000 00000    00000 00000  00000 00000  00000 00000  00000 00000  00000 00000
4518:  00000 00000  00000 00000  00000 00000  00000 00000  00000 00000    00000 00000  00000 00000  00000 00000  00000 00000  00000 00000
4519:  00000 00000  00000 00000  00000 00000  00000 00000  00000 00000    00000 00000  00000 00000  00000 00000  00000 00000  00000 00000
4520:  00000 00000  00000 00000  00000 00000  00000 00000  00000 00000    00000 00000  00000 00000  00000 00000  00000 00000  00000 00000
4521:  00000 00000  00000 00000  00000 00000  00000 00000  00000 00000    00000 00000  00000 00000  00000 00000  00000 00000  00000 00000
4522:  00000 00000  00000 00000  00000 00000  00000 00000  00000 00000    00000 00000  00000 00000  00000 00000  00000 00000  00000 00000
4523:  00000 00000  00000 00000  00000 00000  00000 00000  00000 00000    00000 00000  00000 00000  00000 00000  00000 00000  00000 00000
4524:  00000 00000  00000 00000  00000 00000  00000 00000  00000 00000    00000 00000  00000 00000  00000 00000  00000 00000  00000 00000
4525:  00000 00000  00000 00000  00000 00000  00000 00000  00000 00000    00000 00000  00000 00000  00000 00000  00000 00000  00000 00000
4526:  00000 00000  00000 00000  00000 00000  00000 00000  00000 00000    00000 00000  00000 00000  00000 00000  00000 00000  00000 00000
4527:  00000 00000  00000 00000  00000 00000  00000 00000  00000 00000    00000 00000  00000 00000  00000 00000  00000 00000  00000 00000
4528:  00000 00000  00000 00000  00000 00000  00000 00000  00000 00000    00000 00000  00000 00000  00000 00000  00000 00000  00000 00000
4529:  00000 00000  00000 00000  00000 00000  00000 00000  00000 00000    00000 00000  00000 00000  00000 00000  00000 00000  00000 00000
4530:  00000 00000  00000 00000  00000 00000  00000 00000  00000 00000    00000 00000  00000 00000  00000 00000  00000 00000  00000 00000
4531:  00000 00000  00000 00000  00000 00000  00000 00000  00000 00000    00000 00000  00000 00000  00000 00000  00000 00000  00000 00000
4532:  00000 00000  00000 00000  00000 00000  00000 00000  00000 00000    00000 00000  00000 00000  00000 00000  00000 00000  00000 00000
4533:  00000 00000  00000 00000  00000 00000  00000 00000  00000 00000    00000 00000  00000 00000  00000 00000  00000 00000  00000 00000
4534:  00000 00000  00000 00000  00000 00000  00000 00000  00000 00000    00000 00000  00000 00000  00000 00000  00000 00000  00000 00000
4535:  00000 00000  00000 00000  00000 00000  00000 00000  00000 00000    00000 00000  00000 00000  00000 00000  00000 00000  00000 00000
4536:  00000 00000  00000 00000  00000 00000  00000 00000  00000 00000    00000 00000  00000 00000  00000 00000  00000 00000  00000 00000
4537:  00000 00000  00000 00000  00000 00000  00000 00000  00000 00000    00000 00000  00000 00000  00000 00000  00000 00000  00000 00000
4538:  00000 00000  00000 00000  00000 00000  00000 00000  00000 00000    00000 00000  00000 00000  00000 00000  00000 00000  00000 00000
4539:  00000 00000  00000 00000  00000 00000  00000 00000  00000 00000    00000 00000  00000 00000  00000 00000  00000 00000  00000 00000
4540:  00000 00000  00000 00000  00000 00000  00000 00000  00000 00000    00000 00000  00000 00000  00000 00000  00000 00000  00000 00000
4541:  00000 00000  00000 00000  00000 00000  00000 00000  00000 00000    00000 00000  00000 00000  00000 00000  00000 00000  00000 00000
4542:  00000 00000  00000 00000  00000 00000  00000 00000  00000 00000    00000 00000  00000 00000  00000 00000  00000 00000  00000 00000
4543:  00000 00000  00000 00000  00000 00000  00000 00000  00000 00000    00000 00000  00000 00000  00000 00000  00000 00000  00000 00000
4544:  00000 00000  00000 00000  00000 00000  00000 00000  00000 00000    00000 00000  00000 00000  00000 00000  00000 00000  00000 00000
4545:  00000 00000  00000 00000  00000 00000  00000 00000  00000 00000    00000 00000  00000 00000  00000 00000  00000 00000  00000 00000
4546:  00000 00000  00000 00000  00000 00000  00000 00000  00000 00000    00000 00000  00000 00000  00000 00000  00000 00000  00000 00000
4547:  00000 00000  00000 00000  00000 00000  00000 00000  00000 00000    00000 00000  00000 00000  00000 00000  00000 00000  00000 00000
4548:  00000 00000  00000 00000  00000 00000  00000 00000  00000 00000    00000 00000  00000 00000  00000 00000  00000 00000  00000 00000
4549:  00000 00000  00000 00000  00000 00000  00000 00000  00000 00000    00000 00000  00000 00000  00000 00000  00000 00000  00000 00000
```

```
4550:  00000 00000  00000 00000  00000 00000  00000 00000  00000 00000   00000 00000  00000 00000  00000 00000  00000 00000  00000 00000
4551:  00000 00000  00000 00000  00000 00000  00000 00000  00000 00000   00000 00000  00000 00000  00000 00000  00000 00000  00000 00000
4552:  00000 00000  00000 00000  00000 00000  00000 00000  00000 00000   00000 00000  00000 00000  00000 00000  00000 00000  00000 00000
4553:  00000 00000  00000 00000  00000 00000  00000 00000  00000 00000   00000 00000  00000 00000  00000 00000  00000 00000  00000 00000
4554:  00000 00000  00000 00000  00000 00000  00000 00000  00000 00000   00000 00000  00000 00000  00000 00000  00000 00000  00000 00000
4555:  00000 00000  00000 00000  00000 00000  00000 00000  00000 00000   00000 00000  00000 00000  00000 00000  00000 00000  00000 00000
4556:  00000 00000  00000 00000  00000 00000  00000 00000  00000 00000   00000 00000  00000 00000  00000 00000  00000 00000  00000 00000
4557:  00000 00000  00000 00000  00000 00000  00000 00000  00000 00000   00000 00000  00000 00000  00000 00000  00000 00000  00000 00000
4558:  00000 00000  00000 00000  00000 00000  00000 00000  00000 00000   00000 00000  00000 00000  00000 00000  00000 00000  00000 00000
4559:  00000 00000  00000 00000  00000 00000  00000 00000  00000 00000   00000 00000  00000 00000  00000 00000  00000 00000  00000 00000
4560:  00000 00000  00000 00000  00000 00000  00000 00000  00000 00000   00000 00000  00000 00000  00000 00000  00000 00000  00000 00000
4561:  00000 00000  00000 00000  00000 00000  00000 00000  00000 00000   00000 00000  00000 00000  00000 00000  00000 00000  00000 00000
4562:  00000 00000  00000 00000  00000 00000  00000 00000  00000 00000   00000 00000  00000 00000  00000 00000  00000 00000  00000 00000
4563:  00000 00000  00000 00000  00000 00000  00000 00000  00000 00000   00000 00000  00000 00000  00000 00000  00000 00000  00000 00000
4564:  00000 00000  00000 00000  00000 00000  00000 00000  00000 00000   00000 00000  00000 00000  00000 00000  00000 00000  00000 00000
4565:  00000 00000  00000 00000  00000 00000  00000 00000  00000 00000   00000 00000  00000 00000  00000 00000  00000 00000  00000 00000
4566:  00000 00000  00000 00000  00000 00000  00000 00000  00000 00000   00000 00000  00000 00000  00000 00000  00000 00000  00000 00000
4567:  00000 00000  00000 00000  00000 00000  00000 00000  00000 00000   00000 00000  00000 00000  00000 00000  00000 00000  00000 00000
4568:  00000 00000  00000 00000  00000 00000  00000 00000  00000 00000   00000 00000  00000 00000  00000 00000  00000 00000  00000 00000
4569:  00000 00000  00000 00000  00000 00000  00000 00000  00000 00000   00000 00000  00000 00000  00000 00000  00000 00000  00000 00000
4570:  00000 00000  00000 00000  00000 00000  00000 00000  00000 00000   00000 00000  00000 00000  00000 00000  00000 00000  00000 00000
4571:  00000 00000  00000 00000  00000 00000  00000 00000  00000 00000   00000 00000  00000 00000  00000 00000  00000 00000  00000 00000
4572:  00000 00000  00000 00000  00000 00000  00000 00000  00000 00000   00000 00000  00000 00000  00000 00000  00000 00000  00000 00000
4573:  00000 00000  00000 00000  00000 00000  00000 00000  00000 00000   00000 00000  00000 00000  00000 00000  00000 00000  00000 00000
4574:  00000 00000  00000 00000  00000 00000  00000 00000  00000 00000   00000 00000  00000 00000  00000 00000  00000 00000  00000 00000
4575:  00000 00000  00000 00000  00000 00000  00000 00000  00000 00000   00000 00000  00000 00000  00000 00000  00000 00000  00000 00000
4576:  00000 00000  00000 00000  00000 00000  00000 00000  00000 00000   00000 00000  00000 00000  00000 00000  00000 00000  00000 00000
4577:  00000 00000  00000 00000  00000 00000  00000 00000  00000 00000   00000 00000  00000 00000  00000 00000  00000 00000  00000 00000
4578:  00000 00000  00000 00000  00000 00000  00000 00000  00000 00000   00000 00000  00000 00000  00000 00000  00000 00000  00000 00000
4579:  00000 00000  00000 00000  00000 00000  00000 00000  00000 00000   00000 00000  00000 00000  00000 00000  00000 00000  00000 00000
4580:  00000 00000  00000 00000  00000 00000  00000 00000  00000 00000   00000 00000  00000 00000  00000 00000  00000 00000  00000 00000
4581:  00000 00000  00000 00000  00000 00000  00000 00000  00000 00000   00000 00000  00000 00000  00000 00000  00000 00000  00000 00000
4582:  00000 00000  00000 00000  00000 00000  00000 00000  00000 00000   00000 00000  00000 00000  00000 00000  00000 00000  00000 00000
4583:  00000 00000  00000 00000  00000 00000  00000 00000  00000 00000   00000 00000  00000 00000  00000 00000  00000 00000  00000 00000
4584:  00000 00000  00000 00000  00000 00000  00000 00000  00000 00000   00000 00000  00000 00000  00000 00000  00000 00000  00000 00000
4585:  00000 00000  00000 00000  00000 00000  00000 00000  00000 00000   00000 00000  00000 00000  00000 00000  00000 00000  00000 00000
4586:  00000 00000  00000 00000  00000 00000  00000 00000  00000 00000   00000 00000  00000 00000  00000 00000  00000 00000  00000 00000
4587:  00000 00000  00000 00000  00000 00000  00000 00000  00000 00000   00000 00000  00000 00000  00000 00000  00000 00000  00000 00000
4588:  00000 00000  00000 00000  00000 00000  00000 00000  00000 00000   00000 00000  00000 00000  00000 00000  00000 00000  00000 00000
4589:  00000 00000  00000 00000  00000 00000  00000 00000  00000 00000   00000 00000  00000 00000  00000 00000  00000 00000  00000 00000
4590:  00000 00000  00000 00000  00000 00000  00000 00000  00000 00000   00000 00000  00000 00000  00000 00000  00000 00000  00000 00000
4591:  00000 00000  00000 00000  00000 00000  00000 00000  00000 00000   00000 00000  00000 00000  00000 00000  00000 00000  00000 00000
4592:  00000 00000  00000 00000  00000 00000  00000 00000  00000 00000   00000 00000  00000 00000  00000 00000  00000 00000  00000 00000
4593:  00000 00000  00000 00000  00000 00000  00000 00000  00000 00000   00000 00000  00000 00000  00000 00000  00000 00000  00000 00000
4594:  00000 00000  00000 00000  00000 00000  00000 00000  00000 00000   00000 00000  00000 00000  00000 00000  00000 00000  00000 00000
4595:  00000 00000  00000 00000  00000 00000  00000 00000  00000 00000   00000 00000  00000 00000  00000 00000  00000 00000  00000 00000
4596:  00000 00000  00000 00000  00000 00000  00000 00000  00000 00000   00000 00000  00000 00000  00000 00000  00000 00000  00000 00000
4597:  00000 00000  00000 00000  00000 00000  00000 00000  00000 00000   00000 00000  00000 00000  00000 00000  00000 00000  00000 00000
4598:  00000 00000  00000 00000  00000 00000  00000 00000  00000 00000   00000 00000  00000 00000  00000 00000  00000 00000  00000 00000
4599:  00000 00000  00000 00000  00000 00000  00000 00000  00000 00000   00000 00000  00000 00000  00000 00000  00000 00000  00000 00000
```

```
4600:  00000 00000   00000 00000   00000 00000   00000 00000   00000 00000     00000 00000   00000 00000   00000 00000   00000 00000   00000 00000
4601:  00000 00000   00000 00000   00000 00000   00000 00000   00000 00000     00000 00000   00000 00000   00000 00000   00000 00000   00000 00000
4602:  00000 00000   00000 00000   00000 00000   00000 00000   00000 00000     00000 00000   00000 00000   00000 00000   00000 00000   00000 00000
4603:  00000 00000   00000 00000   00000 00000   00000 00000   00000 00000     00000 00000   00000 00000   00000 00000   00000 00000   00000 00000
4604:  00000 00000   00000 00000   00000 00000   00000 00000   00000 00000     00000 00000   00000 00000   00000 00000   00000 00000   00000 00000
4605:  00000 00000   00000 00000   00000 00000   00000 00000   00000 00000     00000 00000   00000 00000   00000 00000   00000 00000   00000 00000
4606:  00000 00000   00000 00000   00000 00000   00000 00000   00000 00000     00000 00000   00000 00000   00000 00000   00000 00000   00000 00000
4607:  00000 00000   00000 00000   00000 00000   00000 00000   00000 00000     00000 00000   00000 00000   00000 00000   00000 00000   00000 00000
4608:  00000 00000   00000 00000   00000 00000   00000 00000   00000 00000     00000 00000   00000 00000   00000 00000   00000 00000   00000 00000
4609:  00000 00000   00000 00000   00000 00000   00000 00000   00000 00000     00000 00000   00000 00000   00000 00000   00000 00000   00000 00000
4610:  00000 00000   00000 00000   00000 00000   00000 00000   00000 00000     00000 00000   00000 00000   00000 00000   00000 00000   00000 00000
4611:  00000 00000   00000 00000   00000 00000   00000 00000   00000 00000     00000 00000   00000 00000   00000 00000   00000 00000   00000 00000
4612:  00000 00000   00000 00000   00000 00000   00000 00000   00000 00000     00000 00000   00000 00000   00000 00000   00000 00000   00000 00000
4613:  00000 00000   00000 00000   00000 00000   00000 00000   00000 00000     00000 00000   00000 00000   00000 00000   00000 00000   00000 00000
4614:  00000 00000   00000 00000   00000 00000   00000 00000   00000 00000     00000 00000   00000 00000   00000 00000   00000 00000   00000 00000
4615:  00000 00000   00000 00000   00000 00000   00000 00000   00000 00000     00000 00000   00000 00000   00000 00000   00000 00000   00000 00000
4616:  00000 00000   00000 00000   00000 00000   00000 00000   00000 00000     00000 00000   00000 00000   00000 00000   00000 00000   00000 00000
4617:  00000 00000   00000 00000   00000 00000   00000 00000   00000 00000     00000 00000   00000 00000   00000 00000   00000 00000   00000 00000
4618:  00000 00000   00000 00000   00000 00000   00000 00000   00000 00000     00000 00000   00000 00000   00000 00000   00000 00000   00000 00000
4619:  00000 00000   00000 00000   00000 00000   00000 00000   00000 00000     00000 00000   00000 00000   00000 00000   00000 00000   00000 00000
4620:  00000 00000   00000 00000   00000 00000   00000 00000   00000 00000     00000 00000   00000 00000   00000 00000   00000 00000   00000 00000
4621:  00000 00000   00000 00000   00000 00000   00000 00000   00000 00000     00000 00000   00000 00000   00000 00000   00000 00000   00000 00000
4622:  00000 00000   00000 00000   00000 00000   00000 00000   00000 00000     00000 00000   00000 00000   00000 00000   00000 00000   00000 00000
4623:  00000 00000   00000 00000   00000 00000   00000 00000   00000 00000     00000 00000   00000 00000   00000 00000   00000 00000   00000 00000
4624:  00000 00000   00000 00000   00000 00000   00000 00000   00000 00000     00000 00000   00000 00000   00000 00000   00000 00000   00000 00000
4625:  00000 00000   00000 00000   00000 00000   00000 00000   00000 00000     00000 00000   00000 00000   00000 00000   00000 00000   00000 00000
4626:  00000 00000   00000 00000   00000 00000   00000 00000   00000 00000     00000 00000   00000 00000   00000 00000   00000 00000   00000 00000
4627:  00000 00000   00000 00000   00000 00000   00000 00000   00000 00000     00000 00000   00000 00000   00000 00000   00000 00000   00000 00000
4628:  00000 00000   00000 00000   00000 00000   00000 00000   00000 00000     00000 00000   00000 00000   00000 00000   00000 00000   00000 00000
4629:  00000 00000   00000 00000   00000 00000   00000 00000   00000 00000     00000 00000   00000 00000   00000 00000   00000 00000   00000 00000
4630:  00000 00000   00000 00000   00000 00000   00000 00000   00000 00000     00000 00000   00000 00000   00000 00000   00000 00000   00000 00000
4631:  00000 00000   00000 00000   00000 00000   00000 00000   00000 00000     00000 00000   00000 00000   00000 00000   00000 00000   00000 00000
4632:  00000 00000   00000 00000   00000 00000   00000 00000   00000 00000     00000 00000   00000 00000   00000 00000   00000 00000   00000 00000
4633:  00000 00000   00000 00000   00000 00000   00000 00000   00000 00000     00000 00000   00000 00000   00000 00000   00000 00000   00000 00000
4634:  00000 00000   00000 00000   00000 00000   00000 00000   00000 00000     00000 00000   00000 00000   00000 00000   00000 00000   00000 00000
4635:  00000 00000   00000 00000   00000 00000   00000 00000   00000 00000     00000 00000   00000 00000   00000 00000   00000 00000   00000 00000
4636:  00000 00000   00000 00000   00000 00000   00000 00000   00000 00000     00000 00000   00000 00000   00000 00000   00000 00000   00000 00000
4637:  00000 00000   00000 00000   00000 00000   00000 00000   00000 00000     00000 00000   00000 00000   00000 00000   00000 00000   00000 00000
4638:  00000 00000   00000 00000   00000 00000   00000 00000   00000 00000     00000 00000   00000 00000   00000 00000   00000 00000   00000 00000
4639:  00000 00000   00000 00000   00000 00000   00000 00000   00000 00000     00000 00000   00000 00000   00000 00000   00000 00000   00000 00000
4640:  00000 00000   00000 00000   00000 00000   00000 00000   00000 00000     00000 00000   00000 00000   00000 00000   00000 00000   00000 00000
4641:  00000 00000   00000 00000   00000 00000   00000 00000   00000 00000     00000 00000   00000 00000   00000 00000   00000 00000   00000 00000
4642:  00000 00000   00000 00000   00000 00000   00000 00000   00000 00000     00000 00000   00000 00000   00000 00000   00000 00000   00000 00000
4643:  00000 00000   00000 00000   00000 00000   00000 00000   00000 00000     00000 00000   00000 00000   00000 00000   00000 00000   00000 00000
4644:  00000 00000   00000 00000   00000 00000   00000 00000   00000 00000     00000 00000   00000 00000   00000 00000   00000 00000   00000 00000
4645:  00000 00000   00000 00000   00000 00000   00000 00000   00000 00000     00000 00000   00000 00000   00000 00000   00000 00000   00000 00000
4646:  00000 00000   00000 00000   00000 00000   00000 00000   00000 00000     00000 00000   00000 00000   00000 00000   00000 00000   00000 00000
4647:  00000 00000   00000 00000   00000 00000   00000 00000   00000 00000     00000 00000   00000 00000   00000 00000   00000 00000   00000 00000
4648:  00000 00000   00000 00000   00000 00000   00000 00000   00000 00000     00000 00000   00000 00000   00000 00000   00000 00000   00000 00000
4649:  00000 00000   00000 00000   00000 00000   00000 00000   00000 00000     00000 00000   00000 00000   00000 00000   00000 00000   00000 00000
```

```
4650:  00000 00000  00000 00000  00000 00000  00000 00000  00000 00000    00000 00000  00000 00000  00000 00000  00000 00000  00000 00000
4651:  00000 00000  00000 00000  00000 00000  00000 00000  00000 00000    00000 00000  00000 00000  00000 00000  00000 00000  00000 00000
4652:  00000 00000  00000 00000  00000 00000  00000 00000  00000 00000    00000 00000  00000 00000  00000 00000  00000 00000  00000 00000
4653:  00000 00000  00000 00000  00000 00000  00000 00000  00000 00000    00000 00000  00000 00000  00000 00000  00000 00000  00000 00000
4654:  00000 00000  00000 00000  00000 00000  00000 00000  00000 00000    00000 00000  00000 00000  00000 00000  00000 00000  00000 00000
4655:  00000 00000  00000 00000  00000 00000  00000 00000  00000 00000    00000 00000  00000 00000  00000 00000  00000 00000  00000 00000
4656:  00000 00000  00000 00000  00000 00000  00000 00000  00000 00000    00000 00000  00000 00000  00000 00000  00000 00000  00000 00000
4657:  00000 00000  00000 00000  00000 00000  00000 00000  00000 00000    00000 00000  00000 00000  00000 00000  00000 00000  00000 00000
4658:  00000 00000  00000 00000  00000 00000  00000 00000  00000 00000    00000 00000  00000 00000  00000 00000  00000 00000  00000 00000
4659:  00000 00000  00000 00000  00000 00000  00000 00000  00000 00000    00000 00000  00000 00000  00000 00000  00000 00000  00000 00000
4660:  00000 00000  00000 00000  00000 00000  00000 00000  00000 00000    00000 00000  00000 00000  00000 00000  00000 00000  00000 00000
4661:  00000 00000  00000 00000  00000 00000  00000 00000  00000 00000    00000 00000  00000 00000  00000 00000  00000 00000  00000 00000
4662:  00000 00000  00000 00000  00000 00000  00000 00000  00000 00000    00000 00000  00000 00000  00000 00000  00000 00000  00000 00000
4663:  00000 00000  00000 00000  00000 00000  00000 00000  00000 00000    00000 00000  00000 00000  00000 00000  00000 00000  00000 00000
4664:  00000 00000  00000 00000  00000 00000  00000 00000  00000 00000    00000 00000  00000 00000  00000 00000  00000 00000  00000 00000
4665:  00000 00000  00000 00000  00000 00000  00000 00000  00000 00000    00000 00000  00000 00000  00000 00000  00000 00000  00000 00000
4666:  00000 00000  00000 00000  00000 00000  00000 00000  00000 00000    00000 00000  00000 00000  00000 00000  00000 00000  00000 00000
4667:  00000 00000  00000 00000  00000 00000  00000 00000  00000 00000    00000 00000  00000 00000  00000 00000  00000 00000  00000 00000
4668:  00000 00000  00000 00000  00000 00000  00000 00000  00000 00000    00000 00000  00000 00000  00000 00000  00000 00000  00000 00000
4669:  00000 00000  00000 00000  00000 00000  00000 00000  00000 00000    00000 00000  00000 00000  00000 00000  00000 00000  00000 00000
4670:  00000 00000  00000 00000  00000 00000  00000 00000  00000 00000    00000 00000  00000 00000  00000 00000  00000 00000  00000 00000
4671:  00000 00000  00000 00000  00000 00000  00000 00000  00000 00000    00000 00000  00000 00000  00000 00000  00000 00000  00000 00000
4672:  00000 00000  00000 00000  00000 00000  00000 00000  00000 00000    00000 00000  00000 00000  00000 00000  00000 00000  00000 00000
4673:  00000 00000  00000 00000  00000 00000  00000 00000  00000 00000    00000 00000  00000 00000  00000 00000  00000 00000  00000 00000
4674:  00000 00000  00000 00000  00000 00000  00000 00000  00000 00000    00000 00000  00000 00000  00000 00000  00000 00000  00000 00000
4675:  00000 00000  00000 00000  00000 00000  00000 00000  00000 00000    00000 00000  00000 00000  00000 00000  00000 00000  00000 00000
4676:  00000 00000  00000 00000  00000 00000  00000 00000  00000 00000    00000 00000  00000 00000  00000 00000  00000 00000  00000 00000
4677:  00000 00000  00000 00000  00000 00000  00000 00000  00000 00000    00000 00000  00000 00000  00000 00000  00000 00000  00000 00000
4678:  00000 00000  00000 00000  00000 00000  00000 00000  00000 00000    00000 00000  00000 00000  00000 00000  00000 00000  00000 00000
4679:  00000 00000  00000 00000  00000 00000  00000 00000  00000 00000    00000 00000  00000 00000  00000 00000  00000 00000  00000 00000
4680:  00000 00000  00000 00000  00000 00000  00000 00000  00000 00000    00000 00000  00000 00000  00000 00000  00000 00000  00000 00000
4681:  00000 00000  00000 00000  00000 00000  00000 00000  00000 00000    00000 00000  00000 00000  00000 00000  00000 00000  00000 00000
4682:  00000 00000  00000 00000  00000 00000  00000 00000  00000 00000    00000 00000  00000 00000  00000 00000  00000 00000  00000 00000
4683:  00000 00000  00000 00000  00000 00000  00000 00000  00000 00000    00000 00000  00000 00000  00000 00000  00000 00000  00000 00000
4684:  00000 00000  00000 00000  00000 00000  00000 00000  00000 00000    00000 00000  00000 00000  00000 00000  00000 00000  00000 00000
4685:  00000 00000  00000 00000  00000 00000  00000 00000  00000 00000    00000 00000  00000 00000  00000 00000  00000 00000  00000 00000
4686:  00000 00000  00000 00000  00000 00000  00000 00000  00000 00000    00000 00000  00000 00000  00000 00000  00000 00000  00000 00000
4687:  00000 00000  00000 00000  00000 00000  00000 00000  00000 00000    00000 00000  00000 00000  00000 00000  00000 00000  00000 00000
4688:  00000 00000  00000 00000  00000 00000  00000 00000  00000 00000    00000 00000  00000 00000  00000 00000  00000 00000  00000 00000
4689:  00000 00000  00000 00000  00000 00000  00000 00000  00000 00000    00000 00000  00000 00000  00000 00000  00000 00000  00000 00000
4690:  00000 00000  00000 00000  00000 00000  00000 00000  00000 00000    00000 00000  00000 00000  00000 00000  00000 00000  00000 00000
4691:  00000 00000  00000 00000  00000 00000  00000 00000  00000 00000    00000 00000  00000 00000  00000 00000  00000 00000  00000 00000
4692:  00000 00000  00000 00000  00000 00000  00000 00000  00000 00000    00000 00000  00000 00000  00000 00000  00000 00000  00000 00000
4693:  00000 00000  00000 00000  00000 00000  00000 00000  00000 00000    00000 00000  00000 00000  00000 00000  00000 00000  00000 00000
4694:  00000 00000  00000 00000  00000 00000  00000 00000  00000 00000    00000 00000  00000 00000  00000 00000  00000 00000  00000 00000
4695:  00000 00000  00000 00000  00000 00000  00000 00000  00000 00000    00000 00000  00000 00000  00000 00000  00000 00000  00000 00000
4696:  00000 00000  00000 00000  00000 00000  00000 00000  00000 00000    00000 00000  00000 00000  00000 00000  00000 00000  00000 00000
4697:  00000 00000  00000 00000  00000 00000  00000 00000  00000 00000    00000 00000  00000 00000  00000 00000  00000 00000  00000 00000
4698:  00000 00000  00000 00000  00000 00000  00000 00000  00000 00000    00000 00000  00000 00000  00000 00000  00000 00000  00000 00000
4699:  00000 00000  00000 00000  00000 00000  00000 00000  00000 00000    00000 00000  00000 00000  00000 00000  00000 00000  00000 00000
```

```
4700:  00000 00000  00000 00000  00000 00000  00000 00000  00000 00000   00000 00000  00000 00000  00000 00000  00000 00000  00000 00000
4701:  00000 00000  00000 00000  00000 00000  00000 00000  00000 00000   00000 00000  00000 00000  00000 00000  00000 00000  00000 00000
4702:  00000 00000  00000 00000  00000 00000  00000 00000  00000 00000   00000 00000  00000 00000  00000 00000  00000 00000  00000 00000
4703:  00000 00000  00000 00000  00000 00000  00000 00000  00000 00000   00000 00000  00000 00000  00000 00000  00000 00000  00000 00000
4704:  00000 00000  00000 00000  00000 00000  00000 00000  00000 00000   00000 00000  00000 00000  00000 00000  00000 00000  00000 00000
4705:  00000 00000  00000 00000  00000 00000  00000 00000  00000 00000   00000 00000  00000 00000  00000 00000  00000 00000  00000 00000
4706:  00000 00000  00000 00000  00000 00000  00000 00000  00000 00000   00000 00000  00000 00000  00000 00000  00000 00000  00000 00000
4707:  00000 00000  00000 00000  00000 00000  00000 00000  00000 00000   00000 00000  00000 00000  00000 00000  00000 00000  00000 00000
4708:  00000 00000  00000 00000  00000 00000  00000 00000  00000 00000   00000 00000  00000 00000  00000 00000  00000 00000  00000 00000
4709:  00000 00000  00000 00000  00000 00000  00000 00000  00000 00000   00000 00000  00000 00000  00000 00000  00000 00000  00000 00000
4710:  00000 00000  00000 00000  00000 00000  00000 00000  00000 00000   00000 00000  00000 00000  00000 00000  00000 00000  00000 00000
4711:  00000 00000  00000 00000  00000 00000  00000 00000  00000 00000   00000 00000  00000 00000  00000 00000  00000 00000  00000 00000
4712:  00000 00000  00000 00000  00000 00000  00000 00000  00000 00000   00000 00000  00000 00000  00000 00000  00000 00000  00000 00000
4713:  00000 00000  00000 00000  00000 00000  00000 00000  00000 00000   00000 00000  00000 00000  00000 00000  00000 00000  00000 00000
4714:  00000 00000  00000 00000  00000 00000  00000 00000  00000 00000   00000 00000  00000 00000  00000 00000  00000 00000  00000 00000
4715:  00000 00000  00000 00000  00000 00000  00000 00000  00000 00000   00000 00000  00000 00000  00000 00000  00000 00000  00000 00000
4716:  00000 00000  00000 00000  00000 00000  00000 00000  00000 00000   00000 00000  00000 00000  00000 00000  00000 00000  00000 00000
4717:  00000 00000  00000 00000  00000 00000  00000 00000  00000 00000   00000 00000  00000 00000  00000 00000  00000 00000  00000 00000
4718:  00000 00000  00000 00000  00000 00000  00000 00000  00000 00000   00000 00000  00000 00000  00000 00000  00000 00000  00000 00000
4719:  00000 00000  00000 00000  00000 00000  00000 00000  00000 00000   00000 00000  00000 00000  00000 00000  00000 00000  00000 00000
4720:  00000 00000  00000 00000  00000 00000  00000 00000  00000 00000   00000 00000  00000 00000  00000 00000  00000 00000  00000 00000
4721:  00000 00000  00000 00000  00000 00000  00000 00000  00000 00000   00000 00000  00000 00000  00000 00000  00000 00000  00000 00000
4722:  00000 00000  00000 00000  00000 00000  00000 00000  00000 00000   00000 00000  00000 00000  00000 00000  00000 00000  00000 00000
4723:  00000 00000  00000 00000  00000 00000  00000 00000  00000 00000   00000 00000  00000 00000  00000 00000  00000 00000  00000 00000
4724:  00000 00000  00000 00000  00000 00000  00000 00000  00000 00000   00000 00000  00000 00000  00000 00000  00000 00000  00000 00000
4725:  00000 00000  00000 00000  00000 00000  00000 00000  00000 00000   00000 00000  00000 00000  00000 00000  00000 00000  00000 00000
4726:  00000 00000  00000 00000  00000 00000  00000 00000  00000 00000   00000 00000  00000 00000  00000 00000  00000 00000  00000 00000
4727:  00000 00000  00000 00000  00000 00000  00000 00000  00000 00000   00000 00000  00000 00000  00000 00000  00000 00000  00000 00000
4728:  00000 00000  00000 00000  00000 00000  00000 00000  00000 00000   00000 00000  00000 00000  00000 00000  00000 00000  00000 00000
4729:  00000 00000  00000 00000  00000 00000  00000 00000  00000 00000   00000 00000  00000 00000  00000 00000  00000 00000  00000 00000
4730:  00000 00000  00000 00000  00000 00000  00000 00000  00000 00000   00000 00000  00000 00000  00000 00000  00000 00000  00000 00000
4731:  00000 00000  00000 00000  00000 00000  00000 00000  00000 00000   00000 00000  00000 00000  00000 00000  00000 00000  00000 00000
4732:  00000 00000  00000 00000  00000 00000  00000 00000  00000 00000   00000 00000  00000 00000  00000 00000  00000 00000  00000 00000
4733:  00000 00000  00000 00000  00000 00000  00000 00000  00000 00000   00000 00000  00000 00000  00000 00000  00000 00000  00000 00000
4734:  00000 00000  00000 00000  00000 00000  00000 00000  00000 00000   00000 00000  00000 00000  00000 00000  00000 00000  00000 00000
4735:  00000 00000  00000 00000  00000 00000  00000 00000  00000 00000   00000 00000  00000 00000  00000 00000  00000 00000  00000 00000
4736:  00000 00000  00000 00000  00000 00000  00000 00000  00000 00000   00000 00000  00000 00000  00000 00000  00000 00000  00000 00000
4737:  00000 00000  00000 00000  00000 00000  00000 00000  00000 00000   00000 00000  00000 00000  00000 00000  00000 00000  00000 00000
4738:  00000 00000  00000 00000  00000 00000  00000 00000  00000 00000   00000 00000  00000 00000  00000 00000  00000 00000  00000 00000
4739:  00000 00000  00000 00000  00000 00000  00000 00000  00000 00000   00000 00000  00000 00000  00000 00000  00000 00000  00000 00000
4740:  00000 00000  00000 00000  00000 00000  00000 00000  00000 00000   00000 00000  00000 00000  00000 00000  00000 00000  00000 00000
4741:  00000 00000  00000 00000  00000 00000  00000 00000  00000 00000   00000 00000  00000 00000  00000 00000  00000 00000  00000 00000
4742:  00000 00000  00000 00000  00000 00000  00000 00000  00000 00000   00000 00000  00000 00000  00000 00000  00000 00000  00000 00000
4743:  00000 00000  00000 00000  00000 00000  00000 00000  00000 00000   00000 00000  00000 00000  00000 00000  00000 00000  00000 00000
4744:  00000 00000  00000 00000  00000 00000  00000 00000  00000 00000   00000 00000  00000 00000  00000 00000  00000 00000  00000 00000
4745:  00000 00000  00000 00000  00000 00000  00000 00000  00000 00000   00000 00000  00000 00000  00000 00000  00000 00000  00000 00000
4746:  00000 00000  00000 00000  00000 00000  00000 00000  00000 00000   00000 00000  00000 00000  00000 00000  00000 00000  00000 00000
4747:  00000 00000  00000 00000  00000 00000  00000 00000  00000 00000   00000 00000  00000 00000  00000 00000  00000 00000  00000 00000
4748:  00000 00000  00000 00000  00000 00000  00000 00000  00000 00000   00000 00000  00000 00000  00000 00000  00000 00000  00000 00000
4749:  00000 00000  00000 00000  00000 00000  00000 00000  00000 00000   00000 00000  00000 00000  00000 00000  00000 00000  00000 00000
```

```
4750:  00000 00000   00000 00000   00000 00000   00000 00000   00000 00000    00000 00000   00000 00000   00000 00000   00000 00000   00000 00000
4751:  00000 00000   00000 00000   00000 00000   00000 00000   00000 00000    00000 00000   00000 00000   00000 00000   00000 00000   00000 00000
4752:  00000 00000   00000 00000   00000 00000   00000 00000   00000 00000    00000 00000   00000 00000   00000 00000   00000 00000   00000 00000
4753:  00000 00000   00000 00000   00000 00000   00000 00000   00000 00000    00000 00000   00000 00000   00000 00000   00000 00000   00000 00000
4754:  00000 00000   00000 00000   00000 00000   00000 00000   00000 00000    00000 00000   00000 00000   00000 00000   00000 00000   00000 00000
4755:  00000 00000   00000 00000   00000 00000   00000 00000   00000 00000    00000 00000   00000 00000   00000 00000   00000 00000   00000 00000
4756:  00000 00000   00000 00000   00000 00000   00000 00000   00000 00000    00000 00000   00000 00000   00000 00000   00000 00000   00000 00000
4757:  00000 00000   00000 00000   00000 00000   00000 00000   00000 00000    00000 00000   00000 00000   00000 00000   00000 00000   00000 00000
4758:  00000 00000   00000 00000   00000 00000   00000 00000   00000 00000    00000 00000   00000 00000   00000 00000   00000 00000   00000 00000
4759:  00000 00000   00000 00000   00000 00000   00000 00000   00000 00000    00000 00000   00000 00000   00000 00000   00000 00000   00000 00000
4760:  00000 00000   00000 00000   00000 00000   00000 00000   00000 00000    00000 00000   00000 00000   00000 00000   00000 00000   00000 00000
4761:  00000 00000   00000 00000   00000 00000   00000 00000   00000 00000    00000 00000   00000 00000   00000 00000   00000 00000   00000 00000
4762:  00000 00000   00000 00000   00000 00000   00000 00000   00000 00000    00000 00000   00000 00000   00000 00000   00000 00000   00000 00000
4763:  00000 00000   00000 00000   00000 00000   00000 00000   00000 00000    00000 00000   00000 00000   00000 00000   00000 00000   00000 00000
4764:  00000 00000   00000 00000   00000 00000   00000 00000   00000 00000    00000 00000   00000 00000   00000 00000   00000 00000   00000 00000
4765:  00000 00000   00000 00000   00000 00000   00000 00000   00000 00000    00000 00000   00000 00000   00000 00000   00000 00000   00000 00000
4766:  00000 00000   00000 00000   00000 00000   00000 00000   00000 00000    00000 00000   00000 00000   00000 00000   00000 00000   00000 00000
4767:  00000 00000   00000 00000   00000 00000   00000 00000   00000 00000    00000 00000   00000 00000   00000 00000   00000 00000   00000 00000
4768:  00000 00000   00000 00000   00000 00000   00000 00000   00000 00000    00000 00000   00000 00000   00000 00000   00000 00000   00000 00000
4769:  00000 00000   00000 00000   00000 00000   00000 00000   00000 00000    00000 00000   00000 00000   00000 00000   00000 00000   00000 00000
4770:  00000 00000   00000 00000   00000 00000   00000 00000   00000 00000    00000 00000   00000 00000   00000 00000   00000 00000   00000 00000
4771:  00000 00000   00000 00000   00000 00000   00000 00000   00000 00000    00000 00000   00000 00000   00000 00000   00000 00000   00000 00000
4772:  00000 00000   00000 00000   00000 00000   00000 00000   00000 00000    00000 00000   00000 00000   00000 00000   00000 00000   00000 00000
4773:  00000 00000   00000 00000   00000 00000   00000 00000   00000 00000    00000 00000   00000 00000   00000 00000   00000 00000   00000 00000
4774:  00000 00000   00000 00000   00000 00000   00000 00000   00000 00000    00000 00000   00000 00000   00000 00000   00000 00000   00000 00000
4775:  00000 00000   00000 00000   00000 00000   00000 00000   00000 00000    00000 00000   00000 00000   00000 00000   00000 00000   00000 00000
4776:  00000 00000   00000 00000   00000 00000   00000 00000   00000 00000    00000 00000   00000 00000   00000 00000   00000 00000   00000 00000
4777:  00000 00000   00000 00000   00000 00000   00000 00000   00000 00000    00000 00000   00000 00000   00000 00000   00000 00000   00000 00000
4778:  00000 00000   00000 00000   00000 00000   00000 00000   00000 00000    00000 00000   00000 00000   00000 00000   00000 00000   00000 00000
4779:  00000 00000   00000 00000   00000 00000   00000 00000   00000 00000    00000 00000   00000 00000   00000 00000   00000 00000   00000 00000
4780:  00000 00000   00000 00000   00000 00000   00000 00000   00000 00000    00000 00000   00000 00000   00000 00000   00000 00000   00000 00000
4781:  00000 00000   00000 00000   00000 00000   00000 00000   00000 00000    00000 00000   00000 00000   00000 00000   00000 00000   00000 00000
4782:  00000 00000   00000 00000   00000 00000   00000 00000   00000 00000    00000 00000   00000 00000   00000 00000   00000 00000   00000 00000
4783:  00000 00000   00000 00000   00000 00000   00000 00000   00000 00000    00000 00000   00000 00000   00000 00000   00000 00000   00000 00000
4784:  00000 00000   00000 00000   00000 00000   00000 00000   00000 00000    00000 00000   00000 00000   00000 00000   00000 00000   00000 00000
4785:  00000 00000   00000 00000   00000 00000   00000 00000   00000 00000    00000 00000   00000 00000   00000 00000   00000 00000   00000 00000
4786:  00000 00000   00000 00000   00000 00000   00000 00000   00000 00000    00000 00000   00000 00000   00000 00000   00000 00000   00000 00000
4787:  00000 00000   00000 00000   00000 00000   00000 00000   00000 00000    00000 00000   00000 00000   00000 00000   00000 00000   00000 00000
4788:  00000 00000   00000 00000   00000 00000   00000 00000   00000 00000    00000 00000   00000 00000   00000 00000   00000 00000   00000 00000
4789:  00000 00000   00000 00000   00000 00000   00000 00000   00000 00000    00000 00000   00000 00000   00000 00000   00000 00000   00000 00000
4790:  00000 00000   00000 00000   00000 00000   00000 00000   00000 00000    00000 00000   00000 00000   00000 00000   00000 00000   00000 00000
4791:  00000 00000   00000 00000   00000 00000   00000 00000   00000 00000    00000 00000   00000 00000   00000 00000   00000 00000   00000 00000
4792:  00000 00000   00000 00000   00000 00000   00000 00000   00000 00000    00000 00000   00000 00000   00000 00000   00000 00000   00000 00000
4793:  00000 00000   00000 00000   00000 00000   00000 00000   00000 00000    00000 00000   00000 00000   00000 00000   00000 00000   00000 00000
4794:  00000 00000   00000 00000   00000 00000   00000 00000   00000 00000    00000 00000   00000 00000   00000 00000   00000 00000   00000 00000
4795:  00000 00000   00000 00000   00000 00000   00000 00000   00000 00000    00000 00000   00000 00000   00000 00000   00000 00000   00000 00000
4796:  00000 00000   00000 00000   00000 00000   00000 00000   00000 00000    00000 00000   00000 00000   00000 00000   00000 00000   00000 00000
4797:  00000 00000   00000 00000   00000 00000   00000 00000   00000 00000    00000 00000   00000 00000   00000 00000   00000 00000   00000 00000
4798:  00000 00000   00000 00000   00000 00000   00000 00000   00000 00000    00000 00000   00000 00000   00000 00000   00000 00000   00000 00000
4799:  00000 00000   00000 00000   00000 00000   00000 00000   00000 00000    00000 00000   00000 00000   00000 00000   00000 00000   00000 00000
```

```
4800:  00000 00000  00000 00000  00000 00000  00000 00000  00000 00000    00000 00000  00000 00000  00000 00000  00000 00000  00000 00000
4801:  00000 00000  00000 00000  00000 00000  00000 00000  00000 00000    00000 00000  00000 00000  00000 00000  00000 00000  00000 00000
4802:  00000 00000  00000 00000  00000 00000  00000 00000  00000 00000    00000 00000  00000 00000  00000 00000  00000 00000  00000 00000
4803:  00000 00000  00000 00000  00000 00000  00000 00000  00000 00000    00000 00000  00000 00000  00000 00000  00000 00000  00000 00000
4804:  00000 00000  00000 00000  00000 00000  00000 00000  00000 00000    00000 00000  00000 00000  00000 00000  00000 00000  00000 00000
4805:  00000 00000  00000 00000  00000 00000  00000 00000  00000 00000    00000 00000  00000 00000  00000 00000  00000 00000  00000 00000
4806:  00000 00000  00000 00000  00000 00000  00000 00000  00000 00000    00000 00000  00000 00000  00000 00000  00000 00000  00000 00000
4807:  00000 00000  00000 00000  00000 00000  00000 00000  00000 00000    00000 00000  00000 00000  00000 00000  00000 00000  00000 00000
4808:  00000 00000  00000 00000  00000 00000  00000 00000  00000 00000    00000 00000  00000 00000  00000 00000  00000 00000  00000 00000
4809:  00000 00000  00000 00000  00000 00000  00000 00000  00000 00000    00000 00000  00000 00000  00000 00000  00000 00000  00000 00000
4810:  00000 00000  00000 00000  00000 00000  00000 00000  00000 00000    00000 00000  00000 00000  00000 00000  00000 00000  00000 00000
4811:  00000 00000  00000 00000  00000 00000  00000 00000  00000 00000    00000 00000  00000 00000  00000 00000  00000 00000  00000 00000
4812:  00000 00000  00000 00000  00000 00000  00000 00000  00000 00000    00000 00000  00000 00000  00000 00000  00000 00000  00000 00000
4813:  00000 00000  00000 00000  00000 00000  00000 00000  00000 00000    00000 00000  00000 00000  00000 00000  00000 00000  00000 00000
4814:  00000 00000  00000 00000  00000 00000  00000 00000  00000 00000    00000 00000  00000 00000  00000 00000  00000 00000  00000 00000
4815:  00000 00000  00000 00000  00000 00000  00000 00000  00000 00000    00000 00000  00000 00000  00000 00000  00000 00000  00000 00000
4816:  00000 00000  00000 00000  00000 00000  00000 00000  00000 00000    00000 00000  00000 00000  00000 00000  00000 00000  00000 00000
4817:  00000 00000  00000 00000  00000 00000  00000 00000  00000 00000    00000 00000  00000 00000  00000 00000  00000 00000  00000 00000
4818:  00000 00000  00000 00000  00000 00000  00000 00000  00000 00000    00000 00000  00000 00000  00000 00000  00000 00000  00000 00000
4819:  00000 00000  00000 00000  00000 00000  00000 00000  00000 00000    00000 00000  00000 00000  00000 00000  00000 00000  00000 00000
4820:  00000 00000  00000 00000  00000 00000  00000 00000  00000 00000    00000 00000  00000 00000  00000 00000  00000 00000  00000 00000
4821:  00000 00000  00000 00000  00000 00000  00000 00000  00000 00000    00000 00000  00000 00000  00000 00000  00000 00000  00000 00000
4822:  00000 00000  00000 00000  00000 00000  00000 00000  00000 00000    00000 00000  00000 00000  00000 00000  00000 00000  00000 00000
4823:  00000 00000  00000 00000  00000 00000  00000 00000  00000 00000    00000 00000  00000 00000  00000 00000  00000 00000  00000 00000
4824:  00000 00000  00000 00000  00000 00000  00000 00000  00000 00000    00000 00000  00000 00000  00000 00000  00000 00000  00000 00000
4825:  00000 00000  00000 00000  00000 00000  00000 00000  00000 00000    00000 00000  00000 00000  00000 00000  00000 00000  00000 00000
4826:  00000 00000  00000 00000  00000 00000  00000 00000  00000 00000    00000 00000  00000 00000  00000 00000  00000 00000  00000 00000
4827:  00000 00000  00000 00000  00000 00000  00000 00000  00000 00000    00000 00000  00000 00000  00000 00000  00000 00000  00000 00000
4828:  00000 00000  00000 00000  00000 00000  00000 00000  00000 00000    00000 00000  00000 00000  00000 00000  00000 00000  00000 00000
4829:  00000 00000  00000 00000  00000 00000  00000 00000  00000 00000    00000 00000  00000 00000  00000 00000  00000 00000  00000 00000
4830:  00000 00000  00000 00000  00000 00000  00000 00000  00000 00000    00000 00000  00000 00000  00000 00000  00000 00000  00000 00000
4831:  00000 00000  00000 00000  00000 00000  00000 00000  00000 00000    00000 00000  00000 00000  00000 00000  00000 00000  00000 00000
4832:  00000 00000  00000 00000  00000 00000  00000 00000  00000 00000    00000 00000  00000 00000  00000 00000  00000 00000  00000 00000
4833:  00000 00000  00000 00000  00000 00000  00000 00000  00000 00000    00000 00000  00000 00000  00000 00000  00000 00000  00000 00000
4834:  00000 00000  00000 00000  00000 00000  00000 00000  00000 00000    00000 00000  00000 00000  00000 00000  00000 00000  00000 00000
4835:  00000 00000  00000 00000  00000 00000  00000 00000  00000 00000    00000 00000  00000 00000  00000 00000  00000 00000  00000 00000
4836:  00000 00000  00000 00000  00000 00000  00000 00000  00000 00000    00000 00000  00000 00000  00000 00000  00000 00000  00000 00000
4837:  00000 00000  00000 00000  00000 00000  00000 00000  00000 00000    00000 00000  00000 00000  00000 00000  00000 00000  00000 00000
4838:  00000 00000  00000 00000  00000 00000  00000 00000  00000 00000    00000 00000  00000 00000  00000 00000  00000 00000  00000 00000
4839:  00000 00000  00000 00000  00000 00000  00000 00000  00000 00000    00000 00000  00000 00000  00000 00000  00000 00000  00000 00000
4840:  00000 00000  00000 00000  00000 00000  00000 00000  00000 00000    00000 00000  00000 00000  00000 00000  00000 00000  00000 00000
4841:  00000 00000  00000 00000  00000 00000  00000 00000  00000 00000    00000 00000  00000 00000  00000 00000  00000 00000  00000 00000
4842:  00000 00000  00000 00000  00000 00000  00000 00000  00000 00000    00000 00000  00000 00000  00000 00000  00000 00000  00000 00000
4843:  00000 00000  00000 00000  00000 00000  00000 00000  00000 00000    00000 00000  00000 00000  00000 00000  00000 00000  00000 00000
4844:  00000 00000  00000 00000  00000 00000  00000 00000  00000 00000    00000 00000  00000 00000  00000 00000  00000 00000  00000 00000
4845:  00000 00000  00000 00000  00000 00000  00000 00000  00000 00000    00000 00000  00000 00000  00000 00000  00000 00000  00000 00000
4846:  00000 00000  00000 00000  00000 00000  00000 00000  00000 00000    00000 00000  00000 00000  00000 00000  00000 00000  00000 00000
4847:  00000 00000  00000 00000  00000 00000  00000 00000  00000 00000    00000 00000  00000 00000  00000 00000  00000 00000  00000 00000
4848:  00000 00000  00000 00000  00000 00000  00000 00000  00000 00000    00000 00000  00000 00000  00000 00000  00000 00000  00000 00000
4849:  00000 00000  00000 00000  00000 00000  00000 00000  00000 00000    00000 00000  00000 00000  00000 00000  00000 00000  00000 00000
```

```
4850:  00000 00000  00000 00000  00000 00000  00000 00000  00000 00000   00000 00000  00000 00000  00000 00000  00000 00000  00000 00000
4851:  00000 00000  00000 00000  00000 00000  00000 00000  00000 00000   00000 00000  00000 00000  00000 00000  00000 00000  00000 00000
4852:  00000 00000  00000 00000  00000 00000  00000 00000  00000 00000   00000 00000  00000 00000  00000 00000  00000 00000  00000 00000
4853:  00000 00000  00000 00000  00000 00000  00000 00000  00000 00000   00000 00000  00000 00000  00000 00000  00000 00000  00000 00000
4854:  00000 00000  00000 00000  00000 00000  00000 00000  00000 00000   00000 00000  00000 00000  00000 00000  00000 00000  00000 00000
4855:  00000 00000  00000 00000  00000 00000  00000 00000  00000 00000   00000 00000  00000 00000  00000 00000  00000 00000  00000 00000
4856:  00000 00000  00000 00000  00000 00000  00000 00000  00000 00000   00000 00000  00000 00000  00000 00000  00000 00000  00000 00000
4857:  00000 00000  00000 00000  00000 00000  00000 00000  00000 00000   00000 00000  00000 00000  00000 00000  00000 00000  00000 00000
4858:  00000 00000  00000 00000  00000 00000  00000 00000  00000 00000   00000 00000  00000 00000  00000 00000  00000 00000  00000 00000
4859:  00000 00000  00000 00000  00000 00000  00000 00000  00000 00000   00000 00000  00000 00000  00000 00000  00000 00000  00000 00000
4860:  00000 00000  00000 00000  00000 00000  00000 00000  00000 00000   00000 00000  00000 00000  00000 00000  00000 00000  00000 00000
4861:  00000 00000  00000 00000  00000 00000  00000 00000  00000 00000   00000 00000  00000 00000  00000 00000  00000 00000  00000 00000
4862:  00000 00000  00000 00000  00000 00000  00000 00000  00000 00000   00000 00000  00000 00000  00000 00000  00000 00000  00000 00000
4863:  00000 00000  00000 00000  00000 00000  00000 00000  00000 00000   00000 00000  00000 00000  00000 00000  00000 00000  00000 00000
4864:  00000 00000  00000 00000  00000 00000  00000 00000  00000 00000   00000 00000  00000 00000  00000 00000  00000 00000  00000 00000
4865:  00000 00000  00000 00000  00000 00000  00000 00000  00000 00000   00000 00000  00000 00000  00000 00000  00000 00000  00000 00000
4866:  00000 00000  00000 00000  00000 00000  00000 00000  00000 00000   00000 00000  00000 00000  00000 00000  00000 00000  00000 00000
4867:  00000 00000  00000 00000  00000 00000  00000 00000  00000 00000   00000 00000  00000 00000  00000 00000  00000 00000  00000 00000
4868:  00000 00000  00000 00000  00000 00000  00000 00000  00000 00000   00000 00000  00000 00000  00000 00000  00000 00000  00000 00000
4869:  00000 00000  00000 00000  00000 00000  00000 00000  00000 00000   00000 00000  00000 00000  00000 00000  00000 00000  00000 00000
4870:  00000 00000  00000 00000  00000 00000  00000 00000  00000 00000   00000 00000  00000 00000  00000 00000  00000 00000  00000 00000
4871:  00000 00000  00000 00000  00000 00000  00000 00000  00000 00000   00000 00000  00000 00000  00000 00000  00000 00000  00000 00000
4872:  00000 00000  00000 00000  00000 00000  00000 00000  00000 00000   00000 00000  00000 00000  00000 00000  00000 00000  00000 00000
4873:  00000 00000  00000 00000  00000 00000  00000 00000  00000 00000   00000 00000  00000 00000  00000 00000  00000 00000  00000 00000
4874:  00000 00000  00000 00000  00000 00000  00000 00000  00000 00000   00000 00000  00000 00000  00000 00000  00000 00000  00000 00000
4875:  00000 00000  00000 00000  00000 00000  00000 00000  00000 00000   00000 00000  00000 00000  00000 00000  00000 00000  00000 00000
4876:  00000 00000  00000 00000  00000 00000  00000 00000  00000 00000   00000 00000  00000 00000  00000 00000  00000 00000  00000 00000
4877:  00000 00000  00000 00000  00000 00000  00000 00000  00000 00000   00000 00000  00000 00000  00000 00000  00000 00000  00000 00000
4878:  00000 00000  00000 00000  00000 00000  00000 00000  00000 00000   00000 00000  00000 00000  00000 00000  00000 00000  00000 00000
4879:  00000 00000  00000 00000  00000 00000  00000 00000  00000 00000   00000 00000  00000 00000  00000 00000  00000 00000  00000 00000
4880:  00000 00000  00000 00000  00000 00000  00000 00000  00000 00000   00000 00000  00000 00000  00000 00000  00000 00000  00000 00000
4881:  00000 00000  00000 00000  00000 00000  00000 00000  00000 00000   00000 00000  00000 00000  00000 00000  00000 00000  00000 00000
4882:  00000 00000  00000 00000  00000 00000  00000 00000  00000 00000   00000 00000  00000 00000  00000 00000  00000 00000  00000 00000
4883:  00000 00000  00000 00000  00000 00000  00000 00000  00000 00000   00000 00000  00000 00000  00000 00000  00000 00000  00000 00000
4884:  00000 00000  00000 00000  00000 00000  00000 00000  00000 00000   00000 00000  00000 00000  00000 00000  00000 00000  00000 00000
4885:  00000 00000  00000 00000  00000 00000  00000 00000  00000 00000   00000 00000  00000 00000  00000 00000  00000 00000  00000 00000
4886:  00000 00000  00000 00000  00000 00000  00000 00000  00000 00000   00000 00000  00000 00000  00000 00000  00000 00000  00000 00000
4887:  00000 00000  00000 00000  00000 00000  00000 00000  00000 00000   00000 00000  00000 00000  00000 00000  00000 00000  00000 00000
4888:  00000 00000  00000 00000  00000 00000  00000 00000  00000 00000   00000 00000  00000 00000  00000 00000  00000 00000  00000 00000
4889:  00000 00000  00000 00000  00000 00000  00000 00000  00000 00000   00000 00000  00000 00000  00000 00000  00000 00000  00000 00000
4890:  00000 00000  00000 00000  00000 00000  00000 00000  00000 00000   00000 00000  00000 00000  00000 00000  00000 00000  00000 00000
4891:  00000 00000  00000 00000  00000 00000  00000 00000  00000 00000   00000 00000  00000 00000  00000 00000  00000 00000  00000 00000
4892:  00000 00000  00000 00000  00000 00000  00000 00000  00000 00000   00000 00000  00000 00000  00000 00000  00000 00000  00000 00000
4893:  00000 00000  00000 00000  00000 00000  00000 00000  00000 00000   00000 00000  00000 00000  00000 00000  00000 00000  00000 00000
4894:  00000 00000  00000 00000  00000 00000  00000 00000  00000 00000   00000 00000  00000 00000  00000 00000  00000 00000  00000 00000
4895:  00000 00000  00000 00000  00000 00000  00000 00000  00000 00000   00000 00000  00000 00000  00000 00000  00000 00000  00000 00000
4896:  00000 00000  00000 00000  00000 00000  00000 00000  00000 00000   00000 00000  00000 00000  00000 00000  00000 00000  00000 00000
4897:  00000 00000  00000 00000  00000 00000  00000 00000  00000 00000   00000 00000  00000 00000  00000 00000  00000 00000  00000 00000
4898:  00000 00000  00000 00000  00000 00000  00000 00000  00000 00000   00000 00000  00000 00000  00000 00000  00000 00000  00000 00000
4899:  00000 00000  00000 00000  00000 00000  00000 00000  00000 00000   00000 00000  00000 00000  00000 00000  00000 00000  00000 00000
```

```
4900:  00000 00000  00000 00000  00000 00000  00000 00000  00000 00000    00000 00000  00000 00000  00000 00000  00000 00000  00000 00000
4901:  00000 00000  00000 00000  00000 00000  00000 00000  00000 00000    00000 00000  00000 00000  00000 00000  00000 00000  00000 00000
4902:  00000 00000  00000 00000  00000 00000  00000 00000  00000 00000    00000 00000  00000 00000  00000 00000  00000 00000  00000 00000
4903:  00000 00000  00000 00000  00000 00000  00000 00000  00000 00000    00000 00000  00000 00000  00000 00000  00000 00000  00000 00000
4904:  00000 00000  00000 00000  00000 00000  00000 00000  00000 00000    00000 00000  00000 00000  00000 00000  00000 00000  00000 00000
4905:  00000 00000  00000 00000  00000 00000  00000 00000  00000 00000    00000 00000  00000 00000  00000 00000  00000 00000  00000 00000
4906:  00000 00000  00000 00000  00000 00000  00000 00000  00000 00000    00000 00000  00000 00000  00000 00000  00000 00000  00000 00000
4907:  00000 00000  00000 00000  00000 00000  00000 00000  00000 00000    00000 00000  00000 00000  00000 00000  00000 00000  00000 00000
4908:  00000 00000  00000 00000  00000 00000  00000 00000  00000 00000    00000 00000  00000 00000  00000 00000  00000 00000  00000 00000
4909:  00000 00000  00000 00000  00000 00000  00000 00000  00000 00000    00000 00000  00000 00000  00000 00000  00000 00000  00000 00000
4910:  00000 00000  00000 00000  00000 00000  00000 00000  00000 00000    00000 00000  00000 00000  00000 00000  00000 00000  00000 00000
4911:  00000 00000  00000 00000  00000 00000  00000 00000  00000 00000    00000 00000  00000 00000  00000 00000  00000 00000  00000 00000
4912:  00000 00000  00000 00000  00000 00000  00000 00000  00000 00000    00000 00000  00000 00000  00000 00000  00000 00000  00000 00000
4913:  00000 00000  00000 00000  00000 00000  00000 00000  00000 00000    00000 00000  00000 00000  00000 00000  00000 00000  00000 00000
4914:  00000 00000  00000 00000  00000 00000  00000 00000  00000 00000    00000 00000  00000 00000  00000 00000  00000 00000  00000 00000
4915:  00000 00000  00000 00000  00000 00000  00000 00000  00000 00000    00000 00000  00000 00000  00000 00000  00000 00000  00000 00000
4916:  00000 00000  00000 00000  00000 00000  00000 00000  00000 00000    00000 00000  00000 00000  00000 00000  00000 00000  00000 00000
4917:  00000 00000  00000 00000  00000 00000  00000 00000  00000 00000    00000 00000  00000 00000  00000 00000  00000 00000  00000 00000
4918:  00000 00000  00000 00000  00000 00000  00000 00000  00000 00000    00000 00000  00000 00000  00000 00000  00000 00000  00000 00000
4919:  00000 00000  00000 00000  00000 00000  00000 00000  00000 00000    00000 00000  00000 00000  00000 00000  00000 00000  00000 00000
4920:  00000 00000  00000 00000  00000 00000  00000 00000  00000 00000    00000 00000  00000 00000  00000 00000  00000 00000  00000 00000
4921:  00000 00000  00000 00000  00000 00000  00000 00000  00000 00000    00000 00000  00000 00000  00000 00000  00000 00000  00000 00000
4922:  00000 00000  00000 00000  00000 00000  00000 00000  00000 00000    00000 00000  00000 00000  00000 00000  00000 00000  00000 00000
4923:  00000 00000  00000 00000  00000 00000  00000 00000  00000 00000    00000 00000  00000 00000  00000 00000  00000 00000  00000 00000
4924:  00000 00000  00000 00000  00000 00000  00000 00000  00000 00000    00000 00000  00000 00000  00000 00000  00000 00000  00000 00000
4925:  00000 00000  00000 00000  00000 00000  00000 00000  00000 00000    00000 00000  00000 00000  00000 00000  00000 00000  00000 00000
4926:  00000 00000  00000 00000  00000 00000  00000 00000  00000 00000    00000 00000  00000 00000  00000 00000  00000 00000  00000 00000
4927:  00000 00000  00000 00000  00000 00000  00000 00000  00000 00000    00000 00000  00000 00000  00000 00000  00000 00000  00000 00000
4928:  00000 00000  00000 00000  00000 00000  00000 00000  00000 00000    00000 00000  00000 00000  00000 00000  00000 00000  00000 00000
4929:  00000 00000  00000 00000  00000 00000  00000 00000  00000 00000    00000 00000  00000 00000  00000 00000  00000 00000  00000 00000
4930:  00000 00000  00000 00000  00000 00000  00000 00000  00000 00000    00000 00000  00000 00000  00000 00000  00000 00000  00000 00000
4931:  00000 00000  00000 00000  00000 00000  00000 00000  00000 00000    00000 00000  00000 00000  00000 00000  00000 00000  00000 00000
4932:  00000 00000  00000 00000  00000 00000  00000 00000  00000 00000    00000 00000  00000 00000  00000 00000  00000 00000  00000 00000
4933:  00000 00000  00000 00000  00000 00000  00000 00000  00000 00000    00000 00000  00000 00000  00000 00000  00000 00000  00000 00000
4934:  00000 00000  00000 00000  00000 00000  00000 00000  00000 00000    00000 00000  00000 00000  00000 00000  00000 00000  00000 00000
4935:  00000 00000  00000 00000  00000 00000  00000 00000  00000 00000    00000 00000  00000 00000  00000 00000  00000 00000  00000 00000
4936:  00000 00000  00000 00000  00000 00000  00000 00000  00000 00000    00000 00000  00000 00000  00000 00000  00000 00000  00000 00000
4937:  00000 00000  00000 00000  00000 00000  00000 00000  00000 00000    00000 00000  00000 00000  00000 00000  00000 00000  00000 00000
4938:  00000 00000  00000 00000  00000 00000  00000 00000  00000 00000    00000 00000  00000 00000  00000 00000  00000 00000  00000 00000
4939:  00000 00000  00000 00000  00000 00000  00000 00000  00000 00000    00000 00000  00000 00000  00000 00000  00000 00000  00000 00000
4940:  00000 00000  00000 00000  00000 00000  00000 00000  00000 00000    00000 00000  00000 00000  00000 00000  00000 00000  00000 00000
4941:  00000 00000  00000 00000  00000 00000  00000 00000  00000 00000    00000 00000  00000 00000  00000 00000  00000 00000  00000 00000
4942:  00000 00000  00000 00000  00000 00000  00000 00000  00000 00000    00000 00000  00000 00000  00000 00000  00000 00000  00000 00000
4943:  00000 00000  00000 00000  00000 00000  00000 00000  00000 00000    00000 00000  00000 00000  00000 00000  00000 00000  00000 00000
4944:  00000 00000  00000 00000  00000 00000  00000 00000  00000 00000    00000 00000  00000 00000  00000 00000  00000 00000  00000 00000
4945:  00000 00000  00000 00000  00000 00000  00000 00000  00000 00000    00000 00000  00000 00000  00000 00000  00000 00000  00000 00000
4946:  00000 00000  00000 00000  00000 00000  00000 00000  00000 00000    00000 00000  00000 00000  00000 00000  00000 00000  00000 00000
4947:  00000 00000  00000 00000  00000 00000  00000 00000  00000 00000    00000 00000  00000 00000  00000 00000  00000 00000  00000 00000
4948:  00000 00000  00000 00000  00000 00000  00000 00000  00000 00000    00000 00000  00000 00000  00000 00000  00000 00000  00000 00000
4949:  00000 00000  00000 00000  00000 00000  00000 00000  00000 00000    00000 00000  00000 00000  00000 00000  00000 00000  00000 00000
```

```
4950:  00000 00000  00000 00000  00000 00000  00000 00000  00000 00000    00000 00000  00000 00000  00000 00000  00000 00000  00000 00000
4951:  00000 00000  00000 00000  00000 00000  00000 00000  00000 00000    00000 00000  00000 00000  00000 00000  00000 00000  00000 00000
4952:  00000 00000  00000 00000  00000 00000  00000 00000  00000 00000    00000 00000  00000 00000  00000 00000  00000 00000  00000 00000
4953:  00000 00000  00000 00000  00000 00000  00000 00000  00000 00000    00000 00000  00000 00000  00000 00000  00000 00000  00000 00000
4954:  00000 00000  00000 00000  00000 00000  00000 00000  00000 00000    00000 00000  00000 00000  00000 00000  00000 00000  00000 00000
4955:  00000 00000  00000 00000  00000 00000  00000 00000  00000 00000    00000 00000  00000 00000  00000 00000  00000 00000  00000 00000
4956:  00000 00000  00000 00000  00000 00000  00000 00000  00000 00000    00000 00000  00000 00000  00000 00000  00000 00000  00000 00000
4957:  00000 00000  00000 00000  00000 00000  00000 00000  00000 00000    00000 00000  00000 00000  00000 00000  00000 00000  00000 00000
4958:  00000 00000  00000 00000  00000 00000  00000 00000  00000 00000    00000 00000  00000 00000  00000 00000  00000 00000  00000 00000
4959:  00000 00000  00000 00000  00000 00000  00000 00000  00000 00000    00000 00000  00000 00000  00000 00000  00000 00000  00000 00000
4960:  00000 00000  00000 00000  00000 00000  00000 00000  00000 00000    00000 00000  00000 00000  00000 00000  00000 00000  00000 00000
4961:  00000 00000  00000 00000  00000 00000  00000 00000  00000 00000    00000 00000  00000 00000  00000 00000  00000 00000  00000 00000
4962:  00000 00000  00000 00000  00000 00000  00000 00000  00000 00000    00000 00000  00000 00000  00000 00000  00000 00000  00000 00000
4963:  00000 00000  00000 00000  00000 00000  00000 00000  00000 00000    00000 00000  00000 00000  00000 00000  00000 00000  00000 00000
4964:  00000 00000  00000 00000  00000 00000  00000 00000  00000 00000    00000 00000  00000 00000  00000 00000  00000 00000  00000 00000
4965:  00000 00000  00000 00000  00000 00000  00000 00000  00000 00000    00000 00000  00000 00000  00000 00000  00000 00000  00000 00000
4966:  00000 00000  00000 00000  00000 00000  00000 00000  00000 00000    00000 00000  00000 00000  00000 00000  00000 00000  00000 00000
4967:  00000 00000  00000 00000  00000 00000  00000 00000  00000 00000    00000 00000  00000 00000  00000 00000  00000 00000  00000 00000
4968:  00000 00000  00000 00000  00000 00000  00000 00000  00000 00000    00000 00000  00000 00000  00000 00000  00000 00000  00000 00000
4969:  00000 00000  00000 00000  00000 00000  00000 00000  00000 00000    00000 00000  00000 00000  00000 00000  00000 00000  00000 00000
4970:  00000 00000  00000 00000  00000 00000  00000 00000  00000 00000    00000 00000  00000 00000  00000 00000  00000 00000  00000 00000
4971:  00000 00000  00000 00000  00000 00000  00000 00000  00000 00000    00000 00000  00000 00000  00000 00000  00000 00000  00000 00000
4972:  00000 00000  00000 00000  00000 00000  00000 00000  00000 00000    00000 00000  00000 00000  00000 00000  00000 00000  00000 00000
4973:  00000 00000  00000 00000  00000 00000  00000 00000  00000 00000    00000 00000  00000 00000  00000 00000  00000 00000  00000 00000
4974:  00000 00000  00000 00000  00000 00000  00000 00000  00000 00000    00000 00000  00000 00000  00000 00000  00000 00000  00000 00000
4975:  00000 00000  00000 00000  00000 00000  00000 00000  00000 00000    00000 00000  00000 00000  00000 00000  00000 00000  00000 00000
4976:  00000 00000  00000 00000  00000 00000  00000 00000  00000 00000    00000 00000  00000 00000  00000 00000  00000 00000  00000 00000
4977:  00000 00000  00000 00000  00000 00000  00000 00000  00000 00000    00000 00000  00000 00000  00000 00000  00000 00000  00000 00000
4978:  00000 00000  00000 00000  00000 00000  00000 00000  00000 00000    00000 00000  00000 00000  00000 00000  00000 00000  00000 00000
4979:  00000 00000  00000 00000  00000 00000  00000 00000  00000 00000    00000 00000  00000 00000  00000 00000  00000 00000  00000 00000
4980:  00000 00000  00000 00000  00000 00000  00000 00000  00000 00000    00000 00000  00000 00000  00000 00000  00000 00000  00000 00000
4981:  00000 00000  00000 00000  00000 00000  00000 00000  00000 00000    00000 00000  00000 00000  00000 00000  00000 00000  00000 00000
4982:  00000 00000  00000 00000  00000 00000  00000 00000  00000 00000    00000 00000  00000 00000  00000 00000  00000 00000  00000 00000
4983:  00000 00000  00000 00000  00000 00000  00000 00000  00000 00000    00000 00000  00000 00000  00000 00000  00000 00000  00000 00000
4984:  00000 00000  00000 00000  00000 00000  00000 00000  00000 00000    00000 00000  00000 00000  00000 00000  00000 00000  00000 00000
4985:  00000 00000  00000 00000  00000 00000  00000 00000  00000 00000    00000 00000  00000 00000  00000 00000  00000 00000  00000 00000
4986:  00000 00000  00000 00000  00000 00000  00000 00000  00000 00000    00000 00000  00000 00000  00000 00000  00000 00000  00000 00000
4987:  00000 00000  00000 00000  00000 00000  00000 00000  00000 00000    00000 00000  00000 00000  00000 00000  00000 00000  00000 00000
4988:  00000 00000  00000 00000  00000 00000  00000 00000  00000 00000    00000 00000  00000 00000  00000 00000  00000 00000  00000 00000
4989:  00000 00000  00000 00000  00000 00000  00000 00000  00000 00000    00000 00000  00000 00000  00000 00000  00000 00000  00000 00000
4990:  00000 00000  00000 00000  00000 00000  00000 00000  00000 00000    00000 00000  00000 00000  00000 00000  00000 00000  00000 00000
4991:  00000 00000  00000 00000  00000 00000  00000 00000  00000 00000    00000 00000  00000 00000  00000 00000  00000 00000  00000 00000
4992:  00000 00000  00000 00000  00000 00000  00000 00000  00000 00000    00000 00000  00000 00000  00000 00000  00000 00000  00000 00000
4993:  00000 00000  00000 00000  00000 00000  00000 00000  00000 00000    00000 00000  00000 00000  00000 00000  00000 00000  00000 00000
4994:  00000 00000  00000 00000  00000 00000  00000 00000  00000 00000    00000 00000  00000 00000  00000 00000  00000 00000  00000 00000
4995:  00000 00000  00000 00000  00000 00000  00000 00000  00000 00000    00000 00000  00000 00000  00000 00000  00000 00000  00000 00000
4996:  00000 00000  00000 00000  00000 00000  00000 00000  00000 00000    00000 00000  00000 00000  00000 00000  00000 00000  00000 00000
4997:  00000 00000  00000 00000  00000 00000  00000 00000  00000 00000    00000 00000  00000 00000  00000 00000  00000 00000  00000 00000
4998:  00000 00000  00000 00000  00000 00000  00000 00000  00000 00000    00000 00000  00000 00000  00000 00000  00000 00000  00000 00000
4999:  00000 00000  00000 00000  00000 00000  00000 00000  00000 00000    00000 00000  00000 00000  00000 00000  00000 00000  00000 00000
```

```
5000:  00000 00000  00000 00000  00000 00000  00000 00000  00000 00000    00000 00000  00000 00000  00000 00000  00000 00000  00000 00000
5001:  00000 00000  00000 00000  00000 00000  00000 00000  00000 00000    00000 00000  00000 00000  00000 00000  00000 00000  00000 00000
5002:  00000 00000  00000 00000  00000 00000  00000 00000  00000 00000    00000 00000  00000 00000  00000 00000  00000 00000  00000 00000
5003:  00000 00000  00000 00000  00000 00000  00000 00000  00000 00000    00000 00000  00000 00000  00000 00000  00000 00000  00000 00000
5004:  00000 00000  00000 00000  00000 00000  00000 00000  00000 00000    00000 00000  00000 00000  00000 00000  00000 00000  00000 00000
5005:  00000 00000  00000 00000  00000 00000  00000 00000  00000 00000    00000 00000  00000 00000  00000 00000  00000 00000  00000 00000
5006:  00000 00000  00000 00000  00000 00000  00000 00000  00000 00000    00000 00000  00000 00000  00000 00000  00000 00000  00000 00000
5007:  00000 00000  00000 00000  00000 00000  00000 00000  00000 00000    00000 00000  00000 00000  00000 00000  00000 00000  00000 00000
5008:  00000 00000  00000 00000  00000 00000  00000 00000  00000 00000    00000 00000  00000 00000  00000 00000  00000 00000  00000 00000
5009:  00000 00000  00000 00000  00000 00000  00000 00000  00000 00000    00000 00000  00000 00000  00000 00000  00000 00000  00000 00000
5010:  00000 00000  00000 00000  00000 00000  00000 00000  00000 00000    00000 00000  00000 00000  00000 00000  00000 00000  00000 00000
5011:  00000 00000  00000 00000  00000 00000  00000 00000  00000 00000    00000 00000  00000 00000  00000 00000  00000 00000  00000 00000
5012:  00000 00000  00000 00000  00000 00000  00000 00000  00000 00000    00000 00000  00000 00000  00000 00000  00000 00000  00000 00000
5013:  00000 00000  00000 00000  00000 00000  00000 00000  00000 00000    00000 00000  00000 00000  00000 00000  00000 00000  00000 00000
5014:  00000 00000  00000 00000  00000 00000  00000 00000  00000 00000    00000 00000  00000 00000  00000 00000  00000 00000  00000 00000
5015:  00000 00000  00000 00000  00000 00000  00000 00000  00000 00000    00000 00000  00000 00000  00000 00000  00000 00000  00000 00000
5016:  00000 00000  00000 00000  00000 00000  00000 00000  00000 00000    00000 00000  00000 00000  00000 00000  00000 00000  00000 00000
5017:  00000 00000  00000 00000  00000 00000  00000 00000  00000 00000    00000 00000  00000 00000  00000 00000  00000 00000  00000 00000
5018:  00000 00000  00000 00000  00000 00000  00000 00000  00000 00000    00000 00000  00000 00000  00000 00000  00000 00000  00000 00000
5019:  00000 00000  00000 00000  00000 00000  00000 00000  00000 00000    00000 00000  00000 00000  00000 00000  00000 00000  00000 00000
5020:  00000 00000  00000 00000  00000 00000  00000 00000  00000 00000    00000 00000  00000 00000  00000 00000  00000 00000  00000 00000
5021:  00000 00000  00000 00000  00000 00000  00000 00000  00000 00000    00000 00000  00000 00000  00000 00000  00000 00000  00000 00000
5022:  00000 00000  00000 00000  00000 00000  00000 00000  00000 00000    00000 00000  00000 00000  00000 00000  00000 00000  00000 00000
5023:  00000 00000  00000 00000  00000 00000  00000 00000  00000 00000    00000 00000  00000 00000  00000 00000  00000 00000  00000 00000
5024:  00000 00000  00000 00000  00000 00000  00000 00000  00000 00000    00000 00000  00000 00000  00000 00000  00000 00000  00000 00000
5025:  00000 00000  00000 00000  00000 00000  00000 00000  00000 00000    00000 00000  00000 00000  00000 00000  00000 00000  00000 00000
5026:  00000 00000  00000 00000  00000 00000  00000 00000  00000 00000    00000 00000  00000 00000  00000 00000  00000 00000  00000 00000
5027:  00000 00000  00000 00000  00000 00000  00000 00000  00000 00000    00000 00000  00000 00000  00000 00000  00000 00000  00000 00000
5028:  00000 00000  00000 00000  00000 00000  00000 00000  00000 00000    00000 00000  00000 00000  00000 00000  00000 00000  00000 00000
5029:  00000 00000  00000 00000  00000 00000  00000 00000  00000 00000    00000 00000  00000 00000  00000 00000  00000 00000  00000 00000
5030:  00000 00000  00000 00000  00000 00000  00000 00000  00000 00000    00000 00000  00000 00000  00000 00000  00000 00000  00000 00000
5031:  00000 00000  00000 00000  00000 00000  00000 00000  00000 00000    00000 00000  00000 00000  00000 00000  00000 00000  00000 00000
5032:  00000 00000  00000 00000  00000 00000  00000 00000  00000 00000    00000 00000  00000 00000  00000 00000  00000 00000  00000 00000
5033:  00000 00000  00000 00000  00000 00000  00000 00000  00000 00000    00000 00000  00000 00000  00000 00000  00000 00000  00000 00000
5034:  00000 00000  00000 00000  00000 00000  00000 00000  00000 00000    00000 00000  00000 00000  00000 00000  00000 00000  00000 00000
5035:  00000 00000  00000 00000  00000 00000  00000 00000  00000 00000    00000 00000  00000 00000  00000 00000  00000 00000  00000 00000
5036:  00000 00000  00000 00000  00000 00000  00000 00000  00000 00000    00000 00000  00000 00000  00000 00000  00000 00000  00000 00000
5037:  00000 00000  00000 00000  00000 00000  00000 00000  00000 00000    00000 00000  00000 00000  00000 00000  00000 00000  00000 00000
5038:  00000 00000  00000 00000  00000 00000  00000 00000  00000 00000    00000 00000  00000 00000  00000 00000  00000 00000  00000 00000
5039:  00000 00000  00000 00000  00000 00000  00000 00000  00000 00000    00000 00000  00000 00000  00000 00000  00000 00000  00000 00000
5040:  00000 00000  00000 00000  00000 00000  00000 00000  00000 00000    00000 00000  00000 00000  00000 00000  00000 00000  00000 00000
5041:  00000 00000  00000 00000  00000 00000  00000 00000  00000 00000    00000 00000  00000 00000  00000 00000  00000 00000  00000 00000
5042:  00000 00000  00000 00000  00000 00000  00000 00000  00000 00000    00000 00000  00000 00000  00000 00000  00000 00000  00000 00000
5043:  00000 00000  00000 00000  00000 00000  00000 00000  00000 00000    00000 00000  00000 00000  00000 00000  00000 00000  00000 00000
5044:  00000 00000  00000 00000  00000 00000  00000 00000  00000 00000    00000 00000  00000 00000  00000 00000  00000 00000  00000 00000
5045:  00000 00000  00000 00000  00000 00000  00000 00000  00000 00000    00000 00000  00000 00000  00000 00000  00000 00000  00000 00000
5046:  00000 00000  00000 00000  00000 00000  00000 00000  00000 00000    00000 00000  00000 00000  00000 00000  00000 00000  00000 00000
5047:  00000 00000  00000 00000  00000 00000  00000 00000  00000 00000    00000 00000  00000 00000  00000 00000  00000 00000  00000 00000
5048:  00000 00000  00000 00000  00000 00000  00000 00000  00000 00000    00000 00000  00000 00000  00000 00000  00000 00000  00000 00000
5049:  00000 00000  00000 00000  00000 00000  00000 00000  00000 00000    00000 00000  00000 00000  00000 00000  00000 00000  00000 00000
```

```
5050:  00000 00000  00000 00000  00000 00000  00000 00000  00000 00000   00000 00000  00000 00000  00000 00000  00000 00000  00000 00000
5051:  00000 00000  00000 00000  00000 00000  00000 00000  00000 00000   00000 00000  00000 00000  00000 00000  00000 00000  00000 00000
5052:  00000 00000  00000 00000  00000 00000  00000 00000  00000 00000   00000 00000  00000 00000  00000 00000  00000 00000  00000 00000
5053:  00000 00000  00000 00000  00000 00000  00000 00000  00000 00000   00000 00000  00000 00000  00000 00000  00000 00000  00000 00000
5054:  00000 00000  00000 00000  00000 00000  00000 00000  00000 00000   00000 00000  00000 00000  00000 00000  00000 00000  00000 00000
5055:  00000 00000  00000 00000  00000 00000  00000 00000  00000 00000   00000 00000  00000 00000  00000 00000  00000 00000  00000 00000
5056:  00000 00000  00000 00000  00000 00000  00000 00000  00000 00000   00000 00000  00000 00000  00000 00000  00000 00000  00000 00000
5057:  00000 00000  00000 00000  00000 00000  00000 00000  00000 00000   00000 00000  00000 00000  00000 00000  00000 00000  00000 00000
5058:  00000 00000  00000 00000  00000 00000  00000 00000  00000 00000   00000 00000  00000 00000  00000 00000  00000 00000  00000 00000
5059:  00000 00000  00000 00000  00000 00000  00000 00000  00000 00000   00000 00000  00000 00000  00000 00000  00000 00000  00000 00000
5060:  00000 00000  00000 00000  00000 00000  00000 00000  00000 00000   00000 00000  00000 00000  00000 00000  00000 00000  00000 00000
5061:  00000 00000  00000 00000  00000 00000  00000 00000  00000 00000   00000 00000  00000 00000  00000 00000  00000 00000  00000 00000
5062:  00000 00000  00000 00000  00000 00000  00000 00000  00000 00000   00000 00000  00000 00000  00000 00000  00000 00000  00000 00000
5063:  00000 00000  00000 00000  00000 00000  00000 00000  00000 00000   00000 00000  00000 00000  00000 00000  00000 00000  00000 00000
5064:  00000 00000  00000 00000  00000 00000  00000 00000  00000 00000   00000 00000  00000 00000  00000 00000  00000 00000  00000 00000
5065:  00000 00000  00000 00000  00000 00000  00000 00000  00000 00000   00000 00000  00000 00000  00000 00000  00000 00000  00000 00000
5066:  00000 00000  00000 00000  00000 00000  00000 00000  00000 00000   00000 00000  00000 00000  00000 00000  00000 00000  00000 00000
5067:  00000 00000  00000 00000  00000 00000  00000 00000  00000 00000   00000 00000  00000 00000  00000 00000  00000 00000  00000 00000
5068:  00000 00000  00000 00000  00000 00000  00000 00000  00000 00000   00000 00000  00000 00000  00000 00000  00000 00000  00000 00000
5069:  00000 00000  00000 00000  00000 00000  00000 00000  00000 00000   00000 00000  00000 00000  00000 00000  00000 00000  00000 00000
5070:  00000 00000  00000 00000  00000 00000  00000 00000  00000 00000   00000 00000  00000 00000  00000 00000  00000 00000  00000 00000
5071:  00000 00000  00000 00000  00000 00000  00000 00000  00000 00000   00000 00000  00000 00000  00000 00000  00000 00000  00000 00000
5072:  00000 00000  00000 00000  00000 00000  00000 00000  00000 00000   00000 00000  00000 00000  00000 00000  00000 00000  00000 00000
5073:  00000 00000  00000 00000  00000 00000  00000 00000  00000 00000   00000 00000  00000 00000  00000 00000  00000 00000  00000 00000
5074:  00000 00000  00000 00000  00000 00000  00000 00000  00000 00000   00000 00000  00000 00000  00000 00000  00000 00000  00000 00000
5075:  00000 00000  00000 00000  00000 00000  00000 00000  00000 00000   00000 00000  00000 00000  00000 00000  00000 00000  00000 00000
5076:  00000 00000  00000 00000  00000 00000  00000 00000  00000 00000   00000 00000  00000 00000  00000 00000  00000 00000  00000 00000
5077:  00000 00000  00000 00000  00000 00000  00000 00000  00000 00000   00000 00000  00000 00000  00000 00000  00000 00000  00000 00000
5078:  00000 00000  00000 00000  00000 00000  00000 00000  00000 00000   00000 00000  00000 00000  00000 00000  00000 00000  00000 00000
5079:  00000 00000  00000 00000  00000 00000  00000 00000  00000 00000   00000 00000  00000 00000  00000 00000  00000 00000  00000 00000
5080:  00000 00000  00000 00000  00000 00000  00000 00000  00000 00000   00000 00000  00000 00000  00000 00000  00000 00000  00000 00000
5081:  00000 00000  00000 00000  00000 00000  00000 00000  00000 00000   00000 00000  00000 00000  00000 00000  00000 00000  00000 00000
5082:  00000 00000  00000 00000  00000 00000  00000 00000  00000 00000   00000 00000  00000 00000  00000 00000  00000 00000  00000 00000
5083:  00000 00000  00000 00000  00000 00000  00000 00000  00000 00000   00000 00000  00000 00000  00000 00000  00000 00000  00000 00000
5084:  00000 00000  00000 00000  00000 00000  00000 00000  00000 00000   00000 00000  00000 00000  00000 00000  00000 00000  00000 00000
5085:  00000 00000  00000 00000  00000 00000  00000 00000  00000 00000   00000 00000  00000 00000  00000 00000  00000 00000  00000 00000
5086:  00000 00000  00000 00000  00000 00000  00000 00000  00000 00000   00000 00000  00000 00000  00000 00000  00000 00000  00000 00000
5087:  00000 00000  00000 00000  00000 00000  00000 00000  00000 00000   00000 00000  00000 00000  00000 00000  00000 00000  00000 00000
5088:  00000 00000  00000 00000  00000 00000  00000 00000  00000 00000   00000 00000  00000 00000  00000 00000  00000 00000  00000 00000
5089:  00000 00000  00000 00000  00000 00000  00000 00000  00000 00000   00000 00000  00000 00000  00000 00000  00000 00000  00000 00000
5090:  00000 00000  00000 00000  00000 00000  00000 00000  00000 00000   00000 00000  00000 00000  00000 00000  00000 00000  00000 00000
5091:  00000 00000  00000 00000  00000 00000  00000 00000  00000 00000   00000 00000  00000 00000  00000 00000  00000 00000  00000 00000
5092:  00000 00000  00000 00000  00000 00000  00000 00000  00000 00000   00000 00000  00000 00000  00000 00000  00000 00000  00000 00000
5093:  00000 00000  00000 00000  00000 00000  00000 00000  00000 00000   00000 00000  00000 00000  00000 00000  00000 00000  00000 00000
5094:  00000 00000  00000 00000  00000 00000  00000 00000  00000 00000   00000 00000  00000 00000  00000 00000  00000 00000  00000 00000
5095:  00000 00000  00000 00000  00000 00000  00000 00000  00000 00000   00000 00000  00000 00000  00000 00000  00000 00000  00000 00000
5096:  00000 00000  00000 00000  00000 00000  00000 00000  00000 00000   00000 00000  00000 00000  00000 00000  00000 00000  00000 00000
5097:  00000 00000  00000 00000  00000 00000  00000 00000  00000 00000   00000 00000  00000 00000  00000 00000  00000 00000  00000 00000
5098:  00000 00000  00000 00000  00000 00000  00000 00000  00000 00000   00000 00000  00000 00000  00000 00000  00000 00000  00000 00000
5099:  00000 00000  00000 00000  00000 00000  00000 00000  00000 00000   00000 00000  00000 00000  00000 00000  00000 00000  00000 00000
```

```
5100:  00000 00000  00000 00000  00000 00000  00000 00000  00000 00000    00000 00000  00000 00000  00000 00000  00000 00000  00000 00000
5101:  00000 00000  00000 00000  00000 00000  00000 00000  00000 00000    00000 00000  00000 00000  00000 00000  00000 00000  00000 00000
5102:  00000 00000  00000 00000  00000 00000  00000 00000  00000 00000    00000 00000  00000 00000  00000 00000  00000 00000  00000 00000
5103:  00000 00000  00000 00000  00000 00000  00000 00000  00000 00000    00000 00000  00000 00000  00000 00000  00000 00000  00000 00000
5104:  00000 00000  00000 00000  00000 00000  00000 00000  00000 00000    00000 00000  00000 00000  00000 00000  00000 00000  00000 00000
5105:  00000 00000  00000 00000  00000 00000  00000 00000  00000 00000    00000 00000  00000 00000  00000 00000  00000 00000  00000 00000
5106:  00000 00000  00000 00000  00000 00000  00000 00000  00000 00000    00000 00000  00000 00000  00000 00000  00000 00000  00000 00000
5107:  00000 00000  00000 00000  00000 00000  00000 00000  00000 00000    00000 00000  00000 00000  00000 00000  00000 00000  00000 00000
5108:  00000 00000  00000 00000  00000 00000  00000 00000  00000 00000    00000 00000  00000 00000  00000 00000  00000 00000  00000 00000
5109:  00000 00000  00000 00000  00000 00000  00000 00000  00000 00000    00000 00000  00000 00000  00000 00000  00000 00000  00000 00000
5110:  00000 00000  00000 00000  00000 00000  00000 00000  00000 00000    00000 00000  00000 00000  00000 00000  00000 00000  00000 00000
5111:  00000 00000  00000 00000  00000 00000  00000 00000  00000 00000    00000 00000  00000 00000  00000 00000  00000 00000  00000 00000
5112:  00000 00000  00000 00000  00000 00000  00000 00000  00000 00000    00000 00000  00000 00000  00000 00000  00000 00000  00000 00000
5113:  00000 00000  00000 00000  00000 00000  00000 00000  00000 00000    00000 00000  00000 00000  00000 00000  00000 00000  00000 00000
5114:  00000 00000  00000 00000  00000 00000  00000 00000  00000 00000    00000 00000  00000 00000  00000 00000  00000 00000  00000 00000
5115:  00000 00000  00000 00000  00000 00000  00000 00000  00000 00000    00000 00000  00000 00000  00000 00000  00000 00000  00000 00000
5116:  00000 00000  00000 00000  00000 00000  00000 00000  00000 00000    00000 00000  00000 00000  00000 00000  00000 00000  00000 00000
5117:  00000 00000  00000 00000  00000 00000  00000 00000  00000 00000    00000 00000  00000 00000  00000 00000  00000 00000  00000 00000
5118:  00000 00000  00000 00000  00000 00000  00000 00000  00000 00000    00000 00000  00000 00000  00000 00000  00000 00000  00000 00000
5119:  00000 00000  00000 00000  00000 00000  00000 00000  00000 00000    00000 00000  00000 00000  00000 00000  00000 00000  00000 00000
5120:  00000 00000  00000 00000  00000 00000  00000 00000  00000 00000    00000 00000  00000 00000  00000 00000  00000 00000  00000 00000
5121:  00000 00000  00000 00000  00000 00000  00000 00000  00000 00000    00000 00000  00000 00000  00000 00000  00000 00000  00000 00000
5122:  00000 00000  00000 00000  00000 00000  00000 00000  00000 00000    00000 00000  00000 00000  00000 00000  00000 00000  00000 00000
5123:  00000 00000  00000 00000  00000 00000  00000 00000  00000 00000    00000 00000  00000 00000  00000 00000  00000 00000  00000 00000
5124:  00000 00000  00000 00000  00000 00000  00000 00000  00000 00000    00000 00000  00000 00000  00000 00000  00000 00000  00000 00000
5125:  00000 00000  00000 00000  00000 00000  00000 00000  00000 00000    00000 00000  00000 00000  00000 00000  00000 00000  00000 00000
5126:  00000 00000  00000 00000  00000 00000  00000 00000  00000 00000    00000 00000  00000 00000  00000 00000  00000 00000  00000 00000
5127:  00000 00000  00000 00000  00000 00000  00000 00000  00000 00000    00000 00000  00000 00000  00000 00000  00000 00000  00000 00000
5128:  00000 00000  00000 00000  00000 00000  00000 00000  00000 00000    00000 00000  00000 00000  00000 00000  00000 00000  00000 00000
5129:  00000 00000  00000 00000  00000 00000  00000 00000  00000 00000    00000 00000  00000 00000  00000 00000  00000 00000  00000 00000
5130:  00000 00000  00000 00000  00000 00000  00000 00000  00000 00000    00000 00000  00000 00000  00000 00000  00000 00000  00000 00000
5131:  00000 00000  00000 00000  00000 00000  00000 00000  00000 00000    00000 00000  00000 00000  00000 00000  00000 00000  00000 00000
5132:  00000 00000  00000 00000  00000 00000  00000 00000  00000 00000    00000 00000  00000 00000  00000 00000  00000 00000  00000 00000
5133:  00000 00000  00000 00000  00000 00000  00000 00000  00000 00000    00000 00000  00000 00000  00000 00000  00000 00000  00000 00000
5134:  00000 00000  00000 00000  00000 00000  00000 00000  00000 00000    00000 00000  00000 00000  00000 00000  00000 00000  00000 00000
5135:  00000 00000  00000 00000  00000 00000  00000 00000  00000 00000    00000 00000  00000 00000  00000 00000  00000 00000  00000 00000
5136:  00000 00000  00000 00000  00000 00000  00000 00000  00000 00000    00000 00000  00000 00000  00000 00000  00000 00000  00000 00000
5137:  00000 00000  00000 00000  00000 00000  00000 00000  00000 00000    00000 00000  00000 00000  00000 00000  00000 00000  00000 00000
5138:  00000 00000  00000 00000  00000 00000  00000 00000  00000 00000    00000 00000  00000 00000  00000 00000  00000 00000  00000 00000
5139:  00000 00000  00000 00000  00000 00000  00000 00000  00000 00000    00000 00000  00000 00000  00000 00000  00000 00000  00000 00000
5140:  00000 00000  00000 00000  00000 00000  00000 00000  00000 00000    00000 00000  00000 00000  00000 00000  00000 00000  00000 00000
5141:  00000 00000  00000 00000  00000 00000  00000 00000  00000 00000    00000 00000  00000 00000  00000 00000  00000 00000  00000 00000
5142:  00000 00000  00000 00000  00000 00000  00000 00000  00000 00000    00000 00000  00000 00000  00000 00000  00000 00000  00000 00000
5143:  00000 00000  00000 00000  00000 00000  00000 00000  00000 00000    00000 00000  00000 00000  00000 00000  00000 00000  00000 00000
5144:  00000 00000  00000 00000  00000 00000  00000 00000  00000 00000    00000 00000  00000 00000  00000 00000  00000 00000  00000 00000
5145:  00000 00000  00000 00000  00000 00000  00000 00000  00000 00000    00000 00000  00000 00000  00000 00000  00000 00000  00000 00000
5146:  00000 00000  00000 00000  00000 00000  00000 00000  00000 00000    00000 00000  00000 00000  00000 00000  00000 00000  00000 00000
5147:  00000 00000  00000 00000  00000 00000  00000 00000  00000 00000    00000 00000  00000 00000  00000 00000  00000 00000  00000 00000
5148:  00000 00000  00000 00000  00000 00000  00000 00000  00000 00000    00000 00000  00000 00000  00000 00000  00000 00000  00000 00000
5149:  00000 00000  00000 00000  00000 00000  00000 00000  00000 00000    00000 00000  00000 00000  00000 00000  00000 00000  00000 00000
```

```
5150:   00000 00000   00000 00000   00000 00000   00000 00000   00000 00000     00000 00000   00000 00000   00000 00000   00000 00000   00000 00000
5151:   00000 00000   00000 00000   00000 00000   00000 00000   00000 00000     00000 00000   00000 00000   00000 00000   00000 00000   00000 00000
5152:   00000 00000   00000 00000   00000 00000   00000 00000   00000 00000     00000 00000   00000 00000   00000 00000   00000 00000   00000 00000
5153:   00000 00000   00000 00000   00000 00000   00000 00000   00000 00000     00000 00000   00000 00000   00000 00000   00000 00000   00000 00000
5154:   00000 00000   00000 00000   00000 00000   00000 00000   00000 00000     00000 00000   00000 00000   00000 00000   00000 00000   00000 00000
5155:   00000 00000   00000 00000   00000 00000   00000 00000   00000 00000     00000 00000   00000 00000   00000 00000   00000 00000   00000 00000
5156:   00000 00000   00000 00000   00000 00000   00000 00000   00000 00000     00000 00000   00000 00000   00000 00000   00000 00000   00000 00000
5157:   00000 00000   00000 00000   00000 00000   00000 00000   00000 00000     00000 00000   00000 00000   00000 00000   00000 00000   00000 00000
5158:   00000 00000   00000 00000   00000 00000   00000 00000   00000 00000     00000 00000   00000 00000   00000 00000   00000 00000   00000 00000
5159:   00000 00000   00000 00000   00000 00000   00000 00000   00000 00000     00000 00000   00000 00000   00000 00000   00000 00000   00000 00000
5160:   00000 00000   00000 00000   00000 00000   00000 00000   00000 00000     00000 00000   00000 00000   00000 00000   00000 00000   00000 00000
5161:   00000 00000   00000 00000   00000 00000   00000 00000   00000 00000     00000 00000   00000 00000   00000 00000   00000 00000   00000 00000
5162:   00000 00000   00000 00000   00000 00000   00000 00000   00000 00000     00000 00000   00000 00000   00000 00000   00000 00000   00000 00000
5163:   00000 00000   00000 00000   00000 00000   00000 00000   00000 00000     00000 00000   00000 00000   00000 00000   00000 00000   00000 00000
5164:   00000 00000   00000 00000   00000 00000   00000 00000   00000 00000     00000 00000   00000 00000   00000 00000   00000 00000   00000 00000
5165:   00000 00000   00000 00000   00000 00000   00000 00000   00000 00000     00000 00000   00000 00000   00000 00000   00000 00000   00000 00000
5166:   00000 00000   00000 00000   00000 00000   00000 00000   00000 00000     00000 00000   00000 00000   00000 00000   00000 00000   00000 00000
5167:   00000 00000   00000 00000   00000 00000   00000 00000   00000 00000     00000 00000   00000 00000   00000 00000   00000 00000   00000 00000
5168:   00000 00000   00000 00000   00000 00000   00000 00000   00000 00000     00000 00000   00000 00000   00000 00000   00000 00000   00000 00000
5169:   00000 00000   00000 00000   00000 00000   00000 00000   00000 00000     00000 00000   00000 00000   00000 00000   00000 00000   00000 00000
5170:   00000 00000   00000 00000   00000 00000   00000 00000   00000 00000     00000 00000   00000 00000   00000 00000   00000 00000   00000 00000
5171:   00000 00000   00000 00000   00000 00000   00000 00000   00000 00000     00000 00000   00000 00000   00000 00000   00000 00000   00000 00000
5172:   00000 00000   00000 00000   00000 00000   00000 00000   00000 00000     00000 00000   00000 00000   00000 00000   00000 00000   00000 00000
5173:   00000 00000   00000 00000   00000 00000   00000 00000   00000 00000     00000 00000   00000 00000   00000 00000   00000 00000   00000 00000
5174:   00000 00000   00000 00000   00000 00000   00000 00000   00000 00000     00000 00000   00000 00000   00000 00000   00000 00000   00000 00000
5175:   00000 00000   00000 00000   00000 00000   00000 00000   00000 00000     00000 00000   00000 00000   00000 00000   00000 00000   00000 00000
5176:   00000 00000   00000 00000   00000 00000   00000 00000   00000 00000     00000 00000   00000 00000   00000 00000   00000 00000   00000 00000
5177:   00000 00000   00000 00000   00000 00000   00000 00000   00000 00000     00000 00000   00000 00000   00000 00000   00000 00000   00000 00000
5178:   00000 00000   00000 00000   00000 00000   00000 00000   00000 00000     00000 00000   00000 00000   00000 00000   00000 00000   00000 00000
5179:   00000 00000   00000 00000   00000 00000   00000 00000   00000 00000     00000 00000   00000 00000   00000 00000   00000 00000   00000 00000
5180:   00000 00000   00000 00000   00000 00000   00000 00000   00000 00000     00000 00000   00000 00000   00000 00000   00000 00000   00000 00000
5181:   00000 00000   00000 00000   00000 00000   00000 00000   00000 00000     00000 00000   00000 00000   00000 00000   00000 00000   00000 00000
5182:   00000 00000   00000 00000   00000 00000   00000 00000   00000 00000     00000 00000   00000 00000   00000 00000   00000 00000   00000 00000
5183:   00000 00000   00000 00000   00000 00000   00000 00000   00000 00000     00000 00000   00000 00000   00000 00000   00000 00000   00000 00000
5184:   00000 00000   00000 00000   00000 00000   00000 00000   00000 00000     00000 00000   00000 00000   00000 00000   00000 00000   00000 00000
5185:   00000 00000   00000 00000   00000 00000   00000 00000   00000 00000     00000 00000   00000 00000   00000 00000   00000 00000   00000 00000
5186:   00000 00000   00000 00000   00000 00000   00000 00000   00000 00000     00000 00000   00000 00000   00000 00000   00000 00000   00000 00000
5187:   00000 00000   00000 00000   00000 00000   00000 00000   00000 00000     00000 00000   00000 00000   00000 00000   00000 00000   00000 00000
5188:   00000 00000   00000 00000   00000 00000   00000 00000   00000 00000     00000 00000   00000 00000   00000 00000   00000 00000   00000 00000
5189:   00000 00000   00000 00000   00000 00000   00000 00000   00000 00000     00000 00000   00000 00000   00000 00000   00000 00000   00000 00000
5190:   00000 00000   00000 00000   00000 00000   00000 00000   00000 00000     00000 00000   00000 00000   00000 00000   00000 00000   00000 00000
5191:   00000 00000   00000 00000   00000 00000   00000 00000   00000 00000     00000 00000   00000 00000   00000 00000   00000 00000   00000 00000
5192:   00000 00000   00000 00000   00000 00000   00000 00000   00000 00000     00000 00000   00000 00000   00000 00000   00000 00000   00000 00000
5193:   00000 00000   00000 00000   00000 00000   00000 00000   00000 00000     00000 00000   00000 00000   00000 00000   00000 00000   00000 00000
5194:   00000 00000   00000 00000   00000 00000   00000 00000   00000 00000     00000 00000   00000 00000   00000 00000   00000 00000   00000 00000
5195:   00000 00000   00000 00000   00000 00000   00000 00000   00000 00000     00000 00000   00000 00000   00000 00000   00000 00000   00000 00000
5196:   00000 00000   00000 00000   00000 00000   00000 00000   00000 00000     00000 00000   00000 00000   00000 00000   00000 00000   00000 00000
5197:   00000 00000   00000 00000   00000 00000   00000 00000   00000 00000     00000 00000   00000 00000   00000 00000   00000 00000   00000 00000
5198:   00000 00000   00000 00000   00000 00000   00000 00000   00000 00000     00000 00000   00000 00000   00000 00000   00000 00000   00000 00000
5199:   00000 00000   00000 00000   00000 00000   00000 00000   00000 00000     00000 00000   00000 00000   00000 00000   00000 00000   00000 00000
```

```
5200:  00000 00000  00000 00000  00000 00000  00000 00000  00000 00000    00000 00000  00000 00000  00000 00000  00000 00000  00000 00000
5201:  00000 00000  00000 00000  00000 00000  00000 00000  00000 00000    00000 00000  00000 00000  00000 00000  00000 00000  00000 00000
5202:  00000 00000  00000 00000  00000 00000  00000 00000  00000 00000    00000 00000  00000 00000  00000 00000  00000 00000  00000 00000
5203:  00000 00000  00000 00000  00000 00000  00000 00000  00000 00000    00000 00000  00000 00000  00000 00000  00000 00000  00000 00000
5204:  00000 00000  00000 00000  00000 00000  00000 00000  00000 00000    00000 00000  00000 00000  00000 00000  00000 00000  00000 00000
5205:  00000 00000  00000 00000  00000 00000  00000 00000  00000 00000    00000 00000  00000 00000  00000 00000  00000 00000  00000 00000
5206:  00000 00000  00000 00000  00000 00000  00000 00000  00000 00000    00000 00000  00000 00000  00000 00000  00000 00000  00000 00000
5207:  00000 00000  00000 00000  00000 00000  00000 00000  00000 00000    00000 00000  00000 00000  00000 00000  00000 00000  00000 00000
5208:  00000 00000  00000 00000  00000 00000  00000 00000  00000 00000    00000 00000  00000 00000  00000 00000  00000 00000  00000 00000
5209:  00000 00000  00000 00000  00000 00000  00000 00000  00000 00000    00000 00000  00000 00000  00000 00000  00000 00000  00000 00000
5210:  00000 00000  00000 00000  00000 00000  00000 00000  00000 00000    00000 00000  00000 00000  00000 00000  00000 00000  00000 00000
5211:  00000 00000  00000 00000  00000 00000  00000 00000  00000 00000    00000 00000  00000 00000  00000 00000  00000 00000  00000 00000
5212:  00000 00000  00000 00000  00000 00000  00000 00000  00000 00000    00000 00000  00000 00000  00000 00000  00000 00000  00000 00000
5213:  00000 00000  00000 00000  00000 00000  00000 00000  00000 00000    00000 00000  00000 00000  00000 00000  00000 00000  00000 00000
5214:  00000 00000  00000 00000  00000 00000  00000 00000  00000 00000    00000 00000  00000 00000  00000 00000  00000 00000  00000 00000
5215:  00000 00000  00000 00000  00000 00000  00000 00000  00000 00000    00000 00000  00000 00000  00000 00000  00000 00000  00000 00000
5216:  00000 00000  00000 00000  00000 00000  00000 00000  00000 00000    00000 00000  00000 00000  00000 00000  00000 00000  00000 00000
5217:  00000 00000  00000 00000  00000 00000  00000 00000  00000 00000    00000 00000  00000 00000  00000 00000  00000 00000  00000 00000
5218:  00000 00000  00000 00000  00000 00000  00000 00000  00000 00000    00000 00000  00000 00000  00000 00000  00000 00000  00000 00000
5219:  00000 00000  00000 00000  00000 00000  00000 00000  00000 00000    00000 00000  00000 00000  00000 00000  00000 00000  00000 00000
5220:  00000 00000  00000 00000  00000 00000  00000 00000  00000 00000    00000 00000  00000 00000  00000 00000  00000 00000  00000 00000
5221:  00000 00000  00000 00000  00000 00000  00000 00000  00000 00000    00000 00000  00000 00000  00000 00000  00000 00000  00000 00000
5222:  00000 00000  00000 00000  00000 00000  00000 00000  00000 00000    00000 00000  00000 00000  00000 00000  00000 00000  00000 00000
5223:  00000 00000  00000 00000  00000 00000  00000 00000  00000 00000    00000 00000  00000 00000  00000 00000  00000 00000  00000 00000
5224:  00000 00000  00000 00000  00000 00000  00000 00000  00000 00000    00000 00000  00000 00000  00000 00000  00000 00000  00000 00000
5225:  00000 00000  00000 00000  00000 00000  00000 00000  00000 00000    00000 00000  00000 00000  00000 00000  00000 00000  00000 00000
5226:  00000 00000  00000 00000  00000 00000  00000 00000  00000 00000    00000 00000  00000 00000  00000 00000  00000 00000  00000 00000
5227:  00000 00000  00000 00000  00000 00000  00000 00000  00000 00000    00000 00000  00000 00000  00000 00000  00000 00000  00000 00000
5228:  00000 00000  00000 00000  00000 00000  00000 00000  00000 00000    00000 00000  00000 00000  00000 00000  00000 00000  00000 00000
5229:  00000 00000  00000 00000  00000 00000  00000 00000  00000 00000    00000 00000  00000 00000  00000 00000  00000 00000  00000 00000
5230:  00000 00000  00000 00000  00000 00000  00000 00000  00000 00000    00000 00000  00000 00000  00000 00000  00000 00000  00000 00000
5231:  00000 00000  00000 00000  00000 00000  00000 00000  00000 00000    00000 00000  00000 00000  00000 00000  00000 00000  00000 00000
5232:  00000 00000  00000 00000  00000 00000  00000 00000  00000 00000    00000 00000  00000 00000  00000 00000  00000 00000  00000 00000
5233:  00000 00000  00000 00000  00000 00000  00000 00000  00000 00000    00000 00000  00000 00000  00000 00000  00000 00000  00000 00000
5234:  00000 00000  00000 00000  00000 00000  00000 00000  00000 00000    00000 00000  00000 00000  00000 00000  00000 00000  00000 00000
5235:  00000 00000  00000 00000  00000 00000  00000 00000  00000 00000    00000 00000  00000 00000  00000 00000  00000 00000  00000 00000
5236:  00000 00000  00000 00000  00000 00000  00000 00000  00000 00000    00000 00000  00000 00000  00000 00000  00000 00000  00000 00000
5237:  00000 00000  00000 00000  00000 00000  00000 00000  00000 00000    00000 00000  00000 00000  00000 00000  00000 00000  00000 00000
5238:  00000 00000  00000 00000  00000 00000  00000 00000  00000 00000    00000 00000  00000 00000  00000 00000  00000 00000  00000 00000
5239:  00000 00000  00000 00000  00000 00000  00000 00000  00000 00000    00000 00000  00000 00000  00000 00000  00000 00000  00000 00000
5240:  00000 00000  00000 00000  00000 00000  00000 00000  00000 00000    00000 00000  00000 00000  00000 00000  00000 00000  00000 00000
5241:  00000 00000  00000 00000  00000 00000  00000 00000  00000 00000    00000 00000  00000 00000  00000 00000  00000 00000  00000 00000
5242:  00000 00000  00000 00000  00000 00000  00000 00000  00000 00000    00000 00000  00000 00000  00000 00000  00000 00000  00000 00000
5243:  00000 00000  00000 00000  00000 00000  00000 00000  00000 00000    00000 00000  00000 00000  00000 00000  00000 00000  00000 00000
5244:  00000 00000  00000 00000  00000 00000  00000 00000  00000 00000    00000 00000  00000 00000  00000 00000  00000 00000  00000 00000
5245:  00000 00000  00000 00000  00000 00000  00000 00000  00000 00000    00000 00000  00000 00000  00000 00000  00000 00000  00000 00000
5246:  00000 00000  00000 00000  00000 00000  00000 00000  00000 00000    00000 00000  00000 00000  00000 00000  00000 00000  00000 00000
5247:  00000 00000  00000 00000  00000 00000  00000 00000  00000 00000    00000 00000  00000 00000  00000 00000  00000 00000  00000 00000
5248:  00000 00000  00000 00000  00000 00000  00000 00000  00000 00000    00000 00000  00000 00000  00000 00000  00000 00000  00000 00000
5249:  00000 00000  00000 00000  00000 00000  00000 00000  00000 00000    00000 00000  00000 00000  00000 00000  00000 00000  00000 00000
```

```
5250:   00000 00000   00000 00000   00000 00000   00000 00000   00000 00000     00000 00000   00000 00000   00000 00000   00000 00000   00000 00000
5251:   00000 00000   00000 00000   00000 00000   00000 00000   00000 00000     00000 00000   00000 00000   00000 00000   00000 00000   00000 00000
5252:   00000 00000   00000 00000   00000 00000   00000 00000   00000 00000     00000 00000   00000 00000   00000 00000   00000 00000   00000 00000
5253:   00000 00000   00000 00000   00000 00000   00000 00000   00000 00000     00000 00000   00000 00000   00000 00000   00000 00000   00000 00000
5254:   00000 00000   00000 00000   00000 00000   00000 00000   00000 00000     00000 00000   00000 00000   00000 00000   00000 00000   00000 00000
5255:   00000 00000   00000 00000   00000 00000   00000 00000   00000 00000     00000 00000   00000 00000   00000 00000   00000 00000   00000 00000
5256:   00000 00000   00000 00000   00000 00000   00000 00000   00000 00000     00000 00000   00000 00000   00000 00000   00000 00000   00000 00000
5257:   00000 00000   00000 00000   00000 00000   00000 00000   00000 00000     00000 00000   00000 00000   00000 00000   00000 00000   00000 00000
5258:   00000 00000   00000 00000   00000 00000   00000 00000   00000 00000     00000 00000   00000 00000   00000 00000   00000 00000   00000 00000
5259:   00000 00000   00000 00000   00000 00000   00000 00000   00000 00000     00000 00000   00000 00000   00000 00000   00000 00000   00000 00000
5260:   00000 00000   00000 00000   00000 00000   00000 00000   00000 00000     00000 00000   00000 00000   00000 00000   00000 00000   00000 00000
5261:   00000 00000   00000 00000   00000 00000   00000 00000   00000 00000     00000 00000   00000 00000   00000 00000   00000 00000   00000 00000
5262:   00000 00000   00000 00000   00000 00000   00000 00000   00000 00000     00000 00000   00000 00000   00000 00000   00000 00000   00000 00000
5263:   00000 00000   00000 00000   00000 00000   00000 00000   00000 00000     00000 00000   00000 00000   00000 00000   00000 00000   00000 00000
5264:   00000 00000   00000 00000   00000 00000   00000 00000   00000 00000     00000 00000   00000 00000   00000 00000   00000 00000   00000 00000
5265:   00000 00000   00000 00000   00000 00000   00000 00000   00000 00000     00000 00000   00000 00000   00000 00000   00000 00000   00000 00000
5266:   00000 00000   00000 00000   00000 00000   00000 00000   00000 00000     00000 00000   00000 00000   00000 00000   00000 00000   00000 00000
5267:   00000 00000   00000 00000   00000 00000   00000 00000   00000 00000     00000 00000   00000 00000   00000 00000   00000 00000   00000 00000
5268:   00000 00000   00000 00000   00000 00000   00000 00000   00000 00000     00000 00000   00000 00000   00000 00000   00000 00000   00000 00000
5269:   00000 00000   00000 00000   00000 00000   00000 00000   00000 00000     00000 00000   00000 00000   00000 00000   00000 00000   00000 00000
5270:   00000 00000   00000 00000   00000 00000   00000 00000   00000 00000     00000 00000   00000 00000   00000 00000   00000 00000   00000 00000
5271:   00000 00000   00000 00000   00000 00000   00000 00000   00000 00000     00000 00000   00000 00000   00000 00000   00000 00000   00000 00000
5272:   00000 00000   00000 00000   00000 00000   00000 00000   00000 00000     00000 00000   00000 00000   00000 00000   00000 00000   00000 00000
5273:   00000 00000   00000 00000   00000 00000   00000 00000   00000 00000     00000 00000   00000 00000   00000 00000   00000 00000   00000 00000
5274:   00000 00000   00000 00000   00000 00000   00000 00000   00000 00000     00000 00000   00000 00000   00000 00000   00000 00000   00000 00000
5275:   00000 00000   00000 00000   00000 00000   00000 00000   00000 00000     00000 00000   00000 00000   00000 00000   00000 00000   00000 00000
5276:   00000 00000   00000 00000   00000 00000   00000 00000   00000 00000     00000 00000   00000 00000   00000 00000   00000 00000   00000 00000
5277:   00000 00000   00000 00000   00000 00000   00000 00000   00000 00000     00000 00000   00000 00000   00000 00000   00000 00000   00000 00000
5278:   00000 00000   00000 00000   00000 00000   00000 00000   00000 00000     00000 00000   00000 00000   00000 00000   00000 00000   00000 00000
5279:   00000 00000   00000 00000   00000 00000   00000 00000   00000 00000     00000 00000   00000 00000   00000 00000   00000 00000   00000 00000
5280:   00000 00000   00000 00000   00000 00000   00000 00000   00000 00000     00000 00000   00000 00000   00000 00000   00000 00000   00000 00000
5281:   00000 00000   00000 00000   00000 00000   00000 00000   00000 00000     00000 00000   00000 00000   00000 00000   00000 00000   00000 00000
5282:   00000 00000   00000 00000   00000 00000   00000 00000   00000 00000     00000 00000   00000 00000   00000 00000   00000 00000   00000 00000
5283:   00000 00000   00000 00000   00000 00000   00000 00000   00000 00000     00000 00000   00000 00000   00000 00000   00000 00000   00000 00000
5284:   00000 00000   00000 00000   00000 00000   00000 00000   00000 00000     00000 00000   00000 00000   00000 00000   00000 00000   00000 00000
5285:   00000 00000   00000 00000   00000 00000   00000 00000   00000 00000     00000 00000   00000 00000   00000 00000   00000 00000   00000 00000
5286:   00000 00000   00000 00000   00000 00000   00000 00000   00000 00000     00000 00000   00000 00000   00000 00000   00000 00000   00000 00000
5287:   00000 00000   00000 00000   00000 00000   00000 00000   00000 00000     00000 00000   00000 00000   00000 00000   00000 00000   00000 00000
5288:   00000 00000   00000 00000   00000 00000   00000 00000   00000 00000     00000 00000   00000 00000   00000 00000   00000 00000   00000 00000
5289:   00000 00000   00000 00000   00000 00000   00000 00000   00000 00000     00000 00000   00000 00000   00000 00000   00000 00000   00000 00000
5290:   00000 00000   00000 00000   00000 00000   00000 00000   00000 00000     00000 00000   00000 00000   00000 00000   00000 00000   00000 00000
5291:   00000 00000   00000 00000   00000 00000   00000 00000   00000 00000     00000 00000   00000 00000   00000 00000   00000 00000   00000 00000
5292:   00000 00000   00000 00000   00000 00000   00000 00000   00000 00000     00000 00000   00000 00000   00000 00000   00000 00000   00000 00000
5293:   00000 00000   00000 00000   00000 00000   00000 00000   00000 00000     00000 00000   00000 00000   00000 00000   00000 00000   00000 00000
5294:   00000 00000   00000 00000   00000 00000   00000 00000   00000 00000     00000 00000   00000 00000   00000 00000   00000 00000   00000 00000
5295:   00000 00000   00000 00000   00000 00000   00000 00000   00000 00000     00000 00000   00000 00000   00000 00000   00000 00000   00000 00000
5296:   00000 00000   00000 00000   00000 00000   00000 00000   00000 00000     00000 00000   00000 00000   00000 00000   00000 00000   00000 00000
5297:   00000 00000   00000 00000   00000 00000   00000 00000   00000 00000     00000 00000   00000 00000   00000 00000   00000 00000   00000 00000
5298:   00000 00000   00000 00000   00000 00000   00000 00000   00000 00000     00000 00000   00000 00000   00000 00000   00000 00000   00000 00000
5299:   00000 00000   00000 00000   00000 00000   00000 00000   00000 00000     00000 00000   00000 00000   00000 00000   00000 00000   00000 00000
```

```
5300:  00000 00000  00000 00000  00000 00000  00000 00000  00000 00000     00000 00000  00000 00000  00000 00000  00000 00000  00000 00000
5301:  00000 00000  00000 00000  00000 00000  00000 00000  00000 00000     00000 00000  00000 00000  00000 00000  00000 00000  00000 00000
5302:  00000 00000  00000 00000  00000 00000  00000 00000  00000 00000     00000 00000  00000 00000  00000 00000  00000 00000  00000 00000
5303:  00000 00000  00000 00000  00000 00000  00000 00000  00000 00000     00000 00000  00000 00000  00000 00000  00000 00000  00000 00000
5304:  00000 00000  00000 00000  00000 00000  00000 00000  00000 00000     00000 00000  00000 00000  00000 00000  00000 00000  00000 00000
5305:  00000 00000  00000 00000  00000 00000  00000 00000  00000 00000     00000 00000  00000 00000  00000 00000  00000 00000  00000 00000
5306:  00000 00000  00000 00000  00000 00000  00000 00000  00000 00000     00000 00000  00000 00000  00000 00000  00000 00000  00000 00000
5307:  00000 00000  00000 00000  00000 00000  00000 00000  00000 00000     00000 00000  00000 00000  00000 00000  00000 00000  00000 00000
5308:  00000 00000  00000 00000  00000 00000  00000 00000  00000 00000     00000 00000  00000 00000  00000 00000  00000 00000  00000 00000
5309:  00000 00000  00000 00000  00000 00000  00000 00000  00000 00000     00000 00000  00000 00000  00000 00000  00000 00000  00000 00000
5310:  00000 00000  00000 00000  00000 00000  00000 00000  00000 00000     00000 00000  00000 00000  00000 00000  00000 00000  00000 00000
5311:  00000 00000  00000 00000  00000 00000  00000 00000  00000 00000     00000 00000  00000 00000  00000 00000  00000 00000  00000 00000
5312:  00000 00000  00000 00000  00000 00000  00000 00000  00000 00000     00000 00000  00000 00000  00000 00000  00000 00000  00000 00000
5313:  00000 00000  00000 00000  00000 00000  00000 00000  00000 00000     00000 00000  00000 00000  00000 00000  00000 00000  00000 00000
5314:  00000 00000  00000 00000  00000 00000  00000 00000  00000 00000     00000 00000  00000 00000  00000 00000  00000 00000  00000 00000
5315:  00000 00000  00000 00000  00000 00000  00000 00000  00000 00000     00000 00000  00000 00000  00000 00000  00000 00000  00000 00000
5316:  00000 00000  00000 00000  00000 00000  00000 00000  00000 00000     00000 00000  00000 00000  00000 00000  00000 00000  00000 00000
5317:  00000 00000  00000 00000  00000 00000  00000 00000  00000 00000     00000 00000  00000 00000  00000 00000  00000 00000  00000 00000
5318:  00000 00000  00000 00000  00000 00000  00000 00000  00000 00000     00000 00000  00000 00000  00000 00000  00000 00000  00000 00000
5319:  00000 00000  00000 00000  00000 00000  00000 00000  00000 00000     00000 00000  00000 00000  00000 00000  00000 00000  00000 00000
5320:  00000 00000  00000 00000  00000 00000  00000 00000  00000 00000     00000 00000  00000 00000  00000 00000  00000 00000  00000 00000
5321:  00000 00000  00000 00000  00000 00000  00000 00000  00000 00000     00000 00000  00000 00000  00000 00000  00000 00000  00000 00000
5322:  00000 00000  00000 00000  00000 00000  00000 00000  00000 00000     00000 00000  00000 00000  00000 00000  00000 00000  00000 00000
5323:  00000 00000  00000 00000  00000 00000  00000 00000  00000 00000     00000 00000  00000 00000  00000 00000  00000 00000  00000 00000
5324:  00000 00000  00000 00000  00000 00000  00000 00000  00000 00000     00000 00000  00000 00000  00000 00000  00000 00000  00000 00000
5325:  00000 00000  00000 00000  00000 00000  00000 00000  00000 00000     00000 00000  00000 00000  00000 00000  00000 00000  00000 00000
5326:  00000 00000  00000 00000  00000 00000  00000 00000  00000 00000     00000 00000  00000 00000  00000 00000  00000 00000  00000 00000
5327:  00000 00000  00000 00000  00000 00000  00000 00000  00000 00000     00000 00000  00000 00000  00000 00000  00000 00000  00000 00000
5328:  00000 00000  00000 00000  00000 00000  00000 00000  00000 00000     00000 00000  00000 00000  00000 00000  00000 00000  00000 00000
5329:  00000 00000  00000 00000  00000 00000  00000 00000  00000 00000     00000 00000  00000 00000  00000 00000  00000 00000  00000 00000
5330:  00000 00000  00000 00000  00000 00000  00000 00000  00000 00000     00000 00000  00000 00000  00000 00000  00000 00000  00000 00000
5331:  00000 00000  00000 00000  00000 00000  00000 00000  00000 00000     00000 00000  00000 00000  00000 00000  00000 00000  00000 00000
5332:  00000 00000  00000 00000  00000 00000  00000 00000  00000 00000     00000 00000  00000 00000  00000 00000  00000 00000  00000 00000
5333:  00000 00000  00000 00000  00000 00000  00000 00000  00000 00000     00000 00000  00000 00000  00000 00000  00000 00000  00000 00000
5334:  00000 00000  00000 00000  00000 00000  00000 00000  00000 00000     00000 00000  00000 00000  00000 00000  00000 00000  00000 00000
5335:  00000 00000  00000 00000  00000 00000  00000 00000  00000 00000     00000 00000  00000 00000  00000 00000  00000 00000  00000 00000
5336:  00000 00000  00000 00000  00000 00000  00000 00000  00000 00000     00000 00000  00000 00000  00000 00000  00000 00000  00000 00000
5337:  00000 00000  00000 00000  00000 00000  00000 00000  00000 00000     00000 00000  00000 00000  00000 00000  00000 00000  00000 00000
5338:  00000 00000  00000 00000  00000 00000  00000 00000  00000 00000     00000 00000  00000 00000  00000 00000  00000 00000  00000 00000
5339:  00000 00000  00000 00000  00000 00000  00000 00000  00000 00000     00000 00000  00000 00000  00000 00000  00000 00000  00000 00000
5340:  00000 00000  00000 00000  00000 00000  00000 00000  00000 00000     00000 00000  00000 00000  00000 00000  00000 00000  00000 00000
5341:  00000 00000  00000 00000  00000 00000  00000 00000  00000 00000     00000 00000  00000 00000  00000 00000  00000 00000  00000 00000
5342:  00000 00000  00000 00000  00000 00000  00000 00000  00000 00000     00000 00000  00000 00000  00000 00000  00000 00000  00000 00000
5343:  00000 00000  00000 00000  00000 00000  00000 00000  00000 00000     00000 00000  00000 00000  00000 00000  00000 00000  00000 00000
5344:  00000 00000  00000 00000  00000 00000  00000 00000  00000 00000     00000 00000  00000 00000  00000 00000  00000 00000  00000 00000
5345:  00000 00000  00000 00000  00000 00000  00000 00000  00000 00000     00000 00000  00000 00000  00000 00000  00000 00000  00000 00000
5346:  00000 00000  00000 00000  00000 00000  00000 00000  00000 00000     00000 00000  00000 00000  00000 00000  00000 00000  00000 00000
5347:  00000 00000  00000 00000  00000 00000  00000 00000  00000 00000     00000 00000  00000 00000  00000 00000  00000 00000  00000 00000
5348:  00000 00000  00000 00000  00000 00000  00000 00000  00000 00000     00000 00000  00000 00000  00000 00000  00000 00000  00000 00000
5349:  00000 00000  00000 00000  00000 00000  00000 00000  00000 00000     00000 00000  00000 00000  00000 00000  00000 00000  00000 00000
```

```
5350:  00000 00000  00000 00000  00000 00000  00000 00000  00000 00000   00000 00000  00000 00000  00000 00000  00000 00000  00000 00000
5351:  00000 00000  00000 00000  00000 00000  00000 00000  00000 00000   00000 00000  00000 00000  00000 00000  00000 00000  00000 00000
5352:  00000 00000  00000 00000  00000 00000  00000 00000  00000 00000   00000 00000  00000 00000  00000 00000  00000 00000  00000 00000
5353:  00000 00000  00000 00000  00000 00000  00000 00000  00000 00000   00000 00000  00000 00000  00000 00000  00000 00000  00000 00000
5354:  00000 00000  00000 00000  00000 00000  00000 00000  00000 00000   00000 00000  00000 00000  00000 00000  00000 00000  00000 00000
5355:  00000 00000  00000 00000  00000 00000  00000 00000  00000 00000   00000 00000  00000 00000  00000 00000  00000 00000  00000 00000
5356:  00000 00000  00000 00000  00000 00000  00000 00000  00000 00000   00000 00000  00000 00000  00000 00000  00000 00000  00000 00000
5357:  00000 00000  00000 00000  00000 00000  00000 00000  00000 00000   00000 00000  00000 00000  00000 00000  00000 00000  00000 00000
5358:  00000 00000  00000 00000  00000 00000  00000 00000  00000 00000   00000 00000  00000 00000  00000 00000  00000 00000  00000 00000
5359:  00000 00000  00000 00000  00000 00000  00000 00000  00000 00000   00000 00000  00000 00000  00000 00000  00000 00000  00000 00000
5360:  00000 00000  00000 00000  00000 00000  00000 00000  00000 00000   00000 00000  00000 00000  00000 00000  00000 00000  00000 00000
5361:  00000 00000  00000 00000  00000 00000  00000 00000  00000 00000   00000 00000  00000 00000  00000 00000  00000 00000  00000 00000
5362:  00000 00000  00000 00000  00000 00000  00000 00000  00000 00000   00000 00000  00000 00000  00000 00000  00000 00000  00000 00000
5363:  00000 00000  00000 00000  00000 00000  00000 00000  00000 00000   00000 00000  00000 00000  00000 00000  00000 00000  00000 00000
5364:  00000 00000  00000 00000  00000 00000  00000 00000  00000 00000   00000 00000  00000 00000  00000 00000  00000 00000  00000 00000
5365:  00000 00000  00000 00000  00000 00000  00000 00000  00000 00000   00000 00000  00000 00000  00000 00000  00000 00000  00000 00000
5366:  00000 00000  00000 00000  00000 00000  00000 00000  00000 00000   00000 00000  00000 00000  00000 00000  00000 00000  00000 00000
5367:  00000 00000  00000 00000  00000 00000  00000 00000  00000 00000   00000 00000  00000 00000  00000 00000  00000 00000  00000 00000
5368:  00000 00000  00000 00000  00000 00000  00000 00000  00000 00000   00000 00000  00000 00000  00000 00000  00000 00000  00000 00000
5369:  00000 00000  00000 00000  00000 00000  00000 00000  00000 00000   00000 00000  00000 00000  00000 00000  00000 00000  00000 00000
5370:  00000 00000  00000 00000  00000 00000  00000 00000  00000 00000   00000 00000  00000 00000  00000 00000  00000 00000  00000 00000
5371:  00000 00000  00000 00000  00000 00000  00000 00000  00000 00000   00000 00000  00000 00000  00000 00000  00000 00000  00000 00000
5372:  00000 00000  00000 00000  00000 00000  00000 00000  00000 00000   00000 00000  00000 00000  00000 00000  00000 00000  00000 00000
5373:  00000 00000  00000 00000  00000 00000  00000 00000  00000 00000   00000 00000  00000 00000  00000 00000  00000 00000  00000 00000
5374:  00000 00000  00000 00000  00000 00000  00000 00000  00000 00000   00000 00000  00000 00000  00000 00000  00000 00000  00000 00000
5375:  00000 00000  00000 00000  00000 00000  00000 00000  00000 00000   00000 00000  00000 00000  00000 00000  00000 00000  00000 00000
5376:  00000 00000  00000 00000  00000 00000  00000 00000  00000 00000   00000 00000  00000 00000  00000 00000  00000 00000  00000 00000
5377:  00000 00000  00000 00000  00000 00000  00000 00000  00000 00000   00000 00000  00000 00000  00000 00000  00000 00000  00000 00000
5378:  00000 00000  00000 00000  00000 00000  00000 00000  00000 00000   00000 00000  00000 00000  00000 00000  00000 00000  00000 00000
5379:  00000 00000  00000 00000  00000 00000  00000 00000  00000 00000   00000 00000  00000 00000  00000 00000  00000 00000  00000 00000
5380:  00000 00000  00000 00000  00000 00000  00000 00000  00000 00000   00000 00000  00000 00000  00000 00000  00000 00000  00000 00000
5381:  00000 00000  00000 00000  00000 00000  00000 00000  00000 00000   00000 00000  00000 00000  00000 00000  00000 00000  00000 00000
5382:  00000 00000  00000 00000  00000 00000  00000 00000  00000 00000   00000 00000  00000 00000  00000 00000  00000 00000  00000 00000
5383:  00000 00000  00000 00000  00000 00000  00000 00000  00000 00000   00000 00000  00000 00000  00000 00000  00000 00000  00000 00000
5384:  00000 00000  00000 00000  00000 00000  00000 00000  00000 00000   00000 00000  00000 00000  00000 00000  00000 00000  00000 00000
5385:  00000 00000  00000 00000  00000 00000  00000 00000  00000 00000   00000 00000  00000 00000  00000 00000  00000 00000  00000 00000
5386:  00000 00000  00000 00000  00000 00000  00000 00000  00000 00000   00000 00000  00000 00000  00000 00000  00000 00000  00000 00000
5387:  00000 00000  00000 00000  00000 00000  00000 00000  00000 00000   00000 00000  00000 00000  00000 00000  00000 00000  00000 00000
5388:  00000 00000  00000 00000  00000 00000  00000 00000  00000 00000   00000 00000  00000 00000  00000 00000  00000 00000  00000 00000
5389:  00000 00000  00000 00000  00000 00000  00000 00000  00000 00000   00000 00000  00000 00000  00000 00000  00000 00000  00000 00000
5390:  00000 00000  00000 00000  00000 00000  00000 00000  00000 00000   00000 00000  00000 00000  00000 00000  00000 00000  00000 00000
5391:  00000 00000  00000 00000  00000 00000  00000 00000  00000 00000   00000 00000  00000 00000  00000 00000  00000 00000  00000 00000
5392:  00000 00000  00000 00000  00000 00000  00000 00000  00000 00000   00000 00000  00000 00000  00000 00000  00000 00000  00000 00000
5393:  00000 00000  00000 00000  00000 00000  00000 00000  00000 00000   00000 00000  00000 00000  00000 00000  00000 00000  00000 00000
5394:  00000 00000  00000 00000  00000 00000  00000 00000  00000 00000   00000 00000  00000 00000  00000 00000  00000 00000  00000 00000
5395:  00000 00000  00000 00000  00000 00000  00000 00000  00000 00000   00000 00000  00000 00000  00000 00000  00000 00000  00000 00000
5396:  00000 00000  00000 00000  00000 00000  00000 00000  00000 00000   00000 00000  00000 00000  00000 00000  00000 00000  00000 00000
5397:  00000 00000  00000 00000  00000 00000  00000 00000  00000 00000   00000 00000  00000 00000  00000 00000  00000 00000  00000 00000
5398:  00000 00000  00000 00000  00000 00000  00000 00000  00000 00000   00000 00000  00000 00000  00000 00000  00000 00000  00000 00000
5399:  00000 00000  00000 00000  00000 00000  00000 00000  00000 00000   00000 00000  00000 00000  00000 00000  00000 00000  00000 00000
```

```
5400:  00000 00000   00000 00000   00000 00000   00000 00000   00000 00000     00000 00000   00000 00000   00000 00000   00000 00000   00000 00000
5401:  00000 00000   00000 00000   00000 00000   00000 00000   00000 00000     00000 00000   00000 00000   00000 00000   00000 00000   00000 00000
5402:  00000 00000   00000 00000   00000 00000   00000 00000   00000 00000     00000 00000   00000 00000   00000 00000   00000 00000   00000 00000
5403:  00000 00000   00000 00000   00000 00000   00000 00000   00000 00000     00000 00000   00000 00000   00000 00000   00000 00000   00000 00000
5404:  00000 00000   00000 00000   00000 00000   00000 00000   00000 00000     00000 00000   00000 00000   00000 00000   00000 00000   00000 00000
5405:  00000 00000   00000 00000   00000 00000   00000 00000   00000 00000     00000 00000   00000 00000   00000 00000   00000 00000   00000 00000
5406:  00000 00000   00000 00000   00000 00000   00000 00000   00000 00000     00000 00000   00000 00000   00000 00000   00000 00000   00000 00000
5407:  00000 00000   00000 00000   00000 00000   00000 00000   00000 00000     00000 00000   00000 00000   00000 00000   00000 00000   00000 00000
5408:  00000 00000   00000 00000   00000 00000   00000 00000   00000 00000     00000 00000   00000 00000   00000 00000   00000 00000   00000 00000
5409:  00000 00000   00000 00000   00000 00000   00000 00000   00000 00000     00000 00000   00000 00000   00000 00000   00000 00000   00000 00000
5410:  00000 00000   00000 00000   00000 00000   00000 00000   00000 00000     00000 00000   00000 00000   00000 00000   00000 00000   00000 00000
5411:  00000 00000   00000 00000   00000 00000   00000 00000   00000 00000     00000 00000   00000 00000   00000 00000   00000 00000   00000 00000
5412:  00000 00000   00000 00000   00000 00000   00000 00000   00000 00000     00000 00000   00000 00000   00000 00000   00000 00000   00000 00000
5413:  00000 00000   00000 00000   00000 00000   00000 00000   00000 00000     00000 00000   00000 00000   00000 00000   00000 00000   00000 00000
5414:  00000 00000   00000 00000   00000 00000   00000 00000   00000 00000     00000 00000   00000 00000   00000 00000   00000 00000   00000 00000
5415:  00000 00000   00000 00000   00000 00000   00000 00000   00000 00000     00000 00000   00000 00000   00000 00000   00000 00000   00000 00000
5416:  00000 00000   00000 00000   00000 00000   00000 00000   00000 00000     00000 00000   00000 00000   00000 00000   00000 00000   00000 00000
5417:  00000 00000   00000 00000   00000 00000   00000 00000   00000 00000     00000 00000   00000 00000   00000 00000   00000 00000   00000 00000
5418:  00000 00000   00000 00000   00000 00000   00000 00000   00000 00000     00000 00000   00000 00000   00000 00000   00000 00000   00000 00000
5419:  00000 00000   00000 00000   00000 00000   00000 00000   00000 00000     00000 00000   00000 00000   00000 00000   00000 00000   00000 00000
5420:  00000 00000   00000 00000   00000 00000   00000 00000   00000 00000     00000 00000   00000 00000   00000 00000   00000 00000   00000 00000
5421:  00000 00000   00000 00000   00000 00000   00000 00000   00000 00000     00000 00000   00000 00000   00000 00000   00000 00000   00000 00000
5422:  00000 00000   00000 00000   00000 00000   00000 00000   00000 00000     00000 00000   00000 00000   00000 00000   00000 00000   00000 00000
5423:  00000 00000   00000 00000   00000 00000   00000 00000   00000 00000     00000 00000   00000 00000   00000 00000   00000 00000   00000 00000
5424:  00000 00000   00000 00000   00000 00000   00000 00000   00000 00000     00000 00000   00000 00000   00000 00000   00000 00000   00000 00000
5425:  00000 00000   00000 00000   00000 00000   00000 00000   00000 00000     00000 00000   00000 00000   00000 00000   00000 00000   00000 00000
5426:  00000 00000   00000 00000   00000 00000   00000 00000   00000 00000     00000 00000   00000 00000   00000 00000   00000 00000   00000 00000
5427:  00000 00000   00000 00000   00000 00000   00000 00000   00000 00000     00000 00000   00000 00000   00000 00000   00000 00000   00000 00000
5428:  00000 00000   00000 00000   00000 00000   00000 00000   00000 00000     00000 00000   00000 00000   00000 00000   00000 00000   00000 00000
5429:  00000 00000   00000 00000   00000 00000   00000 00000   00000 00000     00000 00000   00000 00000   00000 00000   00000 00000   00000 00000
5430:  00000 00000   00000 00000   00000 00000   00000 00000   00000 00000     00000 00000   00000 00000   00000 00000   00000 00000   00000 00000
5431:  00000 00000   00000 00000   00000 00000   00000 00000   00000 00000     00000 00000   00000 00000   00000 00000   00000 00000   00000 00000
5432:  00000 00000   00000 00000   00000 00000   00000 00000   00000 00000     00000 00000   00000 00000   00000 00000   00000 00000   00000 00000
5433:  00000 00000   00000 00000   00000 00000   00000 00000   00000 00000     00000 00000   00000 00000   00000 00000   00000 00000   00000 00000
5434:  00000 00000   00000 00000   00000 00000   00000 00000   00000 00000     00000 00000   00000 00000   00000 00000   00000 00000   00000 00000
5435:  00000 00000   00000 00000   00000 00000   00000 00000   00000 00000     00000 00000   00000 00000   00000 00000   00000 00000   00000 00000
5436:  00000 00000   00000 00000   00000 00000   00000 00000   00000 00000     00000 00000   00000 00000   00000 00000   00000 00000   00000 00000
5437:  00000 00000   00000 00000   00000 00000   00000 00000   00000 00000     00000 00000   00000 00000   00000 00000   00000 00000   00000 00000
5438:  00000 00000   00000 00000   00000 00000   00000 00000   00000 00000     00000 00000   00000 00000   00000 00000   00000 00000   00000 00000
5439:  00000 00000   00000 00000   00000 00000   00000 00000   00000 00000     00000 00000   00000 00000   00000 00000   00000 00000   00000 00000
5440:  00000 00000   00000 00000   00000 00000   00000 00000   00000 00000     00000 00000   00000 00000   00000 00000   00000 00000   00000 00000
5441:  00000 00000   00000 00000   00000 00000   00000 00000   00000 00000     00000 00000   00000 00000   00000 00000   00000 00000   00000 00000
5442:  00000 00000   00000 00000   00000 00000   00000 00000   00000 00000     00000 00000   00000 00000   00000 00000   00000 00000   00000 00000
5443:  00000 00000   00000 00000   00000 00000   00000 00000   00000 00000     00000 00000   00000 00000   00000 00000   00000 00000   00000 00000
5444:  00000 00000   00000 00000   00000 00000   00000 00000   00000 00000     00000 00000   00000 00000   00000 00000   00000 00000   00000 00000
5445:  00000 00000   00000 00000   00000 00000   00000 00000   00000 00000     00000 00000   00000 00000   00000 00000   00000 00000   00000 00000
5446:  00000 00000   00000 00000   00000 00000   00000 00000   00000 00000     00000 00000   00000 00000   00000 00000   00000 00000   00000 00000
5447:  00000 00000   00000 00000   00000 00000   00000 00000   00000 00000     00000 00000   00000 00000   00000 00000   00000 00000   00000 00000
5448:  00000 00000   00000 00000   00000 00000   00000 00000   00000 00000     00000 00000   00000 00000   00000 00000   00000 00000   00000 00000
5449:  00000 00000   00000 00000   00000 00000   00000 00000   00000 00000     00000 00000   00000 00000   00000 00000   00000 00000   00000 00000
```

```
5450:  00000 00000  00000 00000  00000 00000  00000 00000  00000 00000    00000 00000  00000 00000  00000 00000  00000 00000  00000 00000
5451:  00000 00000  00000 00000  00000 00000  00000 00000  00000 00000    00000 00000  00000 00000  00000 00000  00000 00000  00000 00000
5452:  00000 00000  00000 00000  00000 00000  00000 00000  00000 00000    00000 00000  00000 00000  00000 00000  00000 00000  00000 00000
5453:  00000 00000  00000 00000  00000 00000  00000 00000  00000 00000    00000 00000  00000 00000  00000 00000  00000 00000  00000 00000
5454:  00000 00000  00000 00000  00000 00000  00000 00000  00000 00000    00000 00000  00000 00000  00000 00000  00000 00000  00000 00000
5455:  00000 00000  00000 00000  00000 00000  00000 00000  00000 00000    00000 00000  00000 00000  00000 00000  00000 00000  00000 00000
5456:  00000 00000  00000 00000  00000 00000  00000 00000  00000 00000    00000 00000  00000 00000  00000 00000  00000 00000  00000 00000
5457:  00000 00000  00000 00000  00000 00000  00000 00000  00000 00000    00000 00000  00000 00000  00000 00000  00000 00000  00000 00000
5458:  00000 00000  00000 00000  00000 00000  00000 00000  00000 00000    00000 00000  00000 00000  00000 00000  00000 00000  00000 00000
5459:  00000 00000  00000 00000  00000 00000  00000 00000  00000 00000    00000 00000  00000 00000  00000 00000  00000 00000  00000 00000
5460:  00000 00000  00000 00000  00000 00000  00000 00000  00000 00000    00000 00000  00000 00000  00000 00000  00000 00000  00000 00000
5461:  00000 00000  00000 00000  00000 00000  00000 00000  00000 00000    00000 00000  00000 00000  00000 00000  00000 00000  00000 00000
5462:  00000 00000  00000 00000  00000 00000  00000 00000  00000 00000    00000 00000  00000 00000  00000 00000  00000 00000  00000 00000
5463:  00000 00000  00000 00000  00000 00000  00000 00000  00000 00000    00000 00000  00000 00000  00000 00000  00000 00000  00000 00000
5464:  00000 00000  00000 00000  00000 00000  00000 00000  00000 00000    00000 00000  00000 00000  00000 00000  00000 00000  00000 00000
5465:  00000 00000  00000 00000  00000 00000  00000 00000  00000 00000    00000 00000  00000 00000  00000 00000  00000 00000  00000 00000
5466:  00000 00000  00000 00000  00000 00000  00000 00000  00000 00000    00000 00000  00000 00000  00000 00000  00000 00000  00000 00000
5467:  00000 00000  00000 00000  00000 00000  00000 00000  00000 00000    00000 00000  00000 00000  00000 00000  00000 00000  00000 00000
5468:  00000 00000  00000 00000  00000 00000  00000 00000  00000 00000    00000 00000  00000 00000  00000 00000  00000 00000  00000 00000
5469:  00000 00000  00000 00000  00000 00000  00000 00000  00000 00000    00000 00000  00000 00000  00000 00000  00000 00000  00000 00000
5470:  00000 00000  00000 00000  00000 00000  00000 00000  00000 00000    00000 00000  00000 00000  00000 00000  00000 00000  00000 00000
5471:  00000 00000  00000 00000  00000 00000  00000 00000  00000 00000    00000 00000  00000 00000  00000 00000  00000 00000  00000 00000
5472:  00000 00000  00000 00000  00000 00000  00000 00000  00000 00000    00000 00000  00000 00000  00000 00000  00000 00000  00000 00000
5473:  00000 00000  00000 00000  00000 00000  00000 00000  00000 00000    00000 00000  00000 00000  00000 00000  00000 00000  00000 00000
5474:  00000 00000  00000 00000  00000 00000  00000 00000  00000 00000    00000 00000  00000 00000  00000 00000  00000 00000  00000 00000
5475:  00000 00000  00000 00000  00000 00000  00000 00000  00000 00000    00000 00000  00000 00000  00000 00000  00000 00000  00000 00000
5476:  00000 00000  00000 00000  00000 00000  00000 00000  00000 00000    00000 00000  00000 00000  00000 00000  00000 00000  00000 00000
5477:  00000 00000  00000 00000  00000 00000  00000 00000  00000 00000    00000 00000  00000 00000  00000 00000  00000 00000  00000 00000
5478:  00000 00000  00000 00000  00000 00000  00000 00000  00000 00000    00000 00000  00000 00000  00000 00000  00000 00000  00000 00000
5479:  00000 00000  00000 00000  00000 00000  00000 00000  00000 00000    00000 00000  00000 00000  00000 00000  00000 00000  00000 00000
5480:  00000 00000  00000 00000  00000 00000  00000 00000  00000 00000    00000 00000  00000 00000  00000 00000  00000 00000  00000 00000
5481:  00000 00000  00000 00000  00000 00000  00000 00000  00000 00000    00000 00000  00000 00000  00000 00000  00000 00000  00000 00000
5482:  00000 00000  00000 00000  00000 00000  00000 00000  00000 00000    00000 00000  00000 00000  00000 00000  00000 00000  00000 00000
5483:  00000 00000  00000 00000  00000 00000  00000 00000  00000 00000    00000 00000  00000 00000  00000 00000  00000 00000  00000 00000
5484:  00000 00000  00000 00000  00000 00000  00000 00000  00000 00000    00000 00000  00000 00000  00000 00000  00000 00000  00000 00000
5485:  00000 00000  00000 00000  00000 00000  00000 00000  00000 00000    00000 00000  00000 00000  00000 00000  00000 00000  00000 00000
5486:  00000 00000  00000 00000  00000 00000  00000 00000  00000 00000    00000 00000  00000 00000  00000 00000  00000 00000  00000 00000
5487:  00000 00000  00000 00000  00000 00000  00000 00000  00000 00000    00000 00000  00000 00000  00000 00000  00000 00000  00000 00000
5488:  00000 00000  00000 00000  00000 00000  00000 00000  00000 00000    00000 00000  00000 00000  00000 00000  00000 00000  00000 00000
5489:  00000 00000  00000 00000  00000 00000  00000 00000  00000 00000    00000 00000  00000 00000  00000 00000  00000 00000  00000 00000
5490:  00000 00000  00000 00000  00000 00000  00000 00000  00000 00000    00000 00000  00000 00000  00000 00000  00000 00000  00000 00000
5491:  00000 00000  00000 00000  00000 00000  00000 00000  00000 00000    00000 00000  00000 00000  00000 00000  00000 00000  00000 00000
5492:  00000 00000  00000 00000  00000 00000  00000 00000  00000 00000    00000 00000  00000 00000  00000 00000  00000 00000  00000 00000
5493:  00000 00000  00000 00000  00000 00000  00000 00000  00000 00000    00000 00000  00000 00000  00000 00000  00000 00000  00000 00000
5494:  00000 00000  00000 00000  00000 00000  00000 00000  00000 00000    00000 00000  00000 00000  00000 00000  00000 00000  00000 00000
5495:  00000 00000  00000 00000  00000 00000  00000 00000  00000 00000    00000 00000  00000 00000  00000 00000  00000 00000  00000 00000
5496:  00000 00000  00000 00000  00000 00000  00000 00000  00000 00000    00000 00000  00000 00000  00000 00000  00000 00000  00000 00000
5497:  00000 00000  00000 00000  00000 00000  00000 00000  00000 00000    00000 00000  00000 00000  00000 00000  00000 00000  00000 00000
5498:  00000 00000  00000 00000  00000 00000  00000 00000  00000 00000    00000 00000  00000 00000  00000 00000  00000 00000  00000 00000
5499:  00000 00000  00000 00000  00000 00000  00000 00000  00000 00000    00000 00000  00000 00000  00000 00000  00000 00000  00000 00000
```

```
5500:  00000 00000  00000 00000  00000 00000  00000 00000  00000 00000    00000 00000  00000 00000  00000 00000  00000 00000  00000 00000
5501:  00000 00000  00000 00000  00000 00000  00000 00000  00000 00000    00000 00000  00000 00000  00000 00000  00000 00000  00000 00000
5502:  00000 00000  00000 00000  00000 00000  00000 00000  00000 00000    00000 00000  00000 00000  00000 00000  00000 00000  00000 00000
5503:  00000 00000  00000 00000  00000 00000  00000 00000  00000 00000    00000 00000  00000 00000  00000 00000  00000 00000  00000 00000
5504:  00000 00000  00000 00000  00000 00000  00000 00000  00000 00000    00000 00000  00000 00000  00000 00000  00000 00000  00000 00000
5505:  00000 00000  00000 00000  00000 00000  00000 00000  00000 00000    00000 00000  00000 00000  00000 00000  00000 00000  00000 00000
5506:  00000 00000  00000 00000  00000 00000  00000 00000  00000 00000    00000 00000  00000 00000  00000 00000  00000 00000  00000 00000
5507:  00000 00000  00000 00000  00000 00000  00000 00000  00000 00000    00000 00000  00000 00000  00000 00000  00000 00000  00000 00000
5508:  00000 00000  00000 00000  00000 00000  00000 00000  00000 00000    00000 00000  00000 00000  00000 00000  00000 00000  00000 00000
5509:  00000 00000  00000 00000  00000 00000  00000 00000  00000 00000    00000 00000  00000 00000  00000 00000  00000 00000  00000 00000
5510:  00000 00000  00000 00000  00000 00000  00000 00000  00000 00000    00000 00000  00000 00000  00000 00000  00000 00000  00000 00000
5511:  00000 00000  00000 00000  00000 00000  00000 00000  00000 00000    00000 00000  00000 00000  00000 00000  00000 00000  00000 00000
5512:  00000 00000  00000 00000  00000 00000  00000 00000  00000 00000    00000 00000  00000 00000  00000 00000  00000 00000  00000 00000
5513:  00000 00000  00000 00000  00000 00000  00000 00000  00000 00000    00000 00000  00000 00000  00000 00000  00000 00000  00000 00000
5514:  00000 00000  00000 00000  00000 00000  00000 00000  00000 00000    00000 00000  00000 00000  00000 00000  00000 00000  00000 00000
5515:  00000 00000  00000 00000  00000 00000  00000 00000  00000 00000    00000 00000  00000 00000  00000 00000  00000 00000  00000 00000
5516:  00000 00000  00000 00000  00000 00000  00000 00000  00000 00000    00000 00000  00000 00000  00000 00000  00000 00000  00000 00000
5517:  00000 00000  00000 00000  00000 00000  00000 00000  00000 00000    00000 00000  00000 00000  00000 00000  00000 00000  00000 00000
5518:  00000 00000  00000 00000  00000 00000  00000 00000  00000 00000    00000 00000  00000 00000  00000 00000  00000 00000  00000 00000
5519:  00000 00000  00000 00000  00000 00000  00000 00000  00000 00000    00000 00000  00000 00000  00000 00000  00000 00000  00000 00000
5520:  00000 00000  00000 00000  00000 00000  00000 00000  00000 00000    00000 00000  00000 00000  00000 00000  00000 00000  00000 00000
5521:  00000 00000  00000 00000  00000 00000  00000 00000  00000 00000    00000 00000  00000 00000  00000 00000  00000 00000  00000 00000
5522:  00000 00000  00000 00000  00000 00000  00000 00000  00000 00000    00000 00000  00000 00000  00000 00000  00000 00000  00000 00000
5523:  00000 00000  00000 00000  00000 00000  00000 00000  00000 00000    00000 00000  00000 00000  00000 00000  00000 00000  00000 00000
5524:  00000 00000  00000 00000  00000 00000  00000 00000  00000 00000    00000 00000  00000 00000  00000 00000  00000 00000  00000 00000
5525:  00000 00000  00000 00000  00000 00000  00000 00000  00000 00000    00000 00000  00000 00000  00000 00000  00000 00000  00000 00000
5526:  00000 00000  00000 00000  00000 00000  00000 00000  00000 00000    00000 00000  00000 00000  00000 00000  00000 00000  00000 00000
5527:  00000 00000  00000 00000  00000 00000  00000 00000  00000 00000    00000 00000  00000 00000  00000 00000  00000 00000  00000 00000
5528:  00000 00000  00000 00000  00000 00000  00000 00000  00000 00000    00000 00000  00000 00000  00000 00000  00000 00000  00000 00000
5529:  00000 00000  00000 00000  00000 00000  00000 00000  00000 00000    00000 00000  00000 00000  00000 00000  00000 00000  00000 00000
5530:  00000 00000  00000 00000  00000 00000  00000 00000  00000 00000    00000 00000  00000 00000  00000 00000  00000 00000  00000 00000
5531:  00000 00000  00000 00000  00000 00000  00000 00000  00000 00000    00000 00000  00000 00000  00000 00000  00000 00000  00000 00000
5532:  00000 00000  00000 00000  00000 00000  00000 00000  00000 00000    00000 00000  00000 00000  00000 00000  00000 00000  00000 00000
5533:  00000 00000  00000 00000  00000 00000  00000 00000  00000 00000    00000 00000  00000 00000  00000 00000  00000 00000  00000 00000
5534:  00000 00000  00000 00000  00000 00000  00000 00000  00000 00000    00000 00000  00000 00000  00000 00000  00000 00000  00000 00000
5535:  00000 00000  00000 00000  00000 00000  00000 00000  00000 00000    00000 00000  00000 00000  00000 00000  00000 00000  00000 00000
5536:  00000 00000  00000 00000  00000 00000  00000 00000  00000 00000    00000 00000  00000 00000  00000 00000  00000 00000  00000 00000
5537:  00000 00000  00000 00000  00000 00000  00000 00000  00000 00000    00000 00000  00000 00000  00000 00000  00000 00000  00000 00000
5538:  00000 00000  00000 00000  00000 00000  00000 00000  00000 00000    00000 00000  00000 00000  00000 00000  00000 00000  00000 00000
5539:  00000 00000  00000 00000  00000 00000  00000 00000  00000 00000    00000 00000  00000 00000  00000 00000  00000 00000  00000 00000
5540:  00000 00000  00000 00000  00000 00000  00000 00000  00000 00000    00000 00000  00000 00000  00000 00000  00000 00000  00000 00000
5541:  00000 00000  00000 00000  00000 00000  00000 00000  00000 00000    00000 00000  00000 00000  00000 00000  00000 00000  00000 00000
5542:  00000 00000  00000 00000  00000 00000  00000 00000  00000 00000    00000 00000  00000 00000  00000 00000  00000 00000  00000 00000
5543:  00000 00000  00000 00000  00000 00000  00000 00000  00000 00000    00000 00000  00000 00000  00000 00000  00000 00000  00000 00000
5544:  00000 00000  00000 00000  00000 00000  00000 00000  00000 00000    00000 00000  00000 00000  00000 00000  00000 00000  00000 00000
5545:  00000 00000  00000 00000  00000 00000  00000 00000  00000 00000    00000 00000  00000 00000  00000 00000  00000 00000  00000 00000
5546:  00000 00000  00000 00000  00000 00000  00000 00000  00000 00000    00000 00000  00000 00000  00000 00000  00000 00000  00000 00000
5547:  00000 00000  00000 00000  00000 00000  00000 00000  00000 00000    00000 00000  00000 00000  00000 00000  00000 00000  00000 00000
5548:  00000 00000  00000 00000  00000 00000  00000 00000  00000 00000    00000 00000  00000 00000  00000 00000  00000 00000  00000 00000
5549:  00000 00000  00000 00000  00000 00000  00000 00000  00000 00000    00000 00000  00000 00000  00000 00000  00000 00000  00000 00000
```

```
5550:  00000 00000  00000 00000  00000 00000  00000 00000  00000 00000    00000 00000  00000 00000  00000 00000  00000 00000  00000 00000
5551:  00000 00000  00000 00000  00000 00000  00000 00000  00000 00000    00000 00000  00000 00000  00000 00000  00000 00000  00000 00000
5552:  00000 00000  00000 00000  00000 00000  00000 00000  00000 00000    00000 00000  00000 00000  00000 00000  00000 00000  00000 00000
5553:  00000 00000  00000 00000  00000 00000  00000 00000  00000 00000    00000 00000  00000 00000  00000 00000  00000 00000  00000 00000
5554:  00000 00000  00000 00000  00000 00000  00000 00000  00000 00000    00000 00000  00000 00000  00000 00000  00000 00000  00000 00000
5555:  00000 00000  00000 00000  00000 00000  00000 00000  00000 00000    00000 00000  00000 00000  00000 00000  00000 00000  00000 00000
5556:  00000 00000  00000 00000  00000 00000  00000 00000  00000 00000    00000 00000  00000 00000  00000 00000  00000 00000  00000 00000
5557:  00000 00000  00000 00000  00000 00000  00000 00000  00000 00000    00000 00000  00000 00000  00000 00000  00000 00000  00000 00000
5558:  00000 00000  00000 00000  00000 00000  00000 00000  00000 00000    00000 00000  00000 00000  00000 00000  00000 00000  00000 00000
5559:  00000 00000  00000 00000  00000 00000  00000 00000  00000 00000    00000 00000  00000 00000  00000 00000  00000 00000  00000 00000
5560:  00000 00000  00000 00000  00000 00000  00000 00000  00000 00000    00000 00000  00000 00000  00000 00000  00000 00000  00000 00000
5561:  00000 00000  00000 00000  00000 00000  00000 00000  00000 00000    00000 00000  00000 00000  00000 00000  00000 00000  00000 00000
5562:  00000 00000  00000 00000  00000 00000  00000 00000  00000 00000    00000 00000  00000 00000  00000 00000  00000 00000  00000 00000
5563:  00000 00000  00000 00000  00000 00000  00000 00000  00000 00000    00000 00000  00000 00000  00000 00000  00000 00000  00000 00000
5564:  00000 00000  00000 00000  00000 00000  00000 00000  00000 00000    00000 00000  00000 00000  00000 00000  00000 00000  00000 00000
5565:  00000 00000  00000 00000  00000 00000  00000 00000  00000 00000    00000 00000  00000 00000  00000 00000  00000 00000  00000 00000
5566:  00000 00000  00000 00000  00000 00000  00000 00000  00000 00000    00000 00000  00000 00000  00000 00000  00000 00000  00000 00000
5567:  00000 00000  00000 00000  00000 00000  00000 00000  00000 00000    00000 00000  00000 00000  00000 00000  00000 00000  00000 00000
5568:  00000 00000  00000 00000  00000 00000  00000 00000  00000 00000    00000 00000  00000 00000  00000 00000  00000 00000  00000 00000
5569:  00000 00000  00000 00000  00000 00000  00000 00000  00000 00000    00000 00000  00000 00000  00000 00000  00000 00000  00000 00000
5570:  00000 00000  00000 00000  00000 00000  00000 00000  00000 00000    00000 00000  00000 00000  00000 00000  00000 00000  00000 00000
5571:  00000 00000  00000 00000  00000 00000  00000 00000  00000 00000    00000 00000  00000 00000  00000 00000  00000 00000  00000 00000
5572:  00000 00000  00000 00000  00000 00000  00000 00000  00000 00000    00000 00000  00000 00000  00000 00000  00000 00000  00000 00000
5573:  00000 00000  00000 00000  00000 00000  00000 00000  00000 00000    00000 00000  00000 00000  00000 00000  00000 00000  00000 00000
5574:  00000 00000  00000 00000  00000 00000  00000 00000  00000 00000    00000 00000  00000 00000  00000 00000  00000 00000  00000 00000
5575:  00000 00000  00000 00000  00000 00000  00000 00000  00000 00000    00000 00000  00000 00000  00000 00000  00000 00000  00000 00000
5576:  00000 00000  00000 00000  00000 00000  00000 00000  00000 00000    00000 00000  00000 00000  00000 00000  00000 00000  00000 00000
5577:  00000 00000  00000 00000  00000 00000  00000 00000  00000 00000    00000 00000  00000 00000  00000 00000  00000 00000  00000 00000
5578:  00000 00000  00000 00000  00000 00000  00000 00000  00000 00000    00000 00000  00000 00000  00000 00000  00000 00000  00000 00000
5579:  00000 00000  00000 00000  00000 00000  00000 00000  00000 00000    00000 00000  00000 00000  00000 00000  00000 00000  00000 00000
5580:  00000 00000  00000 00000  00000 00000  00000 00000  00000 00000    00000 00000  00000 00000  00000 00000  00000 00000  00000 00000
5581:  00000 00000  00000 00000  00000 00000  00000 00000  00000 00000    00000 00000  00000 00000  00000 00000  00000 00000  00000 00000
5582:  00000 00000  00000 00000  00000 00000  00000 00000  00000 00000    00000 00000  00000 00000  00000 00000  00000 00000  00000 00000
5583:  00000 00000  00000 00000  00000 00000  00000 00000  00000 00000    00000 00000  00000 00000  00000 00000  00000 00000  00000 00000
5584:  00000 00000  00000 00000  00000 00000  00000 00000  00000 00000    00000 00000  00000 00000  00000 00000  00000 00000  00000 00000
5585:  00000 00000  00000 00000  00000 00000  00000 00000  00000 00000    00000 00000  00000 00000  00000 00000  00000 00000  00000 00000
5586:  00000 00000  00000 00000  00000 00000  00000 00000  00000 00000    00000 00000  00000 00000  00000 00000  00000 00000  00000 00000
5587:  00000 00000  00000 00000  00000 00000  00000 00000  00000 00000    00000 00000  00000 00000  00000 00000  00000 00000  00000 00000
5588:  00000 00000  00000 00000  00000 00000  00000 00000  00000 00000    00000 00000  00000 00000  00000 00000  00000 00000  00000 00000
5589:  00000 00000  00000 00000  00000 00000  00000 00000  00000 00000    00000 00000  00000 00000  00000 00000  00000 00000  00000 00000
5590:  00000 00000  00000 00000  00000 00000  00000 00000  00000 00000    00000 00000  00000 00000  00000 00000  00000 00000  00000 00000
5591:  00000 00000  00000 00000  00000 00000  00000 00000  00000 00000    00000 00000  00000 00000  00000 00000  00000 00000  00000 00000
5592:  00000 00000  00000 00000  00000 00000  00000 00000  00000 00000    00000 00000  00000 00000  00000 00000  00000 00000  00000 00000
5593:  00000 00000  00000 00000  00000 00000  00000 00000  00000 00000    00000 00000  00000 00000  00000 00000  00000 00000  00000 00000
5594:  00000 00000  00000 00000  00000 00000  00000 00000  00000 00000    00000 00000  00000 00000  00000 00000  00000 00000  00000 00000
5595:  00000 00000  00000 00000  00000 00000  00000 00000  00000 00000    00000 00000  00000 00000  00000 00000  00000 00000  00000 00000
5596:  00000 00000  00000 00000  00000 00000  00000 00000  00000 00000    00000 00000  00000 00000  00000 00000  00000 00000  00000 00000
5597:  00000 00000  00000 00000  00000 00000  00000 00000  00000 00000    00000 00000  00000 00000  00000 00000  00000 00000  00000 00000
5598:  00000 00000  00000 00000  00000 00000  00000 00000  00000 00000    00000 00000  00000 00000  00000 00000  00000 00000  00000 00000
5599:  00000 00000  00000 00000  00000 00000  00000 00000  00000 00000    00000 00000  00000 00000  00000 00000  00000 00000  00000 00000
```

```
5600:  00000 00000   00000 00000   00000 00000   00000 00000   00000 00000      00000 00000   00000 00000   00000 00000   00000 00000   00000 00000
5601:  00000 00000   00000 00000   00000 00000   00000 00000   00000 00000      00000 00000   00000 00000   00000 00000   00000 00000   00000 00000
5602:  00000 00000   00000 00000   00000 00000   00000 00000   00000 00000      00000 00000   00000 00000   00000 00000   00000 00000   00000 00000
5603:  00000 00000   00000 00000   00000 00000   00000 00000   00000 00000      00000 00000   00000 00000   00000 00000   00000 00000   00000 00000
5604:  00000 00000   00000 00000   00000 00000   00000 00000   00000 00000      00000 00000   00000 00000   00000 00000   00000 00000   00000 00000
5605:  00000 00000   00000 00000   00000 00000   00000 00000   00000 00000      00000 00000   00000 00000   00000 00000   00000 00000   00000 00000
5606:  00000 00000   00000 00000   00000 00000   00000 00000   00000 00000      00000 00000   00000 00000   00000 00000   00000 00000   00000 00000
5607:  00000 00000   00000 00000   00000 00000   00000 00000   00000 00000      00000 00000   00000 00000   00000 00000   00000 00000   00000 00000
5608:  00000 00000   00000 00000   00000 00000   00000 00000   00000 00000      00000 00000   00000 00000   00000 00000   00000 00000   00000 00000
5609:  00000 00000   00000 00000   00000 00000   00000 00000   00000 00000      00000 00000   00000 00000   00000 00000   00000 00000   00000 00000
5610:  00000 00000   00000 00000   00000 00000   00000 00000   00000 00000      00000 00000   00000 00000   00000 00000   00000 00000   00000 00000
5611:  00000 00000   00000 00000   00000 00000   00000 00000   00000 00000      00000 00000   00000 00000   00000 00000   00000 00000   00000 00000
5612:  00000 00000   00000 00000   00000 00000   00000 00000   00000 00000      00000 00000   00000 00000   00000 00000   00000 00000   00000 00000
5613:  00000 00000   00000 00000   00000 00000   00000 00000   00000 00000      00000 00000   00000 00000   00000 00000   00000 00000   00000 00000
5614:  00000 00000   00000 00000   00000 00000   00000 00000   00000 00000      00000 00000   00000 00000   00000 00000   00000 00000   00000 00000
5615:  00000 00000   00000 00000   00000 00000   00000 00000   00000 00000      00000 00000   00000 00000   00000 00000   00000 00000   00000 00000
5616:  00000 00000   00000 00000   00000 00000   00000 00000   00000 00000      00000 00000   00000 00000   00000 00000   00000 00000   00000 00000
5617:  00000 00000   00000 00000   00000 00000   00000 00000   00000 00000      00000 00000   00000 00000   00000 00000   00000 00000   00000 00000
5618:  00000 00000   00000 00000   00000 00000   00000 00000   00000 00000      00000 00000   00000 00000   00000 00000   00000 00000   00000 00000
5619:  00000 00000   00000 00000   00000 00000   00000 00000   00000 00000      00000 00000   00000 00000   00000 00000   00000 00000   00000 00000
5620:  00000 00000   00000 00000   00000 00000   00000 00000   00000 00000      00000 00000   00000 00000   00000 00000   00000 00000   00000 00000
5621:  00000 00000   00000 00000   00000 00000   00000 00000   00000 00000      00000 00000   00000 00000   00000 00000   00000 00000   00000 00000
5622:  00000 00000   00000 00000   00000 00000   00000 00000   00000 00000      00000 00000   00000 00000   00000 00000   00000 00000   00000 00000
5623:  00000 00000   00000 00000   00000 00000   00000 00000   00000 00000      00000 00000   00000 00000   00000 00000   00000 00000   00000 00000
5624:  00000 00000   00000 00000   00000 00000   00000 00000   00000 00000      00000 00000   00000 00000   00000 00000   00000 00000   00000 00000
5625:  00000 00000   00000 00000   00000 00000   00000 00000   00000 00000      00000 00000   00000 00000   00000 00000   00000 00000   00000 00000
5626:  00000 00000   00000 00000   00000 00000   00000 00000   00000 00000      00000 00000   00000 00000   00000 00000   00000 00000   00000 00000
5627:  00000 00000   00000 00000   00000 00000   00000 00000   00000 00000      00000 00000   00000 00000   00000 00000   00000 00000   00000 00000
5628:  00000 00000   00000 00000   00000 00000   00000 00000   00000 00000      00000 00000   00000 00000   00000 00000   00000 00000   00000 00000
5629:  00000 00000   00000 00000   00000 00000   00000 00000   00000 00000      00000 00000   00000 00000   00000 00000   00000 00000   00000 00000
5630:  00000 00000   00000 00000   00000 00000   00000 00000   00000 00000      00000 00000   00000 00000   00000 00000   00000 00000   00000 00000
5631:  00000 00000   00000 00000   00000 00000   00000 00000   00000 00000      00000 00000   00000 00000   00000 00000   00000 00000   00000 00000
5632:  00000 00000   00000 00000   00000 00000   00000 00000   00000 00000      00000 00000   00000 00000   00000 00000   00000 00000   00000 00000
5633:  00000 00000   00000 00000   00000 00000   00000 00000   00000 00000      00000 00000   00000 00000   00000 00000   00000 00000   00000 00000
5634:  00000 00000   00000 00000   00000 00000   00000 00000   00000 00000      00000 00000   00000 00000   00000 00000   00000 00000   00000 00000
5635:  00000 00000   00000 00000   00000 00000   00000 00000   00000 00000      00000 00000   00000 00000   00000 00000   00000 00000   00000 00000
5636:  00000 00000   00000 00000   00000 00000   00000 00000   00000 00000      00000 00000   00000 00000   00000 00000   00000 00000   00000 00000
5637:  00000 00000   00000 00000   00000 00000   00000 00000   00000 00000      00000 00000   00000 00000   00000 00000   00000 00000   00000 00000
5638:  00000 00000   00000 00000   00000 00000   00000 00000   00000 00000      00000 00000   00000 00000   00000 00000   00000 00000   00000 00000
5639:  00000 00000   00000 00000   00000 00000   00000 00000   00000 00000      00000 00000   00000 00000   00000 00000   00000 00000   00000 00000
5640:  00000 00000   00000 00000   00000 00000   00000 00000   00000 00000      00000 00000   00000 00000   00000 00000   00000 00000   00000 00000
5641:  00000 00000   00000 00000   00000 00000   00000 00000   00000 00000      00000 00000   00000 00000   00000 00000   00000 00000   00000 00000
5642:  00000 00000   00000 00000   00000 00000   00000 00000   00000 00000      00000 00000   00000 00000   00000 00000   00000 00000   00000 00000
5643:  00000 00000   00000 00000   00000 00000   00000 00000   00000 00000      00000 00000   00000 00000   00000 00000   00000 00000   00000 00000
5644:  00000 00000   00000 00000   00000 00000   00000 00000   00000 00000      00000 00000   00000 00000   00000 00000   00000 00000   00000 00000
5645:  00000 00000   00000 00000   00000 00000   00000 00000   00000 00000      00000 00000   00000 00000   00000 00000   00000 00000   00000 00000
5646:  00000 00000   00000 00000   00000 00000   00000 00000   00000 00000      00000 00000   00000 00000   00000 00000   00000 00000   00000 00000
5647:  00000 00000   00000 00000   00000 00000   00000 00000   00000 00000      00000 00000   00000 00000   00000 00000   00000 00000   00000 00000
5648:  00000 00000   00000 00000   00000 00000   00000 00000   00000 00000      00000 00000   00000 00000   00000 00000   00000 00000   00000 00000
5649:  00000 00000   00000 00000   00000 00000   00000 00000   00000 00000      00000 00000   00000 00000   00000 00000   00000 00000   00000 00000
```

```
5650:  00000 00000  00000 00000  00000 00000  00000 00000  00000 00000    00000 00000  00000 00000  00000 00000  00000 00000  00000 00000
5651:  00000 00000  00000 00000  00000 00000  00000 00000  00000 00000    00000 00000  00000 00000  00000 00000  00000 00000  00000 00000
5652:  00000 00000  00000 00000  00000 00000  00000 00000  00000 00000    00000 00000  00000 00000  00000 00000  00000 00000  00000 00000
5653:  00000 00000  00000 00000  00000 00000  00000 00000  00000 00000    00000 00000  00000 00000  00000 00000  00000 00000  00000 00000
5654:  00000 00000  00000 00000  00000 00000  00000 00000  00000 00000    00000 00000  00000 00000  00000 00000  00000 00000  00000 00000
5655:  00000 00000  00000 00000  00000 00000  00000 00000  00000 00000    00000 00000  00000 00000  00000 00000  00000 00000  00000 00000
5656:  00000 00000  00000 00000  00000 00000  00000 00000  00000 00000    00000 00000  00000 00000  00000 00000  00000 00000  00000 00000
5657:  00000 00000  00000 00000  00000 00000  00000 00000  00000 00000    00000 00000  00000 00000  00000 00000  00000 00000  00000 00000
5658:  00000 00000  00000 00000  00000 00000  00000 00000  00000 00000    00000 00000  00000 00000  00000 00000  00000 00000  00000 00000
5659:  00000 00000  00000 00000  00000 00000  00000 00000  00000 00000    00000 00000  00000 00000  00000 00000  00000 00000  00000 00000
5660:  00000 00000  00000 00000  00000 00000  00000 00000  00000 00000    00000 00000  00000 00000  00000 00000  00000 00000  00000 00000
5661:  00000 00000  00000 00000  00000 00000  00000 00000  00000 00000    00000 00000  00000 00000  00000 00000  00000 00000  00000 00000
5662:  00000 00000  00000 00000  00000 00000  00000 00000  00000 00000    00000 00000  00000 00000  00000 00000  00000 00000  00000 00000
5663:  00000 00000  00000 00000  00000 00000  00000 00000  00000 00000    00000 00000  00000 00000  00000 00000  00000 00000  00000 00000
5664:  00000 00000  00000 00000  00000 00000  00000 00000  00000 00000    00000 00000  00000 00000  00000 00000  00000 00000  00000 00000
5665:  00000 00000  00000 00000  00000 00000  00000 00000  00000 00000    00000 00000  00000 00000  00000 00000  00000 00000  00000 00000
5666:  00000 00000  00000 00000  00000 00000  00000 00000  00000 00000    00000 00000  00000 00000  00000 00000  00000 00000  00000 00000
5667:  00000 00000  00000 00000  00000 00000  00000 00000  00000 00000    00000 00000  00000 00000  00000 00000  00000 00000  00000 00000
5668:  00000 00000  00000 00000  00000 00000  00000 00000  00000 00000    00000 00000  00000 00000  00000 00000  00000 00000  00000 00000
5669:  00000 00000  00000 00000  00000 00000  00000 00000  00000 00000    00000 00000  00000 00000  00000 00000  00000 00000  00000 00000
5670:  00000 00000  00000 00000  00000 00000  00000 00000  00000 00000    00000 00000  00000 00000  00000 00000  00000 00000  00000 00000
5671:  00000 00000  00000 00000  00000 00000  00000 00000  00000 00000    00000 00000  00000 00000  00000 00000  00000 00000  00000 00000
5672:  00000 00000  00000 00000  00000 00000  00000 00000  00000 00000    00000 00000  00000 00000  00000 00000  00000 00000  00000 00000
5673:  00000 00000  00000 00000  00000 00000  00000 00000  00000 00000    00000 00000  00000 00000  00000 00000  00000 00000  00000 00000
5674:  00000 00000  00000 00000  00000 00000  00000 00000  00000 00000    00000 00000  00000 00000  00000 00000  00000 00000  00000 00000
5675:  00000 00000  00000 00000  00000 00000  00000 00000  00000 00000    00000 00000  00000 00000  00000 00000  00000 00000  00000 00000
5676:  00000 00000  00000 00000  00000 00000  00000 00000  00000 00000    00000 00000  00000 00000  00000 00000  00000 00000  00000 00000
5677:  00000 00000  00000 00000  00000 00000  00000 00000  00000 00000    00000 00000  00000 00000  00000 00000  00000 00000  00000 00000
5678:  00000 00000  00000 00000  00000 00000  00000 00000  00000 00000    00000 00000  00000 00000  00000 00000  00000 00000  00000 00000
5679:  00000 00000  00000 00000  00000 00000  00000 00000  00000 00000    00000 00000  00000 00000  00000 00000  00000 00000  00000 00000
5680:  00000 00000  00000 00000  00000 00000  00000 00000  00000 00000    00000 00000  00000 00000  00000 00000  00000 00000  00000 00000
5681:  00000 00000  00000 00000  00000 00000  00000 00000  00000 00000    00000 00000  00000 00000  00000 00000  00000 00000  00000 00000
5682:  00000 00000  00000 00000  00000 00000  00000 00000  00000 00000    00000 00000  00000 00000  00000 00000  00000 00000  00000 00000
5683:  00000 00000  00000 00000  00000 00000  00000 00000  00000 00000    00000 00000  00000 00000  00000 00000  00000 00000  00000 00000
5684:  00000 00000  00000 00000  00000 00000  00000 00000  00000 00000    00000 00000  00000 00000  00000 00000  00000 00000  00000 00000
5685:  00000 00000  00000 00000  00000 00000  00000 00000  00000 00000    00000 00000  00000 00000  00000 00000  00000 00000  00000 00000
5686:  00000 00000  00000 00000  00000 00000  00000 00000  00000 00000    00000 00000  00000 00000  00000 00000  00000 00000  00000 00000
5687:  00000 00000  00000 00000  00000 00000  00000 00000  00000 00000    00000 00000  00000 00000  00000 00000  00000 00000  00000 00000
5688:  00000 00000  00000 00000  00000 00000  00000 00000  00000 00000    00000 00000  00000 00000  00000 00000  00000 00000  00000 00000
5689:  00000 00000  00000 00000  00000 00000  00000 00000  00000 00000    00000 00000  00000 00000  00000 00000  00000 00000  00000 00000
5690:  00000 00000  00000 00000  00000 00000  00000 00000  00000 00000    00000 00000  00000 00000  00000 00000  00000 00000  00000 00000
5691:  00000 00000  00000 00000  00000 00000  00000 00000  00000 00000    00000 00000  00000 00000  00000 00000  00000 00000  00000 00000
5692:  00000 00000  00000 00000  00000 00000  00000 00000  00000 00000    00000 00000  00000 00000  00000 00000  00000 00000  00000 00000
5693:  00000 00000  00000 00000  00000 00000  00000 00000  00000 00000    00000 00000  00000 00000  00000 00000  00000 00000  00000 00000
5694:  00000 00000  00000 00000  00000 00000  00000 00000  00000 00000    00000 00000  00000 00000  00000 00000  00000 00000  00000 00000
5695:  00000 00000  00000 00000  00000 00000  00000 00000  00000 00000    00000 00000  00000 00000  00000 00000  00000 00000  00000 00000
5696:  00000 00000  00000 00000  00000 00000  00000 00000  00000 00000    00000 00000  00000 00000  00000 00000  00000 00000  00000 00000
5697:  00000 00000  00000 00000  00000 00000  00000 00000  00000 00000    00000 00000  00000 00000  00000 00000  00000 00000  00000 00000
5698:  00000 00000  00000 00000  00000 00000  00000 00000  00000 00000    00000 00000  00000 00000  00000 00000  00000 00000  00000 00000
5699:  00000 00000  00000 00000  00000 00000  00000 00000  00000 00000    00000 00000  00000 00000  00000 00000  00000 00000  00000 00000
```

```
5700:  00000 00000  00000 00000  00000 00000  00000 00000  00000 00000    00000 00000  00000 00000  00000 00000  00000 00000  00000 00000
5701:  00000 00000  00000 00000  00000 00000  00000 00000  00000 00000    00000 00000  00000 00000  00000 00000  00000 00000  00000 00000
5702:  00000 00000  00000 00000  00000 00000  00000 00000  00000 00000    00000 00000  00000 00000  00000 00000  00000 00000  00000 00000
5703:  00000 00000  00000 00000  00000 00000  00000 00000  00000 00000    00000 00000  00000 00000  00000 00000  00000 00000  00000 00000
5704:  00000 00000  00000 00000  00000 00000  00000 00000  00000 00000    00000 00000  00000 00000  00000 00000  00000 00000  00000 00000
5705:  00000 00000  00000 00000  00000 00000  00000 00000  00000 00000    00000 00000  00000 00000  00000 00000  00000 00000  00000 00000
5706:  00000 00000  00000 00000  00000 00000  00000 00000  00000 00000    00000 00000  00000 00000  00000 00000  00000 00000  00000 00000
5707:  00000 00000  00000 00000  00000 00000  00000 00000  00000 00000    00000 00000  00000 00000  00000 00000  00000 00000  00000 00000
5708:  00000 00000  00000 00000  00000 00000  00000 00000  00000 00000    00000 00000  00000 00000  00000 00000  00000 00000  00000 00000
5709:  00000 00000  00000 00000  00000 00000  00000 00000  00000 00000    00000 00000  00000 00000  00000 00000  00000 00000  00000 00000
5710:  00000 00000  00000 00000  00000 00000  00000 00000  00000 00000    00000 00000  00000 00000  00000 00000  00000 00000  00000 00000
5711:  00000 00000  00000 00000  00000 00000  00000 00000  00000 00000    00000 00000  00000 00000  00000 00000  00000 00000  00000 00000
5712:  00000 00000  00000 00000  00000 00000  00000 00000  00000 00000    00000 00000  00000 00000  00000 00000  00000 00000  00000 00000
5713:  00000 00000  00000 00000  00000 00000  00000 00000  00000 00000    00000 00000  00000 00000  00000 00000  00000 00000  00000 00000
5714:  00000 00000  00000 00000  00000 00000  00000 00000  00000 00000    00000 00000  00000 00000  00000 00000  00000 00000  00000 00000
5715:  00000 00000  00000 00000  00000 00000  00000 00000  00000 00000    00000 00000  00000 00000  00000 00000  00000 00000  00000 00000
5716:  00000 00000  00000 00000  00000 00000  00000 00000  00000 00000    00000 00000  00000 00000  00000 00000  00000 00000  00000 00000
5717:  00000 00000  00000 00000  00000 00000  00000 00000  00000 00000    00000 00000  00000 00000  00000 00000  00000 00000  00000 00000
5718:  00000 00000  00000 00000  00000 00000  00000 00000  00000 00000    00000 00000  00000 00000  00000 00000  00000 00000  00000 00000
5719:  00000 00000  00000 00000  00000 00000  00000 00000  00000 00000    00000 00000  00000 00000  00000 00000  00000 00000  00000 00000
5720:  00000 00000  00000 00000  00000 00000  00000 00000  00000 00000    00000 00000  00000 00000  00000 00000  00000 00000  00000 00000
5721:  00000 00000  00000 00000  00000 00000  00000 00000  00000 00000    00000 00000  00000 00000  00000 00000  00000 00000  00000 00000
5722:  00000 00000  00000 00000  00000 00000  00000 00000  00000 00000    00000 00000  00000 00000  00000 00000  00000 00000  00000 00000
5723:  00000 00000  00000 00000  00000 00000  00000 00000  00000 00000    00000 00000  00000 00000  00000 00000  00000 00000  00000 00000
5724:  00000 00000  00000 00000  00000 00000  00000 00000  00000 00000    00000 00000  00000 00000  00000 00000  00000 00000  00000 00000
5725:  00000 00000  00000 00000  00000 00000  00000 00000  00000 00000    00000 00000  00000 00000  00000 00000  00000 00000  00000 00000
5726:  00000 00000  00000 00000  00000 00000  00000 00000  00000 00000    00000 00000  00000 00000  00000 00000  00000 00000  00000 00000
5727:  00000 00000  00000 00000  00000 00000  00000 00000  00000 00000    00000 00000  00000 00000  00000 00000  00000 00000  00000 00000
5728:  00000 00000  00000 00000  00000 00000  00000 00000  00000 00000    00000 00000  00000 00000  00000 00000  00000 00000  00000 00000
5729:  00000 00000  00000 00000  00000 00000  00000 00000  00000 00000    00000 00000  00000 00000  00000 00000  00000 00000  00000 00000
5730:  00000 00000  00000 00000  00000 00000  00000 00000  00000 00000    00000 00000  00000 00000  00000 00000  00000 00000  00000 00000
5731:  00000 00000  00000 00000  00000 00000  00000 00000  00000 00000    00000 00000  00000 00000  00000 00000  00000 00000  00000 00000
5732:  00000 00000  00000 00000  00000 00000  00000 00000  00000 00000    00000 00000  00000 00000  00000 00000  00000 00000  00000 00000
5733:  00000 00000  00000 00000  00000 00000  00000 00000  00000 00000    00000 00000  00000 00000  00000 00000  00000 00000  00000 00000
5734:  00000 00000  00000 00000  00000 00000  00000 00000  00000 00000    00000 00000  00000 00000  00000 00000  00000 00000  00000 00000
5735:  00000 00000  00000 00000  00000 00000  00000 00000  00000 00000    00000 00000  00000 00000  00000 00000  00000 00000  00000 00000
5736:  00000 00000  00000 00000  00000 00000  00000 00000  00000 00000    00000 00000  00000 00000  00000 00000  00000 00000  00000 00000
5737:  00000 00000  00000 00000  00000 00000  00000 00000  00000 00000    00000 00000  00000 00000  00000 00000  00000 00000  00000 00000
5738:  00000 00000  00000 00000  00000 00000  00000 00000  00000 00000    00000 00000  00000 00000  00000 00000  00000 00000  00000 00000
5739:  00000 00000  00000 00000  00000 00000  00000 00000  00000 00000    00000 00000  00000 00000  00000 00000  00000 00000  00000 00000
5740:  00000 00000  00000 00000  00000 00000  00000 00000  00000 00000    00000 00000  00000 00000  00000 00000  00000 00000  00000 00000
5741:  00000 00000  00000 00000  00000 00000  00000 00000  00000 00000    00000 00000  00000 00000  00000 00000  00000 00000  00000 00000
5742:  00000 00000  00000 00000  00000 00000  00000 00000  00000 00000    00000 00000  00000 00000  00000 00000  00000 00000  00000 00000
5743:  00000 00000  00000 00000  00000 00000  00000 00000  00000 00000    00000 00000  00000 00000  00000 00000  00000 00000  00000 00000
5744:  00000 00000  00000 00000  00000 00000  00000 00000  00000 00000    00000 00000  00000 00000  00000 00000  00000 00000  00000 00000
5745:  00000 00000  00000 00000  00000 00000  00000 00000  00000 00000    00000 00000  00000 00000  00000 00000  00000 00000  00000 00000
5746:  00000 00000  00000 00000  00000 00000  00000 00000  00000 00000    00000 00000  00000 00000  00000 00000  00000 00000  00000 00000
5747:  00000 00000  00000 00000  00000 00000  00000 00000  00000 00000    00000 00000  00000 00000  00000 00000  00000 00000  00000 00000
5748:  00000 00000  00000 00000  00000 00000  00000 00000  00000 00000    00000 00000  00000 00000  00000 00000  00000 00000  00000 00000
5749:  00000 00000  00000 00000  00000 00000  00000 00000  00000 00000    00000 00000  00000 00000  00000 00000  00000 00000  00000 00000
```

```
5750:  00000 00000  00000 00000  00000 00000  00000 00000  00000 00000    00000 00000  00000 00000  00000 00000  00000 00000  00000 00000
5751:  00000 00000  00000 00000  00000 00000  00000 00000  00000 00000    00000 00000  00000 00000  00000 00000  00000 00000  00000 00000
5752:  00000 00000  00000 00000  00000 00000  00000 00000  00000 00000    00000 00000  00000 00000  00000 00000  00000 00000  00000 00000
5753:  00000 00000  00000 00000  00000 00000  00000 00000  00000 00000    00000 00000  00000 00000  00000 00000  00000 00000  00000 00000
5754:  00000 00000  00000 00000  00000 00000  00000 00000  00000 00000    00000 00000  00000 00000  00000 00000  00000 00000  00000 00000
5755:  00000 00000  00000 00000  00000 00000  00000 00000  00000 00000    00000 00000  00000 00000  00000 00000  00000 00000  00000 00000
5756:  00000 00000  00000 00000  00000 00000  00000 00000  00000 00000    00000 00000  00000 00000  00000 00000  00000 00000  00000 00000
5757:  00000 00000  00000 00000  00000 00000  00000 00000  00000 00000    00000 00000  00000 00000  00000 00000  00000 00000  00000 00000
5758:  00000 00000  00000 00000  00000 00000  00000 00000  00000 00000    00000 00000  00000 00000  00000 00000  00000 00000  00000 00000
5759:  00000 00000  00000 00000  00000 00000  00000 00000  00000 00000    00000 00000  00000 00000  00000 00000  00000 00000  00000 00000
5760:  00000 00000  00000 00000  00000 00000  00000 00000  00000 00000    00000 00000  00000 00000  00000 00000  00000 00000  00000 00000
5761:  00000 00000  00000 00000  00000 00000  00000 00000  00000 00000    00000 00000  00000 00000  00000 00000  00000 00000  00000 00000
5762:  00000 00000  00000 00000  00000 00000  00000 00000  00000 00000    00000 00000  00000 00000  00000 00000  00000 00000  00000 00000
5763:  00000 00000  00000 00000  00000 00000  00000 00000  00000 00000    00000 00000  00000 00000  00000 00000  00000 00000  00000 00000
5764:  00000 00000  00000 00000  00000 00000  00000 00000  00000 00000    00000 00000  00000 00000  00000 00000  00000 00000  00000 00000
5765:  00000 00000  00000 00000  00000 00000  00000 00000  00000 00000    00000 00000  00000 00000  00000 00000  00000 00000  00000 00000
5766:  00000 00000  00000 00000  00000 00000  00000 00000  00000 00000    00000 00000  00000 00000  00000 00000  00000 00000  00000 00000
5767:  00000 00000  00000 00000  00000 00000  00000 00000  00000 00000    00000 00000  00000 00000  00000 00000  00000 00000  00000 00000
5768:  00000 00000  00000 00000  00000 00000  00000 00000  00000 00000    00000 00000  00000 00000  00000 00000  00000 00000  00000 00000
5769:  00000 00000  00000 00000  00000 00000  00000 00000  00000 00000    00000 00000  00000 00000  00000 00000  00000 00000  00000 00000
5770:  00000 00000  00000 00000  00000 00000  00000 00000  00000 00000    00000 00000  00000 00000  00000 00000  00000 00000  00000 00000
5771:  00000 00000  00000 00000  00000 00000  00000 00000  00000 00000    00000 00000  00000 00000  00000 00000  00000 00000  00000 00000
5772:  00000 00000  00000 00000  00000 00000  00000 00000  00000 00000    00000 00000  00000 00000  00000 00000  00000 00000  00000 00000
5773:  00000 00000  00000 00000  00000 00000  00000 00000  00000 00000    00000 00000  00000 00000  00000 00000  00000 00000  00000 00000
5774:  00000 00000  00000 00000  00000 00000  00000 00000  00000 00000    00000 00000  00000 00000  00000 00000  00000 00000  00000 00000
5775:  00000 00000  00000 00000  00000 00000  00000 00000  00000 00000    00000 00000  00000 00000  00000 00000  00000 00000  00000 00000
5776:  00000 00000  00000 00000  00000 00000  00000 00000  00000 00000    00000 00000  00000 00000  00000 00000  00000 00000  00000 00000
5777:  00000 00000  00000 00000  00000 00000  00000 00000  00000 00000    00000 00000  00000 00000  00000 00000  00000 00000  00000 00000
5778:  00000 00000  00000 00000  00000 00000  00000 00000  00000 00000    00000 00000  00000 00000  00000 00000  00000 00000  00000 00000
5779:  00000 00000  00000 00000  00000 00000  00000 00000  00000 00000    00000 00000  00000 00000  00000 00000  00000 00000  00000 00000
5780:  00000 00000  00000 00000  00000 00000  00000 00000  00000 00000    00000 00000  00000 00000  00000 00000  00000 00000  00000 00000
5781:  00000 00000  00000 00000  00000 00000  00000 00000  00000 00000    00000 00000  00000 00000  00000 00000  00000 00000  00000 00000
5782:  00000 00000  00000 00000  00000 00000  00000 00000  00000 00000    00000 00000  00000 00000  00000 00000  00000 00000  00000 00000
5783:  00000 00000  00000 00000  00000 00000  00000 00000  00000 00000    00000 00000  00000 00000  00000 00000  00000 00000  00000 00000
5784:  00000 00000  00000 00000  00000 00000  00000 00000  00000 00000    00000 00000  00000 00000  00000 00000  00000 00000  00000 00000
5785:  00000 00000  00000 00000  00000 00000  00000 00000  00000 00000    00000 00000  00000 00000  00000 00000  00000 00000  00000 00000
5786:  00000 00000  00000 00000  00000 00000  00000 00000  00000 00000    00000 00000  00000 00000  00000 00000  00000 00000  00000 00000
5787:  00000 00000  00000 00000  00000 00000  00000 00000  00000 00000    00000 00000  00000 00000  00000 00000  00000 00000  00000 00000
5788:  00000 00000  00000 00000  00000 00000  00000 00000  00000 00000    00000 00000  00000 00000  00000 00000  00000 00000  00000 00000
5789:  00000 00000  00000 00000  00000 00000  00000 00000  00000 00000    00000 00000  00000 00000  00000 00000  00000 00000  00000 00000
5790:  00000 00000  00000 00000  00000 00000  00000 00000  00000 00000    00000 00000  00000 00000  00000 00000  00000 00000  00000 00000
5791:  00000 00000  00000 00000  00000 00000  00000 00000  00000 00000    00000 00000  00000 00000  00000 00000  00000 00000  00000 00000
5792:  00000 00000  00000 00000  00000 00000  00000 00000  00000 00000    00000 00000  00000 00000  00000 00000  00000 00000  00000 00000
5793:  00000 00000  00000 00000  00000 00000  00000 00000  00000 00000    00000 00000  00000 00000  00000 00000  00000 00000  00000 00000
5794:  00000 00000  00000 00000  00000 00000  00000 00000  00000 00000    00000 00000  00000 00000  00000 00000  00000 00000  00000 00000
5795:  00000 00000  00000 00000  00000 00000  00000 00000  00000 00000    00000 00000  00000 00000  00000 00000  00000 00000  00000 00000
5796:  00000 00000  00000 00000  00000 00000  00000 00000  00000 00000    00000 00000  00000 00000  00000 00000  00000 00000  00000 00000
5797:  00000 00000  00000 00000  00000 00000  00000 00000  00000 00000    00000 00000  00000 00000  00000 00000  00000 00000  00000 00000
5798:  00000 00000  00000 00000  00000 00000  00000 00000  00000 00000    00000 00000  00000 00000  00000 00000  00000 00000  00000 00000
5799:  00000 00000  00000 00000  00000 00000  00000 00000  00000 00000    00000 00000  00000 00000  00000 00000  00000 00000  00000 00000
```

```
5800:  00000 00000   00000 00000   00000 00000   00000 00000   00000 00000     00000 00000   00000 00000   00000 00000   00000 00000   00000 00000
5801:  00000 00000   00000 00000   00000 00000   00000 00000   00000 00000     00000 00000   00000 00000   00000 00000   00000 00000   00000 00000
5802:  00000 00000   00000 00000   00000 00000   00000 00000   00000 00000     00000 00000   00000 00000   00000 00000   00000 00000   00000 00000
5803:  00000 00000   00000 00000   00000 00000   00000 00000   00000 00000     00000 00000   00000 00000   00000 00000   00000 00000   00000 00000
5804:  00000 00000   00000 00000   00000 00000   00000 00000   00000 00000     00000 00000   00000 00000   00000 00000   00000 00000   00000 00000
5805:  00000 00000   00000 00000   00000 00000   00000 00000   00000 00000     00000 00000   00000 00000   00000 00000   00000 00000   00000 00000
5806:  00000 00000   00000 00000   00000 00000   00000 00000   00000 00000     00000 00000   00000 00000   00000 00000   00000 00000   00000 00000
5807:  00000 00000   00000 00000   00000 00000   00000 00000   00000 00000     00000 00000   00000 00000   00000 00000   00000 00000   00000 00000
5808:  00000 00000   00000 00000   00000 00000   00000 00000   00000 00000     00000 00000   00000 00000   00000 00000   00000 00000   00000 00000
5809:  00000 00000   00000 00000   00000 00000   00000 00000   00000 00000     00000 00000   00000 00000   00000 00000   00000 00000   00000 00000
5810:  00000 00000   00000 00000   00000 00000   00000 00000   00000 00000     00000 00000   00000 00000   00000 00000   00000 00000   00000 00000
5811:  00000 00000   00000 00000   00000 00000   00000 00000   00000 00000     00000 00000   00000 00000   00000 00000   00000 00000   00000 00000
5812:  00000 00000   00000 00000   00000 00000   00000 00000   00000 00000     00000 00000   00000 00000   00000 00000   00000 00000   00000 00000
5813:  00000 00000   00000 00000   00000 00000   00000 00000   00000 00000     00000 00000   00000 00000   00000 00000   00000 00000   00000 00000
5814:  00000 00000   00000 00000   00000 00000   00000 00000   00000 00000     00000 00000   00000 00000   00000 00000   00000 00000   00000 00000
5815:  00000 00000   00000 00000   00000 00000   00000 00000   00000 00000     00000 00000   00000 00000   00000 00000   00000 00000   00000 00000
5816:  00000 00000   00000 00000   00000 00000   00000 00000   00000 00000     00000 00000   00000 00000   00000 00000   00000 00000   00000 00000
5817:  00000 00000   00000 00000   00000 00000   00000 00000   00000 00000     00000 00000   00000 00000   00000 00000   00000 00000   00000 00000
5818:  00000 00000   00000 00000   00000 00000   00000 00000   00000 00000     00000 00000   00000 00000   00000 00000   00000 00000   00000 00000
5819:  00000 00000   00000 00000   00000 00000   00000 00000   00000 00000     00000 00000   00000 00000   00000 00000   00000 00000   00000 00000
5820:  00000 00000   00000 00000   00000 00000   00000 00000   00000 00000     00000 00000   00000 00000   00000 00000   00000 00000   00000 00000
5821:  00000 00000   00000 00000   00000 00000   00000 00000   00000 00000     00000 00000   00000 00000   00000 00000   00000 00000   00000 00000
5822:  00000 00000   00000 00000   00000 00000   00000 00000   00000 00000     00000 00000   00000 00000   00000 00000   00000 00000   00000 00000
5823:  00000 00000   00000 00000   00000 00000   00000 00000   00000 00000     00000 00000   00000 00000   00000 00000   00000 00000   00000 00000
5824:  00000 00000   00000 00000   00000 00000   00000 00000   00000 00000     00000 00000   00000 00000   00000 00000   00000 00000   00000 00000
5825:  00000 00000   00000 00000   00000 00000   00000 00000   00000 00000     00000 00000   00000 00000   00000 00000   00000 00000   00000 00000
5826:  00000 00000   00000 00000   00000 00000   00000 00000   00000 00000     00000 00000   00000 00000   00000 00000   00000 00000   00000 00000
5827:  00000 00000   00000 00000   00000 00000   00000 00000   00000 00000     00000 00000   00000 00000   00000 00000   00000 00000   00000 00000
5828:  00000 00000   00000 00000   00000 00000   00000 00000   00000 00000     00000 00000   00000 00000   00000 00000   00000 00000   00000 00000
5829:  00000 00000   00000 00000   00000 00000   00000 00000   00000 00000     00000 00000   00000 00000   00000 00000   00000 00000   00000 00000
5830:  00000 00000   00000 00000   00000 00000   00000 00000   00000 00000     00000 00000   00000 00000   00000 00000   00000 00000   00000 00000
5831:  00000 00000   00000 00000   00000 00000   00000 00000   00000 00000     00000 00000   00000 00000   00000 00000   00000 00000   00000 00000
5832:  00000 00000   00000 00000   00000 00000   00000 00000   00000 00000     00000 00000   00000 00000   00000 00000   00000 00000   00000 00000
5833:  00000 00000   00000 00000   00000 00000   00000 00000   00000 00000     00000 00000   00000 00000   00000 00000   00000 00000   00000 00000
5834:  00000 00000   00000 00000   00000 00000   00000 00000   00000 00000     00000 00000   00000 00000   00000 00000   00000 00000   00000 00000
5835:  00000 00000   00000 00000   00000 00000   00000 00000   00000 00000     00000 00000   00000 00000   00000 00000   00000 00000   00000 00000
5836:  00000 00000   00000 00000   00000 00000   00000 00000   00000 00000     00000 00000   00000 00000   00000 00000   00000 00000   00000 00000
5837:  00000 00000   00000 00000   00000 00000   00000 00000   00000 00000     00000 00000   00000 00000   00000 00000   00000 00000   00000 00000
5838:  00000 00000   00000 00000   00000 00000   00000 00000   00000 00000     00000 00000   00000 00000   00000 00000   00000 00000   00000 00000
5839:  00000 00000   00000 00000   00000 00000   00000 00000   00000 00000     00000 00000   00000 00000   00000 00000   00000 00000   00000 00000
5840:  00000 00000   00000 00000   00000 00000   00000 00000   00000 00000     00000 00000   00000 00000   00000 00000   00000 00000   00000 00000
5841:  00000 00000   00000 00000   00000 00000   00000 00000   00000 00000     00000 00000   00000 00000   00000 00000   00000 00000   00000 00000
5842:  00000 00000   00000 00000   00000 00000   00000 00000   00000 00000     00000 00000   00000 00000   00000 00000   00000 00000   00000 00000
5843:  00000 00000   00000 00000   00000 00000   00000 00000   00000 00000     00000 00000   00000 00000   00000 00000   00000 00000   00000 00000
5844:  00000 00000   00000 00000   00000 00000   00000 00000   00000 00000     00000 00000   00000 00000   00000 00000   00000 00000   00000 00000
5845:  00000 00000   00000 00000   00000 00000   00000 00000   00000 00000     00000 00000   00000 00000   00000 00000   00000 00000   00000 00000
5846:  00000 00000   00000 00000   00000 00000   00000 00000   00000 00000     00000 00000   00000 00000   00000 00000   00000 00000   00000 00000
5847:  00000 00000   00000 00000   00000 00000   00000 00000   00000 00000     00000 00000   00000 00000   00000 00000   00000 00000   00000 00000
5848:  00000 00000   00000 00000   00000 00000   00000 00000   00000 00000     00000 00000   00000 00000   00000 00000   00000 00000   00000 00000
5849:  00000 00000   00000 00000   00000 00000   00000 00000   00000 00000     00000 00000   00000 00000   00000 00000   00000 00000   00000 00000
```

```
5850: 00000 00000  00000 00000  00000 00000  00000 00000  00000 00000    00000 00000  00000 00000  00000 00000  00000 00000  00000 00000
5851: 00000 00000  00000 00000  00000 00000  00000 00000  00000 00000    00000 00000  00000 00000  00000 00000  00000 00000  00000 00000
5852: 00000 00000  00000 00000  00000 00000  00000 00000  00000 00000    00000 00000  00000 00000  00000 00000  00000 00000  00000 00000
5853: 00000 00000  00000 00000  00000 00000  00000 00000  00000 00000    00000 00000  00000 00000  00000 00000  00000 00000  00000 00000
5854: 00000 00000  00000 00000  00000 00000  00000 00000  00000 00000    00000 00000  00000 00000  00000 00000  00000 00000  00000 00000
5855: 00000 00000  00000 00000  00000 00000  00000 00000  00000 00000    00000 00000  00000 00000  00000 00000  00000 00000  00000 00000
5856: 00000 00000  00000 00000  00000 00000  00000 00000  00000 00000    00000 00000  00000 00000  00000 00000  00000 00000  00000 00000
5857: 00000 00000  00000 00000  00000 00000  00000 00000  00000 00000    00000 00000  00000 00000  00000 00000  00000 00000  00000 00000
5858: 00000 00000  00000 00000  00000 00000  00000 00000  00000 00000    00000 00000  00000 00000  00000 00000  00000 00000  00000 00000
5859: 00000 00000  00000 00000  00000 00000  00000 00000  00000 00000    00000 00000  00000 00000  00000 00000  00000 00000  00000 00000
5860: 00000 00000  00000 00000  00000 00000  00000 00000  00000 00000    00000 00000  00000 00000  00000 00000  00000 00000  00000 00000
5861: 00000 00000  00000 00000  00000 00000  00000 00000  00000 00000    00000 00000  00000 00000  00000 00000  00000 00000  00000 00000
5862: 00000 00000  00000 00000  00000 00000  00000 00000  00000 00000    00000 00000  00000 00000  00000 00000  00000 00000  00000 00000
5863: 00000 00000  00000 00000  00000 00000  00000 00000  00000 00000    00000 00000  00000 00000  00000 00000  00000 00000  00000 00000
5864: 00000 00000  00000 00000  00000 00000  00000 00000  00000 00000    00000 00000  00000 00000  00000 00000  00000 00000  00000 00000
5865: 00000 00000  00000 00000  00000 00000  00000 00000  00000 00000    00000 00000  00000 00000  00000 00000  00000 00000  00000 00000
5866: 00000 00000  00000 00000  00000 00000  00000 00000  00000 00000    00000 00000  00000 00000  00000 00000  00000 00000  00000 00000
5867: 00000 00000  00000 00000  00000 00000  00000 00000  00000 00000    00000 00000  00000 00000  00000 00000  00000 00000  00000 00000
5868: 00000 00000  00000 00000  00000 00000  00000 00000  00000 00000    00000 00000  00000 00000  00000 00000  00000 00000  00000 00000
5869: 00000 00000  00000 00000  00000 00000  00000 00000  00000 00000    00000 00000  00000 00000  00000 00000  00000 00000  00000 00000
5870: 00000 00000  00000 00000  00000 00000  00000 00000  00000 00000    00000 00000  00000 00000  00000 00000  00000 00000  00000 00000
5871: 00000 00000  00000 00000  00000 00000  00000 00000  00000 00000    00000 00000  00000 00000  00000 00000  00000 00000  00000 00000
5872: 00000 00000  00000 00000  00000 00000  00000 00000  00000 00000    00000 00000  00000 00000  00000 00000  00000 00000  00000 00000
5873: 00000 00000  00000 00000  00000 00000  00000 00000  00000 00000    00000 00000  00000 00000  00000 00000  00000 00000  00000 00000
5874: 00000 00000  00000 00000  00000 00000  00000 00000  00000 00000    00000 00000  00000 00000  00000 00000  00000 00000  00000 00000
5875: 00000 00000  00000 00000  00000 00000  00000 00000  00000 00000    00000 00000  00000 00000  00000 00000  00000 00000  00000 00000
5876: 00000 00000  00000 00000  00000 00000  00000 00000  00000 00000    00000 00000  00000 00000  00000 00000  00000 00000  00000 00000
5877: 00000 00000  00000 00000  00000 00000  00000 00000  00000 00000    00000 00000  00000 00000  00000 00000  00000 00000  00000 00000
5878: 00000 00000  00000 00000  00000 00000  00000 00000  00000 00000    00000 00000  00000 00000  00000 00000  00000 00000  00000 00000
5879: 00000 00000  00000 00000  00000 00000  00000 00000  00000 00000    00000 00000  00000 00000  00000 00000  00000 00000  00000 00000
5880: 00000 00000  00000 00000  00000 00000  00000 00000  00000 00000    00000 00000  00000 00000  00000 00000  00000 00000  00000 00000
5881: 00000 00000  00000 00000  00000 00000  00000 00000  00000 00000    00000 00000  00000 00000  00000 00000  00000 00000  00000 00000
5882: 00000 00000  00000 00000  00000 00000  00000 00000  00000 00000    00000 00000  00000 00000  00000 00000  00000 00000  00000 00000
5883: 00000 00000  00000 00000  00000 00000  00000 00000  00000 00000    00000 00000  00000 00000  00000 00000  00000 00000  00000 00000
5884: 00000 00000  00000 00000  00000 00000  00000 00000  00000 00000    00000 00000  00000 00000  00000 00000  00000 00000  00000 00000
5885: 00000 00000  00000 00000  00000 00000  00000 00000  00000 00000    00000 00000  00000 00000  00000 00000  00000 00000  00000 00000
5886: 00000 00000  00000 00000  00000 00000  00000 00000  00000 00000    00000 00000  00000 00000  00000 00000  00000 00000  00000 00000
5887: 00000 00000  00000 00000  00000 00000  00000 00000  00000 00000    00000 00000  00000 00000  00000 00000  00000 00000  00000 00000
5888: 00000 00000  00000 00000  00000 00000  00000 00000  00000 00000    00000 00000  00000 00000  00000 00000  00000 00000  00000 00000
5889: 00000 00000  00000 00000  00000 00000  00000 00000  00000 00000    00000 00000  00000 00000  00000 00000  00000 00000  00000 00000
5890: 00000 00000  00000 00000  00000 00000  00000 00000  00000 00000    00000 00000  00000 00000  00000 00000  00000 00000  00000 00000
5891: 00000 00000  00000 00000  00000 00000  00000 00000  00000 00000    00000 00000  00000 00000  00000 00000  00000 00000  00000 00000
5892: 00000 00000  00000 00000  00000 00000  00000 00000  00000 00000    00000 00000  00000 00000  00000 00000  00000 00000  00000 00000
5893: 00000 00000  00000 00000  00000 00000  00000 00000  00000 00000    00000 00000  00000 00000  00000 00000  00000 00000  00000 00000
5894: 00000 00000  00000 00000  00000 00000  00000 00000  00000 00000    00000 00000  00000 00000  00000 00000  00000 00000  00000 00000
5895: 00000 00000  00000 00000  00000 00000  00000 00000  00000 00000    00000 00000  00000 00000  00000 00000  00000 00000  00000 00000
5896: 00000 00000  00000 00000  00000 00000  00000 00000  00000 00000    00000 00000  00000 00000  00000 00000  00000 00000  00000 00000
5897: 00000 00000  00000 00000  00000 00000  00000 00000  00000 00000    00000 00000  00000 00000  00000 00000  00000 00000  00000 00000
5898: 00000 00000  00000 00000  00000 00000  00000 00000  00000 00000    00000 00000  00000 00000  00000 00000  00000 00000  00000 00000
5899: 00000 00000  00000 00000  00000 00000  00000 00000  00000 00000    00000 00000  00000 00000  00000 00000  00000 00000  00000 00000
```

```
5900:  00000 00000  00000 00000  00000 00000  00000 00000  00000 00000    00000 00000  00000 00000  00000 00000  00000 00000  00000 00000
5901:  00000 00000  00000 00000  00000 00000  00000 00000  00000 00000    00000 00000  00000 00000  00000 00000  00000 00000  00000 00000
5902:  00000 00000  00000 00000  00000 00000  00000 00000  00000 00000    00000 00000  00000 00000  00000 00000  00000 00000  00000 00000
5903:  00000 00000  00000 00000  00000 00000  00000 00000  00000 00000    00000 00000  00000 00000  00000 00000  00000 00000  00000 00000
5904:  00000 00000  00000 00000  00000 00000  00000 00000  00000 00000    00000 00000  00000 00000  00000 00000  00000 00000  00000 00000
5905:  00000 00000  00000 00000  00000 00000  00000 00000  00000 00000    00000 00000  00000 00000  00000 00000  00000 00000  00000 00000
5906:  00000 00000  00000 00000  00000 00000  00000 00000  00000 00000    00000 00000  00000 00000  00000 00000  00000 00000  00000 00000
5907:  00000 00000  00000 00000  00000 00000  00000 00000  00000 00000    00000 00000  00000 00000  00000 00000  00000 00000  00000 00000
5908:  00000 00000  00000 00000  00000 00000  00000 00000  00000 00000    00000 00000  00000 00000  00000 00000  00000 00000  00000 00000
5909:  00000 00000  00000 00000  00000 00000  00000 00000  00000 00000    00000 00000  00000 00000  00000 00000  00000 00000  00000 00000
5910:  00000 00000  00000 00000  00000 00000  00000 00000  00000 00000    00000 00000  00000 00000  00000 00000  00000 00000  00000 00000
5911:  00000 00000  00000 00000  00000 00000  00000 00000  00000 00000    00000 00000  00000 00000  00000 00000  00000 00000  00000 00000
5912:  00000 00000  00000 00000  00000 00000  00000 00000  00000 00000    00000 00000  00000 00000  00000 00000  00000 00000  00000 00000
5913:  00000 00000  00000 00000  00000 00000  00000 00000  00000 00000    00000 00000  00000 00000  00000 00000  00000 00000  00000 00000
5914:  00000 00000  00000 00000  00000 00000  00000 00000  00000 00000    00000 00000  00000 00000  00000 00000  00000 00000  00000 00000
5915:  00000 00000  00000 00000  00000 00000  00000 00000  00000 00000    00000 00000  00000 00000  00000 00000  00000 00000  00000 00000
5916:  00000 00000  00000 00000  00000 00000  00000 00000  00000 00000    00000 00000  00000 00000  00000 00000  00000 00000  00000 00000
5917:  00000 00000  00000 00000  00000 00000  00000 00000  00000 00000    00000 00000  00000 00000  00000 00000  00000 00000  00000 00000
5918:  00000 00000  00000 00000  00000 00000  00000 00000  00000 00000    00000 00000  00000 00000  00000 00000  00000 00000  00000 00000
5919:  00000 00000  00000 00000  00000 00000  00000 00000  00000 00000    00000 00000  00000 00000  00000 00000  00000 00000  00000 00000
5920:  00000 00000  00000 00000  00000 00000  00000 00000  00000 00000    00000 00000  00000 00000  00000 00000  00000 00000  00000 00000
5921:  00000 00000  00000 00000  00000 00000  00000 00000  00000 00000    00000 00000  00000 00000  00000 00000  00000 00000  00000 00000
5922:  00000 00000  00000 00000  00000 00000  00000 00000  00000 00000    00000 00000  00000 00000  00000 00000  00000 00000  00000 00000
5923:  00000 00000  00000 00000  00000 00000  00000 00000  00000 00000    00000 00000  00000 00000  00000 00000  00000 00000  00000 00000
5924:  00000 00000  00000 00000  00000 00000  00000 00000  00000 00000    00000 00000  00000 00000  00000 00000  00000 00000  00000 00000
5925:  00000 00000  00000 00000  00000 00000  00000 00000  00000 00000    00000 00000  00000 00000  00000 00000  00000 00000  00000 00000
5926:  00000 00000  00000 00000  00000 00000  00000 00000  00000 00000    00000 00000  00000 00000  00000 00000  00000 00000  00000 00000
5927:  00000 00000  00000 00000  00000 00000  00000 00000  00000 00000    00000 00000  00000 00000  00000 00000  00000 00000  00000 00000
5928:  00000 00000  00000 00000  00000 00000  00000 00000  00000 00000    00000 00000  00000 00000  00000 00000  00000 00000  00000 00000
5929:  00000 00000  00000 00000  00000 00000  00000 00000  00000 00000    00000 00000  00000 00000  00000 00000  00000 00000  00000 00000
5930:  00000 00000  00000 00000  00000 00000  00000 00000  00000 00000    00000 00000  00000 00000  00000 00000  00000 00000  00000 00000
5931:  00000 00000  00000 00000  00000 00000  00000 00000  00000 00000    00000 00000  00000 00000  00000 00000  00000 00000  00000 00000
5932:  00000 00000  00000 00000  00000 00000  00000 00000  00000 00000    00000 00000  00000 00000  00000 00000  00000 00000  00000 00000
5933:  00000 00000  00000 00000  00000 00000  00000 00000  00000 00000    00000 00000  00000 00000  00000 00000  00000 00000  00000 00000
5934:  00000 00000  00000 00000  00000 00000  00000 00000  00000 00000    00000 00000  00000 00000  00000 00000  00000 00000  00000 00000
5935:  00000 00000  00000 00000  00000 00000  00000 00000  00000 00000    00000 00000  00000 00000  00000 00000  00000 00000  00000 00000
5936:  00000 00000  00000 00000  00000 00000  00000 00000  00000 00000    00000 00000  00000 00000  00000 00000  00000 00000  00000 00000
5937:  00000 00000  00000 00000  00000 00000  00000 00000  00000 00000    00000 00000  00000 00000  00000 00000  00000 00000  00000 00000
5938:  00000 00000  00000 00000  00000 00000  00000 00000  00000 00000    00000 00000  00000 00000  00000 00000  00000 00000  00000 00000
5939:  00000 00000  00000 00000  00000 00000  00000 00000  00000 00000    00000 00000  00000 00000  00000 00000  00000 00000  00000 00000
5940:  00000 00000  00000 00000  00000 00000  00000 00000  00000 00000    00000 00000  00000 00000  00000 00000  00000 00000  00000 00000
5941:  00000 00000  00000 00000  00000 00000  00000 00000  00000 00000    00000 00000  00000 00000  00000 00000  00000 00000  00000 00000
5942:  00000 00000  00000 00000  00000 00000  00000 00000  00000 00000    00000 00000  00000 00000  00000 00000  00000 00000  00000 00000
5943:  00000 00000  00000 00000  00000 00000  00000 00000  00000 00000    00000 00000  00000 00000  00000 00000  00000 00000  00000 00000
5944:  00000 00000  00000 00000  00000 00000  00000 00000  00000 00000    00000 00000  00000 00000  00000 00000  00000 00000  00000 00000
5945:  00000 00000  00000 00000  00000 00000  00000 00000  00000 00000    00000 00000  00000 00000  00000 00000  00000 00000  00000 00000
5946:  00000 00000  00000 00000  00000 00000  00000 00000  00000 00000    00000 00000  00000 00000  00000 00000  00000 00000  00000 00000
5947:  00000 00000  00000 00000  00000 00000  00000 00000  00000 00000    00000 00000  00000 00000  00000 00000  00000 00000  00000 00000
5948:  00000 00000  00000 00000  00000 00000  00000 00000  00000 00000    00000 00000  00000 00000  00000 00000  00000 00000  00000 00000
5949:  00000 00000  00000 00000  00000 00000  00000 00000  00000 00000    00000 00000  00000 00000  00000 00000  00000 00000  00000 00000
```

```
5950:  00000 00000  00000 00000  00000 00000  00000 00000  00000 00000    00000 00000  00000 00000  00000 00000  00000 00000  00000 00000
5951:  00000 00000  00000 00000  00000 00000  00000 00000  00000 00000    00000 00000  00000 00000  00000 00000  00000 00000  00000 00000
5952:  00000 00000  00000 00000  00000 00000  00000 00000  00000 00000    00000 00000  00000 00000  00000 00000  00000 00000  00000 00000
5953:  00000 00000  00000 00000  00000 00000  00000 00000  00000 00000    00000 00000  00000 00000  00000 00000  00000 00000  00000 00000
5954:  00000 00000  00000 00000  00000 00000  00000 00000  00000 00000    00000 00000  00000 00000  00000 00000  00000 00000  00000 00000
5955:  00000 00000  00000 00000  00000 00000  00000 00000  00000 00000    00000 00000  00000 00000  00000 00000  00000 00000  00000 00000
5956:  00000 00000  00000 00000  00000 00000  00000 00000  00000 00000    00000 00000  00000 00000  00000 00000  00000 00000  00000 00000
5957:  00000 00000  00000 00000  00000 00000  00000 00000  00000 00000    00000 00000  00000 00000  00000 00000  00000 00000  00000 00000
5958:  00000 00000  00000 00000  00000 00000  00000 00000  00000 00000    00000 00000  00000 00000  00000 00000  00000 00000  00000 00000
5959:  00000 00000  00000 00000  00000 00000  00000 00000  00000 00000    00000 00000  00000 00000  00000 00000  00000 00000  00000 00000
5960:  00000 00000  00000 00000  00000 00000  00000 00000  00000 00000    00000 00000  00000 00000  00000 00000  00000 00000  00000 00000
5961:  00000 00000  00000 00000  00000 00000  00000 00000  00000 00000    00000 00000  00000 00000  00000 00000  00000 00000  00000 00000
5962:  00000 00000  00000 00000  00000 00000  00000 00000  00000 00000    00000 00000  00000 00000  00000 00000  00000 00000  00000 00000
5963:  00000 00000  00000 00000  00000 00000  00000 00000  00000 00000    00000 00000  00000 00000  00000 00000  00000 00000  00000 00000
5964:  00000 00000  00000 00000  00000 00000  00000 00000  00000 00000    00000 00000  00000 00000  00000 00000  00000 00000  00000 00000
5965:  00000 00000  00000 00000  00000 00000  00000 00000  00000 00000    00000 00000  00000 00000  00000 00000  00000 00000  00000 00000
5966:  00000 00000  00000 00000  00000 00000  00000 00000  00000 00000    00000 00000  00000 00000  00000 00000  00000 00000  00000 00000
5967:  00000 00000  00000 00000  00000 00000  00000 00000  00000 00000    00000 00000  00000 00000  00000 00000  00000 00000  00000 00000
5968:  00000 00000  00000 00000  00000 00000  00000 00000  00000 00000    00000 00000  00000 00000  00000 00000  00000 00000  00000 00000
5969:  00000 00000  00000 00000  00000 00000  00000 00000  00000 00000    00000 00000  00000 00000  00000 00000  00000 00000  00000 00000
5970:  00000 00000  00000 00000  00000 00000  00000 00000  00000 00000    00000 00000  00000 00000  00000 00000  00000 00000  00000 00000
5971:  00000 00000  00000 00000  00000 00000  00000 00000  00000 00000    00000 00000  00000 00000  00000 00000  00000 00000  00000 00000
5972:  00000 00000  00000 00000  00000 00000  00000 00000  00000 00000    00000 00000  00000 00000  00000 00000  00000 00000  00000 00000
5973:  00000 00000  00000 00000  00000 00000  00000 00000  00000 00000    00000 00000  00000 00000  00000 00000  00000 00000  00000 00000
5974:  00000 00000  00000 00000  00000 00000  00000 00000  00000 00000    00000 00000  00000 00000  00000 00000  00000 00000  00000 00000
5975:  00000 00000  00000 00000  00000 00000  00000 00000  00000 00000    00000 00000  00000 00000  00000 00000  00000 00000  00000 00000
5976:  00000 00000  00000 00000  00000 00000  00000 00000  00000 00000    00000 00000  00000 00000  00000 00000  00000 00000  00000 00000
5977:  00000 00000  00000 00000  00000 00000  00000 00000  00000 00000    00000 00000  00000 00000  00000 00000  00000 00000  00000 00000
5978:  00000 00000  00000 00000  00000 00000  00000 00000  00000 00000    00000 00000  00000 00000  00000 00000  00000 00000  00000 00000
5979:  00000 00000  00000 00000  00000 00000  00000 00000  00000 00000    00000 00000  00000 00000  00000 00000  00000 00000  00000 00000
5980:  00000 00000  00000 00000  00000 00000  00000 00000  00000 00000    00000 00000  00000 00000  00000 00000  00000 00000  00000 00000
5981:  00000 00000  00000 00000  00000 00000  00000 00000  00000 00000    00000 00000  00000 00000  00000 00000  00000 00000  00000 00000
5982:  00000 00000  00000 00000  00000 00000  00000 00000  00000 00000    00000 00000  00000 00000  00000 00000  00000 00000  00000 00000
5983:  00000 00000  00000 00000  00000 00000  00000 00000  00000 00000    00000 00000  00000 00000  00000 00000  00000 00000  00000 00000
5984:  00000 00000  00000 00000  00000 00000  00000 00000  00000 00000    00000 00000  00000 00000  00000 00000  00000 00000  00000 00000
5985:  00000 00000  00000 00000  00000 00000  00000 00000  00000 00000    00000 00000  00000 00000  00000 00000  00000 00000  00000 00000
5986:  00000 00000  00000 00000  00000 00000  00000 00000  00000 00000    00000 00000  00000 00000  00000 00000  00000 00000  00000 00000
5987:  00000 00000  00000 00000  00000 00000  00000 00000  00000 00000    00000 00000  00000 00000  00000 00000  00000 00000  00000 00000
5988:  00000 00000  00000 00000  00000 00000  00000 00000  00000 00000    00000 00000  00000 00000  00000 00000  00000 00000  00000 00000
5989:  00000 00000  00000 00000  00000 00000  00000 00000  00000 00000    00000 00000  00000 00000  00000 00000  00000 00000  00000 00000
5990:  00000 00000  00000 00000  00000 00000  00000 00000  00000 00000    00000 00000  00000 00000  00000 00000  00000 00000  00000 00000
5991:  00000 00000  00000 00000  00000 00000  00000 00000  00000 00000    00000 00000  00000 00000  00000 00000  00000 00000  00000 00000
5992:  00000 00000  00000 00000  00000 00000  00000 00000  00000 00000    00000 00000  00000 00000  00000 00000  00000 00000  00000 00000
5993:  00000 00000  00000 00000  00000 00000  00000 00000  00000 00000    00000 00000  00000 00000  00000 00000  00000 00000  00000 00000
5994:  00000 00000  00000 00000  00000 00000  00000 00000  00000 00000    00000 00000  00000 00000  00000 00000  00000 00000  00000 00000
5995:  00000 00000  00000 00000  00000 00000  00000 00000  00000 00000    00000 00000  00000 00000  00000 00000  00000 00000  00000 00000
5996:  00000 00000  00000 00000  00000 00000  00000 00000  00000 00000    00000 00000  00000 00000  00000 00000  00000 00000  00000 00000
5997:  00000 00000  00000 00000  00000 00000  00000 00000  00000 00000    00000 00000  00000 00000  00000 00000  00000 00000  00000 00000
5998:  00000 00000  00000 00000  00000 00000  00000 00000  00000 00000    00000 00000  00000 00000  00000 00000  00000 00000  00000 00000
5999:  00000 00000  00000 00000  00000 00000  00000 00000  00000 00000    00000 00000  00000 00000  00000 00000  00000 00000  00000 00000
```

```
6000:  00000 00000  00000 00000  00000 00000  00000 00000  00000 00000    00000 00000  00000 00000  00000 00000  00000 00000  00000 00000
6001:  00000 00000  00000 00000  00000 00000  00000 00000  00000 00000    00000 00000  00000 00000  00000 00000  00000 00000  00000 00000
6002:  00000 00000  00000 00000  00000 00000  00000 00000  00000 00000    00000 00000  00000 00000  00000 00000  00000 00000  00000 00000
6003:  00000 00000  00000 00000  00000 00000  00000 00000  00000 00000    00000 00000  00000 00000  00000 00000  00000 00000  00000 00000
6004:  00000 00000  00000 00000  00000 00000  00000 00000  00000 00000    00000 00000  00000 00000  00000 00000  00000 00000  00000 00000
6005:  00000 00000  00000 00000  00000 00000  00000 00000  00000 00000    00000 00000  00000 00000  00000 00000  00000 00000  00000 00000
6006:  00000 00000  00000 00000  00000 00000  00000 00000  00000 00000    00000 00000  00000 00000  00000 00000  00000 00000  00000 00000
6007:  00000 00000  00000 00000  00000 00000  00000 00000  00000 00000    00000 00000  00000 00000  00000 00000  00000 00000  00000 00000
6008:  00000 00000  00000 00000  00000 00000  00000 00000  00000 00000    00000 00000  00000 00000  00000 00000  00000 00000  00000 00000
6009:  00000 00000  00000 00000  00000 00000  00000 00000  00000 00000    00000 00000  00000 00000  00000 00000  00000 00000  00000 00000
6010:  00000 00000  00000 00000  00000 00000  00000 00000  00000 00000    00000 00000  00000 00000  00000 00000  00000 00000  00000 00000
6011:  00000 00000  00000 00000  00000 00000  00000 00000  00000 00000    00000 00000  00000 00000  00000 00000  00000 00000  00000 00000
6012:  00000 00000  00000 00000  00000 00000  00000 00000  00000 00000    00000 00000  00000 00000  00000 00000  00000 00000  00000 00000
6013:  00000 00000  00000 00000  00000 00000  00000 00000  00000 00000    00000 00000  00000 00000  00000 00000  00000 00000  00000 00000
6014:  00000 00000  00000 00000  00000 00000  00000 00000  00000 00000    00000 00000  00000 00000  00000 00000  00000 00000  00000 00000
6015:  00000 00000  00000 00000  00000 00000  00000 00000  00000 00000    00000 00000  00000 00000  00000 00000  00000 00000  00000 00000
6016:  00000 00000  00000 00000  00000 00000  00000 00000  00000 00000    00000 00000  00000 00000  00000 00000  00000 00000  00000 00000
6017:  00000 00000  00000 00000  00000 00000  00000 00000  00000 00000    00000 00000  00000 00000  00000 00000  00000 00000  00000 00000
6018:  00000 00000  00000 00000  00000 00000  00000 00000  00000 00000    00000 00000  00000 00000  00000 00000  00000 00000  00000 00000
6019:  00000 00000  00000 00000  00000 00000  00000 00000  00000 00000    00000 00000  00000 00000  00000 00000  00000 00000  00000 00000
6020:  00000 00000  00000 00000  00000 00000  00000 00000  00000 00000    00000 00000  00000 00000  00000 00000  00000 00000  00000 00000
6021:  00000 00000  00000 00000  00000 00000  00000 00000  00000 00000    00000 00000  00000 00000  00000 00000  00000 00000  00000 00000
6022:  00000 00000  00000 00000  00000 00000  00000 00000  00000 00000    00000 00000  00000 00000  00000 00000  00000 00000  00000 00000
6023:  00000 00000  00000 00000  00000 00000  00000 00000  00000 00000    00000 00000  00000 00000  00000 00000  00000 00000  00000 00000
6024:  00000 00000  00000 00000  00000 00000  00000 00000  00000 00000    00000 00000  00000 00000  00000 00000  00000 00000  00000 00000
6025:  00000 00000  00000 00000  00000 00000  00000 00000  00000 00000    00000 00000  00000 00000  00000 00000  00000 00000  00000 00000
6026:  00000 00000  00000 00000  00000 00000  00000 00000  00000 00000    00000 00000  00000 00000  00000 00000  00000 00000  00000 00000
6027:  00000 00000  00000 00000  00000 00000  00000 00000  00000 00000    00000 00000  00000 00000  00000 00000  00000 00000  00000 00000
6028:  00000 00000  00000 00000  00000 00000  00000 00000  00000 00000    00000 00000  00000 00000  00000 00000  00000 00000  00000 00000
6029:  00000 00000  00000 00000  00000 00000  00000 00000  00000 00000    00000 00000  00000 00000  00000 00000  00000 00000  00000 00000
6030:  00000 00000  00000 00000  00000 00000  00000 00000  00000 00000    00000 00000  00000 00000  00000 00000  00000 00000  00000 00000
6031:  00000 00000  00000 00000  00000 00000  00000 00000  00000 00000    00000 00000  00000 00000  00000 00000  00000 00000  00000 00000
6032:  00000 00000  00000 00000  00000 00000  00000 00000  00000 00000    00000 00000  00000 00000  00000 00000  00000 00000  00000 00000
6033:  00000 00000  00000 00000  00000 00000  00000 00000  00000 00000    00000 00000  00000 00000  00000 00000  00000 00000  00000 00000
6034:  00000 00000  00000 00000  00000 00000  00000 00000  00000 00000    00000 00000  00000 00000  00000 00000  00000 00000  00000 00000
6035:  00000 00000  00000 00000  00000 00000  00000 00000  00000 00000    00000 00000  00000 00000  00000 00000  00000 00000  00000 00000
6036:  00000 00000  00000 00000  00000 00000  00000 00000  00000 00000    00000 00000  00000 00000  00000 00000  00000 00000  00000 00000
6037:  00000 00000  00000 00000  00000 00000  00000 00000  00000 00000    00000 00000  00000 00000  00000 00000  00000 00000  00000 00000
6038:  00000 00000  00000 00000  00000 00000  00000 00000  00000 00000    00000 00000  00000 00000  00000 00000  00000 00000  00000 00000
6039:  00000 00000  00000 00000  00000 00000  00000 00000  00000 00000    00000 00000  00000 00000  00000 00000  00000 00000  00000 00000
6040:  00000 00000  00000 00000  00000 00000  00000 00000  00000 00000    00000 00000  00000 00000  00000 00000  00000 00000  00000 00000
6041:  00000 00000  00000 00000  00000 00000  00000 00000  00000 00000    00000 00000  00000 00000  00000 00000  00000 00000  00000 00000
6042:  00000 00000  00000 00000  00000 00000  00000 00000  00000 00000    00000 00000  00000 00000  00000 00000  00000 00000  00000 00000
6043:  00000 00000  00000 00000  00000 00000  00000 00000  00000 00000    00000 00000  00000 00000  00000 00000  00000 00000  00000 00000
6044:  00000 00000  00000 00000  00000 00000  00000 00000  00000 00000    00000 00000  00000 00000  00000 00000  00000 00000  00000 00000
6045:  00000 00000  00000 00000  00000 00000  00000 00000  00000 00000    00000 00000  00000 00000  00000 00000  00000 00000  00000 00000
6046:  00000 00000  00000 00000  00000 00000  00000 00000  00000 00000    00000 00000  00000 00000  00000 00000  00000 00000  00000 00000
6047:  00000 00000  00000 00000  00000 00000  00000 00000  00000 00000    00000 00000  00000 00000  00000 00000  00000 00000  00000 00000
6048:  00000 00000  00000 00000  00000 00000  00000 00000  00000 00000    00000 00000  00000 00000  00000 00000  00000 00000  00000 00000
6049:  00000 00000  00000 00000  00000 00000  00000 00000  00000 00000    00000 00000  00000 00000  00000 00000  00000 00000  00000 00000
```

```
6050:  00000 00000  00000 00000  00000 00000  00000 00000  00000 00000    00000 00000  00000 00000  00000 00000  00000 00000  00000 00000
6051:  00000 00000  00000 00000  00000 00000  00000 00000  00000 00000    00000 00000  00000 00000  00000 00000  00000 00000  00000 00000
6052:  00000 00000  00000 00000  00000 00000  00000 00000  00000 00000    00000 00000  00000 00000  00000 00000  00000 00000  00000 00000
6053:  00000 00000  00000 00000  00000 00000  00000 00000  00000 00000    00000 00000  00000 00000  00000 00000  00000 00000  00000 00000
6054:  00000 00000  00000 00000  00000 00000  00000 00000  00000 00000    00000 00000  00000 00000  00000 00000  00000 00000  00000 00000
6055:  00000 00000  00000 00000  00000 00000  00000 00000  00000 00000    00000 00000  00000 00000  00000 00000  00000 00000  00000 00000
6056:  00000 00000  00000 00000  00000 00000  00000 00000  00000 00000    00000 00000  00000 00000  00000 00000  00000 00000  00000 00000
6057:  00000 00000  00000 00000  00000 00000  00000 00000  00000 00000    00000 00000  00000 00000  00000 00000  00000 00000  00000 00000
6058:  00000 00000  00000 00000  00000 00000  00000 00000  00000 00000    00000 00000  00000 00000  00000 00000  00000 00000  00000 00000
6059:  00000 00000  00000 00000  00000 00000  00000 00000  00000 00000    00000 00000  00000 00000  00000 00000  00000 00000  00000 00000
6060:  00000 00000  00000 00000  00000 00000  00000 00000  00000 00000    00000 00000  00000 00000  00000 00000  00000 00000  00000 00000
6061:  00000 00000  00000 00000  00000 00000  00000 00000  00000 00000    00000 00000  00000 00000  00000 00000  00000 00000  00000 00000
6062:  00000 00000  00000 00000  00000 00000  00000 00000  00000 00000    00000 00000  00000 00000  00000 00000  00000 00000  00000 00000
6063:  00000 00000  00000 00000  00000 00000  00000 00000  00000 00000    00000 00000  00000 00000  00000 00000  00000 00000  00000 00000
6064:  00000 00000  00000 00000  00000 00000  00000 00000  00000 00000    00000 00000  00000 00000  00000 00000  00000 00000  00000 00000
6065:  00000 00000  00000 00000  00000 00000  00000 00000  00000 00000    00000 00000  00000 00000  00000 00000  00000 00000  00000 00000
6066:  00000 00000  00000 00000  00000 00000  00000 00000  00000 00000    00000 00000  00000 00000  00000 00000  00000 00000  00000 00000
6067:  00000 00000  00000 00000  00000 00000  00000 00000  00000 00000    00000 00000  00000 00000  00000 00000  00000 00000  00000 00000
6068:  00000 00000  00000 00000  00000 00000  00000 00000  00000 00000    00000 00000  00000 00000  00000 00000  00000 00000  00000 00000
6069:  00000 00000  00000 00000  00000 00000  00000 00000  00000 00000    00000 00000  00000 00000  00000 00000  00000 00000  00000 00000
6070:  00000 00000  00000 00000  00000 00000  00000 00000  00000 00000    00000 00000  00000 00000  00000 00000  00000 00000  00000 00000
6071:  00000 00000  00000 00000  00000 00000  00000 00000  00000 00000    00000 00000  00000 00000  00000 00000  00000 00000  00000 00000
6072:  00000 00000  00000 00000  00000 00000  00000 00000  00000 00000    00000 00000  00000 00000  00000 00000  00000 00000  00000 00000
6073:  00000 00000  00000 00000  00000 00000  00000 00000  00000 00000    00000 00000  00000 00000  00000 00000  00000 00000  00000 00000
6074:  00000 00000  00000 00000  00000 00000  00000 00000  00000 00000    00000 00000  00000 00000  00000 00000  00000 00000  00000 00000
6075:  00000 00000  00000 00000  00000 00000  00000 00000  00000 00000    00000 00000  00000 00000  00000 00000  00000 00000  00000 00000
6076:  00000 00000  00000 00000  00000 00000  00000 00000  00000 00000    00000 00000  00000 00000  00000 00000  00000 00000  00000 00000
6077:  00000 00000  00000 00000  00000 00000  00000 00000  00000 00000    00000 00000  00000 00000  00000 00000  00000 00000  00000 00000
6078:  00000 00000  00000 00000  00000 00000  00000 00000  00000 00000    00000 00000  00000 00000  00000 00000  00000 00000  00000 00000
6079:  00000 00000  00000 00000  00000 00000  00000 00000  00000 00000    00000 00000  00000 00000  00000 00000  00000 00000  00000 00000
6080:  00000 00000  00000 00000  00000 00000  00000 00000  00000 00000    00000 00000  00000 00000  00000 00000  00000 00000  00000 00000
6081:  00000 00000  00000 00000  00000 00000  00000 00000  00000 00000    00000 00000  00000 00000  00000 00000  00000 00000  00000 00000
6082:  00000 00000  00000 00000  00000 00000  00000 00000  00000 00000    00000 00000  00000 00000  00000 00000  00000 00000  00000 00000
6083:  00000 00000  00000 00000  00000 00000  00000 00000  00000 00000    00000 00000  00000 00000  00000 00000  00000 00000  00000 00000
6084:  00000 00000  00000 00000  00000 00000  00000 00000  00000 00000    00000 00000  00000 00000  00000 00000  00000 00000  00000 00000
6085:  00000 00000  00000 00000  00000 00000  00000 00000  00000 00000    00000 00000  00000 00000  00000 00000  00000 00000  00000 00000
6086:  00000 00000  00000 00000  00000 00000  00000 00000  00000 00000    00000 00000  00000 00000  00000 00000  00000 00000  00000 00000
6087:  00000 00000  00000 00000  00000 00000  00000 00000  00000 00000    00000 00000  00000 00000  00000 00000  00000 00000  00000 00000
6088:  00000 00000  00000 00000  00000 00000  00000 00000  00000 00000    00000 00000  00000 00000  00000 00000  00000 00000  00000 00000
6089:  00000 00000  00000 00000  00000 00000  00000 00000  00000 00000    00000 00000  00000 00000  00000 00000  00000 00000  00000 00000
6090:  00000 00000  00000 00000  00000 00000  00000 00000  00000 00000    00000 00000  00000 00000  00000 00000  00000 00000  00000 00000
6091:  00000 00000  00000 00000  00000 00000  00000 00000  00000 00000    00000 00000  00000 00000  00000 00000  00000 00000  00000 00000
6092:  00000 00000  00000 00000  00000 00000  00000 00000  00000 00000    00000 00000  00000 00000  00000 00000  00000 00000  00000 00000
6093:  00000 00000  00000 00000  00000 00000  00000 00000  00000 00000    00000 00000  00000 00000  00000 00000  00000 00000  00000 00000
6094:  00000 00000  00000 00000  00000 00000  00000 00000  00000 00000    00000 00000  00000 00000  00000 00000  00000 00000  00000 00000
6095:  00000 00000  00000 00000  00000 00000  00000 00000  00000 00000    00000 00000  00000 00000  00000 00000  00000 00000  00000 00000
6096:  00000 00000  00000 00000  00000 00000  00000 00000  00000 00000    00000 00000  00000 00000  00000 00000  00000 00000  00000 00000
6097:  00000 00000  00000 00000  00000 00000  00000 00000  00000 00000    00000 00000  00000 00000  00000 00000  00000 00000  00000 00000
6098:  00000 00000  00000 00000  00000 00000  00000 00000  00000 00000    00000 00000  00000 00000  00000 00000  00000 00000  00000 00000
6099:  00000 00000  00000 00000  00000 00000  00000 00000  00000 00000    00000 00000  00000 00000  00000 00000  00000 00000  00000 00000
```

```
6100:  00000 00000  00000 00000  00000 00000  00000 00000  00000 00000   00000 00000  00000 00000  00000 00000  00000 00000  00000 00000
6101:  00000 00000  00000 00000  00000 00000  00000 00000  00000 00000   00000 00000  00000 00000  00000 00000  00000 00000  00000 00000
6102:  00000 00000  00000 00000  00000 00000  00000 00000  00000 00000   00000 00000  00000 00000  00000 00000  00000 00000  00000 00000
6103:  00000 00000  00000 00000  00000 00000  00000 00000  00000 00000   00000 00000  00000 00000  00000 00000  00000 00000  00000 00000
6104:  00000 00000  00000 00000  00000 00000  00000 00000  00000 00000   00000 00000  00000 00000  00000 00000  00000 00000  00000 00000
6105:  00000 00000  00000 00000  00000 00000  00000 00000  00000 00000   00000 00000  00000 00000  00000 00000  00000 00000  00000 00000
6106:  00000 00000  00000 00000  00000 00000  00000 00000  00000 00000   00000 00000  00000 00000  00000 00000  00000 00000  00000 00000
6107:  00000 00000  00000 00000  00000 00000  00000 00000  00000 00000   00000 00000  00000 00000  00000 00000  00000 00000  00000 00000
6108:  00000 00000  00000 00000  00000 00000  00000 00000  00000 00000   00000 00000  00000 00000  00000 00000  00000 00000  00000 00000
6109:  00000 00000  00000 00000  00000 00000  00000 00000  00000 00000   00000 00000  00000 00000  00000 00000  00000 00000  00000 00000
6110:  00000 00000  00000 00000  00000 00000  00000 00000  00000 00000   00000 00000  00000 00000  00000 00000  00000 00000  00000 00000
6111:  00000 00000  00000 00000  00000 00000  00000 00000  00000 00000   00000 00000  00000 00000  00000 00000  00000 00000  00000 00000
6112:  00000 00000  00000 00000  00000 00000  00000 00000  00000 00000   00000 00000  00000 00000  00000 00000  00000 00000  00000 00000
6113:  00000 00000  00000 00000  00000 00000  00000 00000  00000 00000   00000 00000  00000 00000  00000 00000  00000 00000  00000 00000
6114:  00000 00000  00000 00000  00000 00000  00000 00000  00000 00000   00000 00000  00000 00000  00000 00000  00000 00000  00000 00000
6115:  00000 00000  00000 00000  00000 00000  00000 00000  00000 00000   00000 00000  00000 00000  00000 00000  00000 00000  00000 00000
6116:  00000 00000  00000 00000  00000 00000  00000 00000  00000 00000   00000 00000  00000 00000  00000 00000  00000 00000  00000 00000
6117:  00000 00000  00000 00000  00000 00000  00000 00000  00000 00000   00000 00000  00000 00000  00000 00000  00000 00000  00000 00000
6118:  00000 00000  00000 00000  00000 00000  00000 00000  00000 00000   00000 00000  00000 00000  00000 00000  00000 00000  00000 00000
6119:  00000 00000  00000 00000  00000 00000  00000 00000  00000 00000   00000 00000  00000 00000  00000 00000  00000 00000  00000 00000
6120:  00000 00000  00000 00000  00000 00000  00000 00000  00000 00000   00000 00000  00000 00000  00000 00000  00000 00000  00000 00000
6121:  00000 00000  00000 00000  00000 00000  00000 00000  00000 00000   00000 00000  00000 00000  00000 00000  00000 00000  00000 00000
6122:  00000 00000  00000 00000  00000 00000  00000 00000  00000 00000   00000 00000  00000 00000  00000 00000  00000 00000  00000 00000
6123:  00000 00000  00000 00000  00000 00000  00000 00000  00000 00000   00000 00000  00000 00000  00000 00000  00000 00000  00000 00000
6124:  00000 00000  00000 00000  00000 00000  00000 00000  00000 00000   00000 00000  00000 00000  00000 00000  00000 00000  00000 00000
6125:  00000 00000  00000 00000  00000 00000  00000 00000  00000 00000   00000 00000  00000 00000  00000 00000  00000 00000  00000 00000
6126:  00000 00000  00000 00000  00000 00000  00000 00000  00000 00000   00000 00000  00000 00000  00000 00000  00000 00000  00000 00000
6127:  00000 00000  00000 00000  00000 00000  00000 00000  00000 00000   00000 00000  00000 00000  00000 00000  00000 00000  00000 00000
6128:  00000 00000  00000 00000  00000 00000  00000 00000  00000 00000   00000 00000  00000 00000  00000 00000  00000 00000  00000 00000
6129:  00000 00000  00000 00000  00000 00000  00000 00000  00000 00000   00000 00000  00000 00000  00000 00000  00000 00000  00000 00000
6130:  00000 00000  00000 00000  00000 00000  00000 00000  00000 00000   00000 00000  00000 00000  00000 00000  00000 00000  00000 00000
6131:  00000 00000  00000 00000  00000 00000  00000 00000  00000 00000   00000 00000  00000 00000  00000 00000  00000 00000  00000 00000
6132:  00000 00000  00000 00000  00000 00000  00000 00000  00000 00000   00000 00000  00000 00000  00000 00000  00000 00000  00000 00000
6133:  00000 00000  00000 00000  00000 00000  00000 00000  00000 00000   00000 00000  00000 00000  00000 00000  00000 00000  00000 00000
6134:  00000 00000  00000 00000  00000 00000  00000 00000  00000 00000   00000 00000  00000 00000  00000 00000  00000 00000  00000 00000
6135:  00000 00000  00000 00000  00000 00000  00000 00000  00000 00000   00000 00000  00000 00000  00000 00000  00000 00000  00000 00000
6136:  00000 00000  00000 00000  00000 00000  00000 00000  00000 00000   00000 00000  00000 00000  00000 00000  00000 00000  00000 00000
6137:  00000 00000  00000 00000  00000 00000  00000 00000  00000 00000   00000 00000  00000 00000  00000 00000  00000 00000  00000 00000
6138:  00000 00000  00000 00000  00000 00000  00000 00000  00000 00000   00000 00000  00000 00000  00000 00000  00000 00000  00000 00000
6139:  00000 00000  00000 00000  00000 00000  00000 00000  00000 00000   00000 00000  00000 00000  00000 00000  00000 00000  00000 00000
6140:  00000 00000  00000 00000  00000 00000  00000 00000  00000 00000   00000 00000  00000 00000  00000 00000  00000 00000  00000 00000
6141:  00000 00000  00000 00000  00000 00000  00000 00000  00000 00000   00000 00000  00000 00000  00000 00000  00000 00000  00000 00000
6142:  00000 00000  00000 00000  00000 00000  00000 00000  00000 00000   00000 00000  00000 00000  00000 00000  00000 00000  00000 00000
6143:  00000 00000  00000 00000  00000 00000  00000 00000  00000 00000   00000 00000  00000 00000  00000 00000  00000 00000  00000 00000
6144:  00000 00000  00000 00000  00000 00000  00000 00000  00000 00000   00000 00000  00000 00000  00000 00000  00000 00000  00000 00000
6145:  00000 00000  00000 00000  00000 00000  00000 00000  00000 00000   00000 00000  00000 00000  00000 00000  00000 00000  00000 00000
6146:  00000 00000  00000 00000  00000 00000  00000 00000  00000 00000   00000 00000  00000 00000  00000 00000  00000 00000  00000 00000
6147:  00000 00000  00000 00000  00000 00000  00000 00000  00000 00000   00000 00000  00000 00000  00000 00000  00000 00000  00000 00000
6148:  00000 00000  00000 00000  00000 00000  00000 00000  00000 00000   00000 00000  00000 00000  00000 00000  00000 00000  00000 00000
6149:  00000 00000  00000 00000  00000 00000  00000 00000  00000 00000   00000 00000  00000 00000  00000 00000  00000 00000  00000 00000
```

```
6150:  00000 00000  00000 00000  00000 00000  00000 00000  00000 00000    00000 00000  00000 00000  00000 00000  00000 00000  00000 00000
6151:  00000 00000  00000 00000  00000 00000  00000 00000  00000 00000    00000 00000  00000 00000  00000 00000  00000 00000  00000 00000
6152:  00000 00000  00000 00000  00000 00000  00000 00000  00000 00000    00000 00000  00000 00000  00000 00000  00000 00000  00000 00000
6153:  00000 00000  00000 00000  00000 00000  00000 00000  00000 00000    00000 00000  00000 00000  00000 00000  00000 00000  00000 00000
6154:  00000 00000  00000 00000  00000 00000  00000 00000  00000 00000    00000 00000  00000 00000  00000 00000  00000 00000  00000 00000
6155:  00000 00000  00000 00000  00000 00000  00000 00000  00000 00000    00000 00000  00000 00000  00000 00000  00000 00000  00000 00000
6156:  00000 00000  00000 00000  00000 00000  00000 00000  00000 00000    00000 00000  00000 00000  00000 00000  00000 00000  00000 00000
6157:  00000 00000  00000 00000  00000 00000  00000 00000  00000 00000    00000 00000  00000 00000  00000 00000  00000 00000  00000 00000
6158:  00000 00000  00000 00000  00000 00000  00000 00000  00000 00000    00000 00000  00000 00000  00000 00000  00000 00000  00000 00000
6159:  00000 00000  00000 00000  00000 00000  00000 00000  00000 00000    00000 00000  00000 00000  00000 00000  00000 00000  00000 00000
6160:  00000 00000  00000 00000  00000 00000  00000 00000  00000 00000    00000 00000  00000 00000  00000 00000  00000 00000  00000 00000
6161:  00000 00000  00000 00000  00000 00000  00000 00000  00000 00000    00000 00000  00000 00000  00000 00000  00000 00000  00000 00000
6162:  00000 00000  00000 00000  00000 00000  00000 00000  00000 00000    00000 00000  00000 00000  00000 00000  00000 00000  00000 00000
6163:  00000 00000  00000 00000  00000 00000  00000 00000  00000 00000    00000 00000  00000 00000  00000 00000  00000 00000  00000 00000
6164:  00000 00000  00000 00000  00000 00000  00000 00000  00000 00000    00000 00000  00000 00000  00000 00000  00000 00000  00000 00000
6165:  00000 00000  00000 00000  00000 00000  00000 00000  00000 00000    00000 00000  00000 00000  00000 00000  00000 00000  00000 00000
6166:  00000 00000  00000 00000  00000 00000  00000 00000  00000 00000    00000 00000  00000 00000  00000 00000  00000 00000  00000 00000
6167:  00000 00000  00000 00000  00000 00000  00000 00000  00000 00000    00000 00000  00000 00000  00000 00000  00000 00000  00000 00000
6168:  00000 00000  00000 00000  00000 00000  00000 00000  00000 00000    00000 00000  00000 00000  00000 00000  00000 00000  00000 00000
6169:  00000 00000  00000 00000  00000 00000  00000 00000  00000 00000    00000 00000  00000 00000  00000 00000  00000 00000  00000 00000
6170:  00000 00000  00000 00000  00000 00000  00000 00000  00000 00000    00000 00000  00000 00000  00000 00000  00000 00000  00000 00000
6171:  00000 00000  00000 00000  00000 00000  00000 00000  00000 00000    00000 00000  00000 00000  00000 00000  00000 00000  00000 00000
6172:  00000 00000  00000 00000  00000 00000  00000 00000  00000 00000    00000 00000  00000 00000  00000 00000  00000 00000  00000 00000
6173:  00000 00000  00000 00000  00000 00000  00000 00000  00000 00000    00000 00000  00000 00000  00000 00000  00000 00000  00000 00000
6174:  00000 00000  00000 00000  00000 00000  00000 00000  00000 00000    00000 00000  00000 00000  00000 00000  00000 00000  00000 00000
6175:  00000 00000  00000 00000  00000 00000  00000 00000  00000 00000    00000 00000  00000 00000  00000 00000  00000 00000  00000 00000
6176:  00000 00000  00000 00000  00000 00000  00000 00000  00000 00000    00000 00000  00000 00000  00000 00000  00000 00000  00000 00000
6177:  00000 00000  00000 00000  00000 00000  00000 00000  00000 00000    00000 00000  00000 00000  00000 00000  00000 00000  00000 00000
6178:  00000 00000  00000 00000  00000 00000  00000 00000  00000 00000    00000 00000  00000 00000  00000 00000  00000 00000  00000 00000
6179:  00000 00000  00000 00000  00000 00000  00000 00000  00000 00000    00000 00000  00000 00000  00000 00000  00000 00000  00000 00000
6180:  00000 00000  00000 00000  00000 00000  00000 00000  00000 00000    00000 00000  00000 00000  00000 00000  00000 00000  00000 00000
6181:  00000 00000  00000 00000  00000 00000  00000 00000  00000 00000    00000 00000  00000 00000  00000 00000  00000 00000  00000 00000
6182:  00000 00000  00000 00000  00000 00000  00000 00000  00000 00000    00000 00000  00000 00000  00000 00000  00000 00000  00000 00000
6183:  00000 00000  00000 00000  00000 00000  00000 00000  00000 00000    00000 00000  00000 00000  00000 00000  00000 00000  00000 00000
6184:  00000 00000  00000 00000  00000 00000  00000 00000  00000 00000    00000 00000  00000 00000  00000 00000  00000 00000  00000 00000
6185:  00000 00000  00000 00000  00000 00000  00000 00000  00000 00000    00000 00000  00000 00000  00000 00000  00000 00000  00000 00000
6186:  00000 00000  00000 00000  00000 00000  00000 00000  00000 00000    00000 00000  00000 00000  00000 00000  00000 00000  00000 00000
6187:  00000 00000  00000 00000  00000 00000  00000 00000  00000 00000    00000 00000  00000 00000  00000 00000  00000 00000  00000 00000
6188:  00000 00000  00000 00000  00000 00000  00000 00000  00000 00000    00000 00000  00000 00000  00000 00000  00000 00000  00000 00000
6189:  00000 00000  00000 00000  00000 00000  00000 00000  00000 00000    00000 00000  00000 00000  00000 00000  00000 00000  00000 00000
6190:  00000 00000  00000 00000  00000 00000  00000 00000  00000 00000    00000 00000  00000 00000  00000 00000  00000 00000  00000 00000
6191:  00000 00000  00000 00000  00000 00000  00000 00000  00000 00000    00000 00000  00000 00000  00000 00000  00000 00000  00000 00000
6192:  00000 00000  00000 00000  00000 00000  00000 00000  00000 00000    00000 00000  00000 00000  00000 00000  00000 00000  00000 00000
6193:  00000 00000  00000 00000  00000 00000  00000 00000  00000 00000    00000 00000  00000 00000  00000 00000  00000 00000  00000 00000
6194:  00000 00000  00000 00000  00000 00000  00000 00000  00000 00000    00000 00000  00000 00000  00000 00000  00000 00000  00000 00000
6195:  00000 00000  00000 00000  00000 00000  00000 00000  00000 00000    00000 00000  00000 00000  00000 00000  00000 00000  00000 00000
6196:  00000 00000  00000 00000  00000 00000  00000 00000  00000 00000    00000 00000  00000 00000  00000 00000  00000 00000  00000 00000
6197:  00000 00000  00000 00000  00000 00000  00000 00000  00000 00000    00000 00000  00000 00000  00000 00000  00000 00000  00000 00000
6198:  00000 00000  00000 00000  00000 00000  00000 00000  00000 00000    00000 00000  00000 00000  00000 00000  00000 00000  00000 00000
6199:  00000 00000  00000 00000  00000 00000  00000 00000  00000 00000    00000 00000  00000 00000  00000 00000  00000 00000  00000 00000
```

```
6200:  00000 00000  00000 00000  00000 00000  00000 00000  00000 00000   00000 00000  00000 00000  00000 00000  00000 00000  00000 00000
6201:  00000 00000  00000 00000  00000 00000  00000 00000  00000 00000   00000 00000  00000 00000  00000 00000  00000 00000  00000 00000
6202:  00000 00000  00000 00000  00000 00000  00000 00000  00000 00000   00000 00000  00000 00000  00000 00000  00000 00000  00000 00000
6203:  00000 00000  00000 00000  00000 00000  00000 00000  00000 00000   00000 00000  00000 00000  00000 00000  00000 00000  00000 00000
6204:  00000 00000  00000 00000  00000 00000  00000 00000  00000 00000   00000 00000  00000 00000  00000 00000  00000 00000  00000 00000
6205:  00000 00000  00000 00000  00000 00000  00000 00000  00000 00000   00000 00000  00000 00000  00000 00000  00000 00000  00000 00000
6206:  00000 00000  00000 00000  00000 00000  00000 00000  00000 00000   00000 00000  00000 00000  00000 00000  00000 00000  00000 00000
6207:  00000 00000  00000 00000  00000 00000  00000 00000  00000 00000   00000 00000  00000 00000  00000 00000  00000 00000  00000 00000
6208:  00000 00000  00000 00000  00000 00000  00000 00000  00000 00000   00000 00000  00000 00000  00000 00000  00000 00000  00000 00000
6209:  00000 00000  00000 00000  00000 00000  00000 00000  00000 00000   00000 00000  00000 00000  00000 00000  00000 00000  00000 00000
6210:  00000 00000  00000 00000  00000 00000  00000 00000  00000 00000   00000 00000  00000 00000  00000 00000  00000 00000  00000 00000
6211:  00000 00000  00000 00000  00000 00000  00000 00000  00000 00000   00000 00000  00000 00000  00000 00000  00000 00000  00000 00000
6212:  00000 00000  00000 00000  00000 00000  00000 00000  00000 00000   00000 00000  00000 00000  00000 00000  00000 00000  00000 00000
6213:  00000 00000  00000 00000  00000 00000  00000 00000  00000 00000   00000 00000  00000 00000  00000 00000  00000 00000  00000 00000
6214:  00000 00000  00000 00000  00000 00000  00000 00000  00000 00000   00000 00000  00000 00000  00000 00000  00000 00000  00000 00000
6215:  00000 00000  00000 00000  00000 00000  00000 00000  00000 00000   00000 00000  00000 00000  00000 00000  00000 00000  00000 00000
6216:  00000 00000  00000 00000  00000 00000  00000 00000  00000 00000   00000 00000  00000 00000  00000 00000  00000 00000  00000 00000
6217:  00000 00000  00000 00000  00000 00000  00000 00000  00000 00000   00000 00000  00000 00000  00000 00000  00000 00000  00000 00000
6218:  00000 00000  00000 00000  00000 00000  00000 00000  00000 00000   00000 00000  00000 00000  00000 00000  00000 00000  00000 00000
6219:  00000 00000  00000 00000  00000 00000  00000 00000  00000 00000   00000 00000  00000 00000  00000 00000  00000 00000  00000 00000
6220:  00000 00000  00000 00000  00000 00000  00000 00000  00000 00000   00000 00000  00000 00000  00000 00000  00000 00000  00000 00000
6221:  00000 00000  00000 00000  00000 00000  00000 00000  00000 00000   00000 00000  00000 00000  00000 00000  00000 00000  00000 00000
6222:  00000 00000  00000 00000  00000 00000  00000 00000  00000 00000   00000 00000  00000 00000  00000 00000  00000 00000  00000 00000
6223:  00000 00000  00000 00000  00000 00000  00000 00000  00000 00000   00000 00000  00000 00000  00000 00000  00000 00000  00000 00000
6224:  00000 00000  00000 00000  00000 00000  00000 00000  00000 00000   00000 00000  00000 00000  00000 00000  00000 00000  00000 00000
6225:  00000 00000  00000 00000  00000 00000  00000 00000  00000 00000   00000 00000  00000 00000  00000 00000  00000 00000  00000 00000
6226:  00000 00000  00000 00000  00000 00000  00000 00000  00000 00000   00000 00000  00000 00000  00000 00000  00000 00000  00000 00000
6227:  00000 00000  00000 00000  00000 00000  00000 00000  00000 00000   00000 00000  00000 00000  00000 00000  00000 00000  00000 00000
6228:  00000 00000  00000 00000  00000 00000  00000 00000  00000 00000   00000 00000  00000 00000  00000 00000  00000 00000  00000 00000
6229:  00000 00000  00000 00000  00000 00000  00000 00000  00000 00000   00000 00000  00000 00000  00000 00000  00000 00000  00000 00000
6230:  00000 00000  00000 00000  00000 00000  00000 00000  00000 00000   00000 00000  00000 00000  00000 00000  00000 00000  00000 00000
6231:  00000 00000  00000 00000  00000 00000  00000 00000  00000 00000   00000 00000  00000 00000  00000 00000  00000 00000  00000 00000
6232:  00000 00000  00000 00000  00000 00000  00000 00000  00000 00000   00000 00000  00000 00000  00000 00000  00000 00000  00000 00000
6233:  00000 00000  00000 00000  00000 00000  00000 00000  00000 00000   00000 00000  00000 00000  00000 00000  00000 00000  00000 00000
6234:  00000 00000  00000 00000  00000 00000  00000 00000  00000 00000   00000 00000  00000 00000  00000 00000  00000 00000  00000 00000
6235:  00000 00000  00000 00000  00000 00000  00000 00000  00000 00000   00000 00000  00000 00000  00000 00000  00000 00000  00000 00000
6236:  00000 00000  00000 00000  00000 00000  00000 00000  00000 00000   00000 00000  00000 00000  00000 00000  00000 00000  00000 00000
6237:  00000 00000  00000 00000  00000 00000  00000 00000  00000 00000   00000 00000  00000 00000  00000 00000  00000 00000  00000 00000
6238:  00000 00000  00000 00000  00000 00000  00000 00000  00000 00000   00000 00000  00000 00000  00000 00000  00000 00000  00000 00000
6239:  00000 00000  00000 00000  00000 00000  00000 00000  00000 00000   00000 00000  00000 00000  00000 00000  00000 00000  00000 00000
6240:  00000 00000  00000 00000  00000 00000  00000 00000  00000 00000   00000 00000  00000 00000  00000 00000  00000 00000  00000 00000
6241:  00000 00000  00000 00000  00000 00000  00000 00000  00000 00000   00000 00000  00000 00000  00000 00000  00000 00000  00000 00000
6242:  00000 00000  00000 00000  00000 00000  00000 00000  00000 00000   00000 00000  00000 00000  00000 00000  00000 00000  00000 00000
6243:  00000 00000  00000 00000  00000 00000  00000 00000  00000 00000   00000 00000  00000 00000  00000 00000  00000 00000  00000 00000
6244:  00000 00000  00000 00000  00000 00000  00000 00000  00000 00000   00000 00000  00000 00000  00000 00000  00000 00000  00000 00000
6245:  00000 00000  00000 00000  00000 00000  00000 00000  00000 00000   00000 00000  00000 00000  00000 00000  00000 00000  00000 00000
6246:  00000 00000  00000 00000  00000 00000  00000 00000  00000 00000   00000 00000  00000 00000  00000 00000  00000 00000  00000 00000
6247:  00000 00000  00000 00000  00000 00000  00000 00000  00000 00000   00000 00000  00000 00000  00000 00000  00000 00000  00000 00000
6248:  00000 00000  00000 00000  00000 00000  00000 00000  00000 00000   00000 00000  00000 00000  00000 00000  00000 00000  00000 00000
6249:  00000 00000  00000 00000  00000 00000  00000 00000  00000 00000   00000 00000  00000 00000  00000 00000  00000 00000  00000 00000
```

6250:	00000 00000	00000 00000	00000 00000	00000 00000	00000 00000	00000 00000	00000 00000	00000 00000	00000 00000	00000 00000
6251:	00000 00000	00000 00000	00000 00000	00000 00000	00000 00000	00000 00000	00000 00000	00000 00000	00000 00000	00000 00000
6252:	00000 00000	00000 00000	00000 00000	00000 00000	00000 00000	00000 00000	00000 00000	00000 00000	00000 00000	00000 00000
6253:	00000 00000	00000 00000	00000 00000	00000 00000	00000 00000	00000 00000	00000 00000	00000 00000	00000 00000	00000 00000
6254:	00000 00000	00000 00000	00000 00000	00000 00000	00000 00000	00000 00000	00000 00000	00000 00000	00000 00000	00000 00000
6255:	00000 00000	00000 00000	00000 00000	00000 00000	00000 00000	00000 00000	00000 00000	00000 00000	00000 00000	00000 00000
6256:	00000 00000	00000 00000	00000 00000	00000 00000	00000 00000	00000 00000	00000 00000	00000 00000	00000 00000	00000 00000
6257:	00000 00000	00000 00000	00000 00000	00000 00000	00000 00000	00000 00000	00000 00000	00000 00000	00000 00000	00000 00000
6258:	00000 00000	00000 00000	00000 00000	00000 00000	00000 00000	00000 00000	00000 00000	00000 00000	00000 00000	00000 00000
6259:	00000 00000	00000 00000	00000 00000	00000 00000	00000 00000	00000 00000	00000 00000	00000 00000	00000 00000	00000 00000
6260:	00000 00000	00000 00000	00000 00000	00000 00000	00000 00000	00000 00000	00000 00000	00000 00000	00000 00000	00000 00000
6261:	00000 00000	00000 00000	00000 00000	00000 00000	00000 00000	00000 00000	00000 00000	00000 00000	00000 00000	00000 00000
6262:	00000 00000	00000 00000	00000 00000	00000 00000	00000 00000	00000 00000	00000 00000	00000 00000	00000 00000	00000 00000
6263:	00000 00000	00000 00000	00000 00000	00000 00000	00000 00000	00000 00000	00000 00000	00000 00000	00000 00000	00000 00000
6264:	00000 00000	00000 00000	00000 00000	00000 00000	00000 00000	00000 00000	00000 00000	00000 00000	00000 00000	00000 00000
6265:	00000 00000	00000 00000	00000 00000	00000 00000	00000 00000	00000 00000	00000 00000	00000 00000	00000 00000	00000 00000
6266:	00000 00000	00000 00000	00000 00000	00000 00000	00000 00000	00000 00000	00000 00000	00000 00000	00000 00000	00000 00000
6267:	00000 00000	00000 00000	00000 00000	00000 00000	00000 00000	00000 00000	00000 00000	00000 00000	00000 00000	00000 00000
6268:	00000 00000	00000 00000	00000 00000	00000 00000	00000 00000	00000 00000	00000 00000	00000 00000	00000 00000	00000 00000
6269:	00000 00000	00000 00000	00000 00000	00000 00000	00000 00000	00000 00000	00000 00000	00000 00000	00000 00000	00000 00000
6270:	00000 00000	00000 00000	00000 00000	00000 00000	00000 00000	00000 00000	00000 00000	00000 00000	00000 00000	00000 00000
6271:	00000 00000	00000 00000	00000 00000	00000 00000	00000 00000	00000 00000	00000 00000	00000 00000	00000 00000	00000 00000
6272:	00000 00000	00000 00000	00000 00000	00000 00000	00000 00000	00000 00000	00000 00000	00000 00000	00000 00000	00000 00000
6273:	00000 00000	00000 00000	00000 00000	00000 00000	00000 00000	00000 00000	00000 00000	00000 00000	00000 00000	00000 00000
6274:	00000 00000	00000 00000	00000 00000	00000 00000	00000 00000	00000 00000	00000 00000	00000 00000	00000 00000	00000 00000
6275:	00000 00000	00000 00000	00000 00000	00000 00000	00000 00000	00000 00000	00000 00000	00000 00000	00000 00000	00000 00000
6276:	00000 00000	00000 00000	00000 00000	00000 00000	00000 00000	00000 00000	00000 00000	00000 00000	00000 00000	00000 00000
6277:	00000 00000	00000 00000	00000 00000	00000 00000	00000 00000	00000 00000	00000 00000	00000 00000	00000 00000	00000 00000
6278:	00000 00000	00000 00000	00000 00000	00000 00000	00000 00000	00000 00000	00000 00000	00000 00000	00000 00000	00000 00000
6279:	00000 00000	00000 00000	00000 00000	00000 00000	00000 00000	00000 00000	00000 00000	00000 00000	00000 00000	00000 00000
6280:	00000 00000	00000 00000	00000 00000	00000 00000	00000 00000	00000 00000	00000 00000	00000 00000	00000 00000	00000 00000
6281:	00000 00000	00000 00000	00000 00000	00000 00000	00000 00000	00000 00000	00000 00000	00000 00000	00000 00000	00000 00000
6282:	00000 00000	00000 00000	00000 00000	00000 00000	00000 00000	00000 00000	00000 00000	00000 00000	00000 00000	00000 00000
6283:	00000 00000	00000 00000	00000 00000	00000 00000	00000 00000	00000 00000	00000 00000	00000 00000	00000 00000	00000 00000
6284:	00000 00000	00000 00000	00000 00000	00000 00000	00000 00000	00000 00000	00000 00000	00000 00000	00000 00000	00000 00000
6285:	00000 00000	00000 00000	00000 00000	00000 00000	00000 00000	00000 00000	00000 00000	00000 00000	00000 00000	00000 00000
6286:	00000 00000	00000 00000	00000 00000	00000 00000	00000 00000	00000 00000	00000 00000	00000 00000	00000 00000	00000 00000
6287:	00000 00000	00000 00000	00000 00000	00000 00000	00000 00000	00000 00000	00000 00000	00000 00000	00000 00000	00000 00000
6288:	00000 00000	00000 00000	00000 00000	00000 00000	00000 00000	00000 00000	00000 00000	00000 00000	00000 00000	00000 00000
6289:	00000 00000	00000 00000	00000 00000	00000 00000	00000 00000	00000 00000	00000 00000	00000 00000	00000 00000	00000 00000
6290:	00000 00000	00000 00000	00000 00000	00000 00000	00000 00000	00000 00000	00000 00000	00000 00000	00000 00000	00000 00000
6291:	00000 00000	00000 00000	00000 00000	00000 00000	00000 00000	00000 00000	00000 00000	00000 00000	00000 00000	00000 00000
6292:	00000 00000	00000 00000	00000 00000	00000 00000	00000 00000	00000 00000	00000 00000	00000 00000	00000 00000	00000 00000
6293:	00000 00000	00000 00000	00000 00000	00000 00000	00000 00000	00000 00000	00000 00000	00000 00000	00000 00000	00000 00000
6294:	00000 00000	00000 00000	00000 00000	00000 00000	00000 00000	00000 00000	00000 00000	00000 00000	00000 00000	00000 00000
6295:	00000 00000	00000 00000	00000 00000	00000 00000	00000 00000	00000 00000	00000 00000	00000 00000	00000 00000	00000 00000
6296:	00000 00000	00000 00000	00000 00000	00000 00000	00000 00000	00000 00000	00000 00000	00000 00000	00000 00000	00000 00000
6297:	00000 00000	00000 00000	00000 00000	00000 00000	00000 00000	00000 00000	00000 00000	00000 00000	00000 00000	00000 00000
6298:	00000 00000	00000 00000	00000 00000	00000 00000	00000 00000	00000 00000	00000 00000	00000 00000	00000 00000	00000 00000
6299:	00000 00000	00000 00000	00000 00000	00000 00000	00000 00000	00000 00000	00000 00000	00000 00000	00000 00000	00000 00000

```
6300:  00000 00000  00000 00000  00000 00000  00000 00000  00000 00000    00000 00000  00000 00000  00000 00000  00000 00000  00000 00000
6301:  00000 00000  00000 00000  00000 00000  00000 00000  00000 00000    00000 00000  00000 00000  00000 00000  00000 00000  00000 00000
6302:  00000 00000  00000 00000  00000 00000  00000 00000  00000 00000    00000 00000  00000 00000  00000 00000  00000 00000  00000 00000
6303:  00000 00000  00000 00000  00000 00000  00000 00000  00000 00000    00000 00000  00000 00000  00000 00000  00000 00000  00000 00000
6304:  00000 00000  00000 00000  00000 00000  00000 00000  00000 00000    00000 00000  00000 00000  00000 00000  00000 00000  00000 00000
6305:  00000 00000  00000 00000  00000 00000  00000 00000  00000 00000    00000 00000  00000 00000  00000 00000  00000 00000  00000 00000
6306:  00000 00000  00000 00000  00000 00000  00000 00000  00000 00000    00000 00000  00000 00000  00000 00000  00000 00000  00000 00000
6307:  00000 00000  00000 00000  00000 00000  00000 00000  00000 00000    00000 00000  00000 00000  00000 00000  00000 00000  00000 00000
6308:  00000 00000  00000 00000  00000 00000  00000 00000  00000 00000    00000 00000  00000 00000  00000 00000  00000 00000  00000 00000
6309:  00000 00000  00000 00000  00000 00000  00000 00000  00000 00000    00000 00000  00000 00000  00000 00000  00000 00000  00000 00000
6310:  00000 00000  00000 00000  00000 00000  00000 00000  00000 00000    00000 00000  00000 00000  00000 00000  00000 00000  00000 00000
6311:  00000 00000  00000 00000  00000 00000  00000 00000  00000 00000    00000 00000  00000 00000  00000 00000  00000 00000  00000 00000
6312:  00000 00000  00000 00000  00000 00000  00000 00000  00000 00000    00000 00000  00000 00000  00000 00000  00000 00000  00000 00000
6313:  00000 00000  00000 00000  00000 00000  00000 00000  00000 00000    00000 00000  00000 00000  00000 00000  00000 00000  00000 00000
6314:  00000 00000  00000 00000  00000 00000  00000 00000  00000 00000    00000 00000  00000 00000  00000 00000  00000 00000  00000 00000
6315:  00000 00000  00000 00000  00000 00000  00000 00000  00000 00000    00000 00000  00000 00000  00000 00000  00000 00000  00000 00000
6316:  00000 00000  00000 00000  00000 00000  00000 00000  00000 00000    00000 00000  00000 00000  00000 00000  00000 00000  00000 00000
6317:  00000 00000  00000 00000  00000 00000  00000 00000  00000 00000    00000 00000  00000 00000  00000 00000  00000 00000  00000 00000
6318:  00000 00000  00000 00000  00000 00000  00000 00000  00000 00000    00000 00000  00000 00000  00000 00000  00000 00000  00000 00000
6319:  00000 00000  00000 00000  00000 00000  00000 00000  00000 00000    00000 00000  00000 00000  00000 00000  00000 00000  00000 00000
6320:  00000 00000  00000 00000  00000 00000  00000 00000  00000 00000    00000 00000  00000 00000  00000 00000  00000 00000  00000 00000
6321:  00000 00000  00000 00000  00000 00000  00000 00000  00000 00000    00000 00000  00000 00000  00000 00000  00000 00000  00000 00000
6322:  00000 00000  00000 00000  00000 00000  00000 00000  00000 00000    00000 00000  00000 00000  00000 00000  00000 00000  00000 00000
6323:  00000 00000  00000 00000  00000 00000  00000 00000  00000 00000    00000 00000  00000 00000  00000 00000  00000 00000  00000 00000
6324:  00000 00000  00000 00000  00000 00000  00000 00000  00000 00000    00000 00000  00000 00000  00000 00000  00000 00000  00000 00000
6325:  00000 00000  00000 00000  00000 00000  00000 00000  00000 00000    00000 00000  00000 00000  00000 00000  00000 00000  00000 00000
6326:  00000 00000  00000 00000  00000 00000  00000 00000  00000 00000    00000 00000  00000 00000  00000 00000  00000 00000  00000 00000
6327:  00000 00000  00000 00000  00000 00000  00000 00000  00000 00000    00000 00000  00000 00000  00000 00000  00000 00000  00000 00000
6328:  00000 00000  00000 00000  00000 00000  00000 00000  00000 00000    00000 00000  00000 00000  00000 00000  00000 00000  00000 00000
6329:  00000 00000  00000 00000  00000 00000  00000 00000  00000 00000    00000 00000  00000 00000  00000 00000  00000 00000  00000 00000
6330:  00000 00000  00000 00000  00000 00000  00000 00000  00000 00000    00000 00000  00000 00000  00000 00000  00000 00000  00000 00000
6331:  00000 00000  00000 00000  00000 00000  00000 00000  00000 00000    00000 00000  00000 00000  00000 00000  00000 00000  00000 00000
6332:  00000 00000  00000 00000  00000 00000  00000 00000  00000 00000    00000 00000  00000 00000  00000 00000  00000 00000  00000 00000
6333:  00000 00000  00000 00000  00000 00000  00000 00000  00000 00000    00000 00000  00000 00000  00000 00000  00000 00000  00000 00000
6334:  00000 00000  00000 00000  00000 00000  00000 00000  00000 00000    00000 00000  00000 00000  00000 00000  00000 00000  00000 00000
6335:  00000 00000  00000 00000  00000 00000  00000 00000  00000 00000    00000 00000  00000 00000  00000 00000  00000 00000  00000 00000
6336:  00000 00000  00000 00000  00000 00000  00000 00000  00000 00000    00000 00000  00000 00000  00000 00000  00000 00000  00000 00000
6337:  00000 00000  00000 00000  00000 00000  00000 00000  00000 00000    00000 00000  00000 00000  00000 00000  00000 00000  00000 00000
6338:  00000 00000  00000 00000  00000 00000  00000 00000  00000 00000    00000 00000  00000 00000  00000 00000  00000 00000  00000 00000
6339:  00000 00000  00000 00000  00000 00000  00000 00000  00000 00000    00000 00000  00000 00000  00000 00000  00000 00000  00000 00000
6340:  00000 00000  00000 00000  00000 00000  00000 00000  00000 00000    00000 00000  00000 00000  00000 00000  00000 00000  00000 00000
6341:  00000 00000  00000 00000  00000 00000  00000 00000  00000 00000    00000 00000  00000 00000  00000 00000  00000 00000  00000 00000
6342:  00000 00000  00000 00000  00000 00000  00000 00000  00000 00000    00000 00000  00000 00000  00000 00000  00000 00000  00000 00000
6343:  00000 00000  00000 00000  00000 00000  00000 00000  00000 00000    00000 00000  00000 00000  00000 00000  00000 00000  00000 00000
6344:  00000 00000  00000 00000  00000 00000  00000 00000  00000 00000    00000 00000  00000 00000  00000 00000  00000 00000  00000 00000
6345:  00000 00000  00000 00000  00000 00000  00000 00000  00000 00000    00000 00000  00000 00000  00000 00000  00000 00000  00000 00000
6346:  00000 00000  00000 00000  00000 00000  00000 00000  00000 00000    00000 00000  00000 00000  00000 00000  00000 00000  00000 00000
6347:  00000 00000  00000 00000  00000 00000  00000 00000  00000 00000    00000 00000  00000 00000  00000 00000  00000 00000  00000 00000
6348:  00000 00000  00000 00000  00000 00000  00000 00000  00000 00000    00000 00000  00000 00000  00000 00000  00000 00000  00000 00000
6349:  00000 00000  00000 00000  00000 00000  00000 00000  00000 00000    00000 00000  00000 00000  00000 00000  00000 00000  00000 00000
```

```
6350:  00000 00000  00000 00000  00000 00000  00000 00000  00000 00000    00000 00000  00000 00000  00000 00000  00000 00000  00000 00000
6351:  00000 00000  00000 00000  00000 00000  00000 00000  00000 00000    00000 00000  00000 00000  00000 00000  00000 00000  00000 00000
6352:  00000 00000  00000 00000  00000 00000  00000 00000  00000 00000    00000 00000  00000 00000  00000 00000  00000 00000  00000 00000
6353:  00000 00000  00000 00000  00000 00000  00000 00000  00000 00000    00000 00000  00000 00000  00000 00000  00000 00000  00000 00000
6354:  00000 00000  00000 00000  00000 00000  00000 00000  00000 00000    00000 00000  00000 00000  00000 00000  00000 00000  00000 00000
6355:  00000 00000  00000 00000  00000 00000  00000 00000  00000 00000    00000 00000  00000 00000  00000 00000  00000 00000  00000 00000
6356:  00000 00000  00000 00000  00000 00000  00000 00000  00000 00000    00000 00000  00000 00000  00000 00000  00000 00000  00000 00000
6357:  00000 00000  00000 00000  00000 00000  00000 00000  00000 00000    00000 00000  00000 00000  00000 00000  00000 00000  00000 00000
6358:  00000 00000  00000 00000  00000 00000  00000 00000  00000 00000    00000 00000  00000 00000  00000 00000  00000 00000  00000 00000
6359:  00000 00000  00000 00000  00000 00000  00000 00000  00000 00000    00000 00000  00000 00000  00000 00000  00000 00000  00000 00000
6360:  00000 00000  00000 00000  00000 00000  00000 00000  00000 00000    00000 00000  00000 00000  00000 00000  00000 00000  00000 00000
6361:  00000 00000  00000 00000  00000 00000  00000 00000  00000 00000    00000 00000  00000 00000  00000 00000  00000 00000  00000 00000
6362:  00000 00000  00000 00000  00000 00000  00000 00000  00000 00000    00000 00000  00000 00000  00000 00000  00000 00000  00000 00000
6363:  00000 00000  00000 00000  00000 00000  00000 00000  00000 00000    00000 00000  00000 00000  00000 00000  00000 00000  00000 00000
6364:  00000 00000  00000 00000  00000 00000  00000 00000  00000 00000    00000 00000  00000 00000  00000 00000  00000 00000  00000 00000
6365:  00000 00000  00000 00000  00000 00000  00000 00000  00000 00000    00000 00000  00000 00000  00000 00000  00000 00000  00000 00000
6366:  00000 00000  00000 00000  00000 00000  00000 00000  00000 00000    00000 00000  00000 00000  00000 00000  00000 00000  00000 00000
6367:  00000 00000  00000 00000  00000 00000  00000 00000  00000 00000    00000 00000  00000 00000  00000 00000  00000 00000  00000 00000
6368:  00000 00000  00000 00000  00000 00000  00000 00000  00000 00000    00000 00000  00000 00000  00000 00000  00000 00000  00000 00000
6369:  00000 00000  00000 00000  00000 00000  00000 00000  00000 00000    00000 00000  00000 00000  00000 00000  00000 00000  00000 00000
6370:  00000 00000  00000 00000  00000 00000  00000 00000  00000 00000    00000 00000  00000 00000  00000 00000  00000 00000  00000 00000
6371:  00000 00000  00000 00000  00000 00000  00000 00000  00000 00000    00000 00000  00000 00000  00000 00000  00000 00000  00000 00000
6372:  00000 00000  00000 00000  00000 00000  00000 00000  00000 00000    00000 00000  00000 00000  00000 00000  00000 00000  00000 00000
6373:  00000 00000  00000 00000  00000 00000  00000 00000  00000 00000    00000 00000  00000 00000  00000 00000  00000 00000  00000 00000
6374:  00000 00000  00000 00000  00000 00000  00000 00000  00000 00000    00000 00000  00000 00000  00000 00000  00000 00000  00000 00000
6375:  00000 00000  00000 00000  00000 00000  00000 00000  00000 00000    00000 00000  00000 00000  00000 00000  00000 00000  00000 00000
6376:  00000 00000  00000 00000  00000 00000  00000 00000  00000 00000    00000 00000  00000 00000  00000 00000  00000 00000  00000 00000
6377:  00000 00000  00000 00000  00000 00000  00000 00000  00000 00000    00000 00000  00000 00000  00000 00000  00000 00000  00000 00000
6378:  00000 00000  00000 00000  00000 00000  00000 00000  00000 00000    00000 00000  00000 00000  00000 00000  00000 00000  00000 00000
6379:  00000 00000  00000 00000  00000 00000  00000 00000  00000 00000    00000 00000  00000 00000  00000 00000  00000 00000  00000 00000
6380:  00000 00000  00000 00000  00000 00000  00000 00000  00000 00000    00000 00000  00000 00000  00000 00000  00000 00000  00000 00000
6381:  00000 00000  00000 00000  00000 00000  00000 00000  00000 00000    00000 00000  00000 00000  00000 00000  00000 00000  00000 00000
6382:  00000 00000  00000 00000  00000 00000  00000 00000  00000 00000    00000 00000  00000 00000  00000 00000  00000 00000  00000 00000
6383:  00000 00000  00000 00000  00000 00000  00000 00000  00000 00000    00000 00000  00000 00000  00000 00000  00000 00000  00000 00000
6384:  00000 00000  00000 00000  00000 00000  00000 00000  00000 00000    00000 00000  00000 00000  00000 00000  00000 00000  00000 00000
6385:  00000 00000  00000 00000  00000 00000  00000 00000  00000 00000    00000 00000  00000 00000  00000 00000  00000 00000  00000 00000
6386:  00000 00000  00000 00000  00000 00000  00000 00000  00000 00000    00000 00000  00000 00000  00000 00000  00000 00000  00000 00000
6387:  00000 00000  00000 00000  00000 00000  00000 00000  00000 00000    00000 00000  00000 00000  00000 00000  00000 00000  00000 00000
6388:  00000 00000  00000 00000  00000 00000  00000 00000  00000 00000    00000 00000  00000 00000  00000 00000  00000 00000  00000 00000
6389:  00000 00000  00000 00000  00000 00000  00000 00000  00000 00000    00000 00000  00000 00000  00000 00000  00000 00000  00000 00000
6390:  00000 00000  00000 00000  00000 00000  00000 00000  00000 00000    00000 00000  00000 00000  00000 00000  00000 00000  00000 00000
6391:  00000 00000  00000 00000  00000 00000  00000 00000  00000 00000    00000 00000  00000 00000  00000 00000  00000 00000  00000 00000
6392:  00000 00000  00000 00000  00000 00000  00000 00000  00000 00000    00000 00000  00000 00000  00000 00000  00000 00000  00000 00000
6393:  00000 00000  00000 00000  00000 00000  00000 00000  00000 00000    00000 00000  00000 00000  00000 00000  00000 00000  00000 00000
6394:  00000 00000  00000 00000  00000 00000  00000 00000  00000 00000    00000 00000  00000 00000  00000 00000  00000 00000  00000 00000
6395:  00000 00000  00000 00000  00000 00000  00000 00000  00000 00000    00000 00000  00000 00000  00000 00000  00000 00000  00000 00000
6396:  00000 00000  00000 00000  00000 00000  00000 00000  00000 00000    00000 00000  00000 00000  00000 00000  00000 00000  00000 00000
6397:  00000 00000  00000 00000  00000 00000  00000 00000  00000 00000    00000 00000  00000 00000  00000 00000  00000 00000  00000 00000
6398:  00000 00000  00000 00000  00000 00000  00000 00000  00000 00000    00000 00000  00000 00000  00000 00000  00000 00000  00000 00000
6399:  00000 00000  00000 00000  00000 00000  00000 00000  00000 00000    00000 00000  00000 00000  00000 00000  00000 00000  00000 00000
```

```
6400:  00000 00000  00000 00000  00000 00000  00000 00000  00000 00000    00000 00000  00000 00000  00000 00000  00000 00000  00000 00000
6401:  00000 00000  00000 00000  00000 00000  00000 00000  00000 00000    00000 00000  00000 00000  00000 00000  00000 00000  00000 00000
6402:  00000 00000  00000 00000  00000 00000  00000 00000  00000 00000    00000 00000  00000 00000  00000 00000  00000 00000  00000 00000
6403:  00000 00000  00000 00000  00000 00000  00000 00000  00000 00000    00000 00000  00000 00000  00000 00000  00000 00000  00000 00000
6404:  00000 00000  00000 00000  00000 00000  00000 00000  00000 00000    00000 00000  00000 00000  00000 00000  00000 00000  00000 00000
6405:  00000 00000  00000 00000  00000 00000  00000 00000  00000 00000    00000 00000  00000 00000  00000 00000  00000 00000  00000 00000
6406:  00000 00000  00000 00000  00000 00000  00000 00000  00000 00000    00000 00000  00000 00000  00000 00000  00000 00000  00000 00000
6407:  00000 00000  00000 00000  00000 00000  00000 00000  00000 00000    00000 00000  00000 00000  00000 00000  00000 00000  00000 00000
6408:  00000 00000  00000 00000  00000 00000  00000 00000  00000 00000    00000 00000  00000 00000  00000 00000  00000 00000  00000 00000
6409:  00000 00000  00000 00000  00000 00000  00000 00000  00000 00000    00000 00000  00000 00000  00000 00000  00000 00000  00000 00000
6410:  00000 00000  00000 00000  00000 00000  00000 00000  00000 00000    00000 00000  00000 00000  00000 00000  00000 00000  00000 00000
6411:  00000 00000  00000 00000  00000 00000  00000 00000  00000 00000    00000 00000  00000 00000  00000 00000  00000 00000  00000 00000
6412:  00000 00000  00000 00000  00000 00000  00000 00000  00000 00000    00000 00000  00000 00000  00000 00000  00000 00000  00000 00000
6413:  00000 00000  00000 00000  00000 00000  00000 00000  00000 00000    00000 00000  00000 00000  00000 00000  00000 00000  00000 00000
6414:  00000 00000  00000 00000  00000 00000  00000 00000  00000 00000    00000 00000  00000 00000  00000 00000  00000 00000  00000 00000
6415:  00000 00000  00000 00000  00000 00000  00000 00000  00000 00000    00000 00000  00000 00000  00000 00000  00000 00000  00000 00000
6416:  00000 00000  00000 00000  00000 00000  00000 00000  00000 00000    00000 00000  00000 00000  00000 00000  00000 00000  00000 00000
6417:  00000 00000  00000 00000  00000 00000  00000 00000  00000 00000    00000 00000  00000 00000  00000 00000  00000 00000  00000 00000
6418:  00000 00000  00000 00000  00000 00000  00000 00000  00000 00000    00000 00000  00000 00000  00000 00000  00000 00000  00000 00000
6419:  00000 00000  00000 00000  00000 00000  00000 00000  00000 00000    00000 00000  00000 00000  00000 00000  00000 00000  00000 00000
6420:  00000 00000  00000 00000  00000 00000  00000 00000  00000 00000    00000 00000  00000 00000  00000 00000  00000 00000  00000 00000
6421:  00000 00000  00000 00000  00000 00000  00000 00000  00000 00000    00000 00000  00000 00000  00000 00000  00000 00000  00000 00000
6422:  00000 00000  00000 00000  00000 00000  00000 00000  00000 00000    00000 00000  00000 00000  00000 00000  00000 00000  00000 00000
6423:  00000 00000  00000 00000  00000 00000  00000 00000  00000 00000    00000 00000  00000 00000  00000 00000  00000 00000  00000 00000
6424:  00000 00000  00000 00000  00000 00000  00000 00000  00000 00000    00000 00000  00000 00000  00000 00000  00000 00000  00000 00000
6425:  00000 00000  00000 00000  00000 00000  00000 00000  00000 00000    00000 00000  00000 00000  00000 00000  00000 00000  00000 00000
6426:  00000 00000  00000 00000  00000 00000  00000 00000  00000 00000    00000 00000  00000 00000  00000 00000  00000 00000  00000 00000
6427:  00000 00000  00000 00000  00000 00000  00000 00000  00000 00000    00000 00000  00000 00000  00000 00000  00000 00000  00000 00000
6428:  00000 00000  00000 00000  00000 00000  00000 00000  00000 00000    00000 00000  00000 00000  00000 00000  00000 00000  00000 00000
6429:  00000 00000  00000 00000  00000 00000  00000 00000  00000 00000    00000 00000  00000 00000  00000 00000  00000 00000  00000 00000
6430:  00000 00000  00000 00000  00000 00000  00000 00000  00000 00000    00000 00000  00000 00000  00000 00000  00000 00000  00000 00000
6431:  00000 00000  00000 00000  00000 00000  00000 00000  00000 00000    00000 00000  00000 00000  00000 00000  00000 00000  00000 00000
6432:  00000 00000  00000 00000  00000 00000  00000 00000  00000 00000    00000 00000  00000 00000  00000 00000  00000 00000  00000 00000
6433:  00000 00000  00000 00000  00000 00000  00000 00000  00000 00000    00000 00000  00000 00000  00000 00000  00000 00000  00000 00000
6434:  00000 00000  00000 00000  00000 00000  00000 00000  00000 00000    00000 00000  00000 00000  00000 00000  00000 00000  00000 00000
6435:  00000 00000  00000 00000  00000 00000  00000 00000  00000 00000    00000 00000  00000 00000  00000 00000  00000 00000  00000 00000
6436:  00000 00000  00000 00000  00000 00000  00000 00000  00000 00000    00000 00000  00000 00000  00000 00000  00000 00000  00000 00000
6437:  00000 00000  00000 00000  00000 00000  00000 00000  00000 00000    00000 00000  00000 00000  00000 00000  00000 00000  00000 00000
6438:  00000 00000  00000 00000  00000 00000  00000 00000  00000 00000    00000 00000  00000 00000  00000 00000  00000 00000  00000 00000
6439:  00000 00000  00000 00000  00000 00000  00000 00000  00000 00000    00000 00000  00000 00000  00000 00000  00000 00000  00000 00000
6440:  00000 00000  00000 00000  00000 00000  00000 00000  00000 00000    00000 00000  00000 00000  00000 00000  00000 00000  00000 00000
6441:  00000 00000  00000 00000  00000 00000  00000 00000  00000 00000    00000 00000  00000 00000  00000 00000  00000 00000  00000 00000
6442:  00000 00000  00000 00000  00000 00000  00000 00000  00000 00000    00000 00000  00000 00000  00000 00000  00000 00000  00000 00000
6443:  00000 00000  00000 00000  00000 00000  00000 00000  00000 00000    00000 00000  00000 00000  00000 00000  00000 00000  00000 00000
6444:  00000 00000  00000 00000  00000 00000  00000 00000  00000 00000    00000 00000  00000 00000  00000 00000  00000 00000  00000 00000
6445:  00000 00000  00000 00000  00000 00000  00000 00000  00000 00000    00000 00000  00000 00000  00000 00000  00000 00000  00000 00000
6446:  00000 00000  00000 00000  00000 00000  00000 00000  00000 00000    00000 00000  00000 00000  00000 00000  00000 00000  00000 00000
6447:  00000 00000  00000 00000  00000 00000  00000 00000  00000 00000    00000 00000  00000 00000  00000 00000  00000 00000  00000 00000
6448:  00000 00000  00000 00000  00000 00000  00000 00000  00000 00000    00000 00000  00000 00000  00000 00000  00000 00000  00000 00000
6449:  00000 00000  00000 00000  00000 00000  00000 00000  00000 00000    00000 00000  00000 00000  00000 00000  00000 00000  00000 00000
```

```
6450: 00000 00000  00000 00000  00000 00000  00000 00000  00000 00000    00000 00000  00000 00000  00000 00000  00000 00000  00000 00000
6451: 00000 00000  00000 00000  00000 00000  00000 00000  00000 00000    00000 00000  00000 00000  00000 00000  00000 00000  00000 00000
6452: 00000 00000  00000 00000  00000 00000  00000 00000  00000 00000    00000 00000  00000 00000  00000 00000  00000 00000  00000 00000
6453: 00000 00000  00000 00000  00000 00000  00000 00000  00000 00000    00000 00000  00000 00000  00000 00000  00000 00000  00000 00000
6454: 00000 00000  00000 00000  00000 00000  00000 00000  00000 00000    00000 00000  00000 00000  00000 00000  00000 00000  00000 00000
6455: 00000 00000  00000 00000  00000 00000  00000 00000  00000 00000    00000 00000  00000 00000  00000 00000  00000 00000  00000 00000
6456: 00000 00000  00000 00000  00000 00000  00000 00000  00000 00000    00000 00000  00000 00000  00000 00000  00000 00000  00000 00000
6457: 00000 00000  00000 00000  00000 00000  00000 00000  00000 00000    00000 00000  00000 00000  00000 00000  00000 00000  00000 00000
6458: 00000 00000  00000 00000  00000 00000  00000 00000  00000 00000    00000 00000  00000 00000  00000 00000  00000 00000  00000 00000
6459: 00000 00000  00000 00000  00000 00000  00000 00000  00000 00000    00000 00000  00000 00000  00000 00000  00000 00000  00000 00000
6460: 00000 00000  00000 00000  00000 00000  00000 00000  00000 00000    00000 00000  00000 00000  00000 00000  00000 00000  00000 00000
6461: 00000 00000  00000 00000  00000 00000  00000 00000  00000 00000    00000 00000  00000 00000  00000 00000  00000 00000  00000 00000
6462: 00000 00000  00000 00000  00000 00000  00000 00000  00000 00000    00000 00000  00000 00000  00000 00000  00000 00000  00000 00000
6463: 00000 00000  00000 00000  00000 00000  00000 00000  00000 00000    00000 00000  00000 00000  00000 00000  00000 00000  00000 00000
6464: 00000 00000  00000 00000  00000 00000  00000 00000  00000 00000    00000 00000  00000 00000  00000 00000  00000 00000  00000 00000
6465: 00000 00000  00000 00000  00000 00000  00000 00000  00000 00000    00000 00000  00000 00000  00000 00000  00000 00000  00000 00000
6466: 00000 00000  00000 00000  00000 00000  00000 00000  00000 00000    00000 00000  00000 00000  00000 00000  00000 00000  00000 00000
6467: 00000 00000  00000 00000  00000 00000  00000 00000  00000 00000    00000 00000  00000 00000  00000 00000  00000 00000  00000 00000
6468: 00000 00000  00000 00000  00000 00000  00000 00000  00000 00000    00000 00000  00000 00000  00000 00000  00000 00000  00000 00000
6469: 00000 00000  00000 00000  00000 00000  00000 00000  00000 00000    00000 00000  00000 00000  00000 00000  00000 00000  00000 00000
6470: 00000 00000  00000 00000  00000 00000  00000 00000  00000 00000    00000 00000  00000 00000  00000 00000  00000 00000  00000 00000
6471: 00000 00000  00000 00000  00000 00000  00000 00000  00000 00000    00000 00000  00000 00000  00000 00000  00000 00000  00000 00000
6472: 00000 00000  00000 00000  00000 00000  00000 00000  00000 00000    00000 00000  00000 00000  00000 00000  00000 00000  00000 00000
6473: 00000 00000  00000 00000  00000 00000  00000 00000  00000 00000    00000 00000  00000 00000  00000 00000  00000 00000  00000 00000
6474: 00000 00000  00000 00000  00000 00000  00000 00000  00000 00000    00000 00000  00000 00000  00000 00000  00000 00000  00000 00000
6475: 00000 00000  00000 00000  00000 00000  00000 00000  00000 00000    00000 00000  00000 00000  00000 00000  00000 00000  00000 00000
6476: 00000 00000  00000 00000  00000 00000  00000 00000  00000 00000    00000 00000  00000 00000  00000 00000  00000 00000  00000 00000
6477: 00000 00000  00000 00000  00000 00000  00000 00000  00000 00000    00000 00000  00000 00000  00000 00000  00000 00000  00000 00000
6478: 00000 00000  00000 00000  00000 00000  00000 00000  00000 00000    00000 00000  00000 00000  00000 00000  00000 00000  00000 00000
6479: 00000 00000  00000 00000  00000 00000  00000 00000  00000 00000    00000 00000  00000 00000  00000 00000  00000 00000  00000 00000
6480: 00000 00000  00000 00000  00000 00000  00000 00000  00000 00000    00000 00000  00000 00000  00000 00000  00000 00000  00000 00000
6481: 00000 00000  00000 00000  00000 00000  00000 00000  00000 00000    00000 00000  00000 00000  00000 00000  00000 00000  00000 00000
6482: 00000 00000  00000 00000  00000 00000  00000 00000  00000 00000    00000 00000  00000 00000  00000 00000  00000 00000  00000 00000
6483: 00000 00000  00000 00000  00000 00000  00000 00000  00000 00000    00000 00000  00000 00000  00000 00000  00000 00000  00000 00000
6484: 00000 00000  00000 00000  00000 00000  00000 00000  00000 00000    00000 00000  00000 00000  00000 00000  00000 00000  00000 00000
6485: 00000 00000  00000 00000  00000 00000  00000 00000  00000 00000    00000 00000  00000 00000  00000 00000  00000 00000  00000 00000
6486: 00000 00000  00000 00000  00000 00000  00000 00000  00000 00000    00000 00000  00000 00000  00000 00000  00000 00000  00000 00000
6487: 00000 00000  00000 00000  00000 00000  00000 00000  00000 00000    00000 00000  00000 00000  00000 00000  00000 00000  00000 00000
6488: 00000 00000  00000 00000  00000 00000  00000 00000  00000 00000    00000 00000  00000 00000  00000 00000  00000 00000  00000 00000
6489: 00000 00000  00000 00000  00000 00000  00000 00000  00000 00000    00000 00000  00000 00000  00000 00000  00000 00000  00000 00000
6490: 00000 00000  00000 00000  00000 00000  00000 00000  00000 00000    00000 00000  00000 00000  00000 00000  00000 00000  00000 00000
6491: 00000 00000  00000 00000  00000 00000  00000 00000  00000 00000    00000 00000  00000 00000  00000 00000  00000 00000  00000 00000
6492: 00000 00000  00000 00000  00000 00000  00000 00000  00000 00000    00000 00000  00000 00000  00000 00000  00000 00000  00000 00000
6493: 00000 00000  00000 00000  00000 00000  00000 00000  00000 00000    00000 00000  00000 00000  00000 00000  00000 00000  00000 00000
6494: 00000 00000  00000 00000  00000 00000  00000 00000  00000 00000    00000 00000  00000 00000  00000 00000  00000 00000  00000 00000
6495: 00000 00000  00000 00000  00000 00000  00000 00000  00000 00000    00000 00000  00000 00000  00000 00000  00000 00000  00000 00000
6496: 00000 00000  00000 00000  00000 00000  00000 00000  00000 00000    00000 00000  00000 00000  00000 00000  00000 00000  00000 00000
6497: 00000 00000  00000 00000  00000 00000  00000 00000  00000 00000    00000 00000  00000 00000  00000 00000  00000 00000  00000 00000
6498: 00000 00000  00000 00000  00000 00000  00000 00000  00000 00000    00000 00000  00000 00000  00000 00000  00000 00000  00000 00000
6499: 00000 00000  00000 00000  00000 00000  00000 00000  00000 00000    00000 00000  00000 00000  00000 00000  00000 00000  00000 00000
```

```
6500:   00000 00000   00000 00000   00000 00000   00000 00000   00000 00000     00000 00000   00000 00000   00000 00000   00000 00000   00000 00000
6501:   00000 00000   00000 00000   00000 00000   00000 00000   00000 00000     00000 00000   00000 00000   00000 00000   00000 00000   00000 00000
6502:   00000 00000   00000 00000   00000 00000   00000 00000   00000 00000     00000 00000   00000 00000   00000 00000   00000 00000   00000 00000
6503:   00000 00000   00000 00000   00000 00000   00000 00000   00000 00000     00000 00000   00000 00000   00000 00000   00000 00000   00000 00000
6504:   00000 00000   00000 00000   00000 00000   00000 00000   00000 00000     00000 00000   00000 00000   00000 00000   00000 00000   00000 00000
6505:   00000 00000   00000 00000   00000 00000   00000 00000   00000 00000     00000 00000   00000 00000   00000 00000   00000 00000   00000 00000
6506:   00000 00000   00000 00000   00000 00000   00000 00000   00000 00000     00000 00000   00000 00000   00000 00000   00000 00000   00000 00000
6507:   00000 00000   00000 00000   00000 00000   00000 00000   00000 00000     00000 00000   00000 00000   00000 00000   00000 00000   00000 00000
6508:   00000 00000   00000 00000   00000 00000   00000 00000   00000 00000     00000 00000   00000 00000   00000 00000   00000 00000   00000 00000
6509:   00000 00000   00000 00000   00000 00000   00000 00000   00000 00000     00000 00000   00000 00000   00000 00000   00000 00000   00000 00000
6510:   00000 00000   00000 00000   00000 00000   00000 00000   00000 00000     00000 00000   00000 00000   00000 00000   00000 00000   00000 00000
6511:   00000 00000   00000 00000   00000 00000   00000 00000   00000 00000     00000 00000   00000 00000   00000 00000   00000 00000   00000 00000
6512:   00000 00000   00000 00000   00000 00000   00000 00000   00000 00000     00000 00000   00000 00000   00000 00000   00000 00000   00000 00000
6513:   00000 00000   00000 00000   00000 00000   00000 00000   00000 00000     00000 00000   00000 00000   00000 00000   00000 00000   00000 00000
6514:   00000 00000   00000 00000   00000 00000   00000 00000   00000 00000     00000 00000   00000 00000   00000 00000   00000 00000   00000 00000
6515:   00000 00000   00000 00000   00000 00000   00000 00000   00000 00000     00000 00000   00000 00000   00000 00000   00000 00000   00000 00000
6516:   00000 00000   00000 00000   00000 00000   00000 00000   00000 00000     00000 00000   00000 00000   00000 00000   00000 00000   00000 00000
6517:   00000 00000   00000 00000   00000 00000   00000 00000   00000 00000     00000 00000   00000 00000   00000 00000   00000 00000   00000 00000
6518:   00000 00000   00000 00000   00000 00000   00000 00000   00000 00000     00000 00000   00000 00000   00000 00000   00000 00000   00000 00000
6519:   00000 00000   00000 00000   00000 00000   00000 00000   00000 00000     00000 00000   00000 00000   00000 00000   00000 00000   00000 00000
6520:   00000 00000   00000 00000   00000 00000   00000 00000   00000 00000     00000 00000   00000 00000   00000 00000   00000 00000   00000 00000
6521:   00000 00000   00000 00000   00000 00000   00000 00000   00000 00000     00000 00000   00000 00000   00000 00000   00000 00000   00000 00000
6522:   00000 00000   00000 00000   00000 00000   00000 00000   00000 00000     00000 00000   00000 00000   00000 00000   00000 00000   00000 00000
6523:   00000 00000   00000 00000   00000 00000   00000 00000   00000 00000     00000 00000   00000 00000   00000 00000   00000 00000   00000 00000
6524:   00000 00000   00000 00000   00000 00000   00000 00000   00000 00000     00000 00000   00000 00000   00000 00000   00000 00000   00000 00000
6525:   00000 00000   00000 00000   00000 00000   00000 00000   00000 00000     00000 00000   00000 00000   00000 00000   00000 00000   00000 00000
6526:   00000 00000   00000 00000   00000 00000   00000 00000   00000 00000     00000 00000   00000 00000   00000 00000   00000 00000   00000 00000
6527:   00000 00000   00000 00000   00000 00000   00000 00000   00000 00000     00000 00000   00000 00000   00000 00000   00000 00000   00000 00000
6528:   00000 00000   00000 00000   00000 00000   00000 00000   00000 00000     00000 00000   00000 00000   00000 00000   00000 00000   00000 00000
6529:   00000 00000   00000 00000   00000 00000   00000 00000   00000 00000     00000 00000   00000 00000   00000 00000   00000 00000   00000 00000
6530:   00000 00000   00000 00000   00000 00000   00000 00000   00000 00000     00000 00000   00000 00000   00000 00000   00000 00000   00000 00000
6531:   00000 00000   00000 00000   00000 00000   00000 00000   00000 00000     00000 00000   00000 00000   00000 00000   00000 00000   00000 00000
6532:   00000 00000   00000 00000   00000 00000   00000 00000   00000 00000     00000 00000   00000 00000   00000 00000   00000 00000   00000 00000
6533:   00000 00000   00000 00000   00000 00000   00000 00000   00000 00000     00000 00000   00000 00000   00000 00000   00000 00000   00000 00000
6534:   00000 00000   00000 00000   00000 00000   00000 00000   00000 00000     00000 00000   00000 00000   00000 00000   00000 00000   00000 00000
6535:   00000 00000   00000 00000   00000 00000   00000 00000   00000 00000     00000 00000   00000 00000   00000 00000   00000 00000   00000 00000
6536:   00000 00000   00000 00000   00000 00000   00000 00000   00000 00000     00000 00000   00000 00000   00000 00000   00000 00000   00000 00000
6537:   00000 00000   00000 00000   00000 00000   00000 00000   00000 00000     00000 00000   00000 00000   00000 00000   00000 00000   00000 00000
6538:   00000 00000   00000 00000   00000 00000   00000 00000   00000 00000     00000 00000   00000 00000   00000 00000   00000 00000   00000 00000
6539:   00000 00000   00000 00000   00000 00000   00000 00000   00000 00000     00000 00000   00000 00000   00000 00000   00000 00000   00000 00000
6540:   00000 00000   00000 00000   00000 00000   00000 00000   00000 00000     00000 00000   00000 00000   00000 00000   00000 00000   00000 00000
6541:   00000 00000   00000 00000   00000 00000   00000 00000   00000 00000     00000 00000   00000 00000   00000 00000   00000 00000   00000 00000
6542:   00000 00000   00000 00000   00000 00000   00000 00000   00000 00000     00000 00000   00000 00000   00000 00000   00000 00000   00000 00000
6543:   00000 00000   00000 00000   00000 00000   00000 00000   00000 00000     00000 00000   00000 00000   00000 00000   00000 00000   00000 00000
6544:   00000 00000   00000 00000   00000 00000   00000 00000   00000 00000     00000 00000   00000 00000   00000 00000   00000 00000   00000 00000
6545:   00000 00000   00000 00000   00000 00000   00000 00000   00000 00000     00000 00000   00000 00000   00000 00000   00000 00000   00000 00000
6546:   00000 00000   00000 00000   00000 00000   00000 00000   00000 00000     00000 00000   00000 00000   00000 00000   00000 00000   00000 00000
6547:   00000 00000   00000 00000   00000 00000   00000 00000   00000 00000     00000 00000   00000 00000   00000 00000   00000 00000   00000 00000
6548:   00000 00000   00000 00000   00000 00000   00000 00000   00000 00000     00000 00000   00000 00000   00000 00000   00000 00000   00000 00000
6549:   00000 00000   00000 00000   00000 00000   00000 00000   00000 00000     00000 00000   00000 00000   00000 00000   00000 00000   00000 00000
```

```
6550:  00000 00000  00000 00000  00000 00000  00000 00000  00000 00000     00000 00000  00000 00000  00000 00000  00000 00000  00000 00000
6551:  00000 00000  00000 00000  00000 00000  00000 00000  00000 00000     00000 00000  00000 00000  00000 00000  00000 00000  00000 00000
6552:  00000 00000  00000 00000  00000 00000  00000 00000  00000 00000     00000 00000  00000 00000  00000 00000  00000 00000  00000 00000
6553:  00000 00000  00000 00000  00000 00000  00000 00000  00000 00000     00000 00000  00000 00000  00000 00000  00000 00000  00000 00000
6554:  00000 00000  00000 00000  00000 00000  00000 00000  00000 00000     00000 00000  00000 00000  00000 00000  00000 00000  00000 00000
6555:  00000 00000  00000 00000  00000 00000  00000 00000  00000 00000     00000 00000  00000 00000  00000 00000  00000 00000  00000 00000
6556:  00000 00000  00000 00000  00000 00000  00000 00000  00000 00000     00000 00000  00000 00000  00000 00000  00000 00000  00000 00000
6557:  00000 00000  00000 00000  00000 00000  00000 00000  00000 00000     00000 00000  00000 00000  00000 00000  00000 00000  00000 00000
6558:  00000 00000  00000 00000  00000 00000  00000 00000  00000 00000     00000 00000  00000 00000  00000 00000  00000 00000  00000 00000
6559:  00000 00000  00000 00000  00000 00000  00000 00000  00000 00000     00000 00000  00000 00000  00000 00000  00000 00000  00000 00000
6560:  00000 00000  00000 00000  00000 00000  00000 00000  00000 00000     00000 00000  00000 00000  00000 00000  00000 00000  00000 00000
6561:  00000 00000  00000 00000  00000 00000  00000 00000  00000 00000     00000 00000  00000 00000  00000 00000  00000 00000  00000 00000
6562:  00000 00000  00000 00000  00000 00000  00000 00000  00000 00000     00000 00000  00000 00000  00000 00000  00000 00000  00000 00000
6563:  00000 00000  00000 00000  00000 00000  00000 00000  00000 00000     00000 00000  00000 00000  00000 00000  00000 00000  00000 00000
6564:  00000 00000  00000 00000  00000 00000  00000 00000  00000 00000     00000 00000  00000 00000  00000 00000  00000 00000  00000 00000
6565:  00000 00000  00000 00000  00000 00000  00000 00000  00000 00000     00000 00000  00000 00000  00000 00000  00000 00000  00000 00000
6566:  00000 00000  00000 00000  00000 00000  00000 00000  00000 00000     00000 00000  00000 00000  00000 00000  00000 00000  00000 00000
6567:  00000 00000  00000 00000  00000 00000  00000 00000  00000 00000     00000 00000  00000 00000  00000 00000  00000 00000  00000 00000
6568:  00000 00000  00000 00000  00000 00000  00000 00000  00000 00000     00000 00000  00000 00000  00000 00000  00000 00000  00000 00000
6569:  00000 00000  00000 00000  00000 00000  00000 00000  00000 00000     00000 00000  00000 00000  00000 00000  00000 00000  00000 00000
6570:  00000 00000  00000 00000  00000 00000  00000 00000  00000 00000     00000 00000  00000 00000  00000 00000  00000 00000  00000 00000
6571:  00000 00000  00000 00000  00000 00000  00000 00000  00000 00000     00000 00000  00000 00000  00000 00000  00000 00000  00000 00000
6572:  00000 00000  00000 00000  00000 00000  00000 00000  00000 00000     00000 00000  00000 00000  00000 00000  00000 00000  00000 00000
6573:  00000 00000  00000 00000  00000 00000  00000 00000  00000 00000     00000 00000  00000 00000  00000 00000  00000 00000  00000 00000
6574:  00000 00000  00000 00000  00000 00000  00000 00000  00000 00000     00000 00000  00000 00000  00000 00000  00000 00000  00000 00000
6575:  00000 00000  00000 00000  00000 00000  00000 00000  00000 00000     00000 00000  00000 00000  00000 00000  00000 00000  00000 00000
6576:  00000 00000  00000 00000  00000 00000  00000 00000  00000 00000     00000 00000  00000 00000  00000 00000  00000 00000  00000 00000
6577:  00000 00000  00000 00000  00000 00000  00000 00000  00000 00000     00000 00000  00000 00000  00000 00000  00000 00000  00000 00000
6578:  00000 00000  00000 00000  00000 00000  00000 00000  00000 00000     00000 00000  00000 00000  00000 00000  00000 00000  00000 00000
6579:  00000 00000  00000 00000  00000 00000  00000 00000  00000 00000     00000 00000  00000 00000  00000 00000  00000 00000  00000 00000
6580:  00000 00000  00000 00000  00000 00000  00000 00000  00000 00000     00000 00000  00000 00000  00000 00000  00000 00000  00000 00000
6581:  00000 00000  00000 00000  00000 00000  00000 00000  00000 00000     00000 00000  00000 00000  00000 00000  00000 00000  00000 00000
6582:  00000 00000  00000 00000  00000 00000  00000 00000  00000 00000     00000 00000  00000 00000  00000 00000  00000 00000  00000 00000
6583:  00000 00000  00000 00000  00000 00000  00000 00000  00000 00000     00000 00000  00000 00000  00000 00000  00000 00000  00000 00000
6584:  00000 00000  00000 00000  00000 00000  00000 00000  00000 00000     00000 00000  00000 00000  00000 00000  00000 00000  00000 00000
6585:  00000 00000  00000 00000  00000 00000  00000 00000  00000 00000     00000 00000  00000 00000  00000 00000  00000 00000  00000 00000
6586:  00000 00000  00000 00000  00000 00000  00000 00000  00000 00000     00000 00000  00000 00000  00000 00000  00000 00000  00000 00000
6587:  00000 00000  00000 00000  00000 00000  00000 00000  00000 00000     00000 00000  00000 00000  00000 00000  00000 00000  00000 00000
6588:  00000 00000  00000 00000  00000 00000  00000 00000  00000 00000     00000 00000  00000 00000  00000 00000  00000 00000  00000 00000
6589:  00000 00000  00000 00000  00000 00000  00000 00000  00000 00000     00000 00000  00000 00000  00000 00000  00000 00000  00000 00000
6590:  00000 00000  00000 00000  00000 00000  00000 00000  00000 00000     00000 00000  00000 00000  00000 00000  00000 00000  00000 00000
6591:  00000 00000  00000 00000  00000 00000  00000 00000  00000 00000     00000 00000  00000 00000  00000 00000  00000 00000  00000 00000
6592:  00000 00000  00000 00000  00000 00000  00000 00000  00000 00000     00000 00000  00000 00000  00000 00000  00000 00000  00000 00000
6593:  00000 00000  00000 00000  00000 00000  00000 00000  00000 00000     00000 00000  00000 00000  00000 00000  00000 00000  00000 00000
6594:  00000 00000  00000 00000  00000 00000  00000 00000  00000 00000     00000 00000  00000 00000  00000 00000  00000 00000  00000 00000
6595:  00000 00000  00000 00000  00000 00000  00000 00000  00000 00000     00000 00000  00000 00000  00000 00000  00000 00000  00000 00000
6596:  00000 00000  00000 00000  00000 00000  00000 00000  00000 00000     00000 00000  00000 00000  00000 00000  00000 00000  00000 00000
6597:  00000 00000  00000 00000  00000 00000  00000 00000  00000 00000     00000 00000  00000 00000  00000 00000  00000 00000  00000 00000
6598:  00000 00000  00000 00000  00000 00000  00000 00000  00000 00000     00000 00000  00000 00000  00000 00000  00000 00000  00000 00000
6599:  00000 00000  00000 00000  00000 00000  00000 00000  00000 00000     00000 00000  00000 00000  00000 00000  00000 00000  00000 00000
```

```
6600:  00000 00000   00000 00000   00000 00000   00000 00000   00000 00000     00000 00000   00000 00000   00000 00000   00000 00000   00000 00000
6601:  00000 00000   00000 00000   00000 00000   00000 00000   00000 00000     00000 00000   00000 00000   00000 00000   00000 00000   00000 00000
6602:  00000 00000   00000 00000   00000 00000   00000 00000   00000 00000     00000 00000   00000 00000   00000 00000   00000 00000   00000 00000
6603:  00000 00000   00000 00000   00000 00000   00000 00000   00000 00000     00000 00000   00000 00000   00000 00000   00000 00000   00000 00000
6604:  00000 00000   00000 00000   00000 00000   00000 00000   00000 00000     00000 00000   00000 00000   00000 00000   00000 00000   00000 00000
6605:  00000 00000   00000 00000   00000 00000   00000 00000   00000 00000     00000 00000   00000 00000   00000 00000   00000 00000   00000 00000
6606:  00000 00000   00000 00000   00000 00000   00000 00000   00000 00000     00000 00000   00000 00000   00000 00000   00000 00000   00000 00000
6607:  00000 00000   00000 00000   00000 00000   00000 00000   00000 00000     00000 00000   00000 00000   00000 00000   00000 00000   00000 00000
6608:  00000 00000   00000 00000   00000 00000   00000 00000   00000 00000     00000 00000   00000 00000   00000 00000   00000 00000   00000 00000
6609:  00000 00000   00000 00000   00000 00000   00000 00000   00000 00000     00000 00000   00000 00000   00000 00000   00000 00000   00000 00000
6610:  00000 00000   00000 00000   00000 00000   00000 00000   00000 00000     00000 00000   00000 00000   00000 00000   00000 00000   00000 00000
6611:  00000 00000   00000 00000   00000 00000   00000 00000   00000 00000     00000 00000   00000 00000   00000 00000   00000 00000   00000 00000
6612:  00000 00000   00000 00000   00000 00000   00000 00000   00000 00000     00000 00000   00000 00000   00000 00000   00000 00000   00000 00000
6613:  00000 00000   00000 00000   00000 00000   00000 00000   00000 00000     00000 00000   00000 00000   00000 00000   00000 00000   00000 00000
6614:  00000 00000   00000 00000   00000 00000   00000 00000   00000 00000     00000 00000   00000 00000   00000 00000   00000 00000   00000 00000
6615:  00000 00000   00000 00000   00000 00000   00000 00000   00000 00000     00000 00000   00000 00000   00000 00000   00000 00000   00000 00000
6616:  00000 00000   00000 00000   00000 00000   00000 00000   00000 00000     00000 00000   00000 00000   00000 00000   00000 00000   00000 00000
6617:  00000 00000   00000 00000   00000 00000   00000 00000   00000 00000     00000 00000   00000 00000   00000 00000   00000 00000   00000 00000
6618:  00000 00000   00000 00000   00000 00000   00000 00000   00000 00000     00000 00000   00000 00000   00000 00000   00000 00000   00000 00000
6619:  00000 00000   00000 00000   00000 00000   00000 00000   00000 00000     00000 00000   00000 00000   00000 00000   00000 00000   00000 00000
6620:  00000 00000   00000 00000   00000 00000   00000 00000   00000 00000     00000 00000   00000 00000   00000 00000   00000 00000   00000 00000
6621:  00000 00000   00000 00000   00000 00000   00000 00000   00000 00000     00000 00000   00000 00000   00000 00000   00000 00000   00000 00000
6622:  00000 00000   00000 00000   00000 00000   00000 00000   00000 00000     00000 00000   00000 00000   00000 00000   00000 00000   00000 00000
6623:  00000 00000   00000 00000   00000 00000   00000 00000   00000 00000     00000 00000   00000 00000   00000 00000   00000 00000   00000 00000
6624:  00000 00000   00000 00000   00000 00000   00000 00000   00000 00000     00000 00000   00000 00000   00000 00000   00000 00000   00000 00000
6625:  00000 00000   00000 00000   00000 00000   00000 00000   00000 00000     00000 00000   00000 00000   00000 00000   00000 00000   00000 00000
6626:  00000 00000   00000 00000   00000 00000   00000 00000   00000 00000     00000 00000   00000 00000   00000 00000   00000 00000   00000 00000
6627:  00000 00000   00000 00000   00000 00000   00000 00000   00000 00000     00000 00000   00000 00000   00000 00000   00000 00000   00000 00000
6628:  00000 00000   00000 00000   00000 00000   00000 00000   00000 00000     00000 00000   00000 00000   00000 00000   00000 00000   00000 00000
6629:  00000 00000   00000 00000   00000 00000   00000 00000   00000 00000     00000 00000   00000 00000   00000 00000   00000 00000   00000 00000
6630:  00000 00000   00000 00000   00000 00000   00000 00000   00000 00000     00000 00000   00000 00000   00000 00000   00000 00000   00000 00000
6631:  00000 00000   00000 00000   00000 00000   00000 00000   00000 00000     00000 00000   00000 00000   00000 00000   00000 00000   00000 00000
6632:  00000 00000   00000 00000   00000 00000   00000 00000   00000 00000     00000 00000   00000 00000   00000 00000   00000 00000   00000 00000
6633:  00000 00000   00000 00000   00000 00000   00000 00000   00000 00000     00000 00000   00000 00000   00000 00000   00000 00000   00000 00000
6634:  00000 00000   00000 00000   00000 00000   00000 00000   00000 00000     00000 00000   00000 00000   00000 00000   00000 00000   00000 00000
6635:  00000 00000   00000 00000   00000 00000   00000 00000   00000 00000     00000 00000   00000 00000   00000 00000   00000 00000   00000 00000
6636:  00000 00000   00000 00000   00000 00000   00000 00000   00000 00000     00000 00000   00000 00000   00000 00000   00000 00000   00000 00000
6637:  00000 00000   00000 00000   00000 00000   00000 00000   00000 00000     00000 00000   00000 00000   00000 00000   00000 00000   00000 00000
6638:  00000 00000   00000 00000   00000 00000   00000 00000   00000 00000     00000 00000   00000 00000   00000 00000   00000 00000   00000 00000
6639:  00000 00000   00000 00000   00000 00000   00000 00000   00000 00000     00000 00000   00000 00000   00000 00000   00000 00000   00000 00000
6640:  00000 00000   00000 00000   00000 00000   00000 00000   00000 00000     00000 00000   00000 00000   00000 00000   00000 00000   00000 00000
6641:  00000 00000   00000 00000   00000 00000   00000 00000   00000 00000     00000 00000   00000 00000   00000 00000   00000 00000   00000 00000
6642:  00000 00000   00000 00000   00000 00000   00000 00000   00000 00000     00000 00000   00000 00000   00000 00000   00000 00000   00000 00000
6643:  00000 00000   00000 00000   00000 00000   00000 00000   00000 00000     00000 00000   00000 00000   00000 00000   00000 00000   00000 00000
6644:  00000 00000   00000 00000   00000 00000   00000 00000   00000 00000     00000 00000   00000 00000   00000 00000   00000 00000   00000 00000
6645:  00000 00000   00000 00000   00000 00000   00000 00000   00000 00000     00000 00000   00000 00000   00000 00000   00000 00000   00000 00000
6646:  00000 00000   00000 00000   00000 00000   00000 00000   00000 00000     00000 00000   00000 00000   00000 00000   00000 00000   00000 00000
6647:  00000 00000   00000 00000   00000 00000   00000 00000   00000 00000     00000 00000   00000 00000   00000 00000   00000 00000   00000 00000
6648:  00000 00000   00000 00000   00000 00000   00000 00000   00000 00000     00000 00000   00000 00000   00000 00000   00000 00000   00000 00000
6649:  00000 00000   00000 00000   00000 00000   00000 00000   00000 00000     00000 00000   00000 00000   00000 00000   00000 00000   00000 00000
```

```
6650:  00000 00000  00000 00000  00000 00000  00000 00000  00000 00000    00000 00000  00000 00000  00000 00000  00000 00000  00000 00000
6651:  00000 00000  00000 00000  00000 00000  00000 00000  00000 00000    00000 00000  00000 00000  00000 00000  00000 00000  00000 00000
6652:  00000 00000  00000 00000  00000 00000  00000 00000  00000 00000    00000 00000  00000 00000  00000 00000  00000 00000  00000 00000
6653:  00000 00000  00000 00000  00000 00000  00000 00000  00000 00000    00000 00000  00000 00000  00000 00000  00000 00000  00000 00000
6654:  00000 00000  00000 00000  00000 00000  00000 00000  00000 00000    00000 00000  00000 00000  00000 00000  00000 00000  00000 00000
6655:  00000 00000  00000 00000  00000 00000  00000 00000  00000 00000    00000 00000  00000 00000  00000 00000  00000 00000  00000 00000
6656:  00000 00000  00000 00000  00000 00000  00000 00000  00000 00000    00000 00000  00000 00000  00000 00000  00000 00000  00000 00000
6657:  00000 00000  00000 00000  00000 00000  00000 00000  00000 00000    00000 00000  00000 00000  00000 00000  00000 00000  00000 00000
6658:  00000 00000  00000 00000  00000 00000  00000 00000  00000 00000    00000 00000  00000 00000  00000 00000  00000 00000  00000 00000
6659:  00000 00000  00000 00000  00000 00000  00000 00000  00000 00000    00000 00000  00000 00000  00000 00000  00000 00000  00000 00000
6660:  00000 00000  00000 00000  00000 00000  00000 00000  00000 00000    00000 00000  00000 00000  00000 00000  00000 00000  00000 00000
6661:  00000 00000  00000 00000  00000 00000  00000 00000  00000 00000    00000 00000  00000 00000  00000 00000  00000 00000  00000 00000
6662:  00000 00000  00000 00000  00000 00000  00000 00000  00000 00000    00000 00000  00000 00000  00000 00000  00000 00000  00000 00000
6663:  00000 00000  00000 00000  00000 00000  00000 00000  00000 00000    00000 00000  00000 00000  00000 00000  00000 00000  00000 00000
6664:  00000 00000  00000 00000  00000 00000  00000 00000  00000 00000    00000 00000  00000 00000  00000 00000  00000 00000  00000 00000
6665:  00000 00000  00000 00000  00000 00000  00000 00000  00000 00000    00000 00000  00000 00000  00000 00000  00000 00000  00000 00000
6666:  00000 00000  00000 00000  00000 00000  00000 00000  00000 00000    00000 00000  00000 00000  00000 00000  00000 00000  00000 00000
6667:  00000 00000  00000 00000  00000 00000  00000 00000  00000 00000    00000 00000  00000 00000  00000 00000  00000 00000  00000 00000
6668:  00000 00000  00000 00000  00000 00000  00000 00000  00000 00000    00000 00000  00000 00000  00000 00000  00000 00000  00000 00000
6669:  00000 00000  00000 00000  00000 00000  00000 00000  00000 00000    00000 00000  00000 00000  00000 00000  00000 00000  00000 00000
6670:  00000 00000  00000 00000  00000 00000  00000 00000  00000 00000    00000 00000  00000 00000  00000 00000  00000 00000  00000 00000
6671:  00000 00000  00000 00000  00000 00000  00000 00000  00000 00000    00000 00000  00000 00000  00000 00000  00000 00000  00000 00000
6672:  00000 00000  00000 00000  00000 00000  00000 00000  00000 00000    00000 00000  00000 00000  00000 00000  00000 00000  00000 00000
6673:  00000 00000  00000 00000  00000 00000  00000 00000  00000 00000    00000 00000  00000 00000  00000 00000  00000 00000  00000 00000
6674:  00000 00000  00000 00000  00000 00000  00000 00000  00000 00000    00000 00000  00000 00000  00000 00000  00000 00000  00000 00000
6675:  00000 00000  00000 00000  00000 00000  00000 00000  00000 00000    00000 00000  00000 00000  00000 00000  00000 00000  00000 00000
6676:  00000 00000  00000 00000  00000 00000  00000 00000  00000 00000    00000 00000  00000 00000  00000 00000  00000 00000  00000 00000
6677:  00000 00000  00000 00000  00000 00000  00000 00000  00000 00000    00000 00000  00000 00000  00000 00000  00000 00000  00000 00000
6678:  00000 00000  00000 00000  00000 00000  00000 00000  00000 00000    00000 00000  00000 00000  00000 00000  00000 00000  00000 00000
6679:  00000 00000  00000 00000  00000 00000  00000 00000  00000 00000    00000 00000  00000 00000  00000 00000  00000 00000  00000 00000
6680:  00000 00000  00000 00000  00000 00000  00000 00000  00000 00000    00000 00000  00000 00000  00000 00000  00000 00000  00000 00000
6681:  00000 00000  00000 00000  00000 00000  00000 00000  00000 00000    00000 00000  00000 00000  00000 00000  00000 00000  00000 00000
6682:  00000 00000  00000 00000  00000 00000  00000 00000  00000 00000    00000 00000  00000 00000  00000 00000  00000 00000  00000 00000
6683:  00000 00000  00000 00000  00000 00000  00000 00000  00000 00000    00000 00000  00000 00000  00000 00000  00000 00000  00000 00000
6684:  00000 00000  00000 00000  00000 00000  00000 00000  00000 00000    00000 00000  00000 00000  00000 00000  00000 00000  00000 00000
6685:  00000 00000  00000 00000  00000 00000  00000 00000  00000 00000    00000 00000  00000 00000  00000 00000  00000 00000  00000 00000
6686:  00000 00000  00000 00000  00000 00000  00000 00000  00000 00000    00000 00000  00000 00000  00000 00000  00000 00000  00000 00000
6687:  00000 00000  00000 00000  00000 00000  00000 00000  00000 00000    00000 00000  00000 00000  00000 00000  00000 00000  00000 00000
6688:  00000 00000  00000 00000  00000 00000  00000 00000  00000 00000    00000 00000  00000 00000  00000 00000  00000 00000  00000 00000
6689:  00000 00000  00000 00000  00000 00000  00000 00000  00000 00000    00000 00000  00000 00000  00000 00000  00000 00000  00000 00000
6690:  00000 00000  00000 00000  00000 00000  00000 00000  00000 00000    00000 00000  00000 00000  00000 00000  00000 00000  00000 00000
6691:  00000 00000  00000 00000  00000 00000  00000 00000  00000 00000    00000 00000  00000 00000  00000 00000  00000 00000  00000 00000
6692:  00000 00000  00000 00000  00000 00000  00000 00000  00000 00000    00000 00000  00000 00000  00000 00000  00000 00000  00000 00000
6693:  00000 00000  00000 00000  00000 00000  00000 00000  00000 00000    00000 00000  00000 00000  00000 00000  00000 00000  00000 00000
6694:  00000 00000  00000 00000  00000 00000  00000 00000  00000 00000    00000 00000  00000 00000  00000 00000  00000 00000  00000 00000
6695:  00000 00000  00000 00000  00000 00000  00000 00000  00000 00000    00000 00000  00000 00000  00000 00000  00000 00000  00000 00000
6696:  00000 00000  00000 00000  00000 00000  00000 00000  00000 00000    00000 00000  00000 00000  00000 00000  00000 00000  00000 00000
6697:  00000 00000  00000 00000  00000 00000  00000 00000  00000 00000    00000 00000  00000 00000  00000 00000  00000 00000  00000 00000
6698:  00000 00000  00000 00000  00000 00000  00000 00000  00000 00000    00000 00000  00000 00000  00000 00000  00000 00000  00000 00000
6699:  00000 00000  00000 00000  00000 00000  00000 00000  00000 00000    00000 00000  00000 00000  00000 00000  00000 00000  00000 00000
```

```
6700:   00000 00000   00000 00000   00000 00000   00000 00000   00000 00000     00000 00000   00000 00000   00000 00000   00000 00000   00000 00000
6701:   00000 00000   00000 00000   00000 00000   00000 00000   00000 00000     00000 00000   00000 00000   00000 00000   00000 00000   00000 00000
6702:   00000 00000   00000 00000   00000 00000   00000 00000   00000 00000     00000 00000   00000 00000   00000 00000   00000 00000   00000 00000
6703:   00000 00000   00000 00000   00000 00000   00000 00000   00000 00000     00000 00000   00000 00000   00000 00000   00000 00000   00000 00000
6704:   00000 00000   00000 00000   00000 00000   00000 00000   00000 00000     00000 00000   00000 00000   00000 00000   00000 00000   00000 00000
6705:   00000 00000   00000 00000   00000 00000   00000 00000   00000 00000     00000 00000   00000 00000   00000 00000   00000 00000   00000 00000
6706:   00000 00000   00000 00000   00000 00000   00000 00000   00000 00000     00000 00000   00000 00000   00000 00000   00000 00000   00000 00000
6707:   00000 00000   00000 00000   00000 00000   00000 00000   00000 00000     00000 00000   00000 00000   00000 00000   00000 00000   00000 00000
6708:   00000 00000   00000 00000   00000 00000   00000 00000   00000 00000     00000 00000   00000 00000   00000 00000   00000 00000   00000 00000
6709:   00000 00000   00000 00000   00000 00000   00000 00000   00000 00000     00000 00000   00000 00000   00000 00000   00000 00000   00000 00000
6710:   00000 00000   00000 00000   00000 00000   00000 00000   00000 00000     00000 00000   00000 00000   00000 00000   00000 00000   00000 00000
6711:   00000 00000   00000 00000   00000 00000   00000 00000   00000 00000     00000 00000   00000 00000   00000 00000   00000 00000   00000 00000
6712:   00000 00000   00000 00000   00000 00000   00000 00000   00000 00000     00000 00000   00000 00000   00000 00000   00000 00000   00000 00000
6713:   00000 00000   00000 00000   00000 00000   00000 00000   00000 00000     00000 00000   00000 00000   00000 00000   00000 00000   00000 00000
6714:   00000 00000   00000 00000   00000 00000   00000 00000   00000 00000     00000 00000   00000 00000   00000 00000   00000 00000   00000 00000
6715:   00000 00000   00000 00000   00000 00000   00000 00000   00000 00000     00000 00000   00000 00000   00000 00000   00000 00000   00000 00000
6716:   00000 00000   00000 00000   00000 00000   00000 00000   00000 00000     00000 00000   00000 00000   00000 00000   00000 00000   00000 00000
6717:   00000 00000   00000 00000   00000 00000   00000 00000   00000 00000     00000 00000   00000 00000   00000 00000   00000 00000   00000 00000
6718:   00000 00000   00000 00000   00000 00000   00000 00000   00000 00000     00000 00000   00000 00000   00000 00000   00000 00000   00000 00000
6719:   00000 00000   00000 00000   00000 00000   00000 00000   00000 00000     00000 00000   00000 00000   00000 00000   00000 00000   00000 00000
6720:   00000 00000   00000 00000   00000 00000   00000 00000   00000 00000     00000 00000   00000 00000   00000 00000   00000 00000   00000 00000
6721:   00000 00000   00000 00000   00000 00000   00000 00000   00000 00000     00000 00000   00000 00000   00000 00000   00000 00000   00000 00000
6722:   00000 00000   00000 00000   00000 00000   00000 00000   00000 00000     00000 00000   00000 00000   00000 00000   00000 00000   00000 00000
6723:   00000 00000   00000 00000   00000 00000   00000 00000   00000 00000     00000 00000   00000 00000   00000 00000   00000 00000   00000 00000
6724:   00000 00000   00000 00000   00000 00000   00000 00000   00000 00000     00000 00000   00000 00000   00000 00000   00000 00000   00000 00000
6725:   00000 00000   00000 00000   00000 00000   00000 00000   00000 00000     00000 00000   00000 00000   00000 00000   00000 00000   00000 00000
6726:   00000 00000   00000 00000   00000 00000   00000 00000   00000 00000     00000 00000   00000 00000   00000 00000   00000 00000   00000 00000
6727:   00000 00000   00000 00000   00000 00000   00000 00000   00000 00000     00000 00000   00000 00000   00000 00000   00000 00000   00000 00000
6728:   00000 00000   00000 00000   00000 00000   00000 00000   00000 00000     00000 00000   00000 00000   00000 00000   00000 00000   00000 00000
6729:   00000 00000   00000 00000   00000 00000   00000 00000   00000 00000     00000 00000   00000 00000   00000 00000   00000 00000   00000 00000
6730:   00000 00000   00000 00000   00000 00000   00000 00000   00000 00000     00000 00000   00000 00000   00000 00000   00000 00000   00000 00000
6731:   00000 00000   00000 00000   00000 00000   00000 00000   00000 00000     00000 00000   00000 00000   00000 00000   00000 00000   00000 00000
6732:   00000 00000   00000 00000   00000 00000   00000 00000   00000 00000     00000 00000   00000 00000   00000 00000   00000 00000   00000 00000
6733:   00000 00000   00000 00000   00000 00000   00000 00000   00000 00000     00000 00000   00000 00000   00000 00000   00000 00000   00000 00000
6734:   00000 00000   00000 00000   00000 00000   00000 00000   00000 00000     00000 00000   00000 00000   00000 00000   00000 00000   00000 00000
6735:   00000 00000   00000 00000   00000 00000   00000 00000   00000 00000     00000 00000   00000 00000   00000 00000   00000 00000   00000 00000
6736:   00000 00000   00000 00000   00000 00000   00000 00000   00000 00000     00000 00000   00000 00000   00000 00000   00000 00000   00000 00000
6737:   00000 00000   00000 00000   00000 00000   00000 00000   00000 00000     00000 00000   00000 00000   00000 00000   00000 00000   00000 00000
6738:   00000 00000   00000 00000   00000 00000   00000 00000   00000 00000     00000 00000   00000 00000   00000 00000   00000 00000   00000 00000
6739:   00000 00000   00000 00000   00000 00000   00000 00000   00000 00000     00000 00000   00000 00000   00000 00000   00000 00000   00000 00000
6740:   00000 00000   00000 00000   00000 00000   00000 00000   00000 00000     00000 00000   00000 00000   00000 00000   00000 00000   00000 00000
6741:   00000 00000   00000 00000   00000 00000   00000 00000   00000 00000     00000 00000   00000 00000   00000 00000   00000 00000   00000 00000
6742:   00000 00000   00000 00000   00000 00000   00000 00000   00000 00000     00000 00000   00000 00000   00000 00000   00000 00000   00000 00000
6743:   00000 00000   00000 00000   00000 00000   00000 00000   00000 00000     00000 00000   00000 00000   00000 00000   00000 00000   00000 00000
6744:   00000 00000   00000 00000   00000 00000   00000 00000   00000 00000     00000 00000   00000 00000   00000 00000   00000 00000   00000 00000
6745:   00000 00000   00000 00000   00000 00000   00000 00000   00000 00000     00000 00000   00000 00000   00000 00000   00000 00000   00000 00000
6746:   00000 00000   00000 00000   00000 00000   00000 00000   00000 00000     00000 00000   00000 00000   00000 00000   00000 00000   00000 00000
6747:   00000 00000   00000 00000   00000 00000   00000 00000   00000 00000     00000 00000   00000 00000   00000 00000   00000 00000   00000 00000
6748:   00000 00000   00000 00000   00000 00000   00000 00000   00000 00000     00000 00000   00000 00000   00000 00000   00000 00000   00000 00000
6749:   00000 00000   00000 00000   00000 00000   00000 00000   00000 00000     00000 00000   00000 00000   00000 00000   00000 00000   00000 00000
```

```
6750:  00000 00000  00000 00000  00000 00000  00000 00000  00000 00000    00000 00000  00000 00000  00000 00000  00000 00000  00000 00000
6751:  00000 00000  00000 00000  00000 00000  00000 00000  00000 00000    00000 00000  00000 00000  00000 00000  00000 00000  00000 00000
6752:  00000 00000  00000 00000  00000 00000  00000 00000  00000 00000    00000 00000  00000 00000  00000 00000  00000 00000  00000 00000
6753:  00000 00000  00000 00000  00000 00000  00000 00000  00000 00000    00000 00000  00000 00000  00000 00000  00000 00000  00000 00000
6754:  00000 00000  00000 00000  00000 00000  00000 00000  00000 00000    00000 00000  00000 00000  00000 00000  00000 00000  00000 00000
6755:  00000 00000  00000 00000  00000 00000  00000 00000  00000 00000    00000 00000  00000 00000  00000 00000  00000 00000  00000 00000
6756:  00000 00000  00000 00000  00000 00000  00000 00000  00000 00000    00000 00000  00000 00000  00000 00000  00000 00000  00000 00000
6757:  00000 00000  00000 00000  00000 00000  00000 00000  00000 00000    00000 00000  00000 00000  00000 00000  00000 00000  00000 00000
6758:  00000 00000  00000 00000  00000 00000  00000 00000  00000 00000    00000 00000  00000 00000  00000 00000  00000 00000  00000 00000
6759:  00000 00000  00000 00000  00000 00000  00000 00000  00000 00000    00000 00000  00000 00000  00000 00000  00000 00000  00000 00000
6760:  00000 00000  00000 00000  00000 00000  00000 00000  00000 00000    00000 00000  00000 00000  00000 00000  00000 00000  00000 00000
6761:  00000 00000  00000 00000  00000 00000  00000 00000  00000 00000    00000 00000  00000 00000  00000 00000  00000 00000  00000 00000
6762:  00000 00000  00000 00000  00000 00000  00000 00000  00000 00000    00000 00000  00000 00000  00000 00000  00000 00000  00000 00000
6763:  00000 00000  00000 00000  00000 00000  00000 00000  00000 00000    00000 00000  00000 00000  00000 00000  00000 00000  00000 00000
6764:  00000 00000  00000 00000  00000 00000  00000 00000  00000 00000    00000 00000  00000 00000  00000 00000  00000 00000  00000 00000
6765:  00000 00000  00000 00000  00000 00000  00000 00000  00000 00000    00000 00000  00000 00000  00000 00000  00000 00000  00000 00000
6766:  00000 00000  00000 00000  00000 00000  00000 00000  00000 00000    00000 00000  00000 00000  00000 00000  00000 00000  00000 00000
6767:  00000 00000  00000 00000  00000 00000  00000 00000  00000 00000    00000 00000  00000 00000  00000 00000  00000 00000  00000 00000
6768:  00000 00000  00000 00000  00000 00000  00000 00000  00000 00000    00000 00000  00000 00000  00000 00000  00000 00000  00000 00000
6769:  00000 00000  00000 00000  00000 00000  00000 00000  00000 00000    00000 00000  00000 00000  00000 00000  00000 00000  00000 00000
6770:  00000 00000  00000 00000  00000 00000  00000 00000  00000 00000    00000 00000  00000 00000  00000 00000  00000 00000  00000 00000
6771:  00000 00000  00000 00000  00000 00000  00000 00000  00000 00000    00000 00000  00000 00000  00000 00000  00000 00000  00000 00000
6772:  00000 00000  00000 00000  00000 00000  00000 00000  00000 00000    00000 00000  00000 00000  00000 00000  00000 00000  00000 00000
6773:  00000 00000  00000 00000  00000 00000  00000 00000  00000 00000    00000 00000  00000 00000  00000 00000  00000 00000  00000 00000
6774:  00000 00000  00000 00000  00000 00000  00000 00000  00000 00000    00000 00000  00000 00000  00000 00000  00000 00000  00000 00000
6775:  00000 00000  00000 00000  00000 00000  00000 00000  00000 00000    00000 00000  00000 00000  00000 00000  00000 00000  00000 00000
6776:  00000 00000  00000 00000  00000 00000  00000 00000  00000 00000    00000 00000  00000 00000  00000 00000  00000 00000  00000 00000
6777:  00000 00000  00000 00000  00000 00000  00000 00000  00000 00000    00000 00000  00000 00000  00000 00000  00000 00000  00000 00000
6778:  00000 00000  00000 00000  00000 00000  00000 00000  00000 00000    00000 00000  00000 00000  00000 00000  00000 00000  00000 00000
6779:  00000 00000  00000 00000  00000 00000  00000 00000  00000 00000    00000 00000  00000 00000  00000 00000  00000 00000  00000 00000
6780:  00000 00000  00000 00000  00000 00000  00000 00000  00000 00000    00000 00000  00000 00000  00000 00000  00000 00000  00000 00000
6781:  00000 00000  00000 00000  00000 00000  00000 00000  00000 00000    00000 00000  00000 00000  00000 00000  00000 00000  00000 00000
6782:  00000 00000  00000 00000  00000 00000  00000 00000  00000 00000    00000 00000  00000 00000  00000 00000  00000 00000  00000 00000
6783:  00000 00000  00000 00000  00000 00000  00000 00000  00000 00000    00000 00000  00000 00000  00000 00000  00000 00000  00000 00000
6784:  00000 00000  00000 00000  00000 00000  00000 00000  00000 00000    00000 00000  00000 00000  00000 00000  00000 00000  00000 00000
6785:  00000 00000  00000 00000  00000 00000  00000 00000  00000 00000    00000 00000  00000 00000  00000 00000  00000 00000  00000 00000
6786:  00000 00000  00000 00000  00000 00000  00000 00000  00000 00000    00000 00000  00000 00000  00000 00000  00000 00000  00000 00000
6787:  00000 00000  00000 00000  00000 00000  00000 00000  00000 00000    00000 00000  00000 00000  00000 00000  00000 00000  00000 00000
6788:  00000 00000  00000 00000  00000 00000  00000 00000  00000 00000    00000 00000  00000 00000  00000 00000  00000 00000  00000 00000
6789:  00000 00000  00000 00000  00000 00000  00000 00000  00000 00000    00000 00000  00000 00000  00000 00000  00000 00000  00000 00000
6790:  00000 00000  00000 00000  00000 00000  00000 00000  00000 00000    00000 00000  00000 00000  00000 00000  00000 00000  00000 00000
6791:  00000 00000  00000 00000  00000 00000  00000 00000  00000 00000    00000 00000  00000 00000  00000 00000  00000 00000  00000 00000
6792:  00000 00000  00000 00000  00000 00000  00000 00000  00000 00000    00000 00000  00000 00000  00000 00000  00000 00000  00000 00000
6793:  00000 00000  00000 00000  00000 00000  00000 00000  00000 00000    00000 00000  00000 00000  00000 00000  00000 00000  00000 00000
6794:  00000 00000  00000 00000  00000 00000  00000 00000  00000 00000    00000 00000  00000 00000  00000 00000  00000 00000  00000 00000
6795:  00000 00000  00000 00000  00000 00000  00000 00000  00000 00000    00000 00000  00000 00000  00000 00000  00000 00000  00000 00000
6796:  00000 00000  00000 00000  00000 00000  00000 00000  00000 00000    00000 00000  00000 00000  00000 00000  00000 00000  00000 00000
6797:  00000 00000  00000 00000  00000 00000  00000 00000  00000 00000    00000 00000  00000 00000  00000 00000  00000 00000  00000 00000
6798:  00000 00000  00000 00000  00000 00000  00000 00000  00000 00000    00000 00000  00000 00000  00000 00000  00000 00000  00000 00000
6799:  00000 00000  00000 00000  00000 00000  00000 00000  00000 00000    00000 00000  00000 00000  00000 00000  00000 00000  00000 00000
```

```
6800:  00000 00000   00000 00000   00000 00000   00000 00000   00000 00000     00000 00000   00000 00000   00000 00000   00000 00000   00000 00000
6801:  00000 00000   00000 00000   00000 00000   00000 00000   00000 00000     00000 00000   00000 00000   00000 00000   00000 00000   00000 00000
6802:  00000 00000   00000 00000   00000 00000   00000 00000   00000 00000     00000 00000   00000 00000   00000 00000   00000 00000   00000 00000
6803:  00000 00000   00000 00000   00000 00000   00000 00000   00000 00000     00000 00000   00000 00000   00000 00000   00000 00000   00000 00000
6804:  00000 00000   00000 00000   00000 00000   00000 00000   00000 00000     00000 00000   00000 00000   00000 00000   00000 00000   00000 00000
6805:  00000 00000   00000 00000   00000 00000   00000 00000   00000 00000     00000 00000   00000 00000   00000 00000   00000 00000   00000 00000
6806:  00000 00000   00000 00000   00000 00000   00000 00000   00000 00000     00000 00000   00000 00000   00000 00000   00000 00000   00000 00000
6807:  00000 00000   00000 00000   00000 00000   00000 00000   00000 00000     00000 00000   00000 00000   00000 00000   00000 00000   00000 00000
6808:  00000 00000   00000 00000   00000 00000   00000 00000   00000 00000     00000 00000   00000 00000   00000 00000   00000 00000   00000 00000
6809:  00000 00000   00000 00000   00000 00000   00000 00000   00000 00000     00000 00000   00000 00000   00000 00000   00000 00000   00000 00000
6810:  00000 00000   00000 00000   00000 00000   00000 00000   00000 00000     00000 00000   00000 00000   00000 00000   00000 00000   00000 00000
6811:  00000 00000   00000 00000   00000 00000   00000 00000   00000 00000     00000 00000   00000 00000   00000 00000   00000 00000   00000 00000
6812:  00000 00000   00000 00000   00000 00000   00000 00000   00000 00000     00000 00000   00000 00000   00000 00000   00000 00000   00000 00000
6813:  00000 00000   00000 00000   00000 00000   00000 00000   00000 00000     00000 00000   00000 00000   00000 00000   00000 00000   00000 00000
6814:  00000 00000   00000 00000   00000 00000   00000 00000   00000 00000     00000 00000   00000 00000   00000 00000   00000 00000   00000 00000
6815:  00000 00000   00000 00000   00000 00000   00000 00000   00000 00000     00000 00000   00000 00000   00000 00000   00000 00000   00000 00000
6816:  00000 00000   00000 00000   00000 00000   00000 00000   00000 00000     00000 00000   00000 00000   00000 00000   00000 00000   00000 00000
6817:  00000 00000   00000 00000   00000 00000   00000 00000   00000 00000     00000 00000   00000 00000   00000 00000   00000 00000   00000 00000
6818:  00000 00000   00000 00000   00000 00000   00000 00000   00000 00000     00000 00000   00000 00000   00000 00000   00000 00000   00000 00000
6819:  00000 00000   00000 00000   00000 00000   00000 00000   00000 00000     00000 00000   00000 00000   00000 00000   00000 00000   00000 00000
6820:  00000 00000   00000 00000   00000 00000   00000 00000   00000 00000     00000 00000   00000 00000   00000 00000   00000 00000   00000 00000
6821:  00000 00000   00000 00000   00000 00000   00000 00000   00000 00000     00000 00000   00000 00000   00000 00000   00000 00000   00000 00000
6822:  00000 00000   00000 00000   00000 00000   00000 00000   00000 00000     00000 00000   00000 00000   00000 00000   00000 00000   00000 00000
6823:  00000 00000   00000 00000   00000 00000   00000 00000   00000 00000     00000 00000   00000 00000   00000 00000   00000 00000   00000 00000
6824:  00000 00000   00000 00000   00000 00000   00000 00000   00000 00000     00000 00000   00000 00000   00000 00000   00000 00000   00000 00000
6825:  00000 00000   00000 00000   00000 00000   00000 00000   00000 00000     00000 00000   00000 00000   00000 00000   00000 00000   00000 00000
6826:  00000 00000   00000 00000   00000 00000   00000 00000   00000 00000     00000 00000   00000 00000   00000 00000   00000 00000   00000 00000
6827:  00000 00000   00000 00000   00000 00000   00000 00000   00000 00000     00000 00000   00000 00000   00000 00000   00000 00000   00000 00000
6828:  00000 00000   00000 00000   00000 00000   00000 00000   00000 00000     00000 00000   00000 00000   00000 00000   00000 00000   00000 00000
6829:  00000 00000   00000 00000   00000 00000   00000 00000   00000 00000     00000 00000   00000 00000   00000 00000   00000 00000   00000 00000
6830:  00000 00000   00000 00000   00000 00000   00000 00000   00000 00000     00000 00000   00000 00000   00000 00000   00000 00000   00000 00000
6831:  00000 00000   00000 00000   00000 00000   00000 00000   00000 00000     00000 00000   00000 00000   00000 00000   00000 00000   00000 00000
6832:  00000 00000   00000 00000   00000 00000   00000 00000   00000 00000     00000 00000   00000 00000   00000 00000   00000 00000   00000 00000
6833:  00000 00000   00000 00000   00000 00000   00000 00000   00000 00000     00000 00000   00000 00000   00000 00000   00000 00000   00000 00000
6834:  00000 00000   00000 00000   00000 00000   00000 00000   00000 00000     00000 00000   00000 00000   00000 00000   00000 00000   00000 00000
6835:  00000 00000   00000 00000   00000 00000   00000 00000   00000 00000     00000 00000   00000 00000   00000 00000   00000 00000   00000 00000
6836:  00000 00000   00000 00000   00000 00000   00000 00000   00000 00000     00000 00000   00000 00000   00000 00000   00000 00000   00000 00000
6837:  00000 00000   00000 00000   00000 00000   00000 00000   00000 00000     00000 00000   00000 00000   00000 00000   00000 00000   00000 00000
6838:  00000 00000   00000 00000   00000 00000   00000 00000   00000 00000     00000 00000   00000 00000   00000 00000   00000 00000   00000 00000
6839:  00000 00000   00000 00000   00000 00000   00000 00000   00000 00000     00000 00000   00000 00000   00000 00000   00000 00000   00000 00000
6840:  00000 00000   00000 00000   00000 00000   00000 00000   00000 00000     00000 00000   00000 00000   00000 00000   00000 00000   00000 00000
6841:  00000 00000   00000 00000   00000 00000   00000 00000   00000 00000     00000 00000   00000 00000   00000 00000   00000 00000   00000 00000
6842:  00000 00000   00000 00000   00000 00000   00000 00000   00000 00000     00000 00000   00000 00000   00000 00000   00000 00000   00000 00000
6843:  00000 00000   00000 00000   00000 00000   00000 00000   00000 00000     00000 00000   00000 00000   00000 00000   00000 00000   00000 00000
6844:  00000 00000   00000 00000   00000 00000   00000 00000   00000 00000     00000 00000   00000 00000   00000 00000   00000 00000   00000 00000
6845:  00000 00000   00000 00000   00000 00000   00000 00000   00000 00000     00000 00000   00000 00000   00000 00000   00000 00000   00000 00000
6846:  00000 00000   00000 00000   00000 00000   00000 00000   00000 00000     00000 00000   00000 00000   00000 00000   00000 00000   00000 00000
6847:  00000 00000   00000 00000   00000 00000   00000 00000   00000 00000     00000 00000   00000 00000   00000 00000   00000 00000   00000 00000
6848:  00000 00000   00000 00000   00000 00000   00000 00000   00000 00000     00000 00000   00000 00000   00000 00000   00000 00000   00000 00000
6849:  00000 00000   00000 00000   00000 00000   00000 00000   00000 00000     00000 00000   00000 00000   00000 00000   00000 00000   00000 00000
```

```
6850:  00000 00000  00000 00000  00000 00000  00000 00000  00000 00000    00000 00000  00000 00000  00000 00000  00000 00000  00000 00000
6851:  00000 00000  00000 00000  00000 00000  00000 00000  00000 00000    00000 00000  00000 00000  00000 00000  00000 00000  00000 00000
6852:  00000 00000  00000 00000  00000 00000  00000 00000  00000 00000    00000 00000  00000 00000  00000 00000  00000 00000  00000 00000
6853:  00000 00000  00000 00000  00000 00000  00000 00000  00000 00000    00000 00000  00000 00000  00000 00000  00000 00000  00000 00000
6854:  00000 00000  00000 00000  00000 00000  00000 00000  00000 00000    00000 00000  00000 00000  00000 00000  00000 00000  00000 00000
6855:  00000 00000  00000 00000  00000 00000  00000 00000  00000 00000    00000 00000  00000 00000  00000 00000  00000 00000  00000 00000
6856:  00000 00000  00000 00000  00000 00000  00000 00000  00000 00000    00000 00000  00000 00000  00000 00000  00000 00000  00000 00000
6857:  00000 00000  00000 00000  00000 00000  00000 00000  00000 00000    00000 00000  00000 00000  00000 00000  00000 00000  00000 00000
6858:  00000 00000  00000 00000  00000 00000  00000 00000  00000 00000    00000 00000  00000 00000  00000 00000  00000 00000  00000 00000
6859:  00000 00000  00000 00000  00000 00000  00000 00000  00000 00000    00000 00000  00000 00000  00000 00000  00000 00000  00000 00000
6860:  00000 00000  00000 00000  00000 00000  00000 00000  00000 00000    00000 00000  00000 00000  00000 00000  00000 00000  00000 00000
6861:  00000 00000  00000 00000  00000 00000  00000 00000  00000 00000    00000 00000  00000 00000  00000 00000  00000 00000  00000 00000
6862:  00000 00000  00000 00000  00000 00000  00000 00000  00000 00000    00000 00000  00000 00000  00000 00000  00000 00000  00000 00000
6863:  00000 00000  00000 00000  00000 00000  00000 00000  00000 00000    00000 00000  00000 00000  00000 00000  00000 00000  00000 00000
6864:  00000 00000  00000 00000  00000 00000  00000 00000  00000 00000    00000 00000  00000 00000  00000 00000  00000 00000  00000 00000
6865:  00000 00000  00000 00000  00000 00000  00000 00000  00000 00000    00000 00000  00000 00000  00000 00000  00000 00000  00000 00000
6866:  00000 00000  00000 00000  00000 00000  00000 00000  00000 00000    00000 00000  00000 00000  00000 00000  00000 00000  00000 00000
6867:  00000 00000  00000 00000  00000 00000  00000 00000  00000 00000    00000 00000  00000 00000  00000 00000  00000 00000  00000 00000
6868:  00000 00000  00000 00000  00000 00000  00000 00000  00000 00000    00000 00000  00000 00000  00000 00000  00000 00000  00000 00000
6869:  00000 00000  00000 00000  00000 00000  00000 00000  00000 00000    00000 00000  00000 00000  00000 00000  00000 00000  00000 00000
6870:  00000 00000  00000 00000  00000 00000  00000 00000  00000 00000    00000 00000  00000 00000  00000 00000  00000 00000  00000 00000
6871:  00000 00000  00000 00000  00000 00000  00000 00000  00000 00000    00000 00000  00000 00000  00000 00000  00000 00000  00000 00000
6872:  00000 00000  00000 00000  00000 00000  00000 00000  00000 00000    00000 00000  00000 00000  00000 00000  00000 00000  00000 00000
6873:  00000 00000  00000 00000  00000 00000  00000 00000  00000 00000    00000 00000  00000 00000  00000 00000  00000 00000  00000 00000
6874:  00000 00000  00000 00000  00000 00000  00000 00000  00000 00000    00000 00000  00000 00000  00000 00000  00000 00000  00000 00000
6875:  00000 00000  00000 00000  00000 00000  00000 00000  00000 00000    00000 00000  00000 00000  00000 00000  00000 00000  00000 00000
6876:  00000 00000  00000 00000  00000 00000  00000 00000  00000 00000    00000 00000  00000 00000  00000 00000  00000 00000  00000 00000
6877:  00000 00000  00000 00000  00000 00000  00000 00000  00000 00000    00000 00000  00000 00000  00000 00000  00000 00000  00000 00000
6878:  00000 00000  00000 00000  00000 00000  00000 00000  00000 00000    00000 00000  00000 00000  00000 00000  00000 00000  00000 00000
6879:  00000 00000  00000 00000  00000 00000  00000 00000  00000 00000    00000 00000  00000 00000  00000 00000  00000 00000  00000 00000
6880:  00000 00000  00000 00000  00000 00000  00000 00000  00000 00000    00000 00000  00000 00000  00000 00000  00000 00000  00000 00000
6881:  00000 00000  00000 00000  00000 00000  00000 00000  00000 00000    00000 00000  00000 00000  00000 00000  00000 00000  00000 00000
6882:  00000 00000  00000 00000  00000 00000  00000 00000  00000 00000    00000 00000  00000 00000  00000 00000  00000 00000  00000 00000
6883:  00000 00000  00000 00000  00000 00000  00000 00000  00000 00000    00000 00000  00000 00000  00000 00000  00000 00000  00000 00000
6884:  00000 00000  00000 00000  00000 00000  00000 00000  00000 00000    00000 00000  00000 00000  00000 00000  00000 00000  00000 00000
6885:  00000 00000  00000 00000  00000 00000  00000 00000  00000 00000    00000 00000  00000 00000  00000 00000  00000 00000  00000 00000
6886:  00000 00000  00000 00000  00000 00000  00000 00000  00000 00000    00000 00000  00000 00000  00000 00000  00000 00000  00000 00000
6887:  00000 00000  00000 00000  00000 00000  00000 00000  00000 00000    00000 00000  00000 00000  00000 00000  00000 00000  00000 00000
6888:  00000 00000  00000 00000  00000 00000  00000 00000  00000 00000    00000 00000  00000 00000  00000 00000  00000 00000  00000 00000
6889:  00000 00000  00000 00000  00000 00000  00000 00000  00000 00000    00000 00000  00000 00000  00000 00000  00000 00000  00000 00000
6890:  00000 00000  00000 00000  00000 00000  00000 00000  00000 00000    00000 00000  00000 00000  00000 00000  00000 00000  00000 00000
6891:  00000 00000  00000 00000  00000 00000  00000 00000  00000 00000    00000 00000  00000 00000  00000 00000  00000 00000  00000 00000
6892:  00000 00000  00000 00000  00000 00000  00000 00000  00000 00000    00000 00000  00000 00000  00000 00000  00000 00000  00000 00000
6893:  00000 00000  00000 00000  00000 00000  00000 00000  00000 00000    00000 00000  00000 00000  00000 00000  00000 00000  00000 00000
6894:  00000 00000  00000 00000  00000 00000  00000 00000  00000 00000    00000 00000  00000 00000  00000 00000  00000 00000  00000 00000
6895:  00000 00000  00000 00000  00000 00000  00000 00000  00000 00000    00000 00000  00000 00000  00000 00000  00000 00000  00000 00000
6896:  00000 00000  00000 00000  00000 00000  00000 00000  00000 00000    00000 00000  00000 00000  00000 00000  00000 00000  00000 00000
6897:  00000 00000  00000 00000  00000 00000  00000 00000  00000 00000    00000 00000  00000 00000  00000 00000  00000 00000  00000 00000
6898:  00000 00000  00000 00000  00000 00000  00000 00000  00000 00000    00000 00000  00000 00000  00000 00000  00000 00000  00000 00000
6899:  00000 00000  00000 00000  00000 00000  00000 00000  00000 00000    00000 00000  00000 00000  00000 00000  00000 00000  00000 00000
```

```
6900:  00000 00000  00000 00000  00000 00000  00000 00000  00000 00000    00000 00000  00000 00000  00000 00000  00000 00000  00000 00000
6901:  00000 00000  00000 00000  00000 00000  00000 00000  00000 00000    00000 00000  00000 00000  00000 00000  00000 00000  00000 00000
6902:  00000 00000  00000 00000  00000 00000  00000 00000  00000 00000    00000 00000  00000 00000  00000 00000  00000 00000  00000 00000
6903:  00000 00000  00000 00000  00000 00000  00000 00000  00000 00000    00000 00000  00000 00000  00000 00000  00000 00000  00000 00000
6904:  00000 00000  00000 00000  00000 00000  00000 00000  00000 00000    00000 00000  00000 00000  00000 00000  00000 00000  00000 00000
6905:  00000 00000  00000 00000  00000 00000  00000 00000  00000 00000    00000 00000  00000 00000  00000 00000  00000 00000  00000 00000
6906:  00000 00000  00000 00000  00000 00000  00000 00000  00000 00000    00000 00000  00000 00000  00000 00000  00000 00000  00000 00000
6907:  00000 00000  00000 00000  00000 00000  00000 00000  00000 00000    00000 00000  00000 00000  00000 00000  00000 00000  00000 00000
6908:  00000 00000  00000 00000  00000 00000  00000 00000  00000 00000    00000 00000  00000 00000  00000 00000  00000 00000  00000 00000
6909:  00000 00000  00000 00000  00000 00000  00000 00000  00000 00000    00000 00000  00000 00000  00000 00000  00000 00000  00000 00000
6910:  00000 00000  00000 00000  00000 00000  00000 00000  00000 00000    00000 00000  00000 00000  00000 00000  00000 00000  00000 00000
6911:  00000 00000  00000 00000  00000 00000  00000 00000  00000 00000    00000 00000  00000 00000  00000 00000  00000 00000  00000 00000
6912:  00000 00000  00000 00000  00000 00000  00000 00000  00000 00000    00000 00000  00000 00000  00000 00000  00000 00000  00000 00000
6913:  00000 00000  00000 00000  00000 00000  00000 00000  00000 00000    00000 00000  00000 00000  00000 00000  00000 00000  00000 00000
6914:  00000 00000  00000 00000  00000 00000  00000 00000  00000 00000    00000 00000  00000 00000  00000 00000  00000 00000  00000 00000
6915:  00000 00000  00000 00000  00000 00000  00000 00000  00000 00000    00000 00000  00000 00000  00000 00000  00000 00000  00000 00000
6916:  00000 00000  00000 00000  00000 00000  00000 00000  00000 00000    00000 00000  00000 00000  00000 00000  00000 00000  00000 00000
6917:  00000 00000  00000 00000  00000 00000  00000 00000  00000 00000    00000 00000  00000 00000  00000 00000  00000 00000  00000 00000
6918:  00000 00000  00000 00000  00000 00000  00000 00000  00000 00000    00000 00000  00000 00000  00000 00000  00000 00000  00000 00000
6919:  00000 00000  00000 00000  00000 00000  00000 00000  00000 00000    00000 00000  00000 00000  00000 00000  00000 00000  00000 00000
6920:  00000 00000  00000 00000  00000 00000  00000 00000  00000 00000    00000 00000  00000 00000  00000 00000  00000 00000  00000 00000
6921:  00000 00000  00000 00000  00000 00000  00000 00000  00000 00000    00000 00000  00000 00000  00000 00000  00000 00000  00000 00000
6922:  00000 00000  00000 00000  00000 00000  00000 00000  00000 00000    00000 00000  00000 00000  00000 00000  00000 00000  00000 00000
6923:  00000 00000  00000 00000  00000 00000  00000 00000  00000 00000    00000 00000  00000 00000  00000 00000  00000 00000  00000 00000
6924:  00000 00000  00000 00000  00000 00000  00000 00000  00000 00000    00000 00000  00000 00000  00000 00000  00000 00000  00000 00000
6925:  00000 00000  00000 00000  00000 00000  00000 00000  00000 00000    00000 00000  00000 00000  00000 00000  00000 00000  00000 00000
6926:  00000 00000  00000 00000  00000 00000  00000 00000  00000 00000    00000 00000  00000 00000  00000 00000  00000 00000  00000 00000
6927:  00000 00000  00000 00000  00000 00000  00000 00000  00000 00000    00000 00000  00000 00000  00000 00000  00000 00000  00000 00000
6928:  00000 00000  00000 00000  00000 00000  00000 00000  00000 00000    00000 00000  00000 00000  00000 00000  00000 00000  00000 00000
6929:  00000 00000  00000 00000  00000 00000  00000 00000  00000 00000    00000 00000  00000 00000  00000 00000  00000 00000  00000 00000
6930:  00000 00000  00000 00000  00000 00000  00000 00000  00000 00000    00000 00000  00000 00000  00000 00000  00000 00000  00000 00000
6931:  00000 00000  00000 00000  00000 00000  00000 00000  00000 00000    00000 00000  00000 00000  00000 00000  00000 00000  00000 00000
6932:  00000 00000  00000 00000  00000 00000  00000 00000  00000 00000    00000 00000  00000 00000  00000 00000  00000 00000  00000 00000
6933:  00000 00000  00000 00000  00000 00000  00000 00000  00000 00000    00000 00000  00000 00000  00000 00000  00000 00000  00000 00000
6934:  00000 00000  00000 00000  00000 00000  00000 00000  00000 00000    00000 00000  00000 00000  00000 00000  00000 00000  00000 00000
6935:  00000 00000  00000 00000  00000 00000  00000 00000  00000 00000    00000 00000  00000 00000  00000 00000  00000 00000  00000 00000
6936:  00000 00000  00000 00000  00000 00000  00000 00000  00000 00000    00000 00000  00000 00000  00000 00000  00000 00000  00000 00000
6937:  00000 00000  00000 00000  00000 00000  00000 00000  00000 00000    00000 00000  00000 00000  00000 00000  00000 00000  00000 00000
6938:  00000 00000  00000 00000  00000 00000  00000 00000  00000 00000    00000 00000  00000 00000  00000 00000  00000 00000  00000 00000
6939:  00000 00000  00000 00000  00000 00000  00000 00000  00000 00000    00000 00000  00000 00000  00000 00000  00000 00000  00000 00000
6940:  00000 00000  00000 00000  00000 00000  00000 00000  00000 00000    00000 00000  00000 00000  00000 00000  00000 00000  00000 00000
6941:  00000 00000  00000 00000  00000 00000  00000 00000  00000 00000    00000 00000  00000 00000  00000 00000  00000 00000  00000 00000
6942:  00000 00000  00000 00000  00000 00000  00000 00000  00000 00000    00000 00000  00000 00000  00000 00000  00000 00000  00000 00000
6943:  00000 00000  00000 00000  00000 00000  00000 00000  00000 00000    00000 00000  00000 00000  00000 00000  00000 00000  00000 00000
6944:  00000 00000  00000 00000  00000 00000  00000 00000  00000 00000    00000 00000  00000 00000  00000 00000  00000 00000  00000 00000
6945:  00000 00000  00000 00000  00000 00000  00000 00000  00000 00000    00000 00000  00000 00000  00000 00000  00000 00000  00000 00000
6946:  00000 00000  00000 00000  00000 00000  00000 00000  00000 00000    00000 00000  00000 00000  00000 00000  00000 00000  00000 00000
6947:  00000 00000  00000 00000  00000 00000  00000 00000  00000 00000    00000 00000  00000 00000  00000 00000  00000 00000  00000 00000
6948:  00000 00000  00000 00000  00000 00000  00000 00000  00000 00000    00000 00000  00000 00000  00000 00000  00000 00000  00000 00000
6949:  00000 00000  00000 00000  00000 00000  00000 00000  00000 00000    00000 00000  00000 00000  00000 00000  00000 00000  00000 00000
```

```
6950:  00000 00000  00000 00000  00000 00000  00000 00000  00000 00000    00000 00000  00000 00000  00000 00000  00000 00000  00000 00000
6951:  00000 00000  00000 00000  00000 00000  00000 00000  00000 00000    00000 00000  00000 00000  00000 00000  00000 00000  00000 00000
6952:  00000 00000  00000 00000  00000 00000  00000 00000  00000 00000    00000 00000  00000 00000  00000 00000  00000 00000  00000 00000
6953:  00000 00000  00000 00000  00000 00000  00000 00000  00000 00000    00000 00000  00000 00000  00000 00000  00000 00000  00000 00000
6954:  00000 00000  00000 00000  00000 00000  00000 00000  00000 00000    00000 00000  00000 00000  00000 00000  00000 00000  00000 00000
6955:  00000 00000  00000 00000  00000 00000  00000 00000  00000 00000    00000 00000  00000 00000  00000 00000  00000 00000  00000 00000
6956:  00000 00000  00000 00000  00000 00000  00000 00000  00000 00000    00000 00000  00000 00000  00000 00000  00000 00000  00000 00000
6957:  00000 00000  00000 00000  00000 00000  00000 00000  00000 00000    00000 00000  00000 00000  00000 00000  00000 00000  00000 00000
6958:  00000 00000  00000 00000  00000 00000  00000 00000  00000 00000    00000 00000  00000 00000  00000 00000  00000 00000  00000 00000
6959:  00000 00000  00000 00000  00000 00000  00000 00000  00000 00000    00000 00000  00000 00000  00000 00000  00000 00000  00000 00000
6960:  00000 00000  00000 00000  00000 00000  00000 00000  00000 00000    00000 00000  00000 00000  00000 00000  00000 00000  00000 00000
6961:  00000 00000  00000 00000  00000 00000  00000 00000  00000 00000    00000 00000  00000 00000  00000 00000  00000 00000  00000 00000
6962:  00000 00000  00000 00000  00000 00000  00000 00000  00000 00000    00000 00000  00000 00000  00000 00000  00000 00000  00000 00000
6963:  00000 00000  00000 00000  00000 00000  00000 00000  00000 00000    00000 00000  00000 00000  00000 00000  00000 00000  00000 00000
6964:  00000 00000  00000 00000  00000 00000  00000 00000  00000 00000    00000 00000  00000 00000  00000 00000  00000 00000  00000 00000
6965:  00000 00000  00000 00000  00000 00000  00000 00000  00000 00000    00000 00000  00000 00000  00000 00000  00000 00000  00000 00000
6966:  00000 00000  00000 00000  00000 00000  00000 00000  00000 00000    00000 00000  00000 00000  00000 00000  00000 00000  00000 00000
6967:  00000 00000  00000 00000  00000 00000  00000 00000  00000 00000    00000 00000  00000 00000  00000 00000  00000 00000  00000 00000
6968:  00000 00000  00000 00000  00000 00000  00000 00000  00000 00000    00000 00000  00000 00000  00000 00000  00000 00000  00000 00000
6969:  00000 00000  00000 00000  00000 00000  00000 00000  00000 00000    00000 00000  00000 00000  00000 00000  00000 00000  00000 00000
6970:  00000 00000  00000 00000  00000 00000  00000 00000  00000 00000    00000 00000  00000 00000  00000 00000  00000 00000  00000 00000
6971:  00000 00000  00000 00000  00000 00000  00000 00000  00000 00000    00000 00000  00000 00000  00000 00000  00000 00000  00000 00000
6972:  00000 00000  00000 00000  00000 00000  00000 00000  00000 00000    00000 00000  00000 00000  00000 00000  00000 00000  00000 00000
6973:  00000 00000  00000 00000  00000 00000  00000 00000  00000 00000    00000 00000  00000 00000  00000 00000  00000 00000  00000 00000
6974:  00000 00000  00000 00000  00000 00000  00000 00000  00000 00000    00000 00000  00000 00000  00000 00000  00000 00000  00000 00000
6975:  00000 00000  00000 00000  00000 00000  00000 00000  00000 00000    00000 00000  00000 00000  00000 00000  00000 00000  00000 00000
6976:  00000 00000  00000 00000  00000 00000  00000 00000  00000 00000    00000 00000  00000 00000  00000 00000  00000 00000  00000 00000
6977:  00000 00000  00000 00000  00000 00000  00000 00000  00000 00000    00000 00000  00000 00000  00000 00000  00000 00000  00000 00000
6978:  00000 00000  00000 00000  00000 00000  00000 00000  00000 00000    00000 00000  00000 00000  00000 00000  00000 00000  00000 00000
6979:  00000 00000  00000 00000  00000 00000  00000 00000  00000 00000    00000 00000  00000 00000  00000 00000  00000 00000  00000 00000
6980:  00000 00000  00000 00000  00000 00000  00000 00000  00000 00000    00000 00000  00000 00000  00000 00000  00000 00000  00000 00000
6981:  00000 00000  00000 00000  00000 00000  00000 00000  00000 00000    00000 00000  00000 00000  00000 00000  00000 00000  00000 00000
6982:  00000 00000  00000 00000  00000 00000  00000 00000  00000 00000    00000 00000  00000 00000  00000 00000  00000 00000  00000 00000
6983:  00000 00000  00000 00000  00000 00000  00000 00000  00000 00000    00000 00000  00000 00000  00000 00000  00000 00000  00000 00000
6984:  00000 00000  00000 00000  00000 00000  00000 00000  00000 00000    00000 00000  00000 00000  00000 00000  00000 00000  00000 00000
6985:  00000 00000  00000 00000  00000 00000  00000 00000  00000 00000    00000 00000  00000 00000  00000 00000  00000 00000  00000 00000
6986:  00000 00000  00000 00000  00000 00000  00000 00000  00000 00000    00000 00000  00000 00000  00000 00000  00000 00000  00000 00000
6987:  00000 00000  00000 00000  00000 00000  00000 00000  00000 00000    00000 00000  00000 00000  00000 00000  00000 00000  00000 00000
6988:  00000 00000  00000 00000  00000 00000  00000 00000  00000 00000    00000 00000  00000 00000  00000 00000  00000 00000  00000 00000
6989:  00000 00000  00000 00000  00000 00000  00000 00000  00000 00000    00000 00000  00000 00000  00000 00000  00000 00000  00000 00000
6990:  00000 00000  00000 00000  00000 00000  00000 00000  00000 00000    00000 00000  00000 00000  00000 00000  00000 00000  00000 00000
6991:  00000 00000  00000 00000  00000 00000  00000 00000  00000 00000    00000 00000  00000 00000  00000 00000  00000 00000  00000 00000
6992:  00000 00000  00000 00000  00000 00000  00000 00000  00000 00000    00000 00000  00000 00000  00000 00000  00000 00000  00000 00000
6993:  00000 00000  00000 00000  00000 00000  00000 00000  00000 00000    00000 00000  00000 00000  00000 00000  00000 00000  00000 00000
6994:  00000 00000  00000 00000  00000 00000  00000 00000  00000 00000    00000 00000  00000 00000  00000 00000  00000 00000  00000 00000
6995:  00000 00000  00000 00000  00000 00000  00000 00000  00000 00000    00000 00000  00000 00000  00000 00000  00000 00000  00000 00000
6996:  00000 00000  00000 00000  00000 00000  00000 00000  00000 00000    00000 00000  00000 00000  00000 00000  00000 00000  00000 00000
6997:  00000 00000  00000 00000  00000 00000  00000 00000  00000 00000    00000 00000  00000 00000  00000 00000  00000 00000  00000 00000
6998:  00000 00000  00000 00000  00000 00000  00000 00000  00000 00000    00000 00000  00000 00000  00000 00000  00000 00000  00000 00000
6999:  00000 00000  00000 00000  00000 00000  00000 00000  00000 00000    00000 00000  00000 00000  00000 00000  00000 00000  00000 00000
```

```
7000:  00000 00000   00000 00000   00000 00000   00000 00000   00000 00000     00000 00000   00000 00000   00000 00000   00000 00000   00000 00000
7001:  00000 00000   00000 00000   00000 00000   00000 00000   00000 00000     00000 00000   00000 00000   00000 00000   00000 00000   00000 00000
7002:  00000 00000   00000 00000   00000 00000   00000 00000   00000 00000     00000 00000   00000 00000   00000 00000   00000 00000   00000 00000
7003:  00000 00000   00000 00000   00000 00000   00000 00000   00000 00000     00000 00000   00000 00000   00000 00000   00000 00000   00000 00000
7004:  00000 00000   00000 00000   00000 00000   00000 00000   00000 00000     00000 00000   00000 00000   00000 00000   00000 00000   00000 00000
7005:  00000 00000   00000 00000   00000 00000   00000 00000   00000 00000     00000 00000   00000 00000   00000 00000   00000 00000   00000 00000
7006:  00000 00000   00000 00000   00000 00000   00000 00000   00000 00000     00000 00000   00000 00000   00000 00000   00000 00000   00000 00000
7007:  00000 00000   00000 00000   00000 00000   00000 00000   00000 00000     00000 00000   00000 00000   00000 00000   00000 00000   00000 00000
7008:  00000 00000   00000 00000   00000 00000   00000 00000   00000 00000     00000 00000   00000 00000   00000 00000   00000 00000   00000 00000
7009:  00000 00000   00000 00000   00000 00000   00000 00000   00000 00000     00000 00000   00000 00000   00000 00000   00000 00000   00000 00000
7010:  00000 00000   00000 00000   00000 00000   00000 00000   00000 00000     00000 00000   00000 00000   00000 00000   00000 00000   00000 00000
7011:  00000 00000   00000 00000   00000 00000   00000 00000   00000 00000     00000 00000   00000 00000   00000 00000   00000 00000   00000 00000
7012:  00000 00000   00000 00000   00000 00000   00000 00000   00000 00000     00000 00000   00000 00000   00000 00000   00000 00000   00000 00000
7013:  00000 00000   00000 00000   00000 00000   00000 00000   00000 00000     00000 00000   00000 00000   00000 00000   00000 00000   00000 00000
7014:  00000 00000   00000 00000   00000 00000   00000 00000   00000 00000     00000 00000   00000 00000   00000 00000   00000 00000   00000 00000
7015:  00000 00000   00000 00000   00000 00000   00000 00000   00000 00000     00000 00000   00000 00000   00000 00000   00000 00000   00000 00000
7016:  00000 00000   00000 00000   00000 00000   00000 00000   00000 00000     00000 00000   00000 00000   00000 00000   00000 00000   00000 00000
7017:  00000 00000   00000 00000   00000 00000   00000 00000   00000 00000     00000 00000   00000 00000   00000 00000   00000 00000   00000 00000
7018:  00000 00000   00000 00000   00000 00000   00000 00000   00000 00000     00000 00000   00000 00000   00000 00000   00000 00000   00000 00000
7019:  00000 00000   00000 00000   00000 00000   00000 00000   00000 00000     00000 00000   00000 00000   00000 00000   00000 00000   00000 00000
7020:  00000 00000   00000 00000   00000 00000   00000 00000   00000 00000     00000 00000   00000 00000   00000 00000   00000 00000   00000 00000
7021:  00000 00000   00000 00000   00000 00000   00000 00000   00000 00000     00000 00000   00000 00000   00000 00000   00000 00000   00000 00000
7022:  00000 00000   00000 00000   00000 00000   00000 00000   00000 00000     00000 00000   00000 00000   00000 00000   00000 00000   00000 00000
7023:  00000 00000   00000 00000   00000 00000   00000 00000   00000 00000     00000 00000   00000 00000   00000 00000   00000 00000   00000 00000
7024:  00000 00000   00000 00000   00000 00000   00000 00000   00000 00000     00000 00000   00000 00000   00000 00000   00000 00000   00000 00000
7025:  00000 00000   00000 00000   00000 00000   00000 00000   00000 00000     00000 00000   00000 00000   00000 00000   00000 00000   00000 00000
7026:  00000 00000   00000 00000   00000 00000   00000 00000   00000 00000     00000 00000   00000 00000   00000 00000   00000 00000   00000 00000
7027:  00000 00000   00000 00000   00000 00000   00000 00000   00000 00000     00000 00000   00000 00000   00000 00000   00000 00000   00000 00000
7028:  00000 00000   00000 00000   00000 00000   00000 00000   00000 00000     00000 00000   00000 00000   00000 00000   00000 00000   00000 00000
7029:  00000 00000   00000 00000   00000 00000   00000 00000   00000 00000     00000 00000   00000 00000   00000 00000   00000 00000   00000 00000
7030:  00000 00000   00000 00000   00000 00000   00000 00000   00000 00000     00000 00000   00000 00000   00000 00000   00000 00000   00000 00000
7031:  00000 00000   00000 00000   00000 00000   00000 00000   00000 00000     00000 00000   00000 00000   00000 00000   00000 00000   00000 00000
7032:  00000 00000   00000 00000   00000 00000   00000 00000   00000 00000     00000 00000   00000 00000   00000 00000   00000 00000   00000 00000
7033:  00000 00000   00000 00000   00000 00000   00000 00000   00000 00000     00000 00000   00000 00000   00000 00000   00000 00000   00000 00000
7034:  00000 00000   00000 00000   00000 00000   00000 00000   00000 00000     00000 00000   00000 00000   00000 00000   00000 00000   00000 00000
7035:  00000 00000   00000 00000   00000 00000   00000 00000   00000 00000     00000 00000   00000 00000   00000 00000   00000 00000   00000 00000
7036:  00000 00000   00000 00000   00000 00000   00000 00000   00000 00000     00000 00000   00000 00000   00000 00000   00000 00000   00000 00000
7037:  00000 00000   00000 00000   00000 00000   00000 00000   00000 00000     00000 00000   00000 00000   00000 00000   00000 00000   00000 00000
7038:  00000 00000   00000 00000   00000 00000   00000 00000   00000 00000     00000 00000   00000 00000   00000 00000   00000 00000   00000 00000
7039:  00000 00000   00000 00000   00000 00000   00000 00000   00000 00000     00000 00000   00000 00000   00000 00000   00000 00000   00000 00000
7040:  00000 00000   00000 00000   00000 00000   00000 00000   00000 00000     00000 00000   00000 00000   00000 00000   00000 00000   00000 00000
7041:  00000 00000   00000 00000   00000 00000   00000 00000   00000 00000     00000 00000   00000 00000   00000 00000   00000 00000   00000 00000
7042:  00000 00000   00000 00000   00000 00000   00000 00000   00000 00000     00000 00000   00000 00000   00000 00000   00000 00000   00000 00000
7043:  00000 00000   00000 00000   00000 00000   00000 00000   00000 00000     00000 00000   00000 00000   00000 00000   00000 00000   00000 00000
7044:  00000 00000   00000 00000   00000 00000   00000 00000   00000 00000     00000 00000   00000 00000   00000 00000   00000 00000   00000 00000
7045:  00000 00000   00000 00000   00000 00000   00000 00000   00000 00000     00000 00000   00000 00000   00000 00000   00000 00000   00000 00000
7046:  00000 00000   00000 00000   00000 00000   00000 00000   00000 00000     00000 00000   00000 00000   00000 00000   00000 00000   00000 00000
7047:  00000 00000   00000 00000   00000 00000   00000 00000   00000 00000     00000 00000   00000 00000   00000 00000   00000 00000   00000 00000
7048:  00000 00000   00000 00000   00000 00000   00000 00000   00000 00000     00000 00000   00000 00000   00000 00000   00000 00000   00000 00000
7049:  00000 00000   00000 00000   00000 00000   00000 00000   00000 00000     00000 00000   00000 00000   00000 00000   00000 00000   00000 00000
```

```
7050:  00000 00000  00000 00000  00000 00000  00000 00000  00000 00000    00000 00000  00000 00000  00000 00000  00000 00000  00000 00000
7051:  00000 00000  00000 00000  00000 00000  00000 00000  00000 00000    00000 00000  00000 00000  00000 00000  00000 00000  00000 00000
7052:  00000 00000  00000 00000  00000 00000  00000 00000  00000 00000    00000 00000  00000 00000  00000 00000  00000 00000  00000 00000
7053:  00000 00000  00000 00000  00000 00000  00000 00000  00000 00000    00000 00000  00000 00000  00000 00000  00000 00000  00000 00000
7054:  00000 00000  00000 00000  00000 00000  00000 00000  00000 00000    00000 00000  00000 00000  00000 00000  00000 00000  00000 00000
7055:  00000 00000  00000 00000  00000 00000  00000 00000  00000 00000    00000 00000  00000 00000  00000 00000  00000 00000  00000 00000
7056:  00000 00000  00000 00000  00000 00000  00000 00000  00000 00000    00000 00000  00000 00000  00000 00000  00000 00000  00000 00000
7057:  00000 00000  00000 00000  00000 00000  00000 00000  00000 00000    00000 00000  00000 00000  00000 00000  00000 00000  00000 00000
7058:  00000 00000  00000 00000  00000 00000  00000 00000  00000 00000    00000 00000  00000 00000  00000 00000  00000 00000  00000 00000
7059:  00000 00000  00000 00000  00000 00000  00000 00000  00000 00000    00000 00000  00000 00000  00000 00000  00000 00000  00000 00000
7060:  00000 00000  00000 00000  00000 00000  00000 00000  00000 00000    00000 00000  00000 00000  00000 00000  00000 00000  00000 00000
7061:  00000 00000  00000 00000  00000 00000  00000 00000  00000 00000    00000 00000  00000 00000  00000 00000  00000 00000  00000 00000
7062:  00000 00000  00000 00000  00000 00000  00000 00000  00000 00000    00000 00000  00000 00000  00000 00000  00000 00000  00000 00000
7063:  00000 00000  00000 00000  00000 00000  00000 00000  00000 00000    00000 00000  00000 00000  00000 00000  00000 00000  00000 00000
7064:  00000 00000  00000 00000  00000 00000  00000 00000  00000 00000    00000 00000  00000 00000  00000 00000  00000 00000  00000 00000
7065:  00000 00000  00000 00000  00000 00000  00000 00000  00000 00000    00000 00000  00000 00000  00000 00000  00000 00000  00000 00000
7066:  00000 00000  00000 00000  00000 00000  00000 00000  00000 00000    00000 00000  00000 00000  00000 00000  00000 00000  00000 00000
7067:  00000 00000  00000 00000  00000 00000  00000 00000  00000 00000    00000 00000  00000 00000  00000 00000  00000 00000  00000 00000
7068:  00000 00000  00000 00000  00000 00000  00000 00000  00000 00000    00000 00000  00000 00000  00000 00000  00000 00000  00000 00000
7069:  00000 00000  00000 00000  00000 00000  00000 00000  00000 00000    00000 00000  00000 00000  00000 00000  00000 00000  00000 00000
7070:  00000 00000  00000 00000  00000 00000  00000 00000  00000 00000    00000 00000  00000 00000  00000 00000  00000 00000  00000 00000
7071:  00000 00000  00000 00000  00000 00000  00000 00000  00000 00000    00000 00000  00000 00000  00000 00000  00000 00000  00000 00000
7072:  00000 00000  00000 00000  00000 00000  00000 00000  00000 00000    00000 00000  00000 00000  00000 00000  00000 00000  00000 00000
7073:  00000 00000  00000 00000  00000 00000  00000 00000  00000 00000    00000 00000  00000 00000  00000 00000  00000 00000  00000 00000
7074:  00000 00000  00000 00000  00000 00000  00000 00000  00000 00000    00000 00000  00000 00000  00000 00000  00000 00000  00000 00000
7075:  00000 00000  00000 00000  00000 00000  00000 00000  00000 00000    00000 00000  00000 00000  00000 00000  00000 00000  00000 00000
7076:  00000 00000  00000 00000  00000 00000  00000 00000  00000 00000    00000 00000  00000 00000  00000 00000  00000 00000  00000 00000
7077:  00000 00000  00000 00000  00000 00000  00000 00000  00000 00000    00000 00000  00000 00000  00000 00000  00000 00000  00000 00000
7078:  00000 00000  00000 00000  00000 00000  00000 00000  00000 00000    00000 00000  00000 00000  00000 00000  00000 00000  00000 00000
7079:  00000 00000  00000 00000  00000 00000  00000 00000  00000 00000    00000 00000  00000 00000  00000 00000  00000 00000  00000 00000
7080:  00000 00000  00000 00000  00000 00000  00000 00000  00000 00000    00000 00000  00000 00000  00000 00000  00000 00000  00000 00000
7081:  00000 00000  00000 00000  00000 00000  00000 00000  00000 00000    00000 00000  00000 00000  00000 00000  00000 00000  00000 00000
7082:  00000 00000  00000 00000  00000 00000  00000 00000  00000 00000    00000 00000  00000 00000  00000 00000  00000 00000  00000 00000
7083:  00000 00000  00000 00000  00000 00000  00000 00000  00000 00000    00000 00000  00000 00000  00000 00000  00000 00000  00000 00000
7084:  00000 00000  00000 00000  00000 00000  00000 00000  00000 00000    00000 00000  00000 00000  00000 00000  00000 00000  00000 00000
7085:  00000 00000  00000 00000  00000 00000  00000 00000  00000 00000    00000 00000  00000 00000  00000 00000  00000 00000  00000 00000
7086:  00000 00000  00000 00000  00000 00000  00000 00000  00000 00000    00000 00000  00000 00000  00000 00000  00000 00000  00000 00000
7087:  00000 00000  00000 00000  00000 00000  00000 00000  00000 00000    00000 00000  00000 00000  00000 00000  00000 00000  00000 00000
7088:  00000 00000  00000 00000  00000 00000  00000 00000  00000 00000    00000 00000  00000 00000  00000 00000  00000 00000  00000 00000
7089:  00000 00000  00000 00000  00000 00000  00000 00000  00000 00000    00000 00000  00000 00000  00000 00000  00000 00000  00000 00000
7090:  00000 00000  00000 00000  00000 00000  00000 00000  00000 00000    00000 00000  00000 00000  00000 00000  00000 00000  00000 00000
7091:  00000 00000  00000 00000  00000 00000  00000 00000  00000 00000    00000 00000  00000 00000  00000 00000  00000 00000  00000 00000
7092:  00000 00000  00000 00000  00000 00000  00000 00000  00000 00000    00000 00000  00000 00000  00000 00000  00000 00000  00000 00000
7093:  00000 00000  00000 00000  00000 00000  00000 00000  00000 00000    00000 00000  00000 00000  00000 00000  00000 00000  00000 00000
7094:  00000 00000  00000 00000  00000 00000  00000 00000  00000 00000    00000 00000  00000 00000  00000 00000  00000 00000  00000 00000
7095:  00000 00000  00000 00000  00000 00000  00000 00000  00000 00000    00000 00000  00000 00000  00000 00000  00000 00000  00000 00000
7096:  00000 00000  00000 00000  00000 00000  00000 00000  00000 00000    00000 00000  00000 00000  00000 00000  00000 00000  00000 00000
7097:  00000 00000  00000 00000  00000 00000  00000 00000  00000 00000    00000 00000  00000 00000  00000 00000  00000 00000  00000 00000
7098:  00000 00000  00000 00000  00000 00000  00000 00000  00000 00000    00000 00000  00000 00000  00000 00000  00000 00000  00000 00000
7099:  00000 00000  00000 00000  00000 00000  00000 00000  00000 00000    00000 00000  00000 00000  00000 00000  00000 00000  00000 00000
```

```
7100:   00000 00000   00000 00000   00000 00000   00000 00000   00000 00000     00000 00000   00000 00000   00000 00000   00000 00000   00000 00000
7101:   00000 00000   00000 00000   00000 00000   00000 00000   00000 00000     00000 00000   00000 00000   00000 00000   00000 00000   00000 00000
7102:   00000 00000   00000 00000   00000 00000   00000 00000   00000 00000     00000 00000   00000 00000   00000 00000   00000 00000   00000 00000
7103:   00000 00000   00000 00000   00000 00000   00000 00000   00000 00000     00000 00000   00000 00000   00000 00000   00000 00000   00000 00000
7104:   00000 00000   00000 00000   00000 00000   00000 00000   00000 00000     00000 00000   00000 00000   00000 00000   00000 00000   00000 00000
7105:   00000 00000   00000 00000   00000 00000   00000 00000   00000 00000     00000 00000   00000 00000   00000 00000   00000 00000   00000 00000
7106:   00000 00000   00000 00000   00000 00000   00000 00000   00000 00000     00000 00000   00000 00000   00000 00000   00000 00000   00000 00000
7107:   00000 00000   00000 00000   00000 00000   00000 00000   00000 00000     00000 00000   00000 00000   00000 00000   00000 00000   00000 00000
7108:   00000 00000   00000 00000   00000 00000   00000 00000   00000 00000     00000 00000   00000 00000   00000 00000   00000 00000   00000 00000
7109:   00000 00000   00000 00000   00000 00000   00000 00000   00000 00000     00000 00000   00000 00000   00000 00000   00000 00000   00000 00000
7110:   00000 00000   00000 00000   00000 00000   00000 00000   00000 00000     00000 00000   00000 00000   00000 00000   00000 00000   00000 00000
7111:   00000 00000   00000 00000   00000 00000   00000 00000   00000 00000     00000 00000   00000 00000   00000 00000   00000 00000   00000 00000
7112:   00000 00000   00000 00000   00000 00000   00000 00000   00000 00000     00000 00000   00000 00000   00000 00000   00000 00000   00000 00000
7113:   00000 00000   00000 00000   00000 00000   00000 00000   00000 00000     00000 00000   00000 00000   00000 00000   00000 00000   00000 00000
7114:   00000 00000   00000 00000   00000 00000   00000 00000   00000 00000     00000 00000   00000 00000   00000 00000   00000 00000   00000 00000
7115:   00000 00000   00000 00000   00000 00000   00000 00000   00000 00000     00000 00000   00000 00000   00000 00000   00000 00000   00000 00000
7116:   00000 00000   00000 00000   00000 00000   00000 00000   00000 00000     00000 00000   00000 00000   00000 00000   00000 00000   00000 00000
7117:   00000 00000   00000 00000   00000 00000   00000 00000   00000 00000     00000 00000   00000 00000   00000 00000   00000 00000   00000 00000
7118:   00000 00000   00000 00000   00000 00000   00000 00000   00000 00000     00000 00000   00000 00000   00000 00000   00000 00000   00000 00000
7119:   00000 00000   00000 00000   00000 00000   00000 00000   00000 00000     00000 00000   00000 00000   00000 00000   00000 00000   00000 00000
7120:   00000 00000   00000 00000   00000 00000   00000 00000   00000 00000     00000 00000   00000 00000   00000 00000   00000 00000   00000 00000
7121:   00000 00000   00000 00000   00000 00000   00000 00000   00000 00000     00000 00000   00000 00000   00000 00000   00000 00000   00000 00000
7122:   00000 00000   00000 00000   00000 00000   00000 00000   00000 00000     00000 00000   00000 00000   00000 00000   00000 00000   00000 00000
7123:   00000 00000   00000 00000   00000 00000   00000 00000   00000 00000     00000 00000   00000 00000   00000 00000   00000 00000   00000 00000
7124:   00000 00000   00000 00000   00000 00000   00000 00000   00000 00000     00000 00000   00000 00000   00000 00000   00000 00000   00000 00000
7125:   00000 00000   00000 00000   00000 00000   00000 00000   00000 00000     00000 00000   00000 00000   00000 00000   00000 00000   00000 00000
7126:   00000 00000   00000 00000   00000 00000   00000 00000   00000 00000     00000 00000   00000 00000   00000 00000   00000 00000   00000 00000
7127:   00000 00000   00000 00000   00000 00000   00000 00000   00000 00000     00000 00000   00000 00000   00000 00000   00000 00000   00000 00000
7128:   00000 00000   00000 00000   00000 00000   00000 00000   00000 00000     00000 00000   00000 00000   00000 00000   00000 00000   00000 00000
7129:   00000 00000   00000 00000   00000 00000   00000 00000   00000 00000     00000 00000   00000 00000   00000 00000   00000 00000   00000 00000
7130:   00000 00000   00000 00000   00000 00000   00000 00000   00000 00000     00000 00000   00000 00000   00000 00000   00000 00000   00000 00000
7131:   00000 00000   00000 00000   00000 00000   00000 00000   00000 00000     00000 00000   00000 00000   00000 00000   00000 00000   00000 00000
7132:   00000 00000   00000 00000   00000 00000   00000 00000   00000 00000     00000 00000   00000 00000   00000 00000   00000 00000   00000 00000
7133:   00000 00000   00000 00000   00000 00000   00000 00000   00000 00000     00000 00000   00000 00000   00000 00000   00000 00000   00000 00000
7134:   00000 00000   00000 00000   00000 00000   00000 00000   00000 00000     00000 00000   00000 00000   00000 00000   00000 00000   00000 00000
7135:   00000 00000   00000 00000   00000 00000   00000 00000   00000 00000     00000 00000   00000 00000   00000 00000   00000 00000   00000 00000
7136:   00000 00000   00000 00000   00000 00000   00000 00000   00000 00000     00000 00000   00000 00000   00000 00000   00000 00000   00000 00000
7137:   00000 00000   00000 00000   00000 00000   00000 00000   00000 00000     00000 00000   00000 00000   00000 00000   00000 00000   00000 00000
7138:   00000 00000   00000 00000   00000 00000   00000 00000   00000 00000     00000 00000   00000 00000   00000 00000   00000 00000   00000 00000
7139:   00000 00000   00000 00000   00000 00000   00000 00000   00000 00000     00000 00000   00000 00000   00000 00000   00000 00000   00000 00000
7140:   00000 00000   00000 00000   00000 00000   00000 00000   00000 00000     00000 00000   00000 00000   00000 00000   00000 00000   00000 00000
7141:   00000 00000   00000 00000   00000 00000   00000 00000   00000 00000     00000 00000   00000 00000   00000 00000   00000 00000   00000 00000
7142:   00000 00000   00000 00000   00000 00000   00000 00000   00000 00000     00000 00000   00000 00000   00000 00000   00000 00000   00000 00000
7143:   00000 00000   00000 00000   00000 00000   00000 00000   00000 00000     00000 00000   00000 00000   00000 00000   00000 00000   00000 00000
7144:   00000 00000   00000 00000   00000 00000   00000 00000   00000 00000     00000 00000   00000 00000   00000 00000   00000 00000   00000 00000
7145:   00000 00000   00000 00000   00000 00000   00000 00000   00000 00000     00000 00000   00000 00000   00000 00000   00000 00000   00000 00000
7146:   00000 00000   00000 00000   00000 00000   00000 00000   00000 00000     00000 00000   00000 00000   00000 00000   00000 00000   00000 00000
7147:   00000 00000   00000 00000   00000 00000   00000 00000   00000 00000     00000 00000   00000 00000   00000 00000   00000 00000   00000 00000
7148:   00000 00000   00000 00000   00000 00000   00000 00000   00000 00000     00000 00000   00000 00000   00000 00000   00000 00000   00000 00000
7149:   00000 00000   00000 00000   00000 00000   00000 00000   00000 00000     00000 00000   00000 00000   00000 00000   00000 00000   00000 00000
```

```
7150:   00000 00000   00000 00000   00000 00000   00000 00000   00000 00000     00000 00000   00000 00000   00000 00000   00000 00000   00000 00000
7151:   00000 00000   00000 00000   00000 00000   00000 00000   00000 00000     00000 00000   00000 00000   00000 00000   00000 00000   00000 00000
7152:   00000 00000   00000 00000   00000 00000   00000 00000   00000 00000     00000 00000   00000 00000   00000 00000   00000 00000   00000 00000
7153:   00000 00000   00000 00000   00000 00000   00000 00000   00000 00000     00000 00000   00000 00000   00000 00000   00000 00000   00000 00000
7154:   00000 00000   00000 00000   00000 00000   00000 00000   00000 00000     00000 00000   00000 00000   00000 00000   00000 00000   00000 00000
7155:   00000 00000   00000 00000   00000 00000   00000 00000   00000 00000     00000 00000   00000 00000   00000 00000   00000 00000   00000 00000
7156:   00000 00000   00000 00000   00000 00000   00000 00000   00000 00000     00000 00000   00000 00000   00000 00000   00000 00000   00000 00000
7157:   00000 00000   00000 00000   00000 00000   00000 00000   00000 00000     00000 00000   00000 00000   00000 00000   00000 00000   00000 00000
7158:   00000 00000   00000 00000   00000 00000   00000 00000   00000 00000     00000 00000   00000 00000   00000 00000   00000 00000   00000 00000
7159:   00000 00000   00000 00000   00000 00000   00000 00000   00000 00000     00000 00000   00000 00000   00000 00000   00000 00000   00000 00000
7160:   00000 00000   00000 00000   00000 00000   00000 00000   00000 00000     00000 00000   00000 00000   00000 00000   00000 00000   00000 00000
7161:   00000 00000   00000 00000   00000 00000   00000 00000   00000 00000     00000 00000   00000 00000   00000 00000   00000 00000   00000 00000
7162:   00000 00000   00000 00000   00000 00000   00000 00000   00000 00000     00000 00000   00000 00000   00000 00000   00000 00000   00000 00000
7163:   00000 00000   00000 00000   00000 00000   00000 00000   00000 00000     00000 00000   00000 00000   00000 00000   00000 00000   00000 00000
7164:   00000 00000   00000 00000   00000 00000   00000 00000   00000 00000     00000 00000   00000 00000   00000 00000   00000 00000   00000 00000
7165:   00000 00000   00000 00000   00000 00000   00000 00000   00000 00000     00000 00000   00000 00000   00000 00000   00000 00000   00000 00000
7166:   00000 00000   00000 00000   00000 00000   00000 00000   00000 00000     00000 00000   00000 00000   00000 00000   00000 00000   00000 00000
7167:   00000 00000   00000 00000   00000 00000   00000 00000   00000 00000     00000 00000   00000 00000   00000 00000   00000 00000   00000 00000
7168:   00000 00000   00000 00000   00000 00000   00000 00000   00000 00000     00000 00000   00000 00000   00000 00000   00000 00000   00000 00000
7169:   00000 00000   00000 00000   00000 00000   00000 00000   00000 00000     00000 00000   00000 00000   00000 00000   00000 00000   00000 00000
7170:   00000 00000   00000 00000   00000 00000   00000 00000   00000 00000     00000 00000   00000 00000   00000 00000   00000 00000   00000 00000
7171:   00000 00000   00000 00000   00000 00000   00000 00000   00000 00000     00000 00000   00000 00000   00000 00000   00000 00000   00000 00000
7172:   00000 00000   00000 00000   00000 00000   00000 00000   00000 00000     00000 00000   00000 00000   00000 00000   00000 00000   00000 00000
7173:   00000 00000   00000 00000   00000 00000   00000 00000   00000 00000     00000 00000   00000 00000   00000 00000   00000 00000   00000 00000
7174:   00000 00000   00000 00000   00000 00000   00000 00000   00000 00000     00000 00000   00000 00000   00000 00000   00000 00000   00000 00000
7175:   00000 00000   00000 00000   00000 00000   00000 00000   00000 00000     00000 00000   00000 00000   00000 00000   00000 00000   00000 00000
7176:   00000 00000   00000 00000   00000 00000   00000 00000   00000 00000     00000 00000   00000 00000   00000 00000   00000 00000   00000 00000
7177:   00000 00000   00000 00000   00000 00000   00000 00000   00000 00000     00000 00000   00000 00000   00000 00000   00000 00000   00000 00000
7178:   00000 00000   00000 00000   00000 00000   00000 00000   00000 00000     00000 00000   00000 00000   00000 00000   00000 00000   00000 00000
7179:   00000 00000   00000 00000   00000 00000   00000 00000   00000 00000     00000 00000   00000 00000   00000 00000   00000 00000   00000 00000
7180:   00000 00000   00000 00000   00000 00000   00000 00000   00000 00000     00000 00000   00000 00000   00000 00000   00000 00000   00000 00000
7181:   00000 00000   00000 00000   00000 00000   00000 00000   00000 00000     00000 00000   00000 00000   00000 00000   00000 00000   00000 00000
7182:   00000 00000   00000 00000   00000 00000   00000 00000   00000 00000     00000 00000   00000 00000   00000 00000   00000 00000   00000 00000
7183:   00000 00000   00000 00000   00000 00000   00000 00000   00000 00000     00000 00000   00000 00000   00000 00000   00000 00000   00000 00000
7184:   00000 00000   00000 00000   00000 00000   00000 00000   00000 00000     00000 00000   00000 00000   00000 00000   00000 00000   00000 00000
7185:   00000 00000   00000 00000   00000 00000   00000 00000   00000 00000     00000 00000   00000 00000   00000 00000   00000 00000   00000 00000
7186:   00000 00000   00000 00000   00000 00000   00000 00000   00000 00000     00000 00000   00000 00000   00000 00000   00000 00000   00000 00000
7187:   00000 00000   00000 00000   00000 00000   00000 00000   00000 00000     00000 00000   00000 00000   00000 00000   00000 00000   00000 00000
7188:   00000 00000   00000 00000   00000 00000   00000 00000   00000 00000     00000 00000   00000 00000   00000 00000   00000 00000   00000 00000
7189:   00000 00000   00000 00000   00000 00000   00000 00000   00000 00000     00000 00000   00000 00000   00000 00000   00000 00000   00000 00000
7190:   00000 00000   00000 00000   00000 00000   00000 00000   00000 00000     00000 00000   00000 00000   00000 00000   00000 00000   00000 00000
7191:   00000 00000   00000 00000   00000 00000   00000 00000   00000 00000     00000 00000   00000 00000   00000 00000   00000 00000   00000 00000
7192:   00000 00000   00000 00000   00000 00000   00000 00000   00000 00000     00000 00000   00000 00000   00000 00000   00000 00000   00000 00000
7193:   00000 00000   00000 00000   00000 00000   00000 00000   00000 00000     00000 00000   00000 00000   00000 00000   00000 00000   00000 00000
7194:   00000 00000   00000 00000   00000 00000   00000 00000   00000 00000     00000 00000   00000 00000   00000 00000   00000 00000   00000 00000
7195:   00000 00000   00000 00000   00000 00000   00000 00000   00000 00000     00000 00000   00000 00000   00000 00000   00000 00000   00000 00000
7196:   00000 00000   00000 00000   00000 00000   00000 00000   00000 00000     00000 00000   00000 00000   00000 00000   00000 00000   00000 00000
7197:   00000 00000   00000 00000   00000 00000   00000 00000   00000 00000     00000 00000   00000 00000   00000 00000   00000 00000   00000 00000
7198:   00000 00000   00000 00000   00000 00000   00000 00000   00000 00000     00000 00000   00000 00000   00000 00000   00000 00000   00000 00000
7199:   00000 00000   00000 00000   00000 00000   00000 00000   00000 00000     00000 00000   00000 00000   00000 00000   00000 00000   00000 00000
```

```
7200:  00000 00000  00000 00000  00000 00000  00000 00000  00000 00000    00000 00000  00000 00000  00000 00000  00000 00000  00000 00000
7201:  00000 00000  00000 00000  00000 00000  00000 00000  00000 00000    00000 00000  00000 00000  00000 00000  00000 00000  00000 00000
7202:  00000 00000  00000 00000  00000 00000  00000 00000  00000 00000    00000 00000  00000 00000  00000 00000  00000 00000  00000 00000
7203:  00000 00000  00000 00000  00000 00000  00000 00000  00000 00000    00000 00000  00000 00000  00000 00000  00000 00000  00000 00000
7204:  00000 00000  00000 00000  00000 00000  00000 00000  00000 00000    00000 00000  00000 00000  00000 00000  00000 00000  00000 00000
7205:  00000 00000  00000 00000  00000 00000  00000 00000  00000 00000    00000 00000  00000 00000  00000 00000  00000 00000  00000 00000
7206:  00000 00000  00000 00000  00000 00000  00000 00000  00000 00000    00000 00000  00000 00000  00000 00000  00000 00000  00000 00000
7207:  00000 00000  00000 00000  00000 00000  00000 00000  00000 00000    00000 00000  00000 00000  00000 00000  00000 00000  00000 00000
7208:  00000 00000  00000 00000  00000 00000  00000 00000  00000 00000    00000 00000  00000 00000  00000 00000  00000 00000  00000 00000
7209:  00000 00000  00000 00000  00000 00000  00000 00000  00000 00000    00000 00000  00000 00000  00000 00000  00000 00000  00000 00000
7210:  00000 00000  00000 00000  00000 00000  00000 00000  00000 00000    00000 00000  00000 00000  00000 00000  00000 00000  00000 00000
7211:  00000 00000  00000 00000  00000 00000  00000 00000  00000 00000    00000 00000  00000 00000  00000 00000  00000 00000  00000 00000
7212:  00000 00000  00000 00000  00000 00000  00000 00000  00000 00000    00000 00000  00000 00000  00000 00000  00000 00000  00000 00000
7213:  00000 00000  00000 00000  00000 00000  00000 00000  00000 00000    00000 00000  00000 00000  00000 00000  00000 00000  00000 00000
7214:  00000 00000  00000 00000  00000 00000  00000 00000  00000 00000    00000 00000  00000 00000  00000 00000  00000 00000  00000 00000
7215:  00000 00000  00000 00000  00000 00000  00000 00000  00000 00000    00000 00000  00000 00000  00000 00000  00000 00000  00000 00000
7216:  00000 00000  00000 00000  00000 00000  00000 00000  00000 00000    00000 00000  00000 00000  00000 00000  00000 00000  00000 00000
7217:  00000 00000  00000 00000  00000 00000  00000 00000  00000 00000    00000 00000  00000 00000  00000 00000  00000 00000  00000 00000
7218:  00000 00000  00000 00000  00000 00000  00000 00000  00000 00000    00000 00000  00000 00000  00000 00000  00000 00000  00000 00000
7219:  00000 00000  00000 00000  00000 00000  00000 00000  00000 00000    00000 00000  00000 00000  00000 00000  00000 00000  00000 00000
7220:  00000 00000  00000 00000  00000 00000  00000 00000  00000 00000    00000 00000  00000 00000  00000 00000  00000 00000  00000 00000
7221:  00000 00000  00000 00000  00000 00000  00000 00000  00000 00000    00000 00000  00000 00000  00000 00000  00000 00000  00000 00000
7222:  00000 00000  00000 00000  00000 00000  00000 00000  00000 00000    00000 00000  00000 00000  00000 00000  00000 00000  00000 00000
7223:  00000 00000  00000 00000  00000 00000  00000 00000  00000 00000    00000 00000  00000 00000  00000 00000  00000 00000  00000 00000
7224:  00000 00000  00000 00000  00000 00000  00000 00000  00000 00000    00000 00000  00000 00000  00000 00000  00000 00000  00000 00000
7225:  00000 00000  00000 00000  00000 00000  00000 00000  00000 00000    00000 00000  00000 00000  00000 00000  00000 00000  00000 00000
7226:  00000 00000  00000 00000  00000 00000  00000 00000  00000 00000    00000 00000  00000 00000  00000 00000  00000 00000  00000 00000
7227:  00000 00000  00000 00000  00000 00000  00000 00000  00000 00000    00000 00000  00000 00000  00000 00000  00000 00000  00000 00000
7228:  00000 00000  00000 00000  00000 00000  00000 00000  00000 00000    00000 00000  00000 00000  00000 00000  00000 00000  00000 00000
7229:  00000 00000  00000 00000  00000 00000  00000 00000  00000 00000    00000 00000  00000 00000  00000 00000  00000 00000  00000 00000
7230:  00000 00000  00000 00000  00000 00000  00000 00000  00000 00000    00000 00000  00000 00000  00000 00000  00000 00000  00000 00000
7231:  00000 00000  00000 00000  00000 00000  00000 00000  00000 00000    00000 00000  00000 00000  00000 00000  00000 00000  00000 00000
7232:  00000 00000  00000 00000  00000 00000  00000 00000  00000 00000    00000 00000  00000 00000  00000 00000  00000 00000  00000 00000
7233:  00000 00000  00000 00000  00000 00000  00000 00000  00000 00000    00000 00000  00000 00000  00000 00000  00000 00000  00000 00000
7234:  00000 00000  00000 00000  00000 00000  00000 00000  00000 00000    00000 00000  00000 00000  00000 00000  00000 00000  00000 00000
7235:  00000 00000  00000 00000  00000 00000  00000 00000  00000 00000    00000 00000  00000 00000  00000 00000  00000 00000  00000 00000
7236:  00000 00000  00000 00000  00000 00000  00000 00000  00000 00000    00000 00000  00000 00000  00000 00000  00000 00000  00000 00000
7237:  00000 00000  00000 00000  00000 00000  00000 00000  00000 00000    00000 00000  00000 00000  00000 00000  00000 00000  00000 00000
7238:  00000 00000  00000 00000  00000 00000  00000 00000  00000 00000    00000 00000  00000 00000  00000 00000  00000 00000  00000 00000
7239:  00000 00000  00000 00000  00000 00000  00000 00000  00000 00000    00000 00000  00000 00000  00000 00000  00000 00000  00000 00000
7240:  00000 00000  00000 00000  00000 00000  00000 00000  00000 00000    00000 00000  00000 00000  00000 00000  00000 00000  00000 00000
7241:  00000 00000  00000 00000  00000 00000  00000 00000  00000 00000    00000 00000  00000 00000  00000 00000  00000 00000  00000 00000
7242:  00000 00000  00000 00000  00000 00000  00000 00000  00000 00000    00000 00000  00000 00000  00000 00000  00000 00000  00000 00000
7243:  00000 00000  00000 00000  00000 00000  00000 00000  00000 00000    00000 00000  00000 00000  00000 00000  00000 00000  00000 00000
7244:  00000 00000  00000 00000  00000 00000  00000 00000  00000 00000    00000 00000  00000 00000  00000 00000  00000 00000  00000 00000
7245:  00000 00000  00000 00000  00000 00000  00000 00000  00000 00000    00000 00000  00000 00000  00000 00000  00000 00000  00000 00000
7246:  00000 00000  00000 00000  00000 00000  00000 00000  00000 00000    00000 00000  00000 00000  00000 00000  00000 00000  00000 00000
7247:  00000 00000  00000 00000  00000 00000  00000 00000  00000 00000    00000 00000  00000 00000  00000 00000  00000 00000  00000 00000
7248:  00000 00000  00000 00000  00000 00000  00000 00000  00000 00000    00000 00000  00000 00000  00000 00000  00000 00000  00000 00000
7249:  00000 00000  00000 00000  00000 00000  00000 00000  00000 00000    00000 00000  00000 00000  00000 00000  00000 00000  00000 00000
```

```
7250:  00000 00000  00000 00000  00000 00000  00000 00000  00000 00000    00000 00000  00000 00000  00000 00000  00000 00000  00000 00000
7251:  00000 00000  00000 00000  00000 00000  00000 00000  00000 00000    00000 00000  00000 00000  00000 00000  00000 00000  00000 00000
7252:  00000 00000  00000 00000  00000 00000  00000 00000  00000 00000    00000 00000  00000 00000  00000 00000  00000 00000  00000 00000
7253:  00000 00000  00000 00000  00000 00000  00000 00000  00000 00000    00000 00000  00000 00000  00000 00000  00000 00000  00000 00000
7254:  00000 00000  00000 00000  00000 00000  00000 00000  00000 00000    00000 00000  00000 00000  00000 00000  00000 00000  00000 00000
7255:  00000 00000  00000 00000  00000 00000  00000 00000  00000 00000    00000 00000  00000 00000  00000 00000  00000 00000  00000 00000
7256:  00000 00000  00000 00000  00000 00000  00000 00000  00000 00000    00000 00000  00000 00000  00000 00000  00000 00000  00000 00000
7257:  00000 00000  00000 00000  00000 00000  00000 00000  00000 00000    00000 00000  00000 00000  00000 00000  00000 00000  00000 00000
7258:  00000 00000  00000 00000  00000 00000  00000 00000  00000 00000    00000 00000  00000 00000  00000 00000  00000 00000  00000 00000
7259:  00000 00000  00000 00000  00000 00000  00000 00000  00000 00000    00000 00000  00000 00000  00000 00000  00000 00000  00000 00000
7260:  00000 00000  00000 00000  00000 00000  00000 00000  00000 00000    00000 00000  00000 00000  00000 00000  00000 00000  00000 00000
7261:  00000 00000  00000 00000  00000 00000  00000 00000  00000 00000    00000 00000  00000 00000  00000 00000  00000 00000  00000 00000
7262:  00000 00000  00000 00000  00000 00000  00000 00000  00000 00000    00000 00000  00000 00000  00000 00000  00000 00000  00000 00000
7263:  00000 00000  00000 00000  00000 00000  00000 00000  00000 00000    00000 00000  00000 00000  00000 00000  00000 00000  00000 00000
7264:  00000 00000  00000 00000  00000 00000  00000 00000  00000 00000    00000 00000  00000 00000  00000 00000  00000 00000  00000 00000
7265:  00000 00000  00000 00000  00000 00000  00000 00000  00000 00000    00000 00000  00000 00000  00000 00000  00000 00000  00000 00000
7266:  00000 00000  00000 00000  00000 00000  00000 00000  00000 00000    00000 00000  00000 00000  00000 00000  00000 00000  00000 00000
7267:  00000 00000  00000 00000  00000 00000  00000 00000  00000 00000    00000 00000  00000 00000  00000 00000  00000 00000  00000 00000
7268:  00000 00000  00000 00000  00000 00000  00000 00000  00000 00000    00000 00000  00000 00000  00000 00000  00000 00000  00000 00000
7269:  00000 00000  00000 00000  00000 00000  00000 00000  00000 00000    00000 00000  00000 00000  00000 00000  00000 00000  00000 00000
7270:  00000 00000  00000 00000  00000 00000  00000 00000  00000 00000    00000 00000  00000 00000  00000 00000  00000 00000  00000 00000
7271:  00000 00000  00000 00000  00000 00000  00000 00000  00000 00000    00000 00000  00000 00000  00000 00000  00000 00000  00000 00000
7272:  00000 00000  00000 00000  00000 00000  00000 00000  00000 00000    00000 00000  00000 00000  00000 00000  00000 00000  00000 00000
7273:  00000 00000  00000 00000  00000 00000  00000 00000  00000 00000    00000 00000  00000 00000  00000 00000  00000 00000  00000 00000
7274:  00000 00000  00000 00000  00000 00000  00000 00000  00000 00000    00000 00000  00000 00000  00000 00000  00000 00000  00000 00000
7275:  00000 00000  00000 00000  00000 00000  00000 00000  00000 00000    00000 00000  00000 00000  00000 00000  00000 00000  00000 00000
7276:  00000 00000  00000 00000  00000 00000  00000 00000  00000 00000    00000 00000  00000 00000  00000 00000  00000 00000  00000 00000
7277:  00000 00000  00000 00000  00000 00000  00000 00000  00000 00000    00000 00000  00000 00000  00000 00000  00000 00000  00000 00000
7278:  00000 00000  00000 00000  00000 00000  00000 00000  00000 00000    00000 00000  00000 00000  00000 00000  00000 00000  00000 00000
7279:  00000 00000  00000 00000  00000 00000  00000 00000  00000 00000    00000 00000  00000 00000  00000 00000  00000 00000  00000 00000
7280:  00000 00000  00000 00000  00000 00000  00000 00000  00000 00000    00000 00000  00000 00000  00000 00000  00000 00000  00000 00000
7281:  00000 00000  00000 00000  00000 00000  00000 00000  00000 00000    00000 00000  00000 00000  00000 00000  00000 00000  00000 00000
7282:  00000 00000  00000 00000  00000 00000  00000 00000  00000 00000    00000 00000  00000 00000  00000 00000  00000 00000  00000 00000
7283:  00000 00000  00000 00000  00000 00000  00000 00000  00000 00000    00000 00000  00000 00000  00000 00000  00000 00000  00000 00000
7284:  00000 00000  00000 00000  00000 00000  00000 00000  00000 00000    00000 00000  00000 00000  00000 00000  00000 00000  00000 00000
7285:  00000 00000  00000 00000  00000 00000  00000 00000  00000 00000    00000 00000  00000 00000  00000 00000  00000 00000  00000 00000
7286:  00000 00000  00000 00000  00000 00000  00000 00000  00000 00000    00000 00000  00000 00000  00000 00000  00000 00000  00000 00000
7287:  00000 00000  00000 00000  00000 00000  00000 00000  00000 00000    00000 00000  00000 00000  00000 00000  00000 00000  00000 00000
7288:  00000 00000  00000 00000  00000 00000  00000 00000  00000 00000    00000 00000  00000 00000  00000 00000  00000 00000  00000 00000
7289:  00000 00000  00000 00000  00000 00000  00000 00000  00000 00000    00000 00000  00000 00000  00000 00000  00000 00000  00000 00000
7290:  00000 00000  00000 00000  00000 00000  00000 00000  00000 00000    00000 00000  00000 00000  00000 00000  00000 00000  00000 00000
7291:  00000 00000  00000 00000  00000 00000  00000 00000  00000 00000    00000 00000  00000 00000  00000 00000  00000 00000  00000 00000
7292:  00000 00000  00000 00000  00000 00000  00000 00000  00000 00000    00000 00000  00000 00000  00000 00000  00000 00000  00000 00000
7293:  00000 00000  00000 00000  00000 00000  00000 00000  00000 00000    00000 00000  00000 00000  00000 00000  00000 00000  00000 00000
7294:  00000 00000  00000 00000  00000 00000  00000 00000  00000 00000    00000 00000  00000 00000  00000 00000  00000 00000  00000 00000
7295:  00000 00000  00000 00000  00000 00000  00000 00000  00000 00000    00000 00000  00000 00000  00000 00000  00000 00000  00000 00000
7296:  00000 00000  00000 00000  00000 00000  00000 00000  00000 00000    00000 00000  00000 00000  00000 00000  00000 00000  00000 00000
7297:  00000 00000  00000 00000  00000 00000  00000 00000  00000 00000    00000 00000  00000 00000  00000 00000  00000 00000  00000 00000
7298:  00000 00000  00000 00000  00000 00000  00000 00000  00000 00000    00000 00000  00000 00000  00000 00000  00000 00000  00000 00000
7299:  00000 00000  00000 00000  00000 00000  00000 00000  00000 00000    00000 00000  00000 00000  00000 00000  00000 00000  00000 00000
```

```
7300:   00000 00000   00000 00000   00000 00000   00000 00000   00000 00000     00000 00000   00000 00000   00000 00000   00000 00000   00000 00000
7301:   00000 00000   00000 00000   00000 00000   00000 00000   00000 00000     00000 00000   00000 00000   00000 00000   00000 00000   00000 00000
7302:   00000 00000   00000 00000   00000 00000   00000 00000   00000 00000     00000 00000   00000 00000   00000 00000   00000 00000   00000 00000
7303:   00000 00000   00000 00000   00000 00000   00000 00000   00000 00000     00000 00000   00000 00000   00000 00000   00000 00000   00000 00000
7304:   00000 00000   00000 00000   00000 00000   00000 00000   00000 00000     00000 00000   00000 00000   00000 00000   00000 00000   00000 00000
7305:   00000 00000   00000 00000   00000 00000   00000 00000   00000 00000     00000 00000   00000 00000   00000 00000   00000 00000   00000 00000
7306:   00000 00000   00000 00000   00000 00000   00000 00000   00000 00000     00000 00000   00000 00000   00000 00000   00000 00000   00000 00000
7307:   00000 00000   00000 00000   00000 00000   00000 00000   00000 00000     00000 00000   00000 00000   00000 00000   00000 00000   00000 00000
7308:   00000 00000   00000 00000   00000 00000   00000 00000   00000 00000     00000 00000   00000 00000   00000 00000   00000 00000   00000 00000
7309:   00000 00000   00000 00000   00000 00000   00000 00000   00000 00000     00000 00000   00000 00000   00000 00000   00000 00000   00000 00000
7310:   00000 00000   00000 00000   00000 00000   00000 00000   00000 00000     00000 00000   00000 00000   00000 00000   00000 00000   00000 00000
7311:   00000 00000   00000 00000   00000 00000   00000 00000   00000 00000     00000 00000   00000 00000   00000 00000   00000 00000   00000 00000
7312:   00000 00000   00000 00000   00000 00000   00000 00000   00000 00000     00000 00000   00000 00000   00000 00000   00000 00000   00000 00000
7313:   00000 00000   00000 00000   00000 00000   00000 00000   00000 00000     00000 00000   00000 00000   00000 00000   00000 00000   00000 00000
7314:   00000 00000   00000 00000   00000 00000   00000 00000   00000 00000     00000 00000   00000 00000   00000 00000   00000 00000   00000 00000
7315:   00000 00000   00000 00000   00000 00000   00000 00000   00000 00000     00000 00000   00000 00000   00000 00000   00000 00000   00000 00000
7316:   00000 00000   00000 00000   00000 00000   00000 00000   00000 00000     00000 00000   00000 00000   00000 00000   00000 00000   00000 00000
7317:   00000 00000   00000 00000   00000 00000   00000 00000   00000 00000     00000 00000   00000 00000   00000 00000   00000 00000   00000 00000
7318:   00000 00000   00000 00000   00000 00000   00000 00000   00000 00000     00000 00000   00000 00000   00000 00000   00000 00000   00000 00000
7319:   00000 00000   00000 00000   00000 00000   00000 00000   00000 00000     00000 00000   00000 00000   00000 00000   00000 00000   00000 00000
7320:   00000 00000   00000 00000   00000 00000   00000 00000   00000 00000     00000 00000   00000 00000   00000 00000   00000 00000   00000 00000
7321:   00000 00000   00000 00000   00000 00000   00000 00000   00000 00000     00000 00000   00000 00000   00000 00000   00000 00000   00000 00000
7322:   00000 00000   00000 00000   00000 00000   00000 00000   00000 00000     00000 00000   00000 00000   00000 00000   00000 00000   00000 00000
7323:   00000 00000   00000 00000   00000 00000   00000 00000   00000 00000     00000 00000   00000 00000   00000 00000   00000 00000   00000 00000
7324:   00000 00000   00000 00000   00000 00000   00000 00000   00000 00000     00000 00000   00000 00000   00000 00000   00000 00000   00000 00000
7325:   00000 00000   00000 00000   00000 00000   00000 00000   00000 00000     00000 00000   00000 00000   00000 00000   00000 00000   00000 00000
7326:   00000 00000   00000 00000   00000 00000   00000 00000   00000 00000     00000 00000   00000 00000   00000 00000   00000 00000   00000 00000
7327:   00000 00000   00000 00000   00000 00000   00000 00000   00000 00000     00000 00000   00000 00000   00000 00000   00000 00000   00000 00000
7328:   00000 00000   00000 00000   00000 00000   00000 00000   00000 00000     00000 00000   00000 00000   00000 00000   00000 00000   00000 00000
7329:   00000 00000   00000 00000   00000 00000   00000 00000   00000 00000     00000 00000   00000 00000   00000 00000   00000 00000   00000 00000
7330:   00000 00000   00000 00000   00000 00000   00000 00000   00000 00000     00000 00000   00000 00000   00000 00000   00000 00000   00000 00000
7331:   00000 00000   00000 00000   00000 00000   00000 00000   00000 00000     00000 00000   00000 00000   00000 00000   00000 00000   00000 00000
7332:   00000 00000   00000 00000   00000 00000   00000 00000   00000 00000     00000 00000   00000 00000   00000 00000   00000 00000   00000 00000
7333:   00000 00000   00000 00000   00000 00000   00000 00000   00000 00000     00000 00000   00000 00000   00000 00000   00000 00000   00000 00000
7334:   00000 00000   00000 00000   00000 00000   00000 00000   00000 00000     00000 00000   00000 00000   00000 00000   00000 00000   00000 00000
7335:   00000 00000   00000 00000   00000 00000   00000 00000   00000 00000     00000 00000   00000 00000   00000 00000   00000 00000   00000 00000
7336:   00000 00000   00000 00000   00000 00000   00000 00000   00000 00000     00000 00000   00000 00000   00000 00000   00000 00000   00000 00000
7337:   00000 00000   00000 00000   00000 00000   00000 00000   00000 00000     00000 00000   00000 00000   00000 00000   00000 00000   00000 00000
7338:   00000 00000   00000 00000   00000 00000   00000 00000   00000 00000     00000 00000   00000 00000   00000 00000   00000 00000   00000 00000
7339:   00000 00000   00000 00000   00000 00000   00000 00000   00000 00000     00000 00000   00000 00000   00000 00000   00000 00000   00000 00000
7340:   00000 00000   00000 00000   00000 00000   00000 00000   00000 00000     00000 00000   00000 00000   00000 00000   00000 00000   00000 00000
7341:   00000 00000   00000 00000   00000 00000   00000 00000   00000 00000     00000 00000   00000 00000   00000 00000   00000 00000   00000 00000
7342:   00000 00000   00000 00000   00000 00000   00000 00000   00000 00000     00000 00000   00000 00000   00000 00000   00000 00000   00000 00000
7343:   00000 00000   00000 00000   00000 00000   00000 00000   00000 00000     00000 00000   00000 00000   00000 00000   00000 00000   00000 00000
7344:   00000 00000   00000 00000   00000 00000   00000 00000   00000 00000     00000 00000   00000 00000   00000 00000   00000 00000   00000 00000
7345:   00000 00000   00000 00000   00000 00000   00000 00000   00000 00000     00000 00000   00000 00000   00000 00000   00000 00000   00000 00000
7346:   00000 00000   00000 00000   00000 00000   00000 00000   00000 00000     00000 00000   00000 00000   00000 00000   00000 00000   00000 00000
7347:   00000 00000   00000 00000   00000 00000   00000 00000   00000 00000     00000 00000   00000 00000   00000 00000   00000 00000   00000 00000
7348:   00000 00000   00000 00000   00000 00000   00000 00000   00000 00000     00000 00000   00000 00000   00000 00000   00000 00000   00000 00000
7349:   00000 00000   00000 00000   00000 00000   00000 00000   00000 00000     00000 00000   00000 00000   00000 00000   00000 00000   00000 00000
```

```
7350: 00000 00000  00000 00000  00000 00000  00000 00000  00000 00000   00000 00000  00000 00000  00000 00000  00000 00000  00000 00000
7351: 00000 00000  00000 00000  00000 00000  00000 00000  00000 00000   00000 00000  00000 00000  00000 00000  00000 00000  00000 00000
7352: 00000 00000  00000 00000  00000 00000  00000 00000  00000 00000   00000 00000  00000 00000  00000 00000  00000 00000  00000 00000
7353: 00000 00000  00000 00000  00000 00000  00000 00000  00000 00000   00000 00000  00000 00000  00000 00000  00000 00000  00000 00000
7354: 00000 00000  00000 00000  00000 00000  00000 00000  00000 00000   00000 00000  00000 00000  00000 00000  00000 00000  00000 00000
7355: 00000 00000  00000 00000  00000 00000  00000 00000  00000 00000   00000 00000  00000 00000  00000 00000  00000 00000  00000 00000
7356: 00000 00000  00000 00000  00000 00000  00000 00000  00000 00000   00000 00000  00000 00000  00000 00000  00000 00000  00000 00000
7357: 00000 00000  00000 00000  00000 00000  00000 00000  00000 00000   00000 00000  00000 00000  00000 00000  00000 00000  00000 00000
7358: 00000 00000  00000 00000  00000 00000  00000 00000  00000 00000   00000 00000  00000 00000  00000 00000  00000 00000  00000 00000
7359: 00000 00000  00000 00000  00000 00000  00000 00000  00000 00000   00000 00000  00000 00000  00000 00000  00000 00000  00000 00000
7360: 00000 00000  00000 00000  00000 00000  00000 00000  00000 00000   00000 00000  00000 00000  00000 00000  00000 00000  00000 00000
7361: 00000 00000  00000 00000  00000 00000  00000 00000  00000 00000   00000 00000  00000 00000  00000 00000  00000 00000  00000 00000
7362: 00000 00000  00000 00000  00000 00000  00000 00000  00000 00000   00000 00000  00000 00000  00000 00000  00000 00000  00000 00000
7363: 00000 00000  00000 00000  00000 00000  00000 00000  00000 00000   00000 00000  00000 00000  00000 00000  00000 00000  00000 00000
7364: 00000 00000  00000 00000  00000 00000  00000 00000  00000 00000   00000 00000  00000 00000  00000 00000  00000 00000  00000 00000
7365: 00000 00000  00000 00000  00000 00000  00000 00000  00000 00000   00000 00000  00000 00000  00000 00000  00000 00000  00000 00000
7366: 00000 00000  00000 00000  00000 00000  00000 00000  00000 00000   00000 00000  00000 00000  00000 00000  00000 00000  00000 00000
7367: 00000 00000  00000 00000  00000 00000  00000 00000  00000 00000   00000 00000  00000 00000  00000 00000  00000 00000  00000 00000
7368: 00000 00000  00000 00000  00000 00000  00000 00000  00000 00000   00000 00000  00000 00000  00000 00000  00000 00000  00000 00000
7369: 00000 00000  00000 00000  00000 00000  00000 00000  00000 00000   00000 00000  00000 00000  00000 00000  00000 00000  00000 00000
7370: 00000 00000  00000 00000  00000 00000  00000 00000  00000 00000   00000 00000  00000 00000  00000 00000  00000 00000  00000 00000
7371: 00000 00000  00000 00000  00000 00000  00000 00000  00000 00000   00000 00000  00000 00000  00000 00000  00000 00000  00000 00000
7372: 00000 00000  00000 00000  00000 00000  00000 00000  00000 00000   00000 00000  00000 00000  00000 00000  00000 00000  00000 00000
7373: 00000 00000  00000 00000  00000 00000  00000 00000  00000 00000   00000 00000  00000 00000  00000 00000  00000 00000  00000 00000
7374: 00000 00000  00000 00000  00000 00000  00000 00000  00000 00000   00000 00000  00000 00000  00000 00000  00000 00000  00000 00000
7375: 00000 00000  00000 00000  00000 00000  00000 00000  00000 00000   00000 00000  00000 00000  00000 00000  00000 00000  00000 00000
7376: 00000 00000  00000 00000  00000 00000  00000 00000  00000 00000   00000 00000  00000 00000  00000 00000  00000 00000  00000 00000
7377: 00000 00000  00000 00000  00000 00000  00000 00000  00000 00000   00000 00000  00000 00000  00000 00000  00000 00000  00000 00000
7378: 00000 00000  00000 00000  00000 00000  00000 00000  00000 00000   00000 00000  00000 00000  00000 00000  00000 00000  00000 00000
7379: 00000 00000  00000 00000  00000 00000  00000 00000  00000 00000   00000 00000  00000 00000  00000 00000  00000 00000  00000 00000
7380: 00000 00000  00000 00000  00000 00000  00000 00000  00000 00000   00000 00000  00000 00000  00000 00000  00000 00000  00000 00000
7381: 00000 00000  00000 00000  00000 00000  00000 00000  00000 00000   00000 00000  00000 00000  00000 00000  00000 00000  00000 00000
7382: 00000 00000  00000 00000  00000 00000  00000 00000  00000 00000   00000 00000  00000 00000  00000 00000  00000 00000  00000 00000
7383: 00000 00000  00000 00000  00000 00000  00000 00000  00000 00000   00000 00000  00000 00000  00000 00000  00000 00000  00000 00000
7384: 00000 00000  00000 00000  00000 00000  00000 00000  00000 00000   00000 00000  00000 00000  00000 00000  00000 00000  00000 00000
7385: 00000 00000  00000 00000  00000 00000  00000 00000  00000 00000   00000 00000  00000 00000  00000 00000  00000 00000  00000 00000
7386: 00000 00000  00000 00000  00000 00000  00000 00000  00000 00000   00000 00000  00000 00000  00000 00000  00000 00000  00000 00000
7387: 00000 00000  00000 00000  00000 00000  00000 00000  00000 00000   00000 00000  00000 00000  00000 00000  00000 00000  00000 00000
7388: 00000 00000  00000 00000  00000 00000  00000 00000  00000 00000   00000 00000  00000 00000  00000 00000  00000 00000  00000 00000
7389: 00000 00000  00000 00000  00000 00000  00000 00000  00000 00000   00000 00000  00000 00000  00000 00000  00000 00000  00000 00000
7390: 00000 00000  00000 00000  00000 00000  00000 00000  00000 00000   00000 00000  00000 00000  00000 00000  00000 00000  00000 00000
7391: 00000 00000  00000 00000  00000 00000  00000 00000  00000 00000   00000 00000  00000 00000  00000 00000  00000 00000  00000 00000
7392: 00000 00000  00000 00000  00000 00000  00000 00000  00000 00000   00000 00000  00000 00000  00000 00000  00000 00000  00000 00000
7393: 00000 00000  00000 00000  00000 00000  00000 00000  00000 00000   00000 00000  00000 00000  00000 00000  00000 00000  00000 00000
7394: 00000 00000  00000 00000  00000 00000  00000 00000  00000 00000   00000 00000  00000 00000  00000 00000  00000 00000  00000 00000
7395: 00000 00000  00000 00000  00000 00000  00000 00000  00000 00000   00000 00000  00000 00000  00000 00000  00000 00000  00000 00000
7396: 00000 00000  00000 00000  00000 00000  00000 00000  00000 00000   00000 00000  00000 00000  00000 00000  00000 00000  00000 00000
7397: 00000 00000  00000 00000  00000 00000  00000 00000  00000 00000   00000 00000  00000 00000  00000 00000  00000 00000  00000 00000
7398: 00000 00000  00000 00000  00000 00000  00000 00000  00000 00000   00000 00000  00000 00000  00000 00000  00000 00000  00000 00000
7399: 00000 00000  00000 00000  00000 00000  00000 00000  00000 00000   00000 00000  00000 00000  00000 00000  00000 00000  00000 00000
```

```
7400:   00000 00000   00000 00000   00000 00000   00000 00000   00000 00000     00000 00000   00000 00000   00000 00000   00000 00000   00000 00000
7401:   00000 00000   00000 00000   00000 00000   00000 00000   00000 00000     00000 00000   00000 00000   00000 00000   00000 00000   00000 00000
7402:   00000 00000   00000 00000   00000 00000   00000 00000   00000 00000     00000 00000   00000 00000   00000 00000   00000 00000   00000 00000
7403:   00000 00000   00000 00000   00000 00000   00000 00000   00000 00000     00000 00000   00000 00000   00000 00000   00000 00000   00000 00000
7404:   00000 00000   00000 00000   00000 00000   00000 00000   00000 00000     00000 00000   00000 00000   00000 00000   00000 00000   00000 00000
7405:   00000 00000   00000 00000   00000 00000   00000 00000   00000 00000     00000 00000   00000 00000   00000 00000   00000 00000   00000 00000
7406:   00000 00000   00000 00000   00000 00000   00000 00000   00000 00000     00000 00000   00000 00000   00000 00000   00000 00000   00000 00000
7407:   00000 00000   00000 00000   00000 00000   00000 00000   00000 00000     00000 00000   00000 00000   00000 00000   00000 00000   00000 00000
7408:   00000 00000   00000 00000   00000 00000   00000 00000   00000 00000     00000 00000   00000 00000   00000 00000   00000 00000   00000 00000
7409:   00000 00000   00000 00000   00000 00000   00000 00000   00000 00000     00000 00000   00000 00000   00000 00000   00000 00000   00000 00000
7410:   00000 00000   00000 00000   00000 00000   00000 00000   00000 00000     00000 00000   00000 00000   00000 00000   00000 00000   00000 00000
7411:   00000 00000   00000 00000   00000 00000   00000 00000   00000 00000     00000 00000   00000 00000   00000 00000   00000 00000   00000 00000
7412:   00000 00000   00000 00000   00000 00000   00000 00000   00000 00000     00000 00000   00000 00000   00000 00000   00000 00000   00000 00000
7413:   00000 00000   00000 00000   00000 00000   00000 00000   00000 00000     00000 00000   00000 00000   00000 00000   00000 00000   00000 00000
7414:   00000 00000   00000 00000   00000 00000   00000 00000   00000 00000     00000 00000   00000 00000   00000 00000   00000 00000   00000 00000
7415:   00000 00000   00000 00000   00000 00000   00000 00000   00000 00000     00000 00000   00000 00000   00000 00000   00000 00000   00000 00000
7416:   00000 00000   00000 00000   00000 00000   00000 00000   00000 00000     00000 00000   00000 00000   00000 00000   00000 00000   00000 00000
7417:   00000 00000   00000 00000   00000 00000   00000 00000   00000 00000     00000 00000   00000 00000   00000 00000   00000 00000   00000 00000
7418:   00000 00000   00000 00000   00000 00000   00000 00000   00000 00000     00000 00000   00000 00000   00000 00000   00000 00000   00000 00000
7419:   00000 00000   00000 00000   00000 00000   00000 00000   00000 00000     00000 00000   00000 00000   00000 00000   00000 00000   00000 00000
7420:   00000 00000   00000 00000   00000 00000   00000 00000   00000 00000     00000 00000   00000 00000   00000 00000   00000 00000   00000 00000
7421:   00000 00000   00000 00000   00000 00000   00000 00000   00000 00000     00000 00000   00000 00000   00000 00000   00000 00000   00000 00000
7422:   00000 00000   00000 00000   00000 00000   00000 00000   00000 00000     00000 00000   00000 00000   00000 00000   00000 00000   00000 00000
7423:   00000 00000   00000 00000   00000 00000   00000 00000   00000 00000     00000 00000   00000 00000   00000 00000   00000 00000   00000 00000
7424:   00000 00000   00000 00000   00000 00000   00000 00000   00000 00000     00000 00000   00000 00000   00000 00000   00000 00000   00000 00000
7425:   00000 00000   00000 00000   00000 00000   00000 00000   00000 00000     00000 00000   00000 00000   00000 00000   00000 00000   00000 00000
7426:   00000 00000   00000 00000   00000 00000   00000 00000   00000 00000     00000 00000   00000 00000   00000 00000   00000 00000   00000 00000
7427:   00000 00000   00000 00000   00000 00000   00000 00000   00000 00000     00000 00000   00000 00000   00000 00000   00000 00000   00000 00000
7428:   00000 00000   00000 00000   00000 00000   00000 00000   00000 00000     00000 00000   00000 00000   00000 00000   00000 00000   00000 00000
7429:   00000 00000   00000 00000   00000 00000   00000 00000   00000 00000     00000 00000   00000 00000   00000 00000   00000 00000   00000 00000
7430:   00000 00000   00000 00000   00000 00000   00000 00000   00000 00000     00000 00000   00000 00000   00000 00000   00000 00000   00000 00000
7431:   00000 00000   00000 00000   00000 00000   00000 00000   00000 00000     00000 00000   00000 00000   00000 00000   00000 00000   00000 00000
7432:   00000 00000   00000 00000   00000 00000   00000 00000   00000 00000     00000 00000   00000 00000   00000 00000   00000 00000   00000 00000
7433:   00000 00000   00000 00000   00000 00000   00000 00000   00000 00000     00000 00000   00000 00000   00000 00000   00000 00000   00000 00000
7434:   00000 00000   00000 00000   00000 00000   00000 00000   00000 00000     00000 00000   00000 00000   00000 00000   00000 00000   00000 00000
7435:   00000 00000   00000 00000   00000 00000   00000 00000   00000 00000     00000 00000   00000 00000   00000 00000   00000 00000   00000 00000
7436:   00000 00000   00000 00000   00000 00000   00000 00000   00000 00000     00000 00000   00000 00000   00000 00000   00000 00000   00000 00000
7437:   00000 00000   00000 00000   00000 00000   00000 00000   00000 00000     00000 00000   00000 00000   00000 00000   00000 00000   00000 00000
7438:   00000 00000   00000 00000   00000 00000   00000 00000   00000 00000     00000 00000   00000 00000   00000 00000   00000 00000   00000 00000
7439:   00000 00000   00000 00000   00000 00000   00000 00000   00000 00000     00000 00000   00000 00000   00000 00000   00000 00000   00000 00000
7440:   00000 00000   00000 00000   00000 00000   00000 00000   00000 00000     00000 00000   00000 00000   00000 00000   00000 00000   00000 00000
7441:   00000 00000   00000 00000   00000 00000   00000 00000   00000 00000     00000 00000   00000 00000   00000 00000   00000 00000   00000 00000
7442:   00000 00000   00000 00000   00000 00000   00000 00000   00000 00000     00000 00000   00000 00000   00000 00000   00000 00000   00000 00000
7443:   00000 00000   00000 00000   00000 00000   00000 00000   00000 00000     00000 00000   00000 00000   00000 00000   00000 00000   00000 00000
7444:   00000 00000   00000 00000   00000 00000   00000 00000   00000 00000     00000 00000   00000 00000   00000 00000   00000 00000   00000 00000
7445:   00000 00000   00000 00000   00000 00000   00000 00000   00000 00000     00000 00000   00000 00000   00000 00000   00000 00000   00000 00000
7446:   00000 00000   00000 00000   00000 00000   00000 00000   00000 00000     00000 00000   00000 00000   00000 00000   00000 00000   00000 00000
7447:   00000 00000   00000 00000   00000 00000   00000 00000   00000 00000     00000 00000   00000 00000   00000 00000   00000 00000   00000 00000
7448:   00000 00000   00000 00000   00000 00000   00000 00000   00000 00000     00000 00000   00000 00000   00000 00000   00000 00000   00000 00000
7449:   00000 00000   00000 00000   00000 00000   00000 00000   00000 00000     00000 00000   00000 00000   00000 00000   00000 00000   00000 00000
```

```
7450: 00000 00000  00000 00000  00000 00000  00000 00000  00000 00000    00000 00000  00000 00000  00000 00000  00000 00000  00000 00000
7451: 00000 00000  00000 00000  00000 00000  00000 00000  00000 00000    00000 00000  00000 00000  00000 00000  00000 00000  00000 00000
7452: 00000 00000  00000 00000  00000 00000  00000 00000  00000 00000    00000 00000  00000 00000  00000 00000  00000 00000  00000 00000
7453: 00000 00000  00000 00000  00000 00000  00000 00000  00000 00000    00000 00000  00000 00000  00000 00000  00000 00000  00000 00000
7454: 00000 00000  00000 00000  00000 00000  00000 00000  00000 00000    00000 00000  00000 00000  00000 00000  00000 00000  00000 00000
7455: 00000 00000  00000 00000  00000 00000  00000 00000  00000 00000    00000 00000  00000 00000  00000 00000  00000 00000  00000 00000
7456: 00000 00000  00000 00000  00000 00000  00000 00000  00000 00000    00000 00000  00000 00000  00000 00000  00000 00000  00000 00000
7457: 00000 00000  00000 00000  00000 00000  00000 00000  00000 00000    00000 00000  00000 00000  00000 00000  00000 00000  00000 00000
7458: 00000 00000  00000 00000  00000 00000  00000 00000  00000 00000    00000 00000  00000 00000  00000 00000  00000 00000  00000 00000
7459: 00000 00000  00000 00000  00000 00000  00000 00000  00000 00000    00000 00000  00000 00000  00000 00000  00000 00000  00000 00000
7460: 00000 00000  00000 00000  00000 00000  00000 00000  00000 00000    00000 00000  00000 00000  00000 00000  00000 00000  00000 00000
7461: 00000 00000  00000 00000  00000 00000  00000 00000  00000 00000    00000 00000  00000 00000  00000 00000  00000 00000  00000 00000
7462: 00000 00000  00000 00000  00000 00000  00000 00000  00000 00000    00000 00000  00000 00000  00000 00000  00000 00000  00000 00000
7463: 00000 00000  00000 00000  00000 00000  00000 00000  00000 00000    00000 00000  00000 00000  00000 00000  00000 00000  00000 00000
7464: 00000 00000  00000 00000  00000 00000  00000 00000  00000 00000    00000 00000  00000 00000  00000 00000  00000 00000  00000 00000
7465: 00000 00000  00000 00000  00000 00000  00000 00000  00000 00000    00000 00000  00000 00000  00000 00000  00000 00000  00000 00000
7466: 00000 00000  00000 00000  00000 00000  00000 00000  00000 00000    00000 00000  00000 00000  00000 00000  00000 00000  00000 00000
7467: 00000 00000  00000 00000  00000 00000  00000 00000  00000 00000    00000 00000  00000 00000  00000 00000  00000 00000  00000 00000
7468: 00000 00000  00000 00000  00000 00000  00000 00000  00000 00000    00000 00000  00000 00000  00000 00000  00000 00000  00000 00000
7469: 00000 00000  00000 00000  00000 00000  00000 00000  00000 00000    00000 00000  00000 00000  00000 00000  00000 00000  00000 00000
7470: 00000 00000  00000 00000  00000 00000  00000 00000  00000 00000    00000 00000  00000 00000  00000 00000  00000 00000  00000 00000
7471: 00000 00000  00000 00000  00000 00000  00000 00000  00000 00000    00000 00000  00000 00000  00000 00000  00000 00000  00000 00000
7472: 00000 00000  00000 00000  00000 00000  00000 00000  00000 00000    00000 00000  00000 00000  00000 00000  00000 00000  00000 00000
7473: 00000 00000  00000 00000  00000 00000  00000 00000  00000 00000    00000 00000  00000 00000  00000 00000  00000 00000  00000 00000
7474: 00000 00000  00000 00000  00000 00000  00000 00000  00000 00000    00000 00000  00000 00000  00000 00000  00000 00000  00000 00000
7475: 00000 00000  00000 00000  00000 00000  00000 00000  00000 00000    00000 00000  00000 00000  00000 00000  00000 00000  00000 00000
7476: 00000 00000  00000 00000  00000 00000  00000 00000  00000 00000    00000 00000  00000 00000  00000 00000  00000 00000  00000 00000
7477: 00000 00000  00000 00000  00000 00000  00000 00000  00000 00000    00000 00000  00000 00000  00000 00000  00000 00000  00000 00000
7478: 00000 00000  00000 00000  00000 00000  00000 00000  00000 00000    00000 00000  00000 00000  00000 00000  00000 00000  00000 00000
7479: 00000 00000  00000 00000  00000 00000  00000 00000  00000 00000    00000 00000  00000 00000  00000 00000  00000 00000  00000 00000
7480: 00000 00000  00000 00000  00000 00000  00000 00000  00000 00000    00000 00000  00000 00000  00000 00000  00000 00000  00000 00000
7481: 00000 00000  00000 00000  00000 00000  00000 00000  00000 00000    00000 00000  00000 00000  00000 00000  00000 00000  00000 00000
7482: 00000 00000  00000 00000  00000 00000  00000 00000  00000 00000    00000 00000  00000 00000  00000 00000  00000 00000  00000 00000
7483: 00000 00000  00000 00000  00000 00000  00000 00000  00000 00000    00000 00000  00000 00000  00000 00000  00000 00000  00000 00000
7484: 00000 00000  00000 00000  00000 00000  00000 00000  00000 00000    00000 00000  00000 00000  00000 00000  00000 00000  00000 00000
7485: 00000 00000  00000 00000  00000 00000  00000 00000  00000 00000    00000 00000  00000 00000  00000 00000  00000 00000  00000 00000
7486: 00000 00000  00000 00000  00000 00000  00000 00000  00000 00000    00000 00000  00000 00000  00000 00000  00000 00000  00000 00000
7487: 00000 00000  00000 00000  00000 00000  00000 00000  00000 00000    00000 00000  00000 00000  00000 00000  00000 00000  00000 00000
7488: 00000 00000  00000 00000  00000 00000  00000 00000  00000 00000    00000 00000  00000 00000  00000 00000  00000 00000  00000 00000
7489: 00000 00000  00000 00000  00000 00000  00000 00000  00000 00000    00000 00000  00000 00000  00000 00000  00000 00000  00000 00000
7490: 00000 00000  00000 00000  00000 00000  00000 00000  00000 00000    00000 00000  00000 00000  00000 00000  00000 00000  00000 00000
7491: 00000 00000  00000 00000  00000 00000  00000 00000  00000 00000    00000 00000  00000 00000  00000 00000  00000 00000  00000 00000
7492: 00000 00000  00000 00000  00000 00000  00000 00000  00000 00000    00000 00000  00000 00000  00000 00000  00000 00000  00000 00000
7493: 00000 00000  00000 00000  00000 00000  00000 00000  00000 00000    00000 00000  00000 00000  00000 00000  00000 00000  00000 00000
7494: 00000 00000  00000 00000  00000 00000  00000 00000  00000 00000    00000 00000  00000 00000  00000 00000  00000 00000  00000 00000
7495: 00000 00000  00000 00000  00000 00000  00000 00000  00000 00000    00000 00000  00000 00000  00000 00000  00000 00000  00000 00000
7496: 00000 00000  00000 00000  00000 00000  00000 00000  00000 00000    00000 00000  00000 00000  00000 00000  00000 00000  00000 00000
7497: 00000 00000  00000 00000  00000 00000  00000 00000  00000 00000    00000 00000  00000 00000  00000 00000  00000 00000  00000 00000
7498: 00000 00000  00000 00000  00000 00000  00000 00000  00000 00000    00000 00000  00000 00000  00000 00000  00000 00000  00000 00000
7499: 00000 00000  00000 00000  00000 00000  00000 00000  00000 00000    00000 00000  00000 00000  00000 00000  00000 00000  00000 00000
```

```
7500:   00000 00000   00000 00000   00000 00000   00000 00000   00000 00000     00000 00000   00000 00000   00000 00000   00000 00000   00000 00000
7501:   00000 00000   00000 00000   00000 00000   00000 00000   00000 00000     00000 00000   00000 00000   00000 00000   00000 00000   00000 00000
7502:   00000 00000   00000 00000   00000 00000   00000 00000   00000 00000     00000 00000   00000 00000   00000 00000   00000 00000   00000 00000
7503:   00000 00000   00000 00000   00000 00000   00000 00000   00000 00000     00000 00000   00000 00000   00000 00000   00000 00000   00000 00000
7504:   00000 00000   00000 00000   00000 00000   00000 00000   00000 00000     00000 00000   00000 00000   00000 00000   00000 00000   00000 00000
7505:   00000 00000   00000 00000   00000 00000   00000 00000   00000 00000     00000 00000   00000 00000   00000 00000   00000 00000   00000 00000
7506:   00000 00000   00000 00000   00000 00000   00000 00000   00000 00000     00000 00000   00000 00000   00000 00000   00000 00000   00000 00000
7507:   00000 00000   00000 00000   00000 00000   00000 00000   00000 00000     00000 00000   00000 00000   00000 00000   00000 00000   00000 00000
7508:   00000 00000   00000 00000   00000 00000   00000 00000   00000 00000     00000 00000   00000 00000   00000 00000   00000 00000   00000 00000
7509:   00000 00000   00000 00000   00000 00000   00000 00000   00000 00000     00000 00000   00000 00000   00000 00000   00000 00000   00000 00000
7510:   00000 00000   00000 00000   00000 00000   00000 00000   00000 00000     00000 00000   00000 00000   00000 00000   00000 00000   00000 00000
7511:   00000 00000   00000 00000   00000 00000   00000 00000   00000 00000     00000 00000   00000 00000   00000 00000   00000 00000   00000 00000
7512:   00000 00000   00000 00000   00000 00000   00000 00000   00000 00000     00000 00000   00000 00000   00000 00000   00000 00000   00000 00000
7513:   00000 00000   00000 00000   00000 00000   00000 00000   00000 00000     00000 00000   00000 00000   00000 00000   00000 00000   00000 00000
7514:   00000 00000   00000 00000   00000 00000   00000 00000   00000 00000     00000 00000   00000 00000   00000 00000   00000 00000   00000 00000
7515:   00000 00000   00000 00000   00000 00000   00000 00000   00000 00000     00000 00000   00000 00000   00000 00000   00000 00000   00000 00000
7516:   00000 00000   00000 00000   00000 00000   00000 00000   00000 00000     00000 00000   00000 00000   00000 00000   00000 00000   00000 00000
7517:   00000 00000   00000 00000   00000 00000   00000 00000   00000 00000     00000 00000   00000 00000   00000 00000   00000 00000   00000 00000
7518:   00000 00000   00000 00000   00000 00000   00000 00000   00000 00000     00000 00000   00000 00000   00000 00000   00000 00000   00000 00000
7519:   00000 00000   00000 00000   00000 00000   00000 00000   00000 00000     00000 00000   00000 00000   00000 00000   00000 00000   00000 00000
7520:   00000 00000   00000 00000   00000 00000   00000 00000   00000 00000     00000 00000   00000 00000   00000 00000   00000 00000   00000 00000
7521:   00000 00000   00000 00000   00000 00000   00000 00000   00000 00000     00000 00000   00000 00000   00000 00000   00000 00000   00000 00000
7522:   00000 00000   00000 00000   00000 00000   00000 00000   00000 00000     00000 00000   00000 00000   00000 00000   00000 00000   00000 00000
7523:   00000 00000   00000 00000   00000 00000   00000 00000   00000 00000     00000 00000   00000 00000   00000 00000   00000 00000   00000 00000
7524:   00000 00000   00000 00000   00000 00000   00000 00000   00000 00000     00000 00000   00000 00000   00000 00000   00000 00000   00000 00000
7525:   00000 00000   00000 00000   00000 00000   00000 00000   00000 00000     00000 00000   00000 00000   00000 00000   00000 00000   00000 00000
7526:   00000 00000   00000 00000   00000 00000   00000 00000   00000 00000     00000 00000   00000 00000   00000 00000   00000 00000   00000 00000
7527:   00000 00000   00000 00000   00000 00000   00000 00000   00000 00000     00000 00000   00000 00000   00000 00000   00000 00000   00000 00000
7528:   00000 00000   00000 00000   00000 00000   00000 00000   00000 00000     00000 00000   00000 00000   00000 00000   00000 00000   00000 00000
7529:   00000 00000   00000 00000   00000 00000   00000 00000   00000 00000     00000 00000   00000 00000   00000 00000   00000 00000   00000 00000
7530:   00000 00000   00000 00000   00000 00000   00000 00000   00000 00000     00000 00000   00000 00000   00000 00000   00000 00000   00000 00000
7531:   00000 00000   00000 00000   00000 00000   00000 00000   00000 00000     00000 00000   00000 00000   00000 00000   00000 00000   00000 00000
7532:   00000 00000   00000 00000   00000 00000   00000 00000   00000 00000     00000 00000   00000 00000   00000 00000   00000 00000   00000 00000
7533:   00000 00000   00000 00000   00000 00000   00000 00000   00000 00000     00000 00000   00000 00000   00000 00000   00000 00000   00000 00000
7534:   00000 00000   00000 00000   00000 00000   00000 00000   00000 00000     00000 00000   00000 00000   00000 00000   00000 00000   00000 00000
7535:   00000 00000   00000 00000   00000 00000   00000 00000   00000 00000     00000 00000   00000 00000   00000 00000   00000 00000   00000 00000
7536:   00000 00000   00000 00000   00000 00000   00000 00000   00000 00000     00000 00000   00000 00000   00000 00000   00000 00000   00000 00000
7537:   00000 00000   00000 00000   00000 00000   00000 00000   00000 00000     00000 00000   00000 00000   00000 00000   00000 00000   00000 00000
7538:   00000 00000   00000 00000   00000 00000   00000 00000   00000 00000     00000 00000   00000 00000   00000 00000   00000 00000   00000 00000
7539:   00000 00000   00000 00000   00000 00000   00000 00000   00000 00000     00000 00000   00000 00000   00000 00000   00000 00000   00000 00000
7540:   00000 00000   00000 00000   00000 00000   00000 00000   00000 00000     00000 00000   00000 00000   00000 00000   00000 00000   00000 00000
7541:   00000 00000   00000 00000   00000 00000   00000 00000   00000 00000     00000 00000   00000 00000   00000 00000   00000 00000   00000 00000
7542:   00000 00000   00000 00000   00000 00000   00000 00000   00000 00000     00000 00000   00000 00000   00000 00000   00000 00000   00000 00000
7543:   00000 00000   00000 00000   00000 00000   00000 00000   00000 00000     00000 00000   00000 00000   00000 00000   00000 00000   00000 00000
7544:   00000 00000   00000 00000   00000 00000   00000 00000   00000 00000     00000 00000   00000 00000   00000 00000   00000 00000   00000 00000
7545:   00000 00000   00000 00000   00000 00000   00000 00000   00000 00000     00000 00000   00000 00000   00000 00000   00000 00000   00000 00000
7546:   00000 00000   00000 00000   00000 00000   00000 00000   00000 00000     00000 00000   00000 00000   00000 00000   00000 00000   00000 00000
7547:   00000 00000   00000 00000   00000 00000   00000 00000   00000 00000     00000 00000   00000 00000   00000 00000   00000 00000   00000 00000
7548:   00000 00000   00000 00000   00000 00000   00000 00000   00000 00000     00000 00000   00000 00000   00000 00000   00000 00000   00000 00000
7549:   00000 00000   00000 00000   00000 00000   00000 00000   00000 00000     00000 00000   00000 00000   00000 00000   00000 00000   00000 00000
```

```
7550:   00000 00000   00000 00000   00000 00000   00000 00000   00000 00000      00000 00000   00000 00000   00000 00000   00000 00000   00000 00000
7551:   00000 00000   00000 00000   00000 00000   00000 00000   00000 00000      00000 00000   00000 00000   00000 00000   00000 00000   00000 00000
7552:   00000 00000   00000 00000   00000 00000   00000 00000   00000 00000      00000 00000   00000 00000   00000 00000   00000 00000   00000 00000
7553:   00000 00000   00000 00000   00000 00000   00000 00000   00000 00000      00000 00000   00000 00000   00000 00000   00000 00000   00000 00000
7554:   00000 00000   00000 00000   00000 00000   00000 00000   00000 00000      00000 00000   00000 00000   00000 00000   00000 00000   00000 00000
7555:   00000 00000   00000 00000   00000 00000   00000 00000   00000 00000      00000 00000   00000 00000   00000 00000   00000 00000   00000 00000
7556:   00000 00000   00000 00000   00000 00000   00000 00000   00000 00000      00000 00000   00000 00000   00000 00000   00000 00000   00000 00000
7557:   00000 00000   00000 00000   00000 00000   00000 00000   00000 00000      00000 00000   00000 00000   00000 00000   00000 00000   00000 00000
7558:   00000 00000   00000 00000   00000 00000   00000 00000   00000 00000      00000 00000   00000 00000   00000 00000   00000 00000   00000 00000
7559:   00000 00000   00000 00000   00000 00000   00000 00000   00000 00000      00000 00000   00000 00000   00000 00000   00000 00000   00000 00000
7560:   00000 00000   00000 00000   00000 00000   00000 00000   00000 00000      00000 00000   00000 00000   00000 00000   00000 00000   00000 00000
7561:   00000 00000   00000 00000   00000 00000   00000 00000   00000 00000      00000 00000   00000 00000   00000 00000   00000 00000   00000 00000
7562:   00000 00000   00000 00000   00000 00000   00000 00000   00000 00000      00000 00000   00000 00000   00000 00000   00000 00000   00000 00000
7563:   00000 00000   00000 00000   00000 00000   00000 00000   00000 00000      00000 00000   00000 00000   00000 00000   00000 00000   00000 00000
7564:   00000 00000   00000 00000   00000 00000   00000 00000   00000 00000      00000 00000   00000 00000   00000 00000   00000 00000   00000 00000
7565:   00000 00000   00000 00000   00000 00000   00000 00000   00000 00000      00000 00000   00000 00000   00000 00000   00000 00000   00000 00000
7566:   00000 00000   00000 00000   00000 00000   00000 00000   00000 00000      00000 00000   00000 00000   00000 00000   00000 00000   00000 00000
7567:   00000 00000   00000 00000   00000 00000   00000 00000   00000 00000      00000 00000   00000 00000   00000 00000   00000 00000   00000 00000
7568:   00000 00000   00000 00000   00000 00000   00000 00000   00000 00000      00000 00000   00000 00000   00000 00000   00000 00000   00000 00000
7569:   00000 00000   00000 00000   00000 00000   00000 00000   00000 00000      00000 00000   00000 00000   00000 00000   00000 00000   00000 00000
7570:   00000 00000   00000 00000   00000 00000   00000 00000   00000 00000      00000 00000   00000 00000   00000 00000   00000 00000   00000 00000
7571:   00000 00000   00000 00000   00000 00000   00000 00000   00000 00000      00000 00000   00000 00000   00000 00000   00000 00000   00000 00000
7572:   00000 00000   00000 00000   00000 00000   00000 00000   00000 00000      00000 00000   00000 00000   00000 00000   00000 00000   00000 00000
7573:   00000 00000   00000 00000   00000 00000   00000 00000   00000 00000      00000 00000   00000 00000   00000 00000   00000 00000   00000 00000
7574:   00000 00000   00000 00000   00000 00000   00000 00000   00000 00000      00000 00000   00000 00000   00000 00000   00000 00000   00000 00000
7575:   00000 00000   00000 00000   00000 00000   00000 00000   00000 00000      00000 00000   00000 00000   00000 00000   00000 00000   00000 00000
7576:   00000 00000   00000 00000   00000 00000   00000 00000   00000 00000      00000 00000   00000 00000   00000 00000   00000 00000   00000 00000
7577:   00000 00000   00000 00000   00000 00000   00000 00000   00000 00000      00000 00000   00000 00000   00000 00000   00000 00000   00000 00000
7578:   00000 00000   00000 00000   00000 00000   00000 00000   00000 00000      00000 00000   00000 00000   00000 00000   00000 00000   00000 00000
7579:   00000 00000   00000 00000   00000 00000   00000 00000   00000 00000      00000 00000   00000 00000   00000 00000   00000 00000   00000 00000
7580:   00000 00000   00000 00000   00000 00000   00000 00000   00000 00000      00000 00000   00000 00000   00000 00000   00000 00000   00000 00000
7581:   00000 00000   00000 00000   00000 00000   00000 00000   00000 00000      00000 00000   00000 00000   00000 00000   00000 00000   00000 00000
7582:   00000 00000   00000 00000   00000 00000   00000 00000   00000 00000      00000 00000   00000 00000   00000 00000   00000 00000   00000 00000
7583:   00000 00000   00000 00000   00000 00000   00000 00000   00000 00000      00000 00000   00000 00000   00000 00000   00000 00000   00000 00000
7584:   00000 00000   00000 00000   00000 00000   00000 00000   00000 00000      00000 00000   00000 00000   00000 00000   00000 00000   00000 00000
7585:   00000 00000   00000 00000   00000 00000   00000 00000   00000 00000      00000 00000   00000 00000   00000 00000   00000 00000   00000 00000
7586:   00000 00000   00000 00000   00000 00000   00000 00000   00000 00000      00000 00000   00000 00000   00000 00000   00000 00000   00000 00000
7587:   00000 00000   00000 00000   00000 00000   00000 00000   00000 00000      00000 00000   00000 00000   00000 00000   00000 00000   00000 00000
7588:   00000 00000   00000 00000   00000 00000   00000 00000   00000 00000      00000 00000   00000 00000   00000 00000   00000 00000   00000 00000
7589:   00000 00000   00000 00000   00000 00000   00000 00000   00000 00000      00000 00000   00000 00000   00000 00000   00000 00000   00000 00000
7590:   00000 00000   00000 00000   00000 00000   00000 00000   00000 00000      00000 00000   00000 00000   00000 00000   00000 00000   00000 00000
7591:   00000 00000   00000 00000   00000 00000   00000 00000   00000 00000      00000 00000   00000 00000   00000 00000   00000 00000   00000 00000
7592:   00000 00000   00000 00000   00000 00000   00000 00000   00000 00000      00000 00000   00000 00000   00000 00000   00000 00000   00000 00000
7593:   00000 00000   00000 00000   00000 00000   00000 00000   00000 00000      00000 00000   00000 00000   00000 00000   00000 00000   00000 00000
7594:   00000 00000   00000 00000   00000 00000   00000 00000   00000 00000      00000 00000   00000 00000   00000 00000   00000 00000   00000 00000
7595:   00000 00000   00000 00000   00000 00000   00000 00000   00000 00000      00000 00000   00000 00000   00000 00000   00000 00000   00000 00000
7596:   00000 00000   00000 00000   00000 00000   00000 00000   00000 00000      00000 00000   00000 00000   00000 00000   00000 00000   00000 00000
7597:   00000 00000   00000 00000   00000 00000   00000 00000   00000 00000      00000 00000   00000 00000   00000 00000   00000 00000   00000 00000
7598:   00000 00000   00000 00000   00000 00000   00000 00000   00000 00000      00000 00000   00000 00000   00000 00000   00000 00000   00000 00000
7599:   00000 00000   00000 00000   00000 00000   00000 00000   00000 00000      00000 00000   00000 00000   00000 00000   00000 00000   00000 00000
```

```
7600:  00000 00000  00000 00000  00000 00000  00000 00000  00000 00000    00000 00000  00000 00000  00000 00000  00000 00000  00000 00000
7601:  00000 00000  00000 00000  00000 00000  00000 00000  00000 00000    00000 00000  00000 00000  00000 00000  00000 00000  00000 00000
7602:  00000 00000  00000 00000  00000 00000  00000 00000  00000 00000    00000 00000  00000 00000  00000 00000  00000 00000  00000 00000
7603:  00000 00000  00000 00000  00000 00000  00000 00000  00000 00000    00000 00000  00000 00000  00000 00000  00000 00000  00000 00000
7604:  00000 00000  00000 00000  00000 00000  00000 00000  00000 00000    00000 00000  00000 00000  00000 00000  00000 00000  00000 00000
7605:  00000 00000  00000 00000  00000 00000  00000 00000  00000 00000    00000 00000  00000 00000  00000 00000  00000 00000  00000 00000
7606:  00000 00000  00000 00000  00000 00000  00000 00000  00000 00000    00000 00000  00000 00000  00000 00000  00000 00000  00000 00000
7607:  00000 00000  00000 00000  00000 00000  00000 00000  00000 00000    00000 00000  00000 00000  00000 00000  00000 00000  00000 00000
7608:  00000 00000  00000 00000  00000 00000  00000 00000  00000 00000    00000 00000  00000 00000  00000 00000  00000 00000  00000 00000
7609:  00000 00000  00000 00000  00000 00000  00000 00000  00000 00000    00000 00000  00000 00000  00000 00000  00000 00000  00000 00000
7610:  00000 00000  00000 00000  00000 00000  00000 00000  00000 00000    00000 00000  00000 00000  00000 00000  00000 00000  00000 00000
7611:  00000 00000  00000 00000  00000 00000  00000 00000  00000 00000    00000 00000  00000 00000  00000 00000  00000 00000  00000 00000
7612:  00000 00000  00000 00000  00000 00000  00000 00000  00000 00000    00000 00000  00000 00000  00000 00000  00000 00000  00000 00000
7613:  00000 00000  00000 00000  00000 00000  00000 00000  00000 00000    00000 00000  00000 00000  00000 00000  00000 00000  00000 00000
7614:  00000 00000  00000 00000  00000 00000  00000 00000  00000 00000    00000 00000  00000 00000  00000 00000  00000 00000  00000 00000
7615:  00000 00000  00000 00000  00000 00000  00000 00000  00000 00000    00000 00000  00000 00000  00000 00000  00000 00000  00000 00000
7616:  00000 00000  00000 00000  00000 00000  00000 00000  00000 00000    00000 00000  00000 00000  00000 00000  00000 00000  00000 00000
7617:  00000 00000  00000 00000  00000 00000  00000 00000  00000 00000    00000 00000  00000 00000  00000 00000  00000 00000  00000 00000
7618:  00000 00000  00000 00000  00000 00000  00000 00000  00000 00000    00000 00000  00000 00000  00000 00000  00000 00000  00000 00000
7619:  00000 00000  00000 00000  00000 00000  00000 00000  00000 00000    00000 00000  00000 00000  00000 00000  00000 00000  00000 00000
7620:  00000 00000  00000 00000  00000 00000  00000 00000  00000 00000    00000 00000  00000 00000  00000 00000  00000 00000  00000 00000
7621:  00000 00000  00000 00000  00000 00000  00000 00000  00000 00000    00000 00000  00000 00000  00000 00000  00000 00000  00000 00000
7622:  00000 00000  00000 00000  00000 00000  00000 00000  00000 00000    00000 00000  00000 00000  00000 00000  00000 00000  00000 00000
7623:  00000 00000  00000 00000  00000 00000  00000 00000  00000 00000    00000 00000  00000 00000  00000 00000  00000 00000  00000 00000
7624:  00000 00000  00000 00000  00000 00000  00000 00000  00000 00000    00000 00000  00000 00000  00000 00000  00000 00000  00000 00000
7625:  00000 00000  00000 00000  00000 00000  00000 00000  00000 00000    00000 00000  00000 00000  00000 00000  00000 00000  00000 00000
7626:  00000 00000  00000 00000  00000 00000  00000 00000  00000 00000    00000 00000  00000 00000  00000 00000  00000 00000  00000 00000
7627:  00000 00000  00000 00000  00000 00000  00000 00000  00000 00000    00000 00000  00000 00000  00000 00000  00000 00000  00000 00000
7628:  00000 00000  00000 00000  00000 00000  00000 00000  00000 00000    00000 00000  00000 00000  00000 00000  00000 00000  00000 00000
7629:  00000 00000  00000 00000  00000 00000  00000 00000  00000 00000    00000 00000  00000 00000  00000 00000  00000 00000  00000 00000
7630:  00000 00000  00000 00000  00000 00000  00000 00000  00000 00000    00000 00000  00000 00000  00000 00000  00000 00000  00000 00000
7631:  00000 00000  00000 00000  00000 00000  00000 00000  00000 00000    00000 00000  00000 00000  00000 00000  00000 00000  00000 00000
7632:  00000 00000  00000 00000  00000 00000  00000 00000  00000 00000    00000 00000  00000 00000  00000 00000  00000 00000  00000 00000
7633:  00000 00000  00000 00000  00000 00000  00000 00000  00000 00000    00000 00000  00000 00000  00000 00000  00000 00000  00000 00000
7634:  00000 00000  00000 00000  00000 00000  00000 00000  00000 00000    00000 00000  00000 00000  00000 00000  00000 00000  00000 00000
7635:  00000 00000  00000 00000  00000 00000  00000 00000  00000 00000    00000 00000  00000 00000  00000 00000  00000 00000  00000 00000
7636:  00000 00000  00000 00000  00000 00000  00000 00000  00000 00000    00000 00000  00000 00000  00000 00000  00000 00000  00000 00000
7637:  00000 00000  00000 00000  00000 00000  00000 00000  00000 00000    00000 00000  00000 00000  00000 00000  00000 00000  00000 00000
7638:  00000 00000  00000 00000  00000 00000  00000 00000  00000 00000    00000 00000  00000 00000  00000 00000  00000 00000  00000 00000
7639:  00000 00000  00000 00000  00000 00000  00000 00000  00000 00000    00000 00000  00000 00000  00000 00000  00000 00000  00000 00000
7640:  00000 00000  00000 00000  00000 00000  00000 00000  00000 00000    00000 00000  00000 00000  00000 00000  00000 00000  00000 00000
7641:  00000 00000  00000 00000  00000 00000  00000 00000  00000 00000    00000 00000  00000 00000  00000 00000  00000 00000  00000 00000
7642:  00000 00000  00000 00000  00000 00000  00000 00000  00000 00000    00000 00000  00000 00000  00000 00000  00000 00000  00000 00000
7643:  00000 00000  00000 00000  00000 00000  00000 00000  00000 00000    00000 00000  00000 00000  00000 00000  00000 00000  00000 00000
7644:  00000 00000  00000 00000  00000 00000  00000 00000  00000 00000    00000 00000  00000 00000  00000 00000  00000 00000  00000 00000
7645:  00000 00000  00000 00000  00000 00000  00000 00000  00000 00000    00000 00000  00000 00000  00000 00000  00000 00000  00000 00000
7646:  00000 00000  00000 00000  00000 00000  00000 00000  00000 00000    00000 00000  00000 00000  00000 00000  00000 00000  00000 00000
7647:  00000 00000  00000 00000  00000 00000  00000 00000  00000 00000    00000 00000  00000 00000  00000 00000  00000 00000  00000 00000
7648:  00000 00000  00000 00000  00000 00000  00000 00000  00000 00000    00000 00000  00000 00000  00000 00000  00000 00000  00000 00000
7649:  00000 00000  00000 00000  00000 00000  00000 00000  00000 00000    00000 00000  00000 00000  00000 00000  00000 00000  00000 00000
```

```
7650:  00000 00000  00000 00000  00000 00000  00000 00000  00000 00000    00000 00000  00000 00000  00000 00000  00000 00000  00000 00000
7651:  00000 00000  00000 00000  00000 00000  00000 00000  00000 00000    00000 00000  00000 00000  00000 00000  00000 00000  00000 00000
7652:  00000 00000  00000 00000  00000 00000  00000 00000  00000 00000    00000 00000  00000 00000  00000 00000  00000 00000  00000 00000
7653:  00000 00000  00000 00000  00000 00000  00000 00000  00000 00000    00000 00000  00000 00000  00000 00000  00000 00000  00000 00000
7654:  00000 00000  00000 00000  00000 00000  00000 00000  00000 00000    00000 00000  00000 00000  00000 00000  00000 00000  00000 00000
7655:  00000 00000  00000 00000  00000 00000  00000 00000  00000 00000    00000 00000  00000 00000  00000 00000  00000 00000  00000 00000
7656:  00000 00000  00000 00000  00000 00000  00000 00000  00000 00000    00000 00000  00000 00000  00000 00000  00000 00000  00000 00000
7657:  00000 00000  00000 00000  00000 00000  00000 00000  00000 00000    00000 00000  00000 00000  00000 00000  00000 00000  00000 00000
7658:  00000 00000  00000 00000  00000 00000  00000 00000  00000 00000    00000 00000  00000 00000  00000 00000  00000 00000  00000 00000
7659:  00000 00000  00000 00000  00000 00000  00000 00000  00000 00000    00000 00000  00000 00000  00000 00000  00000 00000  00000 00000
7660:  00000 00000  00000 00000  00000 00000  00000 00000  00000 00000    00000 00000  00000 00000  00000 00000  00000 00000  00000 00000
7661:  00000 00000  00000 00000  00000 00000  00000 00000  00000 00000    00000 00000  00000 00000  00000 00000  00000 00000  00000 00000
7662:  00000 00000  00000 00000  00000 00000  00000 00000  00000 00000    00000 00000  00000 00000  00000 00000  00000 00000  00000 00000
7663:  00000 00000  00000 00000  00000 00000  00000 00000  00000 00000    00000 00000  00000 00000  00000 00000  00000 00000  00000 00000
7664:  00000 00000  00000 00000  00000 00000  00000 00000  00000 00000    00000 00000  00000 00000  00000 00000  00000 00000  00000 00000
7665:  00000 00000  00000 00000  00000 00000  00000 00000  00000 00000    00000 00000  00000 00000  00000 00000  00000 00000  00000 00000
7666:  00000 00000  00000 00000  00000 00000  00000 00000  00000 00000    00000 00000  00000 00000  00000 00000  00000 00000  00000 00000
7667:  00000 00000  00000 00000  00000 00000  00000 00000  00000 00000    00000 00000  00000 00000  00000 00000  00000 00000  00000 00000
7668:  00000 00000  00000 00000  00000 00000  00000 00000  00000 00000    00000 00000  00000 00000  00000 00000  00000 00000  00000 00000
7669:  00000 00000  00000 00000  00000 00000  00000 00000  00000 00000    00000 00000  00000 00000  00000 00000  00000 00000  00000 00000
7670:  00000 00000  00000 00000  00000 00000  00000 00000  00000 00000    00000 00000  00000 00000  00000 00000  00000 00000  00000 00000
7671:  00000 00000  00000 00000  00000 00000  00000 00000  00000 00000    00000 00000  00000 00000  00000 00000  00000 00000  00000 00000
7672:  00000 00000  00000 00000  00000 00000  00000 00000  00000 00000    00000 00000  00000 00000  00000 00000  00000 00000  00000 00000
7673:  00000 00000  00000 00000  00000 00000  00000 00000  00000 00000    00000 00000  00000 00000  00000 00000  00000 00000  00000 00000
7674:  00000 00000  00000 00000  00000 00000  00000 00000  00000 00000    00000 00000  00000 00000  00000 00000  00000 00000  00000 00000
7675:  00000 00000  00000 00000  00000 00000  00000 00000  00000 00000    00000 00000  00000 00000  00000 00000  00000 00000  00000 00000
7676:  00000 00000  00000 00000  00000 00000  00000 00000  00000 00000    00000 00000  00000 00000  00000 00000  00000 00000  00000 00000
7677:  00000 00000  00000 00000  00000 00000  00000 00000  00000 00000    00000 00000  00000 00000  00000 00000  00000 00000  00000 00000
7678:  00000 00000  00000 00000  00000 00000  00000 00000  00000 00000    00000 00000  00000 00000  00000 00000  00000 00000  00000 00000
7679:  00000 00000  00000 00000  00000 00000  00000 00000  00000 00000    00000 00000  00000 00000  00000 00000  00000 00000  00000 00000
7680:  00000 00000  00000 00000  00000 00000  00000 00000  00000 00000    00000 00000  00000 00000  00000 00000  00000 00000  00000 00000
7681:  00000 00000  00000 00000  00000 00000  00000 00000  00000 00000    00000 00000  00000 00000  00000 00000  00000 00000  00000 00000
7682:  00000 00000  00000 00000  00000 00000  00000 00000  00000 00000    00000 00000  00000 00000  00000 00000  00000 00000  00000 00000
7683:  00000 00000  00000 00000  00000 00000  00000 00000  00000 00000    00000 00000  00000 00000  00000 00000  00000 00000  00000 00000
7684:  00000 00000  00000 00000  00000 00000  00000 00000  00000 00000    00000 00000  00000 00000  00000 00000  00000 00000  00000 00000
7685:  00000 00000  00000 00000  00000 00000  00000 00000  00000 00000    00000 00000  00000 00000  00000 00000  00000 00000  00000 00000
7686:  00000 00000  00000 00000  00000 00000  00000 00000  00000 00000    00000 00000  00000 00000  00000 00000  00000 00000  00000 00000
7687:  00000 00000  00000 00000  00000 00000  00000 00000  00000 00000    00000 00000  00000 00000  00000 00000  00000 00000  00000 00000
7688:  00000 00000  00000 00000  00000 00000  00000 00000  00000 00000    00000 00000  00000 00000  00000 00000  00000 00000  00000 00000
7689:  00000 00000  00000 00000  00000 00000  00000 00000  00000 00000    00000 00000  00000 00000  00000 00000  00000 00000  00000 00000
7690:  00000 00000  00000 00000  00000 00000  00000 00000  00000 00000    00000 00000  00000 00000  00000 00000  00000 00000  00000 00000
7691:  00000 00000  00000 00000  00000 00000  00000 00000  00000 00000    00000 00000  00000 00000  00000 00000  00000 00000  00000 00000
7692:  00000 00000  00000 00000  00000 00000  00000 00000  00000 00000    00000 00000  00000 00000  00000 00000  00000 00000  00000 00000
7693:  00000 00000  00000 00000  00000 00000  00000 00000  00000 00000    00000 00000  00000 00000  00000 00000  00000 00000  00000 00000
7694:  00000 00000  00000 00000  00000 00000  00000 00000  00000 00000    00000 00000  00000 00000  00000 00000  00000 00000  00000 00000
7695:  00000 00000  00000 00000  00000 00000  00000 00000  00000 00000    00000 00000  00000 00000  00000 00000  00000 00000  00000 00000
7696:  00000 00000  00000 00000  00000 00000  00000 00000  00000 00000    00000 00000  00000 00000  00000 00000  00000 00000  00000 00000
7697:  00000 00000  00000 00000  00000 00000  00000 00000  00000 00000    00000 00000  00000 00000  00000 00000  00000 00000  00000 00000
7698:  00000 00000  00000 00000  00000 00000  00000 00000  00000 00000    00000 00000  00000 00000  00000 00000  00000 00000  00000 00000
7699:  00000 00000  00000 00000  00000 00000  00000 00000  00000 00000    00000 00000  00000 00000  00000 00000  00000 00000  00000 00000
```

```
7700:  00000 00000  00000 00000  00000 00000  00000 00000  00000 00000    00000 00000  00000 00000  00000 00000  00000 00000  00000 00000
7701:  00000 00000  00000 00000  00000 00000  00000 00000  00000 00000    00000 00000  00000 00000  00000 00000  00000 00000  00000 00000
7702:  00000 00000  00000 00000  00000 00000  00000 00000  00000 00000    00000 00000  00000 00000  00000 00000  00000 00000  00000 00000
7703:  00000 00000  00000 00000  00000 00000  00000 00000  00000 00000    00000 00000  00000 00000  00000 00000  00000 00000  00000 00000
7704:  00000 00000  00000 00000  00000 00000  00000 00000  00000 00000    00000 00000  00000 00000  00000 00000  00000 00000  00000 00000
7705:  00000 00000  00000 00000  00000 00000  00000 00000  00000 00000    00000 00000  00000 00000  00000 00000  00000 00000  00000 00000
7706:  00000 00000  00000 00000  00000 00000  00000 00000  00000 00000    00000 00000  00000 00000  00000 00000  00000 00000  00000 00000
7707:  00000 00000  00000 00000  00000 00000  00000 00000  00000 00000    00000 00000  00000 00000  00000 00000  00000 00000  00000 00000
7708:  00000 00000  00000 00000  00000 00000  00000 00000  00000 00000    00000 00000  00000 00000  00000 00000  00000 00000  00000 00000
7709:  00000 00000  00000 00000  00000 00000  00000 00000  00000 00000    00000 00000  00000 00000  00000 00000  00000 00000  00000 00000
7710:  00000 00000  00000 00000  00000 00000  00000 00000  00000 00000    00000 00000  00000 00000  00000 00000  00000 00000  00000 00000
7711:  00000 00000  00000 00000  00000 00000  00000 00000  00000 00000    00000 00000  00000 00000  00000 00000  00000 00000  00000 00000
7712:  00000 00000  00000 00000  00000 00000  00000 00000  00000 00000    00000 00000  00000 00000  00000 00000  00000 00000  00000 00000
7713:  00000 00000  00000 00000  00000 00000  00000 00000  00000 00000    00000 00000  00000 00000  00000 00000  00000 00000  00000 00000
7714:  00000 00000  00000 00000  00000 00000  00000 00000  00000 00000    00000 00000  00000 00000  00000 00000  00000 00000  00000 00000
7715:  00000 00000  00000 00000  00000 00000  00000 00000  00000 00000    00000 00000  00000 00000  00000 00000  00000 00000  00000 00000
7716:  00000 00000  00000 00000  00000 00000  00000 00000  00000 00000    00000 00000  00000 00000  00000 00000  00000 00000  00000 00000
7717:  00000 00000  00000 00000  00000 00000  00000 00000  00000 00000    00000 00000  00000 00000  00000 00000  00000 00000  00000 00000
7718:  00000 00000  00000 00000  00000 00000  00000 00000  00000 00000    00000 00000  00000 00000  00000 00000  00000 00000  00000 00000
7719:  00000 00000  00000 00000  00000 00000  00000 00000  00000 00000    00000 00000  00000 00000  00000 00000  00000 00000  00000 00000
7720:  00000 00000  00000 00000  00000 00000  00000 00000  00000 00000    00000 00000  00000 00000  00000 00000  00000 00000  00000 00000
7721:  00000 00000  00000 00000  00000 00000  00000 00000  00000 00000    00000 00000  00000 00000  00000 00000  00000 00000  00000 00000
7722:  00000 00000  00000 00000  00000 00000  00000 00000  00000 00000    00000 00000  00000 00000  00000 00000  00000 00000  00000 00000
7723:  00000 00000  00000 00000  00000 00000  00000 00000  00000 00000    00000 00000  00000 00000  00000 00000  00000 00000  00000 00000
7724:  00000 00000  00000 00000  00000 00000  00000 00000  00000 00000    00000 00000  00000 00000  00000 00000  00000 00000  00000 00000
7725:  00000 00000  00000 00000  00000 00000  00000 00000  00000 00000    00000 00000  00000 00000  00000 00000  00000 00000  00000 00000
7726:  00000 00000  00000 00000  00000 00000  00000 00000  00000 00000    00000 00000  00000 00000  00000 00000  00000 00000  00000 00000
7727:  00000 00000  00000 00000  00000 00000  00000 00000  00000 00000    00000 00000  00000 00000  00000 00000  00000 00000  00000 00000
7728:  00000 00000  00000 00000  00000 00000  00000 00000  00000 00000    00000 00000  00000 00000  00000 00000  00000 00000  00000 00000
7729:  00000 00000  00000 00000  00000 00000  00000 00000  00000 00000    00000 00000  00000 00000  00000 00000  00000 00000  00000 00000
7730:  00000 00000  00000 00000  00000 00000  00000 00000  00000 00000    00000 00000  00000 00000  00000 00000  00000 00000  00000 00000
7731:  00000 00000  00000 00000  00000 00000  00000 00000  00000 00000    00000 00000  00000 00000  00000 00000  00000 00000  00000 00000
7732:  00000 00000  00000 00000  00000 00000  00000 00000  00000 00000    00000 00000  00000 00000  00000 00000  00000 00000  00000 00000
7733:  00000 00000  00000 00000  00000 00000  00000 00000  00000 00000    00000 00000  00000 00000  00000 00000  00000 00000  00000 00000
7734:  00000 00000  00000 00000  00000 00000  00000 00000  00000 00000    00000 00000  00000 00000  00000 00000  00000 00000  00000 00000
7735:  00000 00000  00000 00000  00000 00000  00000 00000  00000 00000    00000 00000  00000 00000  00000 00000  00000 00000  00000 00000
7736:  00000 00000  00000 00000  00000 00000  00000 00000  00000 00000    00000 00000  00000 00000  00000 00000  00000 00000  00000 00000
7737:  00000 00000  00000 00000  00000 00000  00000 00000  00000 00000    00000 00000  00000 00000  00000 00000  00000 00000  00000 00000
7738:  00000 00000  00000 00000  00000 00000  00000 00000  00000 00000    00000 00000  00000 00000  00000 00000  00000 00000  00000 00000
7739:  00000 00000  00000 00000  00000 00000  00000 00000  00000 00000    00000 00000  00000 00000  00000 00000  00000 00000  00000 00000
7740:  00000 00000  00000 00000  00000 00000  00000 00000  00000 00000    00000 00000  00000 00000  00000 00000  00000 00000  00000 00000
7741:  00000 00000  00000 00000  00000 00000  00000 00000  00000 00000    00000 00000  00000 00000  00000 00000  00000 00000  00000 00000
7742:  00000 00000  00000 00000  00000 00000  00000 00000  00000 00000    00000 00000  00000 00000  00000 00000  00000 00000  00000 00000
7743:  00000 00000  00000 00000  00000 00000  00000 00000  00000 00000    00000 00000  00000 00000  00000 00000  00000 00000  00000 00000
7744:  00000 00000  00000 00000  00000 00000  00000 00000  00000 00000    00000 00000  00000 00000  00000 00000  00000 00000  00000 00000
7745:  00000 00000  00000 00000  00000 00000  00000 00000  00000 00000    00000 00000  00000 00000  00000 00000  00000 00000  00000 00000
7746:  00000 00000  00000 00000  00000 00000  00000 00000  00000 00000    00000 00000  00000 00000  00000 00000  00000 00000  00000 00000
7747:  00000 00000  00000 00000  00000 00000  00000 00000  00000 00000    00000 00000  00000 00000  00000 00000  00000 00000  00000 00000
7748:  00000 00000  00000 00000  00000 00000  00000 00000  00000 00000    00000 00000  00000 00000  00000 00000  00000 00000  00000 00000
7749:  00000 00000  00000 00000  00000 00000  00000 00000  00000 00000    00000 00000  00000 00000  00000 00000  00000 00000  00000 00000
```

```
7750:   00000 00000   00000 00000   00000 00000   00000 00000   00000 00000      00000 00000   00000 00000   00000 00000   00000 00000   00000 00000
7751:   00000 00000   00000 00000   00000 00000   00000 00000   00000 00000      00000 00000   00000 00000   00000 00000   00000 00000   00000 00000
7752:   00000 00000   00000 00000   00000 00000   00000 00000   00000 00000      00000 00000   00000 00000   00000 00000   00000 00000   00000 00000
7753:   00000 00000   00000 00000   00000 00000   00000 00000   00000 00000      00000 00000   00000 00000   00000 00000   00000 00000   00000 00000
7754:   00000 00000   00000 00000   00000 00000   00000 00000   00000 00000      00000 00000   00000 00000   00000 00000   00000 00000   00000 00000
7755:   00000 00000   00000 00000   00000 00000   00000 00000   00000 00000      00000 00000   00000 00000   00000 00000   00000 00000   00000 00000
7756:   00000 00000   00000 00000   00000 00000   00000 00000   00000 00000      00000 00000   00000 00000   00000 00000   00000 00000   00000 00000
7757:   00000 00000   00000 00000   00000 00000   00000 00000   00000 00000      00000 00000   00000 00000   00000 00000   00000 00000   00000 00000
7758:   00000 00000   00000 00000   00000 00000   00000 00000   00000 00000      00000 00000   00000 00000   00000 00000   00000 00000   00000 00000
7759:   00000 00000   00000 00000   00000 00000   00000 00000   00000 00000      00000 00000   00000 00000   00000 00000   00000 00000   00000 00000
7760:   00000 00000   00000 00000   00000 00000   00000 00000   00000 00000      00000 00000   00000 00000   00000 00000   00000 00000   00000 00000
7761:   00000 00000   00000 00000   00000 00000   00000 00000   00000 00000      00000 00000   00000 00000   00000 00000   00000 00000   00000 00000
7762:   00000 00000   00000 00000   00000 00000   00000 00000   00000 00000      00000 00000   00000 00000   00000 00000   00000 00000   00000 00000
7763:   00000 00000   00000 00000   00000 00000   00000 00000   00000 00000      00000 00000   00000 00000   00000 00000   00000 00000   00000 00000
7764:   00000 00000   00000 00000   00000 00000   00000 00000   00000 00000      00000 00000   00000 00000   00000 00000   00000 00000   00000 00000
7765:   00000 00000   00000 00000   00000 00000   00000 00000   00000 00000      00000 00000   00000 00000   00000 00000   00000 00000   00000 00000
7766:   00000 00000   00000 00000   00000 00000   00000 00000   00000 00000      00000 00000   00000 00000   00000 00000   00000 00000   00000 00000
7767:   00000 00000   00000 00000   00000 00000   00000 00000   00000 00000      00000 00000   00000 00000   00000 00000   00000 00000   00000 00000
7768:   00000 00000   00000 00000   00000 00000   00000 00000   00000 00000      00000 00000   00000 00000   00000 00000   00000 00000   00000 00000
7769:   00000 00000   00000 00000   00000 00000   00000 00000   00000 00000      00000 00000   00000 00000   00000 00000   00000 00000   00000 00000
7770:   00000 00000   00000 00000   00000 00000   00000 00000   00000 00000      00000 00000   00000 00000   00000 00000   00000 00000   00000 00000
7771:   00000 00000   00000 00000   00000 00000   00000 00000   00000 00000      00000 00000   00000 00000   00000 00000   00000 00000   00000 00000
7772:   00000 00000   00000 00000   00000 00000   00000 00000   00000 00000      00000 00000   00000 00000   00000 00000   00000 00000   00000 00000
7773:   00000 00000   00000 00000   00000 00000   00000 00000   00000 00000      00000 00000   00000 00000   00000 00000   00000 00000   00000 00000
7774:   00000 00000   00000 00000   00000 00000   00000 00000   00000 00000      00000 00000   00000 00000   00000 00000   00000 00000   00000 00000
7775:   00000 00000   00000 00000   00000 00000   00000 00000   00000 00000      00000 00000   00000 00000   00000 00000   00000 00000   00000 00000
7776:   00000 00000   00000 00000   00000 00000   00000 00000   00000 00000      00000 00000   00000 00000   00000 00000   00000 00000   00000 00000
7777:   00000 00000   00000 00000   00000 00000   00000 00000   00000 00000      00000 00000   00000 00000   00000 00000   00000 00000   00000 00000
7778:   00000 00000   00000 00000   00000 00000   00000 00000   00000 00000      00000 00000   00000 00000   00000 00000   00000 00000   00000 00000
7779:   00000 00000   00000 00000   00000 00000   00000 00000   00000 00000      00000 00000   00000 00000   00000 00000   00000 00000   00000 00000
7780:   00000 00000   00000 00000   00000 00000   00000 00000   00000 00000      00000 00000   00000 00000   00000 00000   00000 00000   00000 00000
7781:   00000 00000   00000 00000   00000 00000   00000 00000   00000 00000      00000 00000   00000 00000   00000 00000   00000 00000   00000 00000
7782:   00000 00000   00000 00000   00000 00000   00000 00000   00000 00000      00000 00000   00000 00000   00000 00000   00000 00000   00000 00000
7783:   00000 00000   00000 00000   00000 00000   00000 00000   00000 00000      00000 00000   00000 00000   00000 00000   00000 00000   00000 00000
7784:   00000 00000   00000 00000   00000 00000   00000 00000   00000 00000      00000 00000   00000 00000   00000 00000   00000 00000   00000 00000
7785:   00000 00000   00000 00000   00000 00000   00000 00000   00000 00000      00000 00000   00000 00000   00000 00000   00000 00000   00000 00000
7786:   00000 00000   00000 00000   00000 00000   00000 00000   00000 00000      00000 00000   00000 00000   00000 00000   00000 00000   00000 00000
7787:   00000 00000   00000 00000   00000 00000   00000 00000   00000 00000      00000 00000   00000 00000   00000 00000   00000 00000   00000 00000
7788:   00000 00000   00000 00000   00000 00000   00000 00000   00000 00000      00000 00000   00000 00000   00000 00000   00000 00000   00000 00000
7789:   00000 00000   00000 00000   00000 00000   00000 00000   00000 00000      00000 00000   00000 00000   00000 00000   00000 00000   00000 00000
7790:   00000 00000   00000 00000   00000 00000   00000 00000   00000 00000      00000 00000   00000 00000   00000 00000   00000 00000   00000 00000
7791:   00000 00000   00000 00000   00000 00000   00000 00000   00000 00000      00000 00000   00000 00000   00000 00000   00000 00000   00000 00000
7792:   00000 00000   00000 00000   00000 00000   00000 00000   00000 00000      00000 00000   00000 00000   00000 00000   00000 00000   00000 00000
7793:   00000 00000   00000 00000   00000 00000   00000 00000   00000 00000      00000 00000   00000 00000   00000 00000   00000 00000   00000 00000
7794:   00000 00000   00000 00000   00000 00000   00000 00000   00000 00000      00000 00000   00000 00000   00000 00000   00000 00000   00000 00000
7795:   00000 00000   00000 00000   00000 00000   00000 00000   00000 00000      00000 00000   00000 00000   00000 00000   00000 00000   00000 00000
7796:   00000 00000   00000 00000   00000 00000   00000 00000   00000 00000      00000 00000   00000 00000   00000 00000   00000 00000   00000 00000
7797:   00000 00000   00000 00000   00000 00000   00000 00000   00000 00000      00000 00000   00000 00000   00000 00000   00000 00000   00000 00000
7798:   00000 00000   00000 00000   00000 00000   00000 00000   00000 00000      00000 00000   00000 00000   00000 00000   00000 00000   00000 00000
7799:   00000 00000   00000 00000   00000 00000   00000 00000   00000 00000      00000 00000   00000 00000   00000 00000   00000 00000   00000 00000
```

```
7800:  00000 00000  00000 00000  00000 00000  00000 00000  00000 00000   00000 00000  00000 00000  00000 00000  00000 00000  00000 00000
7801:  00000 00000  00000 00000  00000 00000  00000 00000  00000 00000   00000 00000  00000 00000  00000 00000  00000 00000  00000 00000
7802:  00000 00000  00000 00000  00000 00000  00000 00000  00000 00000   00000 00000  00000 00000  00000 00000  00000 00000  00000 00000
7803:  00000 00000  00000 00000  00000 00000  00000 00000  00000 00000   00000 00000  00000 00000  00000 00000  00000 00000  00000 00000
7804:  00000 00000  00000 00000  00000 00000  00000 00000  00000 00000   00000 00000  00000 00000  00000 00000  00000 00000  00000 00000
7805:  00000 00000  00000 00000  00000 00000  00000 00000  00000 00000   00000 00000  00000 00000  00000 00000  00000 00000  00000 00000
7806:  00000 00000  00000 00000  00000 00000  00000 00000  00000 00000   00000 00000  00000 00000  00000 00000  00000 00000  00000 00000
7807:  00000 00000  00000 00000  00000 00000  00000 00000  00000 00000   00000 00000  00000 00000  00000 00000  00000 00000  00000 00000
7808:  00000 00000  00000 00000  00000 00000  00000 00000  00000 00000   00000 00000  00000 00000  00000 00000  00000 00000  00000 00000
7809:  00000 00000  00000 00000  00000 00000  00000 00000  00000 00000   00000 00000  00000 00000  00000 00000  00000 00000  00000 00000
7810:  00000 00000  00000 00000  00000 00000  00000 00000  00000 00000   00000 00000  00000 00000  00000 00000  00000 00000  00000 00000
7811:  00000 00000  00000 00000  00000 00000  00000 00000  00000 00000   00000 00000  00000 00000  00000 00000  00000 00000  00000 00000
7812:  00000 00000  00000 00000  00000 00000  00000 00000  00000 00000   00000 00000  00000 00000  00000 00000  00000 00000  00000 00000
7813:  00000 00000  00000 00000  00000 00000  00000 00000  00000 00000   00000 00000  00000 00000  00000 00000  00000 00000  00000 00000
7814:  00000 00000  00000 00000  00000 00000  00000 00000  00000 00000   00000 00000  00000 00000  00000 00000  00000 00000  00000 00000
7815:  00000 00000  00000 00000  00000 00000  00000 00000  00000 00000   00000 00000  00000 00000  00000 00000  00000 00000  00000 00000
7816:  00000 00000  00000 00000  00000 00000  00000 00000  00000 00000   00000 00000  00000 00000  00000 00000  00000 00000  00000 00000
7817:  00000 00000  00000 00000  00000 00000  00000 00000  00000 00000   00000 00000  00000 00000  00000 00000  00000 00000  00000 00000
7818:  00000 00000  00000 00000  00000 00000  00000 00000  00000 00000   00000 00000  00000 00000  00000 00000  00000 00000  00000 00000
7819:  00000 00000  00000 00000  00000 00000  00000 00000  00000 00000   00000 00000  00000 00000  00000 00000  00000 00000  00000 00000
7820:  00000 00000  00000 00000  00000 00000  00000 00000  00000 00000   00000 00000  00000 00000  00000 00000  00000 00000  00000 00000
7821:  00000 00000  00000 00000  00000 00000  00000 00000  00000 00000   00000 00000  00000 00000  00000 00000  00000 00000  00000 00000
7822:  00000 00000  00000 00000  00000 00000  00000 00000  00000 00000   00000 00000  00000 00000  00000 00000  00000 00000  00000 00000
7823:  00000 00000  00000 00000  00000 00000  00000 00000  00000 00000   00000 00000  00000 00000  00000 00000  00000 00000  00000 00000
7824:  00000 00000  00000 00000  00000 00000  00000 00000  00000 00000   00000 00000  00000 00000  00000 00000  00000 00000  00000 00000
7825:  00000 00000  00000 00000  00000 00000  00000 00000  00000 00000   00000 00000  00000 00000  00000 00000  00000 00000  00000 00000
7826:  00000 00000  00000 00000  00000 00000  00000 00000  00000 00000   00000 00000  00000 00000  00000 00000  00000 00000  00000 00000
7827:  00000 00000  00000 00000  00000 00000  00000 00000  00000 00000   00000 00000  00000 00000  00000 00000  00000 00000  00000 00000
7828:  00000 00000  00000 00000  00000 00000  00000 00000  00000 00000   00000 00000  00000 00000  00000 00000  00000 00000  00000 00000
7829:  00000 00000  00000 00000  00000 00000  00000 00000  00000 00000   00000 00000  00000 00000  00000 00000  00000 00000  00000 00000
7830:  00000 00000  00000 00000  00000 00000  00000 00000  00000 00000   00000 00000  00000 00000  00000 00000  00000 00000  00000 00000
7831:  00000 00000  00000 00000  00000 00000  00000 00000  00000 00000   00000 00000  00000 00000  00000 00000  00000 00000  00000 00000
7832:  00000 00000  00000 00000  00000 00000  00000 00000  00000 00000   00000 00000  00000 00000  00000 00000  00000 00000  00000 00000
7833:  00000 00000  00000 00000  00000 00000  00000 00000  00000 00000   00000 00000  00000 00000  00000 00000  00000 00000  00000 00000
7834:  00000 00000  00000 00000  00000 00000  00000 00000  00000 00000   00000 00000  00000 00000  00000 00000  00000 00000  00000 00000
7835:  00000 00000  00000 00000  00000 00000  00000 00000  00000 00000   00000 00000  00000 00000  00000 00000  00000 00000  00000 00000
7836:  00000 00000  00000 00000  00000 00000  00000 00000  00000 00000   00000 00000  00000 00000  00000 00000  00000 00000  00000 00000
7837:  00000 00000  00000 00000  00000 00000  00000 00000  00000 00000   00000 00000  00000 00000  00000 00000  00000 00000  00000 00000
7838:  00000 00000  00000 00000  00000 00000  00000 00000  00000 00000   00000 00000  00000 00000  00000 00000  00000 00000  00000 00000
7839:  00000 00000  00000 00000  00000 00000  00000 00000  00000 00000   00000 00000  00000 00000  00000 00000  00000 00000  00000 00000
7840:  00000 00000  00000 00000  00000 00000  00000 00000  00000 00000   00000 00000  00000 00000  00000 00000  00000 00000  00000 00000
7841:  00000 00000  00000 00000  00000 00000  00000 00000  00000 00000   00000 00000  00000 00000  00000 00000  00000 00000  00000 00000
7842:  00000 00000  00000 00000  00000 00000  00000 00000  00000 00000   00000 00000  00000 00000  00000 00000  00000 00000  00000 00000
7843:  00000 00000  00000 00000  00000 00000  00000 00000  00000 00000   00000 00000  00000 00000  00000 00000  00000 00000  00000 00000
7844:  00000 00000  00000 00000  00000 00000  00000 00000  00000 00000   00000 00000  00000 00000  00000 00000  00000 00000  00000 00000
7845:  00000 00000  00000 00000  00000 00000  00000 00000  00000 00000   00000 00000  00000 00000  00000 00000  00000 00000  00000 00000
7846:  00000 00000  00000 00000  00000 00000  00000 00000  00000 00000   00000 00000  00000 00000  00000 00000  00000 00000  00000 00000
7847:  00000 00000  00000 00000  00000 00000  00000 00000  00000 00000   00000 00000  00000 00000  00000 00000  00000 00000  00000 00000
7848:  00000 00000  00000 00000  00000 00000  00000 00000  00000 00000   00000 00000  00000 00000  00000 00000  00000 00000  00000 00000
7849:  00000 00000  00000 00000  00000 00000  00000 00000  00000 00000   00000 00000  00000 00000  00000 00000  00000 00000  00000 00000
```

```
7850:  00000 00000  00000 00000  00000 00000  00000 00000  00000 00000    00000 00000  00000 00000  00000 00000  00000 00000  00000 00000
7851:  00000 00000  00000 00000  00000 00000  00000 00000  00000 00000    00000 00000  00000 00000  00000 00000  00000 00000  00000 00000
7852:  00000 00000  00000 00000  00000 00000  00000 00000  00000 00000    00000 00000  00000 00000  00000 00000  00000 00000  00000 00000
7853:  00000 00000  00000 00000  00000 00000  00000 00000  00000 00000    00000 00000  00000 00000  00000 00000  00000 00000  00000 00000
7854:  00000 00000  00000 00000  00000 00000  00000 00000  00000 00000    00000 00000  00000 00000  00000 00000  00000 00000  00000 00000
7855:  00000 00000  00000 00000  00000 00000  00000 00000  00000 00000    00000 00000  00000 00000  00000 00000  00000 00000  00000 00000
7856:  00000 00000  00000 00000  00000 00000  00000 00000  00000 00000    00000 00000  00000 00000  00000 00000  00000 00000  00000 00000
7857:  00000 00000  00000 00000  00000 00000  00000 00000  00000 00000    00000 00000  00000 00000  00000 00000  00000 00000  00000 00000
7858:  00000 00000  00000 00000  00000 00000  00000 00000  00000 00000    00000 00000  00000 00000  00000 00000  00000 00000  00000 00000
7859:  00000 00000  00000 00000  00000 00000  00000 00000  00000 00000    00000 00000  00000 00000  00000 00000  00000 00000  00000 00000
7860:  00000 00000  00000 00000  00000 00000  00000 00000  00000 00000    00000 00000  00000 00000  00000 00000  00000 00000  00000 00000
7861:  00000 00000  00000 00000  00000 00000  00000 00000  00000 00000    00000 00000  00000 00000  00000 00000  00000 00000  00000 00000
7862:  00000 00000  00000 00000  00000 00000  00000 00000  00000 00000    00000 00000  00000 00000  00000 00000  00000 00000  00000 00000
7863:  00000 00000  00000 00000  00000 00000  00000 00000  00000 00000    00000 00000  00000 00000  00000 00000  00000 00000  00000 00000
7864:  00000 00000  00000 00000  00000 00000  00000 00000  00000 00000    00000 00000  00000 00000  00000 00000  00000 00000  00000 00000
7865:  00000 00000  00000 00000  00000 00000  00000 00000  00000 00000    00000 00000  00000 00000  00000 00000  00000 00000  00000 00000
7866:  00000 00000  00000 00000  00000 00000  00000 00000  00000 00000    00000 00000  00000 00000  00000 00000  00000 00000  00000 00000
7867:  00000 00000  00000 00000  00000 00000  00000 00000  00000 00000    00000 00000  00000 00000  00000 00000  00000 00000  00000 00000
7868:  00000 00000  00000 00000  00000 00000  00000 00000  00000 00000    00000 00000  00000 00000  00000 00000  00000 00000  00000 00000
7869:  00000 00000  00000 00000  00000 00000  00000 00000  00000 00000    00000 00000  00000 00000  00000 00000  00000 00000  00000 00000
7870:  00000 00000  00000 00000  00000 00000  00000 00000  00000 00000    00000 00000  00000 00000  00000 00000  00000 00000  00000 00000
7871:  00000 00000  00000 00000  00000 00000  00000 00000  00000 00000    00000 00000  00000 00000  00000 00000  00000 00000  00000 00000
7872:  00000 00000  00000 00000  00000 00000  00000 00000  00000 00000    00000 00000  00000 00000  00000 00000  00000 00000  00000 00000
7873:  00000 00000  00000 00000  00000 00000  00000 00000  00000 00000    00000 00000  00000 00000  00000 00000  00000 00000  00000 00000
7874:  00000 00000  00000 00000  00000 00000  00000 00000  00000 00000    00000 00000  00000 00000  00000 00000  00000 00000  00000 00000
7875:  00000 00000  00000 00000  00000 00000  00000 00000  00000 00000    00000 00000  00000 00000  00000 00000  00000 00000  00000 00000
7876:  00000 00000  00000 00000  00000 00000  00000 00000  00000 00000    00000 00000  00000 00000  00000 00000  00000 00000  00000 00000
7877:  00000 00000  00000 00000  00000 00000  00000 00000  00000 00000    00000 00000  00000 00000  00000 00000  00000 00000  00000 00000
7878:  00000 00000  00000 00000  00000 00000  00000 00000  00000 00000    00000 00000  00000 00000  00000 00000  00000 00000  00000 00000
7879:  00000 00000  00000 00000  00000 00000  00000 00000  00000 00000    00000 00000  00000 00000  00000 00000  00000 00000  00000 00000
7880:  00000 00000  00000 00000  00000 00000  00000 00000  00000 00000    00000 00000  00000 00000  00000 00000  00000 00000  00000 00000
7881:  00000 00000  00000 00000  00000 00000  00000 00000  00000 00000    00000 00000  00000 00000  00000 00000  00000 00000  00000 00000
7882:  00000 00000  00000 00000  00000 00000  00000 00000  00000 00000    00000 00000  00000 00000  00000 00000  00000 00000  00000 00000
7883:  00000 00000  00000 00000  00000 00000  00000 00000  00000 00000    00000 00000  00000 00000  00000 00000  00000 00000  00000 00000
7884:  00000 00000  00000 00000  00000 00000  00000 00000  00000 00000    00000 00000  00000 00000  00000 00000  00000 00000  00000 00000
7885:  00000 00000  00000 00000  00000 00000  00000 00000  00000 00000    00000 00000  00000 00000  00000 00000  00000 00000  00000 00000
7886:  00000 00000  00000 00000  00000 00000  00000 00000  00000 00000    00000 00000  00000 00000  00000 00000  00000 00000  00000 00000
7887:  00000 00000  00000 00000  00000 00000  00000 00000  00000 00000    00000 00000  00000 00000  00000 00000  00000 00000  00000 00000
7888:  00000 00000  00000 00000  00000 00000  00000 00000  00000 00000    00000 00000  00000 00000  00000 00000  00000 00000  00000 00000
7889:  00000 00000  00000 00000  00000 00000  00000 00000  00000 00000    00000 00000  00000 00000  00000 00000  00000 00000  00000 00000
7890:  00000 00000  00000 00000  00000 00000  00000 00000  00000 00000    00000 00000  00000 00000  00000 00000  00000 00000  00000 00000
7891:  00000 00000  00000 00000  00000 00000  00000 00000  00000 00000    00000 00000  00000 00000  00000 00000  00000 00000  00000 00000
7892:  00000 00000  00000 00000  00000 00000  00000 00000  00000 00000    00000 00000  00000 00000  00000 00000  00000 00000  00000 00000
7893:  00000 00000  00000 00000  00000 00000  00000 00000  00000 00000    00000 00000  00000 00000  00000 00000  00000 00000  00000 00000
7894:  00000 00000  00000 00000  00000 00000  00000 00000  00000 00000    00000 00000  00000 00000  00000 00000  00000 00000  00000 00000
7895:  00000 00000  00000 00000  00000 00000  00000 00000  00000 00000    00000 00000  00000 00000  00000 00000  00000 00000  00000 00000
7896:  00000 00000  00000 00000  00000 00000  00000 00000  00000 00000    00000 00000  00000 00000  00000 00000  00000 00000  00000 00000
7897:  00000 00000  00000 00000  00000 00000  00000 00000  00000 00000    00000 00000  00000 00000  00000 00000  00000 00000  00000 00000
7898:  00000 00000  00000 00000  00000 00000  00000 00000  00000 00000    00000 00000  00000 00000  00000 00000  00000 00000  00000 00000
7899:  00000 00000  00000 00000  00000 00000  00000 00000  00000 00000    00000 00000  00000 00000  00000 00000  00000 00000  00000 00000
```

```
7900:  00000 00000  00000 00000  00000 00000  00000 00000  00000 00000   00000 00000  00000 00000  00000 00000  00000 00000  00000 00000
7901:  00000 00000  00000 00000  00000 00000  00000 00000  00000 00000   00000 00000  00000 00000  00000 00000  00000 00000  00000 00000
7902:  00000 00000  00000 00000  00000 00000  00000 00000  00000 00000   00000 00000  00000 00000  00000 00000  00000 00000  00000 00000
7903:  00000 00000  00000 00000  00000 00000  00000 00000  00000 00000   00000 00000  00000 00000  00000 00000  00000 00000  00000 00000
7904:  00000 00000  00000 00000  00000 00000  00000 00000  00000 00000   00000 00000  00000 00000  00000 00000  00000 00000  00000 00000
7905:  00000 00000  00000 00000  00000 00000  00000 00000  00000 00000   00000 00000  00000 00000  00000 00000  00000 00000  00000 00000
7906:  00000 00000  00000 00000  00000 00000  00000 00000  00000 00000   00000 00000  00000 00000  00000 00000  00000 00000  00000 00000
7907:  00000 00000  00000 00000  00000 00000  00000 00000  00000 00000   00000 00000  00000 00000  00000 00000  00000 00000  00000 00000
7908:  00000 00000  00000 00000  00000 00000  00000 00000  00000 00000   00000 00000  00000 00000  00000 00000  00000 00000  00000 00000
7909:  00000 00000  00000 00000  00000 00000  00000 00000  00000 00000   00000 00000  00000 00000  00000 00000  00000 00000  00000 00000
7910:  00000 00000  00000 00000  00000 00000  00000 00000  00000 00000   00000 00000  00000 00000  00000 00000  00000 00000  00000 00000
7911:  00000 00000  00000 00000  00000 00000  00000 00000  00000 00000   00000 00000  00000 00000  00000 00000  00000 00000  00000 00000
7912:  00000 00000  00000 00000  00000 00000  00000 00000  00000 00000   00000 00000  00000 00000  00000 00000  00000 00000  00000 00000
7913:  00000 00000  00000 00000  00000 00000  00000 00000  00000 00000   00000 00000  00000 00000  00000 00000  00000 00000  00000 00000
7914:  00000 00000  00000 00000  00000 00000  00000 00000  00000 00000   00000 00000  00000 00000  00000 00000  00000 00000  00000 00000
7915:  00000 00000  00000 00000  00000 00000  00000 00000  00000 00000   00000 00000  00000 00000  00000 00000  00000 00000  00000 00000
7916:  00000 00000  00000 00000  00000 00000  00000 00000  00000 00000   00000 00000  00000 00000  00000 00000  00000 00000  00000 00000
7917:  00000 00000  00000 00000  00000 00000  00000 00000  00000 00000   00000 00000  00000 00000  00000 00000  00000 00000  00000 00000
7918:  00000 00000  00000 00000  00000 00000  00000 00000  00000 00000   00000 00000  00000 00000  00000 00000  00000 00000  00000 00000
7919:  00000 00000  00000 00000  00000 00000  00000 00000  00000 00000   00000 00000  00000 00000  00000 00000  00000 00000  00000 00000
7920:  00000 00000  00000 00000  00000 00000  00000 00000  00000 00000   00000 00000  00000 00000  00000 00000  00000 00000  00000 00000
7921:  00000 00000  00000 00000  00000 00000  00000 00000  00000 00000   00000 00000  00000 00000  00000 00000  00000 00000  00000 00000
7922:  00000 00000  00000 00000  00000 00000  00000 00000  00000 00000   00000 00000  00000 00000  00000 00000  00000 00000  00000 00000
7923:  00000 00000  00000 00000  00000 00000  00000 00000  00000 00000   00000 00000  00000 00000  00000 00000  00000 00000  00000 00000
7924:  00000 00000  00000 00000  00000 00000  00000 00000  00000 00000   00000 00000  00000 00000  00000 00000  00000 00000  00000 00000
7925:  00000 00000  00000 00000  00000 00000  00000 00000  00000 00000   00000 00000  00000 00000  00000 00000  00000 00000  00000 00000
7926:  00000 00000  00000 00000  00000 00000  00000 00000  00000 00000   00000 00000  00000 00000  00000 00000  00000 00000  00000 00000
7927:  00000 00000  00000 00000  00000 00000  00000 00000  00000 00000   00000 00000  00000 00000  00000 00000  00000 00000  00000 00000
7928:  00000 00000  00000 00000  00000 00000  00000 00000  00000 00000   00000 00000  00000 00000  00000 00000  00000 00000  00000 00000
7929:  00000 00000  00000 00000  00000 00000  00000 00000  00000 00000   00000 00000  00000 00000  00000 00000  00000 00000  00000 00000
7930:  00000 00000  00000 00000  00000 00000  00000 00000  00000 00000   00000 00000  00000 00000  00000 00000  00000 00000  00000 00000
7931:  00000 00000  00000 00000  00000 00000  00000 00000  00000 00000   00000 00000  00000 00000  00000 00000  00000 00000  00000 00000
7932:  00000 00000  00000 00000  00000 00000  00000 00000  00000 00000   00000 00000  00000 00000  00000 00000  00000 00000  00000 00000
7933:  00000 00000  00000 00000  00000 00000  00000 00000  00000 00000   00000 00000  00000 00000  00000 00000  00000 00000  00000 00000
7934:  00000 00000  00000 00000  00000 00000  00000 00000  00000 00000   00000 00000  00000 00000  00000 00000  00000 00000  00000 00000
7935:  00000 00000  00000 00000  00000 00000  00000 00000  00000 00000   00000 00000  00000 00000  00000 00000  00000 00000  00000 00000
7936:  00000 00000  00000 00000  00000 00000  00000 00000  00000 00000   00000 00000  00000 00000  00000 00000  00000 00000  00000 00000
7937:  00000 00000  00000 00000  00000 00000  00000 00000  00000 00000   00000 00000  00000 00000  00000 00000  00000 00000  00000 00000
7938:  00000 00000  00000 00000  00000 00000  00000 00000  00000 00000   00000 00000  00000 00000  00000 00000  00000 00000  00000 00000
7939:  00000 00000  00000 00000  00000 00000  00000 00000  00000 00000   00000 00000  00000 00000  00000 00000  00000 00000  00000 00000
7940:  00000 00000  00000 00000  00000 00000  00000 00000  00000 00000   00000 00000  00000 00000  00000 00000  00000 00000  00000 00000
7941:  00000 00000  00000 00000  00000 00000  00000 00000  00000 00000   00000 00000  00000 00000  00000 00000  00000 00000  00000 00000
7942:  00000 00000  00000 00000  00000 00000  00000 00000  00000 00000   00000 00000  00000 00000  00000 00000  00000 00000  00000 00000
7943:  00000 00000  00000 00000  00000 00000  00000 00000  00000 00000   00000 00000  00000 00000  00000 00000  00000 00000  00000 00000
7944:  00000 00000  00000 00000  00000 00000  00000 00000  00000 00000   00000 00000  00000 00000  00000 00000  00000 00000  00000 00000
7945:  00000 00000  00000 00000  00000 00000  00000 00000  00000 00000   00000 00000  00000 00000  00000 00000  00000 00000  00000 00000
7946:  00000 00000  00000 00000  00000 00000  00000 00000  00000 00000   00000 00000  00000 00000  00000 00000  00000 00000  00000 00000
7947:  00000 00000  00000 00000  00000 00000  00000 00000  00000 00000   00000 00000  00000 00000  00000 00000  00000 00000  00000 00000
7948:  00000 00000  00000 00000  00000 00000  00000 00000  00000 00000   00000 00000  00000 00000  00000 00000  00000 00000  00000 00000
7949:  00000 00000  00000 00000  00000 00000  00000 00000  00000 00000   00000 00000  00000 00000  00000 00000  00000 00000  00000 00000
```

```
7950:  00000 00000  00000 00000  00000 00000  00000 00000  00000 00000    00000 00000  00000 00000  00000 00000  00000 00000  00000 00000
7951:  00000 00000  00000 00000  00000 00000  00000 00000  00000 00000    00000 00000  00000 00000  00000 00000  00000 00000  00000 00000
7952:  00000 00000  00000 00000  00000 00000  00000 00000  00000 00000    00000 00000  00000 00000  00000 00000  00000 00000  00000 00000
7953:  00000 00000  00000 00000  00000 00000  00000 00000  00000 00000    00000 00000  00000 00000  00000 00000  00000 00000  00000 00000
7954:  00000 00000  00000 00000  00000 00000  00000 00000  00000 00000    00000 00000  00000 00000  00000 00000  00000 00000  00000 00000
7955:  00000 00000  00000 00000  00000 00000  00000 00000  00000 00000    00000 00000  00000 00000  00000 00000  00000 00000  00000 00000
7956:  00000 00000  00000 00000  00000 00000  00000 00000  00000 00000    00000 00000  00000 00000  00000 00000  00000 00000  00000 00000
7957:  00000 00000  00000 00000  00000 00000  00000 00000  00000 00000    00000 00000  00000 00000  00000 00000  00000 00000  00000 00000
7958:  00000 00000  00000 00000  00000 00000  00000 00000  00000 00000    00000 00000  00000 00000  00000 00000  00000 00000  00000 00000
7959:  00000 00000  00000 00000  00000 00000  00000 00000  00000 00000    00000 00000  00000 00000  00000 00000  00000 00000  00000 00000
7960:  00000 00000  00000 00000  00000 00000  00000 00000  00000 00000    00000 00000  00000 00000  00000 00000  00000 00000  00000 00000
7961:  00000 00000  00000 00000  00000 00000  00000 00000  00000 00000    00000 00000  00000 00000  00000 00000  00000 00000  00000 00000
7962:  00000 00000  00000 00000  00000 00000  00000 00000  00000 00000    00000 00000  00000 00000  00000 00000  00000 00000  00000 00000
7963:  00000 00000  00000 00000  00000 00000  00000 00000  00000 00000    00000 00000  00000 00000  00000 00000  00000 00000  00000 00000
7964:  00000 00000  00000 00000  00000 00000  00000 00000  00000 00000    00000 00000  00000 00000  00000 00000  00000 00000  00000 00000
7965:  00000 00000  00000 00000  00000 00000  00000 00000  00000 00000    00000 00000  00000 00000  00000 00000  00000 00000  00000 00000
7966:  00000 00000  00000 00000  00000 00000  00000 00000  00000 00000    00000 00000  00000 00000  00000 00000  00000 00000  00000 00000
7967:  00000 00000  00000 00000  00000 00000  00000 00000  00000 00000    00000 00000  00000 00000  00000 00000  00000 00000  00000 00000
7968:  00000 00000  00000 00000  00000 00000  00000 00000  00000 00000    00000 00000  00000 00000  00000 00000  00000 00000  00000 00000
7969:  00000 00000  00000 00000  00000 00000  00000 00000  00000 00000    00000 00000  00000 00000  00000 00000  00000 00000  00000 00000
7970:  00000 00000  00000 00000  00000 00000  00000 00000  00000 00000    00000 00000  00000 00000  00000 00000  00000 00000  00000 00000
7971:  00000 00000  00000 00000  00000 00000  00000 00000  00000 00000    00000 00000  00000 00000  00000 00000  00000 00000  00000 00000
7972:  00000 00000  00000 00000  00000 00000  00000 00000  00000 00000    00000 00000  00000 00000  00000 00000  00000 00000  00000 00000
7973:  00000 00000  00000 00000  00000 00000  00000 00000  00000 00000    00000 00000  00000 00000  00000 00000  00000 00000  00000 00000
7974:  00000 00000  00000 00000  00000 00000  00000 00000  00000 00000    00000 00000  00000 00000  00000 00000  00000 00000  00000 00000
7975:  00000 00000  00000 00000  00000 00000  00000 00000  00000 00000    00000 00000  00000 00000  00000 00000  00000 00000  00000 00000
7976:  00000 00000  00000 00000  00000 00000  00000 00000  00000 00000    00000 00000  00000 00000  00000 00000  00000 00000  00000 00000
7977:  00000 00000  00000 00000  00000 00000  00000 00000  00000 00000    00000 00000  00000 00000  00000 00000  00000 00000  00000 00000
7978:  00000 00000  00000 00000  00000 00000  00000 00000  00000 00000    00000 00000  00000 00000  00000 00000  00000 00000  00000 00000
7979:  00000 00000  00000 00000  00000 00000  00000 00000  00000 00000    00000 00000  00000 00000  00000 00000  00000 00000  00000 00000
7980:  00000 00000  00000 00000  00000 00000  00000 00000  00000 00000    00000 00000  00000 00000  00000 00000  00000 00000  00000 00000
7981:  00000 00000  00000 00000  00000 00000  00000 00000  00000 00000    00000 00000  00000 00000  00000 00000  00000 00000  00000 00000
7982:  00000 00000  00000 00000  00000 00000  00000 00000  00000 00000    00000 00000  00000 00000  00000 00000  00000 00000  00000 00000
7983:  00000 00000  00000 00000  00000 00000  00000 00000  00000 00000    00000 00000  00000 00000  00000 00000  00000 00000  00000 00000
7984:  00000 00000  00000 00000  00000 00000  00000 00000  00000 00000    00000 00000  00000 00000  00000 00000  00000 00000  00000 00000
7985:  00000 00000  00000 00000  00000 00000  00000 00000  00000 00000    00000 00000  00000 00000  00000 00000  00000 00000  00000 00000
7986:  00000 00000  00000 00000  00000 00000  00000 00000  00000 00000    00000 00000  00000 00000  00000 00000  00000 00000  00000 00000
7987:  00000 00000  00000 00000  00000 00000  00000 00000  00000 00000    00000 00000  00000 00000  00000 00000  00000 00000  00000 00000
7988:  00000 00000  00000 00000  00000 00000  00000 00000  00000 00000    00000 00000  00000 00000  00000 00000  00000 00000  00000 00000
7989:  00000 00000  00000 00000  00000 00000  00000 00000  00000 00000    00000 00000  00000 00000  00000 00000  00000 00000  00000 00000
7990:  00000 00000  00000 00000  00000 00000  00000 00000  00000 00000    00000 00000  00000 00000  00000 00000  00000 00000  00000 00000
7991:  00000 00000  00000 00000  00000 00000  00000 00000  00000 00000    00000 00000  00000 00000  00000 00000  00000 00000  00000 00000
7992:  00000 00000  00000 00000  00000 00000  00000 00000  00000 00000    00000 00000  00000 00000  00000 00000  00000 00000  00000 00000
7993:  00000 00000  00000 00000  00000 00000  00000 00000  00000 00000    00000 00000  00000 00000  00000 00000  00000 00000  00000 00000
7994:  00000 00000  00000 00000  00000 00000  00000 00000  00000 00000    00000 00000  00000 00000  00000 00000  00000 00000  00000 00000
7995:  00000 00000  00000 00000  00000 00000  00000 00000  00000 00000    00000 00000  00000 00000  00000 00000  00000 00000  00000 00000
7996:  00000 00000  00000 00000  00000 00000  00000 00000  00000 00000    00000 00000  00000 00000  00000 00000  00000 00000  00000 00000
7997:  00000 00000  00000 00000  00000 00000  00000 00000  00000 00000    00000 00000  00000 00000  00000 00000  00000 00000  00000 00000
7998:  00000 00000  00000 00000  00000 00000  00000 00000  00000 00000    00000 00000  00000 00000  00000 00000  00000 00000  00000 00000
7999:  00000 00000  00000 00000  00000 00000  00000 00000  00000 00000    00000 00000  00000 00000  00000 00000  00000 00000  00000 00000
```

```
8000:   00000 00000   00000 00000   00000 00000   00000 00000   00000 00000     00000 00000   00000 00000   00000 00000   00000 00000   00000 00000
8001:   00000 00000   00000 00000   00000 00000   00000 00000   00000 00000     00000 00000   00000 00000   00000 00000   00000 00000   00000 00000
8002:   00000 00000   00000 00000   00000 00000   00000 00000   00000 00000     00000 00000   00000 00000   00000 00000   00000 00000   00000 00000
8003:   00000 00000   00000 00000   00000 00000   00000 00000   00000 00000     00000 00000   00000 00000   00000 00000   00000 00000   00000 00000
8004:   00000 00000   00000 00000   00000 00000   00000 00000   00000 00000     00000 00000   00000 00000   00000 00000   00000 00000   00000 00000
8005:   00000 00000   00000 00000   00000 00000   00000 00000   00000 00000     00000 00000   00000 00000   00000 00000   00000 00000   00000 00000
8006:   00000 00000   00000 00000   00000 00000   00000 00000   00000 00000     00000 00000   00000 00000   00000 00000   00000 00000   00000 00000
8007:   00000 00000   00000 00000   00000 00000   00000 00000   00000 00000     00000 00000   00000 00000   00000 00000   00000 00000   00000 00000
8008:   00000 00000   00000 00000   00000 00000   00000 00000   00000 00000     00000 00000   00000 00000   00000 00000   00000 00000   00000 00000
8009:   00000 00000   00000 00000   00000 00000   00000 00000   00000 00000     00000 00000   00000 00000   00000 00000   00000 00000   00000 00000
8010:   00000 00000   00000 00000   00000 00000   00000 00000   00000 00000     00000 00000   00000 00000   00000 00000   00000 00000   00000 00000
8011:   00000 00000   00000 00000   00000 00000   00000 00000   00000 00000     00000 00000   00000 00000   00000 00000   00000 00000   00000 00000
8012:   00000 00000   00000 00000   00000 00000   00000 00000   00000 00000     00000 00000   00000 00000   00000 00000   00000 00000   00000 00000
8013:   00000 00000   00000 00000   00000 00000   00000 00000   00000 00000     00000 00000   00000 00000   00000 00000   00000 00000   00000 00000
8014:   00000 00000   00000 00000   00000 00000   00000 00000   00000 00000     00000 00000   00000 00000   00000 00000   00000 00000   00000 00000
8015:   00000 00000   00000 00000   00000 00000   00000 00000   00000 00000     00000 00000   00000 00000   00000 00000   00000 00000   00000 00000
8016:   00000 00000   00000 00000   00000 00000   00000 00000   00000 00000     00000 00000   00000 00000   00000 00000   00000 00000   00000 00000
8017:   00000 00000   00000 00000   00000 00000   00000 00000   00000 00000     00000 00000   00000 00000   00000 00000   00000 00000   00000 00000
8018:   00000 00000   00000 00000   00000 00000   00000 00000   00000 00000     00000 00000   00000 00000   00000 00000   00000 00000   00000 00000
8019:   00000 00000   00000 00000   00000 00000   00000 00000   00000 00000     00000 00000   00000 00000   00000 00000   00000 00000   00000 00000
8020:   00000 00000   00000 00000   00000 00000   00000 00000   00000 00000     00000 00000   00000 00000   00000 00000   00000 00000   00000 00000
8021:   00000 00000   00000 00000   00000 00000   00000 00000   00000 00000     00000 00000   00000 00000   00000 00000   00000 00000   00000 00000
8022:   00000 00000   00000 00000   00000 00000   00000 00000   00000 00000     00000 00000   00000 00000   00000 00000   00000 00000   00000 00000
8023:   00000 00000   00000 00000   00000 00000   00000 00000   00000 00000     00000 00000   00000 00000   00000 00000   00000 00000   00000 00000
8024:   00000 00000   00000 00000   00000 00000   00000 00000   00000 00000     00000 00000   00000 00000   00000 00000   00000 00000   00000 00000
8025:   00000 00000   00000 00000   00000 00000   00000 00000   00000 00000     00000 00000   00000 00000   00000 00000   00000 00000   00000 00000
8026:   00000 00000   00000 00000   00000 00000   00000 00000   00000 00000     00000 00000   00000 00000   00000 00000   00000 00000   00000 00000
8027:   00000 00000   00000 00000   00000 00000   00000 00000   00000 00000     00000 00000   00000 00000   00000 00000   00000 00000   00000 00000
8028:   00000 00000   00000 00000   00000 00000   00000 00000   00000 00000     00000 00000   00000 00000   00000 00000   00000 00000   00000 00000
8029:   00000 00000   00000 00000   00000 00000   00000 00000   00000 00000     00000 00000   00000 00000   00000 00000   00000 00000   00000 00000
8030:   00000 00000   00000 00000   00000 00000   00000 00000   00000 00000     00000 00000   00000 00000   00000 00000   00000 00000   00000 00000
8031:   00000 00000   00000 00000   00000 00000   00000 00000   00000 00000     00000 00000   00000 00000   00000 00000   00000 00000   00000 00000
8032:   00000 00000   00000 00000   00000 00000   00000 00000   00000 00000     00000 00000   00000 00000   00000 00000   00000 00000   00000 00000
8033:   00000 00000   00000 00000   00000 00000   00000 00000   00000 00000     00000 00000   00000 00000   00000 00000   00000 00000   00000 00000
8034:   00000 00000   00000 00000   00000 00000   00000 00000   00000 00000     00000 00000   00000 00000   00000 00000   00000 00000   00000 00000
8035:   00000 00000   00000 00000   00000 00000   00000 00000   00000 00000     00000 00000   00000 00000   00000 00000   00000 00000   00000 00000
8036:   00000 00000   00000 00000   00000 00000   00000 00000   00000 00000     00000 00000   00000 00000   00000 00000   00000 00000   00000 00000
8037:   00000 00000   00000 00000   00000 00000   00000 00000   00000 00000     00000 00000   00000 00000   00000 00000   00000 00000   00000 00000
8038:   00000 00000   00000 00000   00000 00000   00000 00000   00000 00000     00000 00000   00000 00000   00000 00000   00000 00000   00000 00000
8039:   00000 00000   00000 00000   00000 00000   00000 00000   00000 00000     00000 00000   00000 00000   00000 00000   00000 00000   00000 00000
8040:   00000 00000   00000 00000   00000 00000   00000 00000   00000 00000     00000 00000   00000 00000   00000 00000   00000 00000   00000 00000
8041:   00000 00000   00000 00000   00000 00000   00000 00000   00000 00000     00000 00000   00000 00000   00000 00000   00000 00000   00000 00000
8042:   00000 00000   00000 00000   00000 00000   00000 00000   00000 00000     00000 00000   00000 00000   00000 00000   00000 00000   00000 00000
8043:   00000 00000   00000 00000   00000 00000   00000 00000   00000 00000     00000 00000   00000 00000   00000 00000   00000 00000   00000 00000
8044:   00000 00000   00000 00000   00000 00000   00000 00000   00000 00000     00000 00000   00000 00000   00000 00000   00000 00000   00000 00000
8045:   00000 00000   00000 00000   00000 00000   00000 00000   00000 00000     00000 00000   00000 00000   00000 00000   00000 00000   00000 00000
8046:   00000 00000   00000 00000   00000 00000   00000 00000   00000 00000     00000 00000   00000 00000   00000 00000   00000 00000   00000 00000
8047:   00000 00000   00000 00000   00000 00000   00000 00000   00000 00000     00000 00000   00000 00000   00000 00000   00000 00000   00000 00000
8048:   00000 00000   00000 00000   00000 00000   00000 00000   00000 00000     00000 00000   00000 00000   00000 00000   00000 00000   00000 00000
8049:   00000 00000   00000 00000   00000 00000   00000 00000   00000 00000     00000 00000   00000 00000   00000 00000   00000 00000   00000 00000
```

```
8050:   00000 00000   00000 00000   00000 00000   00000 00000   00000 00000     00000 00000   00000 00000   00000 00000   00000 00000   00000 00000
8051:   00000 00000   00000 00000   00000 00000   00000 00000   00000 00000     00000 00000   00000 00000   00000 00000   00000 00000   00000 00000
8052:   00000 00000   00000 00000   00000 00000   00000 00000   00000 00000     00000 00000   00000 00000   00000 00000   00000 00000   00000 00000
8053:   00000 00000   00000 00000   00000 00000   00000 00000   00000 00000     00000 00000   00000 00000   00000 00000   00000 00000   00000 00000
8054:   00000 00000   00000 00000   00000 00000   00000 00000   00000 00000     00000 00000   00000 00000   00000 00000   00000 00000   00000 00000
8055:   00000 00000   00000 00000   00000 00000   00000 00000   00000 00000     00000 00000   00000 00000   00000 00000   00000 00000   00000 00000
8056:   00000 00000   00000 00000   00000 00000   00000 00000   00000 00000     00000 00000   00000 00000   00000 00000   00000 00000   00000 00000
8057:   00000 00000   00000 00000   00000 00000   00000 00000   00000 00000     00000 00000   00000 00000   00000 00000   00000 00000   00000 00000
8058:   00000 00000   00000 00000   00000 00000   00000 00000   00000 00000     00000 00000   00000 00000   00000 00000   00000 00000   00000 00000
8059:   00000 00000   00000 00000   00000 00000   00000 00000   00000 00000     00000 00000   00000 00000   00000 00000   00000 00000   00000 00000
8060:   00000 00000   00000 00000   00000 00000   00000 00000   00000 00000     00000 00000   00000 00000   00000 00000   00000 00000   00000 00000
8061:   00000 00000   00000 00000   00000 00000   00000 00000   00000 00000     00000 00000   00000 00000   00000 00000   00000 00000   00000 00000
8062:   00000 00000   00000 00000   00000 00000   00000 00000   00000 00000     00000 00000   00000 00000   00000 00000   00000 00000   00000 00000
8063:   00000 00000   00000 00000   00000 00000   00000 00000   00000 00000     00000 00000   00000 00000   00000 00000   00000 00000   00000 00000
8064:   00000 00000   00000 00000   00000 00000   00000 00000   00000 00000     00000 00000   00000 00000   00000 00000   00000 00000   00000 00000
8065:   00000 00000   00000 00000   00000 00000   00000 00000   00000 00000     00000 00000   00000 00000   00000 00000   00000 00000   00000 00000
8066:   00000 00000   00000 00000   00000 00000   00000 00000   00000 00000     00000 00000   00000 00000   00000 00000   00000 00000   00000 00000
8067:   00000 00000   00000 00000   00000 00000   00000 00000   00000 00000     00000 00000   00000 00000   00000 00000   00000 00000   00000 00000
8068:   00000 00000   00000 00000   00000 00000   00000 00000   00000 00000     00000 00000   00000 00000   00000 00000   00000 00000   00000 00000
8069:   00000 00000   00000 00000   00000 00000   00000 00000   00000 00000     00000 00000   00000 00000   00000 00000   00000 00000   00000 00000
8070:   00000 00000   00000 00000   00000 00000   00000 00000   00000 00000     00000 00000   00000 00000   00000 00000   00000 00000   00000 00000
8071:   00000 00000   00000 00000   00000 00000   00000 00000   00000 00000     00000 00000   00000 00000   00000 00000   00000 00000   00000 00000
8072:   00000 00000   00000 00000   00000 00000   00000 00000   00000 00000     00000 00000   00000 00000   00000 00000   00000 00000   00000 00000
8073:   00000 00000   00000 00000   00000 00000   00000 00000   00000 00000     00000 00000   00000 00000   00000 00000   00000 00000   00000 00000
8074:   00000 00000   00000 00000   00000 00000   00000 00000   00000 00000     00000 00000   00000 00000   00000 00000   00000 00000   00000 00000
8075:   00000 00000   00000 00000   00000 00000   00000 00000   00000 00000     00000 00000   00000 00000   00000 00000   00000 00000   00000 00000
8076:   00000 00000   00000 00000   00000 00000   00000 00000   00000 00000     00000 00000   00000 00000   00000 00000   00000 00000   00000 00000
8077:   00000 00000   00000 00000   00000 00000   00000 00000   00000 00000     00000 00000   00000 00000   00000 00000   00000 00000   00000 00000
8078:   00000 00000   00000 00000   00000 00000   00000 00000   00000 00000     00000 00000   00000 00000   00000 00000   00000 00000   00000 00000
8079:   00000 00000   00000 00000   00000 00000   00000 00000   00000 00000     00000 00000   00000 00000   00000 00000   00000 00000   00000 00000
8080:   00000 00000   00000 00000   00000 00000   00000 00000   00000 00000     00000 00000   00000 00000   00000 00000   00000 00000   00000 00000
8081:   00000 00000   00000 00000   00000 00000   00000 00000   00000 00000     00000 00000   00000 00000   00000 00000   00000 00000   00000 00000
8082:   00000 00000   00000 00000   00000 00000   00000 00000   00000 00000     00000 00000   00000 00000   00000 00000   00000 00000   00000 00000
8083:   00000 00000   00000 00000   00000 00000   00000 00000   00000 00000     00000 00000   00000 00000   00000 00000   00000 00000   00000 00000
8084:   00000 00000   00000 00000   00000 00000   00000 00000   00000 00000     00000 00000   00000 00000   00000 00000   00000 00000   00000 00000
8085:   00000 00000   00000 00000   00000 00000   00000 00000   00000 00000     00000 00000   00000 00000   00000 00000   00000 00000   00000 00000
8086:   00000 00000   00000 00000   00000 00000   00000 00000   00000 00000     00000 00000   00000 00000   00000 00000   00000 00000   00000 00000
8087:   00000 00000   00000 00000   00000 00000   00000 00000   00000 00000     00000 00000   00000 00000   00000 00000   00000 00000   00000 00000
8088:   00000 00000   00000 00000   00000 00000   00000 00000   00000 00000     00000 00000   00000 00000   00000 00000   00000 00000   00000 00000
8089:   00000 00000   00000 00000   00000 00000   00000 00000   00000 00000     00000 00000   00000 00000   00000 00000   00000 00000   00000 00000
8090:   00000 00000   00000 00000   00000 00000   00000 00000   00000 00000     00000 00000   00000 00000   00000 00000   00000 00000   00000 00000
8091:   00000 00000   00000 00000   00000 00000   00000 00000   00000 00000     00000 00000   00000 00000   00000 00000   00000 00000   00000 00000
8092:   00000 00000   00000 00000   00000 00000   00000 00000   00000 00000     00000 00000   00000 00000   00000 00000   00000 00000   00000 00000
8093:   00000 00000   00000 00000   00000 00000   00000 00000   00000 00000     00000 00000   00000 00000   00000 00000   00000 00000   00000 00000
8094:   00000 00000   00000 00000   00000 00000   00000 00000   00000 00000     00000 00000   00000 00000   00000 00000   00000 00000   00000 00000
8095:   00000 00000   00000 00000   00000 00000   00000 00000   00000 00000     00000 00000   00000 00000   00000 00000   00000 00000   00000 00000
8096:   00000 00000   00000 00000   00000 00000   00000 00000   00000 00000     00000 00000   00000 00000   00000 00000   00000 00000   00000 00000
8097:   00000 00000   00000 00000   00000 00000   00000 00000   00000 00000     00000 00000   00000 00000   00000 00000   00000 00000   00000 00000
8098:   00000 00000   00000 00000   00000 00000   00000 00000   00000 00000     00000 00000   00000 00000   00000 00000   00000 00000   00000 00000
8099:   00000 00000   00000 00000   00000 00000   00000 00000   00000 00000     00000 00000   00000 00000   00000 00000   00000 00000   00000 00000
```

```
8100:   00000 00000   00000 00000   00000 00000   00000 00000   00000 00000     00000 00000   00000 00000   00000 00000   00000 00000   00000 00000
8101:   00000 00000   00000 00000   00000 00000   00000 00000   00000 00000     00000 00000   00000 00000   00000 00000   00000 00000   00000 00000
8102:   00000 00000   00000 00000   00000 00000   00000 00000   00000 00000     00000 00000   00000 00000   00000 00000   00000 00000   00000 00000
8103:   00000 00000   00000 00000   00000 00000   00000 00000   00000 00000     00000 00000   00000 00000   00000 00000   00000 00000   00000 00000
8104:   00000 00000   00000 00000   00000 00000   00000 00000   00000 00000     00000 00000   00000 00000   00000 00000   00000 00000   00000 00000
8105:   00000 00000   00000 00000   00000 00000   00000 00000   00000 00000     00000 00000   00000 00000   00000 00000   00000 00000   00000 00000
8106:   00000 00000   00000 00000   00000 00000   00000 00000   00000 00000     00000 00000   00000 00000   00000 00000   00000 00000   00000 00000
8107:   00000 00000   00000 00000   00000 00000   00000 00000   00000 00000     00000 00000   00000 00000   00000 00000   00000 00000   00000 00000
8108:   00000 00000   00000 00000   00000 00000   00000 00000   00000 00000     00000 00000   00000 00000   00000 00000   00000 00000   00000 00000
8109:   00000 00000   00000 00000   00000 00000   00000 00000   00000 00000     00000 00000   00000 00000   00000 00000   00000 00000   00000 00000
8110:   00000 00000   00000 00000   00000 00000   00000 00000   00000 00000     00000 00000   00000 00000   00000 00000   00000 00000   00000 00000
8111:   00000 00000   00000 00000   00000 00000   00000 00000   00000 00000     00000 00000   00000 00000   00000 00000   00000 00000   00000 00000
8112:   00000 00000   00000 00000   00000 00000   00000 00000   00000 00000     00000 00000   00000 00000   00000 00000   00000 00000   00000 00000
8113:   00000 00000   00000 00000   00000 00000   00000 00000   00000 00000     00000 00000   00000 00000   00000 00000   00000 00000   00000 00000
8114:   00000 00000   00000 00000   00000 00000   00000 00000   00000 00000     00000 00000   00000 00000   00000 00000   00000 00000   00000 00000
8115:   00000 00000   00000 00000   00000 00000   00000 00000   00000 00000     00000 00000   00000 00000   00000 00000   00000 00000   00000 00000
8116:   00000 00000   00000 00000   00000 00000   00000 00000   00000 00000     00000 00000   00000 00000   00000 00000   00000 00000   00000 00000
8117:   00000 00000   00000 00000   00000 00000   00000 00000   00000 00000     00000 00000   00000 00000   00000 00000   00000 00000   00000 00000
8118:   00000 00000   00000 00000   00000 00000   00000 00000   00000 00000     00000 00000   00000 00000   00000 00000   00000 00000   00000 00000
8119:   00000 00000   00000 00000   00000 00000   00000 00000   00000 00000     00000 00000   00000 00000   00000 00000   00000 00000   00000 00000
8120:   00000 00000   00000 00000   00000 00000   00000 00000   00000 00000     00000 00000   00000 00000   00000 00000   00000 00000   00000 00000
8121:   00000 00000   00000 00000   00000 00000   00000 00000   00000 00000     00000 00000   00000 00000   00000 00000   00000 00000   00000 00000
8122:   00000 00000   00000 00000   00000 00000   00000 00000   00000 00000     00000 00000   00000 00000   00000 00000   00000 00000   00000 00000
8123:   00000 00000   00000 00000   00000 00000   00000 00000   00000 00000     00000 00000   00000 00000   00000 00000   00000 00000   00000 00000
8124:   00000 00000   00000 00000   00000 00000   00000 00000   00000 00000     00000 00000   00000 00000   00000 00000   00000 00000   00000 00000
8125:   00000 00000   00000 00000   00000 00000   00000 00000   00000 00000     00000 00000   00000 00000   00000 00000   00000 00000   00000 00000
8126:   00000 00000   00000 00000   00000 00000   00000 00000   00000 00000     00000 00000   00000 00000   00000 00000   00000 00000   00000 00000
8127:   00000 00000   00000 00000   00000 00000   00000 00000   00000 00000     00000 00000   00000 00000   00000 00000   00000 00000   00000 00000
8128:   00000 00000   00000 00000   00000 00000   00000 00000   00000 00000     00000 00000   00000 00000   00000 00000   00000 00000   00000 00000
8129:   00000 00000   00000 00000   00000 00000   00000 00000   00000 00000     00000 00000   00000 00000   00000 00000   00000 00000   00000 00000
8130:   00000 00000   00000 00000   00000 00000   00000 00000   00000 00000     00000 00000   00000 00000   00000 00000   00000 00000   00000 00000
8131:   00000 00000   00000 00000   00000 00000   00000 00000   00000 00000     00000 00000   00000 00000   00000 00000   00000 00000   00000 00000
8132:   00000 00000   00000 00000   00000 00000   00000 00000   00000 00000     00000 00000   00000 00000   00000 00000   00000 00000   00000 00000
8133:   00000 00000   00000 00000   00000 00000   00000 00000   00000 00000     00000 00000   00000 00000   00000 00000   00000 00000   00000 00000
8134:   00000 00000   00000 00000   00000 00000   00000 00000   00000 00000     00000 00000   00000 00000   00000 00000   00000 00000   00000 00000
8135:   00000 00000   00000 00000   00000 00000   00000 00000   00000 00000     00000 00000   00000 00000   00000 00000   00000 00000   00000 00000
8136:   00000 00000   00000 00000   00000 00000   00000 00000   00000 00000     00000 00000   00000 00000   00000 00000   00000 00000   00000 00000
8137:   00000 00000   00000 00000   00000 00000   00000 00000   00000 00000     00000 00000   00000 00000   00000 00000   00000 00000   00000 00000
8138:   00000 00000   00000 00000   00000 00000   00000 00000   00000 00000     00000 00000   00000 00000   00000 00000   00000 00000   00000 00000
8139:   00000 00000   00000 00000   00000 00000   00000 00000   00000 00000     00000 00000   00000 00000   00000 00000   00000 00000   00000 00000
8140:   00000 00000   00000 00000   00000 00000   00000 00000   00000 00000     00000 00000   00000 00000   00000 00000   00000 00000   00000 00000
8141:   00000 00000   00000 00000   00000 00000   00000 00000   00000 00000     00000 00000   00000 00000   00000 00000   00000 00000   00000 00000
8142:   00000 00000   00000 00000   00000 00000   00000 00000   00000 00000     00000 00000   00000 00000   00000 00000   00000 00000   00000 00000
8143:   00000 00000   00000 00000   00000 00000   00000 00000   00000 00000     00000 00000   00000 00000   00000 00000   00000 00000   00000 00000
8144:   00000 00000   00000 00000   00000 00000   00000 00000   00000 00000     00000 00000   00000 00000   00000 00000   00000 00000   00000 00000
8145:   00000 00000   00000 00000   00000 00000   00000 00000   00000 00000     00000 00000   00000 00000   00000 00000   00000 00000   00000 00000
8146:   00000 00000   00000 00000   00000 00000   00000 00000   00000 00000     00000 00000   00000 00000   00000 00000   00000 00000   00000 00000
8147:   00000 00000   00000 00000   00000 00000   00000 00000   00000 00000     00000 00000   00000 00000   00000 00000   00000 00000   00000 00000
8148:   00000 00000   00000 00000   00000 00000   00000 00000   00000 00000     00000 00000   00000 00000   00000 00000   00000 00000   00000 00000
8149:   00000 00000   00000 00000   00000 00000   00000 00000   00000 00000     00000 00000   00000 00000   00000 00000   00000 00000   00000 00000
```

```
8150:  00000 00000  00000 00000  00000 00000  00000 00000  00000 00000    00000 00000  00000 00000  00000 00000  00000 00000  00000 00000
8151:  00000 00000  00000 00000  00000 00000  00000 00000  00000 00000    00000 00000  00000 00000  00000 00000  00000 00000  00000 00000
8152:  00000 00000  00000 00000  00000 00000  00000 00000  00000 00000    00000 00000  00000 00000  00000 00000  00000 00000  00000 00000
8153:  00000 00000  00000 00000  00000 00000  00000 00000  00000 00000    00000 00000  00000 00000  00000 00000  00000 00000  00000 00000
8154:  00000 00000  00000 00000  00000 00000  00000 00000  00000 00000    00000 00000  00000 00000  00000 00000  00000 00000  00000 00000
8155:  00000 00000  00000 00000  00000 00000  00000 00000  00000 00000    00000 00000  00000 00000  00000 00000  00000 00000  00000 00000
8156:  00000 00000  00000 00000  00000 00000  00000 00000  00000 00000    00000 00000  00000 00000  00000 00000  00000 00000  00000 00000
8157:  00000 00000  00000 00000  00000 00000  00000 00000  00000 00000    00000 00000  00000 00000  00000 00000  00000 00000  00000 00000
8158:  00000 00000  00000 00000  00000 00000  00000 00000  00000 00000    00000 00000  00000 00000  00000 00000  00000 00000  00000 00000
8159:  00000 00000  00000 00000  00000 00000  00000 00000  00000 00000    00000 00000  00000 00000  00000 00000  00000 00000  00000 00000
8160:  00000 00000  00000 00000  00000 00000  00000 00000  00000 00000    00000 00000  00000 00000  00000 00000  00000 00000  00000 00000
8161:  00000 00000  00000 00000  00000 00000  00000 00000  00000 00000    00000 00000  00000 00000  00000 00000  00000 00000  00000 00000
8162:  00000 00000  00000 00000  00000 00000  00000 00000  00000 00000    00000 00000  00000 00000  00000 00000  00000 00000  00000 00000
8163:  00000 00000  00000 00000  00000 00000  00000 00000  00000 00000    00000 00000  00000 00000  00000 00000  00000 00000  00000 00000
8164:  00000 00000  00000 00000  00000 00000  00000 00000  00000 00000    00000 00000  00000 00000  00000 00000  00000 00000  00000 00000
8165:  00000 00000  00000 00000  00000 00000  00000 00000  00000 00000    00000 00000  00000 00000  00000 00000  00000 00000  00000 00000
8166:  00000 00000  00000 00000  00000 00000  00000 00000  00000 00000    00000 00000  00000 00000  00000 00000  00000 00000  00000 00000
8167:  00000 00000  00000 00000  00000 00000  00000 00000  00000 00000    00000 00000  00000 00000  00000 00000  00000 00000  00000 00000
8168:  00000 00000  00000 00000  00000 00000  00000 00000  00000 00000    00000 00000  00000 00000  00000 00000  00000 00000  00000 00000
8169:  00000 00000  00000 00000  00000 00000  00000 00000  00000 00000    00000 00000  00000 00000  00000 00000  00000 00000  00000 00000
8170:  00000 00000  00000 00000  00000 00000  00000 00000  00000 00000    00000 00000  00000 00000  00000 00000  00000 00000  00000 00000
8171:  00000 00000  00000 00000  00000 00000  00000 00000  00000 00000    00000 00000  00000 00000  00000 00000  00000 00000  00000 00000
8172:  00000 00000  00000 00000  00000 00000  00000 00000  00000 00000    00000 00000  00000 00000  00000 00000  00000 00000  00000 00000
8173:  00000 00000  00000 00000  00000 00000  00000 00000  00000 00000    00000 00000  00000 00000  00000 00000  00000 00000  00000 00000
8174:  00000 00000  00000 00000  00000 00000  00000 00000  00000 00000    00000 00000  00000 00000  00000 00000  00000 00000  00000 00000
8175:  00000 00000  00000 00000  00000 00000  00000 00000  00000 00000    00000 00000  00000 00000  00000 00000  00000 00000  00000 00000
8176:  00000 00000  00000 00000  00000 00000  00000 00000  00000 00000    00000 00000  00000 00000  00000 00000  00000 00000  00000 00000
8177:  00000 00000  00000 00000  00000 00000  00000 00000  00000 00000    00000 00000  00000 00000  00000 00000  00000 00000  00000 00000
8178:  00000 00000  00000 00000  00000 00000  00000 00000  00000 00000    00000 00000  00000 00000  00000 00000  00000 00000  00000 00000
8179:  00000 00000  00000 00000  00000 00000  00000 00000  00000 00000    00000 00000  00000 00000  00000 00000  00000 00000  00000 00000
8180:  00000 00000  00000 00000  00000 00000  00000 00000  00000 00000    00000 00000  00000 00000  00000 00000  00000 00000  00000 00000
8181:  00000 00000  00000 00000  00000 00000  00000 00000  00000 00000    00000 00000  00000 00000  00000 00000  00000 00000  00000 00000
8182:  00000 00000  00000 00000  00000 00000  00000 00000  00000 00000    00000 00000  00000 00000  00000 00000  00000 00000  00000 00000
8183:  00000 00000  00000 00000  00000 00000  00000 00000  00000 00000    00000 00000  00000 00000  00000 00000  00000 00000  00000 00000
8184:  00000 00000  00000 00000  00000 00000  00000 00000  00000 00000    00000 00000  00000 00000  00000 00000  00000 00000  00000 00000
8185:  00000 00000  00000 00000  00000 00000  00000 00000  00000 00000    00000 00000  00000 00000  00000 00000  00000 00000  00000 00000
8186:  00000 00000  00000 00000  00000 00000  00000 00000  00000 00000    00000 00000  00000 00000  00000 00000  00000 00000  00000 00000
8187:  00000 00000  00000 00000  00000 00000  00000 00000  00000 00000    00000 00000  00000 00000  00000 00000  00000 00000  00000 00000
8188:  00000 00000  00000 00000  00000 00000  00000 00000  00000 00000    00000 00000  00000 00000  00000 00000  00000 00000  00000 00000
8189:  00000 00000  00000 00000  00000 00000  00000 00000  00000 00000    00000 00000  00000 00000  00000 00000  00000 00000  00000 00000
8190:  00000 00000  00000 00000  00000 00000  00000 00000  00000 00000    00000 00000  00000 00000  00000 00000  00000 00000  00000 00000
8191:  00000 00000  00000 00000  00000 00000  00000 00000  00000 00000    00000 00000  00000 00000  00000 00000  00000 00000  00000 00000
8192:  00000 00000  00000 00000  00000 00000  00000 00000  00000 00000    00000 00000  00000 00000  00000 00000  00000 00000  00000 00000
8193:  00000 00000  00000 00000  00000 00000  00000 00000  00000 00000    00000 00000  00000 00000  00000 00000  00000 00000  00000 00000
8194:  00000 00000  00000 00000  00000 00000  00000 00000  00000 00000    00000 00000  00000 00000  00000 00000  00000 00000  00000 00000
8195:  00000 00000  00000 00000  00000 00000  00000 00000  00000 00000    00000 00000  00000 00000  00000 00000  00000 00000  00000 00000
8196:  00000 00000  00000 00000  00000 00000  00000 00000  00000 00000    00000 00000  00000 00000  00000 00000  00000 00000  00000 00000
8197:  00000 00000  00000 00000  00000 00000  00000 00000  00000 00000    00000 00000  00000 00000  00000 00000  00000 00000  00000 00000
8198:  00000 00000  00000 00000  00000 00000  00000 00000  00000 00000    00000 00000  00000 00000  00000 00000  00000 00000  00000 00000
8199:  00000 00000  00000 00000  00000 00000  00000 00000  00000 00000    00000 00000  00000 00000  00000 00000  00000 00000  00000 00000
```

```
8200:   00000 00000   00000 00000   00000 00000   00000 00000   00000 00000     00000 00000   00000 00000   00000 00000   00000 00000   00000 00000
8201:   00000 00000   00000 00000   00000 00000   00000 00000   00000 00000     00000 00000   00000 00000   00000 00000   00000 00000   00000 00000
8202:   00000 00000   00000 00000   00000 00000   00000 00000   00000 00000     00000 00000   00000 00000   00000 00000   00000 00000   00000 00000
8203:   00000 00000   00000 00000   00000 00000   00000 00000   00000 00000     00000 00000   00000 00000   00000 00000   00000 00000   00000 00000
8204:   00000 00000   00000 00000   00000 00000   00000 00000   00000 00000     00000 00000   00000 00000   00000 00000   00000 00000   00000 00000
8205:   00000 00000   00000 00000   00000 00000   00000 00000   00000 00000     00000 00000   00000 00000   00000 00000   00000 00000   00000 00000
8206:   00000 00000   00000 00000   00000 00000   00000 00000   00000 00000     00000 00000   00000 00000   00000 00000   00000 00000   00000 00000
8207:   00000 00000   00000 00000   00000 00000   00000 00000   00000 00000     00000 00000   00000 00000   00000 00000   00000 00000   00000 00000
8208:   00000 00000   00000 00000   00000 00000   00000 00000   00000 00000     00000 00000   00000 00000   00000 00000   00000 00000   00000 00000
8209:   00000 00000   00000 00000   00000 00000   00000 00000   00000 00000     00000 00000   00000 00000   00000 00000   00000 00000   00000 00000
8210:   00000 00000   00000 00000   00000 00000   00000 00000   00000 00000     00000 00000   00000 00000   00000 00000   00000 00000   00000 00000
8211:   00000 00000   00000 00000   00000 00000   00000 00000   00000 00000     00000 00000   00000 00000   00000 00000   00000 00000   00000 00000
8212:   00000 00000   00000 00000   00000 00000   00000 00000   00000 00000     00000 00000   00000 00000   00000 00000   00000 00000   00000 00000
8213:   00000 00000   00000 00000   00000 00000   00000 00000   00000 00000     00000 00000   00000 00000   00000 00000   00000 00000   00000 00000
8214:   00000 00000   00000 00000   00000 00000   00000 00000   00000 00000     00000 00000   00000 00000   00000 00000   00000 00000   00000 00000
8215:   00000 00000   00000 00000   00000 00000   00000 00000   00000 00000     00000 00000   00000 00000   00000 00000   00000 00000   00000 00000
8216:   00000 00000   00000 00000   00000 00000   00000 00000   00000 00000     00000 00000   00000 00000   00000 00000   00000 00000   00000 00000
8217:   00000 00000   00000 00000   00000 00000   00000 00000   00000 00000     00000 00000   00000 00000   00000 00000   00000 00000   00000 00000
8218:   00000 00000   00000 00000   00000 00000   00000 00000   00000 00000     00000 00000   00000 00000   00000 00000   00000 00000   00000 00000
8219:   00000 00000   00000 00000   00000 00000   00000 00000   00000 00000     00000 00000   00000 00000   00000 00000   00000 00000   00000 00000
8220:   00000 00000   00000 00000   00000 00000   00000 00000   00000 00000     00000 00000   00000 00000   00000 00000   00000 00000   00000 00000
8221:   00000 00000   00000 00000   00000 00000   00000 00000   00000 00000     00000 00000   00000 00000   00000 00000   00000 00000   00000 00000
8222:   00000 00000   00000 00000   00000 00000   00000 00000   00000 00000     00000 00000   00000 00000   00000 00000   00000 00000   00000 00000
8223:   00000 00000   00000 00000   00000 00000   00000 00000   00000 00000     00000 00000   00000 00000   00000 00000   00000 00000   00000 00000
8224:   00000 00000   00000 00000   00000 00000   00000 00000   00000 00000     00000 00000   00000 00000   00000 00000   00000 00000   00000 00000
8225:   00000 00000   00000 00000   00000 00000   00000 00000   00000 00000     00000 00000   00000 00000   00000 00000   00000 00000   00000 00000
8226:   00000 00000   00000 00000   00000 00000   00000 00000   00000 00000     00000 00000   00000 00000   00000 00000   00000 00000   00000 00000
8227:   00000 00000   00000 00000   00000 00000   00000 00000   00000 00000     00000 00000   00000 00000   00000 00000   00000 00000   00000 00000
8228:   00000 00000   00000 00000   00000 00000   00000 00000   00000 00000     00000 00000   00000 00000   00000 00000   00000 00000   00000 00000
8229:   00000 00000   00000 00000   00000 00000   00000 00000   00000 00000     00000 00000   00000 00000   00000 00000   00000 00000   00000 00000
8230:   00000 00000   00000 00000   00000 00000   00000 00000   00000 00000     00000 00000   00000 00000   00000 00000   00000 00000   00000 00000
8231:   00000 00000   00000 00000   00000 00000   00000 00000   00000 00000     00000 00000   00000 00000   00000 00000   00000 00000   00000 00000
8232:   00000 00000   00000 00000   00000 00000   00000 00000   00000 00000     00000 00000   00000 00000   00000 00000   00000 00000   00000 00000
8233:   00000 00000   00000 00000   00000 00000   00000 00000   00000 00000     00000 00000   00000 00000   00000 00000   00000 00000   00000 00000
8234:   00000 00000   00000 00000   00000 00000   00000 00000   00000 00000     00000 00000   00000 00000   00000 00000   00000 00000   00000 00000
8235:   00000 00000   00000 00000   00000 00000   00000 00000   00000 00000     00000 00000   00000 00000   00000 00000   00000 00000   00000 00000
8236:   00000 00000   00000 00000   00000 00000   00000 00000   00000 00000     00000 00000   00000 00000   00000 00000   00000 00000   00000 00000
8237:   00000 00000   00000 00000   00000 00000   00000 00000   00000 00000     00000 00000   00000 00000   00000 00000   00000 00000   00000 00000
8238:   00000 00000   00000 00000   00000 00000   00000 00000   00000 00000     00000 00000   00000 00000   00000 00000   00000 00000   00000 00000
8239:   00000 00000   00000 00000   00000 00000   00000 00000   00000 00000     00000 00000   00000 00000   00000 00000   00000 00000   00000 00000
8240:   00000 00000   00000 00000   00000 00000   00000 00000   00000 00000     00000 00000   00000 00000   00000 00000   00000 00000   00000 00000
8241:   00000 00000   00000 00000   00000 00000   00000 00000   00000 00000     00000 00000   00000 00000   00000 00000   00000 00000   00000 00000
8242:   00000 00000   00000 00000   00000 00000   00000 00000   00000 00000     00000 00000   00000 00000   00000 00000   00000 00000   00000 00000
8243:   00000 00000   00000 00000   00000 00000   00000 00000   00000 00000     00000 00000   00000 00000   00000 00000   00000 00000   00000 00000
8244:   00000 00000   00000 00000   00000 00000   00000 00000   00000 00000     00000 00000   00000 00000   00000 00000   00000 00000   00000 00000
8245:   00000 00000   00000 00000   00000 00000   00000 00000   00000 00000     00000 00000   00000 00000   00000 00000   00000 00000   00000 00000
8246:   00000 00000   00000 00000   00000 00000   00000 00000   00000 00000     00000 00000   00000 00000   00000 00000   00000 00000   00000 00000
8247:   00000 00000   00000 00000   00000 00000   00000 00000   00000 00000     00000 00000   00000 00000   00000 00000   00000 00000   00000 00000
8248:   00000 00000   00000 00000   00000 00000   00000 00000   00000 00000     00000 00000   00000 00000   00000 00000   00000 00000   00000 00000
8249:   00000 00000   00000 00000   00000 00000   00000 00000   00000 00000     00000 00000   00000 00000   00000 00000   00000 00000   00000 00000
```

```
8250:  00000 00000  00000 00000  00000 00000  00000 00000  00000 00000    00000 00000  00000 00000  00000 00000  00000 00000  00000 00000
8251:  00000 00000  00000 00000  00000 00000  00000 00000  00000 00000    00000 00000  00000 00000  00000 00000  00000 00000  00000 00000
8252:  00000 00000  00000 00000  00000 00000  00000 00000  00000 00000    00000 00000  00000 00000  00000 00000  00000 00000  00000 00000
8253:  00000 00000  00000 00000  00000 00000  00000 00000  00000 00000    00000 00000  00000 00000  00000 00000  00000 00000  00000 00000
8254:  00000 00000  00000 00000  00000 00000  00000 00000  00000 00000    00000 00000  00000 00000  00000 00000  00000 00000  00000 00000
8255:  00000 00000  00000 00000  00000 00000  00000 00000  00000 00000    00000 00000  00000 00000  00000 00000  00000 00000  00000 00000
8256:  00000 00000  00000 00000  00000 00000  00000 00000  00000 00000    00000 00000  00000 00000  00000 00000  00000 00000  00000 00000
8257:  00000 00000  00000 00000  00000 00000  00000 00000  00000 00000    00000 00000  00000 00000  00000 00000  00000 00000  00000 00000
8258:  00000 00000  00000 00000  00000 00000  00000 00000  00000 00000    00000 00000  00000 00000  00000 00000  00000 00000  00000 00000
8259:  00000 00000  00000 00000  00000 00000  00000 00000  00000 00000    00000 00000  00000 00000  00000 00000  00000 00000  00000 00000
8260:  00000 00000  00000 00000  00000 00000  00000 00000  00000 00000    00000 00000  00000 00000  00000 00000  00000 00000  00000 00000
8261:  00000 00000  00000 00000  00000 00000  00000 00000  00000 00000    00000 00000  00000 00000  00000 00000  00000 00000  00000 00000
8262:  00000 00000  00000 00000  00000 00000  00000 00000  00000 00000    00000 00000  00000 00000  00000 00000  00000 00000  00000 00000
8263:  00000 00000  00000 00000  00000 00000  00000 00000  00000 00000    00000 00000  00000 00000  00000 00000  00000 00000  00000 00000
8264:  00000 00000  00000 00000  00000 00000  00000 00000  00000 00000    00000 00000  00000 00000  00000 00000  00000 00000  00000 00000
8265:  00000 00000  00000 00000  00000 00000  00000 00000  00000 00000    00000 00000  00000 00000  00000 00000  00000 00000  00000 00000
8266:  00000 00000  00000 00000  00000 00000  00000 00000  00000 00000    00000 00000  00000 00000  00000 00000  00000 00000  00000 00000
8267:  00000 00000  00000 00000  00000 00000  00000 00000  00000 00000    00000 00000  00000 00000  00000 00000  00000 00000  00000 00000
8268:  00000 00000  00000 00000  00000 00000  00000 00000  00000 00000    00000 00000  00000 00000  00000 00000  00000 00000  00000 00000
8269:  00000 00000  00000 00000  00000 00000  00000 00000  00000 00000    00000 00000  00000 00000  00000 00000  00000 00000  00000 00000
8270:  00000 00000  00000 00000  00000 00000  00000 00000  00000 00000    00000 00000  00000 00000  00000 00000  00000 00000  00000 00000
8271:  00000 00000  00000 00000  00000 00000  00000 00000  00000 00000    00000 00000  00000 00000  00000 00000  00000 00000  00000 00000
8272:  00000 00000  00000 00000  00000 00000  00000 00000  00000 00000    00000 00000  00000 00000  00000 00000  00000 00000  00000 00000
8273:  00000 00000  00000 00000  00000 00000  00000 00000  00000 00000    00000 00000  00000 00000  00000 00000  00000 00000  00000 00000
8274:  00000 00000  00000 00000  00000 00000  00000 00000  00000 00000    00000 00000  00000 00000  00000 00000  00000 00000  00000 00000
8275:  00000 00000  00000 00000  00000 00000  00000 00000  00000 00000    00000 00000  00000 00000  00000 00000  00000 00000  00000 00000
8276:  00000 00000  00000 00000  00000 00000  00000 00000  00000 00000    00000 00000  00000 00000  00000 00000  00000 00000  00000 00000
8277:  00000 00000  00000 00000  00000 00000  00000 00000  00000 00000    00000 00000  00000 00000  00000 00000  00000 00000  00000 00000
8278:  00000 00000  00000 00000  00000 00000  00000 00000  00000 00000    00000 00000  00000 00000  00000 00000  00000 00000  00000 00000
8279:  00000 00000  00000 00000  00000 00000  00000 00000  00000 00000    00000 00000  00000 00000  00000 00000  00000 00000  00000 00000
8280:  00000 00000  00000 00000  00000 00000  00000 00000  00000 00000    00000 00000  00000 00000  00000 00000  00000 00000  00000 00000
8281:  00000 00000  00000 00000  00000 00000  00000 00000  00000 00000    00000 00000  00000 00000  00000 00000  00000 00000  00000 00000
8282:  00000 00000  00000 00000  00000 00000  00000 00000  00000 00000    00000 00000  00000 00000  00000 00000  00000 00000  00000 00000
8283:  00000 00000  00000 00000  00000 00000  00000 00000  00000 00000    00000 00000  00000 00000  00000 00000  00000 00000  00000 00000
8284:  00000 00000  00000 00000  00000 00000  00000 00000  00000 00000    00000 00000  00000 00000  00000 00000  00000 00000  00000 00000
8285:  00000 00000  00000 00000  00000 00000  00000 00000  00000 00000    00000 00000  00000 00000  00000 00000  00000 00000  00000 00000
8286:  00000 00000  00000 00000  00000 00000  00000 00000  00000 00000    00000 00000  00000 00000  00000 00000  00000 00000  00000 00000
8287:  00000 00000  00000 00000  00000 00000  00000 00000  00000 00000    00000 00000  00000 00000  00000 00000  00000 00000  00000 00000
8288:  00000 00000  00000 00000  00000 00000  00000 00000  00000 00000    00000 00000  00000 00000  00000 00000  00000 00000  00000 00000
8289:  00000 00000  00000 00000  00000 00000  00000 00000  00000 00000    00000 00000  00000 00000  00000 00000  00000 00000  00000 00000
8290:  00000 00000  00000 00000  00000 00000  00000 00000  00000 00000    00000 00000  00000 00000  00000 00000  00000 00000  00000 00000
8291:  00000 00000  00000 00000  00000 00000  00000 00000  00000 00000    00000 00000  00000 00000  00000 00000  00000 00000  00000 00000
8292:  00000 00000  00000 00000  00000 00000  00000 00000  00000 00000    00000 00000  00000 00000  00000 00000  00000 00000  00000 00000
8293:  00000 00000  00000 00000  00000 00000  00000 00000  00000 00000    00000 00000  00000 00000  00000 00000  00000 00000  00000 00000
8294:  00000 00000  00000 00000  00000 00000  00000 00000  00000 00000    00000 00000  00000 00000  00000 00000  00000 00000  00000 00000
8295:  00000 00000  00000 00000  00000 00000  00000 00000  00000 00000    00000 00000  00000 00000  00000 00000  00000 00000  00000 00000
8296:  00000 00000  00000 00000  00000 00000  00000 00000  00000 00000    00000 00000  00000 00000  00000 00000  00000 00000  00000 00000
8297:  00000 00000  00000 00000  00000 00000  00000 00000  00000 00000    00000 00000  00000 00000  00000 00000  00000 00000  00000 00000
8298:  00000 00000  00000 00000  00000 00000  00000 00000  00000 00000    00000 00000  00000 00000  00000 00000  00000 00000  00000 00000
8299:  00000 00000  00000 00000  00000 00000  00000 00000  00000 00000    00000 00000  00000 00000  00000 00000  00000 00000  00000 00000
```

```
8300:   00000 00000   00000 00000   00000 00000   00000 00000   00000 00000     00000 00000   00000 00000   00000 00000   00000 00000   00000 00000
8301:   00000 00000   00000 00000   00000 00000   00000 00000   00000 00000     00000 00000   00000 00000   00000 00000   00000 00000   00000 00000
8302:   00000 00000   00000 00000   00000 00000   00000 00000   00000 00000     00000 00000   00000 00000   00000 00000   00000 00000   00000 00000
8303:   00000 00000   00000 00000   00000 00000   00000 00000   00000 00000     00000 00000   00000 00000   00000 00000   00000 00000   00000 00000
8304:   00000 00000   00000 00000   00000 00000   00000 00000   00000 00000     00000 00000   00000 00000   00000 00000   00000 00000   00000 00000
8305:   00000 00000   00000 00000   00000 00000   00000 00000   00000 00000     00000 00000   00000 00000   00000 00000   00000 00000   00000 00000
8306:   00000 00000   00000 00000   00000 00000   00000 00000   00000 00000     00000 00000   00000 00000   00000 00000   00000 00000   00000 00000
8307:   00000 00000   00000 00000   00000 00000   00000 00000   00000 00000     00000 00000   00000 00000   00000 00000   00000 00000   00000 00000
8308:   00000 00000   00000 00000   00000 00000   00000 00000   00000 00000     00000 00000   00000 00000   00000 00000   00000 00000   00000 00000
8309:   00000 00000   00000 00000   00000 00000   00000 00000   00000 00000     00000 00000   00000 00000   00000 00000   00000 00000   00000 00000
8310:   00000 00000   00000 00000   00000 00000   00000 00000   00000 00000     00000 00000   00000 00000   00000 00000   00000 00000   00000 00000
8311:   00000 00000   00000 00000   00000 00000   00000 00000   00000 00000     00000 00000   00000 00000   00000 00000   00000 00000   00000 00000
8312:   00000 00000   00000 00000   00000 00000   00000 00000   00000 00000     00000 00000   00000 00000   00000 00000   00000 00000   00000 00000
8313:   00000 00000   00000 00000   00000 00000   00000 00000   00000 00000     00000 00000   00000 00000   00000 00000   00000 00000   00000 00000
8314:   00000 00000   00000 00000   00000 00000   00000 00000   00000 00000     00000 00000   00000 00000   00000 00000   00000 00000   00000 00000
8315:   00000 00000   00000 00000   00000 00000   00000 00000   00000 00000     00000 00000   00000 00000   00000 00000   00000 00000   00000 00000
8316:   00000 00000   00000 00000   00000 00000   00000 00000   00000 00000     00000 00000   00000 00000   00000 00000   00000 00000   00000 00000
8317:   00000 00000   00000 00000   00000 00000   00000 00000   00000 00000     00000 00000   00000 00000   00000 00000   00000 00000   00000 00000
8318:   00000 00000   00000 00000   00000 00000   00000 00000   00000 00000     00000 00000   00000 00000   00000 00000   00000 00000   00000 00000
8319:   00000 00000   00000 00000   00000 00000   00000 00000   00000 00000     00000 00000   00000 00000   00000 00000   00000 00000   00000 00000
8320:   00000 00000   00000 00000   00000 00000   00000 00000   00000 00000     00000 00000   00000 00000   00000 00000   00000 00000   00000 00000
8321:   00000 00000   00000 00000   00000 00000   00000 00000   00000 00000     00000 00000   00000 00000   00000 00000   00000 00000   00000 00000
8322:   00000 00000   00000 00000   00000 00000   00000 00000   00000 00000     00000 00000   00000 00000   00000 00000   00000 00000   00000 00000
8323:   00000 00000   00000 00000   00000 00000   00000 00000   00000 00000     00000 00000   00000 00000   00000 00000   00000 00000   00000 00000
8324:   00000 00000   00000 00000   00000 00000   00000 00000   00000 00000     00000 00000   00000 00000   00000 00000   00000 00000   00000 00000
8325:   00000 00000   00000 00000   00000 00000   00000 00000   00000 00000     00000 00000   00000 00000   00000 00000   00000 00000   00000 00000
8326:   00000 00000   00000 00000   00000 00000   00000 00000   00000 00000     00000 00000   00000 00000   00000 00000   00000 00000   00000 00000
8327:   00000 00000   00000 00000   00000 00000   00000 00000   00000 00000     00000 00000   00000 00000   00000 00000   00000 00000   00000 00000
8328:   00000 00000   00000 00000   00000 00000   00000 00000   00000 00000     00000 00000   00000 00000   00000 00000   00000 00000   00000 00000
8329:   00000 00000   00000 00000   00000 00000   00000 00000   00000 00000     00000 00000   00000 00000   00000 00000   00000 00000   00000 00000
8330:   00000 00000   00000 00000   00000 00000   00000 00000   00000 00000     00000 00000   00000 00000   00000 00000   00000 00000   00000 00000
8331:   00000 00000   00000 00000   00000 00000   00000 00000   00000 00000     00000 00000   00000 00000   00000 00000   00000 00000   00000 00000
8332:   00000 00000   00000 00000   00000 00000   00000 00000   00000 00000     00000 00000   00000 00000   00000 00000   00000 00000   00000 00000
8333:   00000 00000   00000 00000   00000 00000   00000 00000   00000 00000     00000 00000   00000 00000   00000 00000   00000 00000   00000 00000
8334:   00000 00000   00000 00000   00000 00000   00000 00000   00000 00000     00000 00000   00000 00000   00000 00000   00000 00000   00000 00000
8335:   00000 00000   00000 00000   00000 00000   00000 00000   00000 00000     00000 00000   00000 00000   00000 00000   00000 00000   00000 00000
8336:   00000 00000   00000 00000   00000 00000   00000 00000   00000 00000     00000 00000   00000 00000   00000 00000   00000 00000   00000 00000
8337:   00000 00000   00000 00000   00000 00000   00000 00000   00000 00000     00000 00000   00000 00000   00000 00000   00000 00000   00000 00000
8338:   00000 00000   00000 00000   00000 00000   00000 00000   00000 00000     00000 00000   00000 00000   00000 00000   00000 00000   00000 00000
8339:   00000 00000   00000 00000   00000 00000   00000 00000   00000 00000     00000 00000   00000 00000   00000 00000   00000 00000   00000 00000
8340:   00000 00000   00000 00000   00000 00000   00000 00000   00000 00000     00000 00000   00000 00000   00000 00000   00000 00000   00000 00000
8341:   00000 00000   00000 00000   00000 00000   00000 00000   00000 00000     00000 00000   00000 00000   00000 00000   00000 00000   00000 00000
8342:   00000 00000   00000 00000   00000 00000   00000 00000   00000 00000     00000 00000   00000 00000   00000 00000   00000 00000   00000 00000
8343:   00000 00000   00000 00000   00000 00000   00000 00000   00000 00000     00000 00000   00000 00000   00000 00000   00000 00000   00000 00000
8344:   00000 00000   00000 00000   00000 00000   00000 00000   00000 00000     00000 00000   00000 00000   00000 00000   00000 00000   00000 00000
8345:   00000 00000   00000 00000   00000 00000   00000 00000   00000 00000     00000 00000   00000 00000   00000 00000   00000 00000   00000 00000
8346:   00000 00000   00000 00000   00000 00000   00000 00000   00000 00000     00000 00000   00000 00000   00000 00000   00000 00000   00000 00000
8347:   00000 00000   00000 00000   00000 00000   00000 00000   00000 00000     00000 00000   00000 00000   00000 00000   00000 00000   00000 00000
8348:   00000 00000   00000 00000   00000 00000   00000 00000   00000 00000     00000 00000   00000 00000   00000 00000   00000 00000   00000 00000
8349:   00000 00000   00000 00000   00000 00000   00000 00000   00000 00000     00000 00000   00000 00000   00000 00000   00000 00000   00000 00000
```

```
8350:  00000 00000   00000 00000   00000 00000   00000 00000   00000 00000     00000 00000   00000 00000   00000 00000   00000 00000   00000 00000
8351:  00000 00000   00000 00000   00000 00000   00000 00000   00000 00000     00000 00000   00000 00000   00000 00000   00000 00000   00000 00000
8352:  00000 00000   00000 00000   00000 00000   00000 00000   00000 00000     00000 00000   00000 00000   00000 00000   00000 00000   00000 00000
8353:  00000 00000   00000 00000   00000 00000   00000 00000   00000 00000     00000 00000   00000 00000   00000 00000   00000 00000   00000 00000
8354:  00000 00000   00000 00000   00000 00000   00000 00000   00000 00000     00000 00000   00000 00000   00000 00000   00000 00000   00000 00000
8355:  00000 00000   00000 00000   00000 00000   00000 00000   00000 00000     00000 00000   00000 00000   00000 00000   00000 00000   00000 00000
8356:  00000 00000   00000 00000   00000 00000   00000 00000   00000 00000     00000 00000   00000 00000   00000 00000   00000 00000   00000 00000
8357:  00000 00000   00000 00000   00000 00000   00000 00000   00000 00000     00000 00000   00000 00000   00000 00000   00000 00000   00000 00000
8358:  00000 00000   00000 00000   00000 00000   00000 00000   00000 00000     00000 00000   00000 00000   00000 00000   00000 00000   00000 00000
8359:  00000 00000   00000 00000   00000 00000   00000 00000   00000 00000     00000 00000   00000 00000   00000 00000   00000 00000   00000 00000
8360:  00000 00000   00000 00000   00000 00000   00000 00000   00000 00000     00000 00000   00000 00000   00000 00000   00000 00000   00000 00000
8361:  00000 00000   00000 00000   00000 00000   00000 00000   00000 00000     00000 00000   00000 00000   00000 00000   00000 00000   00000 00000
8362:  00000 00000   00000 00000   00000 00000   00000 00000   00000 00000     00000 00000   00000 00000   00000 00000   00000 00000   00000 00000
8363:  00000 00000   00000 00000   00000 00000   00000 00000   00000 00000     00000 00000   00000 00000   00000 00000   00000 00000   00000 00000
8364:  00000 00000   00000 00000   00000 00000   00000 00000   00000 00000     00000 00000   00000 00000   00000 00000   00000 00000   00000 00000
8365:  00000 00000   00000 00000   00000 00000   00000 00000   00000 00000     00000 00000   00000 00000   00000 00000   00000 00000   00000 00000
8366:  00000 00000   00000 00000   00000 00000   00000 00000   00000 00000     00000 00000   00000 00000   00000 00000   00000 00000   00000 00000
8367:  00000 00000   00000 00000   00000 00000   00000 00000   00000 00000     00000 00000   00000 00000   00000 00000   00000 00000   00000 00000
8368:  00000 00000   00000 00000   00000 00000   00000 00000   00000 00000     00000 00000   00000 00000   00000 00000   00000 00000   00000 00000
8369:  00000 00000   00000 00000   00000 00000   00000 00000   00000 00000     00000 00000   00000 00000   00000 00000   00000 00000   00000 00000
8370:  00000 00000   00000 00000   00000 00000   00000 00000   00000 00000     00000 00000   00000 00000   00000 00000   00000 00000   00000 00000
8371:  00000 00000   00000 00000   00000 00000   00000 00000   00000 00000     00000 00000   00000 00000   00000 00000   00000 00000   00000 00000
8372:  00000 00000   00000 00000   00000 00000   00000 00000   00000 00000     00000 00000   00000 00000   00000 00000   00000 00000   00000 00000
8373:  00000 00000   00000 00000   00000 00000   00000 00000   00000 00000     00000 00000   00000 00000   00000 00000   00000 00000   00000 00000
8374:  00000 00000   00000 00000   00000 00000   00000 00000   00000 00000     00000 00000   00000 00000   00000 00000   00000 00000   00000 00000
8375:  00000 00000   00000 00000   00000 00000   00000 00000   00000 00000     00000 00000   00000 00000   00000 00000   00000 00000   00000 00000
8376:  00000 00000   00000 00000   00000 00000   00000 00000   00000 00000     00000 00000   00000 00000   00000 00000   00000 00000   00000 00000
8377:  00000 00000   00000 00000   00000 00000   00000 00000   00000 00000     00000 00000   00000 00000   00000 00000   00000 00000   00000 00000
8378:  00000 00000   00000 00000   00000 00000   00000 00000   00000 00000     00000 00000   00000 00000   00000 00000   00000 00000   00000 00000
8379:  00000 00000   00000 00000   00000 00000   00000 00000   00000 00000     00000 00000   00000 00000   00000 00000   00000 00000   00000 00000
8380:  00000 00000   00000 00000   00000 00000   00000 00000   00000 00000     00000 00000   00000 00000   00000 00000   00000 00000   00000 00000
8381:  00000 00000   00000 00000   00000 00000   00000 00000   00000 00000     00000 00000   00000 00000   00000 00000   00000 00000   00000 00000
8382:  00000 00000   00000 00000   00000 00000   00000 00000   00000 00000     00000 00000   00000 00000   00000 00000   00000 00000   00000 00000
8383:  00000 00000   00000 00000   00000 00000   00000 00000   00000 00000     00000 00000   00000 00000   00000 00000   00000 00000   00000 00000
8384:  00000 00000   00000 00000   00000 00000   00000 00000   00000 00000     00000 00000   00000 00000   00000 00000   00000 00000   00000 00000
8385:  00000 00000   00000 00000   00000 00000   00000 00000   00000 00000     00000 00000   00000 00000   00000 00000   00000 00000   00000 00000
8386:  00000 00000   00000 00000   00000 00000   00000 00000   00000 00000     00000 00000   00000 00000   00000 00000   00000 00000   00000 00000
8387:  00000 00000   00000 00000   00000 00000   00000 00000   00000 00000     00000 00000   00000 00000   00000 00000   00000 00000   00000 00000
8388:  00000 00000   00000 00000   00000 00000   00000 00000   00000 00000     00000 00000   00000 00000   00000 00000   00000 00000   00000 00000
8389:  00000 00000   00000 00000   00000 00000   00000 00000   00000 00000     00000 00000   00000 00000   00000 00000   00000 00000   00000 00000
8390:  00000 00000   00000 00000   00000 00000   00000 00000   00000 00000     00000 00000   00000 00000   00000 00000   00000 00000   00000 00000
8391:  00000 00000   00000 00000   00000 00000   00000 00000   00000 00000     00000 00000   00000 00000   00000 00000   00000 00000   00000 00000
8392:  00000 00000   00000 00000   00000 00000   00000 00000   00000 00000     00000 00000   00000 00000   00000 00000   00000 00000   00000 00000
8393:  00000 00000   00000 00000   00000 00000   00000 00000   00000 00000     00000 00000   00000 00000   00000 00000   00000 00000   00000 00000
8394:  00000 00000   00000 00000   00000 00000   00000 00000   00000 00000     00000 00000   00000 00000   00000 00000   00000 00000   00000 00000
8395:  00000 00000   00000 00000   00000 00000   00000 00000   00000 00000     00000 00000   00000 00000   00000 00000   00000 00000   00000 00000
8396:  00000 00000   00000 00000   00000 00000   00000 00000   00000 00000     00000 00000   00000 00000   00000 00000   00000 00000   00000 00000
8397:  00000 00000   00000 00000   00000 00000   00000 00000   00000 00000     00000 00000   00000 00000   00000 00000   00000 00000   00000 00000
8398:  00000 00000   00000 00000   00000 00000   00000 00000   00000 00000     00000 00000   00000 00000   00000 00000   00000 00000   00000 00000
8399:  00000 00000   00000 00000   00000 00000   00000 00000   00000 00000     00000 00000   00000 00000   00000 00000   00000 00000   00000 00000
```

```
8400:   00000 00000   00000 00000   00000 00000   00000 00000   00000 00000     00000 00000   00000 00000   00000 00000   00000 00000   00000 00000
8401:   00000 00000   00000 00000   00000 00000   00000 00000   00000 00000     00000 00000   00000 00000   00000 00000   00000 00000   00000 00000
8402:   00000 00000   00000 00000   00000 00000   00000 00000   00000 00000     00000 00000   00000 00000   00000 00000   00000 00000   00000 00000
8403:   00000 00000   00000 00000   00000 00000   00000 00000   00000 00000     00000 00000   00000 00000   00000 00000   00000 00000   00000 00000
8404:   00000 00000   00000 00000   00000 00000   00000 00000   00000 00000     00000 00000   00000 00000   00000 00000   00000 00000   00000 00000
8405:   00000 00000   00000 00000   00000 00000   00000 00000   00000 00000     00000 00000   00000 00000   00000 00000   00000 00000   00000 00000
8406:   00000 00000   00000 00000   00000 00000   00000 00000   00000 00000     00000 00000   00000 00000   00000 00000   00000 00000   00000 00000
8407:   00000 00000   00000 00000   00000 00000   00000 00000   00000 00000     00000 00000   00000 00000   00000 00000   00000 00000   00000 00000
8408:   00000 00000   00000 00000   00000 00000   00000 00000   00000 00000     00000 00000   00000 00000   00000 00000   00000 00000   00000 00000
8409:   00000 00000   00000 00000   00000 00000   00000 00000   00000 00000     00000 00000   00000 00000   00000 00000   00000 00000   00000 00000
8410:   00000 00000   00000 00000   00000 00000   00000 00000   00000 00000     00000 00000   00000 00000   00000 00000   00000 00000   00000 00000
8411:   00000 00000   00000 00000   00000 00000   00000 00000   00000 00000     00000 00000   00000 00000   00000 00000   00000 00000   00000 00000
8412:   00000 00000   00000 00000   00000 00000   00000 00000   00000 00000     00000 00000   00000 00000   00000 00000   00000 00000   00000 00000
8413:   00000 00000   00000 00000   00000 00000   00000 00000   00000 00000     00000 00000   00000 00000   00000 00000   00000 00000   00000 00000
8414:   00000 00000   00000 00000   00000 00000   00000 00000   00000 00000     00000 00000   00000 00000   00000 00000   00000 00000   00000 00000
8415:   00000 00000   00000 00000   00000 00000   00000 00000   00000 00000     00000 00000   00000 00000   00000 00000   00000 00000   00000 00000
8416:   00000 00000   00000 00000   00000 00000   00000 00000   00000 00000     00000 00000   00000 00000   00000 00000   00000 00000   00000 00000
8417:   00000 00000   00000 00000   00000 00000   00000 00000   00000 00000     00000 00000   00000 00000   00000 00000   00000 00000   00000 00000
8418:   00000 00000   00000 00000   00000 00000   00000 00000   00000 00000     00000 00000   00000 00000   00000 00000   00000 00000   00000 00000
8419:   00000 00000   00000 00000   00000 00000   00000 00000   00000 00000     00000 00000   00000 00000   00000 00000   00000 00000   00000 00000
8420:   00000 00000   00000 00000   00000 00000   00000 00000   00000 00000     00000 00000   00000 00000   00000 00000   00000 00000   00000 00000
8421:   00000 00000   00000 00000   00000 00000   00000 00000   00000 00000     00000 00000   00000 00000   00000 00000   00000 00000   00000 00000
8422:   00000 00000   00000 00000   00000 00000   00000 00000   00000 00000     00000 00000   00000 00000   00000 00000   00000 00000   00000 00000
8423:   00000 00000   00000 00000   00000 00000   00000 00000   00000 00000     00000 00000   00000 00000   00000 00000   00000 00000   00000 00000
8424:   00000 00000   00000 00000   00000 00000   00000 00000   00000 00000     00000 00000   00000 00000   00000 00000   00000 00000   00000 00000
8425:   00000 00000   00000 00000   00000 00000   00000 00000   00000 00000     00000 00000   00000 00000   00000 00000   00000 00000   00000 00000
8426:   00000 00000   00000 00000   00000 00000   00000 00000   00000 00000     00000 00000   00000 00000   00000 00000   00000 00000   00000 00000
8427:   00000 00000   00000 00000   00000 00000   00000 00000   00000 00000     00000 00000   00000 00000   00000 00000   00000 00000   00000 00000
8428:   00000 00000   00000 00000   00000 00000   00000 00000   00000 00000     00000 00000   00000 00000   00000 00000   00000 00000   00000 00000
8429:   00000 00000   00000 00000   00000 00000   00000 00000   00000 00000     00000 00000   00000 00000   00000 00000   00000 00000   00000 00000
8430:   00000 00000   00000 00000   00000 00000   00000 00000   00000 00000     00000 00000   00000 00000   00000 00000   00000 00000   00000 00000
8431:   00000 00000   00000 00000   00000 00000   00000 00000   00000 00000     00000 00000   00000 00000   00000 00000   00000 00000   00000 00000
8432:   00000 00000   00000 00000   00000 00000   00000 00000   00000 00000     00000 00000   00000 00000   00000 00000   00000 00000   00000 00000
8433:   00000 00000   00000 00000   00000 00000   00000 00000   00000 00000     00000 00000   00000 00000   00000 00000   00000 00000   00000 00000
8434:   00000 00000   00000 00000   00000 00000   00000 00000   00000 00000     00000 00000   00000 00000   00000 00000   00000 00000   00000 00000
8435:   00000 00000   00000 00000   00000 00000   00000 00000   00000 00000     00000 00000   00000 00000   00000 00000   00000 00000   00000 00000
8436:   00000 00000   00000 00000   00000 00000   00000 00000   00000 00000     00000 00000   00000 00000   00000 00000   00000 00000   00000 00000
8437:   00000 00000   00000 00000   00000 00000   00000 00000   00000 00000     00000 00000   00000 00000   00000 00000   00000 00000   00000 00000
8438:   00000 00000   00000 00000   00000 00000   00000 00000   00000 00000     00000 00000   00000 00000   00000 00000   00000 00000   00000 00000
8439:   00000 00000   00000 00000   00000 00000   00000 00000   00000 00000     00000 00000   00000 00000   00000 00000   00000 00000   00000 00000
8440:   00000 00000   00000 00000   00000 00000   00000 00000   00000 00000     00000 00000   00000 00000   00000 00000   00000 00000   00000 00000
8441:   00000 00000   00000 00000   00000 00000   00000 00000   00000 00000     00000 00000   00000 00000   00000 00000   00000 00000   00000 00000
8442:   00000 00000   00000 00000   00000 00000   00000 00000   00000 00000     00000 00000   00000 00000   00000 00000   00000 00000   00000 00000
8443:   00000 00000   00000 00000   00000 00000   00000 00000   00000 00000     00000 00000   00000 00000   00000 00000   00000 00000   00000 00000
8444:   00000 00000   00000 00000   00000 00000   00000 00000   00000 00000     00000 00000   00000 00000   00000 00000   00000 00000   00000 00000
8445:   00000 00000   00000 00000   00000 00000   00000 00000   00000 00000     00000 00000   00000 00000   00000 00000   00000 00000   00000 00000
8446:   00000 00000   00000 00000   00000 00000   00000 00000   00000 00000     00000 00000   00000 00000   00000 00000   00000 00000   00000 00000
8447:   00000 00000   00000 00000   00000 00000   00000 00000   00000 00000     00000 00000   00000 00000   00000 00000   00000 00000   00000 00000
8448:   00000 00000   00000 00000   00000 00000   00000 00000   00000 00000     00000 00000   00000 00000   00000 00000   00000 00000   00000 00000
8449:   00000 00000   00000 00000   00000 00000   00000 00000   00000 00000     00000 00000   00000 00000   00000 00000   00000 00000   00000 00000
```

```
8450:    00000 00000   00000 00000   00000 00000   00000 00000   00000 00000      00000 00000   00000 00000   00000 00000   00000 00000   00000 00000
8451:    00000 00000   00000 00000   00000 00000   00000 00000   00000 00000      00000 00000   00000 00000   00000 00000   00000 00000   00000 00000
8452:    00000 00000   00000 00000   00000 00000   00000 00000   00000 00000      00000 00000   00000 00000   00000 00000   00000 00000   00000 00000
8453:    00000 00000   00000 00000   00000 00000   00000 00000   00000 00000      00000 00000   00000 00000   00000 00000   00000 00000   00000 00000
8454:    00000 00000   00000 00000   00000 00000   00000 00000   00000 00000      00000 00000   00000 00000   00000 00000   00000 00000   00000 00000
8455:    00000 00000   00000 00000   00000 00000   00000 00000   00000 00000      00000 00000   00000 00000   00000 00000   00000 00000   00000 00000
8456:    00000 00000   00000 00000   00000 00000   00000 00000   00000 00000      00000 00000   00000 00000   00000 00000   00000 00000   00000 00000
8457:    00000 00000   00000 00000   00000 00000   00000 00000   00000 00000      00000 00000   00000 00000   00000 00000   00000 00000   00000 00000
8458:    00000 00000   00000 00000   00000 00000   00000 00000   00000 00000      00000 00000   00000 00000   00000 00000   00000 00000   00000 00000
8459:    00000 00000   00000 00000   00000 00000   00000 00000   00000 00000      00000 00000   00000 00000   00000 00000   00000 00000   00000 00000
8460:    00000 00000   00000 00000   00000 00000   00000 00000   00000 00000      00000 00000   00000 00000   00000 00000   00000 00000   00000 00000
8461:    00000 00000   00000 00000   00000 00000   00000 00000   00000 00000      00000 00000   00000 00000   00000 00000   00000 00000   00000 00000
8462:    00000 00000   00000 00000   00000 00000   00000 00000   00000 00000      00000 00000   00000 00000   00000 00000   00000 00000   00000 00000
8463:    00000 00000   00000 00000   00000 00000   00000 00000   00000 00000      00000 00000   00000 00000   00000 00000   00000 00000   00000 00000
8464:    00000 00000   00000 00000   00000 00000   00000 00000   00000 00000      00000 00000   00000 00000   00000 00000   00000 00000   00000 00000
8465:    00000 00000   00000 00000   00000 00000   00000 00000   00000 00000      00000 00000   00000 00000   00000 00000   00000 00000   00000 00000
8466:    00000 00000   00000 00000   00000 00000   00000 00000   00000 00000      00000 00000   00000 00000   00000 00000   00000 00000   00000 00000
8467:    00000 00000   00000 00000   00000 00000   00000 00000   00000 00000      00000 00000   00000 00000   00000 00000   00000 00000   00000 00000
8468:    00000 00000   00000 00000   00000 00000   00000 00000   00000 00000      00000 00000   00000 00000   00000 00000   00000 00000   00000 00000
8469:    00000 00000   00000 00000   00000 00000   00000 00000   00000 00000      00000 00000   00000 00000   00000 00000   00000 00000   00000 00000
8470:    00000 00000   00000 00000   00000 00000   00000 00000   00000 00000      00000 00000   00000 00000   00000 00000   00000 00000   00000 00000
8471:    00000 00000   00000 00000   00000 00000   00000 00000   00000 00000      00000 00000   00000 00000   00000 00000   00000 00000   00000 00000
8472:    00000 00000   00000 00000   00000 00000   00000 00000   00000 00000      00000 00000   00000 00000   00000 00000   00000 00000   00000 00000
8473:    00000 00000   00000 00000   00000 00000   00000 00000   00000 00000      00000 00000   00000 00000   00000 00000   00000 00000   00000 00000
8474:    00000 00000   00000 00000   00000 00000   00000 00000   00000 00000      00000 00000   00000 00000   00000 00000   00000 00000   00000 00000
8475:    00000 00000   00000 00000   00000 00000   00000 00000   00000 00000      00000 00000   00000 00000   00000 00000   00000 00000   00000 00000
8476:    00000 00000   00000 00000   00000 00000   00000 00000   00000 00000      00000 00000   00000 00000   00000 00000   00000 00000   00000 00000
8477:    00000 00000   00000 00000   00000 00000   00000 00000   00000 00000      00000 00000   00000 00000   00000 00000   00000 00000   00000 00000
8478:    00000 00000   00000 00000   00000 00000   00000 00000   00000 00000      00000 00000   00000 00000   00000 00000   00000 00000   00000 00000
8479:    00000 00000   00000 00000   00000 00000   00000 00000   00000 00000      00000 00000   00000 00000   00000 00000   00000 00000   00000 00000
8480:    00000 00000   00000 00000   00000 00000   00000 00000   00000 00000      00000 00000   00000 00000   00000 00000   00000 00000   00000 00000
8481:    00000 00000   00000 00000   00000 00000   00000 00000   00000 00000      00000 00000   00000 00000   00000 00000   00000 00000   00000 00000
8482:    00000 00000   00000 00000   00000 00000   00000 00000   00000 00000      00000 00000   00000 00000   00000 00000   00000 00000   00000 00000
8483:    00000 00000   00000 00000   00000 00000   00000 00000   00000 00000      00000 00000   00000 00000   00000 00000   00000 00000   00000 00000
8484:    00000 00000   00000 00000   00000 00000   00000 00000   00000 00000      00000 00000   00000 00000   00000 00000   00000 00000   00000 00000
8485:    00000 00000   00000 00000   00000 00000   00000 00000   00000 00000      00000 00000   00000 00000   00000 00000   00000 00000   00000 00000
8486:    00000 00000   00000 00000   00000 00000   00000 00000   00000 00000      00000 00000   00000 00000   00000 00000   00000 00000   00000 00000
8487:    00000 00000   00000 00000   00000 00000   00000 00000   00000 00000      00000 00000   00000 00000   00000 00000   00000 00000   00000 00000
8488:    00000 00000   00000 00000   00000 00000   00000 00000   00000 00000      00000 00000   00000 00000   00000 00000   00000 00000   00000 00000
8489:    00000 00000   00000 00000   00000 00000   00000 00000   00000 00000      00000 00000   00000 00000   00000 00000   00000 00000   00000 00000
8490:    00000 00000   00000 00000   00000 00000   00000 00000   00000 00000      00000 00000   00000 00000   00000 00000   00000 00000   00000 00000
8491:    00000 00000   00000 00000   00000 00000   00000 00000   00000 00000      00000 00000   00000 00000   00000 00000   00000 00000   00000 00000
8492:    00000 00000   00000 00000   00000 00000   00000 00000   00000 00000      00000 00000   00000 00000   00000 00000   00000 00000   00000 00000
8493:    00000 00000   00000 00000   00000 00000   00000 00000   00000 00000      00000 00000   00000 00000   00000 00000   00000 00000   00000 00000
8494:    00000 00000   00000 00000   00000 00000   00000 00000   00000 00000      00000 00000   00000 00000   00000 00000   00000 00000   00000 00000
8495:    00000 00000   00000 00000   00000 00000   00000 00000   00000 00000      00000 00000   00000 00000   00000 00000   00000 00000   00000 00000
8496:    00000 00000   00000 00000   00000 00000   00000 00000   00000 00000      00000 00000   00000 00000   00000 00000   00000 00000   00000 00000
8497:    00000 00000   00000 00000   00000 00000   00000 00000   00000 00000      00000 00000   00000 00000   00000 00000   00000 00000   00000 00000
8498:    00000 00000   00000 00000   00000 00000   00000 00000   00000 00000      00000 00000   00000 00000   00000 00000   00000 00000   00000 00000
8499:    00000 00000   00000 00000   00000 00000   00000 00000   00000 00000      00000 00000   00000 00000   00000 00000   00000 00000   00000 00000
```

```
8500:  00000 00000  00000 00000  00000 00000  00000 00000  00000 00000    00000 00000  00000 00000  00000 00000  00000 00000  00000 00000
8501:  00000 00000  00000 00000  00000 00000  00000 00000  00000 00000    00000 00000  00000 00000  00000 00000  00000 00000  00000 00000
8502:  00000 00000  00000 00000  00000 00000  00000 00000  00000 00000    00000 00000  00000 00000  00000 00000  00000 00000  00000 00000
8503:  00000 00000  00000 00000  00000 00000  00000 00000  00000 00000    00000 00000  00000 00000  00000 00000  00000 00000  00000 00000
8504:  00000 00000  00000 00000  00000 00000  00000 00000  00000 00000    00000 00000  00000 00000  00000 00000  00000 00000  00000 00000
8505:  00000 00000  00000 00000  00000 00000  00000 00000  00000 00000    00000 00000  00000 00000  00000 00000  00000 00000  00000 00000
8506:  00000 00000  00000 00000  00000 00000  00000 00000  00000 00000    00000 00000  00000 00000  00000 00000  00000 00000  00000 00000
8507:  00000 00000  00000 00000  00000 00000  00000 00000  00000 00000    00000 00000  00000 00000  00000 00000  00000 00000  00000 00000
8508:  00000 00000  00000 00000  00000 00000  00000 00000  00000 00000    00000 00000  00000 00000  00000 00000  00000 00000  00000 00000
8509:  00000 00000  00000 00000  00000 00000  00000 00000  00000 00000    00000 00000  00000 00000  00000 00000  00000 00000  00000 00000
8510:  00000 00000  00000 00000  00000 00000  00000 00000  00000 00000    00000 00000  00000 00000  00000 00000  00000 00000  00000 00000
8511:  00000 00000  00000 00000  00000 00000  00000 00000  00000 00000    00000 00000  00000 00000  00000 00000  00000 00000  00000 00000
8512:  00000 00000  00000 00000  00000 00000  00000 00000  00000 00000    00000 00000  00000 00000  00000 00000  00000 00000  00000 00000
8513:  00000 00000  00000 00000  00000 00000  00000 00000  00000 00000    00000 00000  00000 00000  00000 00000  00000 00000  00000 00000
8514:  00000 00000  00000 00000  00000 00000  00000 00000  00000 00000    00000 00000  00000 00000  00000 00000  00000 00000  00000 00000
8515:  00000 00000  00000 00000  00000 00000  00000 00000  00000 00000    00000 00000  00000 00000  00000 00000  00000 00000  00000 00000
8516:  00000 00000  00000 00000  00000 00000  00000 00000  00000 00000    00000 00000  00000 00000  00000 00000  00000 00000  00000 00000
8517:  00000 00000  00000 00000  00000 00000  00000 00000  00000 00000    00000 00000  00000 00000  00000 00000  00000 00000  00000 00000
8518:  00000 00000  00000 00000  00000 00000  00000 00000  00000 00000    00000 00000  00000 00000  00000 00000  00000 00000  00000 00000
8519:  00000 00000  00000 00000  00000 00000  00000 00000  00000 00000    00000 00000  00000 00000  00000 00000  00000 00000  00000 00000
8520:  00000 00000  00000 00000  00000 00000  00000 00000  00000 00000    00000 00000  00000 00000  00000 00000  00000 00000  00000 00000
8521:  00000 00000  00000 00000  00000 00000  00000 00000  00000 00000    00000 00000  00000 00000  00000 00000  00000 00000  00000 00000
8522:  00000 00000  00000 00000  00000 00000  00000 00000  00000 00000    00000 00000  00000 00000  00000 00000  00000 00000  00000 00000
8523:  00000 00000  00000 00000  00000 00000  00000 00000  00000 00000    00000 00000  00000 00000  00000 00000  00000 00000  00000 00000
8524:  00000 00000  00000 00000  00000 00000  00000 00000  00000 00000    00000 00000  00000 00000  00000 00000  00000 00000  00000 00000
8525:  00000 00000  00000 00000  00000 00000  00000 00000  00000 00000    00000 00000  00000 00000  00000 00000  00000 00000  00000 00000
8526:  00000 00000  00000 00000  00000 00000  00000 00000  00000 00000    00000 00000  00000 00000  00000 00000  00000 00000  00000 00000
8527:  00000 00000  00000 00000  00000 00000  00000 00000  00000 00000    00000 00000  00000 00000  00000 00000  00000 00000  00000 00000
8528:  00000 00000  00000 00000  00000 00000  00000 00000  00000 00000    00000 00000  00000 00000  00000 00000  00000 00000  00000 00000
8529:  00000 00000  00000 00000  00000 00000  00000 00000  00000 00000    00000 00000  00000 00000  00000 00000  00000 00000  00000 00000
8530:  00000 00000  00000 00000  00000 00000  00000 00000  00000 00000    00000 00000  00000 00000  00000 00000  00000 00000  00000 00000
8531:  00000 00000  00000 00000  00000 00000  00000 00000  00000 00000    00000 00000  00000 00000  00000 00000  00000 00000  00000 00000
8532:  00000 00000  00000 00000  00000 00000  00000 00000  00000 00000    00000 00000  00000 00000  00000 00000  00000 00000  00000 00000
8533:  00000 00000  00000 00000  00000 00000  00000 00000  00000 00000    00000 00000  00000 00000  00000 00000  00000 00000  00000 00000
8534:  00000 00000  00000 00000  00000 00000  00000 00000  00000 00000    00000 00000  00000 00000  00000 00000  00000 00000  00000 00000
8535:  00000 00000  00000 00000  00000 00000  00000 00000  00000 00000    00000 00000  00000 00000  00000 00000  00000 00000  00000 00000
8536:  00000 00000  00000 00000  00000 00000  00000 00000  00000 00000    00000 00000  00000 00000  00000 00000  00000 00000  00000 00000
8537:  00000 00000  00000 00000  00000 00000  00000 00000  00000 00000    00000 00000  00000 00000  00000 00000  00000 00000  00000 00000
8538:  00000 00000  00000 00000  00000 00000  00000 00000  00000 00000    00000 00000  00000 00000  00000 00000  00000 00000  00000 00000
8539:  00000 00000  00000 00000  00000 00000  00000 00000  00000 00000    00000 00000  00000 00000  00000 00000  00000 00000  00000 00000
8540:  00000 00000  00000 00000  00000 00000  00000 00000  00000 00000    00000 00000  00000 00000  00000 00000  00000 00000  00000 00000
8541:  00000 00000  00000 00000  00000 00000  00000 00000  00000 00000    00000 00000  00000 00000  00000 00000  00000 00000  00000 00000
8542:  00000 00000  00000 00000  00000 00000  00000 00000  00000 00000    00000 00000  00000 00000  00000 00000  00000 00000  00000 00000
8543:  00000 00000  00000 00000  00000 00000  00000 00000  00000 00000    00000 00000  00000 00000  00000 00000  00000 00000  00000 00000
8544:  00000 00000  00000 00000  00000 00000  00000 00000  00000 00000    00000 00000  00000 00000  00000 00000  00000 00000  00000 00000
8545:  00000 00000  00000 00000  00000 00000  00000 00000  00000 00000    00000 00000  00000 00000  00000 00000  00000 00000  00000 00000
8546:  00000 00000  00000 00000  00000 00000  00000 00000  00000 00000    00000 00000  00000 00000  00000 00000  00000 00000  00000 00000
8547:  00000 00000  00000 00000  00000 00000  00000 00000  00000 00000    00000 00000  00000 00000  00000 00000  00000 00000  00000 00000
8548:  00000 00000  00000 00000  00000 00000  00000 00000  00000 00000    00000 00000  00000 00000  00000 00000  00000 00000  00000 00000
8549:  00000 00000  00000 00000  00000 00000  00000 00000  00000 00000    00000 00000  00000 00000  00000 00000  00000 00000  00000 00000
```

```
8550:  00000 00000  00000 00000  00000 00000  00000 00000  00000 00000    00000 00000  00000 00000  00000 00000  00000 00000  00000 00000
8551:  00000 00000  00000 00000  00000 00000  00000 00000  00000 00000    00000 00000  00000 00000  00000 00000  00000 00000  00000 00000
8552:  00000 00000  00000 00000  00000 00000  00000 00000  00000 00000    00000 00000  00000 00000  00000 00000  00000 00000  00000 00000
8553:  00000 00000  00000 00000  00000 00000  00000 00000  00000 00000    00000 00000  00000 00000  00000 00000  00000 00000  00000 00000
8554:  00000 00000  00000 00000  00000 00000  00000 00000  00000 00000    00000 00000  00000 00000  00000 00000  00000 00000  00000 00000
8555:  00000 00000  00000 00000  00000 00000  00000 00000  00000 00000    00000 00000  00000 00000  00000 00000  00000 00000  00000 00000
8556:  00000 00000  00000 00000  00000 00000  00000 00000  00000 00000    00000 00000  00000 00000  00000 00000  00000 00000  00000 00000
8557:  00000 00000  00000 00000  00000 00000  00000 00000  00000 00000    00000 00000  00000 00000  00000 00000  00000 00000  00000 00000
8558:  00000 00000  00000 00000  00000 00000  00000 00000  00000 00000    00000 00000  00000 00000  00000 00000  00000 00000  00000 00000
8559:  00000 00000  00000 00000  00000 00000  00000 00000  00000 00000    00000 00000  00000 00000  00000 00000  00000 00000  00000 00000
8560:  00000 00000  00000 00000  00000 00000  00000 00000  00000 00000    00000 00000  00000 00000  00000 00000  00000 00000  00000 00000
8561:  00000 00000  00000 00000  00000 00000  00000 00000  00000 00000    00000 00000  00000 00000  00000 00000  00000 00000  00000 00000
8562:  00000 00000  00000 00000  00000 00000  00000 00000  00000 00000    00000 00000  00000 00000  00000 00000  00000 00000  00000 00000
8563:  00000 00000  00000 00000  00000 00000  00000 00000  00000 00000    00000 00000  00000 00000  00000 00000  00000 00000  00000 00000
8564:  00000 00000  00000 00000  00000 00000  00000 00000  00000 00000    00000 00000  00000 00000  00000 00000  00000 00000  00000 00000
8565:  00000 00000  00000 00000  00000 00000  00000 00000  00000 00000    00000 00000  00000 00000  00000 00000  00000 00000  00000 00000
8566:  00000 00000  00000 00000  00000 00000  00000 00000  00000 00000    00000 00000  00000 00000  00000 00000  00000 00000  00000 00000
8567:  00000 00000  00000 00000  00000 00000  00000 00000  00000 00000    00000 00000  00000 00000  00000 00000  00000 00000  00000 00000
8568:  00000 00000  00000 00000  00000 00000  00000 00000  00000 00000    00000 00000  00000 00000  00000 00000  00000 00000  00000 00000
8569:  00000 00000  00000 00000  00000 00000  00000 00000  00000 00000    00000 00000  00000 00000  00000 00000  00000 00000  00000 00000
8570:  00000 00000  00000 00000  00000 00000  00000 00000  00000 00000    00000 00000  00000 00000  00000 00000  00000 00000  00000 00000
8571:  00000 00000  00000 00000  00000 00000  00000 00000  00000 00000    00000 00000  00000 00000  00000 00000  00000 00000  00000 00000
8572:  00000 00000  00000 00000  00000 00000  00000 00000  00000 00000    00000 00000  00000 00000  00000 00000  00000 00000  00000 00000
8573:  00000 00000  00000 00000  00000 00000  00000 00000  00000 00000    00000 00000  00000 00000  00000 00000  00000 00000  00000 00000
8574:  00000 00000  00000 00000  00000 00000  00000 00000  00000 00000    00000 00000  00000 00000  00000 00000  00000 00000  00000 00000
8575:  00000 00000  00000 00000  00000 00000  00000 00000  00000 00000    00000 00000  00000 00000  00000 00000  00000 00000  00000 00000
8576:  00000 00000  00000 00000  00000 00000  00000 00000  00000 00000    00000 00000  00000 00000  00000 00000  00000 00000  00000 00000
8577:  00000 00000  00000 00000  00000 00000  00000 00000  00000 00000    00000 00000  00000 00000  00000 00000  00000 00000  00000 00000
8578:  00000 00000  00000 00000  00000 00000  00000 00000  00000 00000    00000 00000  00000 00000  00000 00000  00000 00000  00000 00000
8579:  00000 00000  00000 00000  00000 00000  00000 00000  00000 00000    00000 00000  00000 00000  00000 00000  00000 00000  00000 00000
8580:  00000 00000  00000 00000  00000 00000  00000 00000  00000 00000    00000 00000  00000 00000  00000 00000  00000 00000  00000 00000
8581:  00000 00000  00000 00000  00000 00000  00000 00000  00000 00000    00000 00000  00000 00000  00000 00000  00000 00000  00000 00000
8582:  00000 00000  00000 00000  00000 00000  00000 00000  00000 00000    00000 00000  00000 00000  00000 00000  00000 00000  00000 00000
8583:  00000 00000  00000 00000  00000 00000  00000 00000  00000 00000    00000 00000  00000 00000  00000 00000  00000 00000  00000 00000
8584:  00000 00000  00000 00000  00000 00000  00000 00000  00000 00000    00000 00000  00000 00000  00000 00000  00000 00000  00000 00000
8585:  00000 00000  00000 00000  00000 00000  00000 00000  00000 00000    00000 00000  00000 00000  00000 00000  00000 00000  00000 00000
8586:  00000 00000  00000 00000  00000 00000  00000 00000  00000 00000    00000 00000  00000 00000  00000 00000  00000 00000  00000 00000
8587:  00000 00000  00000 00000  00000 00000  00000 00000  00000 00000    00000 00000  00000 00000  00000 00000  00000 00000  00000 00000
8588:  00000 00000  00000 00000  00000 00000  00000 00000  00000 00000    00000 00000  00000 00000  00000 00000  00000 00000  00000 00000
8589:  00000 00000  00000 00000  00000 00000  00000 00000  00000 00000    00000 00000  00000 00000  00000 00000  00000 00000  00000 00000
8590:  00000 00000  00000 00000  00000 00000  00000 00000  00000 00000    00000 00000  00000 00000  00000 00000  00000 00000  00000 00000
8591:  00000 00000  00000 00000  00000 00000  00000 00000  00000 00000    00000 00000  00000 00000  00000 00000  00000 00000  00000 00000
8592:  00000 00000  00000 00000  00000 00000  00000 00000  00000 00000    00000 00000  00000 00000  00000 00000  00000 00000  00000 00000
8593:  00000 00000  00000 00000  00000 00000  00000 00000  00000 00000    00000 00000  00000 00000  00000 00000  00000 00000  00000 00000
8594:  00000 00000  00000 00000  00000 00000  00000 00000  00000 00000    00000 00000  00000 00000  00000 00000  00000 00000  00000 00000
8595:  00000 00000  00000 00000  00000 00000  00000 00000  00000 00000    00000 00000  00000 00000  00000 00000  00000 00000  00000 00000
8596:  00000 00000  00000 00000  00000 00000  00000 00000  00000 00000    00000 00000  00000 00000  00000 00000  00000 00000  00000 00000
8597:  00000 00000  00000 00000  00000 00000  00000 00000  00000 00000    00000 00000  00000 00000  00000 00000  00000 00000  00000 00000
8598:  00000 00000  00000 00000  00000 00000  00000 00000  00000 00000    00000 00000  00000 00000  00000 00000  00000 00000  00000 00000
8599:  00000 00000  00000 00000  00000 00000  00000 00000  00000 00000    00000 00000  00000 00000  00000 00000  00000 00000  00000 00000
```

```
8600:   00000 00000   00000 00000   00000 00000   00000 00000   00000 00000     00000 00000   00000 00000   00000 00000   00000 00000   00000 00000
8601:   00000 00000   00000 00000   00000 00000   00000 00000   00000 00000     00000 00000   00000 00000   00000 00000   00000 00000   00000 00000
8602:   00000 00000   00000 00000   00000 00000   00000 00000   00000 00000     00000 00000   00000 00000   00000 00000   00000 00000   00000 00000
8603:   00000 00000   00000 00000   00000 00000   00000 00000   00000 00000     00000 00000   00000 00000   00000 00000   00000 00000   00000 00000
8604:   00000 00000   00000 00000   00000 00000   00000 00000   00000 00000     00000 00000   00000 00000   00000 00000   00000 00000   00000 00000
8605:   00000 00000   00000 00000   00000 00000   00000 00000   00000 00000     00000 00000   00000 00000   00000 00000   00000 00000   00000 00000
8606:   00000 00000   00000 00000   00000 00000   00000 00000   00000 00000     00000 00000   00000 00000   00000 00000   00000 00000   00000 00000
8607:   00000 00000   00000 00000   00000 00000   00000 00000   00000 00000     00000 00000   00000 00000   00000 00000   00000 00000   00000 00000
8608:   00000 00000   00000 00000   00000 00000   00000 00000   00000 00000     00000 00000   00000 00000   00000 00000   00000 00000   00000 00000
8609:   00000 00000   00000 00000   00000 00000   00000 00000   00000 00000     00000 00000   00000 00000   00000 00000   00000 00000   00000 00000
8610:   00000 00000   00000 00000   00000 00000   00000 00000   00000 00000     00000 00000   00000 00000   00000 00000   00000 00000   00000 00000
8611:   00000 00000   00000 00000   00000 00000   00000 00000   00000 00000     00000 00000   00000 00000   00000 00000   00000 00000   00000 00000
8612:   00000 00000   00000 00000   00000 00000   00000 00000   00000 00000     00000 00000   00000 00000   00000 00000   00000 00000   00000 00000
8613:   00000 00000   00000 00000   00000 00000   00000 00000   00000 00000     00000 00000   00000 00000   00000 00000   00000 00000   00000 00000
8614:   00000 00000   00000 00000   00000 00000   00000 00000   00000 00000     00000 00000   00000 00000   00000 00000   00000 00000   00000 00000
8615:   00000 00000   00000 00000   00000 00000   00000 00000   00000 00000     00000 00000   00000 00000   00000 00000   00000 00000   00000 00000
8616:   00000 00000   00000 00000   00000 00000   00000 00000   00000 00000     00000 00000   00000 00000   00000 00000   00000 00000   00000 00000
8617:   00000 00000   00000 00000   00000 00000   00000 00000   00000 00000     00000 00000   00000 00000   00000 00000   00000 00000   00000 00000
8618:   00000 00000   00000 00000   00000 00000   00000 00000   00000 00000     00000 00000   00000 00000   00000 00000   00000 00000   00000 00000
8619:   00000 00000   00000 00000   00000 00000   00000 00000   00000 00000     00000 00000   00000 00000   00000 00000   00000 00000   00000 00000
8620:   00000 00000   00000 00000   00000 00000   00000 00000   00000 00000     00000 00000   00000 00000   00000 00000   00000 00000   00000 00000
8621:   00000 00000   00000 00000   00000 00000   00000 00000   00000 00000     00000 00000   00000 00000   00000 00000   00000 00000   00000 00000
8622:   00000 00000   00000 00000   00000 00000   00000 00000   00000 00000     00000 00000   00000 00000   00000 00000   00000 00000   00000 00000
8623:   00000 00000   00000 00000   00000 00000   00000 00000   00000 00000     00000 00000   00000 00000   00000 00000   00000 00000   00000 00000
8624:   00000 00000   00000 00000   00000 00000   00000 00000   00000 00000     00000 00000   00000 00000   00000 00000   00000 00000   00000 00000
8625:   00000 00000   00000 00000   00000 00000   00000 00000   00000 00000     00000 00000   00000 00000   00000 00000   00000 00000   00000 00000
8626:   00000 00000   00000 00000   00000 00000   00000 00000   00000 00000     00000 00000   00000 00000   00000 00000   00000 00000   00000 00000
8627:   00000 00000   00000 00000   00000 00000   00000 00000   00000 00000     00000 00000   00000 00000   00000 00000   00000 00000   00000 00000
8628:   00000 00000   00000 00000   00000 00000   00000 00000   00000 00000     00000 00000   00000 00000   00000 00000   00000 00000   00000 00000
8629:   00000 00000   00000 00000   00000 00000   00000 00000   00000 00000     00000 00000   00000 00000   00000 00000   00000 00000   00000 00000
8630:   00000 00000   00000 00000   00000 00000   00000 00000   00000 00000     00000 00000   00000 00000   00000 00000   00000 00000   00000 00000
8631:   00000 00000   00000 00000   00000 00000   00000 00000   00000 00000     00000 00000   00000 00000   00000 00000   00000 00000   00000 00000
8632:   00000 00000   00000 00000   00000 00000   00000 00000   00000 00000     00000 00000   00000 00000   00000 00000   00000 00000   00000 00000
8633:   00000 00000   00000 00000   00000 00000   00000 00000   00000 00000     00000 00000   00000 00000   00000 00000   00000 00000   00000 00000
8634:   00000 00000   00000 00000   00000 00000   00000 00000   00000 00000     00000 00000   00000 00000   00000 00000   00000 00000   00000 00000
8635:   00000 00000   00000 00000   00000 00000   00000 00000   00000 00000     00000 00000   00000 00000   00000 00000   00000 00000   00000 00000
8636:   00000 00000   00000 00000   00000 00000   00000 00000   00000 00000     00000 00000   00000 00000   00000 00000   00000 00000   00000 00000
8637:   00000 00000   00000 00000   00000 00000   00000 00000   00000 00000     00000 00000   00000 00000   00000 00000   00000 00000   00000 00000
8638:   00000 00000   00000 00000   00000 00000   00000 00000   00000 00000     00000 00000   00000 00000   00000 00000   00000 00000   00000 00000
8639:   00000 00000   00000 00000   00000 00000   00000 00000   00000 00000     00000 00000   00000 00000   00000 00000   00000 00000   00000 00000
8640:   00000 00000   00000 00000   00000 00000   00000 00000   00000 00000     00000 00000   00000 00000   00000 00000   00000 00000   00000 00000
8641:   00000 00000   00000 00000   00000 00000   00000 00000   00000 00000     00000 00000   00000 00000   00000 00000   00000 00000   00000 00000
8642:   00000 00000   00000 00000   00000 00000   00000 00000   00000 00000     00000 00000   00000 00000   00000 00000   00000 00000   00000 00000
8643:   00000 00000   00000 00000   00000 00000   00000 00000   00000 00000     00000 00000   00000 00000   00000 00000   00000 00000   00000 00000
8644:   00000 00000   00000 00000   00000 00000   00000 00000   00000 00000     00000 00000   00000 00000   00000 00000   00000 00000   00000 00000
8645:   00000 00000   00000 00000   00000 00000   00000 00000   00000 00000     00000 00000   00000 00000   00000 00000   00000 00000   00000 00000
8646:   00000 00000   00000 00000   00000 00000   00000 00000   00000 00000     00000 00000   00000 00000   00000 00000   00000 00000   00000 00000
8647:   00000 00000   00000 00000   00000 00000   00000 00000   00000 00000     00000 00000   00000 00000   00000 00000   00000 00000   00000 00000
8648:   00000 00000   00000 00000   00000 00000   00000 00000   00000 00000     00000 00000   00000 00000   00000 00000   00000 00000   00000 00000
8649:   00000 00000   00000 00000   00000 00000   00000 00000   00000 00000     00000 00000   00000 00000   00000 00000   00000 00000   00000 00000
```

```
8650:  00000 00000  00000 00000  00000 00000  00000 00000  00000 00000    00000 00000  00000 00000  00000 00000  00000 00000  00000 00000
8651:  00000 00000  00000 00000  00000 00000  00000 00000  00000 00000    00000 00000  00000 00000  00000 00000  00000 00000  00000 00000
8652:  00000 00000  00000 00000  00000 00000  00000 00000  00000 00000    00000 00000  00000 00000  00000 00000  00000 00000  00000 00000
8653:  00000 00000  00000 00000  00000 00000  00000 00000  00000 00000    00000 00000  00000 00000  00000 00000  00000 00000  00000 00000
8654:  00000 00000  00000 00000  00000 00000  00000 00000  00000 00000    00000 00000  00000 00000  00000 00000  00000 00000  00000 00000
8655:  00000 00000  00000 00000  00000 00000  00000 00000  00000 00000    00000 00000  00000 00000  00000 00000  00000 00000  00000 00000
8656:  00000 00000  00000 00000  00000 00000  00000 00000  00000 00000    00000 00000  00000 00000  00000 00000  00000 00000  00000 00000
8657:  00000 00000  00000 00000  00000 00000  00000 00000  00000 00000    00000 00000  00000 00000  00000 00000  00000 00000  00000 00000
8658:  00000 00000  00000 00000  00000 00000  00000 00000  00000 00000    00000 00000  00000 00000  00000 00000  00000 00000  00000 00000
8659:  00000 00000  00000 00000  00000 00000  00000 00000  00000 00000    00000 00000  00000 00000  00000 00000  00000 00000  00000 00000
8660:  00000 00000  00000 00000  00000 00000  00000 00000  00000 00000    00000 00000  00000 00000  00000 00000  00000 00000  00000 00000
8661:  00000 00000  00000 00000  00000 00000  00000 00000  00000 00000    00000 00000  00000 00000  00000 00000  00000 00000  00000 00000
8662:  00000 00000  00000 00000  00000 00000  00000 00000  00000 00000    00000 00000  00000 00000  00000 00000  00000 00000  00000 00000
8663:  00000 00000  00000 00000  00000 00000  00000 00000  00000 00000    00000 00000  00000 00000  00000 00000  00000 00000  00000 00000
8664:  00000 00000  00000 00000  00000 00000  00000 00000  00000 00000    00000 00000  00000 00000  00000 00000  00000 00000  00000 00000
8665:  00000 00000  00000 00000  00000 00000  00000 00000  00000 00000    00000 00000  00000 00000  00000 00000  00000 00000  00000 00000
8666:  00000 00000  00000 00000  00000 00000  00000 00000  00000 00000    00000 00000  00000 00000  00000 00000  00000 00000  00000 00000
8667:  00000 00000  00000 00000  00000 00000  00000 00000  00000 00000    00000 00000  00000 00000  00000 00000  00000 00000  00000 00000
8668:  00000 00000  00000 00000  00000 00000  00000 00000  00000 00000    00000 00000  00000 00000  00000 00000  00000 00000  00000 00000
8669:  00000 00000  00000 00000  00000 00000  00000 00000  00000 00000    00000 00000  00000 00000  00000 00000  00000 00000  00000 00000
8670:  00000 00000  00000 00000  00000 00000  00000 00000  00000 00000    00000 00000  00000 00000  00000 00000  00000 00000  00000 00000
8671:  00000 00000  00000 00000  00000 00000  00000 00000  00000 00000    00000 00000  00000 00000  00000 00000  00000 00000  00000 00000
8672:  00000 00000  00000 00000  00000 00000  00000 00000  00000 00000    00000 00000  00000 00000  00000 00000  00000 00000  00000 00000
8673:  00000 00000  00000 00000  00000 00000  00000 00000  00000 00000    00000 00000  00000 00000  00000 00000  00000 00000  00000 00000
8674:  00000 00000  00000 00000  00000 00000  00000 00000  00000 00000    00000 00000  00000 00000  00000 00000  00000 00000  00000 00000
8675:  00000 00000  00000 00000  00000 00000  00000 00000  00000 00000    00000 00000  00000 00000  00000 00000  00000 00000  00000 00000
8676:  00000 00000  00000 00000  00000 00000  00000 00000  00000 00000    00000 00000  00000 00000  00000 00000  00000 00000  00000 00000
8677:  00000 00000  00000 00000  00000 00000  00000 00000  00000 00000    00000 00000  00000 00000  00000 00000  00000 00000  00000 00000
8678:  00000 00000  00000 00000  00000 00000  00000 00000  00000 00000    00000 00000  00000 00000  00000 00000  00000 00000  00000 00000
8679:  00000 00000  00000 00000  00000 00000  00000 00000  00000 00000    00000 00000  00000 00000  00000 00000  00000 00000  00000 00000
8680:  00000 00000  00000 00000  00000 00000  00000 00000  00000 00000    00000 00000  00000 00000  00000 00000  00000 00000  00000 00000
8681:  00000 00000  00000 00000  00000 00000  00000 00000  00000 00000    00000 00000  00000 00000  00000 00000  00000 00000  00000 00000
8682:  00000 00000  00000 00000  00000 00000  00000 00000  00000 00000    00000 00000  00000 00000  00000 00000  00000 00000  00000 00000
8683:  00000 00000  00000 00000  00000 00000  00000 00000  00000 00000    00000 00000  00000 00000  00000 00000  00000 00000  00000 00000
8684:  00000 00000  00000 00000  00000 00000  00000 00000  00000 00000    00000 00000  00000 00000  00000 00000  00000 00000  00000 00000
8685:  00000 00000  00000 00000  00000 00000  00000 00000  00000 00000    00000 00000  00000 00000  00000 00000  00000 00000  00000 00000
8686:  00000 00000  00000 00000  00000 00000  00000 00000  00000 00000    00000 00000  00000 00000  00000 00000  00000 00000  00000 00000
8687:  00000 00000  00000 00000  00000 00000  00000 00000  00000 00000    00000 00000  00000 00000  00000 00000  00000 00000  00000 00000
8688:  00000 00000  00000 00000  00000 00000  00000 00000  00000 00000    00000 00000  00000 00000  00000 00000  00000 00000  00000 00000
8689:  00000 00000  00000 00000  00000 00000  00000 00000  00000 00000    00000 00000  00000 00000  00000 00000  00000 00000  00000 00000
8690:  00000 00000  00000 00000  00000 00000  00000 00000  00000 00000    00000 00000  00000 00000  00000 00000  00000 00000  00000 00000
8691:  00000 00000  00000 00000  00000 00000  00000 00000  00000 00000    00000 00000  00000 00000  00000 00000  00000 00000  00000 00000
8692:  00000 00000  00000 00000  00000 00000  00000 00000  00000 00000    00000 00000  00000 00000  00000 00000  00000 00000  00000 00000
8693:  00000 00000  00000 00000  00000 00000  00000 00000  00000 00000    00000 00000  00000 00000  00000 00000  00000 00000  00000 00000
8694:  00000 00000  00000 00000  00000 00000  00000 00000  00000 00000    00000 00000  00000 00000  00000 00000  00000 00000  00000 00000
8695:  00000 00000  00000 00000  00000 00000  00000 00000  00000 00000    00000 00000  00000 00000  00000 00000  00000 00000  00000 00000
8696:  00000 00000  00000 00000  00000 00000  00000 00000  00000 00000    00000 00000  00000 00000  00000 00000  00000 00000  00000 00000
8697:  00000 00000  00000 00000  00000 00000  00000 00000  00000 00000    00000 00000  00000 00000  00000 00000  00000 00000  00000 00000
8698:  00000 00000  00000 00000  00000 00000  00000 00000  00000 00000    00000 00000  00000 00000  00000 00000  00000 00000  00000 00000
8699:  00000 00000  00000 00000  00000 00000  00000 00000  00000 00000    00000 00000  00000 00000  00000 00000  00000 00000  00000 00000
```

```
8700:  00000 00000  00000 00000  00000 00000  00000 00000  00000 00000    00000 00000  00000 00000  00000 00000  00000 00000  00000 00000
8701:  00000 00000  00000 00000  00000 00000  00000 00000  00000 00000    00000 00000  00000 00000  00000 00000  00000 00000  00000 00000
8702:  00000 00000  00000 00000  00000 00000  00000 00000  00000 00000    00000 00000  00000 00000  00000 00000  00000 00000  00000 00000
8703:  00000 00000  00000 00000  00000 00000  00000 00000  00000 00000    00000 00000  00000 00000  00000 00000  00000 00000  00000 00000
8704:  00000 00000  00000 00000  00000 00000  00000 00000  00000 00000    00000 00000  00000 00000  00000 00000  00000 00000  00000 00000
8705:  00000 00000  00000 00000  00000 00000  00000 00000  00000 00000    00000 00000  00000 00000  00000 00000  00000 00000  00000 00000
8706:  00000 00000  00000 00000  00000 00000  00000 00000  00000 00000    00000 00000  00000 00000  00000 00000  00000 00000  00000 00000
8707:  00000 00000  00000 00000  00000 00000  00000 00000  00000 00000    00000 00000  00000 00000  00000 00000  00000 00000  00000 00000
8708:  00000 00000  00000 00000  00000 00000  00000 00000  00000 00000    00000 00000  00000 00000  00000 00000  00000 00000  00000 00000
8709:  00000 00000  00000 00000  00000 00000  00000 00000  00000 00000    00000 00000  00000 00000  00000 00000  00000 00000  00000 00000
8710:  00000 00000  00000 00000  00000 00000  00000 00000  00000 00000    00000 00000  00000 00000  00000 00000  00000 00000  00000 00000
8711:  00000 00000  00000 00000  00000 00000  00000 00000  00000 00000    00000 00000  00000 00000  00000 00000  00000 00000  00000 00000
8712:  00000 00000  00000 00000  00000 00000  00000 00000  00000 00000    00000 00000  00000 00000  00000 00000  00000 00000  00000 00000
8713:  00000 00000  00000 00000  00000 00000  00000 00000  00000 00000    00000 00000  00000 00000  00000 00000  00000 00000  00000 00000
8714:  00000 00000  00000 00000  00000 00000  00000 00000  00000 00000    00000 00000  00000 00000  00000 00000  00000 00000  00000 00000
8715:  00000 00000  00000 00000  00000 00000  00000 00000  00000 00000    00000 00000  00000 00000  00000 00000  00000 00000  00000 00000
8716:  00000 00000  00000 00000  00000 00000  00000 00000  00000 00000    00000 00000  00000 00000  00000 00000  00000 00000  00000 00000
8717:  00000 00000  00000 00000  00000 00000  00000 00000  00000 00000    00000 00000  00000 00000  00000 00000  00000 00000  00000 00000
8718:  00000 00000  00000 00000  00000 00000  00000 00000  00000 00000    00000 00000  00000 00000  00000 00000  00000 00000  00000 00000
8719:  00000 00000  00000 00000  00000 00000  00000 00000  00000 00000    00000 00000  00000 00000  00000 00000  00000 00000  00000 00000
8720:  00000 00000  00000 00000  00000 00000  00000 00000  00000 00000    00000 00000  00000 00000  00000 00000  00000 00000  00000 00000
8721:  00000 00000  00000 00000  00000 00000  00000 00000  00000 00000    00000 00000  00000 00000  00000 00000  00000 00000  00000 00000
8722:  00000 00000  00000 00000  00000 00000  00000 00000  00000 00000    00000 00000  00000 00000  00000 00000  00000 00000  00000 00000
8723:  00000 00000  00000 00000  00000 00000  00000 00000  00000 00000    00000 00000  00000 00000  00000 00000  00000 00000  00000 00000
8724:  00000 00000  00000 00000  00000 00000  00000 00000  00000 00000    00000 00000  00000 00000  00000 00000  00000 00000  00000 00000
8725:  00000 00000  00000 00000  00000 00000  00000 00000  00000 00000    00000 00000  00000 00000  00000 00000  00000 00000  00000 00000
8726:  00000 00000  00000 00000  00000 00000  00000 00000  00000 00000    00000 00000  00000 00000  00000 00000  00000 00000  00000 00000
8727:  00000 00000  00000 00000  00000 00000  00000 00000  00000 00000    00000 00000  00000 00000  00000 00000  00000 00000  00000 00000
8728:  00000 00000  00000 00000  00000 00000  00000 00000  00000 00000    00000 00000  00000 00000  00000 00000  00000 00000  00000 00000
8729:  00000 00000  00000 00000  00000 00000  00000 00000  00000 00000    00000 00000  00000 00000  00000 00000  00000 00000  00000 00000
8730:  00000 00000  00000 00000  00000 00000  00000 00000  00000 00000    00000 00000  00000 00000  00000 00000  00000 00000  00000 00000
8731:  00000 00000  00000 00000  00000 00000  00000 00000  00000 00000    00000 00000  00000 00000  00000 00000  00000 00000  00000 00000
8732:  00000 00000  00000 00000  00000 00000  00000 00000  00000 00000    00000 00000  00000 00000  00000 00000  00000 00000  00000 00000
8733:  00000 00000  00000 00000  00000 00000  00000 00000  00000 00000    00000 00000  00000 00000  00000 00000  00000 00000  00000 00000
8734:  00000 00000  00000 00000  00000 00000  00000 00000  00000 00000    00000 00000  00000 00000  00000 00000  00000 00000  00000 00000
8735:  00000 00000  00000 00000  00000 00000  00000 00000  00000 00000    00000 00000  00000 00000  00000 00000  00000 00000  00000 00000
8736:  00000 00000  00000 00000  00000 00000  00000 00000  00000 00000    00000 00000  00000 00000  00000 00000  00000 00000  00000 00000
8737:  00000 00000  00000 00000  00000 00000  00000 00000  00000 00000    00000 00000  00000 00000  00000 00000  00000 00000  00000 00000
8738:  00000 00000  00000 00000  00000 00000  00000 00000  00000 00000    00000 00000  00000 00000  00000 00000  00000 00000  00000 00000
8739:  00000 00000  00000 00000  00000 00000  00000 00000  00000 00000    00000 00000  00000 00000  00000 00000  00000 00000  00000 00000
8740:  00000 00000  00000 00000  00000 00000  00000 00000  00000 00000    00000 00000  00000 00000  00000 00000  00000 00000  00000 00000
8741:  00000 00000  00000 00000  00000 00000  00000 00000  00000 00000    00000 00000  00000 00000  00000 00000  00000 00000  00000 00000
8742:  00000 00000  00000 00000  00000 00000  00000 00000  00000 00000    00000 00000  00000 00000  00000 00000  00000 00000  00000 00000
8743:  00000 00000  00000 00000  00000 00000  00000 00000  00000 00000    00000 00000  00000 00000  00000 00000  00000 00000  00000 00000
8744:  00000 00000  00000 00000  00000 00000  00000 00000  00000 00000    00000 00000  00000 00000  00000 00000  00000 00000  00000 00000
8745:  00000 00000  00000 00000  00000 00000  00000 00000  00000 00000    00000 00000  00000 00000  00000 00000  00000 00000  00000 00000
8746:  00000 00000  00000 00000  00000 00000  00000 00000  00000 00000    00000 00000  00000 00000  00000 00000  00000 00000  00000 00000
8747:  00000 00000  00000 00000  00000 00000  00000 00000  00000 00000    00000 00000  00000 00000  00000 00000  00000 00000  00000 00000
8748:  00000 00000  00000 00000  00000 00000  00000 00000  00000 00000    00000 00000  00000 00000  00000 00000  00000 00000  00000 00000
8749:  00000 00000  00000 00000  00000 00000  00000 00000  00000 00000    00000 00000  00000 00000  00000 00000  00000 00000  00000 00000
```

```
8750:  00000 00000  00000 00000  00000 00000  00000 00000  00000 00000    00000 00000  00000 00000  00000 00000  00000 00000  00000 00000
8751:  00000 00000  00000 00000  00000 00000  00000 00000  00000 00000    00000 00000  00000 00000  00000 00000  00000 00000  00000 00000
8752:  00000 00000  00000 00000  00000 00000  00000 00000  00000 00000    00000 00000  00000 00000  00000 00000  00000 00000  00000 00000
8753:  00000 00000  00000 00000  00000 00000  00000 00000  00000 00000    00000 00000  00000 00000  00000 00000  00000 00000  00000 00000
8754:  00000 00000  00000 00000  00000 00000  00000 00000  00000 00000    00000 00000  00000 00000  00000 00000  00000 00000  00000 00000
8755:  00000 00000  00000 00000  00000 00000  00000 00000  00000 00000    00000 00000  00000 00000  00000 00000  00000 00000  00000 00000
8756:  00000 00000  00000 00000  00000 00000  00000 00000  00000 00000    00000 00000  00000 00000  00000 00000  00000 00000  00000 00000
8757:  00000 00000  00000 00000  00000 00000  00000 00000  00000 00000    00000 00000  00000 00000  00000 00000  00000 00000  00000 00000
8758:  00000 00000  00000 00000  00000 00000  00000 00000  00000 00000    00000 00000  00000 00000  00000 00000  00000 00000  00000 00000
8759:  00000 00000  00000 00000  00000 00000  00000 00000  00000 00000    00000 00000  00000 00000  00000 00000  00000 00000  00000 00000
8760:  00000 00000  00000 00000  00000 00000  00000 00000  00000 00000    00000 00000  00000 00000  00000 00000  00000 00000  00000 00000
8761:  00000 00000  00000 00000  00000 00000  00000 00000  00000 00000    00000 00000  00000 00000  00000 00000  00000 00000  00000 00000
8762:  00000 00000  00000 00000  00000 00000  00000 00000  00000 00000    00000 00000  00000 00000  00000 00000  00000 00000  00000 00000
8763:  00000 00000  00000 00000  00000 00000  00000 00000  00000 00000    00000 00000  00000 00000  00000 00000  00000 00000  00000 00000
8764:  00000 00000  00000 00000  00000 00000  00000 00000  00000 00000    00000 00000  00000 00000  00000 00000  00000 00000  00000 00000
8765:  00000 00000  00000 00000  00000 00000  00000 00000  00000 00000    00000 00000  00000 00000  00000 00000  00000 00000  00000 00000
8766:  00000 00000  00000 00000  00000 00000  00000 00000  00000 00000    00000 00000  00000 00000  00000 00000  00000 00000  00000 00000
8767:  00000 00000  00000 00000  00000 00000  00000 00000  00000 00000    00000 00000  00000 00000  00000 00000  00000 00000  00000 00000
8768:  00000 00000  00000 00000  00000 00000  00000 00000  00000 00000    00000 00000  00000 00000  00000 00000  00000 00000  00000 00000
8769:  00000 00000  00000 00000  00000 00000  00000 00000  00000 00000    00000 00000  00000 00000  00000 00000  00000 00000  00000 00000
8770:  00000 00000  00000 00000  00000 00000  00000 00000  00000 00000    00000 00000  00000 00000  00000 00000  00000 00000  00000 00000
8771:  00000 00000  00000 00000  00000 00000  00000 00000  00000 00000    00000 00000  00000 00000  00000 00000  00000 00000  00000 00000
8772:  00000 00000  00000 00000  00000 00000  00000 00000  00000 00000    00000 00000  00000 00000  00000 00000  00000 00000  00000 00000
8773:  00000 00000  00000 00000  00000 00000  00000 00000  00000 00000    00000 00000  00000 00000  00000 00000  00000 00000  00000 00000
8774:  00000 00000  00000 00000  00000 00000  00000 00000  00000 00000    00000 00000  00000 00000  00000 00000  00000 00000  00000 00000
8775:  00000 00000  00000 00000  00000 00000  00000 00000  00000 00000    00000 00000  00000 00000  00000 00000  00000 00000  00000 00000
8776:  00000 00000  00000 00000  00000 00000  00000 00000  00000 00000    00000 00000  00000 00000  00000 00000  00000 00000  00000 00000
8777:  00000 00000  00000 00000  00000 00000  00000 00000  00000 00000    00000 00000  00000 00000  00000 00000  00000 00000  00000 00000
8778:  00000 00000  00000 00000  00000 00000  00000 00000  00000 00000    00000 00000  00000 00000  00000 00000  00000 00000  00000 00000
8779:  00000 00000  00000 00000  00000 00000  00000 00000  00000 00000    00000 00000  00000 00000  00000 00000  00000 00000  00000 00000
8780:  00000 00000  00000 00000  00000 00000  00000 00000  00000 00000    00000 00000  00000 00000  00000 00000  00000 00000  00000 00000
8781:  00000 00000  00000 00000  00000 00000  00000 00000  00000 00000    00000 00000  00000 00000  00000 00000  00000 00000  00000 00000
8782:  00000 00000  00000 00000  00000 00000  00000 00000  00000 00000    00000 00000  00000 00000  00000 00000  00000 00000  00000 00000
8783:  00000 00000  00000 00000  00000 00000  00000 00000  00000 00000    00000 00000  00000 00000  00000 00000  00000 00000  00000 00000
8784:  00000 00000  00000 00000  00000 00000  00000 00000  00000 00000    00000 00000  00000 00000  00000 00000  00000 00000  00000 00000
8785:  00000 00000  00000 00000  00000 00000  00000 00000  00000 00000    00000 00000  00000 00000  00000 00000  00000 00000  00000 00000
8786:  00000 00000  00000 00000  00000 00000  00000 00000  00000 00000    00000 00000  00000 00000  00000 00000  00000 00000  00000 00000
8787:  00000 00000  00000 00000  00000 00000  00000 00000  00000 00000    00000 00000  00000 00000  00000 00000  00000 00000  00000 00000
8788:  00000 00000  00000 00000  00000 00000  00000 00000  00000 00000    00000 00000  00000 00000  00000 00000  00000 00000  00000 00000
8789:  00000 00000  00000 00000  00000 00000  00000 00000  00000 00000    00000 00000  00000 00000  00000 00000  00000 00000  00000 00000
8790:  00000 00000  00000 00000  00000 00000  00000 00000  00000 00000    00000 00000  00000 00000  00000 00000  00000 00000  00000 00000
8791:  00000 00000  00000 00000  00000 00000  00000 00000  00000 00000    00000 00000  00000 00000  00000 00000  00000 00000  00000 00000
8792:  00000 00000  00000 00000  00000 00000  00000 00000  00000 00000    00000 00000  00000 00000  00000 00000  00000 00000  00000 00000
8793:  00000 00000  00000 00000  00000 00000  00000 00000  00000 00000    00000 00000  00000 00000  00000 00000  00000 00000  00000 00000
8794:  00000 00000  00000 00000  00000 00000  00000 00000  00000 00000    00000 00000  00000 00000  00000 00000  00000 00000  00000 00000
8795:  00000 00000  00000 00000  00000 00000  00000 00000  00000 00000    00000 00000  00000 00000  00000 00000  00000 00000  00000 00000
8796:  00000 00000  00000 00000  00000 00000  00000 00000  00000 00000    00000 00000  00000 00000  00000 00000  00000 00000  00000 00000
8797:  00000 00000  00000 00000  00000 00000  00000 00000  00000 00000    00000 00000  00000 00000  00000 00000  00000 00000  00000 00000
8798:  00000 00000  00000 00000  00000 00000  00000 00000  00000 00000    00000 00000  00000 00000  00000 00000  00000 00000  00000 00000
8799:  00000 00000  00000 00000  00000 00000  00000 00000  00000 00000    00000 00000  00000 00000  00000 00000  00000 00000  00000 00000
```

```
8800:   00000 00000   00000 00000   00000 00000   00000 00000   00000 00000     00000 00000   00000 00000   00000 00000   00000 00000   00000 00000
8801:   00000 00000   00000 00000   00000 00000   00000 00000   00000 00000     00000 00000   00000 00000   00000 00000   00000 00000   00000 00000
8802:   00000 00000   00000 00000   00000 00000   00000 00000   00000 00000     00000 00000   00000 00000   00000 00000   00000 00000   00000 00000
8803:   00000 00000   00000 00000   00000 00000   00000 00000   00000 00000     00000 00000   00000 00000   00000 00000   00000 00000   00000 00000
8804:   00000 00000   00000 00000   00000 00000   00000 00000   00000 00000     00000 00000   00000 00000   00000 00000   00000 00000   00000 00000
8805:   00000 00000   00000 00000   00000 00000   00000 00000   00000 00000     00000 00000   00000 00000   00000 00000   00000 00000   00000 00000
8806:   00000 00000   00000 00000   00000 00000   00000 00000   00000 00000     00000 00000   00000 00000   00000 00000   00000 00000   00000 00000
8807:   00000 00000   00000 00000   00000 00000   00000 00000   00000 00000     00000 00000   00000 00000   00000 00000   00000 00000   00000 00000
8808:   00000 00000   00000 00000   00000 00000   00000 00000   00000 00000     00000 00000   00000 00000   00000 00000   00000 00000   00000 00000
8809:   00000 00000   00000 00000   00000 00000   00000 00000   00000 00000     00000 00000   00000 00000   00000 00000   00000 00000   00000 00000
8810:   00000 00000   00000 00000   00000 00000   00000 00000   00000 00000     00000 00000   00000 00000   00000 00000   00000 00000   00000 00000
8811:   00000 00000   00000 00000   00000 00000   00000 00000   00000 00000     00000 00000   00000 00000   00000 00000   00000 00000   00000 00000
8812:   00000 00000   00000 00000   00000 00000   00000 00000   00000 00000     00000 00000   00000 00000   00000 00000   00000 00000   00000 00000
8813:   00000 00000   00000 00000   00000 00000   00000 00000   00000 00000     00000 00000   00000 00000   00000 00000   00000 00000   00000 00000
8814:   00000 00000   00000 00000   00000 00000   00000 00000   00000 00000     00000 00000   00000 00000   00000 00000   00000 00000   00000 00000
8815:   00000 00000   00000 00000   00000 00000   00000 00000   00000 00000     00000 00000   00000 00000   00000 00000   00000 00000   00000 00000
8816:   00000 00000   00000 00000   00000 00000   00000 00000   00000 00000     00000 00000   00000 00000   00000 00000   00000 00000   00000 00000
8817:   00000 00000   00000 00000   00000 00000   00000 00000   00000 00000     00000 00000   00000 00000   00000 00000   00000 00000   00000 00000
8818:   00000 00000   00000 00000   00000 00000   00000 00000   00000 00000     00000 00000   00000 00000   00000 00000   00000 00000   00000 00000
8819:   00000 00000   00000 00000   00000 00000   00000 00000   00000 00000     00000 00000   00000 00000   00000 00000   00000 00000   00000 00000
8820:   00000 00000   00000 00000   00000 00000   00000 00000   00000 00000     00000 00000   00000 00000   00000 00000   00000 00000   00000 00000
8821:   00000 00000   00000 00000   00000 00000   00000 00000   00000 00000     00000 00000   00000 00000   00000 00000   00000 00000   00000 00000
8822:   00000 00000   00000 00000   00000 00000   00000 00000   00000 00000     00000 00000   00000 00000   00000 00000   00000 00000   00000 00000
8823:   00000 00000   00000 00000   00000 00000   00000 00000   00000 00000     00000 00000   00000 00000   00000 00000   00000 00000   00000 00000
8824:   00000 00000   00000 00000   00000 00000   00000 00000   00000 00000     00000 00000   00000 00000   00000 00000   00000 00000   00000 00000
8825:   00000 00000   00000 00000   00000 00000   00000 00000   00000 00000     00000 00000   00000 00000   00000 00000   00000 00000   00000 00000
8826:   00000 00000   00000 00000   00000 00000   00000 00000   00000 00000     00000 00000   00000 00000   00000 00000   00000 00000   00000 00000
8827:   00000 00000   00000 00000   00000 00000   00000 00000   00000 00000     00000 00000   00000 00000   00000 00000   00000 00000   00000 00000
8828:   00000 00000   00000 00000   00000 00000   00000 00000   00000 00000     00000 00000   00000 00000   00000 00000   00000 00000   00000 00000
8829:   00000 00000   00000 00000   00000 00000   00000 00000   00000 00000     00000 00000   00000 00000   00000 00000   00000 00000   00000 00000
8830:   00000 00000   00000 00000   00000 00000   00000 00000   00000 00000     00000 00000   00000 00000   00000 00000   00000 00000   00000 00000
8831:   00000 00000   00000 00000   00000 00000   00000 00000   00000 00000     00000 00000   00000 00000   00000 00000   00000 00000   00000 00000
8832:   00000 00000   00000 00000   00000 00000   00000 00000   00000 00000     00000 00000   00000 00000   00000 00000   00000 00000   00000 00000
8833:   00000 00000   00000 00000   00000 00000   00000 00000   00000 00000     00000 00000   00000 00000   00000 00000   00000 00000   00000 00000
8834:   00000 00000   00000 00000   00000 00000   00000 00000   00000 00000     00000 00000   00000 00000   00000 00000   00000 00000   00000 00000
8835:   00000 00000   00000 00000   00000 00000   00000 00000   00000 00000     00000 00000   00000 00000   00000 00000   00000 00000   00000 00000
8836:   00000 00000   00000 00000   00000 00000   00000 00000   00000 00000     00000 00000   00000 00000   00000 00000   00000 00000   00000 00000
8837:   00000 00000   00000 00000   00000 00000   00000 00000   00000 00000     00000 00000   00000 00000   00000 00000   00000 00000   00000 00000
8838:   00000 00000   00000 00000   00000 00000   00000 00000   00000 00000     00000 00000   00000 00000   00000 00000   00000 00000   00000 00000
8839:   00000 00000   00000 00000   00000 00000   00000 00000   00000 00000     00000 00000   00000 00000   00000 00000   00000 00000   00000 00000
8840:   00000 00000   00000 00000   00000 00000   00000 00000   00000 00000     00000 00000   00000 00000   00000 00000   00000 00000   00000 00000
8841:   00000 00000   00000 00000   00000 00000   00000 00000   00000 00000     00000 00000   00000 00000   00000 00000   00000 00000   00000 00000
8842:   00000 00000   00000 00000   00000 00000   00000 00000   00000 00000     00000 00000   00000 00000   00000 00000   00000 00000   00000 00000
8843:   00000 00000   00000 00000   00000 00000   00000 00000   00000 00000     00000 00000   00000 00000   00000 00000   00000 00000   00000 00000
8844:   00000 00000   00000 00000   00000 00000   00000 00000   00000 00000     00000 00000   00000 00000   00000 00000   00000 00000   00000 00000
8845:   00000 00000   00000 00000   00000 00000   00000 00000   00000 00000     00000 00000   00000 00000   00000 00000   00000 00000   00000 00000
8846:   00000 00000   00000 00000   00000 00000   00000 00000   00000 00000     00000 00000   00000 00000   00000 00000   00000 00000   00000 00000
8847:   00000 00000   00000 00000   00000 00000   00000 00000   00000 00000     00000 00000   00000 00000   00000 00000   00000 00000   00000 00000
8848:   00000 00000   00000 00000   00000 00000   00000 00000   00000 00000     00000 00000   00000 00000   00000 00000   00000 00000   00000 00000
8849:   00000 00000   00000 00000   00000 00000   00000 00000   00000 00000     00000 00000   00000 00000   00000 00000   00000 00000   00000 00000
```

```
8850:  00000 00000  00000 00000  00000 00000  00000 00000  00000 00000    00000 00000  00000 00000  00000 00000  00000 00000  00000 00000
8851:  00000 00000  00000 00000  00000 00000  00000 00000  00000 00000    00000 00000  00000 00000  00000 00000  00000 00000  00000 00000
8852:  00000 00000  00000 00000  00000 00000  00000 00000  00000 00000    00000 00000  00000 00000  00000 00000  00000 00000  00000 00000
8853:  00000 00000  00000 00000  00000 00000  00000 00000  00000 00000    00000 00000  00000 00000  00000 00000  00000 00000  00000 00000
8854:  00000 00000  00000 00000  00000 00000  00000 00000  00000 00000    00000 00000  00000 00000  00000 00000  00000 00000  00000 00000
8855:  00000 00000  00000 00000  00000 00000  00000 00000  00000 00000    00000 00000  00000 00000  00000 00000  00000 00000  00000 00000
8856:  00000 00000  00000 00000  00000 00000  00000 00000  00000 00000    00000 00000  00000 00000  00000 00000  00000 00000  00000 00000
8857:  00000 00000  00000 00000  00000 00000  00000 00000  00000 00000    00000 00000  00000 00000  00000 00000  00000 00000  00000 00000
8858:  00000 00000  00000 00000  00000 00000  00000 00000  00000 00000    00000 00000  00000 00000  00000 00000  00000 00000  00000 00000
8859:  00000 00000  00000 00000  00000 00000  00000 00000  00000 00000    00000 00000  00000 00000  00000 00000  00000 00000  00000 00000
8860:  00000 00000  00000 00000  00000 00000  00000 00000  00000 00000    00000 00000  00000 00000  00000 00000  00000 00000  00000 00000
8861:  00000 00000  00000 00000  00000 00000  00000 00000  00000 00000    00000 00000  00000 00000  00000 00000  00000 00000  00000 00000
8862:  00000 00000  00000 00000  00000 00000  00000 00000  00000 00000    00000 00000  00000 00000  00000 00000  00000 00000  00000 00000
8863:  00000 00000  00000 00000  00000 00000  00000 00000  00000 00000    00000 00000  00000 00000  00000 00000  00000 00000  00000 00000
8864:  00000 00000  00000 00000  00000 00000  00000 00000  00000 00000    00000 00000  00000 00000  00000 00000  00000 00000  00000 00000
8865:  00000 00000  00000 00000  00000 00000  00000 00000  00000 00000    00000 00000  00000 00000  00000 00000  00000 00000  00000 00000
8866:  00000 00000  00000 00000  00000 00000  00000 00000  00000 00000    00000 00000  00000 00000  00000 00000  00000 00000  00000 00000
8867:  00000 00000  00000 00000  00000 00000  00000 00000  00000 00000    00000 00000  00000 00000  00000 00000  00000 00000  00000 00000
8868:  00000 00000  00000 00000  00000 00000  00000 00000  00000 00000    00000 00000  00000 00000  00000 00000  00000 00000  00000 00000
8869:  00000 00000  00000 00000  00000 00000  00000 00000  00000 00000    00000 00000  00000 00000  00000 00000  00000 00000  00000 00000
8870:  00000 00000  00000 00000  00000 00000  00000 00000  00000 00000    00000 00000  00000 00000  00000 00000  00000 00000  00000 00000
8871:  00000 00000  00000 00000  00000 00000  00000 00000  00000 00000    00000 00000  00000 00000  00000 00000  00000 00000  00000 00000
8872:  00000 00000  00000 00000  00000 00000  00000 00000  00000 00000    00000 00000  00000 00000  00000 00000  00000 00000  00000 00000
8873:  00000 00000  00000 00000  00000 00000  00000 00000  00000 00000    00000 00000  00000 00000  00000 00000  00000 00000  00000 00000
8874:  00000 00000  00000 00000  00000 00000  00000 00000  00000 00000    00000 00000  00000 00000  00000 00000  00000 00000  00000 00000
8875:  00000 00000  00000 00000  00000 00000  00000 00000  00000 00000    00000 00000  00000 00000  00000 00000  00000 00000  00000 00000
8876:  00000 00000  00000 00000  00000 00000  00000 00000  00000 00000    00000 00000  00000 00000  00000 00000  00000 00000  00000 00000
8877:  00000 00000  00000 00000  00000 00000  00000 00000  00000 00000    00000 00000  00000 00000  00000 00000  00000 00000  00000 00000
8878:  00000 00000  00000 00000  00000 00000  00000 00000  00000 00000    00000 00000  00000 00000  00000 00000  00000 00000  00000 00000
8879:  00000 00000  00000 00000  00000 00000  00000 00000  00000 00000    00000 00000  00000 00000  00000 00000  00000 00000  00000 00000
8880:  00000 00000  00000 00000  00000 00000  00000 00000  00000 00000    00000 00000  00000 00000  00000 00000  00000 00000  00000 00000
8881:  00000 00000  00000 00000  00000 00000  00000 00000  00000 00000    00000 00000  00000 00000  00000 00000  00000 00000  00000 00000
8882:  00000 00000  00000 00000  00000 00000  00000 00000  00000 00000    00000 00000  00000 00000  00000 00000  00000 00000  00000 00000
8883:  00000 00000  00000 00000  00000 00000  00000 00000  00000 00000    00000 00000  00000 00000  00000 00000  00000 00000  00000 00000
8884:  00000 00000  00000 00000  00000 00000  00000 00000  00000 00000    00000 00000  00000 00000  00000 00000  00000 00000  00000 00000
8885:  00000 00000  00000 00000  00000 00000  00000 00000  00000 00000    00000 00000  00000 00000  00000 00000  00000 00000  00000 00000
8886:  00000 00000  00000 00000  00000 00000  00000 00000  00000 00000    00000 00000  00000 00000  00000 00000  00000 00000  00000 00000
8887:  00000 00000  00000 00000  00000 00000  00000 00000  00000 00000    00000 00000  00000 00000  00000 00000  00000 00000  00000 00000
8888:  00000 00000  00000 00000  00000 00000  00000 00000  00000 00000    00000 00000  00000 00000  00000 00000  00000 00000  00000 00000
8889:  00000 00000  00000 00000  00000 00000  00000 00000  00000 00000    00000 00000  00000 00000  00000 00000  00000 00000  00000 00000
8890:  00000 00000  00000 00000  00000 00000  00000 00000  00000 00000    00000 00000  00000 00000  00000 00000  00000 00000  00000 00000
8891:  00000 00000  00000 00000  00000 00000  00000 00000  00000 00000    00000 00000  00000 00000  00000 00000  00000 00000  00000 00000
8892:  00000 00000  00000 00000  00000 00000  00000 00000  00000 00000    00000 00000  00000 00000  00000 00000  00000 00000  00000 00000
8893:  00000 00000  00000 00000  00000 00000  00000 00000  00000 00000    00000 00000  00000 00000  00000 00000  00000 00000  00000 00000
8894:  00000 00000  00000 00000  00000 00000  00000 00000  00000 00000    00000 00000  00000 00000  00000 00000  00000 00000  00000 00000
8895:  00000 00000  00000 00000  00000 00000  00000 00000  00000 00000    00000 00000  00000 00000  00000 00000  00000 00000  00000 00000
8896:  00000 00000  00000 00000  00000 00000  00000 00000  00000 00000    00000 00000  00000 00000  00000 00000  00000 00000  00000 00000
8897:  00000 00000  00000 00000  00000 00000  00000 00000  00000 00000    00000 00000  00000 00000  00000 00000  00000 00000  00000 00000
8898:  00000 00000  00000 00000  00000 00000  00000 00000  00000 00000    00000 00000  00000 00000  00000 00000  00000 00000  00000 00000
8899:  00000 00000  00000 00000  00000 00000  00000 00000  00000 00000    00000 00000  00000 00000  00000 00000  00000 00000  00000 00000
```

```
8900:   00000 00000   00000 00000   00000 00000   00000 00000   00000 00000     00000 00000   00000 00000   00000 00000   00000 00000   00000 00000
8901:   00000 00000   00000 00000   00000 00000   00000 00000   00000 00000     00000 00000   00000 00000   00000 00000   00000 00000   00000 00000
8902:   00000 00000   00000 00000   00000 00000   00000 00000   00000 00000     00000 00000   00000 00000   00000 00000   00000 00000   00000 00000
8903:   00000 00000   00000 00000   00000 00000   00000 00000   00000 00000     00000 00000   00000 00000   00000 00000   00000 00000   00000 00000
8904:   00000 00000   00000 00000   00000 00000   00000 00000   00000 00000     00000 00000   00000 00000   00000 00000   00000 00000   00000 00000
8905:   00000 00000   00000 00000   00000 00000   00000 00000   00000 00000     00000 00000   00000 00000   00000 00000   00000 00000   00000 00000
8906:   00000 00000   00000 00000   00000 00000   00000 00000   00000 00000     00000 00000   00000 00000   00000 00000   00000 00000   00000 00000
8907:   00000 00000   00000 00000   00000 00000   00000 00000   00000 00000     00000 00000   00000 00000   00000 00000   00000 00000   00000 00000
8908:   00000 00000   00000 00000   00000 00000   00000 00000   00000 00000     00000 00000   00000 00000   00000 00000   00000 00000   00000 00000
8909:   00000 00000   00000 00000   00000 00000   00000 00000   00000 00000     00000 00000   00000 00000   00000 00000   00000 00000   00000 00000
8910:   00000 00000   00000 00000   00000 00000   00000 00000   00000 00000     00000 00000   00000 00000   00000 00000   00000 00000   00000 00000
8911:   00000 00000   00000 00000   00000 00000   00000 00000   00000 00000     00000 00000   00000 00000   00000 00000   00000 00000   00000 00000
8912:   00000 00000   00000 00000   00000 00000   00000 00000   00000 00000     00000 00000   00000 00000   00000 00000   00000 00000   00000 00000
8913:   00000 00000   00000 00000   00000 00000   00000 00000   00000 00000     00000 00000   00000 00000   00000 00000   00000 00000   00000 00000
8914:   00000 00000   00000 00000   00000 00000   00000 00000   00000 00000     00000 00000   00000 00000   00000 00000   00000 00000   00000 00000
8915:   00000 00000   00000 00000   00000 00000   00000 00000   00000 00000     00000 00000   00000 00000   00000 00000   00000 00000   00000 00000
8916:   00000 00000   00000 00000   00000 00000   00000 00000   00000 00000     00000 00000   00000 00000   00000 00000   00000 00000   00000 00000
8917:   00000 00000   00000 00000   00000 00000   00000 00000   00000 00000     00000 00000   00000 00000   00000 00000   00000 00000   00000 00000
8918:   00000 00000   00000 00000   00000 00000   00000 00000   00000 00000     00000 00000   00000 00000   00000 00000   00000 00000   00000 00000
8919:   00000 00000   00000 00000   00000 00000   00000 00000   00000 00000     00000 00000   00000 00000   00000 00000   00000 00000   00000 00000
8920:   00000 00000   00000 00000   00000 00000   00000 00000   00000 00000     00000 00000   00000 00000   00000 00000   00000 00000   00000 00000
8921:   00000 00000   00000 00000   00000 00000   00000 00000   00000 00000     00000 00000   00000 00000   00000 00000   00000 00000   00000 00000
8922:   00000 00000   00000 00000   00000 00000   00000 00000   00000 00000     00000 00000   00000 00000   00000 00000   00000 00000   00000 00000
8923:   00000 00000   00000 00000   00000 00000   00000 00000   00000 00000     00000 00000   00000 00000   00000 00000   00000 00000   00000 00000
8924:   00000 00000   00000 00000   00000 00000   00000 00000   00000 00000     00000 00000   00000 00000   00000 00000   00000 00000   00000 00000
8925:   00000 00000   00000 00000   00000 00000   00000 00000   00000 00000     00000 00000   00000 00000   00000 00000   00000 00000   00000 00000
8926:   00000 00000   00000 00000   00000 00000   00000 00000   00000 00000     00000 00000   00000 00000   00000 00000   00000 00000   00000 00000
8927:   00000 00000   00000 00000   00000 00000   00000 00000   00000 00000     00000 00000   00000 00000   00000 00000   00000 00000   00000 00000
8928:   00000 00000   00000 00000   00000 00000   00000 00000   00000 00000     00000 00000   00000 00000   00000 00000   00000 00000   00000 00000
8929:   00000 00000   00000 00000   00000 00000   00000 00000   00000 00000     00000 00000   00000 00000   00000 00000   00000 00000   00000 00000
8930:   00000 00000   00000 00000   00000 00000   00000 00000   00000 00000     00000 00000   00000 00000   00000 00000   00000 00000   00000 00000
8931:   00000 00000   00000 00000   00000 00000   00000 00000   00000 00000     00000 00000   00000 00000   00000 00000   00000 00000   00000 00000
8932:   00000 00000   00000 00000   00000 00000   00000 00000   00000 00000     00000 00000   00000 00000   00000 00000   00000 00000   00000 00000
8933:   00000 00000   00000 00000   00000 00000   00000 00000   00000 00000     00000 00000   00000 00000   00000 00000   00000 00000   00000 00000
8934:   00000 00000   00000 00000   00000 00000   00000 00000   00000 00000     00000 00000   00000 00000   00000 00000   00000 00000   00000 00000
8935:   00000 00000   00000 00000   00000 00000   00000 00000   00000 00000     00000 00000   00000 00000   00000 00000   00000 00000   00000 00000
8936:   00000 00000   00000 00000   00000 00000   00000 00000   00000 00000     00000 00000   00000 00000   00000 00000   00000 00000   00000 00000
8937:   00000 00000   00000 00000   00000 00000   00000 00000   00000 00000     00000 00000   00000 00000   00000 00000   00000 00000   00000 00000
8938:   00000 00000   00000 00000   00000 00000   00000 00000   00000 00000     00000 00000   00000 00000   00000 00000   00000 00000   00000 00000
8939:   00000 00000   00000 00000   00000 00000   00000 00000   00000 00000     00000 00000   00000 00000   00000 00000   00000 00000   00000 00000
8940:   00000 00000   00000 00000   00000 00000   00000 00000   00000 00000     00000 00000   00000 00000   00000 00000   00000 00000   00000 00000
8941:   00000 00000   00000 00000   00000 00000   00000 00000   00000 00000     00000 00000   00000 00000   00000 00000   00000 00000   00000 00000
8942:   00000 00000   00000 00000   00000 00000   00000 00000   00000 00000     00000 00000   00000 00000   00000 00000   00000 00000   00000 00000
8943:   00000 00000   00000 00000   00000 00000   00000 00000   00000 00000     00000 00000   00000 00000   00000 00000   00000 00000   00000 00000
8944:   00000 00000   00000 00000   00000 00000   00000 00000   00000 00000     00000 00000   00000 00000   00000 00000   00000 00000   00000 00000
8945:   00000 00000   00000 00000   00000 00000   00000 00000   00000 00000     00000 00000   00000 00000   00000 00000   00000 00000   00000 00000
8946:   00000 00000   00000 00000   00000 00000   00000 00000   00000 00000     00000 00000   00000 00000   00000 00000   00000 00000   00000 00000
8947:   00000 00000   00000 00000   00000 00000   00000 00000   00000 00000     00000 00000   00000 00000   00000 00000   00000 00000   00000 00000
8948:   00000 00000   00000 00000   00000 00000   00000 00000   00000 00000     00000 00000   00000 00000   00000 00000   00000 00000   00000 00000
8949:   00000 00000   00000 00000   00000 00000   00000 00000   00000 00000     00000 00000   00000 00000   00000 00000   00000 00000   00000 00000
```

```
8950:  00000 00000   00000 00000   00000 00000   00000 00000   00000 00000      00000 00000   00000 00000   00000 00000   00000 00000   00000 00000
8951:  00000 00000   00000 00000   00000 00000   00000 00000   00000 00000      00000 00000   00000 00000   00000 00000   00000 00000   00000 00000
8952:  00000 00000   00000 00000   00000 00000   00000 00000   00000 00000      00000 00000   00000 00000   00000 00000   00000 00000   00000 00000
8953:  00000 00000   00000 00000   00000 00000   00000 00000   00000 00000      00000 00000   00000 00000   00000 00000   00000 00000   00000 00000
8954:  00000 00000   00000 00000   00000 00000   00000 00000   00000 00000      00000 00000   00000 00000   00000 00000   00000 00000   00000 00000
8955:  00000 00000   00000 00000   00000 00000   00000 00000   00000 00000      00000 00000   00000 00000   00000 00000   00000 00000   00000 00000
8956:  00000 00000   00000 00000   00000 00000   00000 00000   00000 00000      00000 00000   00000 00000   00000 00000   00000 00000   00000 00000
8957:  00000 00000   00000 00000   00000 00000   00000 00000   00000 00000      00000 00000   00000 00000   00000 00000   00000 00000   00000 00000
8958:  00000 00000   00000 00000   00000 00000   00000 00000   00000 00000      00000 00000   00000 00000   00000 00000   00000 00000   00000 00000
8959:  00000 00000   00000 00000   00000 00000   00000 00000   00000 00000      00000 00000   00000 00000   00000 00000   00000 00000   00000 00000
8960:  00000 00000   00000 00000   00000 00000   00000 00000   00000 00000      00000 00000   00000 00000   00000 00000   00000 00000   00000 00000
8961:  00000 00000   00000 00000   00000 00000   00000 00000   00000 00000      00000 00000   00000 00000   00000 00000   00000 00000   00000 00000
8962:  00000 00000   00000 00000   00000 00000   00000 00000   00000 00000      00000 00000   00000 00000   00000 00000   00000 00000   00000 00000
8963:  00000 00000   00000 00000   00000 00000   00000 00000   00000 00000      00000 00000   00000 00000   00000 00000   00000 00000   00000 00000
8964:  00000 00000   00000 00000   00000 00000   00000 00000   00000 00000      00000 00000   00000 00000   00000 00000   00000 00000   00000 00000
8965:  00000 00000   00000 00000   00000 00000   00000 00000   00000 00000      00000 00000   00000 00000   00000 00000   00000 00000   00000 00000
8966:  00000 00000   00000 00000   00000 00000   00000 00000   00000 00000      00000 00000   00000 00000   00000 00000   00000 00000   00000 00000
8967:  00000 00000   00000 00000   00000 00000   00000 00000   00000 00000      00000 00000   00000 00000   00000 00000   00000 00000   00000 00000
8968:  00000 00000   00000 00000   00000 00000   00000 00000   00000 00000      00000 00000   00000 00000   00000 00000   00000 00000   00000 00000
8969:  00000 00000   00000 00000   00000 00000   00000 00000   00000 00000      00000 00000   00000 00000   00000 00000   00000 00000   00000 00000
8970:  00000 00000   00000 00000   00000 00000   00000 00000   00000 00000      00000 00000   00000 00000   00000 00000   00000 00000   00000 00000
8971:  00000 00000   00000 00000   00000 00000   00000 00000   00000 00000      00000 00000   00000 00000   00000 00000   00000 00000   00000 00000
8972:  00000 00000   00000 00000   00000 00000   00000 00000   00000 00000      00000 00000   00000 00000   00000 00000   00000 00000   00000 00000
8973:  00000 00000   00000 00000   00000 00000   00000 00000   00000 00000      00000 00000   00000 00000   00000 00000   00000 00000   00000 00000
8974:  00000 00000   00000 00000   00000 00000   00000 00000   00000 00000      00000 00000   00000 00000   00000 00000   00000 00000   00000 00000
8975:  00000 00000   00000 00000   00000 00000   00000 00000   00000 00000      00000 00000   00000 00000   00000 00000   00000 00000   00000 00000
8976:  00000 00000   00000 00000   00000 00000   00000 00000   00000 00000      00000 00000   00000 00000   00000 00000   00000 00000   00000 00000
8977:  00000 00000   00000 00000   00000 00000   00000 00000   00000 00000      00000 00000   00000 00000   00000 00000   00000 00000   00000 00000
8978:  00000 00000   00000 00000   00000 00000   00000 00000   00000 00000      00000 00000   00000 00000   00000 00000   00000 00000   00000 00000
8979:  00000 00000   00000 00000   00000 00000   00000 00000   00000 00000      00000 00000   00000 00000   00000 00000   00000 00000   00000 00000
8980:  00000 00000   00000 00000   00000 00000   00000 00000   00000 00000      00000 00000   00000 00000   00000 00000   00000 00000   00000 00000
8981:  00000 00000   00000 00000   00000 00000   00000 00000   00000 00000      00000 00000   00000 00000   00000 00000   00000 00000   00000 00000
8982:  00000 00000   00000 00000   00000 00000   00000 00000   00000 00000      00000 00000   00000 00000   00000 00000   00000 00000   00000 00000
8983:  00000 00000   00000 00000   00000 00000   00000 00000   00000 00000      00000 00000   00000 00000   00000 00000   00000 00000   00000 00000
8984:  00000 00000   00000 00000   00000 00000   00000 00000   00000 00000      00000 00000   00000 00000   00000 00000   00000 00000   00000 00000
8985:  00000 00000   00000 00000   00000 00000   00000 00000   00000 00000      00000 00000   00000 00000   00000 00000   00000 00000   00000 00000
8986:  00000 00000   00000 00000   00000 00000   00000 00000   00000 00000      00000 00000   00000 00000   00000 00000   00000 00000   00000 00000
8987:  00000 00000   00000 00000   00000 00000   00000 00000   00000 00000      00000 00000   00000 00000   00000 00000   00000 00000   00000 00000
8988:  00000 00000   00000 00000   00000 00000   00000 00000   00000 00000      00000 00000   00000 00000   00000 00000   00000 00000   00000 00000
8989:  00000 00000   00000 00000   00000 00000   00000 00000   00000 00000      00000 00000   00000 00000   00000 00000   00000 00000   00000 00000
8990:  00000 00000   00000 00000   00000 00000   00000 00000   00000 00000      00000 00000   00000 00000   00000 00000   00000 00000   00000 00000
8991:  00000 00000   00000 00000   00000 00000   00000 00000   00000 00000      00000 00000   00000 00000   00000 00000   00000 00000   00000 00000
8992:  00000 00000   00000 00000   00000 00000   00000 00000   00000 00000      00000 00000   00000 00000   00000 00000   00000 00000   00000 00000
8993:  00000 00000   00000 00000   00000 00000   00000 00000   00000 00000      00000 00000   00000 00000   00000 00000   00000 00000   00000 00000
8994:  00000 00000   00000 00000   00000 00000   00000 00000   00000 00000      00000 00000   00000 00000   00000 00000   00000 00000   00000 00000
8995:  00000 00000   00000 00000   00000 00000   00000 00000   00000 00000      00000 00000   00000 00000   00000 00000   00000 00000   00000 00000
8996:  00000 00000   00000 00000   00000 00000   00000 00000   00000 00000      00000 00000   00000 00000   00000 00000   00000 00000   00000 00000
8997:  00000 00000   00000 00000   00000 00000   00000 00000   00000 00000      00000 00000   00000 00000   00000 00000   00000 00000   00000 00000
8998:  00000 00000   00000 00000   00000 00000   00000 00000   00000 00000      00000 00000   00000 00000   00000 00000   00000 00000   00000 00000
8999:  00000 00000   00000 00000   00000 00000   00000 00000   00000 00000      00000 00000   00000 00000   00000 00000   00000 00000   00000 00000
```

```
9000:  00000 00000  00000 00000  00000 00000  00000 00000  00000 00000    00000 00000  00000 00000  00000 00000  00000 00000  00000 00000
9001:  00000 00000  00000 00000  00000 00000  00000 00000  00000 00000    00000 00000  00000 00000  00000 00000  00000 00000  00000 00000
9002:  00000 00000  00000 00000  00000 00000  00000 00000  00000 00000    00000 00000  00000 00000  00000 00000  00000 00000  00000 00000
9003:  00000 00000  00000 00000  00000 00000  00000 00000  00000 00000    00000 00000  00000 00000  00000 00000  00000 00000  00000 00000
9004:  00000 00000  00000 00000  00000 00000  00000 00000  00000 00000    00000 00000  00000 00000  00000 00000  00000 00000  00000 00000
9005:  00000 00000  00000 00000  00000 00000  00000 00000  00000 00000    00000 00000  00000 00000  00000 00000  00000 00000  00000 00000
9006:  00000 00000  00000 00000  00000 00000  00000 00000  00000 00000    00000 00000  00000 00000  00000 00000  00000 00000  00000 00000
9007:  00000 00000  00000 00000  00000 00000  00000 00000  00000 00000    00000 00000  00000 00000  00000 00000  00000 00000  00000 00000
9008:  00000 00000  00000 00000  00000 00000  00000 00000  00000 00000    00000 00000  00000 00000  00000 00000  00000 00000  00000 00000
9009:  00000 00000  00000 00000  00000 00000  00000 00000  00000 00000    00000 00000  00000 00000  00000 00000  00000 00000  00000 00000
9010:  00000 00000  00000 00000  00000 00000  00000 00000  00000 00000    00000 00000  00000 00000  00000 00000  00000 00000  00000 00000
9011:  00000 00000  00000 00000  00000 00000  00000 00000  00000 00000    00000 00000  00000 00000  00000 00000  00000 00000  00000 00000
9012:  00000 00000  00000 00000  00000 00000  00000 00000  00000 00000    00000 00000  00000 00000  00000 00000  00000 00000  00000 00000
9013:  00000 00000  00000 00000  00000 00000  00000 00000  00000 00000    00000 00000  00000 00000  00000 00000  00000 00000  00000 00000
9014:  00000 00000  00000 00000  00000 00000  00000 00000  00000 00000    00000 00000  00000 00000  00000 00000  00000 00000  00000 00000
9015:  00000 00000  00000 00000  00000 00000  00000 00000  00000 00000    00000 00000  00000 00000  00000 00000  00000 00000  00000 00000
9016:  00000 00000  00000 00000  00000 00000  00000 00000  00000 00000    00000 00000  00000 00000  00000 00000  00000 00000  00000 00000
9017:  00000 00000  00000 00000  00000 00000  00000 00000  00000 00000    00000 00000  00000 00000  00000 00000  00000 00000  00000 00000
9018:  00000 00000  00000 00000  00000 00000  00000 00000  00000 00000    00000 00000  00000 00000  00000 00000  00000 00000  00000 00000
9019:  00000 00000  00000 00000  00000 00000  00000 00000  00000 00000    00000 00000  00000 00000  00000 00000  00000 00000  00000 00000
9020:  00000 00000  00000 00000  00000 00000  00000 00000  00000 00000    00000 00000  00000 00000  00000 00000  00000 00000  00000 00000
9021:  00000 00000  00000 00000  00000 00000  00000 00000  00000 00000    00000 00000  00000 00000  00000 00000  00000 00000  00000 00000
9022:  00000 00000  00000 00000  00000 00000  00000 00000  00000 00000    00000 00000  00000 00000  00000 00000  00000 00000  00000 00000
9023:  00000 00000  00000 00000  00000 00000  00000 00000  00000 00000    00000 00000  00000 00000  00000 00000  00000 00000  00000 00000
9024:  00000 00000  00000 00000  00000 00000  00000 00000  00000 00000    00000 00000  00000 00000  00000 00000  00000 00000  00000 00000
9025:  00000 00000  00000 00000  00000 00000  00000 00000  00000 00000    00000 00000  00000 00000  00000 00000  00000 00000  00000 00000
9026:  00000 00000  00000 00000  00000 00000  00000 00000  00000 00000    00000 00000  00000 00000  00000 00000  00000 00000  00000 00000
9027:  00000 00000  00000 00000  00000 00000  00000 00000  00000 00000    00000 00000  00000 00000  00000 00000  00000 00000  00000 00000
9028:  00000 00000  00000 00000  00000 00000  00000 00000  00000 00000    00000 00000  00000 00000  00000 00000  00000 00000  00000 00000
9029:  00000 00000  00000 00000  00000 00000  00000 00000  00000 00000    00000 00000  00000 00000  00000 00000  00000 00000  00000 00000
9030:  00000 00000  00000 00000  00000 00000  00000 00000  00000 00000    00000 00000  00000 00000  00000 00000  00000 00000  00000 00000
9031:  00000 00000  00000 00000  00000 00000  00000 00000  00000 00000    00000 00000  00000 00000  00000 00000  00000 00000  00000 00000
9032:  00000 00000  00000 00000  00000 00000  00000 00000  00000 00000    00000 00000  00000 00000  00000 00000  00000 00000  00000 00000
9033:  00000 00000  00000 00000  00000 00000  00000 00000  00000 00000    00000 00000  00000 00000  00000 00000  00000 00000  00000 00000
9034:  00000 00000  00000 00000  00000 00000  00000 00000  00000 00000    00000 00000  00000 00000  00000 00000  00000 00000  00000 00000
9035:  00000 00000  00000 00000  00000 00000  00000 00000  00000 00000    00000 00000  00000 00000  00000 00000  00000 00000  00000 00000
9036:  00000 00000  00000 00000  00000 00000  00000 00000  00000 00000    00000 00000  00000 00000  00000 00000  00000 00000  00000 00000
9037:  00000 00000  00000 00000  00000 00000  00000 00000  00000 00000    00000 00000  00000 00000  00000 00000  00000 00000  00000 00000
9038:  00000 00000  00000 00000  00000 00000  00000 00000  00000 00000    00000 00000  00000 00000  00000 00000  00000 00000  00000 00000
9039:  00000 00000  00000 00000  00000 00000  00000 00000  00000 00000    00000 00000  00000 00000  00000 00000  00000 00000  00000 00000
9040:  00000 00000  00000 00000  00000 00000  00000 00000  00000 00000    00000 00000  00000 00000  00000 00000  00000 00000  00000 00000
9041:  00000 00000  00000 00000  00000 00000  00000 00000  00000 00000    00000 00000  00000 00000  00000 00000  00000 00000  00000 00000
9042:  00000 00000  00000 00000  00000 00000  00000 00000  00000 00000    00000 00000  00000 00000  00000 00000  00000 00000  00000 00000
9043:  00000 00000  00000 00000  00000 00000  00000 00000  00000 00000    00000 00000  00000 00000  00000 00000  00000 00000  00000 00000
9044:  00000 00000  00000 00000  00000 00000  00000 00000  00000 00000    00000 00000  00000 00000  00000 00000  00000 00000  00000 00000
9045:  00000 00000  00000 00000  00000 00000  00000 00000  00000 00000    00000 00000  00000 00000  00000 00000  00000 00000  00000 00000
9046:  00000 00000  00000 00000  00000 00000  00000 00000  00000 00000    00000 00000  00000 00000  00000 00000  00000 00000  00000 00000
9047:  00000 00000  00000 00000  00000 00000  00000 00000  00000 00000    00000 00000  00000 00000  00000 00000  00000 00000  00000 00000
9048:  00000 00000  00000 00000  00000 00000  00000 00000  00000 00000    00000 00000  00000 00000  00000 00000  00000 00000  00000 00000
9049:  00000 00000  00000 00000  00000 00000  00000 00000  00000 00000    00000 00000  00000 00000  00000 00000  00000 00000  00000 00000
```

```
9050:  00000 00000  00000 00000  00000 00000  00000 00000  00000 00000    00000 00000  00000 00000  00000 00000  00000 00000  00000 00000
9051:  00000 00000  00000 00000  00000 00000  00000 00000  00000 00000    00000 00000  00000 00000  00000 00000  00000 00000  00000 00000
9052:  00000 00000  00000 00000  00000 00000  00000 00000  00000 00000    00000 00000  00000 00000  00000 00000  00000 00000  00000 00000
9053:  00000 00000  00000 00000  00000 00000  00000 00000  00000 00000    00000 00000  00000 00000  00000 00000  00000 00000  00000 00000
9054:  00000 00000  00000 00000  00000 00000  00000 00000  00000 00000    00000 00000  00000 00000  00000 00000  00000 00000  00000 00000
9055:  00000 00000  00000 00000  00000 00000  00000 00000  00000 00000    00000 00000  00000 00000  00000 00000  00000 00000  00000 00000
9056:  00000 00000  00000 00000  00000 00000  00000 00000  00000 00000    00000 00000  00000 00000  00000 00000  00000 00000  00000 00000
9057:  00000 00000  00000 00000  00000 00000  00000 00000  00000 00000    00000 00000  00000 00000  00000 00000  00000 00000  00000 00000
9058:  00000 00000  00000 00000  00000 00000  00000 00000  00000 00000    00000 00000  00000 00000  00000 00000  00000 00000  00000 00000
9059:  00000 00000  00000 00000  00000 00000  00000 00000  00000 00000    00000 00000  00000 00000  00000 00000  00000 00000  00000 00000
9060:  00000 00000  00000 00000  00000 00000  00000 00000  00000 00000    00000 00000  00000 00000  00000 00000  00000 00000  00000 00000
9061:  00000 00000  00000 00000  00000 00000  00000 00000  00000 00000    00000 00000  00000 00000  00000 00000  00000 00000  00000 00000
9062:  00000 00000  00000 00000  00000 00000  00000 00000  00000 00000    00000 00000  00000 00000  00000 00000  00000 00000  00000 00000
9063:  00000 00000  00000 00000  00000 00000  00000 00000  00000 00000    00000 00000  00000 00000  00000 00000  00000 00000  00000 00000
9064:  00000 00000  00000 00000  00000 00000  00000 00000  00000 00000    00000 00000  00000 00000  00000 00000  00000 00000  00000 00000
9065:  00000 00000  00000 00000  00000 00000  00000 00000  00000 00000    00000 00000  00000 00000  00000 00000  00000 00000  00000 00000
9066:  00000 00000  00000 00000  00000 00000  00000 00000  00000 00000    00000 00000  00000 00000  00000 00000  00000 00000  00000 00000
9067:  00000 00000  00000 00000  00000 00000  00000 00000  00000 00000    00000 00000  00000 00000  00000 00000  00000 00000  00000 00000
9068:  00000 00000  00000 00000  00000 00000  00000 00000  00000 00000    00000 00000  00000 00000  00000 00000  00000 00000  00000 00000
9069:  00000 00000  00000 00000  00000 00000  00000 00000  00000 00000    00000 00000  00000 00000  00000 00000  00000 00000  00000 00000
9070:  00000 00000  00000 00000  00000 00000  00000 00000  00000 00000    00000 00000  00000 00000  00000 00000  00000 00000  00000 00000
9071:  00000 00000  00000 00000  00000 00000  00000 00000  00000 00000    00000 00000  00000 00000  00000 00000  00000 00000  00000 00000
9072:  00000 00000  00000 00000  00000 00000  00000 00000  00000 00000    00000 00000  00000 00000  00000 00000  00000 00000  00000 00000
9073:  00000 00000  00000 00000  00000 00000  00000 00000  00000 00000    00000 00000  00000 00000  00000 00000  00000 00000  00000 00000
9074:  00000 00000  00000 00000  00000 00000  00000 00000  00000 00000    00000 00000  00000 00000  00000 00000  00000 00000  00000 00000
9075:  00000 00000  00000 00000  00000 00000  00000 00000  00000 00000    00000 00000  00000 00000  00000 00000  00000 00000  00000 00000
9076:  00000 00000  00000 00000  00000 00000  00000 00000  00000 00000    00000 00000  00000 00000  00000 00000  00000 00000  00000 00000
9077:  00000 00000  00000 00000  00000 00000  00000 00000  00000 00000    00000 00000  00000 00000  00000 00000  00000 00000  00000 00000
9078:  00000 00000  00000 00000  00000 00000  00000 00000  00000 00000    00000 00000  00000 00000  00000 00000  00000 00000  00000 00000
9079:  00000 00000  00000 00000  00000 00000  00000 00000  00000 00000    00000 00000  00000 00000  00000 00000  00000 00000  00000 00000
9080:  00000 00000  00000 00000  00000 00000  00000 00000  00000 00000    00000 00000  00000 00000  00000 00000  00000 00000  00000 00000
9081:  00000 00000  00000 00000  00000 00000  00000 00000  00000 00000    00000 00000  00000 00000  00000 00000  00000 00000  00000 00000
9082:  00000 00000  00000 00000  00000 00000  00000 00000  00000 00000    00000 00000  00000 00000  00000 00000  00000 00000  00000 00000
9083:  00000 00000  00000 00000  00000 00000  00000 00000  00000 00000    00000 00000  00000 00000  00000 00000  00000 00000  00000 00000
9084:  00000 00000  00000 00000  00000 00000  00000 00000  00000 00000    00000 00000  00000 00000  00000 00000  00000 00000  00000 00000
9085:  00000 00000  00000 00000  00000 00000  00000 00000  00000 00000    00000 00000  00000 00000  00000 00000  00000 00000  00000 00000
9086:  00000 00000  00000 00000  00000 00000  00000 00000  00000 00000    00000 00000  00000 00000  00000 00000  00000 00000  00000 00000
9087:  00000 00000  00000 00000  00000 00000  00000 00000  00000 00000    00000 00000  00000 00000  00000 00000  00000 00000  00000 00000
9088:  00000 00000  00000 00000  00000 00000  00000 00000  00000 00000    00000 00000  00000 00000  00000 00000  00000 00000  00000 00000
9089:  00000 00000  00000 00000  00000 00000  00000 00000  00000 00000    00000 00000  00000 00000  00000 00000  00000 00000  00000 00000
9090:  00000 00000  00000 00000  00000 00000  00000 00000  00000 00000    00000 00000  00000 00000  00000 00000  00000 00000  00000 00000
9091:  00000 00000  00000 00000  00000 00000  00000 00000  00000 00000    00000 00000  00000 00000  00000 00000  00000 00000  00000 00000
9092:  00000 00000  00000 00000  00000 00000  00000 00000  00000 00000    00000 00000  00000 00000  00000 00000  00000 00000  00000 00000
9093:  00000 00000  00000 00000  00000 00000  00000 00000  00000 00000    00000 00000  00000 00000  00000 00000  00000 00000  00000 00000
9094:  00000 00000  00000 00000  00000 00000  00000 00000  00000 00000    00000 00000  00000 00000  00000 00000  00000 00000  00000 00000
9095:  00000 00000  00000 00000  00000 00000  00000 00000  00000 00000    00000 00000  00000 00000  00000 00000  00000 00000  00000 00000
9096:  00000 00000  00000 00000  00000 00000  00000 00000  00000 00000    00000 00000  00000 00000  00000 00000  00000 00000  00000 00000
9097:  00000 00000  00000 00000  00000 00000  00000 00000  00000 00000    00000 00000  00000 00000  00000 00000  00000 00000  00000 00000
9098:  00000 00000  00000 00000  00000 00000  00000 00000  00000 00000    00000 00000  00000 00000  00000 00000  00000 00000  00000 00000
9099:  00000 00000  00000 00000  00000 00000  00000 00000  00000 00000    00000 00000  00000 00000  00000 00000  00000 00000  00000 00000
```

```
9100:  00000 00000  00000 00000  00000 00000  00000 00000  00000 00000    00000 00000  00000 00000  00000 00000  00000 00000  00000 00000
9101:  00000 00000  00000 00000  00000 00000  00000 00000  00000 00000    00000 00000  00000 00000  00000 00000  00000 00000  00000 00000
9102:  00000 00000  00000 00000  00000 00000  00000 00000  00000 00000    00000 00000  00000 00000  00000 00000  00000 00000  00000 00000
9103:  00000 00000  00000 00000  00000 00000  00000 00000  00000 00000    00000 00000  00000 00000  00000 00000  00000 00000  00000 00000
9104:  00000 00000  00000 00000  00000 00000  00000 00000  00000 00000    00000 00000  00000 00000  00000 00000  00000 00000  00000 00000
9105:  00000 00000  00000 00000  00000 00000  00000 00000  00000 00000    00000 00000  00000 00000  00000 00000  00000 00000  00000 00000
9106:  00000 00000  00000 00000  00000 00000  00000 00000  00000 00000    00000 00000  00000 00000  00000 00000  00000 00000  00000 00000
9107:  00000 00000  00000 00000  00000 00000  00000 00000  00000 00000    00000 00000  00000 00000  00000 00000  00000 00000  00000 00000
9108:  00000 00000  00000 00000  00000 00000  00000 00000  00000 00000    00000 00000  00000 00000  00000 00000  00000 00000  00000 00000
9109:  00000 00000  00000 00000  00000 00000  00000 00000  00000 00000    00000 00000  00000 00000  00000 00000  00000 00000  00000 00000
9110:  00000 00000  00000 00000  00000 00000  00000 00000  00000 00000    00000 00000  00000 00000  00000 00000  00000 00000  00000 00000
9111:  00000 00000  00000 00000  00000 00000  00000 00000  00000 00000    00000 00000  00000 00000  00000 00000  00000 00000  00000 00000
9112:  00000 00000  00000 00000  00000 00000  00000 00000  00000 00000    00000 00000  00000 00000  00000 00000  00000 00000  00000 00000
9113:  00000 00000  00000 00000  00000 00000  00000 00000  00000 00000    00000 00000  00000 00000  00000 00000  00000 00000  00000 00000
9114:  00000 00000  00000 00000  00000 00000  00000 00000  00000 00000    00000 00000  00000 00000  00000 00000  00000 00000  00000 00000
9115:  00000 00000  00000 00000  00000 00000  00000 00000  00000 00000    00000 00000  00000 00000  00000 00000  00000 00000  00000 00000
9116:  00000 00000  00000 00000  00000 00000  00000 00000  00000 00000    00000 00000  00000 00000  00000 00000  00000 00000  00000 00000
9117:  00000 00000  00000 00000  00000 00000  00000 00000  00000 00000    00000 00000  00000 00000  00000 00000  00000 00000  00000 00000
9118:  00000 00000  00000 00000  00000 00000  00000 00000  00000 00000    00000 00000  00000 00000  00000 00000  00000 00000  00000 00000
9119:  00000 00000  00000 00000  00000 00000  00000 00000  00000 00000    00000 00000  00000 00000  00000 00000  00000 00000  00000 00000
9120:  00000 00000  00000 00000  00000 00000  00000 00000  00000 00000    00000 00000  00000 00000  00000 00000  00000 00000  00000 00000
9121:  00000 00000  00000 00000  00000 00000  00000 00000  00000 00000    00000 00000  00000 00000  00000 00000  00000 00000  00000 00000
9122:  00000 00000  00000 00000  00000 00000  00000 00000  00000 00000    00000 00000  00000 00000  00000 00000  00000 00000  00000 00000
9123:  00000 00000  00000 00000  00000 00000  00000 00000  00000 00000    00000 00000  00000 00000  00000 00000  00000 00000  00000 00000
9124:  00000 00000  00000 00000  00000 00000  00000 00000  00000 00000    00000 00000  00000 00000  00000 00000  00000 00000  00000 00000
9125:  00000 00000  00000 00000  00000 00000  00000 00000  00000 00000    00000 00000  00000 00000  00000 00000  00000 00000  00000 00000
9126:  00000 00000  00000 00000  00000 00000  00000 00000  00000 00000    00000 00000  00000 00000  00000 00000  00000 00000  00000 00000
9127:  00000 00000  00000 00000  00000 00000  00000 00000  00000 00000    00000 00000  00000 00000  00000 00000  00000 00000  00000 00000
9128:  00000 00000  00000 00000  00000 00000  00000 00000  00000 00000    00000 00000  00000 00000  00000 00000  00000 00000  00000 00000
9129:  00000 00000  00000 00000  00000 00000  00000 00000  00000 00000    00000 00000  00000 00000  00000 00000  00000 00000  00000 00000
9130:  00000 00000  00000 00000  00000 00000  00000 00000  00000 00000    00000 00000  00000 00000  00000 00000  00000 00000  00000 00000
9131:  00000 00000  00000 00000  00000 00000  00000 00000  00000 00000    00000 00000  00000 00000  00000 00000  00000 00000  00000 00000
9132:  00000 00000  00000 00000  00000 00000  00000 00000  00000 00000    00000 00000  00000 00000  00000 00000  00000 00000  00000 00000
9133:  00000 00000  00000 00000  00000 00000  00000 00000  00000 00000    00000 00000  00000 00000  00000 00000  00000 00000  00000 00000
9134:  00000 00000  00000 00000  00000 00000  00000 00000  00000 00000    00000 00000  00000 00000  00000 00000  00000 00000  00000 00000
9135:  00000 00000  00000 00000  00000 00000  00000 00000  00000 00000    00000 00000  00000 00000  00000 00000  00000 00000  00000 00000
9136:  00000 00000  00000 00000  00000 00000  00000 00000  00000 00000    00000 00000  00000 00000  00000 00000  00000 00000  00000 00000
9137:  00000 00000  00000 00000  00000 00000  00000 00000  00000 00000    00000 00000  00000 00000  00000 00000  00000 00000  00000 00000
9138:  00000 00000  00000 00000  00000 00000  00000 00000  00000 00000    00000 00000  00000 00000  00000 00000  00000 00000  00000 00000
9139:  00000 00000  00000 00000  00000 00000  00000 00000  00000 00000    00000 00000  00000 00000  00000 00000  00000 00000  00000 00000
9140:  00000 00000  00000 00000  00000 00000  00000 00000  00000 00000    00000 00000  00000 00000  00000 00000  00000 00000  00000 00000
9141:  00000 00000  00000 00000  00000 00000  00000 00000  00000 00000    00000 00000  00000 00000  00000 00000  00000 00000  00000 00000
9142:  00000 00000  00000 00000  00000 00000  00000 00000  00000 00000    00000 00000  00000 00000  00000 00000  00000 00000  00000 00000
9143:  00000 00000  00000 00000  00000 00000  00000 00000  00000 00000    00000 00000  00000 00000  00000 00000  00000 00000  00000 00000
9144:  00000 00000  00000 00000  00000 00000  00000 00000  00000 00000    00000 00000  00000 00000  00000 00000  00000 00000  00000 00000
9145:  00000 00000  00000 00000  00000 00000  00000 00000  00000 00000    00000 00000  00000 00000  00000 00000  00000 00000  00000 00000
9146:  00000 00000  00000 00000  00000 00000  00000 00000  00000 00000    00000 00000  00000 00000  00000 00000  00000 00000  00000 00000
9147:  00000 00000  00000 00000  00000 00000  00000 00000  00000 00000    00000 00000  00000 00000  00000 00000  00000 00000  00000 00000
9148:  00000 00000  00000 00000  00000 00000  00000 00000  00000 00000    00000 00000  00000 00000  00000 00000  00000 00000  00000 00000
9149:  00000 00000  00000 00000  00000 00000  00000 00000  00000 00000    00000 00000  00000 00000  00000 00000  00000 00000  00000 00000
```

```
9150:  00000 00000  00000 00000  00000 00000  00000 00000  00000 00000    00000 00000  00000 00000  00000 00000  00000 00000  00000 00000
9151:  00000 00000  00000 00000  00000 00000  00000 00000  00000 00000    00000 00000  00000 00000  00000 00000  00000 00000  00000 00000
9152:  00000 00000  00000 00000  00000 00000  00000 00000  00000 00000    00000 00000  00000 00000  00000 00000  00000 00000  00000 00000
9153:  00000 00000  00000 00000  00000 00000  00000 00000  00000 00000    00000 00000  00000 00000  00000 00000  00000 00000  00000 00000
9154:  00000 00000  00000 00000  00000 00000  00000 00000  00000 00000    00000 00000  00000 00000  00000 00000  00000 00000  00000 00000
9155:  00000 00000  00000 00000  00000 00000  00000 00000  00000 00000    00000 00000  00000 00000  00000 00000  00000 00000  00000 00000
9156:  00000 00000  00000 00000  00000 00000  00000 00000  00000 00000    00000 00000  00000 00000  00000 00000  00000 00000  00000 00000
9157:  00000 00000  00000 00000  00000 00000  00000 00000  00000 00000    00000 00000  00000 00000  00000 00000  00000 00000  00000 00000
9158:  00000 00000  00000 00000  00000 00000  00000 00000  00000 00000    00000 00000  00000 00000  00000 00000  00000 00000  00000 00000
9159:  00000 00000  00000 00000  00000 00000  00000 00000  00000 00000    00000 00000  00000 00000  00000 00000  00000 00000  00000 00000
9160:  00000 00000  00000 00000  00000 00000  00000 00000  00000 00000    00000 00000  00000 00000  00000 00000  00000 00000  00000 00000
9161:  00000 00000  00000 00000  00000 00000  00000 00000  00000 00000    00000 00000  00000 00000  00000 00000  00000 00000  00000 00000
9162:  00000 00000  00000 00000  00000 00000  00000 00000  00000 00000    00000 00000  00000 00000  00000 00000  00000 00000  00000 00000
9163:  00000 00000  00000 00000  00000 00000  00000 00000  00000 00000    00000 00000  00000 00000  00000 00000  00000 00000  00000 00000
9164:  00000 00000  00000 00000  00000 00000  00000 00000  00000 00000    00000 00000  00000 00000  00000 00000  00000 00000  00000 00000
9165:  00000 00000  00000 00000  00000 00000  00000 00000  00000 00000    00000 00000  00000 00000  00000 00000  00000 00000  00000 00000
9166:  00000 00000  00000 00000  00000 00000  00000 00000  00000 00000    00000 00000  00000 00000  00000 00000  00000 00000  00000 00000
9167:  00000 00000  00000 00000  00000 00000  00000 00000  00000 00000    00000 00000  00000 00000  00000 00000  00000 00000  00000 00000
9168:  00000 00000  00000 00000  00000 00000  00000 00000  00000 00000    00000 00000  00000 00000  00000 00000  00000 00000  00000 00000
9169:  00000 00000  00000 00000  00000 00000  00000 00000  00000 00000    00000 00000  00000 00000  00000 00000  00000 00000  00000 00000
9170:  00000 00000  00000 00000  00000 00000  00000 00000  00000 00000    00000 00000  00000 00000  00000 00000  00000 00000  00000 00000
9171:  00000 00000  00000 00000  00000 00000  00000 00000  00000 00000    00000 00000  00000 00000  00000 00000  00000 00000  00000 00000
9172:  00000 00000  00000 00000  00000 00000  00000 00000  00000 00000    00000 00000  00000 00000  00000 00000  00000 00000  00000 00000
9173:  00000 00000  00000 00000  00000 00000  00000 00000  00000 00000    00000 00000  00000 00000  00000 00000  00000 00000  00000 00000
9174:  00000 00000  00000 00000  00000 00000  00000 00000  00000 00000    00000 00000  00000 00000  00000 00000  00000 00000  00000 00000
9175:  00000 00000  00000 00000  00000 00000  00000 00000  00000 00000    00000 00000  00000 00000  00000 00000  00000 00000  00000 00000
9176:  00000 00000  00000 00000  00000 00000  00000 00000  00000 00000    00000 00000  00000 00000  00000 00000  00000 00000  00000 00000
9177:  00000 00000  00000 00000  00000 00000  00000 00000  00000 00000    00000 00000  00000 00000  00000 00000  00000 00000  00000 00000
9178:  00000 00000  00000 00000  00000 00000  00000 00000  00000 00000    00000 00000  00000 00000  00000 00000  00000 00000  00000 00000
9179:  00000 00000  00000 00000  00000 00000  00000 00000  00000 00000    00000 00000  00000 00000  00000 00000  00000 00000  00000 00000
9180:  00000 00000  00000 00000  00000 00000  00000 00000  00000 00000    00000 00000  00000 00000  00000 00000  00000 00000  00000 00000
9181:  00000 00000  00000 00000  00000 00000  00000 00000  00000 00000    00000 00000  00000 00000  00000 00000  00000 00000  00000 00000
9182:  00000 00000  00000 00000  00000 00000  00000 00000  00000 00000    00000 00000  00000 00000  00000 00000  00000 00000  00000 00000
9183:  00000 00000  00000 00000  00000 00000  00000 00000  00000 00000    00000 00000  00000 00000  00000 00000  00000 00000  00000 00000
9184:  00000 00000  00000 00000  00000 00000  00000 00000  00000 00000    00000 00000  00000 00000  00000 00000  00000 00000  00000 00000
9185:  00000 00000  00000 00000  00000 00000  00000 00000  00000 00000    00000 00000  00000 00000  00000 00000  00000 00000  00000 00000
9186:  00000 00000  00000 00000  00000 00000  00000 00000  00000 00000    00000 00000  00000 00000  00000 00000  00000 00000  00000 00000
9187:  00000 00000  00000 00000  00000 00000  00000 00000  00000 00000    00000 00000  00000 00000  00000 00000  00000 00000  00000 00000
9188:  00000 00000  00000 00000  00000 00000  00000 00000  00000 00000    00000 00000  00000 00000  00000 00000  00000 00000  00000 00000
9189:  00000 00000  00000 00000  00000 00000  00000 00000  00000 00000    00000 00000  00000 00000  00000 00000  00000 00000  00000 00000
9190:  00000 00000  00000 00000  00000 00000  00000 00000  00000 00000    00000 00000  00000 00000  00000 00000  00000 00000  00000 00000
9191:  00000 00000  00000 00000  00000 00000  00000 00000  00000 00000    00000 00000  00000 00000  00000 00000  00000 00000  00000 00000
9192:  00000 00000  00000 00000  00000 00000  00000 00000  00000 00000    00000 00000  00000 00000  00000 00000  00000 00000  00000 00000
9193:  00000 00000  00000 00000  00000 00000  00000 00000  00000 00000    00000 00000  00000 00000  00000 00000  00000 00000  00000 00000
9194:  00000 00000  00000 00000  00000 00000  00000 00000  00000 00000    00000 00000  00000 00000  00000 00000  00000 00000  00000 00000
9195:  00000 00000  00000 00000  00000 00000  00000 00000  00000 00000    00000 00000  00000 00000  00000 00000  00000 00000  00000 00000
9196:  00000 00000  00000 00000  00000 00000  00000 00000  00000 00000    00000 00000  00000 00000  00000 00000  00000 00000  00000 00000
9197:  00000 00000  00000 00000  00000 00000  00000 00000  00000 00000    00000 00000  00000 00000  00000 00000  00000 00000  00000 00000
9198:  00000 00000  00000 00000  00000 00000  00000 00000  00000 00000    00000 00000  00000 00000  00000 00000  00000 00000  00000 00000
9199:  00000 00000  00000 00000  00000 00000  00000 00000  00000 00000    00000 00000  00000 00000  00000 00000  00000 00000  00000 00000
```

```
9200:  00000 00000  00000 00000  00000 00000  00000 00000  00000 00000    00000 00000  00000 00000  00000 00000  00000 00000  00000 00000
9201:  00000 00000  00000 00000  00000 00000  00000 00000  00000 00000    00000 00000  00000 00000  00000 00000  00000 00000  00000 00000
9202:  00000 00000  00000 00000  00000 00000  00000 00000  00000 00000    00000 00000  00000 00000  00000 00000  00000 00000  00000 00000
9203:  00000 00000  00000 00000  00000 00000  00000 00000  00000 00000    00000 00000  00000 00000  00000 00000  00000 00000  00000 00000
9204:  00000 00000  00000 00000  00000 00000  00000 00000  00000 00000    00000 00000  00000 00000  00000 00000  00000 00000  00000 00000
9205:  00000 00000  00000 00000  00000 00000  00000 00000  00000 00000    00000 00000  00000 00000  00000 00000  00000 00000  00000 00000
9206:  00000 00000  00000 00000  00000 00000  00000 00000  00000 00000    00000 00000  00000 00000  00000 00000  00000 00000  00000 00000
9207:  00000 00000  00000 00000  00000 00000  00000 00000  00000 00000    00000 00000  00000 00000  00000 00000  00000 00000  00000 00000
9208:  00000 00000  00000 00000  00000 00000  00000 00000  00000 00000    00000 00000  00000 00000  00000 00000  00000 00000  00000 00000
9209:  00000 00000  00000 00000  00000 00000  00000 00000  00000 00000    00000 00000  00000 00000  00000 00000  00000 00000  00000 00000
9210:  00000 00000  00000 00000  00000 00000  00000 00000  00000 00000    00000 00000  00000 00000  00000 00000  00000 00000  00000 00000
9211:  00000 00000  00000 00000  00000 00000  00000 00000  00000 00000    00000 00000  00000 00000  00000 00000  00000 00000  00000 00000
9212:  00000 00000  00000 00000  00000 00000  00000 00000  00000 00000    00000 00000  00000 00000  00000 00000  00000 00000  00000 00000
9213:  00000 00000  00000 00000  00000 00000  00000 00000  00000 00000    00000 00000  00000 00000  00000 00000  00000 00000  00000 00000
9214:  00000 00000  00000 00000  00000 00000  00000 00000  00000 00000    00000 00000  00000 00000  00000 00000  00000 00000  00000 00000
9215:  00000 00000  00000 00000  00000 00000  00000 00000  00000 00000    00000 00000  00000 00000  00000 00000  00000 00000  00000 00000
9216:  00000 00000  00000 00000  00000 00000  00000 00000  00000 00000    00000 00000  00000 00000  00000 00000  00000 00000  00000 00000
9217:  00000 00000  00000 00000  00000 00000  00000 00000  00000 00000    00000 00000  00000 00000  00000 00000  00000 00000  00000 00000
9218:  00000 00000  00000 00000  00000 00000  00000 00000  00000 00000    00000 00000  00000 00000  00000 00000  00000 00000  00000 00000
9219:  00000 00000  00000 00000  00000 00000  00000 00000  00000 00000    00000 00000  00000 00000  00000 00000  00000 00000  00000 00000
9220:  00000 00000  00000 00000  00000 00000  00000 00000  00000 00000    00000 00000  00000 00000  00000 00000  00000 00000  00000 00000
9221:  00000 00000  00000 00000  00000 00000  00000 00000  00000 00000    00000 00000  00000 00000  00000 00000  00000 00000  00000 00000
9222:  00000 00000  00000 00000  00000 00000  00000 00000  00000 00000    00000 00000  00000 00000  00000 00000  00000 00000  00000 00000
9223:  00000 00000  00000 00000  00000 00000  00000 00000  00000 00000    00000 00000  00000 00000  00000 00000  00000 00000  00000 00000
9224:  00000 00000  00000 00000  00000 00000  00000 00000  00000 00000    00000 00000  00000 00000  00000 00000  00000 00000  00000 00000
9225:  00000 00000  00000 00000  00000 00000  00000 00000  00000 00000    00000 00000  00000 00000  00000 00000  00000 00000  00000 00000
9226:  00000 00000  00000 00000  00000 00000  00000 00000  00000 00000    00000 00000  00000 00000  00000 00000  00000 00000  00000 00000
9227:  00000 00000  00000 00000  00000 00000  00000 00000  00000 00000    00000 00000  00000 00000  00000 00000  00000 00000  00000 00000
9228:  00000 00000  00000 00000  00000 00000  00000 00000  00000 00000    00000 00000  00000 00000  00000 00000  00000 00000  00000 00000
9229:  00000 00000  00000 00000  00000 00000  00000 00000  00000 00000    00000 00000  00000 00000  00000 00000  00000 00000  00000 00000
9230:  00000 00000  00000 00000  00000 00000  00000 00000  00000 00000    00000 00000  00000 00000  00000 00000  00000 00000  00000 00000
9231:  00000 00000  00000 00000  00000 00000  00000 00000  00000 00000    00000 00000  00000 00000  00000 00000  00000 00000  00000 00000
9232:  00000 00000  00000 00000  00000 00000  00000 00000  00000 00000    00000 00000  00000 00000  00000 00000  00000 00000  00000 00000
9233:  00000 00000  00000 00000  00000 00000  00000 00000  00000 00000    00000 00000  00000 00000  00000 00000  00000 00000  00000 00000
9234:  00000 00000  00000 00000  00000 00000  00000 00000  00000 00000    00000 00000  00000 00000  00000 00000  00000 00000  00000 00000
9235:  00000 00000  00000 00000  00000 00000  00000 00000  00000 00000    00000 00000  00000 00000  00000 00000  00000 00000  00000 00000
9236:  00000 00000  00000 00000  00000 00000  00000 00000  00000 00000    00000 00000  00000 00000  00000 00000  00000 00000  00000 00000
9237:  00000 00000  00000 00000  00000 00000  00000 00000  00000 00000    00000 00000  00000 00000  00000 00000  00000 00000  00000 00000
9238:  00000 00000  00000 00000  00000 00000  00000 00000  00000 00000    00000 00000  00000 00000  00000 00000  00000 00000  00000 00000
9239:  00000 00000  00000 00000  00000 00000  00000 00000  00000 00000    00000 00000  00000 00000  00000 00000  00000 00000  00000 00000
9240:  00000 00000  00000 00000  00000 00000  00000 00000  00000 00000    00000 00000  00000 00000  00000 00000  00000 00000  00000 00000
9241:  00000 00000  00000 00000  00000 00000  00000 00000  00000 00000    00000 00000  00000 00000  00000 00000  00000 00000  00000 00000
9242:  00000 00000  00000 00000  00000 00000  00000 00000  00000 00000    00000 00000  00000 00000  00000 00000  00000 00000  00000 00000
9243:  00000 00000  00000 00000  00000 00000  00000 00000  00000 00000    00000 00000  00000 00000  00000 00000  00000 00000  00000 00000
9244:  00000 00000  00000 00000  00000 00000  00000 00000  00000 00000    00000 00000  00000 00000  00000 00000  00000 00000  00000 00000
9245:  00000 00000  00000 00000  00000 00000  00000 00000  00000 00000    00000 00000  00000 00000  00000 00000  00000 00000  00000 00000
9246:  00000 00000  00000 00000  00000 00000  00000 00000  00000 00000    00000 00000  00000 00000  00000 00000  00000 00000  00000 00000
9247:  00000 00000  00000 00000  00000 00000  00000 00000  00000 00000    00000 00000  00000 00000  00000 00000  00000 00000  00000 00000
9248:  00000 00000  00000 00000  00000 00000  00000 00000  00000 00000    00000 00000  00000 00000  00000 00000  00000 00000  00000 00000
9249:  00000 00000  00000 00000  00000 00000  00000 00000  00000 00000    00000 00000  00000 00000  00000 00000  00000 00000  00000 00000
```

```
9250:  00000 00000  00000 00000  00000 00000  00000 00000  00000 00000    00000 00000  00000 00000  00000 00000  00000 00000  00000 00000
9251:  00000 00000  00000 00000  00000 00000  00000 00000  00000 00000    00000 00000  00000 00000  00000 00000  00000 00000  00000 00000
9252:  00000 00000  00000 00000  00000 00000  00000 00000  00000 00000    00000 00000  00000 00000  00000 00000  00000 00000  00000 00000
9253:  00000 00000  00000 00000  00000 00000  00000 00000  00000 00000    00000 00000  00000 00000  00000 00000  00000 00000  00000 00000
9254:  00000 00000  00000 00000  00000 00000  00000 00000  00000 00000    00000 00000  00000 00000  00000 00000  00000 00000  00000 00000
9255:  00000 00000  00000 00000  00000 00000  00000 00000  00000 00000    00000 00000  00000 00000  00000 00000  00000 00000  00000 00000
9256:  00000 00000  00000 00000  00000 00000  00000 00000  00000 00000    00000 00000  00000 00000  00000 00000  00000 00000  00000 00000
9257:  00000 00000  00000 00000  00000 00000  00000 00000  00000 00000    00000 00000  00000 00000  00000 00000  00000 00000  00000 00000
9258:  00000 00000  00000 00000  00000 00000  00000 00000  00000 00000    00000 00000  00000 00000  00000 00000  00000 00000  00000 00000
9259:  00000 00000  00000 00000  00000 00000  00000 00000  00000 00000    00000 00000  00000 00000  00000 00000  00000 00000  00000 00000
9260:  00000 00000  00000 00000  00000 00000  00000 00000  00000 00000    00000 00000  00000 00000  00000 00000  00000 00000  00000 00000
9261:  00000 00000  00000 00000  00000 00000  00000 00000  00000 00000    00000 00000  00000 00000  00000 00000  00000 00000  00000 00000
9262:  00000 00000  00000 00000  00000 00000  00000 00000  00000 00000    00000 00000  00000 00000  00000 00000  00000 00000  00000 00000
9263:  00000 00000  00000 00000  00000 00000  00000 00000  00000 00000    00000 00000  00000 00000  00000 00000  00000 00000  00000 00000
9264:  00000 00000  00000 00000  00000 00000  00000 00000  00000 00000    00000 00000  00000 00000  00000 00000  00000 00000  00000 00000
9265:  00000 00000  00000 00000  00000 00000  00000 00000  00000 00000    00000 00000  00000 00000  00000 00000  00000 00000  00000 00000
9266:  00000 00000  00000 00000  00000 00000  00000 00000  00000 00000    00000 00000  00000 00000  00000 00000  00000 00000  00000 00000
9267:  00000 00000  00000 00000  00000 00000  00000 00000  00000 00000    00000 00000  00000 00000  00000 00000  00000 00000  00000 00000
9268:  00000 00000  00000 00000  00000 00000  00000 00000  00000 00000    00000 00000  00000 00000  00000 00000  00000 00000  00000 00000
9269:  00000 00000  00000 00000  00000 00000  00000 00000  00000 00000    00000 00000  00000 00000  00000 00000  00000 00000  00000 00000
9270:  00000 00000  00000 00000  00000 00000  00000 00000  00000 00000    00000 00000  00000 00000  00000 00000  00000 00000  00000 00000
9271:  00000 00000  00000 00000  00000 00000  00000 00000  00000 00000    00000 00000  00000 00000  00000 00000  00000 00000  00000 00000
9272:  00000 00000  00000 00000  00000 00000  00000 00000  00000 00000    00000 00000  00000 00000  00000 00000  00000 00000  00000 00000
9273:  00000 00000  00000 00000  00000 00000  00000 00000  00000 00000    00000 00000  00000 00000  00000 00000  00000 00000  00000 00000
9274:  00000 00000  00000 00000  00000 00000  00000 00000  00000 00000    00000 00000  00000 00000  00000 00000  00000 00000  00000 00000
9275:  00000 00000  00000 00000  00000 00000  00000 00000  00000 00000    00000 00000  00000 00000  00000 00000  00000 00000  00000 00000
9276:  00000 00000  00000 00000  00000 00000  00000 00000  00000 00000    00000 00000  00000 00000  00000 00000  00000 00000  00000 00000
9277:  00000 00000  00000 00000  00000 00000  00000 00000  00000 00000    00000 00000  00000 00000  00000 00000  00000 00000  00000 00000
9278:  00000 00000  00000 00000  00000 00000  00000 00000  00000 00000    00000 00000  00000 00000  00000 00000  00000 00000  00000 00000
9279:  00000 00000  00000 00000  00000 00000  00000 00000  00000 00000    00000 00000  00000 00000  00000 00000  00000 00000  00000 00000
9280:  00000 00000  00000 00000  00000 00000  00000 00000  00000 00000    00000 00000  00000 00000  00000 00000  00000 00000  00000 00000
9281:  00000 00000  00000 00000  00000 00000  00000 00000  00000 00000    00000 00000  00000 00000  00000 00000  00000 00000  00000 00000
9282:  00000 00000  00000 00000  00000 00000  00000 00000  00000 00000    00000 00000  00000 00000  00000 00000  00000 00000  00000 00000
9283:  00000 00000  00000 00000  00000 00000  00000 00000  00000 00000    00000 00000  00000 00000  00000 00000  00000 00000  00000 00000
9284:  00000 00000  00000 00000  00000 00000  00000 00000  00000 00000    00000 00000  00000 00000  00000 00000  00000 00000  00000 00000
9285:  00000 00000  00000 00000  00000 00000  00000 00000  00000 00000    00000 00000  00000 00000  00000 00000  00000 00000  00000 00000
9286:  00000 00000  00000 00000  00000 00000  00000 00000  00000 00000    00000 00000  00000 00000  00000 00000  00000 00000  00000 00000
9287:  00000 00000  00000 00000  00000 00000  00000 00000  00000 00000    00000 00000  00000 00000  00000 00000  00000 00000  00000 00000
9288:  00000 00000  00000 00000  00000 00000  00000 00000  00000 00000    00000 00000  00000 00000  00000 00000  00000 00000  00000 00000
9289:  00000 00000  00000 00000  00000 00000  00000 00000  00000 00000    00000 00000  00000 00000  00000 00000  00000 00000  00000 00000
9290:  00000 00000  00000 00000  00000 00000  00000 00000  00000 00000    00000 00000  00000 00000  00000 00000  00000 00000  00000 00000
9291:  00000 00000  00000 00000  00000 00000  00000 00000  00000 00000    00000 00000  00000 00000  00000 00000  00000 00000  00000 00000
9292:  00000 00000  00000 00000  00000 00000  00000 00000  00000 00000    00000 00000  00000 00000  00000 00000  00000 00000  00000 00000
9293:  00000 00000  00000 00000  00000 00000  00000 00000  00000 00000    00000 00000  00000 00000  00000 00000  00000 00000  00000 00000
9294:  00000 00000  00000 00000  00000 00000  00000 00000  00000 00000    00000 00000  00000 00000  00000 00000  00000 00000  00000 00000
9295:  00000 00000  00000 00000  00000 00000  00000 00000  00000 00000    00000 00000  00000 00000  00000 00000  00000 00000  00000 00000
9296:  00000 00000  00000 00000  00000 00000  00000 00000  00000 00000    00000 00000  00000 00000  00000 00000  00000 00000  00000 00000
9297:  00000 00000  00000 00000  00000 00000  00000 00000  00000 00000    00000 00000  00000 00000  00000 00000  00000 00000  00000 00000
9298:  00000 00000  00000 00000  00000 00000  00000 00000  00000 00000    00000 00000  00000 00000  00000 00000  00000 00000  00000 00000
9299:  00000 00000  00000 00000  00000 00000  00000 00000  00000 00000    00000 00000  00000 00000  00000 00000  00000 00000  00000 00000
```

```
9300:  00000 00000  00000 00000  00000 00000  00000 00000  00000 00000    00000 00000  00000 00000  00000 00000  00000 00000  00000 00000
9301:  00000 00000  00000 00000  00000 00000  00000 00000  00000 00000    00000 00000  00000 00000  00000 00000  00000 00000  00000 00000
9302:  00000 00000  00000 00000  00000 00000  00000 00000  00000 00000    00000 00000  00000 00000  00000 00000  00000 00000  00000 00000
9303:  00000 00000  00000 00000  00000 00000  00000 00000  00000 00000    00000 00000  00000 00000  00000 00000  00000 00000  00000 00000
9304:  00000 00000  00000 00000  00000 00000  00000 00000  00000 00000    00000 00000  00000 00000  00000 00000  00000 00000  00000 00000
9305:  00000 00000  00000 00000  00000 00000  00000 00000  00000 00000    00000 00000  00000 00000  00000 00000  00000 00000  00000 00000
9306:  00000 00000  00000 00000  00000 00000  00000 00000  00000 00000    00000 00000  00000 00000  00000 00000  00000 00000  00000 00000
9307:  00000 00000  00000 00000  00000 00000  00000 00000  00000 00000    00000 00000  00000 00000  00000 00000  00000 00000  00000 00000
9308:  00000 00000  00000 00000  00000 00000  00000 00000  00000 00000    00000 00000  00000 00000  00000 00000  00000 00000  00000 00000
9309:  00000 00000  00000 00000  00000 00000  00000 00000  00000 00000    00000 00000  00000 00000  00000 00000  00000 00000  00000 00000
9310:  00000 00000  00000 00000  00000 00000  00000 00000  00000 00000    00000 00000  00000 00000  00000 00000  00000 00000  00000 00000
9311:  00000 00000  00000 00000  00000 00000  00000 00000  00000 00000    00000 00000  00000 00000  00000 00000  00000 00000  00000 00000
9312:  00000 00000  00000 00000  00000 00000  00000 00000  00000 00000    00000 00000  00000 00000  00000 00000  00000 00000  00000 00000
9313:  00000 00000  00000 00000  00000 00000  00000 00000  00000 00000    00000 00000  00000 00000  00000 00000  00000 00000  00000 00000
9314:  00000 00000  00000 00000  00000 00000  00000 00000  00000 00000    00000 00000  00000 00000  00000 00000  00000 00000  00000 00000
9315:  00000 00000  00000 00000  00000 00000  00000 00000  00000 00000    00000 00000  00000 00000  00000 00000  00000 00000  00000 00000
9316:  00000 00000  00000 00000  00000 00000  00000 00000  00000 00000    00000 00000  00000 00000  00000 00000  00000 00000  00000 00000
9317:  00000 00000  00000 00000  00000 00000  00000 00000  00000 00000    00000 00000  00000 00000  00000 00000  00000 00000  00000 00000
9318:  00000 00000  00000 00000  00000 00000  00000 00000  00000 00000    00000 00000  00000 00000  00000 00000  00000 00000  00000 00000
9319:  00000 00000  00000 00000  00000 00000  00000 00000  00000 00000    00000 00000  00000 00000  00000 00000  00000 00000  00000 00000
9320:  00000 00000  00000 00000  00000 00000  00000 00000  00000 00000    00000 00000  00000 00000  00000 00000  00000 00000  00000 00000
9321:  00000 00000  00000 00000  00000 00000  00000 00000  00000 00000    00000 00000  00000 00000  00000 00000  00000 00000  00000 00000
9322:  00000 00000  00000 00000  00000 00000  00000 00000  00000 00000    00000 00000  00000 00000  00000 00000  00000 00000  00000 00000
9323:  00000 00000  00000 00000  00000 00000  00000 00000  00000 00000    00000 00000  00000 00000  00000 00000  00000 00000  00000 00000
9324:  00000 00000  00000 00000  00000 00000  00000 00000  00000 00000    00000 00000  00000 00000  00000 00000  00000 00000  00000 00000
9325:  00000 00000  00000 00000  00000 00000  00000 00000  00000 00000    00000 00000  00000 00000  00000 00000  00000 00000  00000 00000
9326:  00000 00000  00000 00000  00000 00000  00000 00000  00000 00000    00000 00000  00000 00000  00000 00000  00000 00000  00000 00000
9327:  00000 00000  00000 00000  00000 00000  00000 00000  00000 00000    00000 00000  00000 00000  00000 00000  00000 00000  00000 00000
9328:  00000 00000  00000 00000  00000 00000  00000 00000  00000 00000    00000 00000  00000 00000  00000 00000  00000 00000  00000 00000
9329:  00000 00000  00000 00000  00000 00000  00000 00000  00000 00000    00000 00000  00000 00000  00000 00000  00000 00000  00000 00000
9330:  00000 00000  00000 00000  00000 00000  00000 00000  00000 00000    00000 00000  00000 00000  00000 00000  00000 00000  00000 00000
9331:  00000 00000  00000 00000  00000 00000  00000 00000  00000 00000    00000 00000  00000 00000  00000 00000  00000 00000  00000 00000
9332:  00000 00000  00000 00000  00000 00000  00000 00000  00000 00000    00000 00000  00000 00000  00000 00000  00000 00000  00000 00000
9333:  00000 00000  00000 00000  00000 00000  00000 00000  00000 00000    00000 00000  00000 00000  00000 00000  00000 00000  00000 00000
9334:  00000 00000  00000 00000  00000 00000  00000 00000  00000 00000    00000 00000  00000 00000  00000 00000  00000 00000  00000 00000
9335:  00000 00000  00000 00000  00000 00000  00000 00000  00000 00000    00000 00000  00000 00000  00000 00000  00000 00000  00000 00000
9336:  00000 00000  00000 00000  00000 00000  00000 00000  00000 00000    00000 00000  00000 00000  00000 00000  00000 00000  00000 00000
9337:  00000 00000  00000 00000  00000 00000  00000 00000  00000 00000    00000 00000  00000 00000  00000 00000  00000 00000  00000 00000
9338:  00000 00000  00000 00000  00000 00000  00000 00000  00000 00000    00000 00000  00000 00000  00000 00000  00000 00000  00000 00000
9339:  00000 00000  00000 00000  00000 00000  00000 00000  00000 00000    00000 00000  00000 00000  00000 00000  00000 00000  00000 00000
9340:  00000 00000  00000 00000  00000 00000  00000 00000  00000 00000    00000 00000  00000 00000  00000 00000  00000 00000  00000 00000
9341:  00000 00000  00000 00000  00000 00000  00000 00000  00000 00000    00000 00000  00000 00000  00000 00000  00000 00000  00000 00000
9342:  00000 00000  00000 00000  00000 00000  00000 00000  00000 00000    00000 00000  00000 00000  00000 00000  00000 00000  00000 00000
9343:  00000 00000  00000 00000  00000 00000  00000 00000  00000 00000    00000 00000  00000 00000  00000 00000  00000 00000  00000 00000
9344:  00000 00000  00000 00000  00000 00000  00000 00000  00000 00000    00000 00000  00000 00000  00000 00000  00000 00000  00000 00000
9345:  00000 00000  00000 00000  00000 00000  00000 00000  00000 00000    00000 00000  00000 00000  00000 00000  00000 00000  00000 00000
9346:  00000 00000  00000 00000  00000 00000  00000 00000  00000 00000    00000 00000  00000 00000  00000 00000  00000 00000  00000 00000
9347:  00000 00000  00000 00000  00000 00000  00000 00000  00000 00000    00000 00000  00000 00000  00000 00000  00000 00000  00000 00000
9348:  00000 00000  00000 00000  00000 00000  00000 00000  00000 00000    00000 00000  00000 00000  00000 00000  00000 00000  00000 00000
9349:  00000 00000  00000 00000  00000 00000  00000 00000  00000 00000    00000 00000  00000 00000  00000 00000  00000 00000  00000 00000
```

```
9350:  00000 00000  00000 00000  00000 00000  00000 00000  00000 00000   00000 00000  00000 00000  00000 00000  00000 00000  00000 00000
9351:  00000 00000  00000 00000  00000 00000  00000 00000  00000 00000   00000 00000  00000 00000  00000 00000  00000 00000  00000 00000
9352:  00000 00000  00000 00000  00000 00000  00000 00000  00000 00000   00000 00000  00000 00000  00000 00000  00000 00000  00000 00000
9353:  00000 00000  00000 00000  00000 00000  00000 00000  00000 00000   00000 00000  00000 00000  00000 00000  00000 00000  00000 00000
9354:  00000 00000  00000 00000  00000 00000  00000 00000  00000 00000   00000 00000  00000 00000  00000 00000  00000 00000  00000 00000
9355:  00000 00000  00000 00000  00000 00000  00000 00000  00000 00000   00000 00000  00000 00000  00000 00000  00000 00000  00000 00000
9356:  00000 00000  00000 00000  00000 00000  00000 00000  00000 00000   00000 00000  00000 00000  00000 00000  00000 00000  00000 00000
9357:  00000 00000  00000 00000  00000 00000  00000 00000  00000 00000   00000 00000  00000 00000  00000 00000  00000 00000  00000 00000
9358:  00000 00000  00000 00000  00000 00000  00000 00000  00000 00000   00000 00000  00000 00000  00000 00000  00000 00000  00000 00000
9359:  00000 00000  00000 00000  00000 00000  00000 00000  00000 00000   00000 00000  00000 00000  00000 00000  00000 00000  00000 00000
9360:  00000 00000  00000 00000  00000 00000  00000 00000  00000 00000   00000 00000  00000 00000  00000 00000  00000 00000  00000 00000
9361:  00000 00000  00000 00000  00000 00000  00000 00000  00000 00000   00000 00000  00000 00000  00000 00000  00000 00000  00000 00000
9362:  00000 00000  00000 00000  00000 00000  00000 00000  00000 00000   00000 00000  00000 00000  00000 00000  00000 00000  00000 00000
9363:  00000 00000  00000 00000  00000 00000  00000 00000  00000 00000   00000 00000  00000 00000  00000 00000  00000 00000  00000 00000
9364:  00000 00000  00000 00000  00000 00000  00000 00000  00000 00000   00000 00000  00000 00000  00000 00000  00000 00000  00000 00000
9365:  00000 00000  00000 00000  00000 00000  00000 00000  00000 00000   00000 00000  00000 00000  00000 00000  00000 00000  00000 00000
9366:  00000 00000  00000 00000  00000 00000  00000 00000  00000 00000   00000 00000  00000 00000  00000 00000  00000 00000  00000 00000
9367:  00000 00000  00000 00000  00000 00000  00000 00000  00000 00000   00000 00000  00000 00000  00000 00000  00000 00000  00000 00000
9368:  00000 00000  00000 00000  00000 00000  00000 00000  00000 00000   00000 00000  00000 00000  00000 00000  00000 00000  00000 00000
9369:  00000 00000  00000 00000  00000 00000  00000 00000  00000 00000   00000 00000  00000 00000  00000 00000  00000 00000  00000 00000
9370:  00000 00000  00000 00000  00000 00000  00000 00000  00000 00000   00000 00000  00000 00000  00000 00000  00000 00000  00000 00000
9371:  00000 00000  00000 00000  00000 00000  00000 00000  00000 00000   00000 00000  00000 00000  00000 00000  00000 00000  00000 00000
9372:  00000 00000  00000 00000  00000 00000  00000 00000  00000 00000   00000 00000  00000 00000  00000 00000  00000 00000  00000 00000
9373:  00000 00000  00000 00000  00000 00000  00000 00000  00000 00000   00000 00000  00000 00000  00000 00000  00000 00000  00000 00000
9374:  00000 00000  00000 00000  00000 00000  00000 00000  00000 00000   00000 00000  00000 00000  00000 00000  00000 00000  00000 00000
9375:  00000 00000  00000 00000  00000 00000  00000 00000  00000 00000   00000 00000  00000 00000  00000 00000  00000 00000  00000 00000
9376:  00000 00000  00000 00000  00000 00000  00000 00000  00000 00000   00000 00000  00000 00000  00000 00000  00000 00000  00000 00000
9377:  00000 00000  00000 00000  00000 00000  00000 00000  00000 00000   00000 00000  00000 00000  00000 00000  00000 00000  00000 00000
9378:  00000 00000  00000 00000  00000 00000  00000 00000  00000 00000   00000 00000  00000 00000  00000 00000  00000 00000  00000 00000
9379:  00000 00000  00000 00000  00000 00000  00000 00000  00000 00000   00000 00000  00000 00000  00000 00000  00000 00000  00000 00000
9380:  00000 00000  00000 00000  00000 00000  00000 00000  00000 00000   00000 00000  00000 00000  00000 00000  00000 00000  00000 00000
9381:  00000 00000  00000 00000  00000 00000  00000 00000  00000 00000   00000 00000  00000 00000  00000 00000  00000 00000  00000 00000
9382:  00000 00000  00000 00000  00000 00000  00000 00000  00000 00000   00000 00000  00000 00000  00000 00000  00000 00000  00000 00000
9383:  00000 00000  00000 00000  00000 00000  00000 00000  00000 00000   00000 00000  00000 00000  00000 00000  00000 00000  00000 00000
9384:  00000 00000  00000 00000  00000 00000  00000 00000  00000 00000   00000 00000  00000 00000  00000 00000  00000 00000  00000 00000
9385:  00000 00000  00000 00000  00000 00000  00000 00000  00000 00000   00000 00000  00000 00000  00000 00000  00000 00000  00000 00000
9386:  00000 00000  00000 00000  00000 00000  00000 00000  00000 00000   00000 00000  00000 00000  00000 00000  00000 00000  00000 00000
9387:  00000 00000  00000 00000  00000 00000  00000 00000  00000 00000   00000 00000  00000 00000  00000 00000  00000 00000  00000 00000
9388:  00000 00000  00000 00000  00000 00000  00000 00000  00000 00000   00000 00000  00000 00000  00000 00000  00000 00000  00000 00000
9389:  00000 00000  00000 00000  00000 00000  00000 00000  00000 00000   00000 00000  00000 00000  00000 00000  00000 00000  00000 00000
9390:  00000 00000  00000 00000  00000 00000  00000 00000  00000 00000   00000 00000  00000 00000  00000 00000  00000 00000  00000 00000
9391:  00000 00000  00000 00000  00000 00000  00000 00000  00000 00000   00000 00000  00000 00000  00000 00000  00000 00000  00000 00000
9392:  00000 00000  00000 00000  00000 00000  00000 00000  00000 00000   00000 00000  00000 00000  00000 00000  00000 00000  00000 00000
9393:  00000 00000  00000 00000  00000 00000  00000 00000  00000 00000   00000 00000  00000 00000  00000 00000  00000 00000  00000 00000
9394:  00000 00000  00000 00000  00000 00000  00000 00000  00000 00000   00000 00000  00000 00000  00000 00000  00000 00000  00000 00000
9395:  00000 00000  00000 00000  00000 00000  00000 00000  00000 00000   00000 00000  00000 00000  00000 00000  00000 00000  00000 00000
9396:  00000 00000  00000 00000  00000 00000  00000 00000  00000 00000   00000 00000  00000 00000  00000 00000  00000 00000  00000 00000
9397:  00000 00000  00000 00000  00000 00000  00000 00000  00000 00000   00000 00000  00000 00000  00000 00000  00000 00000  00000 00000
9398:  00000 00000  00000 00000  00000 00000  00000 00000  00000 00000   00000 00000  00000 00000  00000 00000  00000 00000  00000 00000
9399:  00000 00000  00000 00000  00000 00000  00000 00000  00000 00000   00000 00000  00000 00000  00000 00000  00000 00000  00000 00000
```

```
9400:  00000 00000  00000 00000  00000 00000  00000 00000  00000 00000    00000 00000  00000 00000  00000 00000  00000 00000  00000 00000
9401:  00000 00000  00000 00000  00000 00000  00000 00000  00000 00000    00000 00000  00000 00000  00000 00000  00000 00000  00000 00000
9402:  00000 00000  00000 00000  00000 00000  00000 00000  00000 00000    00000 00000  00000 00000  00000 00000  00000 00000  00000 00000
9403:  00000 00000  00000 00000  00000 00000  00000 00000  00000 00000    00000 00000  00000 00000  00000 00000  00000 00000  00000 00000
9404:  00000 00000  00000 00000  00000 00000  00000 00000  00000 00000    00000 00000  00000 00000  00000 00000  00000 00000  00000 00000
9405:  00000 00000  00000 00000  00000 00000  00000 00000  00000 00000    00000 00000  00000 00000  00000 00000  00000 00000  00000 00000
9406:  00000 00000  00000 00000  00000 00000  00000 00000  00000 00000    00000 00000  00000 00000  00000 00000  00000 00000  00000 00000
9407:  00000 00000  00000 00000  00000 00000  00000 00000  00000 00000    00000 00000  00000 00000  00000 00000  00000 00000  00000 00000
9408:  00000 00000  00000 00000  00000 00000  00000 00000  00000 00000    00000 00000  00000 00000  00000 00000  00000 00000  00000 00000
9409:  00000 00000  00000 00000  00000 00000  00000 00000  00000 00000    00000 00000  00000 00000  00000 00000  00000 00000  00000 00000
9410:  00000 00000  00000 00000  00000 00000  00000 00000  00000 00000    00000 00000  00000 00000  00000 00000  00000 00000  00000 00000
9411:  00000 00000  00000 00000  00000 00000  00000 00000  00000 00000    00000 00000  00000 00000  00000 00000  00000 00000  00000 00000
9412:  00000 00000  00000 00000  00000 00000  00000 00000  00000 00000    00000 00000  00000 00000  00000 00000  00000 00000  00000 00000
9413:  00000 00000  00000 00000  00000 00000  00000 00000  00000 00000    00000 00000  00000 00000  00000 00000  00000 00000  00000 00000
9414:  00000 00000  00000 00000  00000 00000  00000 00000  00000 00000    00000 00000  00000 00000  00000 00000  00000 00000  00000 00000
9415:  00000 00000  00000 00000  00000 00000  00000 00000  00000 00000    00000 00000  00000 00000  00000 00000  00000 00000  00000 00000
9416:  00000 00000  00000 00000  00000 00000  00000 00000  00000 00000    00000 00000  00000 00000  00000 00000  00000 00000  00000 00000
9417:  00000 00000  00000 00000  00000 00000  00000 00000  00000 00000    00000 00000  00000 00000  00000 00000  00000 00000  00000 00000
9418:  00000 00000  00000 00000  00000 00000  00000 00000  00000 00000    00000 00000  00000 00000  00000 00000  00000 00000  00000 00000
9419:  00000 00000  00000 00000  00000 00000  00000 00000  00000 00000    00000 00000  00000 00000  00000 00000  00000 00000  00000 00000
9420:  00000 00000  00000 00000  00000 00000  00000 00000  00000 00000    00000 00000  00000 00000  00000 00000  00000 00000  00000 00000
9421:  00000 00000  00000 00000  00000 00000  00000 00000  00000 00000    00000 00000  00000 00000  00000 00000  00000 00000  00000 00000
9422:  00000 00000  00000 00000  00000 00000  00000 00000  00000 00000    00000 00000  00000 00000  00000 00000  00000 00000  00000 00000
9423:  00000 00000  00000 00000  00000 00000  00000 00000  00000 00000    00000 00000  00000 00000  00000 00000  00000 00000  00000 00000
9424:  00000 00000  00000 00000  00000 00000  00000 00000  00000 00000    00000 00000  00000 00000  00000 00000  00000 00000  00000 00000
9425:  00000 00000  00000 00000  00000 00000  00000 00000  00000 00000    00000 00000  00000 00000  00000 00000  00000 00000  00000 00000
9426:  00000 00000  00000 00000  00000 00000  00000 00000  00000 00000    00000 00000  00000 00000  00000 00000  00000 00000  00000 00000
9427:  00000 00000  00000 00000  00000 00000  00000 00000  00000 00000    00000 00000  00000 00000  00000 00000  00000 00000  00000 00000
9428:  00000 00000  00000 00000  00000 00000  00000 00000  00000 00000    00000 00000  00000 00000  00000 00000  00000 00000  00000 00000
9429:  00000 00000  00000 00000  00000 00000  00000 00000  00000 00000    00000 00000  00000 00000  00000 00000  00000 00000  00000 00000
9430:  00000 00000  00000 00000  00000 00000  00000 00000  00000 00000    00000 00000  00000 00000  00000 00000  00000 00000  00000 00000
9431:  00000 00000  00000 00000  00000 00000  00000 00000  00000 00000    00000 00000  00000 00000  00000 00000  00000 00000  00000 00000
9432:  00000 00000  00000 00000  00000 00000  00000 00000  00000 00000    00000 00000  00000 00000  00000 00000  00000 00000  00000 00000
9433:  00000 00000  00000 00000  00000 00000  00000 00000  00000 00000    00000 00000  00000 00000  00000 00000  00000 00000  00000 00000
9434:  00000 00000  00000 00000  00000 00000  00000 00000  00000 00000    00000 00000  00000 00000  00000 00000  00000 00000  00000 00000
9435:  00000 00000  00000 00000  00000 00000  00000 00000  00000 00000    00000 00000  00000 00000  00000 00000  00000 00000  00000 00000
9436:  00000 00000  00000 00000  00000 00000  00000 00000  00000 00000    00000 00000  00000 00000  00000 00000  00000 00000  00000 00000
9437:  00000 00000  00000 00000  00000 00000  00000 00000  00000 00000    00000 00000  00000 00000  00000 00000  00000 00000  00000 00000
9438:  00000 00000  00000 00000  00000 00000  00000 00000  00000 00000    00000 00000  00000 00000  00000 00000  00000 00000  00000 00000
9439:  00000 00000  00000 00000  00000 00000  00000 00000  00000 00000    00000 00000  00000 00000  00000 00000  00000 00000  00000 00000
9440:  00000 00000  00000 00000  00000 00000  00000 00000  00000 00000    00000 00000  00000 00000  00000 00000  00000 00000  00000 00000
9441:  00000 00000  00000 00000  00000 00000  00000 00000  00000 00000    00000 00000  00000 00000  00000 00000  00000 00000  00000 00000
9442:  00000 00000  00000 00000  00000 00000  00000 00000  00000 00000    00000 00000  00000 00000  00000 00000  00000 00000  00000 00000
9443:  00000 00000  00000 00000  00000 00000  00000 00000  00000 00000    00000 00000  00000 00000  00000 00000  00000 00000  00000 00000
9444:  00000 00000  00000 00000  00000 00000  00000 00000  00000 00000    00000 00000  00000 00000  00000 00000  00000 00000  00000 00000
9445:  00000 00000  00000 00000  00000 00000  00000 00000  00000 00000    00000 00000  00000 00000  00000 00000  00000 00000  00000 00000
9446:  00000 00000  00000 00000  00000 00000  00000 00000  00000 00000    00000 00000  00000 00000  00000 00000  00000 00000  00000 00000
9447:  00000 00000  00000 00000  00000 00000  00000 00000  00000 00000    00000 00000  00000 00000  00000 00000  00000 00000  00000 00000
9448:  00000 00000  00000 00000  00000 00000  00000 00000  00000 00000    00000 00000  00000 00000  00000 00000  00000 00000  00000 00000
9449:  00000 00000  00000 00000  00000 00000  00000 00000  00000 00000    00000 00000  00000 00000  00000 00000  00000 00000  00000 00000
```

```
9450:  00000 00000  00000 00000  00000 00000  00000 00000  00000 00000    00000 00000  00000 00000  00000 00000  00000 00000  00000 00000
9451:  00000 00000  00000 00000  00000 00000  00000 00000  00000 00000    00000 00000  00000 00000  00000 00000  00000 00000  00000 00000
9452:  00000 00000  00000 00000  00000 00000  00000 00000  00000 00000    00000 00000  00000 00000  00000 00000  00000 00000  00000 00000
9453:  00000 00000  00000 00000  00000 00000  00000 00000  00000 00000    00000 00000  00000 00000  00000 00000  00000 00000  00000 00000
9454:  00000 00000  00000 00000  00000 00000  00000 00000  00000 00000    00000 00000  00000 00000  00000 00000  00000 00000  00000 00000
9455:  00000 00000  00000 00000  00000 00000  00000 00000  00000 00000    00000 00000  00000 00000  00000 00000  00000 00000  00000 00000
9456:  00000 00000  00000 00000  00000 00000  00000 00000  00000 00000    00000 00000  00000 00000  00000 00000  00000 00000  00000 00000
9457:  00000 00000  00000 00000  00000 00000  00000 00000  00000 00000    00000 00000  00000 00000  00000 00000  00000 00000  00000 00000
9458:  00000 00000  00000 00000  00000 00000  00000 00000  00000 00000    00000 00000  00000 00000  00000 00000  00000 00000  00000 00000
9459:  00000 00000  00000 00000  00000 00000  00000 00000  00000 00000    00000 00000  00000 00000  00000 00000  00000 00000  00000 00000
9460:  00000 00000  00000 00000  00000 00000  00000 00000  00000 00000    00000 00000  00000 00000  00000 00000  00000 00000  00000 00000
9461:  00000 00000  00000 00000  00000 00000  00000 00000  00000 00000    00000 00000  00000 00000  00000 00000  00000 00000  00000 00000
9462:  00000 00000  00000 00000  00000 00000  00000 00000  00000 00000    00000 00000  00000 00000  00000 00000  00000 00000  00000 00000
9463:  00000 00000  00000 00000  00000 00000  00000 00000  00000 00000    00000 00000  00000 00000  00000 00000  00000 00000  00000 00000
9464:  00000 00000  00000 00000  00000 00000  00000 00000  00000 00000    00000 00000  00000 00000  00000 00000  00000 00000  00000 00000
9465:  00000 00000  00000 00000  00000 00000  00000 00000  00000 00000    00000 00000  00000 00000  00000 00000  00000 00000  00000 00000
9466:  00000 00000  00000 00000  00000 00000  00000 00000  00000 00000    00000 00000  00000 00000  00000 00000  00000 00000  00000 00000
9467:  00000 00000  00000 00000  00000 00000  00000 00000  00000 00000    00000 00000  00000 00000  00000 00000  00000 00000  00000 00000
9468:  00000 00000  00000 00000  00000 00000  00000 00000  00000 00000    00000 00000  00000 00000  00000 00000  00000 00000  00000 00000
9469:  00000 00000  00000 00000  00000 00000  00000 00000  00000 00000    00000 00000  00000 00000  00000 00000  00000 00000  00000 00000
9470:  00000 00000  00000 00000  00000 00000  00000 00000  00000 00000    00000 00000  00000 00000  00000 00000  00000 00000  00000 00000
9471:  00000 00000  00000 00000  00000 00000  00000 00000  00000 00000    00000 00000  00000 00000  00000 00000  00000 00000  00000 00000
9472:  00000 00000  00000 00000  00000 00000  00000 00000  00000 00000    00000 00000  00000 00000  00000 00000  00000 00000  00000 00000
9473:  00000 00000  00000 00000  00000 00000  00000 00000  00000 00000    00000 00000  00000 00000  00000 00000  00000 00000  00000 00000
9474:  00000 00000  00000 00000  00000 00000  00000 00000  00000 00000    00000 00000  00000 00000  00000 00000  00000 00000  00000 00000
9475:  00000 00000  00000 00000  00000 00000  00000 00000  00000 00000    00000 00000  00000 00000  00000 00000  00000 00000  00000 00000
9476:  00000 00000  00000 00000  00000 00000  00000 00000  00000 00000    00000 00000  00000 00000  00000 00000  00000 00000  00000 00000
9477:  00000 00000  00000 00000  00000 00000  00000 00000  00000 00000    00000 00000  00000 00000  00000 00000  00000 00000  00000 00000
9478:  00000 00000  00000 00000  00000 00000  00000 00000  00000 00000    00000 00000  00000 00000  00000 00000  00000 00000  00000 00000
9479:  00000 00000  00000 00000  00000 00000  00000 00000  00000 00000    00000 00000  00000 00000  00000 00000  00000 00000  00000 00000
9480:  00000 00000  00000 00000  00000 00000  00000 00000  00000 00000    00000 00000  00000 00000  00000 00000  00000 00000  00000 00000
9481:  00000 00000  00000 00000  00000 00000  00000 00000  00000 00000    00000 00000  00000 00000  00000 00000  00000 00000  00000 00000
9482:  00000 00000  00000 00000  00000 00000  00000 00000  00000 00000    00000 00000  00000 00000  00000 00000  00000 00000  00000 00000
9483:  00000 00000  00000 00000  00000 00000  00000 00000  00000 00000    00000 00000  00000 00000  00000 00000  00000 00000  00000 00000
9484:  00000 00000  00000 00000  00000 00000  00000 00000  00000 00000    00000 00000  00000 00000  00000 00000  00000 00000  00000 00000
9485:  00000 00000  00000 00000  00000 00000  00000 00000  00000 00000    00000 00000  00000 00000  00000 00000  00000 00000  00000 00000
9486:  00000 00000  00000 00000  00000 00000  00000 00000  00000 00000    00000 00000  00000 00000  00000 00000  00000 00000  00000 00000
9487:  00000 00000  00000 00000  00000 00000  00000 00000  00000 00000    00000 00000  00000 00000  00000 00000  00000 00000  00000 00000
9488:  00000 00000  00000 00000  00000 00000  00000 00000  00000 00000    00000 00000  00000 00000  00000 00000  00000 00000  00000 00000
9489:  00000 00000  00000 00000  00000 00000  00000 00000  00000 00000    00000 00000  00000 00000  00000 00000  00000 00000  00000 00000
9490:  00000 00000  00000 00000  00000 00000  00000 00000  00000 00000    00000 00000  00000 00000  00000 00000  00000 00000  00000 00000
9491:  00000 00000  00000 00000  00000 00000  00000 00000  00000 00000    00000 00000  00000 00000  00000 00000  00000 00000  00000 00000
9492:  00000 00000  00000 00000  00000 00000  00000 00000  00000 00000    00000 00000  00000 00000  00000 00000  00000 00000  00000 00000
9493:  00000 00000  00000 00000  00000 00000  00000 00000  00000 00000    00000 00000  00000 00000  00000 00000  00000 00000  00000 00000
9494:  00000 00000  00000 00000  00000 00000  00000 00000  00000 00000    00000 00000  00000 00000  00000 00000  00000 00000  00000 00000
9495:  00000 00000  00000 00000  00000 00000  00000 00000  00000 00000    00000 00000  00000 00000  00000 00000  00000 00000  00000 00000
9496:  00000 00000  00000 00000  00000 00000  00000 00000  00000 00000    00000 00000  00000 00000  00000 00000  00000 00000  00000 00000
9497:  00000 00000  00000 00000  00000 00000  00000 00000  00000 00000    00000 00000  00000 00000  00000 00000  00000 00000  00000 00000
9498:  00000 00000  00000 00000  00000 00000  00000 00000  00000 00000    00000 00000  00000 00000  00000 00000  00000 00000  00000 00000
9499:  00000 00000  00000 00000  00000 00000  00000 00000  00000 00000    00000 00000  00000 00000  00000 00000  00000 00000  00000 00000
```

```
9500:   00000 00000   00000 00000   00000 00000   00000 00000   00000 00000     00000 00000   00000 00000   00000 00000   00000 00000   00000 00000
9501:   00000 00000   00000 00000   00000 00000   00000 00000   00000 00000     00000 00000   00000 00000   00000 00000   00000 00000   00000 00000
9502:   00000 00000   00000 00000   00000 00000   00000 00000   00000 00000     00000 00000   00000 00000   00000 00000   00000 00000   00000 00000
9503:   00000 00000   00000 00000   00000 00000   00000 00000   00000 00000     00000 00000   00000 00000   00000 00000   00000 00000   00000 00000
9504:   00000 00000   00000 00000   00000 00000   00000 00000   00000 00000     00000 00000   00000 00000   00000 00000   00000 00000   00000 00000
9505:   00000 00000   00000 00000   00000 00000   00000 00000   00000 00000     00000 00000   00000 00000   00000 00000   00000 00000   00000 00000
9506:   00000 00000   00000 00000   00000 00000   00000 00000   00000 00000     00000 00000   00000 00000   00000 00000   00000 00000   00000 00000
9507:   00000 00000   00000 00000   00000 00000   00000 00000   00000 00000     00000 00000   00000 00000   00000 00000   00000 00000   00000 00000
9508:   00000 00000   00000 00000   00000 00000   00000 00000   00000 00000     00000 00000   00000 00000   00000 00000   00000 00000   00000 00000
9509:   00000 00000   00000 00000   00000 00000   00000 00000   00000 00000     00000 00000   00000 00000   00000 00000   00000 00000   00000 00000
9510:   00000 00000   00000 00000   00000 00000   00000 00000   00000 00000     00000 00000   00000 00000   00000 00000   00000 00000   00000 00000
9511:   00000 00000   00000 00000   00000 00000   00000 00000   00000 00000     00000 00000   00000 00000   00000 00000   00000 00000   00000 00000
9512:   00000 00000   00000 00000   00000 00000   00000 00000   00000 00000     00000 00000   00000 00000   00000 00000   00000 00000   00000 00000
9513:   00000 00000   00000 00000   00000 00000   00000 00000   00000 00000     00000 00000   00000 00000   00000 00000   00000 00000   00000 00000
9514:   00000 00000   00000 00000   00000 00000   00000 00000   00000 00000     00000 00000   00000 00000   00000 00000   00000 00000   00000 00000
9515:   00000 00000   00000 00000   00000 00000   00000 00000   00000 00000     00000 00000   00000 00000   00000 00000   00000 00000   00000 00000
9516:   00000 00000   00000 00000   00000 00000   00000 00000   00000 00000     00000 00000   00000 00000   00000 00000   00000 00000   00000 00000
9517:   00000 00000   00000 00000   00000 00000   00000 00000   00000 00000     00000 00000   00000 00000   00000 00000   00000 00000   00000 00000
9518:   00000 00000   00000 00000   00000 00000   00000 00000   00000 00000     00000 00000   00000 00000   00000 00000   00000 00000   00000 00000
9519:   00000 00000   00000 00000   00000 00000   00000 00000   00000 00000     00000 00000   00000 00000   00000 00000   00000 00000   00000 00000
9520:   00000 00000   00000 00000   00000 00000   00000 00000   00000 00000     00000 00000   00000 00000   00000 00000   00000 00000   00000 00000
9521:   00000 00000   00000 00000   00000 00000   00000 00000   00000 00000     00000 00000   00000 00000   00000 00000   00000 00000   00000 00000
9522:   00000 00000   00000 00000   00000 00000   00000 00000   00000 00000     00000 00000   00000 00000   00000 00000   00000 00000   00000 00000
9523:   00000 00000   00000 00000   00000 00000   00000 00000   00000 00000     00000 00000   00000 00000   00000 00000   00000 00000   00000 00000
9524:   00000 00000   00000 00000   00000 00000   00000 00000   00000 00000     00000 00000   00000 00000   00000 00000   00000 00000   00000 00000
9525:   00000 00000   00000 00000   00000 00000   00000 00000   00000 00000     00000 00000   00000 00000   00000 00000   00000 00000   00000 00000
9526:   00000 00000   00000 00000   00000 00000   00000 00000   00000 00000     00000 00000   00000 00000   00000 00000   00000 00000   00000 00000
9527:   00000 00000   00000 00000   00000 00000   00000 00000   00000 00000     00000 00000   00000 00000   00000 00000   00000 00000   00000 00000
9528:   00000 00000   00000 00000   00000 00000   00000 00000   00000 00000     00000 00000   00000 00000   00000 00000   00000 00000   00000 00000
9529:   00000 00000   00000 00000   00000 00000   00000 00000   00000 00000     00000 00000   00000 00000   00000 00000   00000 00000   00000 00000
9530:   00000 00000   00000 00000   00000 00000   00000 00000   00000 00000     00000 00000   00000 00000   00000 00000   00000 00000   00000 00000
9531:   00000 00000   00000 00000   00000 00000   00000 00000   00000 00000     00000 00000   00000 00000   00000 00000   00000 00000   00000 00000
9532:   00000 00000   00000 00000   00000 00000   00000 00000   00000 00000     00000 00000   00000 00000   00000 00000   00000 00000   00000 00000
9533:   00000 00000   00000 00000   00000 00000   00000 00000   00000 00000     00000 00000   00000 00000   00000 00000   00000 00000   00000 00000
9534:   00000 00000   00000 00000   00000 00000   00000 00000   00000 00000     00000 00000   00000 00000   00000 00000   00000 00000   00000 00000
9535:   00000 00000   00000 00000   00000 00000   00000 00000   00000 00000     00000 00000   00000 00000   00000 00000   00000 00000   00000 00000
9536:   00000 00000   00000 00000   00000 00000   00000 00000   00000 00000     00000 00000   00000 00000   00000 00000   00000 00000   00000 00000
9537:   00000 00000   00000 00000   00000 00000   00000 00000   00000 00000     00000 00000   00000 00000   00000 00000   00000 00000   00000 00000
9538:   00000 00000   00000 00000   00000 00000   00000 00000   00000 00000     00000 00000   00000 00000   00000 00000   00000 00000   00000 00000
9539:   00000 00000   00000 00000   00000 00000   00000 00000   00000 00000     00000 00000   00000 00000   00000 00000   00000 00000   00000 00000
9540:   00000 00000   00000 00000   00000 00000   00000 00000   00000 00000     00000 00000   00000 00000   00000 00000   00000 00000   00000 00000
9541:   00000 00000   00000 00000   00000 00000   00000 00000   00000 00000     00000 00000   00000 00000   00000 00000   00000 00000   00000 00000
9542:   00000 00000   00000 00000   00000 00000   00000 00000   00000 00000     00000 00000   00000 00000   00000 00000   00000 00000   00000 00000
9543:   00000 00000   00000 00000   00000 00000   00000 00000   00000 00000     00000 00000   00000 00000   00000 00000   00000 00000   00000 00000
9544:   00000 00000   00000 00000   00000 00000   00000 00000   00000 00000     00000 00000   00000 00000   00000 00000   00000 00000   00000 00000
9545:   00000 00000   00000 00000   00000 00000   00000 00000   00000 00000     00000 00000   00000 00000   00000 00000   00000 00000   00000 00000
9546:   00000 00000   00000 00000   00000 00000   00000 00000   00000 00000     00000 00000   00000 00000   00000 00000   00000 00000   00000 00000
9547:   00000 00000   00000 00000   00000 00000   00000 00000   00000 00000     00000 00000   00000 00000   00000 00000   00000 00000   00000 00000
9548:   00000 00000   00000 00000   00000 00000   00000 00000   00000 00000     00000 00000   00000 00000   00000 00000   00000 00000   00000 00000
9549:   00000 00000   00000 00000   00000 00000   00000 00000   00000 00000     00000 00000   00000 00000   00000 00000   00000 00000   00000 00000
```

```
9550: 00000 00000  00000 00000  00000 00000  00000 00000  00000 00000    00000 00000  00000 00000  00000 00000  00000 00000  00000 00000
9551: 00000 00000  00000 00000  00000 00000  00000 00000  00000 00000    00000 00000  00000 00000  00000 00000  00000 00000  00000 00000
9552: 00000 00000  00000 00000  00000 00000  00000 00000  00000 00000    00000 00000  00000 00000  00000 00000  00000 00000  00000 00000
9553: 00000 00000  00000 00000  00000 00000  00000 00000  00000 00000    00000 00000  00000 00000  00000 00000  00000 00000  00000 00000
9554: 00000 00000  00000 00000  00000 00000  00000 00000  00000 00000    00000 00000  00000 00000  00000 00000  00000 00000  00000 00000
9555: 00000 00000  00000 00000  00000 00000  00000 00000  00000 00000    00000 00000  00000 00000  00000 00000  00000 00000  00000 00000
9556: 00000 00000  00000 00000  00000 00000  00000 00000  00000 00000    00000 00000  00000 00000  00000 00000  00000 00000  00000 00000
9557: 00000 00000  00000 00000  00000 00000  00000 00000  00000 00000    00000 00000  00000 00000  00000 00000  00000 00000  00000 00000
9558: 00000 00000  00000 00000  00000 00000  00000 00000  00000 00000    00000 00000  00000 00000  00000 00000  00000 00000  00000 00000
9559: 00000 00000  00000 00000  00000 00000  00000 00000  00000 00000    00000 00000  00000 00000  00000 00000  00000 00000  00000 00000
9560: 00000 00000  00000 00000  00000 00000  00000 00000  00000 00000    00000 00000  00000 00000  00000 00000  00000 00000  00000 00000
9561: 00000 00000  00000 00000  00000 00000  00000 00000  00000 00000    00000 00000  00000 00000  00000 00000  00000 00000  00000 00000
9562: 00000 00000  00000 00000  00000 00000  00000 00000  00000 00000    00000 00000  00000 00000  00000 00000  00000 00000  00000 00000
9563: 00000 00000  00000 00000  00000 00000  00000 00000  00000 00000    00000 00000  00000 00000  00000 00000  00000 00000  00000 00000
9564: 00000 00000  00000 00000  00000 00000  00000 00000  00000 00000    00000 00000  00000 00000  00000 00000  00000 00000  00000 00000
9565: 00000 00000  00000 00000  00000 00000  00000 00000  00000 00000    00000 00000  00000 00000  00000 00000  00000 00000  00000 00000
9566: 00000 00000  00000 00000  00000 00000  00000 00000  00000 00000    00000 00000  00000 00000  00000 00000  00000 00000  00000 00000
9567: 00000 00000  00000 00000  00000 00000  00000 00000  00000 00000    00000 00000  00000 00000  00000 00000  00000 00000  00000 00000
9568: 00000 00000  00000 00000  00000 00000  00000 00000  00000 00000    00000 00000  00000 00000  00000 00000  00000 00000  00000 00000
9569: 00000 00000  00000 00000  00000 00000  00000 00000  00000 00000    00000 00000  00000 00000  00000 00000  00000 00000  00000 00000
9570: 00000 00000  00000 00000  00000 00000  00000 00000  00000 00000    00000 00000  00000 00000  00000 00000  00000 00000  00000 00000
9571: 00000 00000  00000 00000  00000 00000  00000 00000  00000 00000    00000 00000  00000 00000  00000 00000  00000 00000  00000 00000
9572: 00000 00000  00000 00000  00000 00000  00000 00000  00000 00000    00000 00000  00000 00000  00000 00000  00000 00000  00000 00000
9573: 00000 00000  00000 00000  00000 00000  00000 00000  00000 00000    00000 00000  00000 00000  00000 00000  00000 00000  00000 00000
9574: 00000 00000  00000 00000  00000 00000  00000 00000  00000 00000    00000 00000  00000 00000  00000 00000  00000 00000  00000 00000
9575: 00000 00000  00000 00000  00000 00000  00000 00000  00000 00000    00000 00000  00000 00000  00000 00000  00000 00000  00000 00000
9576: 00000 00000  00000 00000  00000 00000  00000 00000  00000 00000    00000 00000  00000 00000  00000 00000  00000 00000  00000 00000
9577: 00000 00000  00000 00000  00000 00000  00000 00000  00000 00000    00000 00000  00000 00000  00000 00000  00000 00000  00000 00000
9578: 00000 00000  00000 00000  00000 00000  00000 00000  00000 00000    00000 00000  00000 00000  00000 00000  00000 00000  00000 00000
9579: 00000 00000  00000 00000  00000 00000  00000 00000  00000 00000    00000 00000  00000 00000  00000 00000  00000 00000  00000 00000
9580: 00000 00000  00000 00000  00000 00000  00000 00000  00000 00000    00000 00000  00000 00000  00000 00000  00000 00000  00000 00000
9581: 00000 00000  00000 00000  00000 00000  00000 00000  00000 00000    00000 00000  00000 00000  00000 00000  00000 00000  00000 00000
9582: 00000 00000  00000 00000  00000 00000  00000 00000  00000 00000    00000 00000  00000 00000  00000 00000  00000 00000  00000 00000
9583: 00000 00000  00000 00000  00000 00000  00000 00000  00000 00000    00000 00000  00000 00000  00000 00000  00000 00000  00000 00000
9584: 00000 00000  00000 00000  00000 00000  00000 00000  00000 00000    00000 00000  00000 00000  00000 00000  00000 00000  00000 00000
9585: 00000 00000  00000 00000  00000 00000  00000 00000  00000 00000    00000 00000  00000 00000  00000 00000  00000 00000  00000 00000
9586: 00000 00000  00000 00000  00000 00000  00000 00000  00000 00000    00000 00000  00000 00000  00000 00000  00000 00000  00000 00000
9587: 00000 00000  00000 00000  00000 00000  00000 00000  00000 00000    00000 00000  00000 00000  00000 00000  00000 00000  00000 00000
9588: 00000 00000  00000 00000  00000 00000  00000 00000  00000 00000    00000 00000  00000 00000  00000 00000  00000 00000  00000 00000
9589: 00000 00000  00000 00000  00000 00000  00000 00000  00000 00000    00000 00000  00000 00000  00000 00000  00000 00000  00000 00000
9590: 00000 00000  00000 00000  00000 00000  00000 00000  00000 00000    00000 00000  00000 00000  00000 00000  00000 00000  00000 00000
9591: 00000 00000  00000 00000  00000 00000  00000 00000  00000 00000    00000 00000  00000 00000  00000 00000  00000 00000  00000 00000
9592: 00000 00000  00000 00000  00000 00000  00000 00000  00000 00000    00000 00000  00000 00000  00000 00000  00000 00000  00000 00000
9593: 00000 00000  00000 00000  00000 00000  00000 00000  00000 00000    00000 00000  00000 00000  00000 00000  00000 00000  00000 00000
9594: 00000 00000  00000 00000  00000 00000  00000 00000  00000 00000    00000 00000  00000 00000  00000 00000  00000 00000  00000 00000
9595: 00000 00000  00000 00000  00000 00000  00000 00000  00000 00000    00000 00000  00000 00000  00000 00000  00000 00000  00000 00000
9596: 00000 00000  00000 00000  00000 00000  00000 00000  00000 00000    00000 00000  00000 00000  00000 00000  00000 00000  00000 00000
9597: 00000 00000  00000 00000  00000 00000  00000 00000  00000 00000    00000 00000  00000 00000  00000 00000  00000 00000  00000 00000
9598: 00000 00000  00000 00000  00000 00000  00000 00000  00000 00000    00000 00000  00000 00000  00000 00000  00000 00000  00000 00000
9599: 00000 00000  00000 00000  00000 00000  00000 00000  00000 00000    00000 00000  00000 00000  00000 00000  00000 00000  00000 00000
```

```
9600:  00000 00000  00000 00000  00000 00000  00000 00000  00000 00000   00000 00000  00000 00000  00000 00000  00000 00000  00000 00000
9601:  00000 00000  00000 00000  00000 00000  00000 00000  00000 00000   00000 00000  00000 00000  00000 00000  00000 00000  00000 00000
9602:  00000 00000  00000 00000  00000 00000  00000 00000  00000 00000   00000 00000  00000 00000  00000 00000  00000 00000  00000 00000
9603:  00000 00000  00000 00000  00000 00000  00000 00000  00000 00000   00000 00000  00000 00000  00000 00000  00000 00000  00000 00000
9604:  00000 00000  00000 00000  00000 00000  00000 00000  00000 00000   00000 00000  00000 00000  00000 00000  00000 00000  00000 00000
9605:  00000 00000  00000 00000  00000 00000  00000 00000  00000 00000   00000 00000  00000 00000  00000 00000  00000 00000  00000 00000
9606:  00000 00000  00000 00000  00000 00000  00000 00000  00000 00000   00000 00000  00000 00000  00000 00000  00000 00000  00000 00000
9607:  00000 00000  00000 00000  00000 00000  00000 00000  00000 00000   00000 00000  00000 00000  00000 00000  00000 00000  00000 00000
9608:  00000 00000  00000 00000  00000 00000  00000 00000  00000 00000   00000 00000  00000 00000  00000 00000  00000 00000  00000 00000
9609:  00000 00000  00000 00000  00000 00000  00000 00000  00000 00000   00000 00000  00000 00000  00000 00000  00000 00000  00000 00000
9610:  00000 00000  00000 00000  00000 00000  00000 00000  00000 00000   00000 00000  00000 00000  00000 00000  00000 00000  00000 00000
9611:  00000 00000  00000 00000  00000 00000  00000 00000  00000 00000   00000 00000  00000 00000  00000 00000  00000 00000  00000 00000
9612:  00000 00000  00000 00000  00000 00000  00000 00000  00000 00000   00000 00000  00000 00000  00000 00000  00000 00000  00000 00000
9613:  00000 00000  00000 00000  00000 00000  00000 00000  00000 00000   00000 00000  00000 00000  00000 00000  00000 00000  00000 00000
9614:  00000 00000  00000 00000  00000 00000  00000 00000  00000 00000   00000 00000  00000 00000  00000 00000  00000 00000  00000 00000
9615:  00000 00000  00000 00000  00000 00000  00000 00000  00000 00000   00000 00000  00000 00000  00000 00000  00000 00000  00000 00000
9616:  00000 00000  00000 00000  00000 00000  00000 00000  00000 00000   00000 00000  00000 00000  00000 00000  00000 00000  00000 00000
9617:  00000 00000  00000 00000  00000 00000  00000 00000  00000 00000   00000 00000  00000 00000  00000 00000  00000 00000  00000 00000
9618:  00000 00000  00000 00000  00000 00000  00000 00000  00000 00000   00000 00000  00000 00000  00000 00000  00000 00000  00000 00000
9619:  00000 00000  00000 00000  00000 00000  00000 00000  00000 00000   00000 00000  00000 00000  00000 00000  00000 00000  00000 00000
9620:  00000 00000  00000 00000  00000 00000  00000 00000  00000 00000   00000 00000  00000 00000  00000 00000  00000 00000  00000 00000
9621:  00000 00000  00000 00000  00000 00000  00000 00000  00000 00000   00000 00000  00000 00000  00000 00000  00000 00000  00000 00000
9622:  00000 00000  00000 00000  00000 00000  00000 00000  00000 00000   00000 00000  00000 00000  00000 00000  00000 00000  00000 00000
9623:  00000 00000  00000 00000  00000 00000  00000 00000  00000 00000   00000 00000  00000 00000  00000 00000  00000 00000  00000 00000
9624:  00000 00000  00000 00000  00000 00000  00000 00000  00000 00000   00000 00000  00000 00000  00000 00000  00000 00000  00000 00000
9625:  00000 00000  00000 00000  00000 00000  00000 00000  00000 00000   00000 00000  00000 00000  00000 00000  00000 00000  00000 00000
9626:  00000 00000  00000 00000  00000 00000  00000 00000  00000 00000   00000 00000  00000 00000  00000 00000  00000 00000  00000 00000
9627:  00000 00000  00000 00000  00000 00000  00000 00000  00000 00000   00000 00000  00000 00000  00000 00000  00000 00000  00000 00000
9628:  00000 00000  00000 00000  00000 00000  00000 00000  00000 00000   00000 00000  00000 00000  00000 00000  00000 00000  00000 00000
9629:  00000 00000  00000 00000  00000 00000  00000 00000  00000 00000   00000 00000  00000 00000  00000 00000  00000 00000  00000 00000
9630:  00000 00000  00000 00000  00000 00000  00000 00000  00000 00000   00000 00000  00000 00000  00000 00000  00000 00000  00000 00000
9631:  00000 00000  00000 00000  00000 00000  00000 00000  00000 00000   00000 00000  00000 00000  00000 00000  00000 00000  00000 00000
9632:  00000 00000  00000 00000  00000 00000  00000 00000  00000 00000   00000 00000  00000 00000  00000 00000  00000 00000  00000 00000
9633:  00000 00000  00000 00000  00000 00000  00000 00000  00000 00000   00000 00000  00000 00000  00000 00000  00000 00000  00000 00000
9634:  00000 00000  00000 00000  00000 00000  00000 00000  00000 00000   00000 00000  00000 00000  00000 00000  00000 00000  00000 00000
9635:  00000 00000  00000 00000  00000 00000  00000 00000  00000 00000   00000 00000  00000 00000  00000 00000  00000 00000  00000 00000
9636:  00000 00000  00000 00000  00000 00000  00000 00000  00000 00000   00000 00000  00000 00000  00000 00000  00000 00000  00000 00000
9637:  00000 00000  00000 00000  00000 00000  00000 00000  00000 00000   00000 00000  00000 00000  00000 00000  00000 00000  00000 00000
9638:  00000 00000  00000 00000  00000 00000  00000 00000  00000 00000   00000 00000  00000 00000  00000 00000  00000 00000  00000 00000
9639:  00000 00000  00000 00000  00000 00000  00000 00000  00000 00000   00000 00000  00000 00000  00000 00000  00000 00000  00000 00000
9640:  00000 00000  00000 00000  00000 00000  00000 00000  00000 00000   00000 00000  00000 00000  00000 00000  00000 00000  00000 00000
9641:  00000 00000  00000 00000  00000 00000  00000 00000  00000 00000   00000 00000  00000 00000  00000 00000  00000 00000  00000 00000
9642:  00000 00000  00000 00000  00000 00000  00000 00000  00000 00000   00000 00000  00000 00000  00000 00000  00000 00000  00000 00000
9643:  00000 00000  00000 00000  00000 00000  00000 00000  00000 00000   00000 00000  00000 00000  00000 00000  00000 00000  00000 00000
9644:  00000 00000  00000 00000  00000 00000  00000 00000  00000 00000   00000 00000  00000 00000  00000 00000  00000 00000  00000 00000
9645:  00000 00000  00000 00000  00000 00000  00000 00000  00000 00000   00000 00000  00000 00000  00000 00000  00000 00000  00000 00000
9646:  00000 00000  00000 00000  00000 00000  00000 00000  00000 00000   00000 00000  00000 00000  00000 00000  00000 00000  00000 00000
9647:  00000 00000  00000 00000  00000 00000  00000 00000  00000 00000   00000 00000  00000 00000  00000 00000  00000 00000  00000 00000
9648:  00000 00000  00000 00000  00000 00000  00000 00000  00000 00000   00000 00000  00000 00000  00000 00000  00000 00000  00000 00000
9649:  00000 00000  00000 00000  00000 00000  00000 00000  00000 00000   00000 00000  00000 00000  00000 00000  00000 00000  00000 00000
```

```
9650:   00000 00000   00000 00000   00000 00000   00000 00000   00000 00000     00000 00000   00000 00000   00000 00000   00000 00000   00000 00000
9651:   00000 00000   00000 00000   00000 00000   00000 00000   00000 00000     00000 00000   00000 00000   00000 00000   00000 00000   00000 00000
9652:   00000 00000   00000 00000   00000 00000   00000 00000   00000 00000     00000 00000   00000 00000   00000 00000   00000 00000   00000 00000
9653:   00000 00000   00000 00000   00000 00000   00000 00000   00000 00000     00000 00000   00000 00000   00000 00000   00000 00000   00000 00000
9654:   00000 00000   00000 00000   00000 00000   00000 00000   00000 00000     00000 00000   00000 00000   00000 00000   00000 00000   00000 00000
9655:   00000 00000   00000 00000   00000 00000   00000 00000   00000 00000     00000 00000   00000 00000   00000 00000   00000 00000   00000 00000
9656:   00000 00000   00000 00000   00000 00000   00000 00000   00000 00000     00000 00000   00000 00000   00000 00000   00000 00000   00000 00000
9657:   00000 00000   00000 00000   00000 00000   00000 00000   00000 00000     00000 00000   00000 00000   00000 00000   00000 00000   00000 00000
9658:   00000 00000   00000 00000   00000 00000   00000 00000   00000 00000     00000 00000   00000 00000   00000 00000   00000 00000   00000 00000
9659:   00000 00000   00000 00000   00000 00000   00000 00000   00000 00000     00000 00000   00000 00000   00000 00000   00000 00000   00000 00000
9660:   00000 00000   00000 00000   00000 00000   00000 00000   00000 00000     00000 00000   00000 00000   00000 00000   00000 00000   00000 00000
9661:   00000 00000   00000 00000   00000 00000   00000 00000   00000 00000     00000 00000   00000 00000   00000 00000   00000 00000   00000 00000
9662:   00000 00000   00000 00000   00000 00000   00000 00000   00000 00000     00000 00000   00000 00000   00000 00000   00000 00000   00000 00000
9663:   00000 00000   00000 00000   00000 00000   00000 00000   00000 00000     00000 00000   00000 00000   00000 00000   00000 00000   00000 00000
9664:   00000 00000   00000 00000   00000 00000   00000 00000   00000 00000     00000 00000   00000 00000   00000 00000   00000 00000   00000 00000
9665:   00000 00000   00000 00000   00000 00000   00000 00000   00000 00000     00000 00000   00000 00000   00000 00000   00000 00000   00000 00000
9666:   00000 00000   00000 00000   00000 00000   00000 00000   00000 00000     00000 00000   00000 00000   00000 00000   00000 00000   00000 00000
9667:   00000 00000   00000 00000   00000 00000   00000 00000   00000 00000     00000 00000   00000 00000   00000 00000   00000 00000   00000 00000
9668:   00000 00000   00000 00000   00000 00000   00000 00000   00000 00000     00000 00000   00000 00000   00000 00000   00000 00000   00000 00000
9669:   00000 00000   00000 00000   00000 00000   00000 00000   00000 00000     00000 00000   00000 00000   00000 00000   00000 00000   00000 00000
9670:   00000 00000   00000 00000   00000 00000   00000 00000   00000 00000     00000 00000   00000 00000   00000 00000   00000 00000   00000 00000
9671:   00000 00000   00000 00000   00000 00000   00000 00000   00000 00000     00000 00000   00000 00000   00000 00000   00000 00000   00000 00000
9672:   00000 00000   00000 00000   00000 00000   00000 00000   00000 00000     00000 00000   00000 00000   00000 00000   00000 00000   00000 00000
9673:   00000 00000   00000 00000   00000 00000   00000 00000   00000 00000     00000 00000   00000 00000   00000 00000   00000 00000   00000 00000
9674:   00000 00000   00000 00000   00000 00000   00000 00000   00000 00000     00000 00000   00000 00000   00000 00000   00000 00000   00000 00000
9675:   00000 00000   00000 00000   00000 00000   00000 00000   00000 00000     00000 00000   00000 00000   00000 00000   00000 00000   00000 00000
9676:   00000 00000   00000 00000   00000 00000   00000 00000   00000 00000     00000 00000   00000 00000   00000 00000   00000 00000   00000 00000
9677:   00000 00000   00000 00000   00000 00000   00000 00000   00000 00000     00000 00000   00000 00000   00000 00000   00000 00000   00000 00000
9678:   00000 00000   00000 00000   00000 00000   00000 00000   00000 00000     00000 00000   00000 00000   00000 00000   00000 00000   00000 00000
9679:   00000 00000   00000 00000   00000 00000   00000 00000   00000 00000     00000 00000   00000 00000   00000 00000   00000 00000   00000 00000
9680:   00000 00000   00000 00000   00000 00000   00000 00000   00000 00000     00000 00000   00000 00000   00000 00000   00000 00000   00000 00000
9681:   00000 00000   00000 00000   00000 00000   00000 00000   00000 00000     00000 00000   00000 00000   00000 00000   00000 00000   00000 00000
9682:   00000 00000   00000 00000   00000 00000   00000 00000   00000 00000     00000 00000   00000 00000   00000 00000   00000 00000   00000 00000
9683:   00000 00000   00000 00000   00000 00000   00000 00000   00000 00000     00000 00000   00000 00000   00000 00000   00000 00000   00000 00000
9684:   00000 00000   00000 00000   00000 00000   00000 00000   00000 00000     00000 00000   00000 00000   00000 00000   00000 00000   00000 00000
9685:   00000 00000   00000 00000   00000 00000   00000 00000   00000 00000     00000 00000   00000 00000   00000 00000   00000 00000   00000 00000
9686:   00000 00000   00000 00000   00000 00000   00000 00000   00000 00000     00000 00000   00000 00000   00000 00000   00000 00000   00000 00000
9687:   00000 00000   00000 00000   00000 00000   00000 00000   00000 00000     00000 00000   00000 00000   00000 00000   00000 00000   00000 00000
9688:   00000 00000   00000 00000   00000 00000   00000 00000   00000 00000     00000 00000   00000 00000   00000 00000   00000 00000   00000 00000
9689:   00000 00000   00000 00000   00000 00000   00000 00000   00000 00000     00000 00000   00000 00000   00000 00000   00000 00000   00000 00000
9690:   00000 00000   00000 00000   00000 00000   00000 00000   00000 00000     00000 00000   00000 00000   00000 00000   00000 00000   00000 00000
9691:   00000 00000   00000 00000   00000 00000   00000 00000   00000 00000     00000 00000   00000 00000   00000 00000   00000 00000   00000 00000
9692:   00000 00000   00000 00000   00000 00000   00000 00000   00000 00000     00000 00000   00000 00000   00000 00000   00000 00000   00000 00000
9693:   00000 00000   00000 00000   00000 00000   00000 00000   00000 00000     00000 00000   00000 00000   00000 00000   00000 00000   00000 00000
9694:   00000 00000   00000 00000   00000 00000   00000 00000   00000 00000     00000 00000   00000 00000   00000 00000   00000 00000   00000 00000
9695:   00000 00000   00000 00000   00000 00000   00000 00000   00000 00000     00000 00000   00000 00000   00000 00000   00000 00000   00000 00000
9696:   00000 00000   00000 00000   00000 00000   00000 00000   00000 00000     00000 00000   00000 00000   00000 00000   00000 00000   00000 00000
9697:   00000 00000   00000 00000   00000 00000   00000 00000   00000 00000     00000 00000   00000 00000   00000 00000   00000 00000   00000 00000
9698:   00000 00000   00000 00000   00000 00000   00000 00000   00000 00000     00000 00000   00000 00000   00000 00000   00000 00000   00000 00000
9699:   00000 00000   00000 00000   00000 00000   00000 00000   00000 00000     00000 00000   00000 00000   00000 00000   00000 00000   00000 00000
```

```
9700:   00000 00000  00000 00000  00000 00000  00000 00000  00000 00000    00000 00000  00000 00000  00000 00000  00000 00000  00000 00000
9701:   00000 00000  00000 00000  00000 00000  00000 00000  00000 00000    00000 00000  00000 00000  00000 00000  00000 00000  00000 00000
9702:   00000 00000  00000 00000  00000 00000  00000 00000  00000 00000    00000 00000  00000 00000  00000 00000  00000 00000  00000 00000
9703:   00000 00000  00000 00000  00000 00000  00000 00000  00000 00000    00000 00000  00000 00000  00000 00000  00000 00000  00000 00000
9704:   00000 00000  00000 00000  00000 00000  00000 00000  00000 00000    00000 00000  00000 00000  00000 00000  00000 00000  00000 00000
9705:   00000 00000  00000 00000  00000 00000  00000 00000  00000 00000    00000 00000  00000 00000  00000 00000  00000 00000  00000 00000
9706:   00000 00000  00000 00000  00000 00000  00000 00000  00000 00000    00000 00000  00000 00000  00000 00000  00000 00000  00000 00000
9707:   00000 00000  00000 00000  00000 00000  00000 00000  00000 00000    00000 00000  00000 00000  00000 00000  00000 00000  00000 00000
9708:   00000 00000  00000 00000  00000 00000  00000 00000  00000 00000    00000 00000  00000 00000  00000 00000  00000 00000  00000 00000
9709:   00000 00000  00000 00000  00000 00000  00000 00000  00000 00000    00000 00000  00000 00000  00000 00000  00000 00000  00000 00000
9710:   00000 00000  00000 00000  00000 00000  00000 00000  00000 00000    00000 00000  00000 00000  00000 00000  00000 00000  00000 00000
9711:   00000 00000  00000 00000  00000 00000  00000 00000  00000 00000    00000 00000  00000 00000  00000 00000  00000 00000  00000 00000
9712:   00000 00000  00000 00000  00000 00000  00000 00000  00000 00000    00000 00000  00000 00000  00000 00000  00000 00000  00000 00000
9713:   00000 00000  00000 00000  00000 00000  00000 00000  00000 00000    00000 00000  00000 00000  00000 00000  00000 00000  00000 00000
9714:   00000 00000  00000 00000  00000 00000  00000 00000  00000 00000    00000 00000  00000 00000  00000 00000  00000 00000  00000 00000
9715:   00000 00000  00000 00000  00000 00000  00000 00000  00000 00000    00000 00000  00000 00000  00000 00000  00000 00000  00000 00000
9716:   00000 00000  00000 00000  00000 00000  00000 00000  00000 00000    00000 00000  00000 00000  00000 00000  00000 00000  00000 00000
9717:   00000 00000  00000 00000  00000 00000  00000 00000  00000 00000    00000 00000  00000 00000  00000 00000  00000 00000  00000 00000
9718:   00000 00000  00000 00000  00000 00000  00000 00000  00000 00000    00000 00000  00000 00000  00000 00000  00000 00000  00000 00000
9719:   00000 00000  00000 00000  00000 00000  00000 00000  00000 00000    00000 00000  00000 00000  00000 00000  00000 00000  00000 00000
9720:   00000 00000  00000 00000  00000 00000  00000 00000  00000 00000    00000 00000  00000 00000  00000 00000  00000 00000  00000 00000
9721:   00000 00000  00000 00000  00000 00000  00000 00000  00000 00000    00000 00000  00000 00000  00000 00000  00000 00000  00000 00000
9722:   00000 00000  00000 00000  00000 00000  00000 00000  00000 00000    00000 00000  00000 00000  00000 00000  00000 00000  00000 00000
9723:   00000 00000  00000 00000  00000 00000  00000 00000  00000 00000    00000 00000  00000 00000  00000 00000  00000 00000  00000 00000
9724:   00000 00000  00000 00000  00000 00000  00000 00000  00000 00000    00000 00000  00000 00000  00000 00000  00000 00000  00000 00000
9725:   00000 00000  00000 00000  00000 00000  00000 00000  00000 00000    00000 00000  00000 00000  00000 00000  00000 00000  00000 00000
9726:   00000 00000  00000 00000  00000 00000  00000 00000  00000 00000    00000 00000  00000 00000  00000 00000  00000 00000  00000 00000
9727:   00000 00000  00000 00000  00000 00000  00000 00000  00000 00000    00000 00000  00000 00000  00000 00000  00000 00000  00000 00000
9728:   00000 00000  00000 00000  00000 00000  00000 00000  00000 00000    00000 00000  00000 00000  00000 00000  00000 00000  00000 00000
9729:   00000 00000  00000 00000  00000 00000  00000 00000  00000 00000    00000 00000  00000 00000  00000 00000  00000 00000  00000 00000
9730:   00000 00000  00000 00000  00000 00000  00000 00000  00000 00000    00000 00000  00000 00000  00000 00000  00000 00000  00000 00000
9731:   00000 00000  00000 00000  00000 00000  00000 00000  00000 00000    00000 00000  00000 00000  00000 00000  00000 00000  00000 00000
9732:   00000 00000  00000 00000  00000 00000  00000 00000  00000 00000    00000 00000  00000 00000  00000 00000  00000 00000  00000 00000
9733:   00000 00000  00000 00000  00000 00000  00000 00000  00000 00000    00000 00000  00000 00000  00000 00000  00000 00000  00000 00000
9734:   00000 00000  00000 00000  00000 00000  00000 00000  00000 00000    00000 00000  00000 00000  00000 00000  00000 00000  00000 00000
9735:   00000 00000  00000 00000  00000 00000  00000 00000  00000 00000    00000 00000  00000 00000  00000 00000  00000 00000  00000 00000
9736:   00000 00000  00000 00000  00000 00000  00000 00000  00000 00000    00000 00000  00000 00000  00000 00000  00000 00000  00000 00000
9737:   00000 00000  00000 00000  00000 00000  00000 00000  00000 00000    00000 00000  00000 00000  00000 00000  00000 00000  00000 00000
9738:   00000 00000  00000 00000  00000 00000  00000 00000  00000 00000    00000 00000  00000 00000  00000 00000  00000 00000  00000 00000
9739:   00000 00000  00000 00000  00000 00000  00000 00000  00000 00000    00000 00000  00000 00000  00000 00000  00000 00000  00000 00000
9740:   00000 00000  00000 00000  00000 00000  00000 00000  00000 00000    00000 00000  00000 00000  00000 00000  00000 00000  00000 00000
9741:   00000 00000  00000 00000  00000 00000  00000 00000  00000 00000    00000 00000  00000 00000  00000 00000  00000 00000  00000 00000
9742:   00000 00000  00000 00000  00000 00000  00000 00000  00000 00000    00000 00000  00000 00000  00000 00000  00000 00000  00000 00000
9743:   00000 00000  00000 00000  00000 00000  00000 00000  00000 00000    00000 00000  00000 00000  00000 00000  00000 00000  00000 00000
9744:   00000 00000  00000 00000  00000 00000  00000 00000  00000 00000    00000 00000  00000 00000  00000 00000  00000 00000  00000 00000
9745:   00000 00000  00000 00000  00000 00000  00000 00000  00000 00000    00000 00000  00000 00000  00000 00000  00000 00000  00000 00000
9746:   00000 00000  00000 00000  00000 00000  00000 00000  00000 00000    00000 00000  00000 00000  00000 00000  00000 00000  00000 00000
9747:   00000 00000  00000 00000  00000 00000  00000 00000  00000 00000    00000 00000  00000 00000  00000 00000  00000 00000  00000 00000
9748:   00000 00000  00000 00000  00000 00000  00000 00000  00000 00000    00000 00000  00000 00000  00000 00000  00000 00000  00000 00000
9749:   00000 00000  00000 00000  00000 00000  00000 00000  00000 00000    00000 00000  00000 00000  00000 00000  00000 00000  00000 00000
```

```
9750:  00000 00000  00000 00000  00000 00000  00000 00000  00000 00000   00000 00000  00000 00000  00000 00000  00000 00000  00000 00000
9751:  00000 00000  00000 00000  00000 00000  00000 00000  00000 00000   00000 00000  00000 00000  00000 00000  00000 00000  00000 00000
9752:  00000 00000  00000 00000  00000 00000  00000 00000  00000 00000   00000 00000  00000 00000  00000 00000  00000 00000  00000 00000
9753:  00000 00000  00000 00000  00000 00000  00000 00000  00000 00000   00000 00000  00000 00000  00000 00000  00000 00000  00000 00000
9754:  00000 00000  00000 00000  00000 00000  00000 00000  00000 00000   00000 00000  00000 00000  00000 00000  00000 00000  00000 00000
9755:  00000 00000  00000 00000  00000 00000  00000 00000  00000 00000   00000 00000  00000 00000  00000 00000  00000 00000  00000 00000
9756:  00000 00000  00000 00000  00000 00000  00000 00000  00000 00000   00000 00000  00000 00000  00000 00000  00000 00000  00000 00000
9757:  00000 00000  00000 00000  00000 00000  00000 00000  00000 00000   00000 00000  00000 00000  00000 00000  00000 00000  00000 00000
9758:  00000 00000  00000 00000  00000 00000  00000 00000  00000 00000   00000 00000  00000 00000  00000 00000  00000 00000  00000 00000
9759:  00000 00000  00000 00000  00000 00000  00000 00000  00000 00000   00000 00000  00000 00000  00000 00000  00000 00000  00000 00000
9760:  00000 00000  00000 00000  00000 00000  00000 00000  00000 00000   00000 00000  00000 00000  00000 00000  00000 00000  00000 00000
9761:  00000 00000  00000 00000  00000 00000  00000 00000  00000 00000   00000 00000  00000 00000  00000 00000  00000 00000  00000 00000
9762:  00000 00000  00000 00000  00000 00000  00000 00000  00000 00000   00000 00000  00000 00000  00000 00000  00000 00000  00000 00000
9763:  00000 00000  00000 00000  00000 00000  00000 00000  00000 00000   00000 00000  00000 00000  00000 00000  00000 00000  00000 00000
9764:  00000 00000  00000 00000  00000 00000  00000 00000  00000 00000   00000 00000  00000 00000  00000 00000  00000 00000  00000 00000
9765:  00000 00000  00000 00000  00000 00000  00000 00000  00000 00000   00000 00000  00000 00000  00000 00000  00000 00000  00000 00000
9766:  00000 00000  00000 00000  00000 00000  00000 00000  00000 00000   00000 00000  00000 00000  00000 00000  00000 00000  00000 00000
9767:  00000 00000  00000 00000  00000 00000  00000 00000  00000 00000   00000 00000  00000 00000  00000 00000  00000 00000  00000 00000
9768:  00000 00000  00000 00000  00000 00000  00000 00000  00000 00000   00000 00000  00000 00000  00000 00000  00000 00000  00000 00000
9769:  00000 00000  00000 00000  00000 00000  00000 00000  00000 00000   00000 00000  00000 00000  00000 00000  00000 00000  00000 00000
9770:  00000 00000  00000 00000  00000 00000  00000 00000  00000 00000   00000 00000  00000 00000  00000 00000  00000 00000  00000 00000
9771:  00000 00000  00000 00000  00000 00000  00000 00000  00000 00000   00000 00000  00000 00000  00000 00000  00000 00000  00000 00000
9772:  00000 00000  00000 00000  00000 00000  00000 00000  00000 00000   00000 00000  00000 00000  00000 00000  00000 00000  00000 00000
9773:  00000 00000  00000 00000  00000 00000  00000 00000  00000 00000   00000 00000  00000 00000  00000 00000  00000 00000  00000 00000
9774:  00000 00000  00000 00000  00000 00000  00000 00000  00000 00000   00000 00000  00000 00000  00000 00000  00000 00000  00000 00000
9775:  00000 00000  00000 00000  00000 00000  00000 00000  00000 00000   00000 00000  00000 00000  00000 00000  00000 00000  00000 00000
9776:  00000 00000  00000 00000  00000 00000  00000 00000  00000 00000   00000 00000  00000 00000  00000 00000  00000 00000  00000 00000
9777:  00000 00000  00000 00000  00000 00000  00000 00000  00000 00000   00000 00000  00000 00000  00000 00000  00000 00000  00000 00000
9778:  00000 00000  00000 00000  00000 00000  00000 00000  00000 00000   00000 00000  00000 00000  00000 00000  00000 00000  00000 00000
9779:  00000 00000  00000 00000  00000 00000  00000 00000  00000 00000   00000 00000  00000 00000  00000 00000  00000 00000  00000 00000
9780:  00000 00000  00000 00000  00000 00000  00000 00000  00000 00000   00000 00000  00000 00000  00000 00000  00000 00000  00000 00000
9781:  00000 00000  00000 00000  00000 00000  00000 00000  00000 00000   00000 00000  00000 00000  00000 00000  00000 00000  00000 00000
9782:  00000 00000  00000 00000  00000 00000  00000 00000  00000 00000   00000 00000  00000 00000  00000 00000  00000 00000  00000 00000
9783:  00000 00000  00000 00000  00000 00000  00000 00000  00000 00000   00000 00000  00000 00000  00000 00000  00000 00000  00000 00000
9784:  00000 00000  00000 00000  00000 00000  00000 00000  00000 00000   00000 00000  00000 00000  00000 00000  00000 00000  00000 00000
9785:  00000 00000  00000 00000  00000 00000  00000 00000  00000 00000   00000 00000  00000 00000  00000 00000  00000 00000  00000 00000
9786:  00000 00000  00000 00000  00000 00000  00000 00000  00000 00000   00000 00000  00000 00000  00000 00000  00000 00000  00000 00000
9787:  00000 00000  00000 00000  00000 00000  00000 00000  00000 00000   00000 00000  00000 00000  00000 00000  00000 00000  00000 00000
9788:  00000 00000  00000 00000  00000 00000  00000 00000  00000 00000   00000 00000  00000 00000  00000 00000  00000 00000  00000 00000
9789:  00000 00000  00000 00000  00000 00000  00000 00000  00000 00000   00000 00000  00000 00000  00000 00000  00000 00000  00000 00000
9790:  00000 00000  00000 00000  00000 00000  00000 00000  00000 00000   00000 00000  00000 00000  00000 00000  00000 00000  00000 00000
9791:  00000 00000  00000 00000  00000 00000  00000 00000  00000 00000   00000 00000  00000 00000  00000 00000  00000 00000  00000 00000
9792:  00000 00000  00000 00000  00000 00000  00000 00000  00000 00000   00000 00000  00000 00000  00000 00000  00000 00000  00000 00000
9793:  00000 00000  00000 00000  00000 00000  00000 00000  00000 00000   00000 00000  00000 00000  00000 00000  00000 00000  00000 00000
9794:  00000 00000  00000 00000  00000 00000  00000 00000  00000 00000   00000 00000  00000 00000  00000 00000  00000 00000  00000 00000
9795:  00000 00000  00000 00000  00000 00000  00000 00000  00000 00000   00000 00000  00000 00000  00000 00000  00000 00000  00000 00000
9796:  00000 00000  00000 00000  00000 00000  00000 00000  00000 00000   00000 00000  00000 00000  00000 00000  00000 00000  00000 00000
9797:  00000 00000  00000 00000  00000 00000  00000 00000  00000 00000   00000 00000  00000 00000  00000 00000  00000 00000  00000 00000
9798:  00000 00000  00000 00000  00000 00000  00000 00000  00000 00000   00000 00000  00000 00000  00000 00000  00000 00000  00000 00000
9799:  00000 00000  00000 00000  00000 00000  00000 00000  00000 00000   00000 00000  00000 00000  00000 00000  00000 00000  00000 00000
```

```
9800:  00000 00000  00000 00000  00000 00000  00000 00000  00000 00000    00000 00000  00000 00000  00000 00000  00000 00000  00000 00000
9801:  00000 00000  00000 00000  00000 00000  00000 00000  00000 00000    00000 00000  00000 00000  00000 00000  00000 00000  00000 00000
9802:  00000 00000  00000 00000  00000 00000  00000 00000  00000 00000    00000 00000  00000 00000  00000 00000  00000 00000  00000 00000
9803:  00000 00000  00000 00000  00000 00000  00000 00000  00000 00000    00000 00000  00000 00000  00000 00000  00000 00000  00000 00000
9804:  00000 00000  00000 00000  00000 00000  00000 00000  00000 00000    00000 00000  00000 00000  00000 00000  00000 00000  00000 00000
9805:  00000 00000  00000 00000  00000 00000  00000 00000  00000 00000    00000 00000  00000 00000  00000 00000  00000 00000  00000 00000
9806:  00000 00000  00000 00000  00000 00000  00000 00000  00000 00000    00000 00000  00000 00000  00000 00000  00000 00000  00000 00000
9807:  00000 00000  00000 00000  00000 00000  00000 00000  00000 00000    00000 00000  00000 00000  00000 00000  00000 00000  00000 00000
9808:  00000 00000  00000 00000  00000 00000  00000 00000  00000 00000    00000 00000  00000 00000  00000 00000  00000 00000  00000 00000
9809:  00000 00000  00000 00000  00000 00000  00000 00000  00000 00000    00000 00000  00000 00000  00000 00000  00000 00000  00000 00000
9810:  00000 00000  00000 00000  00000 00000  00000 00000  00000 00000    00000 00000  00000 00000  00000 00000  00000 00000  00000 00000
9811:  00000 00000  00000 00000  00000 00000  00000 00000  00000 00000    00000 00000  00000 00000  00000 00000  00000 00000  00000 00000
9812:  00000 00000  00000 00000  00000 00000  00000 00000  00000 00000    00000 00000  00000 00000  00000 00000  00000 00000  00000 00000
9813:  00000 00000  00000 00000  00000 00000  00000 00000  00000 00000    00000 00000  00000 00000  00000 00000  00000 00000  00000 00000
9814:  00000 00000  00000 00000  00000 00000  00000 00000  00000 00000    00000 00000  00000 00000  00000 00000  00000 00000  00000 00000
9815:  00000 00000  00000 00000  00000 00000  00000 00000  00000 00000    00000 00000  00000 00000  00000 00000  00000 00000  00000 00000
9816:  00000 00000  00000 00000  00000 00000  00000 00000  00000 00000    00000 00000  00000 00000  00000 00000  00000 00000  00000 00000
9817:  00000 00000  00000 00000  00000 00000  00000 00000  00000 00000    00000 00000  00000 00000  00000 00000  00000 00000  00000 00000
9818:  00000 00000  00000 00000  00000 00000  00000 00000  00000 00000    00000 00000  00000 00000  00000 00000  00000 00000  00000 00000
9819:  00000 00000  00000 00000  00000 00000  00000 00000  00000 00000    00000 00000  00000 00000  00000 00000  00000 00000  00000 00000
9820:  00000 00000  00000 00000  00000 00000  00000 00000  00000 00000    00000 00000  00000 00000  00000 00000  00000 00000  00000 00000
9821:  00000 00000  00000 00000  00000 00000  00000 00000  00000 00000    00000 00000  00000 00000  00000 00000  00000 00000  00000 00000
9822:  00000 00000  00000 00000  00000 00000  00000 00000  00000 00000    00000 00000  00000 00000  00000 00000  00000 00000  00000 00000
9823:  00000 00000  00000 00000  00000 00000  00000 00000  00000 00000    00000 00000  00000 00000  00000 00000  00000 00000  00000 00000
9824:  00000 00000  00000 00000  00000 00000  00000 00000  00000 00000    00000 00000  00000 00000  00000 00000  00000 00000  00000 00000
9825:  00000 00000  00000 00000  00000 00000  00000 00000  00000 00000    00000 00000  00000 00000  00000 00000  00000 00000  00000 00000
9826:  00000 00000  00000 00000  00000 00000  00000 00000  00000 00000    00000 00000  00000 00000  00000 00000  00000 00000  00000 00000
9827:  00000 00000  00000 00000  00000 00000  00000 00000  00000 00000    00000 00000  00000 00000  00000 00000  00000 00000  00000 00000
9828:  00000 00000  00000 00000  00000 00000  00000 00000  00000 00000    00000 00000  00000 00000  00000 00000  00000 00000  00000 00000
9829:  00000 00000  00000 00000  00000 00000  00000 00000  00000 00000    00000 00000  00000 00000  00000 00000  00000 00000  00000 00000
9830:  00000 00000  00000 00000  00000 00000  00000 00000  00000 00000    00000 00000  00000 00000  00000 00000  00000 00000  00000 00000
9831:  00000 00000  00000 00000  00000 00000  00000 00000  00000 00000    00000 00000  00000 00000  00000 00000  00000 00000  00000 00000
9832:  00000 00000  00000 00000  00000 00000  00000 00000  00000 00000    00000 00000  00000 00000  00000 00000  00000 00000  00000 00000
9833:  00000 00000  00000 00000  00000 00000  00000 00000  00000 00000    00000 00000  00000 00000  00000 00000  00000 00000  00000 00000
9834:  00000 00000  00000 00000  00000 00000  00000 00000  00000 00000    00000 00000  00000 00000  00000 00000  00000 00000  00000 00000
9835:  00000 00000  00000 00000  00000 00000  00000 00000  00000 00000    00000 00000  00000 00000  00000 00000  00000 00000  00000 00000
9836:  00000 00000  00000 00000  00000 00000  00000 00000  00000 00000    00000 00000  00000 00000  00000 00000  00000 00000  00000 00000
9837:  00000 00000  00000 00000  00000 00000  00000 00000  00000 00000    00000 00000  00000 00000  00000 00000  00000 00000  00000 00000
9838:  00000 00000  00000 00000  00000 00000  00000 00000  00000 00000    00000 00000  00000 00000  00000 00000  00000 00000  00000 00000
9839:  00000 00000  00000 00000  00000 00000  00000 00000  00000 00000    00000 00000  00000 00000  00000 00000  00000 00000  00000 00000
9840:  00000 00000  00000 00000  00000 00000  00000 00000  00000 00000    00000 00000  00000 00000  00000 00000  00000 00000  00000 00000
9841:  00000 00000  00000 00000  00000 00000  00000 00000  00000 00000    00000 00000  00000 00000  00000 00000  00000 00000  00000 00000
9842:  00000 00000  00000 00000  00000 00000  00000 00000  00000 00000    00000 00000  00000 00000  00000 00000  00000 00000  00000 00000
9843:  00000 00000  00000 00000  00000 00000  00000 00000  00000 00000    00000 00000  00000 00000  00000 00000  00000 00000  00000 00000
9844:  00000 00000  00000 00000  00000 00000  00000 00000  00000 00000    00000 00000  00000 00000  00000 00000  00000 00000  00000 00000
9845:  00000 00000  00000 00000  00000 00000  00000 00000  00000 00000    00000 00000  00000 00000  00000 00000  00000 00000  00000 00000
9846:  00000 00000  00000 00000  00000 00000  00000 00000  00000 00000    00000 00000  00000 00000  00000 00000  00000 00000  00000 00000
9847:  00000 00000  00000 00000  00000 00000  00000 00000  00000 00000    00000 00000  00000 00000  00000 00000  00000 00000  00000 00000
9848:  00000 00000  00000 00000  00000 00000  00000 00000  00000 00000    00000 00000  00000 00000  00000 00000  00000 00000  00000 00000
9849:  00000 00000  00000 00000  00000 00000  00000 00000  00000 00000    00000 00000  00000 00000  00000 00000  00000 00000  00000 00000
```

```
9850:  00000 00000  00000 00000  00000 00000  00000 00000  00000 00000    00000 00000  00000 00000  00000 00000  00000 00000  00000 00000
9851:  00000 00000  00000 00000  00000 00000  00000 00000  00000 00000    00000 00000  00000 00000  00000 00000  00000 00000  00000 00000
9852:  00000 00000  00000 00000  00000 00000  00000 00000  00000 00000    00000 00000  00000 00000  00000 00000  00000 00000  00000 00000
9853:  00000 00000  00000 00000  00000 00000  00000 00000  00000 00000    00000 00000  00000 00000  00000 00000  00000 00000  00000 00000
9854:  00000 00000  00000 00000  00000 00000  00000 00000  00000 00000    00000 00000  00000 00000  00000 00000  00000 00000  00000 00000
9855:  00000 00000  00000 00000  00000 00000  00000 00000  00000 00000    00000 00000  00000 00000  00000 00000  00000 00000  00000 00000
9856:  00000 00000  00000 00000  00000 00000  00000 00000  00000 00000    00000 00000  00000 00000  00000 00000  00000 00000  00000 00000
9857:  00000 00000  00000 00000  00000 00000  00000 00000  00000 00000    00000 00000  00000 00000  00000 00000  00000 00000  00000 00000
9858:  00000 00000  00000 00000  00000 00000  00000 00000  00000 00000    00000 00000  00000 00000  00000 00000  00000 00000  00000 00000
9859:  00000 00000  00000 00000  00000 00000  00000 00000  00000 00000    00000 00000  00000 00000  00000 00000  00000 00000  00000 00000
9860:  00000 00000  00000 00000  00000 00000  00000 00000  00000 00000    00000 00000  00000 00000  00000 00000  00000 00000  00000 00000
9861:  00000 00000  00000 00000  00000 00000  00000 00000  00000 00000    00000 00000  00000 00000  00000 00000  00000 00000  00000 00000
9862:  00000 00000  00000 00000  00000 00000  00000 00000  00000 00000    00000 00000  00000 00000  00000 00000  00000 00000  00000 00000
9863:  00000 00000  00000 00000  00000 00000  00000 00000  00000 00000    00000 00000  00000 00000  00000 00000  00000 00000  00000 00000
9864:  00000 00000  00000 00000  00000 00000  00000 00000  00000 00000    00000 00000  00000 00000  00000 00000  00000 00000  00000 00000
9865:  00000 00000  00000 00000  00000 00000  00000 00000  00000 00000    00000 00000  00000 00000  00000 00000  00000 00000  00000 00000
9866:  00000 00000  00000 00000  00000 00000  00000 00000  00000 00000    00000 00000  00000 00000  00000 00000  00000 00000  00000 00000
9867:  00000 00000  00000 00000  00000 00000  00000 00000  00000 00000    00000 00000  00000 00000  00000 00000  00000 00000  00000 00000
9868:  00000 00000  00000 00000  00000 00000  00000 00000  00000 00000    00000 00000  00000 00000  00000 00000  00000 00000  00000 00000
9869:  00000 00000  00000 00000  00000 00000  00000 00000  00000 00000    00000 00000  00000 00000  00000 00000  00000 00000  00000 00000
9870:  00000 00000  00000 00000  00000 00000  00000 00000  00000 00000    00000 00000  00000 00000  00000 00000  00000 00000  00000 00000
9871:  00000 00000  00000 00000  00000 00000  00000 00000  00000 00000    00000 00000  00000 00000  00000 00000  00000 00000  00000 00000
9872:  00000 00000  00000 00000  00000 00000  00000 00000  00000 00000    00000 00000  00000 00000  00000 00000  00000 00000  00000 00000
9873:  00000 00000  00000 00000  00000 00000  00000 00000  00000 00000    00000 00000  00000 00000  00000 00000  00000 00000  00000 00000
9874:  00000 00000  00000 00000  00000 00000  00000 00000  00000 00000    00000 00000  00000 00000  00000 00000  00000 00000  00000 00000
9875:  00000 00000  00000 00000  00000 00000  00000 00000  00000 00000    00000 00000  00000 00000  00000 00000  00000 00000  00000 00000
9876:  00000 00000  00000 00000  00000 00000  00000 00000  00000 00000    00000 00000  00000 00000  00000 00000  00000 00000  00000 00000
9877:  00000 00000  00000 00000  00000 00000  00000 00000  00000 00000    00000 00000  00000 00000  00000 00000  00000 00000  00000 00000
9878:  00000 00000  00000 00000  00000 00000  00000 00000  00000 00000    00000 00000  00000 00000  00000 00000  00000 00000  00000 00000
9879:  00000 00000  00000 00000  00000 00000  00000 00000  00000 00000    00000 00000  00000 00000  00000 00000  00000 00000  00000 00000
9880:  00000 00000  00000 00000  00000 00000  00000 00000  00000 00000    00000 00000  00000 00000  00000 00000  00000 00000  00000 00000
9881:  00000 00000  00000 00000  00000 00000  00000 00000  00000 00000    00000 00000  00000 00000  00000 00000  00000 00000  00000 00000
9882:  00000 00000  00000 00000  00000 00000  00000 00000  00000 00000    00000 00000  00000 00000  00000 00000  00000 00000  00000 00000
9883:  00000 00000  00000 00000  00000 00000  00000 00000  00000 00000    00000 00000  00000 00000  00000 00000  00000 00000  00000 00000
9884:  00000 00000  00000 00000  00000 00000  00000 00000  00000 00000    00000 00000  00000 00000  00000 00000  00000 00000  00000 00000
9885:  00000 00000  00000 00000  00000 00000  00000 00000  00000 00000    00000 00000  00000 00000  00000 00000  00000 00000  00000 00000
9886:  00000 00000  00000 00000  00000 00000  00000 00000  00000 00000    00000 00000  00000 00000  00000 00000  00000 00000  00000 00000
9887:  00000 00000  00000 00000  00000 00000  00000 00000  00000 00000    00000 00000  00000 00000  00000 00000  00000 00000  00000 00000
9888:  00000 00000  00000 00000  00000 00000  00000 00000  00000 00000    00000 00000  00000 00000  00000 00000  00000 00000  00000 00000
9889:  00000 00000  00000 00000  00000 00000  00000 00000  00000 00000    00000 00000  00000 00000  00000 00000  00000 00000  00000 00000
9890:  00000 00000  00000 00000  00000 00000  00000 00000  00000 00000    00000 00000  00000 00000  00000 00000  00000 00000  00000 00000
9891:  00000 00000  00000 00000  00000 00000  00000 00000  00000 00000    00000 00000  00000 00000  00000 00000  00000 00000  00000 00000
9892:  00000 00000  00000 00000  00000 00000  00000 00000  00000 00000    00000 00000  00000 00000  00000 00000  00000 00000  00000 00000
9893:  00000 00000  00000 00000  00000 00000  00000 00000  00000 00000    00000 00000  00000 00000  00000 00000  00000 00000  00000 00000
9894:  00000 00000  00000 00000  00000 00000  00000 00000  00000 00000    00000 00000  00000 00000  00000 00000  00000 00000  00000 00000
9895:  00000 00000  00000 00000  00000 00000  00000 00000  00000 00000    00000 00000  00000 00000  00000 00000  00000 00000  00000 00000
9896:  00000 00000  00000 00000  00000 00000  00000 00000  00000 00000    00000 00000  00000 00000  00000 00000  00000 00000  00000 00000
9897:  00000 00000  00000 00000  00000 00000  00000 00000  00000 00000    00000 00000  00000 00000  00000 00000  00000 00000  00000 00000
9898:  00000 00000  00000 00000  00000 00000  00000 00000  00000 00000    00000 00000  00000 00000  00000 00000  00000 00000  00000 00000
9899:  00000 00000  00000 00000  00000 00000  00000 00000  00000 00000    00000 00000  00000 00000  00000 00000  00000 00000  00000 00000
```

```
9900: 00000 00000  00000 00000  00000 00000  00000 00000  00000 00000    00000 00000  00000 00000  00000 00000  00000 00000  00000 00000
9901: 00000 00000  00000 00000  00000 00000  00000 00000  00000 00000    00000 00000  00000 00000  00000 00000  00000 00000  00000 00000
9902: 00000 00000  00000 00000  00000 00000  00000 00000  00000 00000    00000 00000  00000 00000  00000 00000  00000 00000  00000 00000
9903: 00000 00000  00000 00000  00000 00000  00000 00000  00000 00000    00000 00000  00000 00000  00000 00000  00000 00000  00000 00000
9904: 00000 00000  00000 00000  00000 00000  00000 00000  00000 00000    00000 00000  00000 00000  00000 00000  00000 00000  00000 00000
9905: 00000 00000  00000 00000  00000 00000  00000 00000  00000 00000    00000 00000  00000 00000  00000 00000  00000 00000  00000 00000
9906: 00000 00000  00000 00000  00000 00000  00000 00000  00000 00000    00000 00000  00000 00000  00000 00000  00000 00000  00000 00000
9907: 00000 00000  00000 00000  00000 00000  00000 00000  00000 00000    00000 00000  00000 00000  00000 00000  00000 00000  00000 00000
9908: 00000 00000  00000 00000  00000 00000  00000 00000  00000 00000    00000 00000  00000 00000  00000 00000  00000 00000  00000 00000
9909: 00000 00000  00000 00000  00000 00000  00000 00000  00000 00000    00000 00000  00000 00000  00000 00000  00000 00000  00000 00000
9910: 00000 00000  00000 00000  00000 00000  00000 00000  00000 00000    00000 00000  00000 00000  00000 00000  00000 00000  00000 00000
9911: 00000 00000  00000 00000  00000 00000  00000 00000  00000 00000    00000 00000  00000 00000  00000 00000  00000 00000  00000 00000
9912: 00000 00000  00000 00000  00000 00000  00000 00000  00000 00000    00000 00000  00000 00000  00000 00000  00000 00000  00000 00000
9913: 00000 00000  00000 00000  00000 00000  00000 00000  00000 00000    00000 00000  00000 00000  00000 00000  00000 00000  00000 00000
9914: 00000 00000  00000 00000  00000 00000  00000 00000  00000 00000    00000 00000  00000 00000  00000 00000  00000 00000  00000 00000
9915: 00000 00000  00000 00000  00000 00000  00000 00000  00000 00000    00000 00000  00000 00000  00000 00000  00000 00000  00000 00000
9916: 00000 00000  00000 00000  00000 00000  00000 00000  00000 00000    00000 00000  00000 00000  00000 00000  00000 00000  00000 00000
9917: 00000 00000  00000 00000  00000 00000  00000 00000  00000 00000    00000 00000  00000 00000  00000 00000  00000 00000  00000 00000
9918: 00000 00000  00000 00000  00000 00000  00000 00000  00000 00000    00000 00000  00000 00000  00000 00000  00000 00000  00000 00000
9919: 00000 00000  00000 00000  00000 00000  00000 00000  00000 00000    00000 00000  00000 00000  00000 00000  00000 00000  00000 00000
9920: 00000 00000  00000 00000  00000 00000  00000 00000  00000 00000    00000 00000  00000 00000  00000 00000  00000 00000  00000 00000
9921: 00000 00000  00000 00000  00000 00000  00000 00000  00000 00000    00000 00000  00000 00000  00000 00000  00000 00000  00000 00000
9922: 00000 00000  00000 00000  00000 00000  00000 00000  00000 00000    00000 00000  00000 00000  00000 00000  00000 00000  00000 00000
9923: 00000 00000  00000 00000  00000 00000  00000 00000  00000 00000    00000 00000  00000 00000  00000 00000  00000 00000  00000 00000
9924: 00000 00000  00000 00000  00000 00000  00000 00000  00000 00000    00000 00000  00000 00000  00000 00000  00000 00000  00000 00000
9925: 00000 00000  00000 00000  00000 00000  00000 00000  00000 00000    00000 00000  00000 00000  00000 00000  00000 00000  00000 00000
9926: 00000 00000  00000 00000  00000 00000  00000 00000  00000 00000    00000 00000  00000 00000  00000 00000  00000 00000  00000 00000
9927: 00000 00000  00000 00000  00000 00000  00000 00000  00000 00000    00000 00000  00000 00000  00000 00000  00000 00000  00000 00000
9928: 00000 00000  00000 00000  00000 00000  00000 00000  00000 00000    00000 00000  00000 00000  00000 00000  00000 00000  00000 00000
9929: 00000 00000  00000 00000  00000 00000  00000 00000  00000 00000    00000 00000  00000 00000  00000 00000  00000 00000  00000 00000
9930: 00000 00000  00000 00000  00000 00000  00000 00000  00000 00000    00000 00000  00000 00000  00000 00000  00000 00000  00000 00000
9931: 00000 00000  00000 00000  00000 00000  00000 00000  00000 00000    00000 00000  00000 00000  00000 00000  00000 00000  00000 00000
9932: 00000 00000  00000 00000  00000 00000  00000 00000  00000 00000    00000 00000  00000 00000  00000 00000  00000 00000  00000 00000
9933: 00000 00000  00000 00000  00000 00000  00000 00000  00000 00000    00000 00000  00000 00000  00000 00000  00000 00000  00000 00000
9934: 00000 00000  00000 00000  00000 00000  00000 00000  00000 00000    00000 00000  00000 00000  00000 00000  00000 00000  00000 00000
9935: 00000 00000  00000 00000  00000 00000  00000 00000  00000 00000    00000 00000  00000 00000  00000 00000  00000 00000  00000 00000
9936: 00000 00000  00000 00000  00000 00000  00000 00000  00000 00000    00000 00000  00000 00000  00000 00000  00000 00000  00000 00000
9937: 00000 00000  00000 00000  00000 00000  00000 00000  00000 00000    00000 00000  00000 00000  00000 00000  00000 00000  00000 00000
9938: 00000 00000  00000 00000  00000 00000  00000 00000  00000 00000    00000 00000  00000 00000  00000 00000  00000 00000  00000 00000
9939: 00000 00000  00000 00000  00000 00000  00000 00000  00000 00000    00000 00000  00000 00000  00000 00000  00000 00000  00000 00000
9940: 00000 00000  00000 00000  00000 00000  00000 00000  00000 00000    00000 00000  00000 00000  00000 00000  00000 00000  00000 00000
9941: 00000 00000  00000 00000  00000 00000  00000 00000  00000 00000    00000 00000  00000 00000  00000 00000  00000 00000  00000 00000
9942: 00000 00000  00000 00000  00000 00000  00000 00000  00000 00000    00000 00000  00000 00000  00000 00000  00000 00000  00000 00000
9943: 00000 00000  00000 00000  00000 00000  00000 00000  00000 00000    00000 00000  00000 00000  00000 00000  00000 00000  00000 00000
9944: 00000 00000  00000 00000  00000 00000  00000 00000  00000 00000    00000 00000  00000 00000  00000 00000  00000 00000  00000 00000
9945: 00000 00000  00000 00000  00000 00000  00000 00000  00000 00000    00000 00000  00000 00000  00000 00000  00000 00000  00000 00000
9946: 00000 00000  00000 00000  00000 00000  00000 00000  00000 00000    00000 00000  00000 00000  00000 00000  00000 00000  00000 00000
9947: 00000 00000  00000 00000  00000 00000  00000 00000  00000 00000    00000 00000  00000 00000  00000 00000  00000 00000  00000 00000
9948: 00000 00000  00000 00000  00000 00000  00000 00000  00000 00000    00000 00000  00000 00000  00000 00000  00000 00000  00000 00000
9949: 00000 00000  00000 00000  00000 00000  00000 00000  00000 00000    00000 00000  00000 00000  00000 00000  00000 00000  00000 00000
```

```
9950: 0000000000 0000000000 0000000000 0000000000 0000000000  0000000000 0000000000 0000000000 0000000000 0000000000
9951: 0000000000 0000000000 0000000000 0000000000 0000000000  0000000000 0000000000 0000000000 0000000000 0000000000
9952: 0000000000 0000000000 0000000000 0000000000 0000000000  0000000000 0000000000 0000000000 0000000000 0000000000
9953: 0000000000 0000000000 0000000000 0000000000 0000000000  0000000000 0000000000 0000000000 0000000000 0000000000
9954: 0000000000 0000000000 0000000000 0000000000 0000000000  0000000000 0000000000 0000000000 0000000000 0000000000
9955: 0000000000 0000000000 0000000000 0000000000 0000000000  0000000000 0000000000 0000000000 0000000000 0000000000
9956: 0000000000 0000000000 0000000000 0000000000 0000000000  0000000000 0000000000 0000000000 0000000000 0000000000
9957: 0000000000 0000000000 0000000000 0000000000 0000000000  0000000000 0000000000 0000000000 0000000000 0000000000
9958: 0000000000 0000000000 0000000000 0000000000 0000000000  0000000000 0000000000 0000000000 0000000000 0000000000
9959: 0000000000 0000000000 0000000000 0000000000 0000000000  0000000000 0000000000 0000000000 0000000000 0000000000
9960: 0000000000 0000000000 0000000000 0000000000 0000000000  0000000000 0000000000 0000000000 0000000000 0000000000
9961: 0000000000 0000000000 0000000000 0000000000 0000000000  0000000000 0000000000 0000000000 0000000000 0000000000
9962: 0000000000 0000000000 0000000000 0000000000 0000000000  0000000000 0000000000 0000000000 0000000000 0000000000
9963: 0000000000 0000000000 0000000000 0000000000 0000000000  0000000000 0000000000 0000000000 0000000000 0000000000
9964: 0000000000 0000000000 0000000000 0000000000 0000000000  0000000000 0000000000 0000000000 0000000000 0000000000
9965: 0000000000 0000000000 0000000000 0000000000 0000000000  0000000000 0000000000 0000000000 0000000000 0000000000
9966: 0000000000 0000000000 0000000000 0000000000 0000000000  0000000000 0000000000 0000000000 0000000000 0000000000
9967: 0000000000 0000000000 0000000000 0000000000 0000000000  0000000000 0000000000 0000000000 0000000000 0000000000
9968: 0000000000 0000000000 0000000000 0000000000 0000000000  0000000000 0000000000 0000000000 0000000000 0000000000
9969: 0000000000 0000000000 0000000000 0000000000 0000000000  0000000000 0000000000 0000000000 0000000000 0000000000
9970: 0000000000 0000000000 0000000000 0000000000 0000000000  0000000000 0000000000 0000000000 0000000000 0000000000
9971: 0000000000 0000000000 0000000000 0000000000 0000000000  0000000000 0000000000 0000000000 0000000000 0000000000
9972: 0000000000 0000000000 0000000000 0000000000 0000000000  0000000000 0000000000 0000000000 0000000000 0000000000
9973: 0000000000 0000000000 0000000000 0000000000 0000000000  0000000000 0000000000 0000000000 0000000000 0000000000
9974: 0000000000 0000000000 0000000000 0000000000 0000000000  0000000000 0000000000 0000000000 0000000000 0000000000
9975: 0000000000 0000000000 0000000000 0000000000 0000000000  0000000000 0000000000 0000000000 0000000000 0000000000
9976: 0000000000 0000000000 0000000000 0000000000 0000000000  0000000000 0000000000 0000000000 0000000000 0000000000
9977: 0000000000 0000000000 0000000000 0000000000 0000000000  0000000000 0000000000 0000000000 0000000000 0000000000
9978: 0000000000 0000000000 0000000000 0000000000 0000000000  0000000000 0000000000 0000000000 0000000000 0000000000
9979: 0000000000 0000000000 0000000000 0000000000 0000000000  0000000000 0000000000 0000000000 0000000000 0000000000
9980: 0000000000 0000000000 0000000000 0000000000 0000000000  0000000000 0000000000 0000000000 0000000000 0000000000
9981: 0000000000 0000000000 0000000000 0000000000 0000000000  0000000000 0000000000 0000000000 0000000000 0000000000
9982: 0000000000 0000000000 0000000000 0000000000 0000000000  0000000000 0000000000 0000000000 0000000000 0000000000
9983: 0000000000 0000000000 0000000000 0000000000 0000000000  0000000000 0000000000 0000000000 0000000000 0000000000
9984: 0000000000 0000000000 0000000000 0000000000 0000000000  0000000000 0000000000 0000000000 0000000000 0000000000
9985: 0000000000 0000000000 0000000000 0000000000 0000000000  0000000000 0000000000 0000000000 0000000000 0000000000
9986: 0000000000 0000000000 0000000000 0000000000 0000000000  0000000000 0000000000 0000000000 0000000000 0000000000
9987: 0000000000 0000000000 0000000000 0000000000 0000000000  0000000000 0000000000 0000000000 0000000000 0000000000
9988: 0000000000 0000000000 0000000000 0000000000 0000000000  0000000000 0000000000 0000000000 0000000000 0000000000
9989: 0000000000 0000000000 0000000000 0000000000 0000000000  0000000000 0000000000 0000000000 0000000000 0000000000
9990: 0000000000 0000000000 0000000000 0000000000 0000000000  0000000000 0000000000 0000000000 0000000000 0000000000
9991: 0000000000 0000000000 0000000000 0000000000 0000000000  0000000000 0000000000 0000000000 0000000000 0000000000
9992: 0000000000 0000000000 0000000000 0000000000 0000000000  0000000000 0000000000 0000000000 0000000000 0000000000
9993: 0000000000 0000000000 0000000000 0000000000 0000000000  0000000000 0000000000 0000000000 0000000000 0000000000
9994: 0000000000 0000000000 0000000000 0000000000 0000000000  0000000000 0000000000 0000000000 0000000000 0000000000
9995: 0000000000 0000000000 0000000000 0000000000 0000000000  0000000000 0000000000 0000000000 0000000000 0000000000
9996: 0000000000 0000000000 0000000000 0000000000 0000000000  0000000000 0000000000 0000000000 0000000000 0000000000
9997: 0000000000 0000000000 0000000000 0000000000 0000000000  0000000000 0000000000 0000000000 0000000000 0000000000
9998: 0000000000 0000000000 0000000000 0000000000 0000000000  0000000000 0000000000 0000000000 0000000000 0000000000
9999: 0000000000 0000000000 0000000000 0000000000 0000000000  0000000000 0000000000 0000000000 0000000000 0000000000
```

A Million Zeros

www.ingramcontent.com/pod-product-compliance
Lightning Source LLC
Chambersburg PA
CBHW081456190326
41458CB00015B/5265